Noel C. W. Beadle

The Vegetation of Australia

The Vegetation of Australia

Noel C. W. Beadle
Foundation Professor of Botany
University of New England, Armidale, New South Wales, Australia

With 416 Figures and 91 Tables

Cambridge University Press
Cambridge
London   New York   New Rochelle   Melbourne   Sydney

Published by the Press Syndicate of the University of Cambridge
The Pitt Building, Trumpington Street, Cambridge CB2 1RP
32 East, 57th Street, New York, NY 10022, USA
296 Beaconsfield Parade, Middle Park, Melbourne 3206, Australia

© Gustav Fischer Verlag Stuttgart 1981
(ISBN 3-437-30313-9)
(published as Vol. IV of the Series "Vegetationsmonographien der einzelnen Großräume",
edited by Heinrich Walter and S.-W. Breckle)

First published 1981

Designed and printed in West Germany by Passavia Druckerei GmbH Passau

In Continental Europe including the Soviet Union only available through
Gustav Fischer Verlag, Stuttgart

This work is a coproduction of Gustav Fischer Verlag, Stuttgart (Bundesrepublik Deutschland)
and Cambridge University Press, Cambridge UK

British Library Cataloguing in Publication Data

Beadle, Noel C.W.
The vegetation of Australia
1. Botany – Australia
1. Title
581.9′94   QK431

ISBN 0-521-24195-2

Library of Congress Catalog Card Number
81′2662

Preface of the editors

This is the first account of the vegetation of the whole of Australia – a continent which has many peculiarities and which has not received sufficient attention in the general field of vegetation science. The 'Australian Plate' separated from the supercontinent, Gondwanaland in the relatively recent past when its biota commenced an isolated development. Because of this the Australian flora is so different from all other continents that one regards it as a separate floral kingdom.

The author has presented an analysis of the flora, its origin, development and composition. Endemism is particularly marked. It is unique in that the genera *Eucalyptus* (about 450 good species) and *Acacia* (about 700 mostly phyllodineous species) dominate almost all Australian forests and woody communities outside the tropical rain forests.

The flora of the woody communities in the wetter areas is mainly xeromorphic, the taxa having evolved as adaptations to the sandy soils of low fertility (i.e. they are peinomorphic). In contrast to the Holarctic where one finds species-poor plant communities on soils of low fertility, the Australian xeromorphic communities are floristically rich; indeed, it is not unusual to find upward of 100 species in heaths especially in south-western Australia. These xeromorphic species belong mainly to the families Myrtaceae, Proteaceae, Epacridaceae, Papilionaceae (Tribe Podalyrieae), Rutaceae, Casuarinaceae and the genus *Acacia*.

The vegetation is described in seventeen chapters. One chapter is devoted to the rainforests and their remnants in the semi-arid and arid regions. These forests were derived partly from ancient Gondwanaland taxa (especially in the south-east) and partly from Asian taxa which have invaded the continent by distance dispersal in more recent times, either directly or via New Guinea.

Eight chapters are devoted to the *Eucalyptus* forests, woodlands and mallee communities, which cover much of the wetter parts of the continent from the lowlands to the tree-line on the highest mountains in the south-east. They are replaced in these wetter areas by rainforests on the more fertile soils and by scrubs and heaths on the soils of lowest fertility.

In the semi-arid and arid zones, the eucalypts are replaced in most areas either by species of *Acacia* (especially *A. harpophylla* – Brigalow, and *A. aneura* – Mulga), by *Casuarina* spp., by species of *Atriplex* (Saltbush) and *Maireana* (Bluebush), or by grassland.

The Australian grasslands are unusual in so far as the species which dominate the driest sandy areas from hummocks. In the case of *Triodia* and *Plectrachne* (Porcupine Grasses or "Spinifex"), the plants have needle-like leaves, whereas *Zygochloa paradoxa* has caducous leaves and forms shrub-like hummocks up to 2 m high. All three genera are endemic. Tussock grasslands also occur, among them the Mitchell Grass Downs dominated by *Astrebla* spp. (endemic) and occurring in semi-arid to arid areas on clays, *Dichanthium* spp. in the tropics and subtropics, *Stipa* and *Danthonia* in the south, and *Poa* in cool to cold climates. Smaller areas are dominated by other grasses and each community is described in some detail.

The variety of landscapes produced in the Holarctic by the different physiognomy of needle-leaved and broad-leaved forests is absent from Australia. However, many species of conifers do occur, but mostly as subdominants, only. A relationship between the Northern Hemisphere and the Australian floras is seen through the occurrence of many bipolar taxa in alpine Australia, among them *Poa*, *Ranunculus*, *Caltha* and *Veronica*. The Australian high-altitude flora is closely related also to the cold-climate floras of New Zealand and Southern America.

The ecologist and the botanist and the vegetation scientist from the Northern hemisphere often has to learn anew in Australia. Insectivorous plants grow not only on mires. *Drosera* and *Utricularia* can be ephemerals or geophytes in forests or dry heaths.

The book contains 378 photographs of plant communities, 90 tables of data and 38 distribution maps, including a vegetation map of the continent, showing the distributions of c. 90 plant communities. It contains a wealth of information on the flora and vegetation and should be of particular value to all botanists, ecologists, and plant geographers, both in Australia and elsewhere, even if they have no special knowledge of the often unique Australian conditions.

Stuttgart, September 1980

H. Walter
S.-W. Breckle

Preface

In 1964, at the International Botanical Congress held in Edinburgh, Scotland, Professor Heinrich Walter suggested that I should write this book. At first I rejected the idea on the grounds that I did not know enough about the vegetation of Australia. However, on my return to Australia I reconsidered the project and decided to make a start, especially since I had already written several sections on the arid zone flora and vegetation, which up until that time had been my special interest.

I was soon to find that my original estimate of my knowledge of the continent was correct and that it would be necessary for me to visit many parts of Australia which I had not seen. These included the accessible parts of the west and north, as well as much of the east, especially the section south of the Tropic, and western Tasmania. Consequently, over the past fourteen years I have made numerous excursions which have been fitted in between teaching at the University.

A vegetation map is provided (Fig. 6.1) which locates about 90 plant alliances. This has been constructed partly from published works, which are quoted in the text and listed in the bibliography, and partly from my own transects (chiefly south of the Tropic in the east). This map supplements the recently published map of Carnahan (1976) based on structural form and which has been very useful to me.

I am most grateful to Professor Walter for inviting me to write the book. Furthermore, I should like to thank him for his stimulating discussions during my stay in Germany in 1961, and for his continued support for this project, and to compliment him on the high standard he has set for his associates and students. Also, my thanks are extented to Professor S.-W. Breckle (Universität Bielefeld) who assisted Professor Walter with the editing. Originally I expected the book to be a joint effort with some of my colleagues but, when no assistance was forthcoming, I decided to continue alone, though at times the task appeared too difficult for one man. Two sections have been supplied by others, namely, "Rainforest of north-eastern Australia" (Chap. 7, Sect. 7) by G.C. Stocker and B.P.M. Hyland (C.S.I.R.O., Division of Forest Research, Atherton, Queensland), and "Orchidaceae" (Chap. 4, Section 5.9.4) by B.J. Wallace (Botany Department, University of New England).

Several people have assisted by reading parts of the manuscript, namely, my sister, Miss Lois D. Beadle, Drs R.D.B. Whalley and G.P. Guymer, Messrs B.J. Wallace, J.B. Williams, R.A. Boyd and Mrs Dorothy Wheeler.

I am especially indebted to my friends in the departments of Geophysics and Geology, for many helpful discussions on continental drift, general geological problems and fossils, namely, Professor R. Green, Dr R.E. Gould and Professor H.J. Harrington. I thank also Mr G. Wray (Department of Botany) for printing the photographs, Mr M.J. Roach (Department of Geography) for drawing the smaller maps and Mr G. ter Hedde for photographing and printing the vegetation map (Fig. 6.1).

About a dozen of the photographs are not my own but have been supplied by others (as acknowledged below the photos); these people are: the late Mr R.H. Anderson (Fig. 7.14), Dr J.S. Beard (16.29), Mr M.I.T. Brooker (14.20), Dr Winifred Curtis (7.12), Mr. D.B. Foreman (23.19), Mr D. Gibson (8.14), Dr G.P. Guymer (7.32), Dr (Mrs) Betsy Jackes (13.2), Mrs Madeleine O. Rankin (22.3), Mr B.J. Wallace (2.11) and Mr G. Wray (2.11 and 20.6).

During my travels I received assistance, especially with plant identication and transport, in several centres and I should like to thank those people concerned both for their botanical assistance and for their hospitality. In Darwin, Mr G. Brown, Superintendent of Parks, Gardens and Reserves, Mr N. Byrnes (now at Brisbane), and Miss Jenny Harman. At Broome, Mr F. Lullfitz. At Perth, at the University of Western Australia, Emeritus Professor B.J. Grieve and his secretary, Mrs K. Holland, Miss Alison Baird, Mr G.G. Smith, Drs S. James, B. Lamont, W.A. Loneragan, N. Malajczug; at the State Herbarium, Messrs P.B. Wilson, B.R. Maselin and A.S. George. At Manjimup, W.A. Messrs J.J. Havel, O.W. Loneragan and P. Skinner. In Hobart, Professor W. Jackson and his wife Vanny, Drs Winifred Curtis, J. Townrow and Mr A.B. Mount. In Sydney, Dr L.A.S. Johnson and his staff.

Several of my colleagues have given valuable assistance by driving vehicles to enable me to make continuous transect records, and of these I thank particularly Mr N. Petrasz (Department of Geology) who accompanied me around Australia in 1972.

The work was financed almost entirely by the University of New England, with some support for field work from the Rural Credits Development Fund. Over a number of years I received grants from the latter Fund to support my research in the interior, and this greatly extended my knowledge of the eastern segment of the arid zone, where I worked in my first position after graduation as an officer of the Soil Conservation Service of New South Wales. In 1972, a grant from the Fund subsidized a long trip around Australia, north of the Tropic and in Western Australia.

Typing the manuscript has been a tedious job and several secretaries have made contributions over the years, especially Miss Helene Dawson who typed most of the final draft.

The nomenclature used is that in local floras or lists. For Northern Territory, Chippendale (1971); Western Australia, Beard (1970); South Australia and the arid zone, Black (1943–57) and Eichler (1965); eastern New South Wales, Beadle et al (1972) and Beadle and Beadle (1970–76) (using the terminology accepted at the National Herbarium, Sydney); Victoria, Willis (1970–72); Tasmania, Curtis (1963–67) and Curtis and Morris (1975). Bentham's Flora Australiensis (1863–78) has been used for taxa not included in the other floras, especially for tropical Queensland.

Two other groups of people must be recognized as active helpers. Firstly those who have published previously and whose names are included in the bibliography. Of these I pay tribute especially to the late Dr. N.H. Speck whose unpublished Ph.D. thesis on the vegetation of south-west Western Australia was made available to me by Professor Grieve. Also, the list of plant communities by Specht, Roe and Boughton (1974) made my task a lot easier. The second group includes those diligent persons (mostly anonymous) in the Dixson Library at the University of New England who unfailingly secured for me any literature needed.

Finally I should like to thank also the publishers, especially Mr B. Gaebler, for their attention to the production of the book and the arrangement of the text.

Armidale 1980   Noel C. W. Beadle

Contents

1. The Australian Environment

1.	Location of Australia and its Origin	1
2.	Topography: Rivers	5
2.1.	Topography	5
2.2.	Rivers	6
3.	Some Comments on the Geomorphology and Geology	7
4.	Climate	11
4.1.	General	11
4.2.	Solar Radiation	12
4.3.	Temperature	14
4.4.	Precipitation	14
4.4.1.	Mean Annual Rainfall and Seasonal Incidence	14
4.4.2.	Explanation of the Seasonal Rainfall Patterns	15
4.4.3.	Rainfall Reliability	17
4.4.4.	Dew and Fog	18
4.5.	Natural Irrigation	19
4.6.	Winds	19
4.6.1.	General	19
4.6.2.	Cyclic Salt	20
4.7.	Growth Regimes (Climate types)	20
5.	Soils	22
5.1.	Soil Parent Materials	22
5.2.	Weathering and the Development of the Soil Profile	23
5.2.1.	General	23
5.2.2.	Soil Profile Development under Conditions of Impeded Drainage. Laterite	23
5.3.	The Gilgai Phenomenon	30
5.4.	Soil Classification	31
5.5.	Relationship between Soil Group and Vegetation	31
5.6.	Individual Soil Properties and their Impact on the Flora and Vegetation	31
6.	Fire	32

2. The Flora

1.	Introduction	33
2.	The Vascular Flora	33
2.1.	Pteridophytes	33
2.1.1.	The Indigenous Flora	33
2.1.2.	Occurrence of Pteridophytes as Community Dominants	37
2.1.3.	Changes in the Abundance of Pteridophytes as a Result of Man's Activities	38
2.2.	Gymnosperms	41
2.2.1.	The Indigenous Flora	41
2.2.2.	Habitats of Gymnosperms	41
2.2.3.	The Naturalized Flora	44
2.3.	Angiosperms	44
2.3.1.	General	44
2.3.2.	The Indigenous Flora	44
2.3.3.	The Endemic Families and Genera	49

2.3.4.	Primitive Angiosperm Families	51
2.3.5.	Heterotrophic Angiosperms	52
2.3.6.	The Naturalized Flora	56

3. Origins of the Vascular Flora

1.	Introduction	57
2.	Pteridophytes and Gymnosperms	57
2.1.	Pre-Cretaceous Flora	57
2.2.	Cretaceous Flora	58
3.	Cretaceous Angiosperm Flora	59
4.	Early Tertiary Angiosperm Floras of Australia and New Zealand	60
5.	Floristic Relationships of the Modern Australian Flora with other Continents and Islands	62
5.1.	General	62
5.2.	Africa – Australia – South America Relationships	64
6.	History of the Angiosperms in Australia	65
7.	Distance Dispersal of Plants	68
7.1.	General	68
7.2.	New Guinea – Australia Relationships	68
7.3.	Australia – Asia Relationships	69
7.4.	Australia – New Zealand Relationships	72

4. Development of the Modern Flora

1.	Introduction	73
2.	Tertiary and Quaternary Conditions and Events Relevant to Plant Distribution and Dispersal	73
2.1.	The Coast-line and Topography	73
2.2.	Soil Parent Materials	74
2.3.	Climate	74
3.	History of the Early Tertiary Floras	75
3.1.	History of the *Nothofagus* Assemblage	75
3.2.	History of the Broad-leaved Rainforest Assemblage	75
3.3.	History of the Herbaceous Cold-climate Assemblage	77
4.	The Endemic Families and Genera in Relation to Habitat	78
4.1.	The Endemic Families	78
4.2.	The Endemic Genera	78
5.	Comments on Selected Families	80
5.1.	General	80
5.2.	Myrtaceae	80
5.2.1.	The Subfamilies and Genera	80
5.2.2.	*Eucalyptus*	82
5.3.	Proteaceae	84
5.4.	Rutaceae	85
5.5.	Papilionaceae and Mimosaceae *(Acacia)*	85
5.6.	Euphorbiaceae	88
5.7.	Epacridaceae and Ericaceae	88
5.8.	Casuarinaceae	89
5.9.	Some Monocotyledonous Families	89
5.9.1.	Liliales	89
5.9.2.	Cyperaceae	91

5.9.3.	Restionaceae	92
5.9.4.	Orchidaceae (section supplied by B. J. Wallace)	92
6.	Xeromorphy and Sclerophylly	93
6.1.	General	93
6.2.	Some Characters of Xeromorphs	94
6.3.	Occurrence of Xeromorphy in Families and Genera	95
6.4.	Geographic Distribution of the Xeromorphic Flora in Wetter Areas	95
6.5.	Habitat Factors	96
7.	History of the Xeromorphic Flora	96

5. The Flora of the Arid Zone

1.	Introduction	100
2.	The Flora	100
2.1.	General	100
2.2.	The Vascular Flora	101
2.2.1.	Pteridophytes and Gymnosperms	101
2.2.2.	Angiosperms	101
2.2.3.	The Naturalized Flora	102
3.	Distribution of the Angiosperm Flora	103
3.1.	General	103
3.2.	Species-Abundance in Various Habitats	104
3.3.	Distributions of the More Abundant Taxa	105
3.4.	Factors Affecting the Distribution of the Flora	105
3.4.1.	Mean Annual Rainfall	105
3.4.2.	Seasonal Rainfall Incidence	105
3.4.3.	Edaphic Factors	106
3.5.	Disjunctions	106
4.	Origins of the Angiosperm Flora	107
4.1.	General	107
4.2.	Taxa Derived from the Rainforests	109
4.3.	Taxa Derived from the Wet Xeromorphic Assemblages	109
4.4.	Taxa Derived from Temperate Genera	110
4.5.	Taxa Derived from the Littoral Zone	111
4.6.	Unclassified Genera	111
5.	The Endemic Genera	111
6.	Some Vegetative Features of Arid Zone Plants	114
6.1.	Life Forms	114
6.2.	Leaves	116
7.	Reproduction	117
7.1.	General	118
7.2.	Fruits	118
7.3.	Seeds; Germination; Longevity	118
7.4.	Dispersal of Fruits and Seeds	120
7.5.	Vegetative Reproduction	121

6. The Plant Communities
Classification: Factors Affecting the Patterning of the Communities, and the Distribution of the Communities in the Continent

1.	Introduction	122
2.	Classification of the Communities	123

2.1.	General: Life-forms; Leaf-size Classes; Structural Units; Succession	123
2.2.	Floristic Units	123
3.	Factors Affecting the Patterning of the Communities	123
3.1.	General	123
3.2.	Historic Factors	125
3.3.	Climate	125
3.3.1.	Rainfall	125
3.3.2.	Temperature	126
3.4.	Edaphic Factors	126
3.4.1.	General	126
3.4.2.	Soil Depth	126
3.4.3.	Soil Nutrients	127
3.4.4.	Soil Salinity	127
3.4.5.	Soil Texture	127
3.4.6.	Soil Waterlogging	128
3.4.7.	Soil Acidity/Alkalinity	129
3.4.8.	Toxic Substances	129
3.4.8.1.	Gypsum	129
3.4.8.2.	Heavy Metals	130
3.4.8.3.	Serpentines	130
4.	Distribution of the Communities in the Continent	130
4.1.	General	130
4.2.	Region 1: Wetter Tropics	131
4.3.	Region 2: Eastern Coastal Lowlands	131
4.4.	Region 3: Eastern Inland Lowlands	132
4.5.	Region 4: Eastern Highlands	132
4.6.	Region 5: Tasmania	133
4.7.	Region 6: The Mallee	133
4.8.	Region 7: South-western Australia	134
4.9.	Region 8: Semi-arid and Arid Areas	134

7. The Rainforests

1.	Introduction	136
2.	Factors Governing the Distributions of the Rainforests	136
2.1.	Mean Annual Rainfall and Seasonal Incidence	136
2.2.	Temperature	136
2.3.	Edaphic Factors	136
2.4.	Margins of Rainforests	137
3.	Characteristics of Rainforest Plants	138
3.1.	Leaf Characters	138
3.2.	Life Forms	138
3.2.1.	General	138
3.2.2.	Woody Plants (Phanerophytes)	139
3.2.3.	Lianas	140
3.2.4.	Epiphytes	140
4.	The Vascular Flora	142
4.1.	Pteridophytes	142
4.2.	Gymnosperms	142
4.3.	Angiosperms	142
5.	The Alliances	143
6.	Alliances South of the Tropic in Wetter Areas	143
6.1.	General	143

6.2.	The *Nothofagus* Alliances	143
6.2.1.	General	143
6.2.2.	*Nothofagus cunninghamii* Alliance	143
6.2.2.1.	Communities in Tasmania	151
6.2.2.2.	Communities in Victoria	152
6.2.3.	Some Related Communities	153
6.2.4.	*Nothofagus gunnii* Alliance	154
6.2.5.	*Nothofagus moorei* Alliance	154
6.3.	Alliances Dominated by Cunoniaceae	156
6.3.1.	General	156
6.3.2.	*Anodopetalum biglandulosum* Alliance	156
6.3.3.	*Ceratopetalum apetalum* Alliance	157
6.3.3.1.	General	157
6.3.3.2.	*C. apetalum – Doryphora sassafras* Suballiance	157
6.3.3.3.	*Doryphora sassafras – Acacia melanoxylon* Suballiance	158
6.3.3.4.	*C. apetalum – Acmena smithii – Tristania laurina* Suballiance	158
6.3.3.5.	*C. apetalum – Diploglottis australis* Suballiance	161
6.3.3.6.	*C. apetalum – Schizomeria ovata* Suballiance	161
6.3.3.7.	*C. apetalum – Argyrodendron* spp. Suballiance	161
6.3.4.	*Schizomeria ovata – Doryphora sassafras – Ackama paniculata – Cryptocarya glaucescens* Alliance	161
6.4.	*Argyrodendron* spp. Alliance	162
6.5.	*Drypetes australasica – Araucaria* spp. *– Brachychiton discolor – Flindersia* spp. Alliance	164
6.6.	Littoral Rainforests	166
6.6.1.	General	166
6.6.2.	*Cupaniopsis anacardioides* Alliance	166
6.6.3.	*Tristania conferta* Alliance	166
7.	Rainforests of Tropical North-eastern Australia (by G. C. Stocker and B. P. M. Hyland)	167
7.1.	Distribution – Past and Present	167
7.2.	Floristic Composition	167
7.3.	Characteristics of some Representative Forest Types	170
8.	Semi-Evergreen Vine Forests (Monsoon Forests)	179
8.1.	General	179
8.2.	Associations	180
8.3.	The Effects of Fire	181
8.4.	Littoral Rainforests	181
9.	Rainforest-derived Communities in the Semi-Arid and Arid Zones	181
9.1.	General	181
9.2.	*Cadellia pentastylis* Alliance	183
9.3.	*Brachychiton* spp. Alliance	183
9.3.1.	General	183
9.3.2.	*B. rupestris* Suballiance	184
9.3.3.	*B. australis* Suballiance	184
9.3.4.	*B. grandiflorus* Suballiance	185
9.4.	Mixed Stands	185
9.5.	*Erythrophleum chlorostachys* Alliance	186
9.6.	*Adansonia gregorii* Alliance	186
9.7.	*Terminalia* spp. *– Bauhinia cunninghamii* Alliance	187
9.7.1.	General	187
9.7.2.	*Terminalia* spp. Suballiances	187
9.7.3.	*Bauhinia cunninghamii* Suballiance	188
9.8.	*Macropteranthes* spp. Alliance	189

9.9.	*Excoecaria latifolia* Alliance	189
9.10.	*Geijera parviflora – Flindersia maculosa – Heterodendrum oleifolium* Alliance	190
9.10.1.	General	190
9.10.2.	Associations Containing *Geijera parviflora*	190
9.10.3.	Associations Containing *Flindersia maculosa*	191
9.10.4.	Associations Containing *Heterodendrum oleifolium*	191
9.11.	*Atalaya hemiglauca – Grevillea striata – Ventilago viminalis* Alliance	191
9.12.	Other Species	193
10.	Gallery Rainforests	193
10.1.	General	193
10.2.	Communities between the Clarence R. and the Tropic	194
10.3.	Communities North of the Tropic in the East	194
10.4.	Communities in the Drier Tropics (N.T.)	195

8. Eucalyptus Communities of the Tropics

1.	Introduction	196
2.	Descriptions of the Alliances	197
2.1.	*Eucalyptus tetrodonta – E. miniata – E. polycarpa* (sens. lat.) Alliance	197
2.1.1.	*E. tetrodonta – E. miniata* Suballiance	198
2.1.2.	*E. tetrodonta – E. polycarpa* Suballiance	200
2.1.3.	*E. polycarpa – E. apodophylla* Suballiance	202
2.1.4.	*E. polycarpa – E. tessellaris* Suballiance	202
2.2.	*E. tectifica – E. confertiflora – E. grandifolia* Alliance	204
2.2.1.	*E. tectifica – E. confertiflora – E. foelschiana* Suballiance	204
2.2.2.	*E. grandifolia* Suballiance	208
2.2.3.	*E. latifolia* Suballiance	209
2.3.	*E. ptychocarpa* Alliance	209
2.4.	*E. alba* Alliance	210
2.5.	*E. phoenicea* Alliance	211
2.5.1.	*E. phoenicea – E. ferruginea* Suballiance	212
2.5.2.	*Eucalyptus* spp. – *Calytrix – Triodia* Edaphic Complex	213
2.6.	*E. papuana* Alliance	214
2.7.	*E. argillacea – E. terminalis* Alliance	216
2.8.	*E. dichromophloia* Alliance	217
2.8.1.	*E. dichromophloia* Suballiance	218
2.8.2.	*E. dichromophloia – E. herbertiana – E. collina* Suballiance	219
2.8.3.	*E. dichromophloia – E. zygophylla – Acacia* ssp. Suballiance	220
2.8.4.	*E. aspera* Suballiance	221
2.9.	*E. pruinosa* Alliance	222
2.10.	*E. brevifolia* Alliance	223

9. The Tall Eucalyptus Forests of the Eastern Coastal Lowlands Mostly on Soils of Higher Fertility

1.	Introduction	225
2.	Descriptions of the Alliances	225
2.1.	*Eucalyptus cloëziana* Alliance	225
2.2.	*E. grandis* Alliance	229
2.3.	*E. saligna* Alliance	229
2.4.	*E. botryoides* Alliance	232
2.5.	*E. resinifera – E. acmenoides – E. propinqua* Alliance	232

2.6.	*E. tereticornis* Alliance	233
2.6.1.	*E. tereticornis* Suballiance	233
2.6.2.	*E. amplifolia* Suballiance	234
2.6.3.	*E. seeana* Suballiance	234
2.7.	*E. pilularis* Alliance	234
2.7.1.	*E. pilularis* Suballiance	235
2.7.2.	*E. pilularis* – *E. microcorys* Suballiance	236
2.7.3.	*E. pilularis* – *E. intermedia* – *E. siderophloia* Suballiance	237
2.7.4.	*E. pyrocarpa* Suballiance	237
2.7.5.	*E. pilularis* – *E. saligna* – *E. paniculata* Suballiance	237
2.7.6.	*E. pilularis* – *E. gummifera* Suballiance	238
2.7.7.	*E. pilularis* – *E. piperita* Suballiance	239
2.7.8.	*E. pilularis* – *Angophora costata* Suballiance	239
2.8.	*E. maculata* Alliance	239
2.9.	*E. robusta* Alliance	240
2.10.	*E. elata* Alliance	241

10. Eucalyptus Woodlands and Forests on Soils of Low Fertility Chiefly on the Eastern Coastal Lowlands

1.	Introduction	242
2.	Descriptions of the Alliances	243
2.1.	*Eucalyptus intermedia* – *E. acmenoides* – *E. signata* – *E. nigra* Alliance	243
2.1.1.	*E. intermedia* – *E. acmenoides* Suballiance	243
2.1.2.	*E. signata* – *E. intermedia* – *E. nigra* Suballiance	244
2.2.	*E. baileyana* – *E. planchoniana* – *E. bancroftii* Alliance	246
2.3.	*E. gummifera* – *E. racemosa* – *E. sieberi* Alliance	247
2.3.1.	General: Climate; Soil Parent Materials; Soils; Topography	247
2.3.2.	The Flora	249
2.3.3.	The Plant Communities: General	251
2.3.4.	Communities in Rocky Terrain and in Swamps	251
2.3.5.	*Eucalyptus* Communities: General	253
2.3.5.1.	*E. gummifera* – *E. racemosa* – *Angophora costata* Suballiance	253
2.3.5.2.	*E. sieberi* – *E. piperita* – *E. racemosa* Suballiance	253
2.3.5.3.	*E. eximia* – *E. punctata* Suballiance	255
2.3.5.4.	The Mallee Species and Communities	257
2.4.	*E. globoidea* Alliance	258
2.5.	*E. baxteri* Alliance	259
2.6.	*E. amygdalina* Alliance	261
2.7.	*E. nitida* Alliance	263

11. Eucalyptus Communities of the Cooler Climates of the Eastern Highlands, Lowland Victoria and Tasmania

1.	Introduction	265
2.	Alliances Occurring from the Northern Tablelands to Tasmania	267
2.1.	*Eucalyptus pauciflora* Alliance	267
2.1.1.	*E. pauciflora* Suballiance	267
2.1.2.	*E. stellulata* Suballiance	269
2.1.3.	*E. niphophila* Suballiance	271
2.1.4.	*E. moorei* – *E. glaucescens* Suballiance	271
2.2.	*E. delegatensis* – *E. dalrympleana* Alliance	271
2.3.	*E. obliqua* – *E. fastigata* Alliance	274

2.4.	*E. viminalis* – *E. rubida* – *E. huberana* Alliance	275
2.4.1.	*E. viminalis* – *E. rubida* Suballiance	275
2.4.2.	*E. huberana* Suballiance	278
2.5.	*E. radiata* Alliance	279
3.	Other Alliances Confined to the Mainland	279
3.1.	*E. mannifera* Alliance	279
3.2.	*E. macrorhyncha* – *E. rossii* Alliance	280
3.3.	*E. cypellocarpa* – *E. muellerana* – *E. maidenii* Alliance	282
4.	Alliances Confined to the Northern Tablelands	283
4.1.	*E. laevopinea* – *E. caliginosa* – *E. youmanii* Alliance	283
4.1.1.	General	283
4.1.2.	*E. laevopinea* Suballiance	284
4.1.3.	*E. caliginosa* Suballiance	284
4.1.4.	*E. youmanii* Suballiance	286
4.2.	*E. campanulata* Alliance	287
4.3.	*E. andrewsii* Alliance	287
4.4.	*E. nova-anglica* Alliance	288
5.	Alliances Occurring Mainly in Victoria and Tasmania	289
5.1.	*E. ovata* Alliance	289
5.2.	*E. regnans* Alliance	290
5.3.	*E. bicostata* – *E. globulus* Alliance	293
6.	Alliances Restricted to Tasmania	293
6.1.	General	293
6.2.	*E. coccifera* – *E. subcrenulata* Alliance	293
6.3.	*E. gunnii* Alliance	296
6.4.	*E. rodwayi* Alliance	297

12. The Ironbark Forests and Woodlands of the East (Cape York Peninsula to Victoria)

1.	Introduction	298
2.	Descriptions of the Alliances	299
2.1.	*Eucalyptus cullenii* Alliance	299
2.2.	*E. drepanophylla* – *E. crebra* Alliance	299
2.2.1.	General	299
2.2.2.	*E. drepanophylla* Suballiance	301
2.2.3.	*E. crebra* Suballiance	302
2.2.4.	*Angophora costata* – *Eucalyptus* spp. Suballiance	304
2.3.	*E. melanophloia* Alliance	306
2.3.1.	*E. melanophloia* Suballiance	306
2.3.2.	*E. shirleyi* Suballiance	308
2.3.3.	*E. similis* Suballiance	308
2.3.4.	*Angophora melanoxylon* Suballiance	309
2.4.	*E. sideroxylon* Alliance	310
2.4.1.	General	310
2.4.2.	*E. sideroxylon* – *E. dealbata* Suballiance	312
2.4.3.	*E. fibrosa* spp. *nubila* Suballiance	313

13. The Box Woodlands of the East and South-East

1.	Introduction	315
2.	Descriptions of the Alliances	317
2.1.	*Eucalyptus leptophleba* – *E. microneura* – *E. normantonensis* Alliance	317

2.1.1.	*E. leptophleba* Suballiance	317
2.1.2.	*E. microneura* Suballiance	319
2.1.3.	*E. normantonensis* Suballiance	319
2.2.	*E. orgadophila* Alliance	320
2.3.	*E. intertexta* Alliance	320
2.3.1.	*E. intertexta – Acacia* Suballiance	321
2.3.2.	*E. intertexta – Callitris "glauca"* Suballiance	322
2.4.	*E. populnea* Alliance	322
2.4.1.	General	322
2.4.2.	*E. brownii* Suballiance	324
2.4.3.	*E. populnea* Suballiance	325
2.4.4.	*E. populnea – Casuarina cristata* Suballiance	325
2.4.5.	*E. populnea – Eremophila mitchellii* Suballiance	325
2.4.6.	*E. populnea – Callitris glauca* Suballiance	325
2.4.7.	*E. populnea – Geijera parviflora – Flindersia maculosa* Suballiance	327
2.4.8.	*E. populnea – Acacia* spp. Suballiance	327
2.5.	*E. melliodora – E. blakelyi* Alliance	328
2.6.	*E. moluccana* Alliance	330
2.7.	*E. albens* Alliance	331
2.8.	*E. woollsiana* Alliance	334
2.9.	*E. pilligaensis* Alliance	335
2.10.	*E. dawsonii* Alliance	336
2.11.	*E. leucoxylon – E. fasciculosa* Alliance	336
2.12.	*E. odorata – E. porosa* Alliance	338
2.13.	*E. cladocalyx* Alliance	339

14. The Mallee and Marlock Communities

1.	Introduction	340
1.1.	General	340
1.2.	Distribution; Climate; Soils	340
1.3.	Community Structure	341
1.4.	The Species of *Eucalyptus*	341
2.	Descriptions of the Alliances	343
2.1.	General	343
2.2.	Mallee and marlock alliances on deep soil profiles	343
2.2.1.	*Eucalyptus diversifolia* Alliance	343
2.2.2.	*E. incrassata – E. foecunda* Alliance	344
2.2.3.	*E. socialis – E. dumosa – E. gracilis – E. oleosa* Alliance	345
2.2.4.	*E. viridis* Alliance	349
2.2.5.	*E. redunca – E. uncinata* Alliance	350
2.2.6.	*E. platypus* Alliance	351
2.2.7.	*E. nutans – E. gardneri* Alliance	352
2.2.8.	*E. cooperana* Alliance	352
2.2.9.	*E. eremophila* Alliance	352
2.2.10.	*E. sheathiana – E. loxophleba – E. oleosa* Alliance	354
2.2.11.	*E. burracoppinensis – Casuarina acutivalvis* Alliance	355
2.2.12.	*E. pyriformis – E. oldfieldii – E. leptopoda* Alliance	355
2.2.13.	*E. concinna* Alliance	358
2.2.14.	*E. dongarrensis – E. oraria* Alliance	358
2.3.	Mallee-heaths	359
2.3.1.	General	359
2.3.2.	*E. tetragona – E. incrassata* Alliance	359

2.3.3.	Species on the West Coast and Inland	361
2.4.	Communities on Rocky Outcrops in South-western Western Australia	361
2.4.1.	General	361
2.4.2.	*Eucalyptus preissiana* Alliance	362
2.4.3.	Other *Eucalyptus* Species	362
2.5.	*Eucalyptus* Species in the Arid Zone	363

15. Eucalyptus Forests and Woodlands in the South-West

1.	Introduction	368
2.	Descriptions of the Alliances	370
2.1.	*Eucalyptus diversicolor* Alliance	370
2.2.	*E. jacksonii* Alliance	372
2.3.	*E. marginata – E. calophylla* Alliance	372
2.3.1.	Forest Stands of *E. marginata* and *E. calophylla*	375
2.3.2.	Woodland Stands of *E. marginata* and *E. calophylla*	376
2.3.3.	Stunted Communities	376
2.3.4.	The Effect of Fire	376
2.3.5.	Other Associations	377
2.3.5.1.	*E. megacarpa* Association	377
2.3.5.2.	*E. patens* Association	377
2.4.	*E. wandoo – E. accedens – E. astringens* Alliance	378
2.4.1.	*E. wandoo* Suballiance	378
2.4.2.	*E. accedens* Suballiance	380
2.4.3.	*E. astringens* Suballiance	380
2.5.	*E. gomphocephala* Alliance	380
2.6.	*E. erythrocorys* Alliance	381
2.7.	*E. loxophleba* Alliance	382
2.8.	*E. cornuta* Alliance	382
2.9.	*E. occidentalis* Alliance	383
2.10.	*E. salmonophloia – E. salubris* Alliance	384
2.11.	*E. lesouefii – E. dundasii* Alliance	387
2.11.1.	*E. lesouefii* Suballiance	388
2.11.2.	*E. dundasii* Suballiance	388

16. Heaths, Banksia Scrubs and Related Communities on the Lowlands

1.	Introduction	390
1.1.	Definitions	390
1.2.	Location and Habitats	391
1.3.	Flora	398
2.	Communities in the Tropics	398
2.1.	General	398
2.2.	*Fenzlia – Melaleuca – Leptospermum – Sinoga* Alliance	398
2.3.	"High Mountain Scrub"	399
2.4.	Communities in Northern Territory	399
3.	Communities South of the Tropic in the East	400
3.1.	General	400
3.2.	Communities on Coastal Sands	400
3.2.1.	General	400
3.2.2.	*Banksia serratifolia* Alliance	400

3.2.3.	*B. aspleniifolia* Alliance	401
3.2.4.	*B. robur* Alliance	401
3.2.5.	*B. ericifolia* Alliance	401
3.2.6.	*B. marginata* Alliance	402
3.2.7.	*Leptospermum flavescens – L. attenuatum* Alliance	405
3.2.8.	*L. liversidgei* Alliance	405
3.2.9.	*L. myrsinoides* Alliance	405
3.2.10.	*Leptospermum* Communities in Tasmania	405
3.2.11.	*Acacia mucronata – Phebalium squameum* Alliance	407
3.2.12.	Sedge – Heaths	407
3.3.	Headland Heaths	408
3.3.1.	General	408
3.3.2.	*Casuarina distyla – Jacksonia stackhousii* Alliance	408
3.3.3.	*C. littoralis – Banksia aspleniifolia* Alliance	408
3.4.	Heaths, Scrubs and Thickets on Rocky Outcrops	409
3.4.1.	General	409
3.4.2.	On Glasshouse Mountains, Queensland	409
3.4.3.	On Trachyte Exposures, Warrumbungle Mountains	410
3.4.4.	On Triassic Sandstones, Central New South Wales	411
3.4.4.1.	General	411
3.4.4.2.	*Casuarina distyla* Alliance	411
3.4.4.3.	*C. nana* Alliance	411
3.4.5.	On Granite Exposures on Wilson's Promontory	413
4.	Communities on South Australia and Western Victoria	413
4.1.	General	413
4.2.	*Xanthorrhoea australis – Banksia ornata – Hakea rostrata – Casuarina paludosa* Alliance	413
4.3.	*Xanthorrhoea australis – Banksia ornata – Casuarina pusilla* Alliance	414
5.	Communities in South-western Western Australia	414
5.1.	General	414
5.1.1.	Location and Habitats	414
5.1.2.	Soil Parent Materials and Soils	416
5.1.3.	Flora	416
5.2.	Communities on the Western Coastal Plain	416
5.2.1.	General	416
5.2.2.	*Banksia menziesii – B. attenuata* Alliance	418
5.2.3.	*B. menziesii – B. attenuata – Casuarina – fraserana Eucalyptus todtiana* Alliance	418
5.2.4.	*Banksia* Scrub – heaths	418
5.2.5.	The Heaths	419
5.2.5.1.	General	419
5.2.5.2.	Heaths on Limestone	420
5.2.5.3.	Heaths on Sandplain	421
5.2.5.4.	Heaths on Laterite	421
5.2.6.	*Actinostrobus arenarius* Alliance	422
5.2.7.	*Banksia ashbyi* Alliance	424
5.2.8.	Thickets Dominated by Species of *Acacia, Melaleuca* and *Casuarina*	425
5.2.8.1.	*Acacia rostellifera* Thickets	425
5.2.8.2.	*Melaleuca cardiophylla* Thickets	425
5.2.8.3.	Other Species of *Melaleuca*	425
5.2.8.4.	*Melaleuca megacephala – Hakea pycnoneura* Thickets	425
5.2.8.5.	*Acacia – Casuarina – Melaleuca* Thickets	425
5.3.	Communities on the Southern Coastal Plain and Inland	425
5.3.1.	General	425

5.3.2.	The Coastal Heaths and Scrub-Heaths	426
5.3.3.	The Inland Heaths and Scrub-Heaths	427
5.4.	Communities on the Coastal Sands along the Wettest Part of the Coast	428
5.4.1.	General	428
5.4.2.	Heaths on Headlands	430
5.4.3.	*Agonis flexuosa* Alliance	430
5.4.4.	Communities Dominated by *Banksia* spp.	433
5.4.5.	*Acacia decipiens* Alliance	433
5.5.	Heaths and Related Communities on Ranges and Hills	434
5.5.1.	General	434
5.5.2.	Colonization of the Rock Surfaces	434
5.5.3.	Heaths on Headlands or on Exposed Hilltops Close to the Ocean	435
5.5.4.	Heaths on Rocky Outcrops in Wetter Areas	436
5.5.5.	Heaths on Rocky Outcrops in Drier Areas	436
5.6.	Heaths in the Arid Zone	436

17. The Alpine Communities

1.	Introduction	438
1.1.	General	438
1.2.	Climate	438
1.3.	Topography	438
1.4.	Soil Parent Materials and Soils	440
2.	Flora	441
3.	Vegetation	445
3.1.	Life-forms	445
3.2.	Structural Forms of Vegetation and their Interrelationships	446
4.	Descriptions of the Alliances	447
4.1.	The Fellfields	447
4.1.1.	General	447
4.1.2.	Communities of Rock Ledges and Clefts	448
4.1.3.	Open Communities	449
4.1.4.	Cushion Mosaics	450
4.2.	The Herbfields	451
4.2.1.	General	451
4.2.2.	*Plantago muelleri – Montia australasica* Alliance	451
4.2.3.	*Celmisia longifolia – Poa* Alliance	451
4.3.	The Grasslands	452
4.4.	The Heaths and Shrublands	452
4.4.1.	General	452
4.4.2.	Communities in Well Drained Habitats on Plateaux	452
4.4.2.1.	Communities on the Mainland	452
4.4.2.2.	Communities in Tasmania	452
4.4.3.	Wet Heaths	454
4.5.	Communities in Free Water and in Waterlogged Areas	455
4.5.1.	General	455
4.5.2.	Aquatic Communities	455
4.5.3.	Fens (Sedgelands)	456
4.5.3.1.	*Carex gaudichaudiana* Alliance	456
4.5.3.2.	*Astelia alpina* Alliance	456
4.5.3.3.	*Carpha – Uncinnia – Oreobolis* Alliance	456
4.5.4.	Bogs	457
4.5.4.1.	General	457

4.5.4.2.	*Carex gaudichaudiana – Sphagnum cristatum* Alliance	457
4.5.4.3.	*Epacris paludosa – Sphagnum cristatum* Alliance	459
4.5.4.4.	*E. breviflora – Blindia robusta* Alliance	459

18. Acacia and Casuarina Communities of the Arid and Semi-Arid Zones

1.	Introduction	461
2.	Communities Dominated by *Acacia*	462
2.1.	General	462
2.2.	*A. shirleyi* Alliance	463
2.3.	*A. harpophylla* Alliance	464
2.4.	*A. pendula* Alliance	466
2.5.	*A. cambagei* Alliance	467
2.6.	*A. aneura* Alliance	469
2.6.1.	General; Autecology of *A. aneura*	469
2.6.2.	Community Structure, Associations and Subassociations	472
2.6.3.	The Effects of Grazing; Degeneration and Secondary Successions	476
2.7.	*A. catenulata* Alliance	477
2.8.	*A. excelsa* Alliance	478
2.9.	*A. pachycarpa* (= *A. ancistrocarpa*) Alliance	479
2.10.	*A. translucens* Alliance	481
2.11.	*A. sclerosperma – A. tetragonophylla – Eremophila pterocarpa – Atriplex* spp. Edaphic Complex	481
2.12.	*A. ramulosa – A. linophylla* Alliance	482
2.13.	*A. acuminata* Alliance	483
2.14.	*A. xiphophylla* Alliance	483
2.15.	*A. grasbyi* Alliance	483
2.16.	*A. calcicola* Alliance	485
2.17.	*A. sowdenii – A. loderi* Alliance	485
2.17.1.	*A. sowdenii* Suballiance	486
2.17.2.	*A. loderi* Suballiance	486
2.18	*A. ligulata* Alliance	487
3.	Communities Dominated by *Casuarina*	487
3.1.	General	487
3.2.	*C. luehmannii* Alliance	487
3.3.	*C. cristata* Alliance	489
3.4.	*C. decaisneana* Alliance	492

19. The Halophytic Shrublands

1.	Introduction	493
1.1.	Definitions	493
1.2.	The Chenopodiaceae in Australia	493
1.3.	Economic Importance	494
1.4.	Leaf Anatomy and Physiology of Some Chenopodiaceae	495
2.	Descriptions of the Alliances	496
2.1.	Alliances Dominated by *Atriplex* spp.	496
2.1.1.	General	496
2.1.2.	*A. nummularia* Alliance	496
2.1.3.	*A. vesicaria* Alliance	497
2.1.3.1.	*A. vesicaria* Suballiance (on deep clays in the east)	499
2.1.3.2.	*A. vesicaria – Bassia* spp. Suballiance	501
2.1.3.3.	*A. vesicaria – Ixiolaena leptolepis* Suballiance	503

2.1.3.4.	*A. rhagodioides* Suballiance	504
2.1.3.5.	*A. hymenotheca* Suballiance	504
2.1.3.6.	*A. vesicaria* – *Maireana astrotricha* Suballiance	504
2.2.	Alliances Dominated by *Maireana (Kochia)* spp.	506
2.2.1.	General	506
2.2.2.	*M. sedifolia* Alliance	506
2.2.3.	*M. astrotricha* Alliance	508
2.2.4.	*M. pyramidata* Alliance	508

20. The Natural Grasslands and Savannahs

1.	Introduction	510
2.	The Grass Flora	510
3.	Habits of Australian Grasses	512
4.	Distribution of the Grasses and Grasslands in Australia.	514
5.	Descriptions of the Alliances	515
5.1.	Littoral Grasslands	515
5.1.1.	*Spinifex* spp. Alliance	515
5.1.2.	*Sporobolus virginicus* Alliance	516
5.1.3.	*Xerochloa* spp. Alliance	517
5.1.4.	*Aristida hirta* – *A. superpendens* Alliance	517
5.1.5.	*Themeda australis* on Headlands	517
5.2.	Grasslands in Fresh-water Aquatic Habitats or on Soils which are Periodically Flooded	517
5.2.1.	*Oryza australiensis* Alliance	519
5.2.2.	*Phragmites* spp. Alliance	519
5.2.2.1.	*P. karka* Suballiance	519
5.2.2.2.	*P. australis* Suballiance	520
5.2.3.	*Themeda australis* – *Eriachne burkittii* Alliance	520
5.2.4.	*Chionachne cyathopoda* Alliance	520
5.2.5.	*Pseudoraphis spinescens* Alliance	520
5.2.6.	*Leptochloa digitata* Alliance	520
5.2.7.	*Eragrostis australasica* Alliance	521
5.2.8.	Other Species of Limited Occurrence	521
5.3.	Grasslands Occurring on Fine-textured Soils mainly in the Semi-arid Zone	522
5.3.1.	Distributions of the Grasslands and their Interrelationships	522
5.3.2.	*Dichanthium* spp. Alliance	522
5.3.2.1.	*D. tenuiculum* Suballiance	523
5.3.2.2.	*D. sericeum* Suballiance	524
5.3.3.	*Astrebla* spp. Alliance	524
5.3.3.1.	General	524
5.3.3.2.	*Astrebla squarrosa* Suballiance	525
5.3.3.3.	*A. elymoides* Suballiance	525
5.3.3.4.	*A. lappaceae* Suballiance	526
5.3.3.5.	*A. pectinata* Suballiance	527
5.3.3.6.	Effects of Grazing on the *Astrebla* Grasslands	527
5.3.4.	*Eragrostis xerophila* Alliance	528
5.3.5.	*E. setifolia* Alliance	528
5.3.6.	*E. dielsii* Alliance	528
5.3.7.	The *Stipa* Grasslands	528
5.3.7.1.	General	528
5.3.7.2.	*Stipa aristiglumis* Alliance	529
5.3.7.3.	*S. scabra* – *S. bigeniculata* Alliance	530
5.3.7.4.	*S. falcata* – *Danthonia caespitosa* Disclimax Grassland	530

5.3.7.5.	Other Species of *Stipa*	530
5.4.	Grasslands and Savannahs of the Arid Zone	532
5.4.1.	General	532
5.4.2.	*Plectrachne* and *Triodia* Alliances – "Spinifex Country"	533
5.4.2.1.	*Plectrachne schinzii* Alliance	534
5.4.2.2.	*Triodia pungens* Alliance	535
5.4.2.3.	*Triodia basedowii* Alliance	537
5.4.2.4.	*Triodia irritans* var. *irritans* Alliance	538
5.4.2.5.	*Triodia mitchellii* Savannahs	539
5.4.3.	*Zygochloa paradoxa* Alliance	539
5.4.4.	The Ephemeral Grasslands	540
5.4.4.1.	General	540
5.4.4.2.	*Sporobolus australasicus* – *Enneapogon* spp. Grassland	540
5.4.4.3.	*Aristida* spp. Grasslands	540
5.4.4.4.	*Enneapogon avenaceus* Grasslands	541
5.4.4.5.	Other Species	542
5.5.	The *Poa* Grasslands	542
5.5.1.	General	542
5.5.2.	*Poa hiemata* Alliance	543
5.5.3.	*Poa labillardieri* Alliance	544
5.5.3.1.	Associations on the Mainland	544
5.5.3.2.	Associations in Tasmania	544
5.5.4.	*Poa poiformis* Alliance	548

21. Communities of the Inland Watercourses, Flood-Plains and Discharge Areas

1.	Introduction	549
1.1.	The Watercourses	549
1.2.	Water-tables and Underground Waters	550
1.3.	Discharge Areas	550
1.3.1.	Playas	550
1.3.2.	Wadis, Minor Watercourses and Fan-Deltas	551
2.	The Plant Communities	551
2.1.	General	551
2.2.	Aquatic Communities	551
2.3.	"Swamp" Communities	552
2.3.1.	*Phragmites australis* Alliance	552
2.3.2.	*Typha domingensis* Alliance	552
2.3.3.	*Marsilea drummondii* Alliance	552
2.3.4.	*Muehlenbeckia cunninghamii* Alliance	554
2.3.5.	*Eleocharis pallens* Alliance	554
2.3.6.	*Chenopodium auricomum* Alliance	555
2.3.7.	*Eremophila maculata* Alliance	555
2.3.8.	*Eragrostis australasica* Alliance	556
2.4.	Communities on Clays in Flooded Areas, Mainly Channel Country and Playas	556
2.4.1.	Ephemeral Communities in the Channel Country	556
2.4.2.	Communities on and around Playas	557
2.4.2.1.	General	557
2.4.2.2.	*Arthrocnemum* spp. – *Pachycornia tenuis* Alliance	558
2.4.2.3.	*Frankenia* spp. Alliance	559
2.4.2.4.	Communities of the Subsaline Zone	559
2.4.2.5.	Communities on Upper Beaches	559
2.4.2.6.	Communities on Sandy Areas Surrounding Playas	560
2.5.	*Eucalyptus* Communities Fringing Watercourses and on Flood-plains	561

2.5.1.	*Eucalyptus camaldulensis* Alliance	561
2.5.2.	*E. rudis* Alliance	564
2.5.3.	*E. microtheca* Alliance	565
2.5.4.	*E. largiflorens* Alliance	567
2.6.	Communities of Minor Watercourses and other Small Irrigated Areas	568
2.6.1.	Minor Watercourses	568
2.6.2.	Semi-permanent Rock Holes and Deep Gorges	569
2.6.3.	Rock-crevice and Boulder Communities	569

22. Communities in Fresh or Brackish Water, Mainly on the Coastal Lowlands, Including Lagoons, Lakes, Rivers, Swamps and Flooded Areas

1.	Introduction	570
1.1.	General	570
1.2.	The Flora	570
2.	Communities in the Tropics	571
2.1.	General	571
2.2.	*Nymphaea gigantea* – *Nelumbo nucifera* Alliance	573
2.3.	Grassland – Sedgeland Communities	574
2.4.	*Pandanus* spp. Alliance	574
2.5.	*Livistona humilis* Alliance	575
2.6.	*Tristania lactiflua* – *Grevillea pteridifolia* – *Banksia dentata* Alliance	576
2.7.	Communities Dominated by *Melaleuca*	577
2.7.1.	General	577
2.7.2.	*Melaleuca leucadendron* Alliance	577
2.7.3.	*M. viridiflora* Alliance	579
2.7.4.	*M. minutiflora* Alliance	580
3.	Communities in the East from South-eastern Queensland to South Australia and Tasmania	580
3.1.	Communities of Coastal Brackish Lakes and Estuaries	580
3.2.	Communities of Fresh Water Lakes and Lagoons	582
3.3.	The Sedgelands	584
3.3.1.	General	584
3.3.2.	*Baumea juncea* Alliance	587
3.3.3.	*Calorophus minor* – *Leptocarpus tenax* Alliance	587
3.3.4.	*Gymnoschoenus sphaerocephalus* Alliance	588
3.3.5.	*Gahnia trifida* – *G. filum* Alliance	592
3.3.6.	*Lomandra dura* – *L. effusa* Alliance	593
3.4.	Communities Associated with Rivers	593
3.4.1.	General	593
4.	Communities in South-western Western Australia	594
4.1.	General	594
4.2.	Communities in Permanent and Semi-permanent Fresh Water	595
4.3.	Sedgelands	595
4.3.1.	General	595
4.3.2.	*Leptocarpus aristatus* Alliance	595
4.3.3.	*Evandra* – *Anarthria* – *Lyginia* spp. Alliance	595
4.4.	Communities Dominated by *Melaleuca*	597
4.4.1.	*Melaleuca raphiophylla* Alliance	597
4.4.2.	*M. preissiana* Alliance	597
4.4.3.	Other Species of *Melaleuca*	597
4.5.	Communities Dominated by *Banksia*	597

23. Communities of the Littoral Zone

1.	Introduction	598
1.1.	General; Ocean Currents; Climates; Tides; Sea Water	598
1.2.	Habitats	599
1.3.	Flora	599
2.	Communities on Rocky Outcrops and Headlands	601
3.	Communities on Sand Dunes	602
3.1.	General	602
3.2.	Zonation of the Communities	605
4.	Communities on Mudflats	611
4.1.	General	611
4.2.	The Sea-grasses and Marine Meadows	611
4.2.1.	General	611
4.2.2.	Communities in the North-east	613
4.2.3.	Communities in the South	613
4.3.	Mangroves and Mangrove Communities (Mangals)	615
4.3.1.	General	615
4.3.2.	Fruits, Seeds and Germination	616
4.3.3.	Tolerance to Flooding; Root Systems; Reactions to Salinity	617
4.3.4.	The Mangals: General	619
4.3.5.	Zonations and Alliances in the Tropics	621
4.3.5.1.	*Sonneratia caseolaris* Alliance	621
4.3.5.2.	*Avicennia marina* var. *resinifera* Alliance	621
4.3.5.3.	*Rhizophora* spp. Alliance	621
4.3.5.4.	*Bruguiera* spp. Alliance	621
4.3.5.5.	*Ceriops tagal* Alliance	622
4.3.5.6.	The Inner Zone (Landward Fringe)	622
4.3.5.7.	*Nypa fruticans* Alliance	624
4.3.6.	Zonations and Alliances South of the Tropic	624
4.3.6.1.	General	624
4.3.6.2.	*Avicennia marina* var. *australasica* Alliance	625
4.4.	Mangrove Islands	627
4.5.	Communities Adjoining Mangroves on the Landward Side	628
4.6.	The Samphire, Sedgeland and Grassland Communities	628
4.6.1.	General	628
4.6.2.	Communities in the Tropics	628
4.6.3.	Communities South of the Tropic	631
5.	Communities and Flora of Small Islands	632
5.1.	General	632
5.2.	Torres Strait Islands	633
5.3.	Coral Cays	635
5.4.	Fraser Island (a sand island)	635
5.5.	The Five Islands (rocky)	636
5.6.	Granite Islands in South-eastern Victoria	636
5.7.	Kangaroo Island	636
5.8.	Coastal Islands near Fremantle	637
5.9.	Dorre Island	637

References ... 639

Index ... 657

Abbreviations

A
A.C.T.   Australian Capital Territory
Afr.   Africa
alt.   altitude
Amer.   America
approx.   approximately
Austr.   Australia
av.   average

B
B.P.   before present

C
c.   about
Centr.   Central
cf.   compare
Chap.   Chapter
cm   centimetre
cm^2 square cm
cm^{-2} per square cm
cosmopol.   cosmopolitan
C.S.I.R.O.   Commonwealth Scientific and Industrial Research Organization

D
diam.   diameter
dil.   dilute

E
E.   east
Ed.   Editor
Edn   edition
e.g.   for example
equiv.   equivalent
espec.   especially
et al.   and others
etc.   et cetera
Eur.   Europe
excl.   excluding

F
Fig.   Figure

H
ha   hectare(s)
Hemisph.   Hemisphere

I
ibid   in the same place (as in a book or journal)
i.e.   that is
Is.   Island

K
km   kilometre
km^2 square km

L
Lat.   Latitude
Long.   Longitude

M
m   metre
m^2 square m
Madagas.   Madagascar
Malay.   Malaysia(n)
max.   maximum
min.   minimum
mm   millimetre
mm^2 square mm
Mt.   Mount(ain)

N
N.   north(ern)
N.Cal.   New Caledonia
n.d.   no date
NE   north-east
N.G.   New Guinea
No(s)   number(s)
N.S.W.   New South Wales
N.T.   Northern Territory
NW   north-west
N.Z.   New Zealand

P
p.   page
Pacif.   Pacific (Ocean)
Philipp.   Philippines
Polynes.   Polynesia
pp.   pages
ppm   parts per million
Publ.   Publisher

Q
Q. or Qld   Queensland
q.v.   which see

R
R.   River

S
S.   south(ern)
S.A.   South Australia
S.Afr.   South Africa
SE   south-east
Ser.   Series
sp.   species (singular)
spp.   species (plural)
ssp.   subspecies
Str.   Strait
syn.   synonym

T
Tas.	Tasmania
temp.	temperature

U
Univ.	University
unpubl.	unpublished
U.S.	United States of America

V
v.	versus
var.	variety
Vic.	Victoria
viz.	namely

W
W.	west(ern)
W.A.	Western Australia
W. Afr.	West Africa
wt.	weight

=	equal
<	less than
>	greater than
°C	degrees Celsius
%	percentage
*	introduced
#	endemic

1. The Australian Environment

1. Location of Australia and its Origin

Australia presently lies between lat. 10°34′ S. (Thursday Is.) and 43°40′ S. (Maatsuyker Is.) and together with New Guinea, Tasmania and a large number of smaller islands on the continental shelf (Fig. 1.1) is part of one of the earth's crustal plates.

Geological and geophysical data collected over the past couple of decades indicate that prior to the Triassic Period (195–225 million years ago, see Table 1.1), the world's crustal plates were connected to form one vast landmass (Pangea) which rifted to form two supercontinents, Laurasia (comprising North America and Eurasia, excluding India), and Gondwanaland (comprising Africa, India, Australia, South America and Antarctica) (Fig. 1.2). The two supercontinents again fragmented, the fragments moving apart to form the

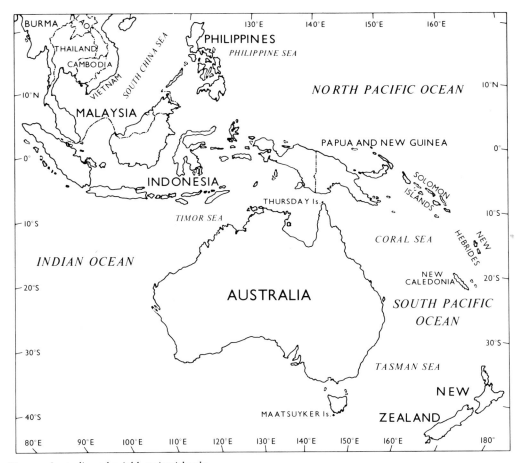

Fig. 1.1. Australia and neighbouring islands.

Table 1.1. Geological time-scale.
At left: Eras, Periods, and Epochs from Pre-Cambrian to the Cainozoic Era.
At right: Epochs of the Cretaceous, Tertiary and Quaternary Periods. Ages in millions of years.

Era	Period	Epoch	Age (millions) of years	Era	Epoch	Age (millions) of years
CAINOZOIC	Quaternary	Holocene	0–3	QUATERNARY	Holocene	0
		Pleistocene			Pleistocene	
	Tertiary	Pliocene	3–7	TERTIARY	Pliocene	10
		Miocene	7–26		Miocene	20
		Oligocene	26–38		Oligocene	30, 40
		Eocene	38–53		Eocene	
		Palaeocene	53–65		Palaeocene	50, 60
MESOZOIC	Cretaceous		65–136	UPPER CRETACEOUS	Maestrichtian	70
	Jurassic		136–195		Campanian	
	Triassic		195–225		Santonian	80
PALAEOZOIC	Permian		225–280		Coniacian	90
	Carboniferous		280–345		Turonian	
	Devonian		345–395		Cenomanian	100
	Silurian		395–440	LOWER CRETACEOUS	Albian	
	Ordovician		440–500		Aptian	110
	Cambrian		500–570		Barremian	120
PRE-CAMBRIAN	Proterozoic		570–2,400		Hauterivian	
	Archaean		2,400–>3,000		Valanginian	130
					Ryazanian	140

present-day continents. This theory of continental drift, first contemplated in the 17th Century, was promoted by Alfred Wegener in 1922, but the speculation lacked convincing data until the work on palaeomagnetism was developed. The historical development of this work is documented in Tarling and Tarling (1971).

As far as Australia is concerned, the events leading to the formation of the modern continent can be summarized in the reconstruction of Gondwanaland. The rifting of the supercontinent commenced in the Triassic (Green, 1975) (see Table 1.1), separating West Gondwana (South America and Africa) from East Gondwana by the formation of the Indian Ocean rift, and, by the Cretaceous (Sutton, 1970), the dis-

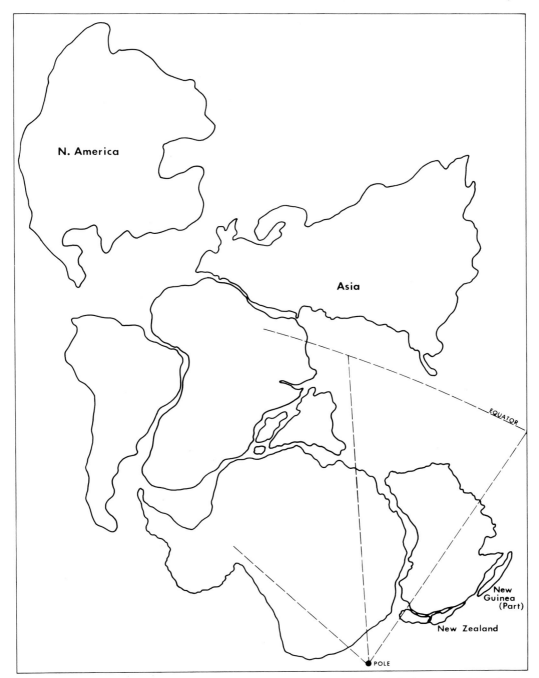

Fig. 1.2. Sketch map showing the relative positions of the land masses and islands which constituted Gondwanaland in the early Cretaceous (c. 120 million years ago). North America and Asia shown in approximate positions. Redrawn, with some modifications, from Smith et al. in Hughes (1973). New Guinea and New Zealand located by R. T. Green and H. J. Harrington (personal communication).

persion of the East Gondwana plate is clearly under way.

Following the separation of Africa and South America as a unit, India separated from the southern landmass and moved rapidly northward (10 cm per year), finally to impinge upon Asia. During the Tertiary, the various continents continued to separate and the intervening oceans continued to widen.

The latitudinal position of the Australian plate and its relationships with Antarctica, New Guinea and New Zealand, as well as its distance from the Asian plate, are important for the interpretation of the late Cretaceous and early Tertiary floras of Australia.

The location of Australia relative to the South Pole is uncertain and various positions have been postulated by geophysicists, including the Pole close to the south-east coast of Australia in the Jurassic – Cretaceous. This implies a northern movement of the continent during the Cretaceous and Tertiary, possibly through 30° degrees of latitude.

The distances between Asia and Australia are also uncertain; biological data, chiefly the absence of placental animals in Australia, indicates an age-old ocean barrier between the two continents.

New Zealand, once part of Gondwanaland, is thought to have separated from West Antarctica in the early Tertiary (late Palaeocene, c. 63 million years ago – Harrington, 1975). It moved to its present position in the early Tertiary, whereas Australia continued its northerly drift throughout the whole of the Tertiary. New Caledonia consists of continental crust and is considered to be a portion of Gondwanaland, which is confirmed by the composition of its flora.

New Guinea, though represented on some

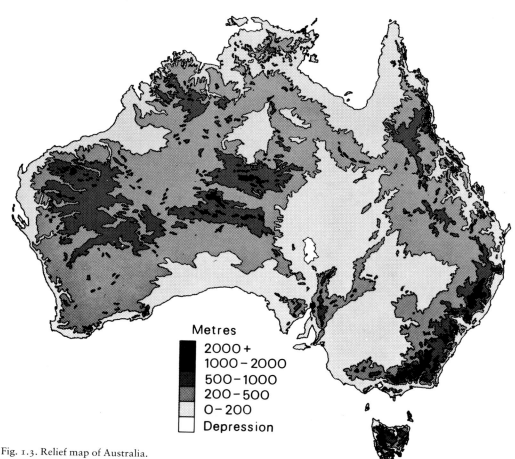

Fig. 1.3. Relief map of Australia.

maps (see Smith et al, in Hughes, 1973) as a fixed landmass relative to the Australian continent, is thought by others (Green and Pitt, 1967; Harrington, 1975) to have developed from two segments, the more southerly of which (Papua) once lay under the ocean along the north-east coast of Queensland. This segment of land separated from the Australian coast and rotated through 90° in an anticlockwise direction relative to Australia to occupy its present position, where it impinged upon a chain of volanic islands now incorporated in New Guinea. The geologically composite island of New Guinea consequently did not exist until the mid-Tertiary; this nullifies or diminishes the significance of this island as a migration route for plants or animals during the Cretaceous and early Tertiary.

In addition to the northern migration of the Australian plate possibly through 30°, the plate has rotated in an anticlockwise direction relative to the Mesozoic latitudinal zonations (Green, 1975). Thus, the western part of the continent, and particularly the south-western "corner", formerly occupied a more northerly position relative to the rest of the continent. The south-east part of the continent (Victoria and Tasmania) has always lain in the higher latitudes. These observations have bearing on the former climates of the various parts of the continent and consequently on the flora, as discussed in Chap. 3.

The account given above indicates that Australia and New Guinea, still lying on the same plate, have been isolated from the other continents and from New Zealand since the early Tertiary. Modern Australia is characterized by general crustal stability, though it is subjected to minor earthquakes. There are no active volcanoes, the nearest being in New Guinea.

2. Topography: Rivers

2.1. Topography

Most of the continent lies at altitudes below 300 m and a considerable area lies below 200 m (Fig. 1.3). A small area around Lake Eyre lies below sea level. These low-lying areas correspond approximately with the one-time location of vast inland lakes and seas which have slowly been filled with sediment (Fig. 1.4). A large proportion of the inland is covered by sand deposits which have been blown into dunes in the past, though most of these are now stabilized by vegetation. The locations of the main dune systems are given in Fig. 1.5. The monotony of this flat inland landscape is occasionally broken by emergent low ridges or isolated hills which are the residuals of ancient ranges or plateaux (Fig. 1.6).

Mountainous country occurs on the mainland only in the east, especially the south-east where several mountain tops and plateaux exceed 2000 m in height (Fig. 1.7), the highest being

Fig. 1.4. Part of the Lake Eyre basin. Lake Callabonna in the distance on the left.

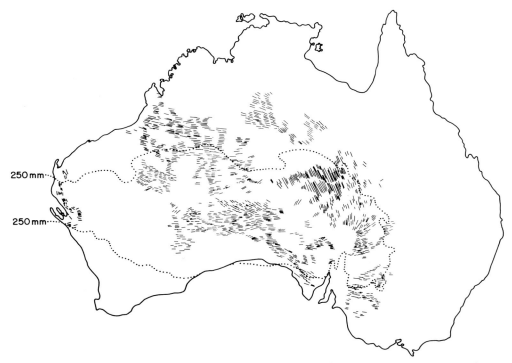

Fig. 1.5. The inland sand dune systems. Redrawn from Mabbutt and Sullivan, using unpublished data of Ruxton and Loeffler, in Moore (1970).

Mt. Kosciusko standing at 2216 m. Rugged mountainous terrain occurs also in Tasmania, though the highest peak, Mt. Ossa, stands only 1607 m above sea level. None of the mountains is high enough to support permanent snowfields.

On the mainland the mountainous area in the south-east lies at the extremity of the eastern Great Divide, which extends northward close to the ocean for the full length of the continent. This range, which is a tableland over much of its length, determines the pattern of rivers in the eastern half of the continent. Higher land occurs also in the south-west, the north-west and the centre, where the hills represent the oldest (Precambrian) parts of the continent.

These elevated areas determine the direction of flow of the rivers in these areas.

2.2. Rivers

The Australian rivers (Fig. 1.8), in comparison with those in other continents, carry relatively small volumes of water. Indeed, most of the longest do not flow the year round and many of them in times of drought are no more than a chain of elongated waterholes (Fig. 1.9). Since rainfall is highest along the eastern coastal lowlands and on the adjacent range, which constitutes a watershed in the east, the largest rivers occur in this area. The easterly flowing rivers travel over relatively short distances into the Pacific Ocean. Many of these rivers are antecedent streams which have cut their way through the mountains, forming deep gorges. Others, especially those with sources on lower parts of the dividing range, meander across broad valleys.

Water which falls west of the eastern Great Divide flows westward, and the rivers drain 3 main basins. In the north, a number of slow-flowing streams empty into the Gulf of Carpentaria. The rivers in central Queensland, supplemented by water from the north and some from the west, drain into Lake Eyre, which lies c. 14 m below sea-level. These rivers are irregular in their flow and spread over vast areas of flat (Channel) country which is described in

Fig. 1.6. Two landmarks of the inland. Above: Ayers Rock, N.T.; the rock is conglomerate. Below: Mt. Conner, N.T.; the mesa is capped with silica, probably the remnants of an ancient laterite profile.

some detail in Chap. 21. The Murray-Darling system in the south is fed from the eastern and south-eastern highlands and the combined waters ultimately reach the ocean at Encounter Bay.

The western and central segments of the arid zone and the Nullarbor Plain are completely devoid of rivers, or even watercourses, this being due partly to the low rainfall, and, in the sandy central and north-western segments, to the high permeability of the sands from which there is no runoff. These sandy areas, however, show evidence of prior streams.

3. Some Comments on the Geomorphology and Geology

The notes below are provided as a basis for the understanding of the origins of the modern flora and ecosystems. They are based mainly on the works of David and Browne (1950), Noakes (1966) and Brown et al. (1968).

Structurally the continent is made up of 3 major compartments (Mabbutt and Sullivan,

Fig. 1.7. The Kosciusko highlands, N.S.W. View from Mt. Kosciusko. *Eucalyptus niphophila* in the foreground.

Fig. 1.8. The main river systems.

Fig. 1.9. Darling R. south of Menindee, N.S.W. showing the dry bed. The trees are *Eucalyptus camaldulensis* (River Red Gum). Photo taken October, 1940.

1970): the stable western platform (Precambrian shield); the central basin of gently warped, relatively young sedimentary rocks; the eastern uplands. The Precambrian shield occupies approximately the western half of the continent, and through the ages this has been the most stable part, especially the south-west. This area is dominated by a vast basement of Archaean granite and gneiss, and it is the only part of the continent which has never lain under the sea. The original distribution of the Archaean rocks is obscure, but Noakes suggests that they extended eastward from the known outcrops and that the crust was broken into huge blocks which sank, later to be buried by sediments. At all events, the east and particularly the southeast lay under water in a trough, the Tasman Geosyncline. Precambrian rocks occur at the surface today in the west, with smaller occurrences in Northern Territory, northern Queensland, South Australia and extending into western New South Wales, and in western Tasmania. The rocks are mainly metamorphics, including gneisses, schists and slates, but sedimentary rocks also occur, especially sandstones, shales, dolomite and greywackes.

The building of the eastern half of the continent, which involved the filling of the Tasman Geosyncline (Harrington, 1974), proceeded over a period of over 400 million years (Cambrian to Jurassic) by sedimentation, intrusions and vulcanism. Sedimentation was most active during the Permian and Triassic. Permian deposits have a maximum depth of 7000 m and some of these contain coal built up from gymnospermic and pteridophyte material. The rocks are mainly limestones, sandstones, conglomerates and greywackes. The intrusions, occurring mainly in late Silurian and Devonian, were largely granites. Vulcanism occurred during all the geological periods, the lavas being mostly acidic but some were basic, including, for example, the Cambrian basalts in the north-west and the Permian basalts in the south-east, of which only tiny remnants remain today.

The Jurassic has special significance botanically through the origin and early diversification of the angiosperms. Geologically the Australian Jurassic is noted for widespread crustal sagging between the stable western shield and the eastern highlands. Depression of the continent, mainly in the east, led to the development of swamps and lakes in an area corresponding to the Great Artesian Basin. Extensive sediments were laid down, including some coal, the deposits being up to 2400 m deep. Igneous rocks were injected into the Permian and Triassic sediments, especially dolerite in Tasmania.

The crustal sagging initiated during the Jurassic continued during the Cretaceous and two additional basins were added in the south, the Eucla and Murray basins. Marine shales, sandstones, calcareous sediments and some conglomerates were laid down in all these basins. Towards the end of the Lower Cretaceous much of the continent began to rise, but the Great Artesian Basin persisted as a lake during the Upper Cretaceous and into the Tertiary.

The Cretaceous has considerable significance botanically. During this period the angiosperms had differentiated into a very large number of families most of which were probably widely distributed across Gondwanaland, and probably Laurasia as well. Also, Gondwanaland fragmented and the southern continents drifted apart, leaving Australia, possibly still attached to Antarctica, as an isolated landmass at the end of the Cretaceous. The separation of Australia and Antarctica at the end of the Cretaceous or very early Tertiary (H. J. Harrington, personal communication) established Australia as an isolated continent.

During the Tertiary the continent was considerably modified by its elevation, additions of sedimentary rocks along parts of the coast line, and through volcanic activity. General uplift of the continent, especially in the south, formed a connection between Tasmania and the mainland. The mainland was effectively drained, so that seas and fresh water lakes became progressively smaller during the Tertiary. Local orogenies involved the elevation of the eastern Great Divide (Kosciusko uplift), especially in the south, and the elevation of the Flinders Range in South Australia. At the same time, local subsidences occurred, one to sever Tasmania from the mainland, the other to form the Gulf of Carpentaria. The two stretches of water formed by these subsidences present biological barriers to migration and the disjunction of some plant taxa.

Additions to the coast occurred mainly in the south-west and south. Large areas of marine limestone were elevated along the west coast and in the Eucla (Nullarbor) region and Murray Basins. The Nullarbor limestones are of special significance because they have provided an abundant source of lime which has been blown eastward and deposited over relatively large areas, where the lime has had considerable effects on the modern soils.

Volcanic activity occurred throughout the Tertiary. It was confined to the east, with very minor activity in the south-west. The lava was chiefly basalt and in places covered considerable areas, though much of it has now been eroded away. Many basalt dykes and minor intrusions also occurred. The basalts are high in plant nutrients and produce fertile soils which, in suitable climates, support the richest of the Australian rainforests. Also, other types of lava were produced, probably all in the Lower Tertiary and many of the volcanoes can be located today by the residual volcanic plugs (Fig. 1.10).

Lastly, during the Pleistocene glaciation, falls in sea level, amounting to a maximum of c. 100 m, periodically connected both Tasmania and New Guinea with the mainland. This has considerable significance for the migration of plants and animals from islands to the mainland or vice versa, including the migration of aboriginal man from the mainland to Tasmania.

Fig. 1.10. Glasshouse Mountains – volcanic plugs of trachyte. South-eastern Queensland.

4. Climate

4.1. General

The climatic analysis presented here is designed to quantify those factors which are important in determining the distribution of the natural plant communities. Water and temperature are dealt with in more detail than the other factors, since these two determine the vegetation zones and growth patterns. Other factors are mentioned only cursorily here, but in some cases are treated in more detail in later chapters when they have some special effect locally. The quantitative data which have been used in the compilation of graphs and maps are those published by the Bureau of Meteorology.

In most cases mean monthly and mean annual figures are used, and mention is made here of the possible misleading impressions that mean values may give, as discussed in appropriate places below. However, for vegetation studies, far more than for agricultural applications, mean values, particularly means of annual rainfall, provide more reliable parameters for correlation with vegetation than they do for crops. This is due to the fact that the component species of natural plant communities are adapted to the environment, including the vagaries of the climate particularly the highly variable rainfall, to a much greater extent than are crop plants, for which climatic analyses are commonly designed (Fitzpatrick and Nix, 1970; Leeper, 1970). Crop plants are mainly introduced species with short life-cycles, many being

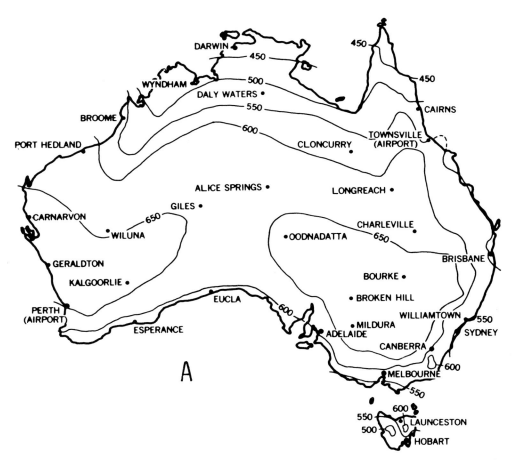

Fig. 1.11. Average distribution of total radiation in calories cm^{-2} per day. A. for January; B. for July. Data from Bureau of Meteorology, Bull. No. 1 (1965). Redrawn from Fitzpatrick and Nix, in Moore (1970).

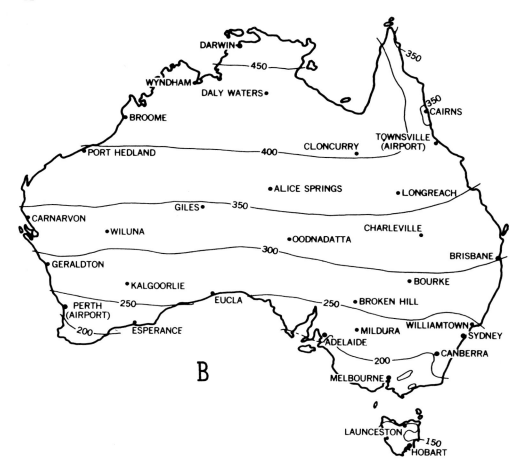

annuals with seasonal growth, in contrast to the native, perennial, long-lived species which have evolved mechanisms for combatting the extremes of the climate.

4.2. Solar Radiation

The amount of energy received at any given point on the earth's surface depends on the latitude, the season, and the cloud cover. As the sun makes its apparent movement to the Tropic of Capricorn in mid-summer (Australian conditions) day-length increases to almost 13 hours; when the sun is directly overhead at noon, the daily energy-input at sea-level at the Tropic increases to c. 660 cal. cm^{-2}. South of the Tropic day-length increases still further to c. 15$^{1}/_{2}$ hours in Tasmania (Hobart). Though the sun's rays pass obliquely through the atmosphere, when more energy is absorbed by the atmosphere, the increased day-length compensates for the energy absorption, with the result that solar radiation is more or less constant from the equator to c. lat. 43$^{1}/_{2}$°S., the southern tip of Tasmania. Consequently the whole continent would receive approximately the same amount of solar energy in mid-summer, if the skies were always cloudless.

The solar energy pattern for the continent is given in Fig. 1.11. This shows a reduction in sunshine in the north in the summer (Fig. 1.11A), which is due to the high incidence of cloud, and a similar screening effect by cloud is seen along the east and south coasts and in Tasmania. The reason for the two highest sunshine areas not meeting in Central Australia is due to the passage of cloud over this area during mid-summer.

With the return of the sun to the north after

the summer solstice (21st December), days become shorter in the Southern Hemisphere and the sun's rays become increasingly oblique until 21st June. In mid-winter day-length north of the Tropic is c. 11 hours and in Tasmania c. 9 hours (Hobart). Thus the receipt of solar energy is greatly reduced in the south by the position of the sun, and the input of solar energy is further decreased by the regular clouding of the skies (Fig. 1.11B). However, the clouds, in addition to bringing rain, have a blanketing effect on the land and tend to conserve heat and to avert frosts by reducing heat-radiation to the upper atmosphere.

These data explain the temperature patterns over most of the continent, excluding higher altitudes (see below), but they do not separate the effects of visible light from those associated with heat. Summer light régimes are clearly adequate over the whole continent, but the visible light component might well be deficient in the south in winter in areas where temperature is not limiting growth. That this possible inadequacy in light is connected with cloud is suggested by the retardation of crop-growth during long periods of cloudy weather when temperatures are adequate. Natural communities are likely to respond in a similar manner. However, there is little evidence to indicate that light intensity alone determines the distribution or zonation of communities in any area, though day-length possibly plays some part in controlling flowering of some taxa.

The effects of reduced light intensity in determining specialized plant communities are occasionally seen in intensely shaded areas such as

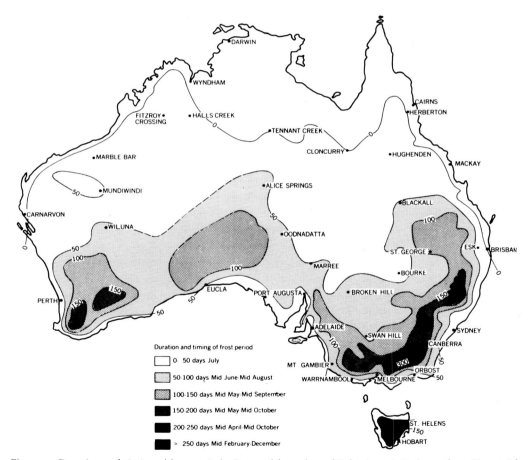

Fig. 1.12. Duration and timing of frost periods. Prepared from data of Foley (1945). Redrawn from Fitzpatrick and Nix, in Moore (1970).

deep ravines or caves. Aspect differences such as south-facing slopes with lower illumination as opposed to north-facing slopes with high illumination often lead to differences in community structure or floristics, but these differences can usually be interpreted by differences in exposure to wind or in temperature, rather than light.

The most noticeable effects of reduced light intensity are seen within communities, especially rainforests, where the density of the tree-canopy determines the microclimate and the understory species.

4.3. Temperature

Temperatures over most of the continent are relatively high, though a considerable range of temperatures occurs. The highest summer maximum ever recorded is 53°C for Cloncurry in western Queensland; the lowest minimum is −22°C at Charlotte's Pass (Kosciusko area) in New South Wales. Mean summer maximum temperatures vary between 33° and 13°C, the highest summer temperatures being recorded in the tropical segment of the arid/semi-arid zone. At Marble Bar, for example, the mean maximum summer temperatures lie around 40°C for 6 months (October to March). In the wetter tropics mean summer temperatures are somewhat lower (32–35°C in the west and 28–31°C in the east).

Highest maxima over the whole continent are fairly uniform as a result of the occasional movement of hot, dry air-masses from the arid zone to the south and east. Almost all stations, including Hobart, have recorded a highest maximum of 40°C or above, excepting the higher stations on the south-eastern highlands (and probably the mountain plateaux of Tasmania where there are no recording stations).

The isotherms run mostly from west to east, so that mean summer temperatures decrease from the Tropic southward, with the coolest areas in the south-east at higher altitudes.

Minimum temperatures are more important with regard to plant distribution. The littoral zone and the wetter tropics are frost-free or almost so (Fig. 1.12). Inland, however, subzero temperatures are recorded in the tropics, especially at higher altitudes, and the severity of frosts increases towards the south, especially in the east. Mean minimum temperatures are most misleading as illustrated by the following comments on the Armidale (lat. 30°31′ S., altitude 980 m) winter climate: The mean minimum temperature for the coldest month (July) is 1.4°C. However, the lowest minimum is −10°C and, although this is not experienced each year, temperatures between −5° and −8°C are recorded on many occasions each winter and a frost could occur on any day between 21st March and 4th December.

There are no permanent snowfields either on the mainland or in Tasmania, but snow may lie for several months at higher altitudes on the south-eastern highlands and mountain plateaux of Tasmania where snow-drifts may sometimes persist on south-facing slopes throughout the summer.

Frost damage to plant communities is unusual but may occur with a very early frost, presumably before plant organs have become "hardened" by slowly diminishing temperatures, or with late frosts in the spring which may destroy young foliage. The destruction of mature trees by exceptionally low temperatures, experienced before a blanket of snow has fallen, has been reported in subalpine woodlands. Also, in the arid zone young seedlings may be destroyed by the occasional heavy frost (temperature c. −5°C) sometimes occurring in mid-winter.

4.4. Precipitation

4.4.1. Mean Annual Rainfall and Seasonal Incidence

The isohyets run more or less parallel with the coast (Fig. 1.13). The recorded extremes of mean annual rainfall are 3479 mm at Innisfail (eastern Queensland), and c. 125 mm at William Creek, South Australia. The following zones of mean annual rainfall are defined: the *arid zone* includes that area which receives a mean annual rainfall of less than 250 mm; the *semi-arid zone* is the area bounded by the 250 and 500 mm isohyets; areas receiving a mean annual rainfall exceeding 500 mm are referred to as *humid* or *wetter*.

Seasonal rainfall patterns vary with latitude. In general terms, summer rainfall predominates in the north, and winter rainfall in the south, except along the east coast and adjoining hills and mountains, where the rainfall is more or less

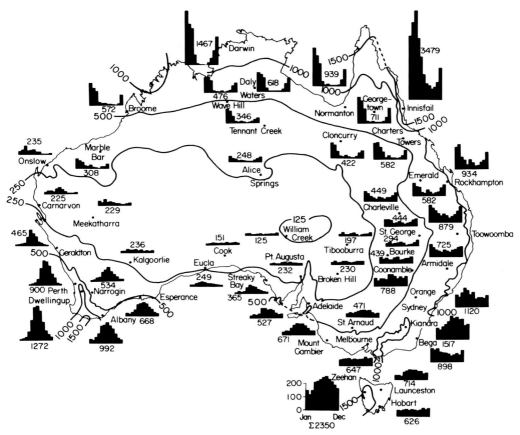

Fig. 1.13. Isohyets and diagrams showing the mean monthly rainfall of selected recording stations. For vertical scale, see Zeehan, Tasmania. The mean annual rainfall is recorded under each diagram.

evenly distributed throughout the year. The seasonal distribution is indicated for selected recording stations in Fig. 1.13. Rainfall in almost every area is very variable, the reliability of both annual and monthly total decreasing from wet to arid. Consequently mean values, especially in the arid zone, give a distorted picture of the effectiveness of the rainfall.

4.4.2. Explanation of the Seasonal Rainfall Patterns

Rainfall over the continent is governed mainly by latitude which determines the paths of the low pressure systems over the continent.

In the north the climate is dominated during the summer by the monsoonal low-pressure systems induced by the heating of the tropical part of the continent. The low-pressure cells lead to precipitation, which is highest in the north and least at the Tropic; sometimes the rain extends further south, but the amounts are small, though possibly significant in the arid interior. In the winter the tropical area is dominated by high pressure cells and the rainfall is scant. Consequently the northern part of the continent experiences a predominantly summer rainfall, with dry winters, the extremes of wet and dry occurring in the extreme north.

In addition to the monsoonal rains, tropical cyclones (hurricanes) of considerable intensity strike the land mostly on the north-western and north-eastern coasts. These are described by Gentilli (1972) as intense tropical lows involving large-scale mass-rising of air. Once they have reached great intensity (wind velocities up to 150 km per hour) their centres move more rapidly than those of tropical depressions. He estimates that their frequency is about one every 2 years and that their contribution to the rainfall may be 100–150 mm in an annual total of 800 to 1000 mm. In the west, tropical cyclones commonly strike the coast at Onslow (with an arid climate – Fig. 1.13), and move in a north-easterly direction along the coast, but they sometimes move inland and deposit rain in the arid zone, and their effects sometimes extend to the southern coast. The paths of cyclones are unpredictable. Slatyer (1962), in his analysis of the climate of the Alice Springs area, reports that here cyclones move in from either Queensland or the Darwin area; their frequency is less than one per year and they deposit 100–200 mm of rain in periods of 24–48 hours.

In the south the position is reversed. Rain occurs mainly in the winter when the climate is dominated by the temperate low-pressure systems. These strike the continent between latitudes 30° and 34° S. depositing rain especially in the south, the amount diminishing along a gradient from south to north (Gentilli, 1972). Thus the arid zone receives less rain from these depressions than do the more southerly tracts of land. The latitude at which the depressions strike the coast determines the amount of rainfall. When the depressions strike at c. 30° S. (between Geraldton and Perth) rainfall is high and the arid zone receives some rain. On the other hand, when the depressions strike the land at c. 34° S. (along the south coast in the west) the rain extends only for short distances inland and a drought occurs, being increasingly intense inland. Both the shape of the coast-line and the occurrence of hills or mountains influence the amount of rain which falls. Land which projects into the ocean receives more rain than does the land adjacent to broad gulfs. Thus, the southwest of Western Australia and Eyre Peninsula, South Australia, have a higher rainfall than the Nullarbor region which adjoins the Great Australian Bight.

Elevated land induces rain by orographic precipitation, when the water-laden air-masses are deflected upward and cool, the water vapour condensing and falling as rain. Thus, the hills east of Perth receive a higher rainfall than the lowlands to the west and the Flinders Range in South Australia extends the semi-arid zone northward into the arid zone (Fig. 1.13). The effects are greater still when the air-masses reach the highlands of Victoria, e.g., at Mt. Buffalo (altitude c. 1350 m), winter precipitation (rain and snow) reaches the high value of 1060 mm for the 5 months May to September. The same occurs on the west coast of Tasmania, where the winter rainfall (5 months) on the lowlands (Zeehan, altitude 175 m) is 1186 mm, while that at Lake Margaret on the highlands c. 20 km east of Zeehan) it is 1745 mm for the same period.

The arid zone lies in an area which is too far south to benefit from the monsoonal rains and too far north and/or too flat in the south to benefit from the southern rains. Most of the precipitation comes from convectional thunderstorms, especially in the northern segment, which is somewhat wetter than the southern segment. Tropical cyclones make an occasional significant contribution to the water supply.

The low ranges in the arid zone have some effect on precipitation. Some of these reach an altitude of c. 1600 m and convectional air currents, greatest around the ranges, induce local thunderstorms more frequently than on the plains. Rain-shadows may occur on one side of the range; for example a summer rain-shadow occurs in the Charlotte Waters area, south of the McDonnell Ranges (Slatyer, 1962).

Along the east coast and adjacent tablelands, the rainfall is higher than that to the west, which is due partly to orographic precipitation associated with the ranges and partly to the additional moisture brought to the continent in winds blowing from the east, the moisture being derived from the Pacific Ocean.

The hills and mountains in the east lie close to the ocean. In the tropics the highest occur around lat. 16–17° S. (west of Cairns – Mt. Bartle Frère standing at 1602 m). The tropical strip of coast and adjacent mountains receives moisture from the trade winds, especially in summer, and this intensifies the pronounced summer rainfall in this region (see Innisfail – Fig. 1.13). Since the trade winds blow throughout the year, a small but significant winter rainfall is also experienced (c.f. Darwin – Fig. 1.13).

South of the Tropic, rain is brought to the coast and adjacent highlands by low pressure systems developed over the Pacific Ocean. The rain occurs mainly in summer, with a smaller winter component. The shape of the coastline and the height of the highlands both affect the amount of precipitation. Coastal land which projects into the ocean usually receives a high rainfall, while the higher mountains receive more rain than low hills as a result of orographic precipitation. Thus around the Tropic, where the Great Divide is low, precipitation inland is low and the semi-arid zone extends eastwards close to the ocean, there being a steep rainfall gradient over the narrow wet coastal belt (Fig. 1.13).

At about lat. 27° S. the range becomes progressively higher towards the south and the total annual rainfall is high, with a highly significant summer component, especially close to the coast and on the higher ranges. A low gap in the range (and lower rainfall) occurs around lat. 32° S., between Cassilis and the Hunter Valley, through which the western semi-arid vegetation has invaded the coastal lowlands.

Along the south-east coast the summer rainfall often equals or even exceeds winter precipitation, e.g. Bega. This coastal strip lies in a rainshadow with respect to the winter rainfall moving from the west, since the rain clouds have shed much or most of their moisture on the highlands. The summer component on this southern strip of coast, together with the warmer winter temperatures derived from the warm East Australian Current, which flows past this part of the coast, accounts for the southern extension of the subtropical climate along the coast, an important factor in the distribution of some plant communities (see Rainforests, Chap. 7).

Rain-shadows occur on the coastal strip and within mountainous country inland. They are areas in which rainfall is low because moisture-laden air-masses are deflected upward by the usually steep adjacent topography; this results in low precipitation on the lowlands and higher precipitation on the uplands. Rain-shadow areas usually support a type of vegetation which is unusually dry for the general area.

4.4.3. Rainfall Reliability

Australia has been described as a "land of droughts and flooding rains" and the variability of the rainfall was the concern of early settlers long before meteorological records were kept. However, the reactions of natural vegetation to both droughts and floods are quite different from man's reactions to these phenomena. Man becomes conscious of drought when water is in short supply for crops or domestic stock. On the other hand, natural vegetation is adapted to the variable seasonal and annual extremes of rainfall and few cases of loss of species or permanent damage to a plant community caused by drought can be quoted, except in areas where communities had previously been altered by white man or domestic stock.

Similarly, floods which devastate man's property over a few days do not damage natural vegetation, except in places where silt is deposited thickly; in such cases it is probable that man's disturbance of the natural vegetation has led to the intensification of the flood and to accelerated soil erosion in the uplands. On the contrary, there are many plant communities which owe their existence to flood-waters, especially some of the woody communities in the arid zone (see 4.5, below).

Generally speaking, winter rain is more reliable than summer rain. Consequently the wetter, more southerly parts of the continent, especially Tasmania, have the most uniform rainfall patterns, the percentage deviation from the annual mean being 15–20% (Leeper, 1970). In the wetter parts of the mainland the percentage deviation varies between 20 and 30%, the annual rainfall varying from less than half to more than twice the mean annual figure. In drier areas this variation increases, so that in the arid zone the annual totals may be as low as one-quarter of the mean and as high as 3 times the mean (Fig. 1.14). Actually, the lowest annual rainfall recorded is zero, experienced in the de Grey district of Western Australia (east of Port Hedland), where a drought occurred with 17 consecutive months without rain, from July 1923 to December 1924 (Dwyer, 1957).

There appears to be no relationship between the extremely low or extremely high values of one year with either the previous or the following year, i.e., an extremely dry year may be followed by an extremely wet year and *vice versa*. The latter is perhaps surprising since one may expect that the high soil moisture levels produced during a wet year could add to local precipitation to the east during the following months or year. This does not occur, nor do the extensive flood-waters which occasionally stretch over vast areas of the eastern segment of the arid zone have any significant effect on local precipitation.

Just as annual rainfall totals vary so also do the monthly totals. This erratic variation is illustrated by the data in Fig. 1.14 and Table 1.2. The perennial species are adapted to these variations in rainfall, so that the effects of wet and

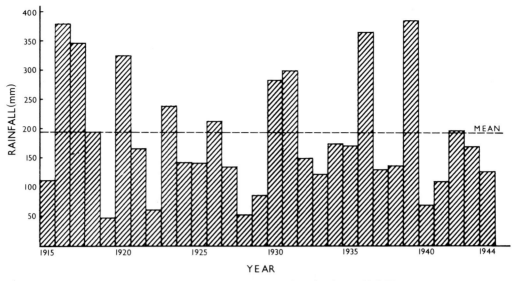

Fig. 1.14. Annual rainfall for 30 successive years (1915–1944) for Tibooburra, N.S.W.

dry months, seasons or years are reflected chiefly through the annual species (see Chap. 5).

The effects of drought are more severe in the wetter areas, where species are less drought resistant, than in the arid zone. For example, the drought of 1965 was responsible for the death of eucalypts in some woodland communities on shallow, stony soils on the southern highlands of New South Wales (Pook et al., 1966), and similar deaths of trees have been recorded on the Northern Tablelands by Boyd (1976).

4.4.4. Dew and Fog

Dews are deposited in most parts of the continent, especially in the south, including the arid zone, where a large diurnal temperature range is experienced. The heaviest dews occur when soil moisture conditions are adequate, but light dews may occur when plants are water-deficient, e.g., during drought periods in most areas, and regularly during the cooler months in the arid zone. The amount of water deposited by dew in a single night probably does not exceed 0.3 mm and is unlikely to come into contact with roots (Warren, et al., 1962).

It is probable that some plants in the arid zone absorb water from dew, but whether the amount absorbed is significant in prolonging the life of the leaf or the plant has not been demonstrated. Wood (1925) has measured increases in leaf weight as a result of exposure of the leaves of Chenopodiaceae to high humidities and regards leaf-absorption as significant in arid areas. However, there is some doubt as to

Table 1.2. Monthly rainfall (mm) for 3 individual years for Tibooburra, N.S.W. 1919 was the driest year for the 30-year period 1915–1944. 1918 represents the year which most closely approaches the mean for the 30-year period, and 1939 was the wettest year for the same period. The mean monthly and mean annual values are given on the lowest line.

	J	F	M	A	M	J	J	A	S	O	N	D	Total for year
1919	3	7	0	3	14	0	0	12	1	0	1	6	47
1918	60	0	2	79	25	0	0	13	10	6	0	0	195
1939	32	97	67	57	38	32	15	7	0	2	44	0	391
Mean values	18	25	18	13	15	19	13	10	10	15	22	22	193

whether the water enters the living cell or is held hygroscopically in salt crystallised on the outsides of the leaves (see Chap. 19).

Absorption of liquid water by leaves is controlled partly or even entirely by the wettability of the leaf surface. Many Australian plants have wettable leaves, which can be observed during rain. On the other hand many species have unwettable surfaces, especially when young, which are produced by: i. Aggregations of wax rods outside the epidermis, as in many glaucous leaves, e.g., some species of *Eucalyptus* (Hallam and Chambers, 1970). ii. Certain types of hairy surfaces; the hairs, which are simple or vesicular, either repel water or hold air and prevent the entry of water films among the hairs. iii. Very shiny, smooth cuticles. These occur on a variety of plants and are most apparent in young leaves, e.g. *Angophora* and most *Eucalyptus* spp. They occur on leaves of all sizes, some of the largest being the water-repellant upper surfaces of leaves of some Araceae growing in rainforests.

Unwettable leaf surfaces often become wettable with age. The wax rods of young glaucous leaves usually drop off, hairs become more distant and cuticles become thinner and possibly minutely undulate as the leaves expand, so that the chances of water absorption through the leaf increase with age.

Fogs are of local occurrence only, and are most common in hilly country where they sometimes lie in valleys in the mornings. On mountain tops fog and low cloud often envelop the vegetation and deposit water on leaves in sufficient quantities to cause dripping from the leaf tips. Fogs, by reducing transpiration, and liquid water deposited on leaves or on the soil doubtless have some significance in the water balance of some communities, but no quantitative data can be presented.

4.5. Natural Irrigation

In most parts of the continent, even in the arid zone, but excluding flat areas or very sandy ones, run-off water from the more elevated areas irrigates rock-clefts, the lower slopes and valley floors. In addition to the extra moisture provided in the lower areas, soil fertility levels are commonly increased downslope by organic matter carried in the run-off water. This natural irrigation accounts for the occurrence of taller or more luxuriant communities than the environment supports in upland sites, e.g., ribbons of rainforest along streams which flow through xeromorphic *Eucalyptus* forest.

The most extensive areas of communities which are sustained by natural irrigation occur in the semi-arid and arid zones. Much of the precipitation (rain and snow) falling westward of the watershed of the eastern highlands is channelled into rivers which flow inland, and in times of high flood vast areas of flat land adjacent to the rivers are flooded. This additional water has significant effects on the vegetation (discussed in Chap. 21).

4.6. Winds

4.6.1. General

As well as carrying water, either in the form of vapour or liquid, winds transport a variety of solid objects, both organic and inorganic. Many large groups of plants are dependent on air movements for pollination, notably the Gramineae, Cyperaceae, *Casuarina* and the gymnosperms. Again, many plants with light seeds or fruits are disseminated by the wind, and the wide distribution of many inland species across the arid zone in Australia is probably the result of wind dispersal of the disseminules.

Strong winds have injurious effects on vegetation by causing direct mechanical damage through the breaking of aerial parts or the uprooting of trees, especially during cyclones which may permanently control the vegetation, as in Cyclone Vine Forests (see Chap. 7). The most destructive cyclone on record is Cyclone Tracey which struck Darwin on Christmas morning, 1974 and destroyed more than half the city; the cyclone destroyed or mutilated vegetation over an area of c. 460 km^2 (Stocker 1976). Also, damage by abrasion caused by the hurling of sand grains or ice crystals against leaves and stems is a common occurrence in areas exposed to high wind velocities, especially along the coast, inland, or at high altitudes. Plants growing in such exposed habitats are commonly adapted to high wind velocities and have developed prostrate habits. Or, if the plants grow erect they become deformed and wind-shorn. Other particles carried in winds are soluble salts (discussed below) and soil particles transported during dust and sandstorms.

4.6.2. Cyclic Salt

Salt solution, picked up as spray from the ocean, carried inland by winds and brought to the earth in rain, is referred to as cyclic salt. A few quantitative estimates of cyclic salt are available (Table 1.3). The salt is estimated as chloride and expressed as sodium chloride, but since it is derived from the ocean it contains essential plant nutrients, especially calcium, magnesium, potassium and sulphate.

Along the coast the amounts of salt deposited are seemingly high, but the actual concentration of salt in the rain approximates 40 ppm (0.004% sodium chloride solution) which could not have any deleterious effects on plants; furthermore no salt accumulation could occur under conditions of free drainage.

Hutton (1976) using both Victorian and other data indicates that the amount of chloride deposited decreases with distance from the ocean. Inland, the position is complicated by the possibility of salt being lifted by the wind from the beds of salt lakes or from soils. This may account for the big difference between the two figures for Roma in Table 1.3.

The interest in cyclic salt lies in the theory that some soil profiles have developed as a result of solonization, a process involving the downward movement of clay rendered more mobile in the profile by adsorbed sodium ions deposited inland as cyclic salt.

Though there is much evidence to support the view that the saltpans in the arid zone are derived mainly from salts carried in streams (Chap. 21), cyclic salt also may have made some contribution to these salt deposits both directly and in local surface drainage waters.

4.7. Growth Régimes (Climate-types)

The interaction of water supply and temperature determines the distribution of the flora, the patterning of the plant communities and the vegetative growth patterns of the communities. Five main climatic regions can be distinguished, which are distinct at their extremes but merge one with another. They are discussed in relation to vegetation-patterning in Chap. 6. The climatic diagrams (Fig. 1.15), constructed according to the scheme of Walter and Lieth (1967), serve to illustrate rapidly the interaction of rainfall and temperature (Abscissa: Months, S. Hemisphere July–June; Ordinate: one division = 10°C or 20 mm rain). When the rainfall curve lies above the temperature curve this particular climate the relatively humid period of the year, indicated on the diagram by longitudinal hatching. Rain in excess of 100 mm is indicated in solid black at a scale reduced to one-tenth. Conversely, when the temperature curve lies above the moisture curve, the area between the two curves indicates the relatively arid period of the year (stippled areas on the diagram). The intensity and duration of the frosty period are shown under the abscissa: in black, the month with mean daily minimum below 0°C (see Hotham Heights); hatched, the month with absolute minimum below 0°C (frost may occur). Other data are: Above Station altitude in m, number of observation years (left) and yearly mean temperature and rainfall (right); below left side-mean daily minimum of the coldest month and minimum, in the mid-mean duration of frost free season in days. A large climatic diagram-map of Australia (with more than 200 diagrams)

Table 1.3. Cyclic salt, estimated as chloride and expressed as NaCl, in kg per hectare per annum.

Station, state	Latitude	Distance from ocean km	Mean annual rainfall mm	Salt as NaCl kg/ha/annum	Source of data
Geraldton, W.A.	28°45'	<1	464	195	Teakle (1937)
Esperance, W.A.	33°50'	<1	668	290	Teakle (1937)
Cape Bridgewater, Vic.	38°58'	<1	703	288	Hutton and Leslie (1958)
Perth, W.A.	31°57'	c. 20	899	340	Teakle (1937)
Dookie, Vic.	36°23'	c. 230	533	8.7	Hutton and Leslie (1958)
Katherine, N.T.	14°28'	c. 300	887	2.9	Wetselaar and Hutton (1963)
Roma, Qld.	26°32'	c. 430	511	10.6 and 59.1	Brunnick (1909)
Mundiwindi, W.A.	23°52'	c. 540	264	10.8	Teakle (1937)

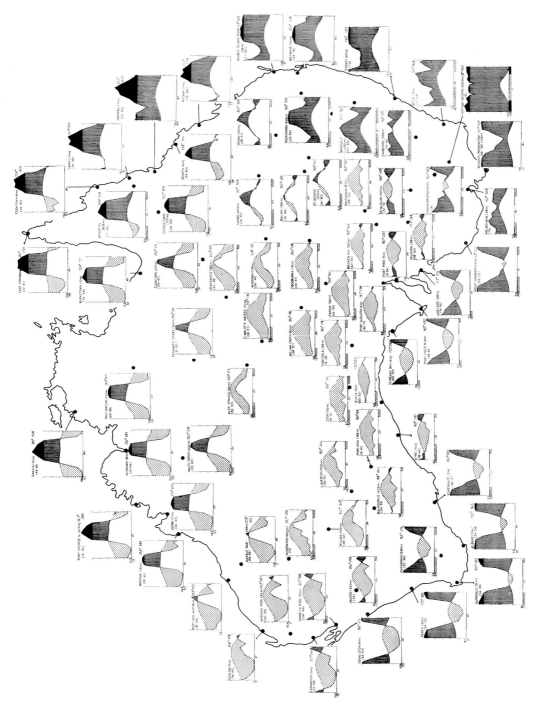

Fig. 1.15. Climatic diagrams for selected stations. For explanation, see text. Supplied by Prof. Dr. H. Walter.

was published in Supplement Vol. X of these Vegetation Monographs: Walter-Harnickell-Mueller-Dombois: Climate diagram Maps of the Individual continents (Springer-Verlag, New York 1975).

The diagrams are most appropriate for indicating the climate-type and are especially useful for finding homoclimates in the different parts of the world. However, for the semi-arid and the arid zone in particular, the diagrams cannot indicate the growing season since rainfall in this region is so unreliable that mean monthly values obscure the rainfall potential.

5. Soils

5.1. Soil Parent Materials

The parent materials are classified into 4 categories: i. rocks; ii. alluvial, colluvial and aeolian deposits (unconsolidated materials); iii. fossil soil profiles (laterites); iv. organic accumulations built up by plants. Rocks form the parent materials of soils mainly in the more elevated areas and some of their properties relevant to vegetation are discussed in the next section. Unconsolidated parent materials occur mostly at lower elevations, mainly along rivers as alluviums, or as aeolian sands in the arid interior, the sands being derived from sedimentary rocks or from reweathered laterite profiles; these materials cover at least half the continent. Fossil laterite profiles likewise cover large areas. Organic accumulations cover very small areas in alpine and subalpine tracts, with still smaller patches in some coastal areas.

The continent contains much the same variety of rocks as do the other continents, including igneous, sedimentary and metamorphic types. In Australia there is a high proportion of sedimentary rocks, including sandstones, greywackes, limestones and shales, which were laid down in vast basins or seas, the particles comprising the sediments being derived directly or indirectly from igneous rocks. Of the igneous rocks, granites and related types which are classed as "acid" and contain 65% or more of free silica (quartz) are the most abundant, and give rise to sandy soils or sandy alluviums. In contrast, basalts which are "basic" and contain little or no free quartz give rise to fine-textured soils, often with a high proportion of ferric hydroxide. Soil texture is important in delimiting certain communities, as discussed in Chap. 6.

The chemical composition of rocks determines soil fertility levels and in this respect phosphate is probably the most important in Australian soils. Igneous rocks contain apatite, a phosphatic mineral, the amount varying considerably from one rock type to another (Joplin, 1963). For example the phosphorus contents of most granites lie within the range 90–300 ppm P (extremes c. 30–1200 ppm), in contrast to basalts where the common values are mostly 1200–3500 ppm P (extremes c. 30–7000 ppm).

It follows that soils derived from basalts are likely to be more fertile with respect to phosphate than soils derived from granites. The same applies to alluviums and to sedimentary rocks built up from particles washed or blown from these soils, though transport of particles either by wind or water is likely to change the composition of the soil ingredients, mainly through sorting the particles according to their sizes. Thus, under the influence of water in wetter climates and of wind in drier areas, the lighter soil particles, with which plant nutrients are associated, are likely to be removed and the coarser particles left behind. The general low nutrient status of Australian soils, most of which are derived from sedimentary rocks or from alluvial or aeolian deposits, is probably the result of one or even several cycles of soil formation, erosion, particle-sorting and deposition of coarser fractions and loss of the finer fractions (Stewart, 1959; Beadle, 1962b). However, the finer fractions are in some cases deposited as alluvium, particularly along the main river systems in the east, where soils of relatively high fertility occur.

Fertility levels in sedimentary rocks cover much the same ranges as in the igneous rocks. In general terms the lowest levels are recorded in pale coloured coarse sediments, the highest levels in dark coloured fine-grained sediments; limestones, because they commonly contain dead marine organisms, are usually of potentially high fertility, though in wet climates they commonly lose their nutrients rapidly as a result of leaching. Successive seams of sedimentary rock from the one deposit often vary considerably in nutrient status, especially with respect to phosphate.

The significance of soil phosphorus in the development of xeromorphy is discussed in

Chap. 4, and the role of soil phosphate in delimiting plant communities, especially in wetter areas in the south-east is dealt with in Chap. 6.

5.2. Weathering and the Development of the Soil Profile

5.2.1. General

The processes involved in the weathering of rock minerals and in the development of the soil profile (Table 1.4), especially under conditions of free drainage, are well documented elsewhere and are not repeated here (see Leeper, 1957).

5.2.2. Soil Profile Development under Conditions of Impeded Drainage. Laterite

Waterlogging of the soil has several significant consequences, among them the accumulation of organic matter as a result of slow oxidation through lack of oxygen (see Table 1.4, Groups 42, 43), and the production of anaerobic conditions under which certain substances are reduced. For example, methane (marsh gas – CH_4) is produced in some bogs and sulphates can be reduced to sulphides, especially H_2S. Also, the anaerobic reduction of iron from the ferric to the ferrous condition sometimes occurs and is detected by yellow-green or bluish mottling in the waterlogged subsoil. The mottled material is known as gley and its presence is used to identify some Groups of soils (Table 1.4, Nos. 15 and 36). This reduction, on a vast scale, appears to be responsible for the widespread production of laterite during the Tertiary.

The fossil laterites which once covered most, if not all, of the continent appear to have been produced under conditions of seasonal waterlogging. The profiles, some of them 30–50 m deep, are usually differentiated into 3 horizons. The surface horizon consists of ferruginous material, often concreted and several metres deep. This overlies a mottled zone consisting of ferric hydroxide segregates forming mosaics with white clay, usually kaolin. Below this there is a layer of white kaolin. A fourth and lower-most horizon, of quartzite is sometimes present.

Laterite has been studied by many authors, including Whitehouse (1940), Prescott and Pendleton (1952), Beadle and Burges (1953), Jessup (1960), Mabbutt (1965), Prider (1966) and many others. The mechanism of formation is not completely understood. The profiles probably formed under warm, humid climatic condition, the soil being waterlogged for at least part of the year. The soils supported vegetation which supplied organic matter to the profile as a source of energy for micro-organisms which appear to be involved in the process of reduction of iron, either directly or through the production of reducing substances. Under these conditions, the soluble ferrous ions diffused in the soil and at the surface were oxidized to insoluble ferric ions. This repeated migration of ferrous ions from below upwards shifted the iron from the lower to the upper layers. During the weathering process desilicification of primary rock minerals proceeded at the same time, if the laterite was forming on an igneous rock. The soluble silicates so formed apparently moved downward in the profile and in the wettest climates the silicates were apparently removed by deep percolation. However, in some areas, possibly in regions of lesser rainfall, free silica was precipitated at or near the base of the profile (Whitehouse, 1940). In a few areas, e.g. White Cliffs (N.S.W.) and Coober Pedy (S.A.) some silica is deposited as opal, possibly under some special conditions of filtration through the clay (Jones and Segnit, 1966). Siliceous layers (e.g. quartzites) referred to as silcrete or grey billy are found only in what is now the arid zone. The material is either grey or reddish and the reasons for the differences are not known, though they may be associated with the chemical composition of the materials in which the secondary silica was deposited.

Since any type of material may form laterite under conditions of waterlogging, the composition of the laterite profiles varied. Parent materials high in free quartz, such as granite and sandstones, produced laterites with a high free quartz content, the quartz occurring in all horizons, being cemented by ferric hydroxide in the surface ferruginous zone, or mixed with kaolin in lower horizons. Basalts, on the other hand, in which free quartz is low or absent produced laterites with little or no free quartz. These variations, the results of differences in both parent material and the then climatic conditions, enable

Table 1.4. *Great Soil Groups in order of degree of profile development and degree of leaching, as defined and described by Stace et al. (1968). See also Hubbard (1970).*

I. No profile differentiation. Nos. 2–6 classed as Regosols, which are deeper soils formed from transported rock particles which may or may not have been weathered chemically. In all soils, the only horizon development is the possible accumulation of organic matter in the surface soil.	1. LITHOSOLS. Stony and gravelly soils; texture varies from sand to clay loam; rock fragments usually present; particles produced by physical weathering; chemical weathering minimal.	Elevated rocky or stony areas. Arid to humid. Many communities.
	2. SOLONCHAKS. Highly saline; sands to clays, the clays often granular and cracking into polygons on drying; surface sometimes salt-encrusted. Formed from sediments and may be layered.	Near the sea, or in arid zone espec. playas. Halophytic communities.
	3. ALLUVIAL SOILS. Formed on alluvium too young to show horizon development. Texture, stoniness, depth, colour and carbonate-content vary. May be deposits over other soil profiles – complete or truncated.	All parts of continent, usually associated with rivers, fan-formations. Vegetation various.
	4. CALCAREOUS SANDS. Siliceous sands enriched with calcium carbonate from organisms; loosely coherent when wet, loose when dry.	Mainly littoral zone.
	5. SILICEOUS SANDS. Quartz sand without or with a coating of hydrated ferric hydroxide, hence white, yellow, yellow-brown to red. Acid throughout. Formed by leaching, e.g. of limestones or sandstones, or from ancient laterites in drier areas.	Coast, or inland as dunes.
	6. EARTHY SANDS. Sandy profiles, yellow or red, the sand grains coated with clay which causes weak cohesion. Some eluviation of clay possible. Some contain ironstone gravel. Acid. Possibly derived from ancient laterites. Related to Nos. 5, 25 and 26.	Semi-arid areas with xeromorphic communities, or arid areas with hummock grassland.
II. Minimal Profile Development. Soils exhibit little or no evidence of leaching, either because they occur in drier climates, or the high clay content retards water movement through the profile. Horizon development indicated by downward movement of either calcium carbonate and/or clay.	7. GREY, BROWN AND RED CALCAREOUS SOILS. Shallow (to c. 40 cm), weakly structured loams to light clays with fine carbonates throughout (pH c. 9). Surface horizon shows slight eluviation of clay. Grey-brown or red. Organic matter very low. Often with gilgais.	Arid zone. Support halophytic shrublands.
	8. DESERT LOAMS. Moderate texture contrast. A horizon to 10 cm deep, neutral to alkaline, light brown to red-brown. Surface often strewn with polished stones. B. horizon with added clay, alkaline. Derived mainly from medium to fine-grained sedimentary rocks.	Arid zone; alluvial plains and associated stony uplands (gibber downs). Support halophytic shrublands.
	9. RED AND BROWN HARDPAN SOILS. Shallow to moderately deep. A horizon non-saline, acid to neutral, sometimes strewn with polished stones. B horizon indurated with silica cementation and clay, massive, sometimes with gravel. Probably derived from clayey materials from ancient laterites.	Arid zone; restricted to central W. A. and north-eastern S. A. Support *Acacia* shrublands.

	10. GREY, BROWN AND RED CLAYS. Deep clays, usually paler with depth. Surface soils granular or self-mulching, acid or alkaline, cracking, often with gilgais. Subsoil blocky, sometimes with carbonate or gypsum, acid or alkaline. Derived mostly from alluviums.	Acid to humid; along watercourses and floodplains. Support grasslands in driest areas, woodlands of *Casuarina*, *Acacia* and *Eucalyptus* elsewhere.
III. Dark Soils. Clayey soils with a relatively high organic content. Texture profile developed as a result of eluviation of clay from A horizon. Occur between 500 and 1000 mm isohyets. The 4 groups are closely related.	11. BLACK EARTHS. Heavy clays with fairly uniform texture profile, dark grey to very dark brown to black. A horizon with c. 40% clay (c.f. Chernozem), neutral to alkaline. Lower horizons with secondary calcium carbonate (c.f. Rendzina). Crack deeply on drying. Developed mainly on alluvium.	Tas. to north Qld. High fertility. Support grasslands or *Eucalyptus* woodland with a grassy understory.
	12. RENDZINAS. Shallow soils developed on limestones or marls. Texture profile as No. 11. Surface neutral to alkaline. Subsoil alkaline, with fragments of parent material (c.f. No. 11).	Small areas in eastern States. High fertility. Vegetation as No. 11.
	13. CHERNOZEMS. Resemble No. 11, but differ in having only c. 20% clay in topsoil which is mildly acid. Lower B horizon with secondary calcium carbonate. Formed on basic alluviums or rocks.	Small areas in eastern States, supporting *Eucalyptus* woodlands with a grassy understory.
	14. PRAIRIE SOILS. A horizon light clay or loam, dark grey-brown, brown or black, acid. B horizon more clayey; carbonate absent; acid to mildly alkaline. Grade with No. 11.	Coastal and subcoastal areas from southern Qld to Tas. and S.A. Support *Eucalyptus* forest or woodland.
	15. WIESENBODEN (Meadow Soils). Related to Nos. 11 and 13. but have gley, formed under conditions of waterlogging, deep in profile. Clay varies from 40–80%; pH 6.0–8.0.	Same areas as No. 11. Support *Eucalyptus* spp. tolerant of waterlogging, e.g. *E. ovata*, *E. tereticornis*, or *Melaleuca* spp.
IV. Mildly Leached Soils. Texture-contrast soils, mostly with brown or grey brown A horizons and darker brown B horizons.	16. SOLONETZ. Sandy or loamy A horizon abruptly differentiated from clayey B horizon. A horizon (to 10 cm deep) neutral to slightly alkaline. B horizon very compact resulting in poor drainage, sometimes mottled, the clay dominated by sodium and magnesium; pH high (to 9.0). Related to No. 17 but are less leached.	Semi-arid regions in south-east and south-west. Support *Eucalyptus* woodlands.
	17. SOLODIZED SOLONETZ AND SOLODIC SOILS. Resemble No. 16 but are more leached. A and B horizons abruptly differentiated. A horizon sandy, up to 60 cm deep, with a bleached A2, acid. B horizon columnar to blocky, very alkaline below. Derived from various siliceous parent materials.	Widespread in all States in areas with mean annual rainfall of 400–1000 mm. Support *Eucalyptus* and *Acacia* woodlands.
	18. SOLOTHS (Solods). Resemble No. 17 but are acid throughout the profile.	Occupy small areas in the same districts as No. 17. Vegetation as No. 17.

19. SOLONIZED BROWN SOILS. Contain abundant calcium carbonate. Profile sandy throughout. A horizon deep (on crests of dunes), or shallow (in troughs), with or without carbonate, neutral or slightly alkaline, brown to red-brown. B horizon with much carbonate, alkaline. Derived from calcareous and saline aeolian sands.

Semi-arid and arid south in west and east. Support mallee, or woodlands (chiefly *Casuarina cristata*). Small areas on lunettes on eastern sides of playas.

20. RED-BROWN EARTHS. A horizon loamy, the A1 with organic matter overlying paler red-brown A2, acid. B horizon with added clay, prismatic to blocky, calcium carbonate present, sometimes gypsum below. Derived mainly from siliceous parent materials. Usually with high base status.

Mainly south of Tropic between 400 and 630 mm isohyets where they support *Eucalyptus* woodlands. Also a bleached form occurs in arid western W. A. near the Tropic, with open *Acacia* scrub.

21. NON-CALCIC BROWN SOILS. Similar to No. 20 but with no A2 horizon and no calcium carbonate in the B horizon. Profile slightly acid to slightly alkaline. Derived from siliceous parent materials.

Semi-arid south-east and south-west. Support *Eucalyptus* woodlands.

22. CHOCOLATE SOILS. Acid, friable clay loams with some horizon differentiation; some leaching but high in bases. 3 sub-groups: reddish, normal and grey. Derived from basic parent materials. High fertility.

Mainly south-east, Qld to Vic. between 500 and 1250 mm isohyets. Support *Eucalyptus* woodlands and forests with dense grass understories.

23. BROWN EARTHS (Brown Forest Soils). Little profile differentiation; yellowish, reddish or dark brown throughout. Light loams to clays. Surface acid to neutral; pH and bases increase with depth. Derived from shales and calcareous materials.

Humid climates from S. A. to northern N. S. W. High fertility and support *Eucalyptus* forest with mesomorphic understory, or rainforests.

V. Soils with predominantly sesquioxidic clay minerals. Other than having uniform colour profile (excepting the surface with accumulated organic matter), the 7 Soil Groups have little in common. No. 24 occurs in the arid zone, the soils containing calcium carbonate; the skeleton is siliceous sand, the ferric hydroxide coating the sand grains. At the other extreme, No. 30 occurs in humid areas, the soils being derived mainly from basalt; the soil skeleton is ferric hydroxide mixed with kaolin.

24. CALCAREOUS RED EARTHS. Soil uniformly sandy, clay increasing slightly with depth. Calcium carbonate present throughout; profile alkaline. Derived possibly from red sand from ancient laterites with calcium carbonate added from another source.

Arid zone. Support *Acacia* scrubs, *Casuarina cristata* woodlands, or halophytic shrublands.

25. RED EARTHS. Red throughout except for surface organic matter; contain some clay (c.f. No. 6 which they resemble); structure weak crumby to blocky. Acid. Derived from siliceous materials or laterites. Manganiferous or ferruginous nodules sometimes present.

Widely distributed in northern half of continent, from arid zone to wetter tropics. Support *Acacia* woodlands (drier) to *Eucalyptus* forests (wetter).

26. YELLOW EARTHS. Similar to No. 25, but yellow to yellow-brown. Acid. Derived from sandstones or other siliceous materials or from leached lateritic remnants. May contain ferruginous nodules deep in profile.

Subhumid to humid areas, central N. S. W. to Cape York; south-west and north-west W. A. Low fertility and usually support xeromorphic communities.

27. TERRA ROSSA SOILS. Shallow red or red-brown soils developed on highly calcareous materials. Surface grey, with some organic matter. Neutral to slightly alkaline, with fragments of parent material in lower layers.

Uncommon; close to coast in south-east and Tas. Support *Eucalyptus* communities including some types of mallee.

28. EUCHROZEMS. Red, clayey, strongly structured; derived from basalt. A horizon dark with organic matter, mildly leached, slightly acid. Clay increases with depth. Resemble No. 30 but have higher base saturation and more compact subsoil.

Small patches in the east and tropics between 630–750 mm isohyets. Support *Eucalyptus* woodlands often with some residual rainforest species.

29. XANTHOZEMS. Yellow, clayey, friable, strongly structured, with moderate horizon differentiation, acid, high fertility. Derived from basalts, phyllites and schists. Yellow counterparts of No. 30.

Subhumid and moister regions in the east in summer rainfall zone. Support drier types of rainforest or grassy *Eucalyptus* woodlands.

30. KRASNOZEMS (Red Loams). Deep, red or red-brown; high organic matter in surface. Acid (pH 4.5–5.5); base saturation c. 50%. Derived from basalts or basic rocks, except in tropics where they are formed on intermediate rocks. Some tropical forms with gibbsite ($Al_2O_3 \cdot H_2O$) in surface layer.

Mainly east from north Qld to Tas. with mean annual rainfall 850–3700 mm. Support rainforests or tall *Eucalyptus* forests with a mesomorphic understory.

VI. Mildly to strongly acid and highly differentiated: Podzols and Podzolic Soils. Characterized by strong eluviation of clay and sesquioxides from the A horizon and their deposition in the B horizon. The soils occur in wet climates where some or most of the rain passes through the soil profile. The extent of horizon differentiation is determined by the proportion of clay + sesquioxide to sand (quartz), which is determined by the parent material. The amount of eluviation is controlled by the interaction of the amount of percolating water, the soil ingredients (as above) which also determine the porosity and drainage. The amount of organic matter in the profile and its location are also used to identify Groups. The Gleyed Podzolic Soils are formed under water-logged conditions. Profiles acid throughout, rarely neutral; calcium carbonate never present.

31. GREY-BROWN PODZOLIC SOILS. A horizon grey to grey-brown, medium textured; A2 distinct but not bleached. B horizon yellow to yellow brown, with blocky structure and high clay. Acid, except neutral in lower B. Base saturation 50–80%. Derived mainly from siliceous rocks or dolerite.

East and south-east from southern Qld to S. A. Mean annual rainfall 500–1000 mm. Support *Eucalyptus* forests and woodlands.

32. RED PODZOLIC SOILS. A1 grey-brown; A2 light grey to yellowish; both sandy loam to clay loam. B horizon red brown to red, sandy clay to heavy clay. Profile acid; base saturation high to low. Derived mainly from intermediate rocks.

Humid east and north. Support *Eucalyptus* forests and woodlands, rarely rainforests.

33. YELLOW PODZOLIC SOILS. A horizon greyish or brownish; A2 paler; both horizons sand to clay loam. B horizon yellow-brown, friable, clayey. Profile acid; base saturation medium to low. Derived from siliceous materials.

Mainly humid east. Support *Eucalyptus* forests and woodlands; *Nothofagus* forest in Tas.

34. BROWN PODZOLIC SOILS. Profiles brownish to yellowish throughout. A horizon lacks a well defined A2 which is stained with organic matter washed from the A1. B horizon with some eluviated clay. Derived from various materials – acid to basic.

More humid areas of Tas., highlands of Vic. and Kosciusko region, south-western W. A. Support tall *Eucalyptus* forests or grassy forests and woodlands or heaths.

35. LATERITIC PODZOLIC SOILS. Presumably derived from old laterite profiles under conditions of free drainage. See 44. below. A horizon sandy; B horizon mottled yellow-brown to red clay. Pisolitic or massive ironstone present at base of A2 and part of B horizons. Mottled horizon and white clay form the base of the profile. Acid throughout.

Many parts of continent with mean annual rainfall exceeding 500 mm. Support *Eucalyptus* forest or woodland.

36. GLEYED PODZOLIC SOILS. Profile of the podzolic type with mottlings from A2 downward which are the result of waterlogging – rust or ochreous root tracings and spottings. Ferrous iron usually detectable. Acid throughout.

Areas subject to waterlogging, mainly humid south-east. Support species adapted to some waterlogging, e.g. *Eucalyptus ovata* and *Melaleuca* spp.

37. PODZOLS. Strongly differentiated profiles. A1 dark, with organic matter; A2 white. B yellow to red to black, with eluviated clay, sesquioxide and organic matter. Derived from sandy parent materials.

High rainfall areas, chiefly near coast. Low fertility soils, supporting xeromorphic woodlands, scrubs or heaths.

38. HUMUS PODZOLS. Resemble No. 37., but have a dark coloured (sometimes black) B horizon formed as an accumulation at the level of watertable. B horizon may be mottled and A2 is light grey rather than white. Formed on sandy parent materials.

Higher rainfall areas, chiefly near the coast in areas subject to waterlogging. Support sedgelands, wet heath or low scrubs with *Banksia* or *Melaleuca* dominant.

39. PEATY PODZOLS. Profile surmounted by a thick layer of brown or black peat up to 25 cm deep. Below, a layer of humified peat. These 2 layers regarded as A0 horizon. Lower A0 grades into organic, sandy clay loam (A1) which overlies sandy material containing much organic matter. Very acid (pH 4–5).

Mainly south-west Tas. in perhumid cool-cold climates. Support sedgelands, sometimes with emergent eucalypts.

VII. Dominated by organic matter.
Profile with much organic matter in the topsoil, obscuring the effects of other components. Formed under wet conditions, some with waterlogging.

40. ALPINE HUMUS SOILS. Organic matter well humified and incorporated in the mineral soil. Horizons merge. A horizon c. 25 cm deep, grey, brown to black, friable. Lower layers brown to yellow brown, merging to grey. High fertility. Derived from acid or basic rocks.

Highlands of south-east above 800 m and Tas. above 1000 m; also eastern Vic. at 400 m. Support subalpine woodlands, heath, or grasslands.

41. HUMIC GLEYS. Similar to No. 36 but with much organic matter intimately incorporated in the dark A horizon. Acid to neutral, some subsaline.

East coast behind littoral zone or valley plains along rivers. Support *Eucalyptus* swamp forests, or *Melaleuca* or *Casuarina* spp.

42. NEUTRAL TO ALKALINE PEATS. Organic matter accumulated under influence of alkaline ground-water. From c. 25 cm to 1–2 m deep; black and granular to dark brown and fibrous; may contain lenses of shells or soft carbonates.

Coastal areas, south-east S.A., north-west Tas. and possibly other coastal areas in south-east. Support sedgelands, sedge-heath or *Melaleuca* spp.

	43. ACID PEATS. Dark-coloured to black; organic matter with or without fibrous plant remains, some partly decomposed. Profiles up to c. 1 m deep, waterlogged permanently or regularly. Developed over any type of rock.	High mountains and plateaux, N.S.W., Vic. and Tas. Minor occurrences near coast. Support sedgelands, heath and sometimes *Eucalyptus* spp. *Sphagnum* moss present in some.
VIII. Fossil Laterite Residuals.	44. IRONSTONE GRAVEL SOILS. Gravelly soils sometimes with a sandy matrix; massive ferruginous laterite commonly present and sometimes part of the mottled zone below. See No. 35.	Extensive areas in south-west W.A. and smaller areas in north-west W.A., N.T., arid zone and the east. Vegetation various, but always xeromorphic.

a classification in 3 subgroups, ferruginous, bauxitic and manganiferous (Hallsworth and Costin, 1953).

The time of formation of laterite cannot be fixed precisely. Many authors suggest Miocene as the main period of formation and it is probable that two or even more periods of formation occurred, perhaps as late as the Holocene. Accurate dating is difficult, but Mabbutt (1965) states that in Central Australia the laterites are overlaid by beds containing mammalian fossils which indicate a Miocene to mid-Pliocene age for these laterites. Also, Dury and Langford-Smith (1969), using potassium-argon isotope dating techniques on basalts formed over laterite profiles in the Warrumbungle Mountains (N.S.W.), indicate that the basalts are about 14 million years old, which must be the minimum age of the laterites in this area, i.e., mid-Miocene.

The climatic and topographic conditions needed for the formation of the laterites have certainly changed since they were formed. The ancient laterite surfaces have been dissected and drained, and most of the profiles have been destroyed or altered. Destruction involves complete removal of the profiles to expose underlying rocks, or truncation to any level to expose the mottled zone, or the kaolin zone, or the basal silica zone. The last, being very resistant to weathering sometimes caps flat-topped residuals (Fig. 1.6), and some of this material, fragmented and washed downhill and polished by wind-abrasion is now strewn on the surfaces of some soils as stones, known as gibbers.

Alteration to complete or truncated profiles occurred under conditions of free drainage in relatively wet climates. The process involved the downward movement of iron, as in podzolization. Since most of the ancient laterites were derived from siliceous parent materials (e.g., gneiss, granite, sandstone), the removal of the ferric hydroxide from the surface layers has produced a sandy layer which now lies above the residue of the ferruginous and/or mottled-horizons; the sand, which is often white as a result of leaching, is sometimes mixed with ironstone gravel. Such soils are classified as Lateritic Podzols (Table 1.4, No. 35). Further, it is apparent that during the process of reweathering, but before the sand contained within the ferruginous zone of the laterite had been washed free of iron hydroxide, the newly formed surface sandy layer was washed downhill and distributed as sand sheets (Mulcahy, 1960). The sand grains are variously coloured, from red and red-brown (least leached) to pale brown or yellow (more leached). These sands, either loose or weakly coherent, are the parent materials for some of the modern soils, e.g., Siliceous Sands and Earthy Sands (Table 1.4, Nos. 5 and 6) and the Red and Yellow Earths (Nos. 25 and 26).

The amount of laterite formed on basalt or other parent materials low in quartz was apparently relatively small but some residual profiles occur in the east and it is probable that the Euchrozems (Table 1.4, No. 28) owe their development, in part at least, to the residual lateritic material derived from basalt (Hallsworth et al., 1953).

The fossil laterites, as they occur today, are soils of low fertility, which is due partly to the fixation of phosphate by iron or aluminium as insoluble phosphate. A factor contributing to their low fertility is the compactness and rock-like nature of the ferruginous zone. This compactness is the result of drying and loss of water by the colloids. It is probable that when the laterites were forming and before the profiles were drained, the lateritic soils exhibited a much higher fertility, if only because of the relative softness of the material and the more effective exploration of the medium by roots.

Fertility levels in the ancient laterites can only be conjectured and it is likely that they varied considerably; for example those derived from basalt were likely to be more fertile than those derived from granite or sandstone. Losses of nutrients by leaching were likely to be smaller under conditions of poor drainage than under conditions of free drainage, though Wild (1958) postulates a loss of phosphate during the period of laterite formation.

That a loss of nutrients is likely to have occurred after the break-up of the laterites and during the processes of reweathering and redistribution of the lateritic remnants is deduced from the following: Firstly, in the ancient laterites, desilicification of the primary minerals had occurred and the macronutrients (except phosphorus) were therefore in a mobile form. Consequently, when the laterites were drained and reweathered, removal of the nutrients (except phosphorus) by leaching was likely to be rapid. Secondly, as far as phosphorus is concerned, this element is rendered insoluble by precipitation as ferric phosphate and its concentration in the ferruginous layer of laterites is relatively high (c. 500–1500 ppm P) in comparison with levels in the kaolin zone (30–50 ppm P). Consequently the presence of iron in a profile tends to conserve phosphate, though the phosphorus is not readily available to plants. This is demonstrated by the fact that phosphorus content is usually related to the amount of ferric hydroxide on the surface of the sand grains; thus red sands have phosphorus contents of 100–250 ppm P, white sands from Lateritic Podzols 15–30 ppm, and yellow sands have values intermediate between these two.

5.3. The Gilgai Phenomenon

Many Australian soils with a high clay content (Table 1.4, especially Nos. 7 and 10) occurring under a mean annual rainfall of 150 to 1500 mm with regular or intermittent periods of drought, exhibit the gilgai phenomenon, which in simple terms is described as undulations in the soil surface identified by hummocks and hollows. The phenomenon is a natural and permanent feature of certain soils which cannot be altered by cultivation or levelling (Jensen, 1914). Gilgais have considerable effects on the patterning of microcommunities within major associations, partly through the redistribution of rain water, the hollows filling with water, and partly through differences in the soil chemical properties of the hummocks and hollows.

Gilgais have been studied by many authors, most recently by Hallsworth and Beckmann (1969) and Paton (1974), and they appear to result from the differential movement of blocks of soil, i.e. the upward movement of some blocks (which become mounds), and the downward movement of other blocks which become depressions. In some cases mounds and depressions are contiguous. In other cases, mounds and depressions are separated by a stretch of flat or concave surface (shelf), so that the sequence from high to low is mound, shelf, depression (= melon-hole or crab-hole). Within the depression sink-holes may occur, these being small, usually angular holes extending into the subsoil.

In many cases the A and B horizons remain in position, though the depth of the A horizon on mounds may be reduced following the washing of surface soil aggregates downward into the depression. However, in some cases the B horizon lies at the surface of the mound (usually readily identified by the presence of calcium carbonate), which has a granular surface, and has been referred to as a "puff".

The forces which initiated the vertical movements of the soil blocks are not known precisely. It is usually considered that the upward movement of soil is caused by the differential swelling of the A and B horizons. Since the B horizon contains a higher proportion of sodium-dominated clay with an expanding lattice, and is also more clayey than the A horizon, it swells more than the A horizon on wetting and rises towards the surface. Once the B horizon becomes elevated, it remains in this position as a mound. Some circulation of soil by erosion of the mound by water or by the falling of soil aggregates down cracks (chiefly in the depression) may occur. Soil material washed or falling into the subsoil adds to the volume of the subsoil, so that pressures in the subsoil are increased when the soil is wetted and upward movement of soil results. The early wetting of the subsoil by water-flow down cracks may also be a factor contributing to the upward movement of soil, in so far as the B horizon may commence to swell before the A horizon is fully wetted. On the other hand, Paton (1974) ascribes the upward movements to "differential loadings", i.e., the wet, plastic subsoil is pushed to the surface in compensation for the downward pressures exerted by more solid and less plastic blocks of soils.

5.4. Soil Classification

The first system for the classification of Australian soils was based on the geological nature of the soil parent material (Jensen, 1914). The concept of Great Soil Groups devised by Russian pedologists was introduced into Australia by Prescott (1931), who discussed also the genesis of the soils, and he was the first to recognize the difficulties in soil classification imposed by the occurrence over much of Australia of ancient landscapes dominated by fossil laterite. Noteworthy contributions to soil classification were made by Stephens (1953, 1964) who used the various profile characters based on horizon development, and the physical and chemical properties of the horizons to identify Soil Groups. This system with some modifications and regroupings is still in use and the various Great Soil Groups are listed in Table 1.4 under 7 major headings (Stace et al., 1968).

The most recent development in soil mapping is that of Northcote (1965) and Northcote et al. (various publications; for reference, see Stace et al. 1968). His "factual key for the recognition of Australian soils" is based on specific soil properties of the profiles. This detailed classification which refines the units in Stace et al. (1968) covers the whole of the continent mapped on a scale 1:2,000,000.

5.5. Relationship between Soil Group and Vegetation

Although soil properties commonly control the floristics and structure of vegetation, rarely can a Soil Group be used to indicate a specific plant community. This is due partly to the fact that the Soil Groups, identified by the morphology of the profile, commonly extend over a range of climates, especially when the soil parent material is derived from the fossil laterites. For example, the Red Earths extend from the arid zone to the wet tropics and support communities as diverse as *Acacia* woodlands and tall *Eucalyptus* forest. However, subdivision of the Soil Group into smaller categories would show a closer relationship between soil and vegetation.

A second factor militating against a close relationship between Soil Group and vegetation is the adaptation of dominant species to a wide range of soil properties. Thus a tree species, such as *Eucalyptus populnea*, which grows on sands or clays with a pH range of c. 5.5–8 occurs on several Soil Groups. However, if the understory species of the *E. populnea* woodlands are considered and several subgroups of communities are recognized, a correlation between soil and vegetation is apparent (see Chap. 13 for details).

In many cases single soil properties appear to delimit both the floristics and structure of communities. These are dealt with in the next section.

5.6. Individual Soil Properties and their Impact on the Flora and Vegetation

Individual soil properties can sometimes be identified as the main or even controlling factor in the evolution of plant taxa, particularly with respect to adaptation to large excesses or minimal supplies of certain materials, or in determining the local patterning of plant communities. In most cases, however, soil properties are commonly determined by interactions between soil parent material and the other environmental variables. Moreover, single soil properties frequently have different effects in different environments. Consequently individual soil properties and interactions will be discussed after the flora and plant communities have been defined (see Chap. 6). Items to be considered are: soil depth; soil waterlogging; soil texture; soil fertility; soil reaction (pH); salinity; excesses of essential elements, such as calcium in the form of gypsum ($CaSO_4 \cdot 2H_2O$), and magnesium in serpentines; toxic substances, especially heavy metals.

6. Fire

Fires caused by lightning have probably always been an important ecological factor, and most Australian plant communities are adjusted to periodic burning. The incidence of fires has probably increased since the arrival of man. The aborigines, who migrated to the continent c. 50,000 years ago, lit fires partly to herd native animals and partly to produce new grass shoots to attract these animals. Also, according to early explorers, they rarely, if ever, extinguished their cooking fires which may possibly have spread to the adjacent vegetation. White man likewise has burned extensively, though purposeful burning has decreased with the passage of time (Whalley, 1970). On the other hand, accidental conflagrations are periodic and most of these are initiated by humans. When data are available, the effects of fire on specific plant communities are given in the chapters dealing with the vegetation.

2. The Flora

1. Introduction

The Australian flora is usually regarded as unique in the world, chiefly because it contains a high proportion of endemic genera and partly because two genera, *Eucalyptus* and *Acacia*, with a total of over 1000 species, dominate the vegetation over most of the continent. However, this uniqueness applies mainly to the genera and species of angiosperms and gymnosperms.

In this chapter only the vascular flora is considered. Nevertheless, the other groups of organisms are ecologically important and some mention is made of some of these in later chapters, when the organisms have some special significance in specific plant communities. For example, mosses are so abundant in certain rainforests that they are used to designate a type of rainforest (Chap. 7). Also, mosses and sometimes liverworts and lichens are often community dominants on bare rock faces or they form societies on forest floors. Some algae are mentioned as soil organisms, especially *Anabaena* and *Nostoc*, which are known to fix atmospheric nitrogen.

Micro-organisms likewise are ecologically essential for such processes as the decomposition of organic matter and the cycling of nutrients, and some are parasitic. Some are concerned with the fixation of atmospheric nitrogen, especially rhizobia in association with a legume, the actinomycete forming nodules on the roots of some species of *Casuarina* and the free-living, saprophytic *Azotobacter*, *Beijerinckia* and *Clostridium*. Many Australian plants have mycorrhiza in their roots, including bryophytes, pteridophytes, gymnosperms and angiosperms.

The angiosperms dominate most plant communities, and in this chapter data on numbers of taxa are presented; other items such as origins in the continent and community – dominants are dealt with in succeeding chapters.

The pteridophytes and gymnosperms are given a somewhat different treatment below, since most of the taxa are not community dominants and are consequently not featured in the community descriptions, except for their possible inclusion in lists of subsidiary species. For this reason some additional ecological notes are included below for those taxa which cannot be adequately dealt with under the climax communities.

2. The Vascular Flora

2.1. Pteridophytes

2.1.1. The Indigenous Flora

The pteridophyte flora consists of 99 genera with c. 300 species (Table 2.1). They represent c. 40% of the world's total pteridophyte genera (247, Copeland, 1947) and 3% of the world's total species (c. 10,000). Only four of the Australian genera are endemic and three of these are monotypic. Most of the genera are found on all other continents. 25 genera have restricted distribution, occurring in Asia, the southern continents or on islands in the Indian or south Pacific Oceans.

Most of the genera and species occur in the tropics, mainly in the rainforests (Chap. 7). However, the pteridophytes are widely distributed in the continent and are found in most communities, mainly as subsidiary species in habitats ranging from aquatic to arid. As subsidiary species they rarely become so prominent that they are used to classify a community.

The pteridophytes exhibit a variety of habits. While most grow in soil, some are aquatic and many are epiphytic. Those growing in soil have either creeping rhizomes or caudices, the latter sometimes reaching a height of several metres, as in the tree-ferns (Fig. 2.1). The aquatics include a heterosporous group, known as the Water Ferns (Marsileaceae and Azollaceae), and a few homosporous ferns are also aquatic. The epiphytes are found mainly in the rainforests

Table 2.1. The Australian pteridophyte flora and the world's flora for the same genera. #indicates endemic in Australia Families 1–14 arranged in probable evolutionary sequence. Water Ferns (Families 35, 36) at end. Other families (15–34, advanced ferns with vertical anulus) arranged alphabetically according to family. Sources of data: World, Willis and Airy Shaw (1973); Australia, Wakefield (1955), Tindale (1961a, b, 1963, 1965), Smith (1966), Willis (1970), Croxall (1975).

	WORLD spp.	Distribution outside Australia	AUSTRALIA spp.	Distribution in Australia
PSILOPSIDA				
1. Psilotaceae				
Psilotum	3	Trop. & Subtrop.	2	Widespread, wetter areas
2. Tmesipteridaceae				
Tmesipteris	10	Polynes.; N.Z.	5	E. Qld to Tas.
LYCOPSIDA				
3. Lycopodiaceae				
Lycopodium	450	Trop & Temp.	8	Widespread, wetter areas
Phylloglossum	1	N.Z.	1	Vic.; Tas.; S.A.; SW. W.A.
4. Selaginellaceae				
Selaginella	600	Mainly Trop.	9	Widespread, wetter areas
5. Isoëtaceae				
Isoëtes	75	Trop. & Temp.	5	Temp.; Centr. Aust. (isolated)
PTEROPSIDA				
6. Ophioglossaceae				
Botrychium	36	Cosmopol.	2	SE. Qld to Tas.
Helminthostachys	1	SE. Asia; N.G.	1	Trop.
Ophioglossum	56	Trop. & Temp.	c. 6	Widespread
7. Angiopteridaceae				
Angiopteris	c. 100	Asia; Polynes.; Madagas.	1	E. Qld; NE. N.S.W.
8. Marattiaceae				
Marattia	60	Trop., N.Z.	2	E. Qld
9. Osmundaceae				
Leptopteris	7	N.G.; Pacif.	1	E. Qld; E. N.S.W.
Todea	1	S. Afr.; N.Z.	1	E. Qld to Tas.; S.A.
10. Schizeaceae				
Lygodium	3	Trop. & Subtrop.	3	Trop.; E. Qld; E. N.S.W.
Schizaea	c. 35	N. Amer.; S. continents	4	E. Qld to Tas.; wetter S.A.
11. Gleicheniaceae				
Dicranopteris	10	Trop.	2	Trop.; E. Qld; E. N.S.W.
Diplopterygium	20	Trop. & Subtrop.	1	N. Qld
Gleichenia	10	Malays.; Afr.; Pacif.	5	E. Qld to Tas.; S.A.; W.A.
Sticherus	100	Trop. & S. Hemisph.	3	E. Qld to Tas
12. Hymenophyllaceae				
Aptopteris	1	N.Z.	1	Tas.
Cephalomanes	10	India; Madagas.; Pacif.	1	E. Qld; E. N.S.W.
Crepidomanes	12	E. Afr. Is.; Pacif.	5	E. Qld
Didymoglossum	20	Amer., Afr., Ceylon	1	NE. Qld
Gonocormus	c. 6	Afr. to Polynes.	2	E. Qld; E. N.S.W.
Hymenophyllum	25	N. & S. Temp.	19	E. Qld to Tas.
Macroglena	12	E. Afr. Is.; Pacif.	2	E. Qld to Vic.
Microgonium	12	Afr.; Asia; Amer.	4	E. Qld
Microtrichomanes	10	S. Hemisph.	2	E. Qld; E. N.S.W.
Pleuromanes	3	Ceylon to Tahiti	1	NE. Qld
Polyphlebium	1	N.Z.	1	E. Qld; E. N.S.W.
Reediella	2	Pacif.; N.G.	2	NE. Qld
Selenodesmium	10	India; Pacif.	2	E. Qld; E. N.S.W.

		WORLD		AUSTRALIA	
		spp.	Distribution outside Australia	spp.	Distribution in Australia
	Sphaerocionium	63	Mainly Trop.; N.Z.	1	E. N.S.W.
	Trichomanes	25	Trop. & Subtrop.	2	NE. Qld
13. Cyatheaceae					
	Cyathea	800	Trop. & Subtrop.	9	E. Qld to Tas.
14. Dicksoniaceae					
	Culcita	9	Trop. Amer.; Pacif.	2	E. Qld to Tas.
	Dicksonia	30	Malays.; Amer.; Pacif.	2	E. Qld to Tas.; S.A.
15. Adiantaceae					
	Adiantum	200	Cosmopol.	9	Widespread, wetter areas
16. Aspidiaceae					
	Arachniodes	50	Trop. & Subtrop.	1	E. Qld; E. N.S.W.
	Lastreopsis	35	Trop. & Subtrop.	11	E. Qld to Tas.
	Polystichum	175	Cosmopol.	4	E. Qld to Tas.
	Tectaria	200	Trop.	1	E. Qld
17. Aspleniaceae					
	Asplenium	700	Cosmopol.	19	Widespread
	Pleurosorus	3	N. Afr.; Spain; Chile	1	Widespread
18. Athyriaceae					
	Athyrium	180	Cosmopol.	2	E. Qld to Tas.
	Cystopteris	18	Cosmopol. (cold)	1	SE. to Tas. (cold)
	Diplazium	400	Trop. & N. Temp.	7	E. Qld; E. N.S.W.
19. Blechnaceae					
	Blechnum	200	Cosmopol.	13	Trop.; E. Qld to Tas.
	Doodia	10	Ceylon; N.Z.; Hawaii	3	E. Qld to Tas.
	# *Pteridoblechnum*	1	Endemic to Aust.	1	N. Qld
	Stenochlaena	5	Afr.; Asia; Pacif.	1	Trop.
20. Davalliaceae					
	Davallia	40	Asia; Pacif.	4	E. Qld to Tas.
	Humata	50	Asia; Madagas.; Pacif.	1	NE. Qld
	Rumohra	6	Cosmopol.	1	E. Qld to Tas.
21. Dennstaedtiaceae					
	Dennstaedtia	70	Trop.	1	E. Qld to Vic.
	Histiopteris	10	Warm S. Hemisph.	1	E. Qld to Tas.; S.A.; N.T.
	Hypolepis	45	Trop. & Subtrop.	5	E. Qld to Tas.
	Microlepia	45	Old World Trop.; Japan	1	NE. Qld
	Oenotrichia	4	N.G.; New Cal.	1	NE. Qld
	Pteridium	6	Cosmopol.	1	Wetter Temp.
22. Dipteridaceae					
	Dipteris	8	Asia; Polynes.	1	N. Qld
23. Grammitidaceae					
	Ctenopteris	200	Trop.; N.Z.	4	E. N.S.W. to Tas.
	Grammitis	154	Trop.; N.Z.; Chile	4	E. Qld to Tas.
24. Gymnogrammaceae					
	Anogramma	7	N. & S. Temp.	1	S. N.S.W. to Tas.; S.A.; SW. W.A.
	Gymnogramma	5	Amer.; Asia	1	Arid S.A.; N.T.
	# *Paraceterach*	2	Endemic to Aust.	2	Trop. and Subtrop.
	# *Platyzoma*	1	Endemic to Aust.	1	Trop.
25. Lindsaeaceae					
	Lindsaea	200	Trop.; S. Afr.; N.Z.	13	Widespread, wetter areas
26. Lomariopsidaceae					
	Bolbitis	85	Trop.	1	N. Qld
	Elaphoglossum	400	Trop. & Subtrop.	1	N. Qld.
	Teratophyllum	9	Trop.; Malays.; Pacif.	1	N. Qld
27. Oleandraceae					
	Arthropteris	20	Afr. to N.Z.	4	E. Qld; E. N.S.W.

	WORLD		AUSTRALIA	
	spp.	Distribution outside Australia	spp.	Distribution in Australia
Nephrolepis	30	Mainly Trop.	1	Trop., E. N.S.W.; Centr. Aust.
Oleandra	40	Trop.	1	N. Qld
28. Parkeriaceae				
Ceratopteris	2	Trop. & Subtrop.	1	Trop.
29. Plagiogyriaceae				
Plagiogyria	36	E. Asia; Amer.	1	E. Qld
30. Polypodiaceae				
Belvisia	15	Afr.; Asia; Polynes.	1	E. Qld; E. N.S.W.
Colysis	30	Afr. to N.G.	2	E. Qld
Dictymia	4	N. Cal.; Fiji	1	E. Qld; E. N.S.W.
Drynaria	20	Afr.; Asia; Pacif.	3	E. Qld; E. N.S.W.
Microsorium	60	Asia; Polynes.	6	Trop., E. Qld to Vic.
Platycerium	17	Malays.; Afr.; Pacif.	4	E. Qld. to Vic.
Pyrrosia	100	Asia; Afr.; N.Z.	4	E. Qld; E. N.S.W.
31. Pteridaceae				
Acrostichum	3	Trop.	1	Trop. to N. N.S.W.
Pteris	c. 280	Cosmopol.	7	Widespread, wetter areas
32. Sinopteridaceae				
Cheilanthes	180	Trop. to Temp.	c. 9	Widespread
Dryopteris	35	Trop. & Subtrop.	1	E. Qld
# *Neurosoria*	1	Endemic to Aust.	1	Qld; N.T.
Pellaea	80	Trop. & Subtrop.	1	E. Qld
33. Thelypteridaceae				
Ampelopteris	1	Trop.	1	E. Qld; E. N.S.W.
Cyclosorus	300	Trop. & Subtrop.	9	Widespread, wetter areas
Thelypteris	500	Cosmopol.	1	E. Qld
34. Vittariaceae				
Antrophyum	40	Trop. & Subtrop.	1	E. Qld
Monogramma	2	Mascarene Is., Malays.	1	E. Qld
Vaginularia	6	Ceylon to Fiji	1	E. Qld
Vittaria	50	Trop. & Subtrop.	1	E. Qld; NE. N.S.W.
35. Marsileaceae				
Marsilea	60	Trop. to Temp.	6	Widespread
Pilularia	6	N. & S. Temp.	1	Temp.
36. Azollaceae				
Azolla	6	Trop. & Subtrop.	2	Widespread

and belong to several families, from primitive to advanced, e.g., *Psilotum complanatum, P. nudum, Tmesipteris* spp. (occurring almost entirely on the trunks of tree-ferns), *Ophioglossum pendulum, Asplenium nidus* and *Platycerium* spp. (Chap. 7). The epiphytic pteridophytes sometimes form small micro-communities in the branches of trees in rainforests or on fallen logs or rocks, and in more open areas in *Eucalyptus* forests they occasionally form similar communities. Commonly, a single species constitutes a community and sometimes a single plant, but usually more than one species occur including other pteridophytes or angiosperms; these usually become established in the debris collected or produced by the epiphyte. The production of debris by the epiphyte itself is a characteristic of a few ferns, e.g. *Drynaria* and *Platycerium*, which produce both fertile and sterile fronds; the latter, known as "nest" leaves, die and accumulate at the back of the green fronds.

Several pteridophytes occur as dominants over small areas, usually under extreme conditions of the environment. The main occurrences, classified according to habitat, are dealt with in the next section.

Fig. 2.1. *Dicksonia antarctica*. Trees behind are *Eucalyptus cypellocarpa*. Understory species: *Doryphora sassafras* and *Hedycarya angustifolia*. Mt. Irvine (Blue Mts), N.S.W.

2.1.2. Occurrence of Pteridophytes as Community Dominants

i. **In Aquatic Habitats:** Throughout the continent, but mainly in the south, free water surfaces are often covered by species of *Azolla* (*A. filiculoides* and *A. pinnata*). The plants propagate rapidly by vegetative means and spread over still waters.

Submerged aquatic communities of pteridophytes include masses of *Ceratopteris thalictroides* along some of the tropical rivers, and communities of *Isoëtes*, usually in the southern half of the continent. *I. gunnii*, endemic in Tasmania, occurs in alpine lakes, submerged to a depth of about 1 m. *I. elatior*, also confined to Tasmania, is found along water-courses (Wakefield, 1955). The two mainland species, *I. drummondii* and *I. humilior*, are both widespread in the continent, sometimes forming carpets just below the water surface.

ii. **In Swamps:** In swampy areas, from the coast to the interior, members of the Marsileaceae, *Marsilea* and *Pilularia*, are often common or even dominant. *Marsilea* is found throughout the continent in permanently or temporarily wet situations. When growing in pools of water or in streams, the petioles of the leaves elongate and the four leaflets float on the water surface; this is the usual habit for *M. angustifolia* and *M. mutica* which are confined to wetter climates. In drier regions the plants usually form a sub-aquatic community and as the water evaporates from the soil *Marsilea* becomes a terrestrial plant, forming a herbaceous sward, until the sporocarps mature and the leaves wither. The most noteworthy species is *M. drummondii* (Chap. 21). *Pilularia novae-hollandiae* forms small societies in freshwater swamps along the wetter coast in the south. The communities are possibly ephemeral, and are often overlooked because of the sedge-like nature of the foliage.

Other pteridophytes which are prominent or dominant in freshwater swamps include species of *Lycopodium*, *Selaginella* and *Gleichenia*. Ferns which occur in saline or subsaline swampy areas along the coast are *Acrostichum speciosum* and *Blechnum indicum*. The former is of fairly common occurrence in some of the mangrove communities (Chap. 23). *B. indicum* occurs mainly along the east coast, sometimes becoming dominant at the edges of subsaline swamps (Chap. 22); it may form the herbaceous layer of subsaline communities dominated by species of *Melaleuca*.

iii. **In rocky situations:** Microcommunities of pteridophytes occur among rocks in both wet and dry situations, and even on saline cliffs facing the ocean (*Asplenium obtusatum*). Many of the species are referred to as "rock ferns".

In the wetter areas, bare rock-faces are commonly covered with small communities of pteridophytes which root in clefts or along bedding seams in sedimentary rocks. The most common species are *Lycopodium cernuum* (Fig. 2.2) and *Gleichenia dicarpa*, and less frequently *L. laterale*. A few angiosperms are usual in these communities, especially species of *Drosera*.

In drier situations, but in forested areas mostly in the south, the following species form the nucleus of microcommunities among rocks: *Adiantum aethiopicum*, *Anagramma leptophylla*, *Asplenium flabellifolium*, *Cheilanthes tenuifolia*, *Culcita dubia*, *Ophioglossum lusitanicum*, *Pleurosorus rutifolius*, *Polystichum* spp., *Psilotum nudum* and *Todea barbara* (Fig.

Fig. 2.2. *Lycopodium cernuum* on a rock face of sandstone in a road cutting. Some *Gleichenia dicarpa* and *Blechnum minus* also present. Royal National Park, south of Sydney, N.S.W.

2.3). Even in the arid zone pteridophyte communities occur on rocky outcrops (poikilohydric spp.).

Of the ferns mentioned above *Anogramma* is unique in that the gametophyte is perennial, whereas the sporophyte is annual.

iv. **In caves and ravines:** In these shaded habitats light intensity is apparently too low for the growth of angiosperms, so that the vegetation is commonly dominated by bryophytes, often liverworts. In the mats formed by these non-vascular plants ferns sometimes occur as emergents, of which *Leptopteris fraseri* is the most noteworthy. This species has filmy fronds up to 2 m long; the plants require high humidities and often occur under or near waterfalls. Also, *Schizaea rupestre,* which forms turf-like swards over the wet soil surface is often present, commonly at the mouths of caves.

2.1.3. Changes in the Abundance of Pteridophytes as a Result of Man's Activities

By clearing land for agriculture man has considerably reduced the pteridophyte flora in the wetter areas, possibly with the local extinction of many species, particularly epiphytes and members of the Hymenophyllaceae, most of which require very humid conditions. On the other hand, a few pteridophytes have undoubtedly increased in abundance as a result of man's activities:

i. **In road cuttings.** Along roads and cuttings, pteridophytes are often the first vascular plants to colonize bare soil or rock ledges, notably *Lycopodium cernuum* (Fig. 2.2), *Gleichenia dicarpa* (Fig. 2.4) and *Culcita dubia*. In drains along roads in forested areas, ferns are usually prominent, e.g. *Adiantum* spp., *Culcita dubia, Cyathea australis, Doodia aspera* and *Pellaea falcata*.

Fig. 2.3. Community of *Todea barbara* (larger fronds) with *Blechnum minus* below. Road cutting through sandstone. Royal National Park, south of Sydney, N.S.W.

Fig. 2.4. *Gleichenia dicarpa* on a road-cutting. Point Lookout, near Armidale, N.S.W.

ii. **After fires.** Fires sometimes increase the density of some ferns in the herbaceous stratum, often to the exclusion of angiosperms. In *Eucalyptus* communities, the dominant trees are rarely killed by fire and the regenerating community consists, in some cases, of a tree layer with a continuous fern layer below. The most common fern which assumes dominance is bracken, *Pteridium esculentum*. A similar condition is sometimes encountered with *Culcita dubia*. Tree-ferns (*Cyathea* and *Dicksonia*) are fairly fire-resistant and not uncommonly become more prominent, at least temporarily, in forest communities which have been burned.

iii. **Invasion into cleared land.** Forest land cleared for grazing, less commonly for cropping, is sometimes invaded by bracken *(Pteridium esculentum)*. Invasion occurs mainly on the soils of higher fertility and may be so complete that all other plants are excluded. Invasion in most cases appears to be the result of lateral spread of the bracken rhizomes, but in some cases the patchy distribution of bracken in paddocks suggests that new clones have developed as the result of spore dispersal followed by the establishment of a gametophyte and a new sporophyte which initiates a new clone.

Table 2.2. Australian gymnosperm genera and the distribution of the genera elsewhere in the world. + indicates presence. − indicates absence. # indicates endemic in Australia. Sources of data: Guillaumin (1948); Curtis (1956); Burbidge (1963); Dallimore and Jackson (1966); Willis and Airy Shaw (1973); Womersley (1969); Duffy (1972).

	WORLD								AUSTRALIA	
	spp.	S. Amer.	N.Z.	N.G.	SE. Asia	Afr.	N. Cal.	Elsewhere	spp.	Distribution
CYCADALES										
Cycadaceae										
Cycas	20	−	−	+	+	+	+	Madagas., Polynes.	7	Trop.
Zamiaceae										
# *Bowenia*	2	−	−	−	−	−	−	−	2	E. Qld
# *Lepidozamia*	2	−	−	−	−	−	−	−	2	E. Qld, E. N.S.W.
Macrozamia	14	−	−	+	−	−	−	−	14	E. Qld, E. N.S.W. SW. W.A., Cent. Aust.
CONIFERALES										
Araucariaceae										
Agathis	20	−	+	+	+	−	+	−	3	E. Qld
Araucaria	18	+	−	+	−	−	+	Pacif. Is.	2	E. Qld; NE. N.S.W.
Cupressaceae										
# *Actinostrobus*	3	−	−	−	−	−	−	−	3	SW. W.A.
Callitris	18	−	−	−	−	−	+	−	16	Widespread
# *Diselma*	1	−	−	−	−	−	−	−	1	Tas.
Podocarpaceae										
Dacrydium	25	+	+	+	+	−	+	Fiji	1	Tas.
# *Microcachrys*	1	−	−	−	−	−	−	−	1	Tas.
# *Microstrobus*	2	−	−	−	−	−	−	−	2	N.S.W., Tas.
Phyllocladus	7	−	+	+	+	−	−	Philipp.	1	Tas.
Podocarpus	100	+	+	+	+	+	+	Japan	5	E. Qld to Tas., SW. W.A.
Taxodiaceae										
# *Athrotaxis*	3	−	−	−	−	−	−	−	3	Tas.

Other genera in New Guinea: *Libocedrus* (also N.Z. and N.Cal.), *Papuacedrus* (also Moluccas) (both Cupressaceae)

2.2. Gymnosperms

2.2.1. The Indigenous Flora

The gymnosperm flora consists of 15 genera and 62 spp., which represents a large proportion of the world's gymnosperms (c. 75 genera and 700 spp.). Seven genera are endemic (Table 2.2). The genera are mostly confined to the Southern Hemisphere with few extensions north of the equator, notably *Podocarpus*. This southern assemblage of genera is different from the Northern Hemisphere assemblage.

Extensive forests of gymnosperms, comparable in area or height with those in the Northern Hemisphere, do not occur in Australia, though small areas of forest and woodland do occur. The habitats of the genera are outlined in the next section.

2.2.2. Habitats of Gymnosperms

i. The cycads are found mainly as understory species in forests. *Cycas* (Fig. 2.5) occurs in the tropical *Eucalyptus* forests, *Bowenia* in the tropical rainforests. *Lepidozamia* usually occurs at the margins of rainforests and *Macrozamia* in *Eucalyptus* forests and woodlands, sometimes becoming locally dominant on hills (Fig. 2.6) where the soil is too shallow to support trees. The relic occurrence of *Macrozamia macdonnellii* (Chap. 5) in Central Australia gives evidence of a former more extensive distribution of this genus.

ii. *Agathis* and *Araucaria* have restricted distributions in the east and are confined to the rainforests (Chap. 7).

iii. *Actinostrobus* spp. are usually emergents in heaths on sands in Western Australia (Chap.

Fig. 2.5. *Cycas media*, with male cone. Near Darwin, N.T.

Fig. 2.6. *Macrozamia moorei*. Near Springsure, Qld.

Fig. 2.7. Mature female cones of *Callitris "glauca"*.

16). *Callitris* (Fig. 2.7), the largest genus, is widespread, growing usually on sandy soils or rocky outcrops. *C. intratropica* occurs in the tropics sometimes in pure stands or in association with eucalyptus (Chap. 8); it provides useful timber. *C. columellaris* exists in two forms, one near the coast in the east (referred to as *C. columellaris*), the other inland and referred to as *C. "glauca"*. The inland form, which provides useful timber, is found almost from east to west in the southern half of the continent, mostly in association with eucalypts (Chap. 13) but also in the arid zone on sandhills (Chaps 5, 18). *C. macleayana* is found mainly in rainforests. *C. preissei* occurs mainly in the mallee (Chap. 14). The remaining species occur mainly as subdominants in *Eucalyptus* communities. *Diselma* is confined to subalpine Tasmania (Chap. 17).

iv. Of the Podocarpaceae, three genera are confined to Tasmania. *Dacrydium* and *Phyllocladus* are constituents of the rainforests (Chap. 7), and *Microcachrys* forms shrubberies on alpine plateaux (Chap. 17). *Microstrobus niphophilus* also occurs in alpine shrubberies in Tasmania, while the other species, *M. fitzgeraldii*, a shrub, is found only in the Blue Mts west of Sydney on wet rocks in the range of spray of waterfalls.

Podocarpus occurs in a variety of habitats. *P. amarus* and *P. elatus*, trees c. 24 m and 40 m high respectively, occur in the eastern rainforests, the former being confined to the tropics, the latter extending to lat. 35° S. *P. drouyniana* and *P. spinulosus* are shrubs 1–2 m high occurring in *Eucalyptus* forests, the former in Western Australia, the latter in the east. *P. lawrencii* (= *P. alpina*) is a prostrate shrub in alpine areas (Chap. 17). It may form a small tree at lower altitudes in eastern Victoria (Chap. 7).

v. *Athrotaxis*, with two common species, forms forests in wet situations (Fig. 2.8) in montane and subalpine Tasmania, often around lakes (Chap. 17) or on valley slopes, sometimes in association with *Nothofagus cunninghamii* (Chap. 7).

Fig. 2.8. *Athrotaxis selaginoides* edging a river. *Eucalyptus coccifera* behind. Near Lake Fenton, Tas.

2.2.3. The Naturalized Flora

Many Northern Hemisphere conifers are grown as ornamentals or for windbreaks, and *Pinus radiata* (Monterey Pine or Radiata Pine), native to Monterey County, coastal California, is grown in plantations for timber. This species has become naturalized in a few areas, e.g., near Sydney on sandy soils. The growth of the pine is dependent on a mycorrhizal association and lack of the appropriate fungus in Australian soils presumably prevents invasion of this species into soils which are otherwise suitable for its establishment.

2.3. Angiosperms

2.3.1. General

The angiosperms dominate the Australian flora and they exhibit the range of size, family diversification, modes of nutrition and life-form as do the assemblages on other continents. The smallest is the tiny, rootless, floating aquatic plant, *Wolffia arrhiza*, c. 1 mm long, the largest is *Eucalyptus regnans* which sometimes exceeds 100 m in height.

The numbers of angiosperm taxa in Australia in relation to the world's totals are given in Table 2.3, and a list of the families and numbers of genera in Table 2.4. The flora contains a relatively high proportion of primitive angiosperm families and of heterotropic plants, which are enumerated below but are not mentioned elsewhere in the text, unless they have some special significance in a specific plant community.

2.3.2. The Indigenous Flora

Of the 411 families in the world (Hutchinson, 1959), 221 occur in Australia (Table 2.3). The 190 families which do not occur are mostly small; 184 contain fewer than 20 genera and most of these are confined to the Americas. Only 5 families with more than 20 genera are absent from Australia, viz., Bromeliaceae, Cactaceae (both chiefly American), Marantaceae, Papaveraceae and Saxifragaceae (the last represented in Australia by the Southern Hemisphere counterpart, Escalloniaceae). These families are absent from New Guinea also, except Marantaceae (7 genera in New Guinea). The angiosperm flora is therefore not distinctive at the family level.

Families with the largest number of genera in Australia are Gramineae, Compositae, Papilionaceae and Orchidaceae. These are the four largest families in the world. However, of the large families, the one with the highest proportion of the world's total genera in Australia is Myrtaceae (Fig. 2.9). This family has the largest number of species in Australia and consequently the highest proportion of the world's species for individual families (Fig. 2.10). These 2 graphs indicate that Australia contains also a high proportion of the world's genera and species of Chenopodiaceae and Rutaceae.

Smaller families (20–100 genera in the world) with a strong representation in Australia are Cunoniaceae, Cyperaceae, Epacridaceae, Restionaceae, Proteaceae and Sterculiaceae. There are several of the smallest families (1–20 genera) which have their greatest representation in Australia: Casuarinaceae, Eupomatiaceae, Goodeniaceae, Himantandraceae, Myoporaceae, Pittosporaceae, Stackhousiaceae, Stylidiaceae, and the Monocotyledons, Centrolepidaceae, Philydraceae and Xanthorrhoeaceae. There are 12 endemic families (see 2.3.3 below).

With 1,686 genera (13.5% of the world's total) Australia, which covers c. 5% of the world's land surface, possibly has the expected number of genera, since many genera occur on

Table 2.3. The Australian angiosperm flora, classified according to the families recognized by Hutchinson (1959). Data from Burbidge (1963).

	Families	Genera	Species
Dicotyledons: Australia	173	1,250	10,500
Monocotyledons: Australia	49	436	2,600
Total: Australia	221	1,686	13,100
Total: World	411	13,000	300,000

Table 2.4. The Australian angiosperm flora (genera and species) and the world's flora for the same families. # indicates endemic to Australia Sources of data: World, Hutchinson (1959, 1964), Willis and Airy Shaw (1973); Australia, Burbidge (1960, 1963).

	WORLD			AUSTRALIA			
	Gen.	spp.	Distribution	Gen.	spp.	Distribution	Endemic Gen.
DICOTYLEDONS							
Acanthaceae	250	2500	Mainly Trop.	15	27	Trop. to arid	2
Actinidiaceae	3	350	Trop.; SE. Asia; Amer.	1	1	NE. Qld	0
Aegicerataceae†	1	2	India to N.G.	1	1	Coast: N.T.; Qld; N.S.W.	0
Aizoaceae†	130	1200	Afr.; Asia; Amer.	8	26	Littoral; arid	3
#Akaniaceae	1	1	Endemic to Aust.	1	1	Central E. coast	1
Alangiaceae	2	20	Trop.	1	5	E. coast: Qld; N.S.W.	0
Amaranthaceae	65	850	Trop. to Temp.	9	139	Widespread	2
Anacardiaceae	60	600	Trop. to Temp.	7	12	Trop. N.T.; Qld; N.S.W.	2
Annonaceae	120	2100	Trop.	12	21	Trop. N.T.; Qld; N.S.W.	3
Apocynaceae	180	1500	Mainly Trop.	11	66	Trop. Qld; arid; 1 Tas.	1
Aquifoliaceae	5	500	Trop. to Temp.	2	2	Trop.	0
Aquilariaceae	6	18	Afr.; Asia; N. Cal.	1	1	NE. Qld	0
Araliaceae	55	700	Trop.; some Temp.	12	32	Trop. to SE.; Tas.	3
Aristolochiaceae	10	400	Trop. to Temp.	1	9	Trop. N.T.; Qld; N.S.W.	0
Asclepiadaceae	130	2000	Trop. & Subtrop.	13	66	Mainly Trop. & arid.	2
#Austrobaileyaceae	1	2	Endemic to Aust.	1	2	NE. Qld	1
Balanophoraceae	18	120	Trop.	1	1	NE. Qld	0
Balanopsidaceae	2	8	N. Cal.	1	1	NE. Qld	0
#Baueraceae	1	3	Endemic to Aust.	1	3	SE.; Tas.	1
Bignoniaceae	120	650	Mainly Trop.	5	10	E.; arid	0
Bombacaceae	20	180	Trop., espec. Amer.	3	3	Trop.	0
Boraginaceae	100	2000	Trop. to Temp.	9	47	Widespread	2
#Brunoniaceae	1	1	Endemic to Aust.	1	1	Mainly S.; arid	1
Burseraceae	16	500	Trop.	2	4	Trop. to E. Qld, N.S.W.	0
Byblidaceae	2	4	S. Afr.	1	2	Trop.; SW. W.A.	1
Cabombaceae	2	7	Cosmopol. excl. Eur.	1	1	E. Qld; N.S.W.	0
Caesalpiniaceae	152	2800	Trop. to Temp.	15	70	Trop.; arid	2
Callitrichaceae	1	25	Cosmopol.	1	10	Mainly S. & E. coast	0
Campanulaceae	50	1500	Temp.; Subtrop.	3	15	Widespread	0
Capparidaceae	30	650	Trop. to Temp.	6	25	Widespread	3
Caprifoliaceae	12	450	Mainly N. Temp.	1	2	E. Temp. Qld to Tas.	0
Capusiaceae†	1	6	SE. Asia; Philipp.	1	3	E. Qld	0
Cardiopteridaceae	1	3	SE. Asia	1	1	NE. Qld	0
Caryophyllaceae	70	1750	Cosmopol.	6	29	Mainly Temp.	0
Casuarinaceae	2	65	SE. Asia; Pacif. Is.	1	30	General	0
Celastraceae	40	600	Trop. to Temp.	11	22	Trop. 1 SW.W.A.	5
#Cephalotaceae	1	1	Endemic to Aust.	1	1	SW. W.A.	1
Ceratophyllaceae	1	10	Cosmopol.	1	3	E. Qld to Vic.; S.A.	0
Chenopodiaceae	102	1400	Cosmopol.	21	226	Littoral; arid	11
#Cloanthaceae†	10	66	Endemic to Aust.	10	66	Widespread	10
Clusiaceae†	35	550	Mainly Trop.	4	13	Trop.	0
Cochlospermaceae	2	25	Trop.	1	3	Trop.	0
Combretaceae	19	600	Trop. & Subtrop.	3	34	Trop.; E. Qld	1
Compositae	900	13000	Cosmopol.	104	705	Widespread	55
Connaraceae	25	200	Trop.	2	2	NE. Qld	0
Convolvulaceae	55	1650	Trop. to Temp.	18	66	Widespread	1
Cornaceae	12	100	Mainly Temp.	1	1	N. coast N.S.W.	0
Corynocarpaceae	1	5	SW. Pacif.	1	1	NE. Qld	0
Crassulaceae	35	1500	Cosmopol., espec. Afr.	1	10	E. to W. in S.	0
Cruciferae	375	3200	Cosmopol.	24	85	Temp.; arid	15

	WORLD			AUSTRALIA			
	Gen.	spp.	Distribution	Gen.	spp.	Distribution	Endemic Gen.
Cucurbitaceae	110	640	Mainly Trop.	12	26	Trop.; arid	0
Cunoniaceae	26	250	S. Trop. & Subtrop.	14	23	NE. Qld to Tas.; 1 W.A.	4
Cuscutaceae†	1	170	Cosmop.	1	4	Temp.	0
Dichapetalaceae	5	200	Trop.	1	1	NE. Qld	0
Dilleniaceae	10	400	Trop. & Subtrop.	4	107	Widespread, excl. arid	1
Donatiaceae	1	2	Mainly Subantarctic	1	1	Tas.	0
Droseraceae	4	105	Trop. to Temp.	1	71	Widespread	0
Ebenaceae	7	500	Trop. espec. SE. Asia	1	6	Trop. to E. N.S.W.	0
Ehretiaceae†	13	400	Trop. & Subtrop.	4	25	Trop.; arid	1
Elaeagnaceae	3	50	Mainly N. Hemisph.	1	1	NE. Qld	0
Elaeocarpaceae	12	350	Trop. & Subtrop.	5	33	E., Qld to Tas.	1
Elatinaceae	2	40	Trop. to Temp.	2	7	Widespread	0
Epacridaceae	30	400	SE. Asia to S. Amer.	28	332	Mainly humid S.	21
#Eremosynaceae	1	1	Endemic to Aust.	1	1	SW. W.A.	1
Ericaceae	50	1350	Almost cosmopol.	5	9	NE. Qld to Tas.	1
Erythroxylaceae	2	250	Trop.	1	3	Trop. N.T., Qld; E. N.S.W.	0
Escalloniaceae	18	160	Mainly S. Hemisph.	7	20	Humid E.	4
Eucryphiaceae	1	5	S. Amer.	1	2	E. N.S.W.; Tas.	0
Euphorbiaceae	300	5000	Cosmopol.	47	290	Widespread	15
Eupomatiaceae	1	2	New Guinea	1	2	E. Qld to Vic.	0
Fagaceae	8	900	N. & S. Temp., excl. Afr.	1	3	SE. Qld to Tas.	0
Flacourtiaceae	93	1000	Trop. & Subtrop.	7	13	E. Qld, N. coast. N.S.W.	2
Frankeniaceae	4	90	Trop. to Temp.	1	49	Littoral; arid	0
Gentianaceae	80	900	Cosmopol.	4	5	Temp., espec. SE.	0
Geraniaceae	5	750	Cosmopol.	3	12	Temp.; S. arid	0
Gesneriaceae	120	2000	Trop. & Subtrop.	5	5	E. Qld to Vic.	1
Goodeniaceae	14	310	SE. Asia to S. Amer.	13	300	Widespread	8
#Gyrostemonaceae	5	15	Endemic to Aust.	5	15	Widespread	5
Haloragaceae	6	120	Cosmopol.	5	90	Widespread	2
Hamamelidaceae	22	80	Trop. & Subtrop.	2	2	NE. Qld	2
Helleboraceae†	27	850	Mainly N. Temp.	1	2	N.S.W. to Tas.	0
Hernandiaceae	4	50	Trop.	2	4	Trop.	1
Himantandraceae	1	2	N.G.; Moluccas	1	1	NE. Qld	0
Hippocrateaceae	18	220	Trop. & Subtrop.	3	4	NE. Qld	0
Hydrophyllaceae	18	250	Cosmopol.	1	2	Trop.	0
Hypericaceae	8	450	Trop. to Temp.	2	3	NE. Qld to Temp.	0
Icacinaceae	58	400	Trop.	5	6	E: Qld, N.S.W.	1
#Idiospermaceae	1	1	Endemic to Aust.	1	1	NE. Qld.	1
Illecebraceae†	10	100	Mainly arid	1	5	Mainly Temp.	0
Labiatae	180	3500	Cosmopol.	21	183	Widespread	6
Lauraceae	32	2500	Trop. & Subtrop.	6	65	E: Qld, N.S.W.	0
Lecythidaceae	15	325	Trop., espec. Amer.	2	3	Trop.	0
Lentibulariaceae	4	170	Cosmopol.	2	39	Widespread, excl. arid	0
Linaceae	12	290	Cosmopol.	2	3	Temp. to E. Qld.	0
Lobeliaceae	20	450	Trop. to Temp.	5	36	Widespread	0
Loganiaceae	7	130	Trop. to Temp.	2	21	Temp.	0
Loranthaceae	65	900	Trop. to Temp.	10	62	Widespread, excl. Tas.	6
Lythraceae	25	550	Cosmopol.	7	25	Widespread	0
Malpighiaceae	60	800	Trop. espec. S. Amer.	2	2	NE. Qld.	0
Malvaceae	75	1000	Trop. to Temp.	16	144	Widespread	4
Melastomataceae	240	3000	Trop. & Subtrop.	5	9	Trop. to E. coast N.S.W.	0
Meliaceae	50	1400	Trop. & Subtrop.	14	53	Mainly Trop.; arid	3
Menispermaceae	65	350	Trop. & Subtrop.	13	21	Trop. to E. coast N.S.W.	5
Menyanthaceae	5	33	Mainly Temp.	3	19	Mainly Temp.	0
Mimosaceae	55	2800	Trop. & Subtrop.; arid	10	c.750	Widespread	0

	WORLD			AUSTRALIA			
	Gen.	spp.	Distribution	Gen.	spp.	Distribution	Endemic Gen.
Monimiaceae	32	350	S. Hemisph.	10	21	E. coast	5
Moraceae	53	1400	Mainly Trop. & Subtrop.	7	40	Trop. to coast N.S.W.	0
Myoporaceae	4	130	Asia; Afr.; Pacif.	2	120	Widespread	1
Myristicaceae	18	300	Trop., espec. Asia	2	2	Trop.	0
Myrsinaceae	35	1000	Trop. & Subtrop.	6	24	Trop.; E. Qld to Vic.	0
Myrtaceae	100	3000	Trop. to Temp.	47	1300	Widespread	33
Nepenthaceae	2	68	Mainly Asia; N. Cal.	1	1	NE. Qld	0
Nyctaginaceae	30	290	Trop. & Subtrop.	4	6	Mainly Trop. & arid	0
Nymphaeaceae	6	90	Almost cosmopol.	2	7	Trop. to coast N.S.W.	0
Ochnaceae	40	600	Mainly Trop.	1	1	NE. Qld	0
Olacaceae	25	250	Mainly Trop.	2	7	Widespread	0
Oleaceae	29	600	Cosmopol., espec. Asia	6	23	Trop.; E. Qld to Vic.	0
Onagraceae	21	640	Trop. to cool Temp.	4	16	Widespread	0
Opiliaceae	8	60	Trop., espec. Asia	2	2	Trop.	0
Orobanchaceae	14	180	Mainly Temp. Eurasia	1	1	Temp.	0
Oxalidaceae	4	900	Trop. to Temp.	1	2	Temp.	0
Papilionaceae	482	12000	Cosmopol.	86	846	Widespread	37
Passifloraceae	12	600	Trop. to Temp.	2	11	Trop.; E. Qld to Vic.	0
Pedaliaceae	12	50	Trop. Asia, Afr.	2	5	Trop.; arid	1
Periplocaceae	45	200	Trop. to warm Temp.	1	1	Trop.	0
Petiveraceae†	11	96	Afr.; Amer.; N. Cal.	1	1	E. Qld; N.S.W.	1
Piperaceae	4	2000	Trop.	2	14	Trop.; E. Qld; N.S.W.	0
Pittosporaceae	9	200	Afr. to Pacif.	9	49	Mainly Temp; arid	6
Plantaginaceae	3	270	Cosmopol.	1	14	Widespread	0
Plumbaginaceae	10	500	Cosmopol.	3	5	Trop.; littoral; arid	0
Podostemaceae	45	130	Mainly Trop.	1	1	E. Qld	0
Polygalaceae	12	800	Almost cosmopol.	4	38	Widespread	1
Polygonaceae	40	800	Mainly N. Temp.	3	34	Widespread	0
Portulacaceae	19	350	Cosmopol.	6	48	Widespread	0
Potaliaceae†	4	70	Trop.	1	4	Trop.	0
Primulaceae	20	1000	Cosmopol.	2	5	Temp.	0
Proteaceae	62	1050	Afr.; Asia to S. Amer.	37	840	Widespread	28
Rafflesiaceae	8	50	Trop.	1	1	SW. W.A.	0
Ranunculaceae	22	1040	Cosmopol.	3	39	Temp.	0
Rhamnaceae	58	900	Cosmopol.	16	157	Widespread	7
Rhizophoraceae	16	120	Trop.	5	8	Trop. to N. coast N.S.W.	0
Rosaceae	124	3375	Cosmopol.	7	15	Mainly Temp.	1
Rubiaceae	500	6000	Cosmopol.	38	195	Widespread	5
Rutaceae	150	900	Trop. to Temp.	36	324	Widespread	20
Santalaceae	30	400	Trop. to Temp.	9	50	Temp. & arid	5
Sapindaceae	150	2000	Trop. & Subtrop.	25	144	Widespread	7
Sapotaceae	75	800	Trop.	9	47	Trop. to coast N.S.W.	2
Sauraviaceae	1	300	Trop. Asia, Amer.	1	1	NE. Qld	0
Scrophulariaceae	220	3000	Cosmopol.	28	107	Widespread	2
Simaroubaceae	20	120	Trop. & Subtrop.	8	9	E. Trop. to N. N.S.W.	3
Solanaceae	90	2200	Cosmopol.	8	127	Widespread	3
Sonneratiaceae	2	7	E. Afr. to Polynes.	1	2	Trop.	0
Spigeliaceae†	3	90	Afr.; Asia; S. Amer.	2	37	Widespread excl. arid	0
Stackhousiaceae	2	25	SE. Asia to N.Z.	2	23	Widespread	1
Sterculiaceae	60	700	Mainly Trop.	21	152	Widespread	10
Strychnaceae†	6	250	Trop. & Subtrop.	1	4	Trop.	0
Stylidiaceae	5	150	SE. Asia; Pacif.; S. Amer.	4	120	Widespread excl. arid	1
Symplocaceae	2	300	Trop. & Subtrop.	1	7	E. Qld; N.S.W.	0
Theaceae	16	500	Trop. & Subtrop.	1	1	NE. Qld	0
Thymelaeaceae	50	500	Almost cosmopol.	6	89	Widespread	1

	WORLD			AUSTRALIA			
	Gen.	spp.	Distribution	Gen.	spp.	Distribution	Endemic Gen.
Tiliaceae	50	450	Trop. to Temp.	6	58	Trop.; N. arid	0
#Tremandraceae	3	28	Endemic to Aust.	3	28	Temp.	3
Trimeniaceae†	4	15	Afr.; SE. Asia to Pacif.	1	1	N. coast N.S.W.	1
Ulmaceae	15	200	Trop. to Temp.	2	7	Trop. to E. N.S.W.	0
Umbelliferae	275	2850	Cosmopol.	23	156	Temp.; arid	9
Urticaceae	45	550	Trop. to Temp.	9	16	Widespread	0
Verbenaceae	75	3000	Trop. to Temp.	14	41	Widespread	3
Violaceae	22	900	Cosmopol.	5	19	Mainly Temp.	1
Viscaceae	11	450	Trop. & Subtrop.	3	11	Mainly SE. & S.	0
Vitaceae	12	700	Trop. & Subtrop.	6	22	E. coast; 1 SW.W.A.	1
Winteraceae	8	120	SE. Asia to S. Amer.	2	8	E.: Qld to Tas.	0
Zygophyllaceae	25	240	Trop. & Subtrop.	4	33	Arid & dry Temp.	0

MONOCOTYLEDONS

	WORLD			AUSTRALIA			
	Gen.	spp.	Distribution	Gen.	spp.	Distribution	Endemic Gen.
Agavaceae	19	520	Drier Trop. & Subtrop.	3	7	E. coast Qld. N.S.W.	1
Alismataceae	13	90	Trop. to Temp.	2	5	Widespread	0
Amaryllidaceae	85	1100	Trop. & Subtrop.	3	18	Trop.; E. coast; arid	1
Aponogetonaceae	1	30	Trop. Afr.; Asia	1	1	Trop. to E. coast N.S.W.	0
Apostasiaceae	3	20	SE. Asia, N.G.	1	1	NE. Qld	0
Araceae	115	2000	Trop. & Subtrop.	8	16	Trop. to E. coast N.S.W.	1
Burmanniaceae	11	100	Trop. to Temp.	1	2	Trop. & Subtrop.	0
Butomaceae	5	8	Trop. to Temp.	1	1	E. Qld	0
#Cartonemataceae	1	7	Endemic to Aust.	1	7	Trop.; SW. W.A.	1
Centrolepidaceae	7	40	SE. Asia; Pacif.; S. Amer.	7	35	Mainly Temp.	4
Commelinaceae	38	500	Trop. & Subtrop.	6	17	Trop.; E. coast; SW. W.A.	0
Cyperaceae	95	4000	Cosmopol.	42	530	Widespread	12
Dioscoreaceae	6	750	Mainly Trop.	1	3	Trop.; E. coast; SW. W.A.	0
Eriocaulaceae	13	1150	Trop. to Temp.	1	19	Widespread	0
Flagellariaceae	3	10	Trop.	2	2	Trop. to E. coast N.S.W.	0
Gramineae	620	10000	Cosmopol.	40	860	Widespread	32
Haemadoraceae	16	90	S. Afr.; Trop. Amer.	7	76	E. coast; SW. W.A.	6
Hydrocharitaceae	16	80	Trop. to Temp.	9	14	Widespread	1
Hypoxidaceae	5	95	Mainly S. Hemisph.	4	8	Trop.; E., Qld to Tas.	0
Iridaceae	70	800	Subtrop. & Temp.	5	30	E. coast; S.A.; SW. W.A.	2
Juncaceae	9	400	Mainly Temp. S. Hemisph.	2	23	Mainly Temp.	0
Juncaginaceae	4	25	Almost cosmopol.	3	15	Widespread	2
Lemnaceae	2	27	Almost cosmopol.	2	5	Widespread	0
Liliaceae	180	3000	Cosmopol.	33	140	Widespread	22
Musaceae	2	42	Trop.	1	4	NE. Qld	0
Najadaceae	1	50	Cosmopol.	1	5	Trop. to E. coast N.S.W.	0
Orchidaceae	735	20000	Almost cosmopol.	91	550	Widespread, excl. arid	25
Palmae	217	2500	Trop. & Subtrop.	19	41	Trop.; E. Qld to Vic.	3
Pandanaceae	3	700	Trop. excl. Amer.	2	17	Mainly Trop.	0
#Petermanniaceae	1	1	Endemic to Aust.	1	1	E. coast Qld, N.S.W.	1
Philesiaceae	7	9	S. Amer.; Afr.; N.Z.	3	3	Trop.; E. Qld to Vic.	0
Philydraceae	4	5	Malays.	4	4	E. coast; SW. W.A.	0
Pontederiaceae	7	30	Mainly Trop.	1	2	Trop.	0
Posidoniaceae	1	2	Mediterranean; marine	1	2	Mainly S.	0
Potamogetonaceae	2	100	Cosmopol.	1	20	Widespread	0
Restionaceae	29	320	Afr.; Asia; S. Amer.	17	90	E. coast; SW. W.A.	13
Roxburghiaceae†	3	30	SE. Asia; N. Amer.	1	2	Trop.	0
Ruppiaceae	1	2	Temp.; Subtrop.; marine	1	2	Temp. & Subtrop.	0
Smilacaceae	4	375	Trop. to Temp.	2	8	Trop.; E. Qld to Vic.	0
Sparganiaceae	1	20	Mainly N. Temp.	1	2	E. coast, Qld to Vic.	0
Taccaceae	2	31	Trop.	1	3	Trop.	0

| | WORLD | | | AUSTRALIA | | | Endemic |
	Gen.	spp.	Distribution	Gen.	spp.	Distribution	Gen.
Thismiaceae	10	44	Trop.	1	1	Vic.; Tas.	0
Triuridaceae	4	30	Trop.	1	1	NE. Qld	0
Typhaceae	1	20	Cosmopol.	1	2	Widespread	0
Xanthorrhoeaceae	8	66	N.G.; N. Cal.	8	58	S. Qld to Tas.; SW. W.A.	7
Xyridaceae	2	250	SE. Asia; Amer.; Afr.	1	15	Trop. to Temp.	0
Zannichelliaceae	6	20	Cosmopol.	4	12	Widespread	1
Zingiberaceae	45	700	Trop. & Subtrop.	7	10	N. & E. Qld; N.S.W.	0
Zosteraceae	2	10	Both Hemisph.; marine	1	2	S.A.; Vic.; N.S.W.; Tas.	0

† Aegiceratacea: sometimes included under Myrsinaceae
Aizoaceae: includes Molluginaceae
Capusiaceae: also known as Siphonodontaceae
Cloanthaceae: sometimes included under Verbenaceae
Clusiaceae: also known as Guttiferae
Cuscutaceae: sometimes included under Scrophulariaceae
Ehretiaceae: sometimes included under Boraginaceae
Eremosynaceae: sometimes included under Saxifragaceae
Helleboraceae: sometimes included under Ranunculaceae
Illecebraceae: sometimes included under Caryophyllaceae
Petiveraceae: sometimes included under Phytolaccaceae
Potaliaceae: sometimes included under Loganiaceae
Roxburghiaceae: also known as Stemonaceae
Spigeliaceae: sometimes included under Loganiaceae
Strychnaceae: sometimes included under Loganiaceae
Trimeniaceae: sometimes included under Monimiaceae

more than one continent and some are cosmopolitan. The largest Australian genera are *Acacia* and *Eucalyptus*, which dominate the vegetation over vast areas. Genera containing more than 50 species are listed in Table 2.5. Many of these genera are endemic (see 2.3.3, below), and only a few of them (*Acacia*, *Helichrysum* etc.) are represented abundantly outside Australia.

At the generic level, the flora is closely related to that of New Guinea and south-east Asia in the north, and with New Zealand and South America in the south, which enables the identification of elements (see Chap. 3).

The number of species in the continent (c. 13,000) representing only 4.3% of the world's total species, might be regarded as low in view of the fact that the continent occupies 5% of the world's land surface. About 75% of the species are endemic. Those which are not endemic occur mainly in the rainforests, in aquatic habitats or in the littoral zone.

2.3.3. The Endemic Families and Genera

There are 12 endemic families (marked # in Table 2.4). These are all small, containing a total of 27 genera and 126 species. The number of endemic families is possibly the expected number for the area. For comparison, the Americas, covering c. 5 times the area of Australia, contain 49 endemic families.

There are 538 endemic genera which constitute 32% of the indigenous genera (1,686); they occur in 92 families (Table 2.4). In addition, c. 100 genera are near-endemic and probably originated in Australia, e.g. *Eucalyptus*, *Melaleuca* (see Table 2.5); one or a few species of such genera occur either on the islands close to Australia, mainly New Guinea, New Caledonia, and New Zealand, or on the Asian mainland. Endemism is dealt with in more detail in Chap. 4.

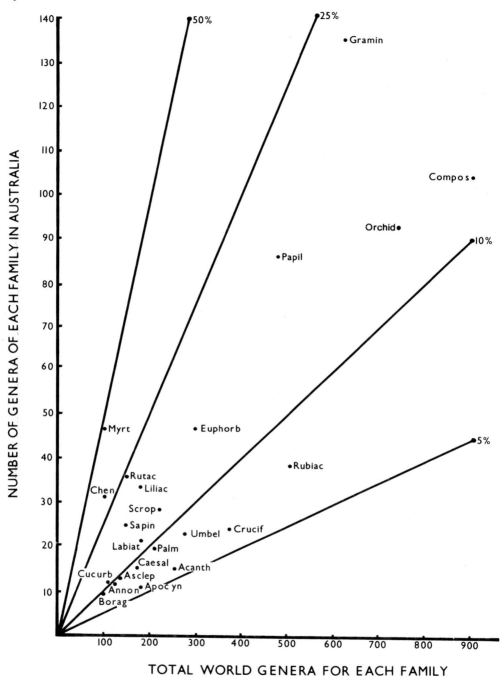

Fig. 2.9. Proportions of the world's genera occurring in Australia for each of the families containing more than 100 genera in the world.

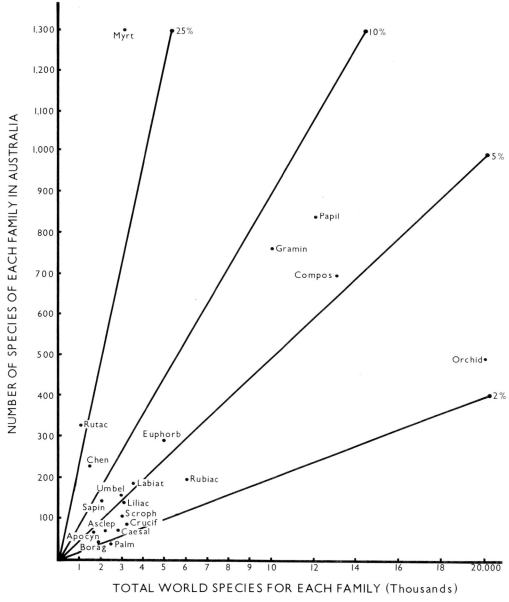

Fig. 2.10. Proportions of the world's species occurring in Australia for each of the families containing more than 100 genera in the world.

2.3.4. Primitive Angiosperm Families

Several primitive families of angiosperms occur in Australia (Table 2.6 and Fig. 2.11) and their presence, together with the occurrence of other primitive families in the west-Pacific region, has been used by some authors to postulate that the angiosperms originated in this general region, a view discounted by fossil evidence (Chap. 3). Nearly all of these families occur in the rainforests, the notable exception being *Hibbertia* (Dilleniaceae), though a few

Table 2.5. Indigenous genera with more than 50 species. Endemic genera are marked #.
Data from Burbidge (1963), Willis and Airy Shaw (1973).

Genus	Family	Number of Australian species	Distribution outside Australia	Total world species
Acacia	Mimosaceae	c. 750	Trop. & Subtrop., espec. Afr.	c. 950
Eucalyptus	Myrtaceae	c. 450	N. G. to Malays.; Philipp.	c. 450
Grevillea	Proteaceae	250	N. G. to Malays.	260
# *Hakea*	Proteaceae	140		140
Leucopogon	Epacridaceae	140	Malays.; N.Z.	150
Melaleuca	Myrtaceae	140	N. G.; SE. Asia	141
Goodenia	Goodeniaceae	120	N. G.; SE. Asia	121
Stylidium	Stylidiaceae	110	SE. Asia	114
# *Eremophila*	Myoporaceae	110		110
Hibbertia	Dilleniaceae	100	N. G.; N. Cal.; Madagas.	103
Helichrysum	Compositae	100	All continents, excl. Amer.	500
Ptilotus	Amaranthaceae	100	Malays.	101
# *Pultenaea*	Papilionaceae	100		100
Prasophyllum	Orchidaceae	85	N. Z.	85
Cyperus	Cyperaceae	82	Widespread. Trop. to Temp.	550
Fimbristylis	Cyperaceae	80	Widespread in Trop.	300
Olearia	Compositae	80	N. G.; N. Z.	100
Pimelea	Thymelaeaceae	80	Timor; N. Z.	85
Schoenus	Cyperaceae	80	N. Z.; Malays.; N. Temp.	100
Solanum	Solanaceae	80	Trop. to Temp.	1700
Baeckea	Myrtaceae	70	SE. Asia; SW. Pacif.	72
# *Boronia*	Rutaceae	70		70
Caladenia	Orchidaceae	70	Indonesia; N. Z.; N. Cal.	75
# *Daviesia*	Papilionaceae	70		70
Drosera	Droseraceae	70	Trop. to Temp.	100
Haloragis	Haloragaceae	70	SE. Asia; N. Z.; S. Amer.	76
Scaevola	Goodeniaceae	70	Trop., mainly sea coasts	80–100
# *Prostanthera*	Labiatae	65		65
Pterostylis	Orchidaceae	60	N. G.; N. Cal.; N. Z.	70
Brachycome	Compositae	52	N. G.; N.Z.; N. Cal.	60
Helipterum	Compositae	60	S. Afr.	90
Phyllanthus	Euphorbiaceae	60	Trop. to Temp.	600
# *Dampiera*	Goodeniaceae	58		58
# *Dryandra*	Proteaceae	56		56
Dendrobium	Orchidaceae	55	SE. Asia; N. G.; SW. Pacif.	c. 1000
Dodonaea	Sapindaceae	55	Trop. & Subtrop.	60
Stipa	Gramineae	55	Trop. to Temp.	300
Aristida	Gramineae	50	Subtrop. to Temp.	350
# *Bossiaea*	Papilionaceae	50		50
# *Calytrix*	Myrtaceae	50		50
Swainsona	Papilionaceae	50	N. Z.	51
# *Verticordia*	Myrtaceae	50		50

species of this genus grow in rainforest margins. Four of the families are confined to Australia and New Guinea. The significance of the primitive families with respect to the origins and evolution of the modern flora is discussed in Chap. 4.

2.3.5. Heterotrophic Angiosperms

A large number of heterotrophic plants occur in the continent, including holoparasites and saprophytes which lack chlorophyll, carnivorous plants, and stem or root hemiparasites. They

Table 2.6. Primitive woody angiosperm families in Australia. Families arranged in the order of Hutchinson (1959). Data from Airy Shaw (1973), Burbidge (1963) and Womersley (1969). #indicates endemic to Australia. (x) indicates confined to Australia and New Guinea. Occurrences outside Australia and New Guinea given in last column.

Family	Number of Genera in World	Australia	N.G.	Genera in Australia	Genera occurring in both Australia and N.G., and occurrences elsewhere
Winteraceae	8	2	4	Bubbia, Tasmannia	Bubbia (SW. Pacif.), Tasmannia (SE. Asia; N. Cal.)
Himantandraceae	1	1	1	Himantandra	Himantandra (x)
Annonaceae	120	13	31	#Ancana, Cananga, Fissistigma, #Fitzalania, #Haplostichanthus, Mitrephora, Polyalthia, Popowia, Rauwenhoffia, Saccopetalum, Unona, Uvaria	Cananga (SE. Asia), Fissistigma (Afr.; Asia), Mitrephora (Trop. Asia), Polyalthia (Afr.; Asia; Pacif.), Popowia (Afr.; Asia), Rauwenhoffia (SE. Asia), Unona (Afr., Asia), Uvaria (Afr.; Asia)
Eupomatiaceae	1	1	1	Eupomatia	Eupomatia (x)
Monimiaceae	32	9	11	#Atherosperma, #Daphnandra, #Doryphora, Hedycarya, Levieria, Steganthera, #Tetrasynandra, Wilkiea	Daphnandra (x), Hedycarya (W. Pacif.), Levieria (x), Palmeria (SE. Asia), Steganthera (Celebes), Wilkiea (x)
Austrobaileyaceae	1	1	1	Austrobaileya	Austrobaileya (x)
Lauraceae	32	6	14	Beilschmiedia, Cinnamomum, Cryptocarya, Endiandra, Litsea, Neolitsea	Beilschmiedia (Pantrop.), Cinnamomum (E. Asia), Cryptocarya (Trop., excl. Afr.), Endiandra (SE. Asia; W. Pacif.), Litsea (Asia; Amer.) Neolitsea (SE. Asia)
Trimeniaceae	4	1	2	Piptocalyx	Piptocalyx (x)
Hernandiaceae	4	3	3	Gyrocarpus, Hernandia #Valvanthera	Gyrocarpus (Pantrop.), Hernandia (Pantrop.)
Myristicaceae	18	2	4	Horsfieldia, Myristica	Horsfieldia (SE. Asia), Myristica (Afr.; Asia)
Dilleniaceae	10	4	3	Dillenia, Hibbertia #Pachynema, Tetracera	Dillenia (SE. Asia; SW. Pacif.), Hibbertia (Madagas.; N. Cal.), Tetracera (Pantrop.)
#Idiospermaceae	1	1	0	#Idiospermum	

Fig. 2.11. Two primitive angiosperms. Left, *Idiospermum australiense*. The flowers, c. 25–30 mm diam. have spirally arranged red perianth-segments in c. 4 rows. The stamens are visible in the centre and are best seen in the uppermost flower (right) which has lost its perianth. The free carpels are enclosed in the receptacle. Photo: B. J. Wallace. Centre and right, *Eupomatia laurina*. Bud (note operculum) and flowers shown in central photograph. Flowers are 20–25 mm diam. The fertile stamens (reflexed) are outermost and in c. 3 rows, spirally arranged. The petal-like structures (erect) are inserted inside the stamens and are interpreted as staminodes. At right: aggregate fruits which are 15–20 mm diam., and consist of a fleshy receptacle enclosing a number of seeds embedded in pulp. Photo: George Wray.

Table 2.7. Heterotrophic angiosperms indigenous in Australia, arranged according to method of nutrition. Data from Bentham (1863–1878), Black (1943–1957), Burbidge (1963), Barlow (1966), Beard (1970). #indicates endemic in Australia. Number of Australian species in brackets.

I. HOLOPARASITES WITHOUT CHLOROPHYLL, LIVING ON ROOTS OF OTHER PLANTS		
BALANOPHORACEAE	*Balanophora* (1)	Herb. Subterranean except inflorescence. Rainforests. NE. Qld; SE. Asia; SW. Pacif.
OROBANCHACEAE Broom Rape	*Orobanche* (2)	Herbs to c. 40 cm. SW. W. A. to S. N. S. W. All continents
RAFFLESIACEAE	*Pilostyles* (1)	Herb. Parasitic on *Daviesia*. SW. W. A.; Amer.; Afr.; Asia
II. SAPROPHYTES WITHOUT CHLOROPHYLL		
ORCHIDACEAE (for additional spp., see text)	*Aphyllorchis* (2)	Herbs. Trop.; SE. Asia
	#*Cryptanthemis* (1)	Herb. NE. N.S.W.
	Didymoplexis (1)	Herb. Trop.; Afr.; Asia; Pacif.
	Epipogium (1)	Herb. NE. Qld; NE. N.S.W.; Afr.; Asia; Eur.
	Galeola (2)	Climbing herbs. E. Qld; E. N.S.W.; E. Asia; SW. Pacif.
	Gastrodia (2)	Herbs. SE. Qld and wetter S.; Asia; N. Z.
	#*Rhizanthella* (1)	Herb. SW. W. A.
THISMIACEAE	*Thismia* (1)	SE. and Tas.; Trop. Asia
III. CARNIVOROUS PLANTS WITH PITCHERS		
#CEPHALOTACEAE	#*Cephalotus* (1)	Herb. Swamps. SW. W. A.
NEPENTHACEAE	*Nepenthes* (1)	Herb. NE. Qld; SE. Asia; New Caled.; Madagascar

IV. CARNIVOROUS PLANTS WITH STICKY HAIRS OR THE LEAF-BLADE MODIFIED AS A TRAP.

BYBLIDACEAE	# *Byblis* (2)	Herbs with stick hairs. One sp. SW. W. A., one Trop. N.T.
DROSERACEAE Sundews	*Aldrovanda* (1)	Aquatic herb, with leaf-blade modified as a trap. Trop. N.T.; Qld; Eur.; Asia
	Drosera (65)	Herbs with sticky hairs. Widespread in wetter areas, mostly SW. W. A., rare in arid zone. All continents

V. CARNIVOROUS PLANTS WITH MINUTE BLADDERS (TRAPS) ON FINELY DISSECTED SUBMERGED OR SUBTERRANEAN LEAVES

LENTIBULARIACEAE	*Polypompholyx* (3)	Herbs. Swamps, wet rocks. SW. W. A.; S. A.; Vic.; Tas.; S. Amer.
	Utricularia (37)	Herbs, one sp. twining. Swamps and free water. Widespread in wetter areas. All continents

VI. STEM AND ROOT HEMIPARASITES (except *Cuscuta*)

CASSYTHACEAE Devil's Twine	*Cassytha* (12)	Stems filiform; rootless, except seedling. Widespread near coast, mainland and Tas.; Afr.; Asia
CUSCUTACEAE	*Cuscuta* (4 + 4 introd.)	Stems filiform; rootless, except seedling. Chlorophyll absent. S. half of Aust. All continents
LORANTHACEAE Mistletoes	# *Atkinsonia* (1)	Shrub; root parasite. Blue Mts, N.S.W.
	# *Nuytsia* (1)	Tree; root parasite. SW. W. A.
	Amyema (32)	Stem parasites. Mainland. SE. Asia; Pacif.
	Amylotheca (4)	Stem parasites. N.S.W.; Qld; N.G.; Pacif.
	# *Benthamina* (1)	Stem parasite. N.S.W.; Qld
	Decaisnina (6)	Stem parasites. Trop.; N.G.; W. Pacif.
	Dendrophthoe (6)	Stem parasites. N.S.W.; Qld; Afr.; Asia
	# *Diplatia* (3)	Stem parasites. N.S.W.; Qld; N.T.
	# *Lysiana* (6)	Stem parasites. Mainland
	Muellerina (4)	Stem parasites. Qld; N.S.W.; Vic.
VISCACEAE Mistletoes	*Korthalsella* (2)	Stem parasites on rainforest trees, Qld; N.S.W.; Afr.; Asia; Pacif.
	Notothixos (4)	Parasitic on other mistletoes. Qld; N.S.W.; Vic.; E. Asia
	Viscum (5)	Stem parasites. Qld; N.S.W.; Afr.; Asia; Eur.
OLACACEAE	*Olax* (5)	Shrubs, root parasites. Widespread in wetter areas. Afr.; Asia
SANTALACEAE All root parasites	# *Anthobolus* (5)	Shrubs. Mainly semi-arid and arid
	# *Choretrum* (6)	Shrubs. SE. Qld and wetter SE.
	# *Eucarya* (*Fusanus*) (5)	Shrubs or small trees. S. half of Aust., mainly semiarid and arid
	Exocarpos (9)	Shrubs and trees. Widespread; rainforests to arid. Philipp.; Pacif.; N. Z.
	Henslowia (1)	Tree. NE. Qld; SE. Asia
	# *Leptomeria* (16)	Shrubs. Mainly S., wet to arid
	# *Omphacomeria* (1)	Shrub. SE. N.S.W.; Vic.
	Santalum (3)	Shrubs and tree. Mainly N. & arid; SE. Asia
	# *Spirogardnera* (1)	Shrub. SW. W. A.
	Thesium (1)	Herb. SE. Qld to Tas.; Afr.; Asia; S. Amer.
SCROPHULARIACEAE some root parasites)	*Euphrasia* (12–15,	Herbs. SE. and Tas.; SW. W. A.; Cool Temp., both Hemisph.

Table 2.8. Numbers of angiosperm genera naturalized in Australia.

Family	Genera
Compositae	87
Gramineae	66
Cruciferae	30
Papilionaceae	30
Iridaceae	22
Scrophulariaceae	21
Caryophyllaceae	18
Umbelliferae	16
Labiatae	14
Liliaceae	13
Solanaceae	11
Boraginaceae	10
Rosaceae	10
Other families (87)	192
Total	540

occur in several families and are listed in Table 2.7, which provides also notes on habit and distribution.

Only one family is endemic but several genera, especially among the hemiparasites, are confined to Australia. Many of the taxa have restricted habitats, notably carnivorous plants occurring in free water or in swamps. On the other hand, the hemiparasites, especially the mistletoes, are widely distributed, occurring even in the driest areas, provided trees or shrubs are present (Chap. 5). Very many of these plants have succulent fruits and are dispersed by birds. A general account of some of the heterotrophic plants in Australia is provided in McLuckie and McKee (1954).

In addition to the taxa listed in Table 2.7, there are a few additional genera of Orchidaceae which are mostly autotrophic but which contain one or two saprophytic species. These are: *Calochilus* (1 saprophytic sp. – 7 autotrophic), *Cryptostylis* (1–5), *Dipodium* (2–2), *Prasophyllum* (2–100) and *Eulophia* (2–4), (B. J. Wallace, personal communication).

2.3.6. The Naturalized Flora

About 540 genera belonging to 100 families have become naturalized (Burbidge, 1963). The number of species belonging to these genera, together with introduced species belonging to genera which are native to Australia, is about 1,500. A list of the families containing most of the introduced genera is provided in Table 2.8. Many of the plants were introduced for the production of food or as ornamentals. They were initially grown in fields or gardens, from which they escaped, to become established in the natural vegetation. Some of these are now regarded as weeds. Many species were introduced for pasturage, particularly grasses and legumes. Accidental introduction, for example as seed impurities in grain, accounts for most of the weed species, many of which belong to the Compositae.

3. Origins of the Vascular Flora

1. Introduction

When Australia became isolated as a result of the fragmentation of Gondwanaland (Chap. 1), it supported a considerable vascular flora which occurred elsewhere on Gondwanaland and consisted of pteridophytes, gymnosperms and angiosperms. The effects of continental drift in dividing the flora and fauna of the world as they existed in the Cretaceous and the effects of isolation on the various groups of organisms have been discussed by many authors, e.g., Raven and Axelrod (1974) and Smith (1974) for the plants, and Keast (1973) for the animals. Consequently only the main plant taxa relevant to Australia are discussed below. Continental drift cannot explain the distribution of many taxa, especially those which exhibit major disjunctions (Thorne 1972), and in particular bipolar taxa, some of which are represented in Australia and are mentioned below.

This explanation for the origins of the earliest flora of Australia, particularly for the angiosperms, replaces the "invasion theories" which emanated from Hooker's (1859) classification of the angiosperm flora into 3 "elements", viz., Indo-Malayan, the autochthonous (Australian) and the Antarctic elements. Though these elements cannot be defined "with exactitude" (Gardner, 1942) they nevertheless indicate the close floristic relationships between the Asian flora in the north and the New Zealand – South American flora in the south. Hooker also recognized the occurrence in Australia of cosmopolitan taxa, many of them bipolar.

The recognition of elements led later botanists to assume that the autochthonous (Australian) flora, the origins of which could not be explained, was invaded from the north by tropical taxa and from the south by a cool climate assemblage typified by *Nothofagus*. The invasions were assumed to have occurred across land connections during the Cretaceous or Early Tertiary. Possible land connections to the north were readily devised; for the south, attenuated "isthmian connections" between the southern continents had to be postulated. These explanations are now obsolete, though distance dispersal must be invoked to explain the presence of certain taxa in the continent, so that Hooker's "elements" and the invasion theories are not so aberrant as it might at first appear. Two other items need modification or explanation. Firstly, the "Indo-Malayan element" appears to have been derived from two segments, one of Gondwanaland origin, the other through recent additions from Asia. Secondly, the autochthonous element is a secondary development in the continent, having been derived mainly from the original Gondwanaland flora.

The history of the flora prior to the evolution of the angiosperms, is not dealt with in detail, but a brief account of the sequence of pteridophytes and gymnosperms is given, and the vascular flora of the Cretaceous is discussed in as much detail as the fossil record permits.

2. Pteridophytes and Gymnosperms

2.1. Pre-Cretaceous Flora

The first vascular plants are recorded from the Devonian (Gould, 1976) and included Psilopsida (*Hedeia* and *Yarravia*), *Zosterophyllum australianum*, one of the group which were probably the precursors of the lycopods, and Lycopsida (*Baragwanathia* and, later, *Leptophloeum*). Through the ages, additional groups of pteridophytes evolved, viz., the Sphenopsida (now extinct in Australia) and the Pteropsida (Table 2.1). The first seed plants were pteridosperms (now extinct in the world) of which the Glossopterideae were the most significant, since the existence of plants of this group in India and the four southern continents has been used to support the one-time existence of Gondwanaland (Plumstead, 1962; Melville, 1966). The pteridosperms were replaced by the cycads and, later, the Ginkgoales (now extinct in Australia), and finally the conifers evolved (Table 2.2). Gould (1976) indicates that the Australian megafossil floras of the pre-Cretaceous show a "cycle of cosmopolitan and endemic forms", the Glossopterideae being the significant group confined to the south. Furthermore, the fossil record indicates that all the main orders of pteridophytes (except the advanced ferns with a vertical annulus, see 7.3, below) and of the gymnosperms were established in the Australian segment of Gondwanaland before the Cretaceous.

2.2. Cretaceous Flora

A summary of the Australian and New Zealand Cretaceous floras is given in Table 3.1 and 3.2. Only fossils which can be referred to recognized botanical taxa have been included. Leaf fragments and vegetative shoots assigned to such "genera" as *Phyllites* and *Taxites* have been omitted and also spores of unknown taxonomic affinities.

The Cretaceous pteridophytes and gymnosperms amounted to several tens of genera and possibly a few hundred species. Most of the genera occurred elsewhere in the world. Dettmann's (1963) spore-records for south-eastern Australia number 110 species in 20 genera, the great majority of which are pteridophytes or gymnosperms. A large number of the spores belong to species and genera which occur in rocks of similar age in Western Australia, Canada and Siberia. The megafossils indicate the presence of lycopods, osmundaceous ferns, and araucarian and podocarpaceous gymnosperms.

Several of the groups well represented in the Australian Cretaceous have become extinct in the world. In Australia the pteridosperms were represented by some 15 genera, the Bennettitales by 3 or 4. Also, the extant genera *Equisetum* and *Ginkgo*, widely distributed in the Australian Cretaceous, have become extinct in Australia. Both these genera persisted until the early Tertiary (Deane, 1925; Slade, 1964).

Among the pteridophytes, the Lycopsida and many of the Pteropsida are well represented in the Cretaceous and presumably have had a continuous existence in Australia from the early Cretaceous to the present.

For the conifers, the data in Table 3.2 indicate that *Araucaria*, *Microcachrys* and *Podocarpus* have had a continuous existence in Australia from the beginning of the Cretaceous and that these genera occurred also in New Zealand during the Cretaceous. Of interest is the extinction in New Zealand of three genera recorded for the Cretaceous, *Araucaria*, *Athrotaxis* and *Microcachrys* which are still present in Australia. These is no Australian Cretaceous record for Cupressaceae, or for *Athrotaxis*.

Table 3.1. Australian Cretaceous pteridophytes. M = megafossil; S = spore. For living genera, see Table 2.1. Sources of data: Walkom (1915, 1917, 1918, 1919); Jones and de Jersey (1947); Medwell (1952, 1954); Kenley (1954); Dettmann (1963); Playford (1965); Playford and Dettmann (1965); Helby (1966); Playford and Cornelius (1967); Douglas (1973); Gould (1976).

	Cretaceous Early	Late
LYCOPSIDA		
Lycopodiaceae	M S	S
Selaginellaceae	S	S
Isoëtaceae	M	M
SPHENOPSIDA		
Equisetaceae	M	M
PTEROPSIDA		
Ophioglossaceae	S?	S?
Marattiaceae	M	
Osmundaceae	M S	M S
Schizaeaceae	S	S
Gleicheniaceae	M S	M S
Hymenophyllaceae	S	M? S
Cyatheaceae	S	S
Dicksoniaceae	M S	M S
Marsileaceae	S?	S?

Table 3.2. Australian and New Zealand Cretaceous conifers. M = megafossil; P = pollen. Sources of data, see text and Table 3.1.

	Australia Cretaceous Early	Australia Cretaceous Late	New Zealand Cretaceous Early	New Zealand Cretaceous Late
Agathis			M	M
Araucariacites	M P	M P	M P	M P
Athrotaxis				M?
Dacrydium	P	P		P
Microcachryidites	P	P	P	P
Phyllocladus		P	M	M
Podocarpus	M P	M P	M P	M P

3. Cretaceous Angiosperm Flora

In Australia, the first suspected angiosperm megafossils (three broad-leaved plants, two of which were referred to *Celastrophyllum*) were recorded by Walkom (1919) in Cenomanian rocks from Styx R., western Queensland. However, failure by Cookson (1964) to recover fossil angiosperm pollen from the same beds has thrown some doubt on the identity of these fossils.

The recovery of fossil pollen from Late Cretaceous rocks (Cookson, 1961; Cookson and Balme, 1962; Dettmann and Playford, 1968; Dettmann, 1973) indicates the presence of angiosperms in the continent during the Albian in the north (Bathurst and Melville Islands), the Great Artesian Basin area, and in the south-east and south-west. The affinities of the pollen grains are uncertain but probably both dicotyledons and monocotyledons existed, the total number of species recovered being 20. The pollen assemblages are fairly uniform from north to south, but with a greater diversity in the north.

The first identifications with modern taxa were made by Dettmann and Playford (1968) for three species from Senonian strata from the Otway Basin, Victoria. These species are: *Nothofagidites senectus,* morphologically comparable with the pollen of the *Nothofagus brassii* group (Cookson and Pike, 1955); *Proteacidites scaboratus,* which is described for New Zealand by Couper (1960), where it is restricted to the Upper Senonian and Danian sediments and is recorded also for the lower Paleocene of Victoria; *Proteacidites amolosexinus,* a new species so far recorded only for Australia. Other pollen grains are recorded which may be angiospermic but which cannot be related to modern genera. The sediments contain also spores of pteridophytes and gymnosperms. According to Dettmann and Playford (1968), the spore-assemblages of the late Albian – early Senonian in eastern Australia have few specific features in common with European, equatorial African and eastern North American microfloras of the same age, but some similarities can be detected with western North America, U.S.S.R. (particularly Siberia) and New Zealand. Martin (1977) has recorded *Ilex* pollen in Turonian rocks in the south-east, where it predates the first appearance of *Nothofagus*.

A much larger number of pollen species have been recorded for New Zealand (Couper, 1953, 1960). In addition to *Nothofagus* and Proteaceae the most significant records are those for Myrtaceae (Table 3.3). For Antarctica there is only a single angiosperm record for the Cretaceous (Senonian), viz., *Nothofagidites* pollen (Cranwell, 1964). The South American fossil angiosperm flora described by Menendez (1969) from leaf impressions of possible Cenomanian and Turonian age contains several taxa which relate it to the Australian Cretaceous flora, though the relationship is no closer than that shown by the modern floras.

The fossil plants recovered from the Cretaceous have been used by several authors to deduce climates of the period. Antarctica, or part of it, was warm enough to support forest vegetation. Apart from *Nothofagus,* this continent supported *Araucaria* and *Podocarpus,* and, during earlier periods, the taxa recorded for the other southern continents, including a *Glossopteris* flora, Bennettitales and Cycadales (Wace, 1965). For Australia, Dettmann and Playford (1968) suggest that the Early Cretaceous was cool and climatically uniform over the southern part of the continent, as indicated by the presence of *Microcachrydites*. During the Senonian, when the angiosperms proliferated, the presence of the *brassii* type pollen of *Nothofagus* in Vic-

Table 3.3. Australian and New Zealand Cretaceous angiosperms. M = megafossil; P = pollen. Sources of data, see text.

	Australia Cretaceous Early Late		New Zealand Cretaceous Early Late	
Aquifoliaceae *(Ilex)*	P			
Caryophyllaceae	P			P
Cruciferae				P
Cyperaceae			P	P
Fagaceae:				
Nothofagus		P		P
Liliaceae				P (3 spp.
Leptospermum type				P
Metrosideros type				P
Olacaceae				P
Proteaceae		P (3 spp.)		P (12 spp.)
Unidentified		P		
Dicotyledons		M? P	P	P (16 spp.)
Monocotyledons	P			P (2 spp.)

toria indicates warm temperate conditions in the south. This type of pollen is found in successively younger rocks from south to north [Eocene in south-east Queensland (Harris, 1965) and Pliocene in New Guinea (Cookson and Pike, 1955)].

4. Early Tertiary Angiosperm Floras of Australia and New Zealand

Fossil angiosperm leaves from the early Tertiary were first described by Ettinghausen (1883, 1866 – English translation by Etheridge, 1888 and 1895). Ettinghausen regarded this flora as part of a cosmopolitan type and related it to the Northern Hemisphere flora. Many of his identifications have proved incorrect.

This flora, which provides the earliest record in Australia of what appears to be broad-leaved rainforest communities, has been referred to as the *"Cinnamomum"* flora. Revisions of Ettinghausen's identification led to the suggestion that in the Eocene and Lower Oligocene the following families were represented in this flora: Cunoniaceae, Elaeocarpaceae, Fagaceae (*Nothofagus*), Lauraceae, Meliaceae, Moraceae, Myrtaceae, Palmae, Proteaceae, Sapindaceae, Sterculiaceae (Deane, 1902 a, b, 1925 a, b; Chapman, 1937). The assemblage, with some variations in composition, has been recorded from Tasmania, south-west Western Australia, Central Australia and in many places near the east coast from Victoria to Queensland.

More information is supplied from fossil pollen (Table 3.4) and there is some agreement between the pollen records and the suggested families based on leaf impressions.

The assemblages from different parts of the continent vary considerably in specific composition, and in some cases an assembly appears to be inconsistent with modern plant-assemblages. Various authors have suggested that the Early Tertiary communities were altitudinally zoned and that plant-parts from different altitudes were washed into basins where the fragments were fossilized together.

The wide distribution of *Nothofagus* and Proteaceae is possibly indicative of the dominance of these two taxa over large areas. The most northerly assemblage is reported from the mid-

Table 3.4. Palaeocene and Eocene angiosperm and Gnetales pollen records for Australia and New Zealand. f. = fossil genus; other genera are extant. Sources of data: See text, and Harris (1965), Martin (1973), Stover and Evans (1973).

Family (or Order)	Genus	Aust.	N.Z.
Agavaceae	*Phormium*		+
Araliaceae			+
Barringtoniaceae	*Barringtonia*	+	
Casuarinaceae	*Casuarina*	+	+
Chloranthaceae	*Ascarina*		+
Cruciferae			+
Cyperaceae		+	+
Epacridaceae	*Dracophyllum*		+
Ericales		+	
Escalloniaceae	*Quintinia*	+	
Fagaceae	*Nothofagus*	+	+
Gnetales	*Ephedra*	+	+
Haloragaceae	f. *Gunnerites*	+	
Liliaceae	f. *Liliacidites*	+	+
Loranthaceae	*Elytranthe*		+
Malvaceae	*Hohera*		+
	f. *Malvacipollis*	+	
	Plagianthus		+
Meliaceae	*Dysoxylum*		+
Myrtaceae	*Eucalyptus?*	+	
	Leptospermum		+
	Metrosideros		+
	f. *Myrtaceidites*	6 spp.	
Olacaceae	f. *Anacolosidites*	+	+
Palmae	*Nypa*	+	
	Rhopalostylis		+
Proteaceae	f. *Banksieidites*	+	
	f. *Beaupreaidites*	+	+
	Knightia		+
	f. *Proteacidites*	18 spp.	16 spp.
Rhizophoraceae		+	
Santalaceae	f. *Santalumidites*	+	
Sapindaceae	f. *Cupanieidites*	+	+
Sonneratiaceae	*Sonneratia*	+	
Sparganiaceae		+	
Tiliaceae	f. *Tilliaepollenites*	+	
Verbenaceae	*Avicennia*	+	

Eocene of Central Australia, near Alice Springs. It contains *Nothofagus,* Proteaceae, Cyperaceae, Haloragaceae, Liliaceae, Myrtaceae and Sapotaceae, plus the conifers *Araucaria, Microcachrys, Phyllocladus* and Cupressaceae (Kemp, 1976).

In the south-west of Western Australia, McWhae et al. (1958) record *Nothofagus* with Proteaceae, Moraceae and a palm, together with *Araucaria.* In the same region, Churchill (1973) records three mangroves, *Avicennia* (which still occurs in that region), and Rhizophoraceae and Sonneratiaceae from Mid-to Late-Eocene sediments, together with the fresh water mangrove, *Barringtonia,* and the palm *Nypa.* The last four taxa are now confined to the tropics.

Of interest is the recording of pollen of *Casuarina* in both Australia and New Zealand in the Eocene. The genus has been recorded as a megafossil in Miocene rocks from Patagonia by Frenguelli (1943).

Since there are no Early Tertiary fossil records for the now-tropical part of the continent, it is impossible to state whether a latitudinal zonation of the flora existed in the Tertiary, as it does today. If latitudinal and/or altitudinal zonations occurred, they may possibly be reconstructed through the composition of modern assemblages, expressed in two broad groups, a *Nothofagus* – Proteaceae – Cunoniaceae etc. assemblage and one containing a variety of broad-leaved plants referred to above as the *"Cinnamomum"* flora. This item is discussed in more detail below.

Two extra-Australian fossil floras which have bearing on the Australian flora and inter-continental floristic relations come from the Ninetyeast Ridge and Antarctica.

The Ninetyeast Ridge lies in the Indian Ocean (Fig. 3.1) and represents submerged islands. Pollen analyses of cores from deep sea drillings of Palaeocene and Oligocene age have identified the following taxa (Kemp and Harris, 1975): Casuarinaceae, Chloranthaceae (*Ascarina* – extant in south-east Asia to west Pacific and New Zealand), Compositae (Tubuliflorae), Didymelaceae (now confined to Madagascar), Fagaceae *(Nothofagus),* Haloragaceae *(Gunnera),* Loranthaceae, Moraceae/Urticaceae, Myrtaceae, Palmae, Proteaceae, Restionaceae, Sapindaceae (Cupanieae), Sapotaceae, Sparganiaceae, together with the conifers, *Araucaria* and Podocarpaceae. This assemblage is undoubtedly of Gondwanaland origin and is

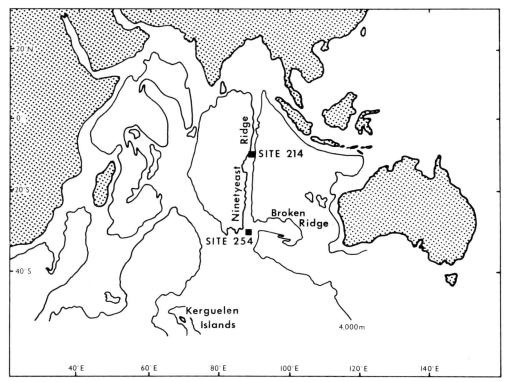

Fig. 3.1. Sketch map of the Indian Ocean showing the Ninetyeast Ridge and related features, and the positions of Deep sea Drilling Project sites 214 and 254. Redrawn from Kemp and Harris (1975).

Table 3.5. Tertiary angiosperm fossil records from Seymour Island, Antarctica Peninsula (c. 64°S., 56°W.) (Wace, 1965), and from the Antarctica ice shelves (Kemp, 1972, and Kemp and Barrett, 1975). Pollen records indicated by P. Genera with living species in italics.

Family	Genus	Present distribution of living genera
Cyperaceae	Scirpitis	
Aquifoliaceae	Iliciphyllum	
Cruciferae	P	
Cunoniaceae	*Caldcluvia*	Chile
Fagaceae	*Nothofagus* (5 spp.)	Aust., N. G., N. Cal., N. Z., S. Amer.
Lauraceae	Lauriphyllum	
Leguminosae	Leguminosites	
Loranthaceae	P	
Melastomataceae	Miconiiphyllum	
Monimiaceae	*Laurelia*	N. Z., Chile
	Mollinedia	Centr. & Trop. S. Amer.
Myricaceae	*Myrica*	All continents, except Aust.; absent N. Z.
Myrtaceae	P	
Onagraceae	? *Fuchsia* P	N. Z., Tahiti, Centr. & S. Amer.
Palmae		
Proteaceae	*Knightia* P	N. Cal., N. Z.
	Lomatia	Aust., Chile
	Proteacidites P (8 spp.)	
Winteraceae	? *Drimys*	Aust., Borneo, N. G., N. Cal., N. Z., S. Amer. (Now *Tasmannia* in Aust. and *Pseudowintera* in N. Z.)

closely related to the early Australian assemblages.

The Antarctica flora for the Tertiary is listed in Table 3.5. This flora shows some relationships with the Australian-New Zealand flora of the early Tertiary but much stronger relationships with the flora of South America. Antarctica appears to have acted as a bridge between South America and Australia up to the time of separation of the continents (see Fig. 1.2).

Many of the taxa recorded as fossils in the Early Tertiary occur also in the modern flora of the southern continents, but not in the Northern Hemisphere, e.g., *Nothofagus,* Proteaceae, Escalloniaceae and the identifiable genera of Myrtaceae. It is probable that only a small fraction of the total Early Tertiary flora has been recovered in fossil form and that many other taxa occurred in Gondwanaland and were shared by the southern continents after the supercontinent fragmented. These taxa also are likely to have persisted into the modern floras of the southern continents.

5. Floristic Relationships of the Modern Australian Flora with other Continents and Islands

5.1. General

Since only about one-third of the Australian angiosperm genera are endemic it follows that a high proportion of the genera exist on other continents. Very many of these occur in Asia, many are pantropical, and some are cosmopolitan. On the other hand, there are many genera and a few families which have restricted ranges, most being confined to the southern continents. This last group of taxa, which occur today mainly in the cooler parts of Australia, probably existed on Gondwanaland, and are dealt with in this section. Some of the more northerly taxa may also have existed on Gondwanaland, though many appear to be recent migrants from Asia. These taxa occur in the modern rainforests and an attempt is made in Section 7, below, to

Table 3.6. Distributions of the genera of those families which are confined to the Southern Hemisphere (a few with minor extensions into the Northern Hemisphere). Sources of data: Hooker (1872–1897); Allan (1961); Burbridge (1963); Womersley (1969); Moore and Edgar (1970); Willis and Airy Shaw (1973).

	Numbers of genera in							Genera shared between continents
	World	Afr.	India	Aust.	N.G.	N.Z.	S.Amer.	
Aponogetonaceae	1	1	1	1	1	–	–	*Aponogeton*
Byblidaceae	2	1	0	1	1	–	–	None
Centrolepidaceae	7	0	0	7	2	3	1	*Gaimardia* (Aust.; S. Amer.)
Cunoniaceae	27	3	0	14	12	2	4	*Weinmannia* (Aust.; S. Amer.)†
Donatiaceae	1	0	0	1	0	1	1	*Donatia* (Aust.; S. Amer.)
Epacridaceae	30	0	1	28	5	5	1	None
Escalloniaceae	23	4	2	7	5	3	4	None
Eucryphiaceae	1	0	0	1	0	0	1	*Eucryphia* (Aust.; S. Amer.)
Goodeniaceae	14	0	1	13	5	2	1	*Selliera* (Aust.; S. Amer.)
								Scaevola (Aust.; India)
Monimiaceae	34	6	0	10	11	2	9	None
Myoporaceae	4	1	0	2	1	1	0	*Myoporum* (Aust.; SE. Asia)
Philesiaceae	7	1	0	3	2	1	4	*Luzuriaga* (Aust.; S. Amer.)
Pittosporaceae	9	1	1	9	3	1	–	*Pittosporum* (Aust.; Afr.; SE. Asia)
Proteaceae	62	14	1	38	8	2	8	*Gevuina, Lomatia, Oreocallis, Orites* (Aust.; S. Amer.)
								Helicia (Asia; Aust.)
Restionaceae	29	15	0	17	2	3	1	*Restio* (Aust.; S. Afr.)
								Leptocarpus (Aust.; S. Amer.)
Stylidiaceae	5	0	1	4	1	3	2	*Stylidium* (Aust.; India)
								Forstera, Phyllachne (Aust.; S. Amer.)
Trimeniaceae	4	1	0	1	2	0	0	None
Winteraceae	8	0	0	2	4	1	1	*Tasmannia* (Aust.; SE. Asia)

† Now *Pseudoweinmannia* in Australia.

differentiate between ancient Gondwanaland lineages and late Tertiary-Quaternary migrants.

5.2. Africa – Australia – South America Relationships

Eighteen families are confined to or have their maximum development in the three southern continents. These are listed in Table 3.6, which shows also those genera which are shared between continents. These shared genera indicate a closer relationship between Australia and South America than between Australia and Africa.

It is difficult to assess the significance of the shared genera because of the possibility of distance dispersal in recent times. For example, those families which are apparently Australian (Centrolepidaceae, Epacridaceae, Goodeniaceae, Pittosporaceae and Stylidiaceae) have probably been carried from Australia to other islands and continents by ocean currents, winds or birds. Also, it is noteworthy that the African species of *Restio* are not closely allied to the Australian species referred to the same genus (Cutler, 1966).

The families which have the most significance in intercontinental floristic relationships are the woody ones, viz., Cunoniaceae, Escalloniaceae, Eucryphiaceae, Monimiaceae, Proteaceae and Winteraceae, and the herbaceous Restionaceae. These families were probably established in Gondwanaland. The paucity or absence of these families in India supports the view that India was the first segment of Gondwanaland to separate and it appears not to have carried with it any of the southern families, except, possibly, Escalloniaceae and Proteaceae, the last family being represented in India by *Heliciopsis* which does not occur east of Wallace's Line. India may have carried some other Gondwanaland taxa to Asia (Schuster, 1972) but these cannot be identified. They could have included some members of the *Cinnamomum* flora.

In addition to the taxa listed in Table 3.6, the relationship between Australia and Africa is strengthened by the restriction to these two continents of *Adansonia* (possibly a very recent introduction into Australia in the north-west, Chap. 7), four Liliaceae (*Bulbine, Caesia, Iphigenia* and *Wurmbea*) and *Helipterum*

(Compositae). In contrast, the relationship between the floras of Australia and South America is strengthened by the occurrence of c. 40 genera which are confined to the cooler or colder parts of the Southern Hemisphere (Table 3.7); some of these have minor extensions to the north. Of these genera, only *Acaena, Gunnera* and *Haloragis* occur in the African region. Many of the genera occur on the islands in the Pacific Ocean, and on the subantarctic islands (see 7.4 below).

Of greatest significance with respect to the floristic relationships among the southern continents is *Nothofagus* (Hooker, 1859; Gordon, 1949; Burbidge, 1960; Darlington, 1965). The genus is thought to have originated in the Southern Hemisphere, except by Darlington who postulates an Asian origin followed by migration to the south in the Cretaceous. The genus is readily identified by its pollen. Three pollen types are recognised, which identify groups of species, *brassii, menziesii* and *fusca*. The *brassii* type is found in species with a higher temperature-tolerance (warm temperate); species with this type of pollen occur today only in New Guinea and New Caledonia, but *brassii* type pollen has been recorded in fossil form from the Australian Cretaceous (Dettmann and Playford, 1968) and from Cretaceous and Tertiary New Zealand up to the Pliocene (Couper, 1960) and from South America (Darlington, 1965). The *brassii* group of species has been replaced in all these areas by the more cold-resistant *menziesii* and *fusca* groups.

Evidence which supports the one-time contact of the *Nothofagus* communities of the world is the regular occurrence in different countries, with different species of *Nothofagus*, of the same associated organisms, viz., the parasitic ascomycetous fungus *Cyttaria* (except in New Guinea and New Caledonia where the species of *Nothofagus* are of the *brassii* type), the same moss family (Dicnemonaceae), with the same group of animals – the primitive, wingless, sucking bugs of the family Peloridiidae dependent on the wet moss (Evans, 1941).

The flora of Australia is closely related to both the New Guinea and New Zealand floras and some of the genera shared among these landmasses are listed in Table 3.6. Since the close relations are probably the result of distance dispersal in relatively recent times, detailed discussion is reserved for Section 7, below.

6. History of the Angiosperms in Australia

The centre of origin of the angiosperms is not known, but the early occurrence of megafossils and monosulcate pollen (with one furrow) in the Northern Hemisphere might suggest an origin in the Northern Hemisphere or the tropical segment of Gondwanaland. Raven and Axelrod (1974) suggest a rapid diversification of the angiosperms in West Gondwanaland, followed by migrations throughout Laurasia and Gondwanaland.

Muller (1970), using extensive data on fossil angiosperm pollens postulates an origin of the angiosperms in the Jurassic as the Magnolian type, the pollen of which cannot be distinguished from some gymnosperms, and that many of the modern genera evolved in the Turonian and Senonian (Table 1.1). He states that from the Jurassic to the top of the Albian the primitive angiosperms colonized the major landmasses without much further local radiation, which suggests the absence of major barriers to dispersal. The earliest pollen types have been recorded from Eurasia, Africa, the Americas and Australia.

In the Cenomanian the angiosperms were still in the minority, but new pollen types became conspicuous. In the Turonian and Senonian the angiosperms became dominant, and in the Senonian, there are sufficient pollen records to enable a differentiation of floras on a geographic basis, thus: i. Eastern North America and western Europe; ii. Eastern Siberia and western North America; iii. Malaysia (which is distinct from Australia); iv. Australia – New Zealand – Antarctica; v. Central Atlantic (northern part of South America and Gold Coast of Africa) (Muller 1970). This division into botanical provinces in the Senonian suggests the isolation of the continents before that epoch.

Wherever the centre of origin and diversification, it is probable that a few tens of millions of years had elapsed before the woody angiosperms reached the Australian segment of Gondwanaland. Furthermore, the last portion of "Australia" to receive the migrant flora would have been the south-east, because of its remote geographic position at that time (see Fig. 1.2). Consequently the first fossil records for the south-east (*Nothofagus* and Proteaceae) were not necessarily the first woody plants in the continent.

The families which migrated into "Australia" were both primitive (Table 2.6) and advanced. The significance of the primitive families is difficult to assess. Most of them occur in the modern rainforests. Four of them are confined to Australia–New Guinea and must be regarded as descendants of ancient Gondwanaland lineages (Himantadraceae, Eupomatiaceae, Austrobaileyaceae and Idiospermaceae). All genera of these families are rare and have restricted areas. They appear to have played no part in the evolution of new taxa, all genera being monotypic, except *Eupomatia*, with 2 species. The Annonaceae, Hernandiaceae and Myristicaceae appear to be recent migrants from Asia (see 7.3 below).

The remaining families, Winteraceae, Monimiaceae, Lauraceae and Dilleniaceae, probably made up part of the very early flora, the first two being constituents of the cooler rainforests dominated by *Nothofagus* (as today), the Lauraceae occurring in the broad-leaved rainforests (*Cinnamomum* flora). The habitats of the Dilleniaceae are vague, especially since *Hibbertia*, with c. 100 species, has an extensive secondary development on low fertility soils (see Chap. 4).

Other families in the first rainforests were Fagaceae *(Nothofagus)* and Proteaceae, both recorded for the Late Cretaceous and several other families (see next paragraph) recorded first for the Early Tertiary probably occurred in the Cretaceous. It is these families which form a significant part of the modern rainforests, and several of them diversified in Australia to produce the vast assemblage of xeromorphic taxa enumerated in Chap. 4 (chiefly Myrtaceae, Rutaceae, Epacridaceae and Proteaceae).

It appears that two large assemblages of rainforest plants existed in the Late Cretaceous and Early Tertiary: i. In warmer areas: Lauraceae, Meliaceae, Moraceae, Myrtaceae, Palmae, Proteaceae, Sapindaceae and Sterculiaceae, and probably others, including the primitive angiosperm families Eupomatiaceae, Himantandraceae and Idiospermaceae. The conifers *Araucaria, Agathis* and *Podocarpus* occurred in this assemblage. ii. In cooler areas: Fagaceae *(Nothofagus)*, Cunoniaceae, Epacridaceae, Escalloniaceae, Eucryphiaceae, Monimiaceae, Proteaceae and Winteraceae, plus the conifers *Athrotaxis, Dacrydium* and *Phyllocladus*.

It is possible that some families occurred in both assemblages, especially Elaeocarpaceae

Table 3.7. Data on angiosperm genera which are confined to South America, New Zealand, Australia and some Subantarctic islands (some with extensions to the north, as indicated). Absence indicated by –. Sources of data: Philippi (1881); Guillaumin (1948); Hoogland (1958); Allan (1961); Burbidge (1963); Wace (1965); Willis and Airy Shaw (1966); Moore and Edgar (1970); Thorne (1972); Green and Walton (1975).

		Number of species in					Occurrence elsewhere
		World	S. Amer.	N. Z.	Aust.	N. G.	
DICOTYLEDONS							
Campanulaceae	Hypsela	5	2	1	2	–	Trop. Asia
	Pratia	35	2	5	13	2	Antarctica & subantarctic Is.
Caryophyllaceae	Colobanthus	20	5	13	2	–	
Compositae	Abrotanella	20	c.9	8	3	2	SE. Asia, Japan, N. Cal.
	Lagenophora	30	3	6	3	3	North-western N. Amer.
	Microseris	14	1	1	1	–	
Donatiaceae	Donatia	2	1	1	1	–	
Ericaceae	Gaultheria	200	5	8	2	6	North-western N. Amer.; E. Asia Centr. Amer.
	Pernettya	30	8	2	2	–	New Hebrides
Elaeocarpaceae	Aristotelia	12	2	2	5	1	
Eucryphiaceae	Eucryphia	6	3	–	3	–	
Fagaceae	Nothofagus	35	9	4	3	17	N. Cal.
Goodeniaceae	Selliera	2	1	1	1	–	
Haloragaceae	Gunnera	40	15	10	1	1	Java; Costa Rica; S. Afr.
	Haloragis	76	1	7	70	10	Madagascar; N. Cal.
Polygonaceae	Muehlenbeckia	c.20	2	5	10	5	N. Cal.
Proteaceae	Geuina	3	1	–	1	1	Malays.
	Lomatia	12	4	–	8	–	
	Oreocallis	5	2	–	2	1	Malays.
	Orites	9	1	–	8	–	
Rhamnaceae	Discaria	c.13	11	1	1	–	
Rosaceae	Acaena	c.100	c.90	14	2	1	S. Afr.; California; many islands
Rubiaceae	Coprosma	90	2	45	2	8	Indonesia; Pacific & subantarctic Is.
	Nertera	12	1	6	2	2	Philipp.; Malays.
Scrophulariaceae	Ourisia	c.20	12	10	1	–	
Stylidiaceae	Forstera	5	1	4	1	–	
	Phyllachne	4	1	3	1	–	
Thymeleaeceae	Drapetes	c.8	1	5	1	1	Malays.
Umbelliferae	Lilaeopsis	20	11	3	3	–	Western N. Amer.
	Oreomyrrhis	23	2	3	7	6	Mexico; Borneo; Formosa
	Schizeilema	18	2	11	1	–	

		70	3	6	25	
Winteraceae	Drimys†					Malays.; Borneo
MONOCOTYLEDONS						
Centrolepidaceae	Gaimardia	2	1	1	2	
Cyperaceae	Carpha	11	4	1	1	Mascarene Is.; Japan; N. Cal.
	Oreobolus	10	2	3	2	Borneo
	Uncinia	45	18	14	6	Centr. Amer.; N. Cal.
Gramineae	Amphibromus	9	1	4	—	
Iridaceae	Libertia	10	3	5	1	
Liliaceae	Astelia	25	1	2	1	Pacific & Mascarene Is.
Restionaceae	Leptocarpus	15	1	12	2	SE. Asia

† Now *Pseudowintera* (in New Zealand) and *Tasmannia* (in Australia).

and Proteaceae, and that the two types of rainforest mixed when in contact either in latitudinal or altitudinal zonations.

In addition, at least two other types of communities and floral assemblages occurred in the Early Tertiary, viz., mangrove swamps in the littoral zone and probably fresh water swamps containing Cyperaceae and Sparganiaceae, at least.

Two genera are mentioned in more detail because of their importance in the continent, viz., *Nothofagus* and *Acacia*.

Nothofagus is closely related taxonomically with the Northern Hemisphere *Fagus*. It must be assumed that the ancient fagalian lineage diversified in the Early Cretaceous and that *Nothofagus* differentiated in eastern Gondwanaland in the same areas as the unique assemblage of conifers (*Dacrydium* etc.) occurred. Like these conifers, *Nothofagus* has never been present in Africa.

Acacia possibly had its origin in Gondwanaland, which is deduced from the following. The genus, with over 900 species, occurs on all continents, but with its greatest development in Australia. *Acacia* was divided by Bentham (1875) into 6 sections, only 4 of which are represented in Australia, the Gummiferae (6 spp.), Botryocephalae (32), Pulchellae (14) and Phyllodineae (c. 700). The first three sections are bipinnate, the last phyllodineous. The Gummiferae is regarded as the primitive section and it occurs extensively outside Australia, especially Africa. The other three sections are confined to Australia, excepting two or three phyllodineous species which have apparently migrated to some Pacific islands. It is possible that the ancient *Acacia* stock (Gummiferae) occurred in Gondwanaland and that Australia retained a couple of taxa when the supercontinent fragmented. These primitive taxa gave rise to the other sections in Australia, the phyllodineous types, which have been most successful, being the last to develop. The Australian Gummiferae now occur in the tropics (except *A. farnesiana,* possibly a recent introduction), and it is probable that diversification proceeded from north to south (Andrews, 1914), initially at a very slow rate. This is consistent with the fact that fossil pollen of *Acacia* has not been recovered in Victoria from sediments older than the Pliocene (Cookson, 1954). Additional information on *Acacia* in Australia is provided in Chap. 4.

7. Distance Dispersal of Plants

7.1. General

There is much evidence to indicate that distance dispersal is a powerful factor in the spread of plants throughout the world. The agents for dispersal are usually defined as winds, ocean currents and birds. The effectiveness of the three cannot be assessed, especially for by-gone times. While winds and ocean currents may have been as effective in the distant past as they are today, the same cannot be said for birds. Indeed, the capacity for flight and the migratory habits of Cretaceous and Early Tertiary birds are not known.

There are few quantitative guide-lines discriminating between distance dispersal and migration overland. Apparent morphological seed or fruit-dispersing mechanisms are usually taken into account, such as succulent fruits, appendages for air-transport or buoyancy in water. A useful guide is the apparent community-assemblage, assumed to have travelled overland, as illustrated by the *Nothofagus* assemblage (see 5.2, above), as opposed to the random accumulation of taxa. The latter is illustrated by genera and species common to Australia and New Zealand, the taxa belonging to a number of communities (see 7.4, below).

The best positive evidence for distance dispersal comes from the colonization of islands by plants. For the southern taxa, the floras of the subantarctic islands provide strong evidence for the distance dispersal of some taxa (Taylor, 1955; Wace, 1965), probably by birds. These islands are known to have been glaciated during the last Ice Age and must have been colonized during the past 15,000 years.

Man has possibly been responsible for the introduction of some species into Northern Australia. The aborigines migrated from Asia and may have brought seeds with them, either inadvertently or purposely. Also, the Malaysian Baijini settled in Arnhem Land and cultivated rice. They departed before the arrival in the same area of the Macassans from Celebes in the early 16th Century (Stocker, 1968). *Acacia farnesiana*, a member of the primitive Section Gummiferae is one of the species possibly introduced recently by man.

The significance of distance dispersal can be evaluated only through comparisons of the floras of land masses isolated by water barriers and this is done in the sections below, viz., for some of the islands and south-east Asia in the north, and for New Zealand, the subantarctic islands and temperate South America in the south. The relationships with southern Africa are so slight that a detailed analysis is not warranted.

7.2. New Guinea – Australia Relationships

Data on the New Guinea flora in comparison with that of Australia are given in Table 3.8. Noteworthy features of the New Guinea flora are the absence of endemic families, the relatively low proportion of endemic genera and the high proportion of endemic species. These characters are consistent with the relative youthfulness of the flora; the island emerged from the ocean in the late Oligocene to early Miocene.

Good (1960) classified the New Guinea flora into 8 categories: i. Widespread, and often predominantly temperate; ii. Pantropical; iii. Palaeotropical; iv. Genera of the Asiatic-Australian and American tropical sectors only; v. Indomalaysian genera; vi. Australian genera; vii. Remaining non-endemic genera; viii. Endemic genera. The Australian genera could likewise be

Table 3.8. Composition of the New Guinea flora (Good, 1960; Hoogland, 1972) in comparison with the Australian flora.

	Families Total	Endemic	Genera Total	Endemic	%	Species Total	Endemic	%
New Guinea	c. 217	0	1,501	116	8	9,250	8,500	92
Australia	221	11	1,686	538	32	13,100	c. 10,200	c. 80

classed under the same categories, except that some of the genera regarded as Australian by Good have been regarded as New Zealand–South American by the writer, and some (e.g., *Acacia* and *Drosera*) are regarded as more or less world-wide, though they are exceptionally well represented in Australia, as Good indicates. Alternatively, the New Guinea flora could be classified under the same "elements" as occur in Australia, but the elements occur in different proportions.

The New Guinea and Australian floras are closely related both at the family and generic levels (Womersley, 1969). Hoogland (in Walker, 1972) states that 893 genera occur in both New Guinea and Australia. New Guinea lacks two of the Southern Hemisphere families, Donatiaceae and Eucryphiaceae, but the others are well represented there (Table 3.6). New Guinea contains a few tropical families not found in Australia, e.g., Dipterocarpaceae (8 genera in New Guinea, occurring mainly in the north and west of the island) and Marantaceae (6 genera). Families which occur in New Guinea and New Zealand but not in Australia are Chloranthaceae *(Ascarina)* and Coriariaceae *(Coriaria)*.

7.3. Australia – Asia Relationships

A very large number of genera and species occur both on the Asian mainland and in Australia. Most of these, especially rainforest taxa, occur also in New Guinea. Blesser (in Specht and Mountford, 1958) states that 620 of the 2220 species which occur in Northern Territory north of lat. 15° S occur outside Australia, mainly in Malaysia and India in either the monsoon rainforests or in littoral communities. Some of the species are of obvious Asian origin and some are Australian. Consequently, an interchange of species between the two continents has occurred.

The interchange of taxa between Australia and Asia has probably been active since the Mid-Tertiary and the rate of interchange is likely to have increased as the Australian plate drifted towards Asia. The emergence of New Guinea in the Mid-Tertiary provided an additional stepping stone for migrations, especially when it was connected to Australia during the Pleistocene glaciation (see Walker, 1972).

Obvious angiosperm migrants from Asia into Australia are those genera which are represented in Australia by one or a few species, in Asia by dozens or even hundreds of species, e.g., genera of the primitive families Annonaceae, Hernandiaceae and Myristicaceae (Table 2.6), and the two Ericaceae, *Agapetes* and *Rhododendron* which occur on Mt. Bellenden Ker, north Queensland. Also, there are some species of undoubted Asian genera which extend from Asia through New Guinea to Australia, e.g., *Ceiba malabaricum* (Bombacaceae) and *Derris scandens* (Papilionaceae). Another significant segment of the flora which possibly migrated from Asia is the group of advanced ferns with a vertical annulus (Chap. 2, Table 2.1, families 15–34), which have not been recorded as fossils in any Australian rocks. There are Northern

Table 3.9. World distributions of the genera of selected families occurring in the Australian rainforests. Sources of data: Burbidge (1963); Womersley (1969); Guillaumin (1948); Willis and Airy-Shaw (1973).

Family	Number of genera in Australian rainforests	Endemic in Australia	Distribution outside Australia			
			N.G.	N. Cal.	SE. Asia	Pantropical or Africa only
Euphorbiaceae	31	5	25	13	24	16
Lauraceae	6	0	6	4	6	3
Meliaceae	16	2	14	4	14	3
Myrtaceae	15	4	12	8	8	1
Proteaceae	19	14	5	3	1	0
Rutaceae	16	5	11	9	8	2
Sapindaceae	20	5	15	6	11	2
Sterculiaceae	3	0	3	2	1	1

Hemisphere fossil records for this group of ferns, though the identifications are based on vegetative rather than sporangial characters, e.g., *Acrostichum* from the Eocene in England, and Blechnaceae from the European Oligocene (Harland et al., 1967). Assuming that these records are correct and that this group of ferns is monophyletic, they must have migrated into Australia from the Northern Hemisphere during the Tertiary.

Table 3.10. Distribution of species of selected genera of families in Table 3.9. Genera arranged in order of species-abundance in Australia. Sources of data: World, Willis and Airy Shaw (1973); New Guinea, E.E. Henty (personal communitication); New Caledonia, Guillaumin (1948), Australia, Burbidge (1963).

	World spp.	Distribution outside Australia (spp. in brackets)	Australia spp.
EUPHORBIACEAE			
Glochidion	300	Afr.; Asia; Amer.; N.G. (65); N.Cal. (7)	13
Baloghia	13	N.Cal. (9)	4
Drypetes	200	Afr.; Asia; N.G. (5); N.Cal. (1)	3
Tragia	100	Trop. & Subtrop. (not N.G.)	1
LAURACEAE			
Endiandra	80	SE. Asia; Polynesia; N.G. (37); N.Cal. (5)	26
Cinnamomum	250	Indo-Malay.; E. Asia; N.G. (20); N.Cal. (2)	4
Litsea	400	Asia; Amer.; N.G. (60); N.Cal. (12)	4
MELIACEAE			
Dysoxylum	200	SE. Asia; N.G. (48); N.Cal. (25)	14
Aglaia	c. 300	SE. Asia; N.G. (90); N.Cal. (1)	3
Chisocheton	100	Indo-Malay.; N.G. (21)	1
Toona	15	SE. Asia; N.G. (1)	1
MYRTACEAE			
Myrtoideae			
Syzygium	500	Afr.; Asia; N.G. (138); N.Cal. (27)	c. 60
Rhodamnia	20	Asia; N.G. (12); N.Cal. (1)	7
Leptospermoideae			
Tristania	50	SE. Asia; N.G. (6); N.Cal. (6)	14
Xanthostemon	45	E. Asia; Philipp.; N.G. (5); N.Cal. (37)	5
RUTACEAE			
Flindersia	22	Moluccas; N.G. (5); N.Cal. (1)	14
Melicope	70	SE. Asia; N.G. (23); N.Cal. (11)	12
Geijera	7	N.G. (1); N.Cal. (3)	3
Halfordia	4	N.G. (2); N.Cal. (1)	2
Glycosmis	60	Malay.; N.G. (3); N.Cal. (1)	1
SAPINDACEAE			
Alectryon	15	Malay.; N.G. (7); N.Cal. (1)	9
Cupaniopsis	60	Polynesia; N.G. (18); N.Cal. (21)	5
Mischocarpus	25	Malay.; N.G. (9)	5
Guoia	70	Malay.; N.G. (20); N.Cal. (9)	3
Erioglossum	1	E. Asia; N.G. (1)	1
STERCULIACEAE			
Brachychiton	18	N.G. (2)	17
Argyrodendron	4	N.G. (1)	4
Sterculia	150	Trop. (general); N.G. (25); N.Cal. (5)	4

Migrations of taxa from Australia to Asia have likewise occurred, as evidenced, for example, by the occurrence of *Baeckea* in China, and *Scaevola* and *Stylidium* in India (Table 3.6). However, the possibility of the migration into Asia from Australia of rainforest genera of Gondwanaland origin has not previously been entertained, chiefly because the northern Australian rainforest species have been regarded as derivatives of migrant taxa from Asia. In order to determine the region of origin of some genera present in the Australian rainforests in the north, an analysis of 8 families has been made. The families selected are those which were named as constituents of the "*Cinnamomum*" flora. The Euphorbiaceae has been added, partly because of its abundance in the Australian rainforests and partly because of its strong development as xeromorphic taxa outside the rainforests, which suggests a long residency in the continent.

The data in Table 3.9 indicate that the genera of the Proteaceae are Australian, one genus *(Helicia)* having migrated into Asia. Genera of the Myrtaceae, Rutaceae, Sapindaceae and Sterculiaceae appear also to belong to the Australia–New Guinea region and these are presumably of Gondwanaland origin; some have migrated to Asia. On the other hand, the Euphorbiaceae, Lauraceae and Meliaceae could well be of Asian origin, though an Australian origin for some of the widely distributed genera is possible. Indeed, it is possible that some of the ancient lineages may have been carried from Gondwanaland on

Table 3.11. Species which occur in both Australia and New Zealand. (M) = species is confined to the Australian mainland; (T) = species is confined to Tasmania. Other species occur both on the mainland and Tasmania.

DICOTYLEDONS

Caryophyllaceae	*Colobanthus affinis; C. apetalus*
Compositae	*Craspedia uniflora*
	Vittadinia australis
Epacridaceae	*Leucopogon juniperinus; L. parviflorus*
	Pentachondra pumila
Ericaceae	*Gaultheria depressa* (T)
Euphorbiaceae	*Poranthera microphylla*
Gentianaceae	*Liparophyllum gunnii* (T)
Goodeniaceae	*Selliera radicans*
Haloragaceae	*Haloragis micrantha*
Nyctaginaceae	*Pisonia umbellifera* (M)
Rhamnaceae	*Pomaderris apetala; P. phylicifolia*
Rubiaceae	*Coprosma pumila* (T)
Stylidiaceae	*Phyllachne colensoi* (T)

MONOCOTYLEDONS

Centrolepidaceae	*Centrolepis strigosa*
Cyperaceae	*Carpha alpina*
	Lepidosperma laterale; L. filiforme
	Oreobolus pumilio
	Uncinia compacta; U. riparia; U. tenella
Gramineae	*Dichelachne crinita; D. sciurea*
	Echinopogon ovatus
	Microlaena stipoides
Iridaceae	*Libertia pulchella*
Liliaceae	*Herpolirion novae-zealandiae*
Orchidaceae	*Caleana minor*
	Chiloglottis formicifera (M)
	Orthoceras strictum (M)
	Pterostylis mutica; P. plumosa
	Thelymitra ixioides; T. longifolia; T. venosa (M)
Restionaceae	*Hypolaena lateriflora*
Zannichelliaceae	*Lepilaena preissei* (M)

India, later to re-enter Australia. More positive information comes from the distribution of the species of the various genera.

The distributions of the species of 89 genera belonging to 7 of the families in Table 3.9, excluding endemic genera and Proteaceae, has been done and the distribution patterns of selected genera are given in Table 3.10. The data show that some genera do not occur in Asia, but are restricted to Australia, New Guinea and New Caledonia, especially Rutaceae, Sterculiaceae, Euphorbiaceae and Sapindaceae; these genera have probably been derived from Gondwanaland lineages. Conversely, several genera, with many species in Asia and a few in Australia appear to be of Asian origin. Of the 89 genera considered, 60 appear to be of extra-Australian origin and most of these have probably migrated into Australia from Asia.

A further comment on the Meliaceae and Lauraceae is relevant. In addition to their probable occurrence in the *"Cinnamomum"* flora, the pollen of *Dysoxylum* has been recorded in the early Tertiary flora of New Zealand (Table 3.4) and the modern distribution of *Endiandra* suggests a Gondwanaland origin. Consequently, it is probable that interchange of rainforest taxa between Asia and Australia continued throughout the Tertiary, though the flow from north to south exceeded the flow from south to north. It follows that the "Indo-Malayan element" must refer to only a segment of the modern rainforest assemblage and it does not include those genera which were derived from the ancient Gondwanaland segment.

7.4. Australia – New Zealand Relationships

Some of the taxa common to the modern floras of Australia and New Zealand are listed in Table 3.6 and 3.7. Some of these appear to establish an age-old relationship of the floras through Gondwanaland occurrences. Others, especially those which occur on oceanic islands, have probably been distributed throughout the Southern Hemisphere by distance dispersal. Also, there are many bipolar taxa which have probably likewise been distributed by distance dispersal, e.g. *Agrostis, Apium, Carex, Drosera* and many others.

In addition to the taxa referred to in the last paragraph, there are 57 genera belonging to 30 families which are more or less restricted to Australia and New Zealand. About half of these are indicated in Table 3.11. Only one of the genera *(Cassinia)* is represented in Africa. Many of the genera occur on Pacific islands, but none occurs in South America. The various distributions suggest an interchange of taxa between Australia and New Zealand in relatively recent times. The centres of origin of the genera might possibly be deduced from the abundance of species in the two countries and, if this is so, transport of taxa has occurred in both directions. Furthermore, the variety of habitats represented suggests random transport of taxa from several plant communities. Also, there are c. 40 species which occur in both Australia and New Zealand, which are listed in Table 3.11. These also support distance dispersal, possibly in quite recent times.

Many of the taxa included in the tables in this section occur in the treeless, alpine communities of both Australia and New Zealand. When this alpine flora originated cannot be determined with the data available, but it seems improbable that it existed on Gondwanaland. This follows from the fact that extremely few of the genera occur in Australia, New Zealand and South America only, and that most of the genera occur on islands, some of which were under ice during the Pleistocene glaciation.

4. Development of the Modern Flora

1 Introduction

After separating from Antarctica and New Zealand, probably in the Palaeocene, Australia was isolated and Tasmania was attached to the mainland by a land bridge. The flora at that time possibly occurred in two main assemblages, *Nothofagus* and its associates occupying cooler areas, and a broad-leaved rainforest assemblage occurring in warmer areas. Fresh water swamps probably existed and mangroves occurred along part of the coast. A cold climate, herbaceous assemblage may have existed in the coldest parts in the south, especially the south-east.

During the Tertiary vast changes occurred in the climate and the soil parent materials, which stimulated the evolution of new taxa. About 630 new genera evolved, many of these being xeromorphic, including *Eucalyptus,* which supplanted the rainforests over much or most of the continent. These xeromorphic taxa constitute the second major phase in the history of the flora.

During the Mid-Tertiary, the *Eucalyptus* communities probably extended from east to west across the continent and they were interspersed with the contracting rainforests, the latter occupying soils of higher fertility. Communities dominated by conifers possibly occurred as well.

The third and final stage in the general history of the flora was caused by the development of the semi-arid and arid zones. The reduction in rainfall fragmented the *Eucalyptus* communities, almost eliminating them from the driest areas, where they were replaced mainly by *Acacia* and grasses.

In addition to internal floristic changes which resulted from local adaptation of the Gondwanaland flora to new habitats, a large number of taxa were brought to the continent by distance dispersal from the north and the south-east (discussed in Chap 3) and interchange of taxa between Australia and Antarctica may have occurred before the latter became ice-bound.

This chapter deals with the changes in the Tertiary environment and the attendant evolution of xeromorphic taxa, with special reference to those families and genera which are abundant in the Australian flora. The development of aridity and the arid zone flora are dealt with in Chap. 5.

2. Tertiary and Quaternary Conditions and Events Relevant to Plant Distribution and Dispersal

2.1 The Coast-line and Topography

At the end of the Cretaceous the Australian coast-line was more or less as it is today and the relatively flat topography of the Cretaceous probably persisted until the end of the Eocene (David and Browne, 1950). Some minor changes to the coast-line occurred during the Tertiary. The first resulted from the inundations along the Great Australian Bight and part of the west coast, where extensive limestones were laid down and later elevated. The largest of these limestone areas is the Nullarbor Plain. Also, the southern portion of the Murray River was inundated at this time, but without extensive deposition of limestone.

As the continent drifted northward towards Asia, the land which is now New Guinea began to emerge from its position along the north-eastern coast of Queensland and moved northward to its present position (see Chap 1).

Towards the end of the Miocene uplifting occurred in the south-east with the elevation of the Flinders Range in South Australia and the further elevation of the eastern Great Divide, particularly in the south. The uplift ended in the early Pleistocene. During the Pliocene a block fault isolated Tasmania with the formation of Bass Strait, the average depth of which is 60–70 m.

The Pleistocene glaciation, by lowering sea level, resulted in the formation of a land bridge to New Guinea and one to Tasmania, at least periodically (see Walker, 1972).

2.2 Soil Parent Materials

In the Oligocene volcanic activity occurred in the east, from Queensland to Tasmania which was then attached to the mainland. The lava was mostly basalt, high in plant nutrients, including phosphate. It is mainly on these basalt flows, now largely removed by erosion, that most of the modern rainforests occur. Judging from the shapes and sizes of the extinct volcanoes, some of the cones probably reached a height of 2,500 metres. The volcanic activity may possibly have continued until the end of the Tertiary. Basaltic flows occurred also in the south-west but only fragments of these remain today.

The formation of laterite, much of which still remains today, was one of the main pedological events of the Tertiary. Its formation is described in Chap. 1. The formation of laterite implies wet, warm climates over most, if not all of the continent, possibly in the Miocene.

The laterite profiles were probably first dissected at the end of the Miocene when orographic uplifts occurred and streams became more active. The presence of fossil laterite on the tops of the eastern Great Divide is used to indicate the development of the laterite in the Miocene and its fragmentation at the end of this epoch. In many areas the complete laterite profiles were removed. In others the profiles were truncated, so that the kaolin horizon lies at the surface in some cases or, by deeper truncation, the basal silica horizon caps the modern surface, sometimes persisting on table-top ridges in the arid zone and referred to as mesas (Chap. 1). The sands derived from the weathered laterites are of low fertility, and they were spread over much of the continent, adding to the area of low fertility sandy soils derived from such rocks as sandstones and coarse granites. These low-fertility media favoured the expansion of the xeromorphic communities and the contraction of the rainforests.

2.3. Climate

Climatic changes were inevitable as the continent moved northward into the tropics. Other variations in climate were brought about by topographic changes in the continent and probably also by variations in world climates, determined, for example, by the extent of the ice sheets at the poles, which have an effect also on sea level.

It is usually considered that the early Tertiary was warm and wet, suggested by oxygen isotope data (Dorman, 1966) and the occurrence across the continent in the south of fossil plants indicative of at least a wet climate. The warm, wet period possibly continued over the whole continent into the Miocene, as suggested by the wisespread formation of laterite.

Climatic events of the Pliocene and Quaternary have been discussed by many writers (Hills, 1939; Whitehouse, 1940; Browne, 1945; Crocker and Wood, 1947; David and Browne, 1950 and Jessup, 1961). The various schemes differ considerably in detail which is due, in part, to the fact that they represent local chronologies, and in some cases such chronologies have been extended to the whole of the continent. In no case is the time of development of the semi-arid and arid zones indicated, items which are all important for a discussion on the history of the flora. Further to this, these chronologies were postulated before the acceptance of continental drift.

During the Early Tertiary when the continent lay at lower latitudes, a wet climate is likely to have prevailed over the entire area, with the wettest conditions in the east, partly as a result of rainfall derived from the Pacific Ocean (as today, see Chap. 1) and partly because of the presence of the Great Divide in the east.

As the continent moved northward towards the Tropic, rotating in an anti-clockwise direction, the north-western coast would have come under the influence of the low rainfall belt, when the modern semi-arid zone was initiated. It is probable that such a latitudinal position was reached towards the end of the Miocene. As the continent moved northward, the semi-arid and arid zones moved southward, and the southern belt of wet-climate country in the south became progressively narrower, except in the east which remained wet, though probably with climatic fluctuations. As the continent assumed its present orientation and moved into the tropics, rainfall along the northern coast line increased.

It is therefore probable that the modern arid zone commenced to develop c. 15 million years ago, which is consistent with the large changes which occurred in the inland flora, involving the evolution of c. 100 endemic genera, a few of them with 50–100 species.

The Pliocene uplift in the east produced or intensified cold habitats, thus favouring the expansion of the cold climate flora. The uplift probably increased rainfall in the south-east, accentuating the rainfall gradient from the south-east to the inland, which was becoming drier.

The Pleistocene glaciation had some effects on the landscapes and flora. The effects of glaciation were localized mainly on the southern highlands in the south-east of the mainland and in Tasmania (Browne, 1945; MacPhail and Peterson, 1975). Widespread extinction and migration of floras, comparable with those in the Northern Hemisphere, did not occur.

The more recent work of Bowler et al. (1976) indicates that during the past 60,000 years climates in the east, at least, have changed considerably. Their data were collected from north-east Queensland to Tasmania, South Australia and coastal Western Australia, and include pollen analyses and estimated lake-levels.

They indicate that between 60,000 and 40,000 years B.P. the areas were drier than at present, becoming drier and colder, maximum glaciation occurring between 25,000 and 15,000 years B.P. Desert dune-building took place at the same time. The period possibly corresponds with Crocker and Wood's (1947) "recent arid period" in the south, dated at c. 6,000 years B.P. Ice began to retreat from the Kosciusko area before 20,000 years B.P. and from Tasmania 15,000 years B.P. "In the last 10,000 years climate has been relatively stable, although there are some indications that temperature and rainfall were marginally higher than now between 8,000 and 5,000 B.P." The pollen analyses of Kershaw (1970) of lake sediments in north-eastern Queensland, and of Dodson (1974) and Dodson and Wilson (1975) of sediments from Victorian lakes indicate fluctuations in rainfall during the past 20,000 years.

Although a continuous record for the climate of Australia, or for any smaller area, cannot be presented, it is clear that since the Miocene when the laterites commenced to disintegrate, there have been vast changes in the climate which, in some areas, notably the arid zone, brought about enormous floristic and vegetation changes. Also, in wetter areas lesser climatic changes produced expansions and contractions of plant communities, which have been responsible for the isolation of patches of vegetation; these have led to disjunctions in the modern flora and vegetational patterns. Some of these are discussed in Chaps. 5 and 6, especially as they affect the problem of community classification.

3. History of the Early Tertiary Floras

3.1. History of the *Nothofagus* Assemblage

In the Early Tertiary *Nothofagus* was associated regularly with the conifers *Dacrydium*, #*Microcachrys*, *Phyllocladus* and probably *Podocarpus*. In addition, several other angiosperm families probably occurred with *Nothofagus*, as they do today, though they have not been recorded as fossils; these are Cunoniaceae, Epacridaceae, Escalloniaceae, Eucryphiaceae, Proteaceae, Winteraceae, and possibly some herbaceous taxa, such as Cyperaceae. The exact distribution of the forests is not known, either in terms of latitude or altitude, but they appear to have extended across southern Australia and possibly well into northern Queensland where they provided New Guinea with part of its first angiosperm flora. The *Nothofagus* forests occurred in the same districts as the broad-leaved rainforests, the latter occurring probably in warmer lowland sites, the *Nothofagus* forests on the cooler uplands (Cookson and Duigan, 1950). Xeromorphic communities existed in the Early Tertiary in the same districts and possibly occupied soils of lower fertility.

During the Tertiary the *Nothofagus* assemblage contracted to the east, *Nothofagus* itself being found today only in Tasmania and in a few isolated localities in the south-east (Chap. 7). The times when *Nothofagus* disappeared from the north, centre and west cannot be fixed, but some data are available for the south-east.

The palynological studies of Martin (1973a, b, c, 1974) have established that the *Nothofa-*

gus-conifer flora persisted in the Lachlan Valley until the end of the Pliocene or even early Pleistocene. Martin's samples, collected from bores in the Ivanhoe district (the area now receives a mean annual rainfall of 276 mm) supplied almost 100 species of spores and pollens probably from the Eocene-Oligocene. The assemblage contained all 3 *Nothofagus* pollen types (*brassii, fusca* and *menziesii*), together with 11 species of Proteaceae, *Casuarina,* both succulent-fruited and dry-fruited Myrtaceae, Escalloniaceae (*Quintinia,* found today with *Nothofagus*), Sapindaceae, the gymnosperm genera listed above and *Ephedra* (now extinct in Australia), together with an unidentified Ericales and several monocotyledonous families. The *Nothofagus brassii* type is indicative of a warm climate, whereas the *fusca* and *menziesii* types indicate cooler climates, so that the three species in the one assemblage suggests an altitudinal zonation somewhere in the district.

A second set of samples analysed by Martin from the Cowra-Forbes district (mean annual rainfall now c. 500 mm), ascribed to the late Pliocene or early Pleistocene, shows a similar assemblage, but with the absence of the *Nothofagus brassii* pollen, suggesting a cooler climate. Additional taxa recorded are Cupressaceae, *Acacia, Drimys, Haloragis,* Compositae and more Myrtaceae. Furthermore, Martin was able to separate assemblages into phases, viz., i. a Myrtaceae-*Casuarina* phase (oldest), ii. a *Nothofagus-Cyathea* phase, and iii. a second Myrtaceae-*Casuarina* phase (youngest). In addition, in some of the profiles the *Nothofagus* phase is absent and in some Compositae and Gramineae increase towards the top of some sequences, probably indicating the development of an herbaceous ground cover. In these areas today Myrtaceae (especially *Eucalyptus*) and *Casuarina* are both abundant, but *Nothofagus* does not occur, nor is it found towards the east on the highlands. Whether the pollens were accumulated *in situ* or carried westward in the river from the highlands is not known for certain. Martin (1974) has also recorded in the same area 5 pollens of Euphorbiaceae which match modern genera.

It must be assumed that the ancient genera in this assemblage were unable to adapt to the changing environments of the Tertiary, i.e., to lower rainfall and/or lower soil fertility levels. *Nothofagus* and its associates are still confined to the cooler rainforests, *Nothofagus* dominating some stands, Cunoniaceae-Monimiaceae other stands (Chap. 7). Possibly some endemic genera evolved (see 4, below).

Adaptations to low fertility were relatively few, the most significant being in the Epacridaceae and possibly some Proteaceae (see 5.7 and 5.3, below). Also, one species of *Ceratopetalum (C. gummiferum)* occurs on soils of low fertility, mainly in the Sydney area.

3.2. History of the Broad-leaved Rainforest Assemblage

The broad-leaved rainforest assemblage (*Cinnamonum* flora) which occurred in the Eocene constituted a type of rainforest and appears to have been widely distributed in the continent. The assemblage contained Lauraceae, Meliaceae, Moraceae, Myrtaceae, Palmae, Proteaceae, Sapindaceae, Sterculiaceae and other families (see Chap. 3). The conifers, *Agathis, Araucaria* and *Podocarpus* were constituents of the forests.

The broad-leaved rainforests possibly reached their maximum development in the Mid-Tertiary when the laterites were forming. In the later Tertiary they started to contract, partly because of the spreading of low fertility soil parent materials produced by the weathering of the laterites, and partly as a result of the decrease in rainfall in the north-west which culminated in the extensive semi-arid and arid zones. Rejuvenation of the rainforests in the east occurred following volcanic activity and the addition of basalts. Today, the rainforests have contracted to small areas along the east coast and adjacent ranges, and still smaller areas along the north coast (Chap. 7). They disappeared completely from the south-west, leaving no significant relicts, and in this area low soil fertility levels may be responsible for their disappearance (Beadle, 1966), rather than climatic change.

Although the ancient forests contracted to small areas, remnants of the forests are to be seen over most of the continent, especially in the arid zone, e.g., palms in Central Australia and a large number of arid adapted trees of rainforest genera, of which *Flindersia, Geijera* (Rutaceae), # *Owenia* (Meliaceae) and *Brachychiton* (Sterculiaceae) only are mentioned here. A more extensive list and discussion are provided in Chap 5. A relatively large number of endemic genera evolved in the various families within the rainforests (see 4, below).

The identity of this ancient Gondwanaland flora has been obscured by additions of taxa from south-east Asia, many of them via New Guinea (see Chap. 3).

A most significant development, possibly initiated in the early Tertiary, or even in the Cretaceous, was the adaptation of several ancient lineages of certain families to low fertility soil conditions, resulting in the evolution of new genotypes (mostly new genera) which exhibited xeromorphic characters. These changes occurred extensively in the Myrtaceae, Proteaceae and Rutaceae, with a lesser development in some other families, and are discussed in Sections 5 and 6, below.

3.3. History of the Herbaceous Cold-climate Assemblage

The existence of a cold-climate herbaceous flora on Gondwanaland has been discounted (Chap. 3). However, that such a flora was established in the Tertiary is suggested by the fact that most of the cold-climate herbaceous taxa do not occur in the modern *Nothofagus* forests. Also, several of the taxa occur across Australia in the south and some of the taxa are abundant in the arid zone.

The taxa concerned are most of those listed in Chap. 3, occurring in New Zealand and South America, though many of these could well have been dispersed around the lower latitudes by distance dispersal during the late Tertiary or even the Quaternary. Other taxa include bipolar genera, especially those belonging to the usually temperate families Caryophyllaceae, Compositae, Cruciferae, Cyperaceae and Umbelliferae, though all of these families have some development in the tropics. Some of these bipolar genera have been studied and it has been established that the species-groups of one Hemisphere are not closely related to the species groups of the other Hemisphere, e.g., *Caltha* (Hill, 1918), and *Ranunculus* (Briggs, 1959). This suggests long isolation of the groups, but whether this isolation dates back to the Cretaceous cannot be stated.

The elevation of the southern highlands in the Pliocene undoubtedly extended the area available for cold-climate assemblages (Smith-White, 1959), and after this elevation occurred many more taxa of Australia-New Zealand origin have been added to the flora.

Remnants of the cold climate flora are seen in parts of Western and South Australia and in the arid zone, and it is possible that at some time during the Tertiary the hills of Central Australia (which area is known to have supported *Nothofagus*) may have harboured a cold climate flora which has diversified as annual taxa in the arid zone (see Chap. 5).

To illustrate the widespread occurrence of a family usually regarded as temperate, the Umbelliferae is selected. This family, though represented in the tropics in Australia by *Hydrocotyle* and *Trachymene,* has its greatest development in the south. Of the 9 endemic genera, only 3 are confined to cool or cold climates (#*Dichosciadium,* monotypic; #*Diplaspis* and #*Oschatzia,* each with 2 species). The remaining genera occur mainly across the southern half of the continent, some on soils of low fertility, some in the arid interior, and a few in both these habitats. The largest of the endemic genera are #*Platysace* (22 species) and #*Xanthosia* (20). The former is an aberrant genus of the family, being woody, and the leaves are simple. #*Xanthosia* on the other hand is closely allied to *Actinotus,* which seems to have played an important role in the evolution of the Australian genera. In Australia 17 species of *Actinotus* occur; there is one species in New Caledonia. In Australia *Actinotus* (Fig. 4.1) is found on the

Fig. 4.1. Two inflorescences of *Actinotus helianthi* (Umbelliferae). The inflorescence, including bracts, is c. 5 cm diam. The bracts are white, sometimes tinged with green, and resemble flannel in texture. Near Manildra, N.S.W.

low fertility soils and extends into the arid regions. Species of this genus appear to have provided the stock for the evolution of some of the monotypic endemic genera of the south west and the arid regions (# *Chlaenosciadium*, # *Homalosciadium* and # *Neosciadium*).

4. The Endemic Families and Genera in Relation to Habitat

4.1 The Endemic Families

Twelve families are endemic (Table 4.1). These are all small, containing a total of 27 genera and 126 spp. With the exception of Tremandraceae all the families have been classed from time to time under other families as subfamilies or tribes. Three families, each with a single genus (two of them monotypic) are confined to the rainforests. The remainder occur chiefly on the soils of low fertility or in the arid zone, the two habitats which have produced the most extensive evolutionary changes in the flora. Cephalotaceae contains the single monotypic genus # *Cephalotus*, a pitcher plant.

4.2. The Endemic Genera

Generic evolution has occurred in 92 families (Table 2.4). The family with the largest number of endemic genera is Compositae (55). Of the larger families, those with the highest proportions of endemic genera are Proteaceae, Epacridaceae, Restionaceae, Myrtaceae, Liliaceae, Chenopodiaceae and Cruciferae (Fig. 4.2).

These are 538 endemic genera occurring in a variety of habitats, which are classified here under four main headings: i. Rainforests; ii. Soils of low fertility; iii. Arid and semi-arid; iv. Cold climates and Tasmania. The numbers of genera confined to each habitat are given in Table 4.2. A large number of genera extend over two or more habitats and these have not been classified; about one-third of these are herbaceous and have no habitat preferences, while the remaining 54 genera are woody and are found on the soils of low fertility and in the arid zone.

The 86 endemic genera in the rainforests occur in 36 families. Of these, some occur in or are derived from lineages which existed in the southern *Nothofagus* assemblage, though many of the genera have migrated to the north, some to New Guinea (when they are not regarded as endemic). The main developments have occurred in Cunoniaceae (# *Anodopetalum*, # *Callicoma*, # *Calycomis*, # *Pseudoweinmannia*), Escalloniaceae (# *Abrophyllum*, # *Anopterus*,

Table 4.1. The endemic families. Habitat is defined in broad ecological terms. Monocotyledons are indicated by (M).

Family	Number of Genera	Number of Monotypic Genera	Total Number of Species	Habitat
Chloanthaceae	10	2	66	low fertility soils in humid areas; arid
Gyrostemonaceae	5	1	15	rainforests (margins); low fertility soils in humid areas; arid
Tremandraceae	3	1	28	mainly low fertility soils in humid areas
Cartonemataceae (M)	1	0	7	mainly low fertility soils is humid areas; one sp. on Aru Is.
Baueraceae	1	0	3	mainly low fertility soils in humid areas
Austrobaileyaceae	1	1	1	rainforests (east)
Akaniaceae	1	1	1	rainforests (east)
Brunoniaceae	1	1	1	arid to humid on low fertility soils
Cephalotaceae	1	1	1	low fertility, wet soils (SW. W.A.)
Eremosynaceae	1	1	1	low fertility soils (SW. W.A.)
Idiospermaceae	1	1	1	rainforests (NE. Qld)
Petermanniaceae (M)	1	1	1	rainforests (east)

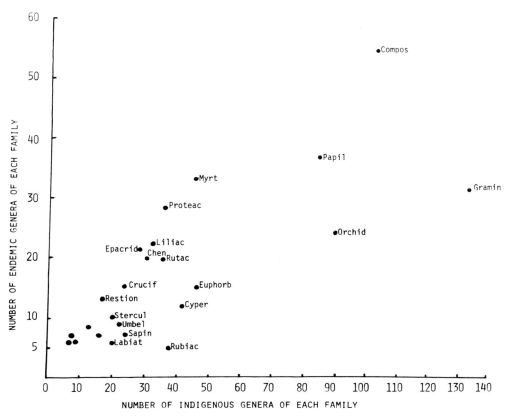

Fig. 4.2. Proportions of endemic to indigenous genera for those families which contain 6 or more endemic genera. The 5 points without names are for: Goodeniaceae 8/13; Rhamnaceae 7/16; Xanthorrhoeceae 7/8; Loranthaceae 6/10; Pittosporaceae 6/9; Haemodoraceae 6/7.

Table 4.2. Number of endemic angiosperm genera confined to specific habitats. Genera which occur in 2 or more of the habitats indicated are unclassified. All Tasmanian endemic genera have been classed under "Cold".

	Rainforests (Soils of higher fertility, humid climates)	Soils of low fertility, humid climates	Arid and semi-arid	Cold climates (mainly subalpine and alpine)	Unclassified	Total
Dicotyledons	76	136	94	17	90	413
Monocotyledons	10	58	13	2	42	125
Total	86	194	107	19	132	538

#*Cuttsia*, #*Tetracarpaea*), and Proteaceae (Table 4.4). Many of these genera are dominant or prominent in the rainforests in the south-east (Chap. 7).

The rainforests in the north have likewise produced many endemic genera, notably in Proteaceae (14 endemic genera – Table 4.4), Orchidaceae (7), Menispermaceae (5), Rutaceae (5), Sapindaceae (5), Annonaceae (3), Liliaceae (3), Palmae (3), Myrtaceae (2 – Table 4.3). It is probable that many more genera evolved in the northern rainforests of Australia and migrated to New Caledonia or south-east Asia as discussed in Chap. 3.

The soils of low fertility in wetter climates support the richest endemic flora. The genera belong mainly to the Papilionaceae (30 genera), Myrtaceae (23 – Table 4.3), Epacridaceae (19), Liliaceae (16), Proteaceae (17 – Table 4.4), Rutaceae (13), Restionaceae (12), Cyperaceae (9), Euphorbiaceae (8), Goodeniaceae (8), Haemodoraceae (6), Xanthorrhoeaceae (6). These genera and the near-endemic genera, such as *Eucalyptus, Melaleuca, Baeckea* and some others which have migrated out of Australia characterize what is usually referred to as the Australian element. Almost all of the genera are xeromorphic, and many are sclerophyllous. Some of these families are discussed in more detail in Sect. 5, below.

The 107 genera (52 monotypic) occurring in the semi-arid and arid zones belong to 31 families. Endemic genera occur most abundantly in the Compositae (23 genera), Cruciferae (14), Chenopodiaceae (13), Gramineae (12) and Capparidaceae (3). However, the largest endemic genera belong to smaller families: *Eremophila* with 120 species (Myoporaceae) and *Ptilotus* c. 100 species (Amaranthaceae). This flora is discussed in more detail in Chap. 5.

The 19 cold climate endemic genera belong to 11 families, with the largest development in the Compositae *(# Ewartia, # Nablobium, # Paranrennaria, # Pterigopappus)*, Umbelliferae *(# Dichosciadium, # Diplaspis, # Oschatzia)* and Epacridaceae *(# Prionotes, # Richea).*

5. Comments on Selected Families

5.1. General

Those families in which xeromorphy and sclerophylly developed during the Tertiary are dealt with in some detail in this section, viz., Myrtaceae, Proteaceae, Rutaceae, Papilionaceae, Euphorbiaceae, Epacridaceae, Casuarinaceae and *Acacia,* members of which constitute the xeromorphic communities dominated mainly by *Eucalyptus,* and the heaths. Understory species in these communities include several monocotyledonous families, some of which are dealt with here, viz., the Liliaceae and related families, Cyperaceae, Restionaceae and part of the Orchidaceae.

Other large families are dealt with more appropriately elsewhere. The Chenopodiaceae (Chap. 19), Compositae and Cruciferae are most abundant in the arid zone (Chap. 5) and the Gramineae are discussed with the grasslands (Chap. 20).

5.2. Myrtaceae

5.2.1. The Subfamilies and Genera

All three subfamilies of Myrtaceae, Myrtoideae (fruit a berry), Leptospermoideae (fruit a capsule) and Chamelaucioideae (fruit a nut) occur in Australia (Bentham, 1866). The Myrtoideae are widely distributed in the world (especially South America) and are regarded as the oldest of the three subfamilies (Andrews, 1913). Members of this subfamily occur in Australia mainly in rainforests of in rainforest-margins (Table 4.3).

The Leptospermoideae are represented in Australian by 26 genera, most of which are endemic and the remainder may possibly have originated in the Australian region. A few genera of obvious Australian origin have migrated northward, e.g.: *Eucalyptus* to Timor and the Philippines; *Melaleuca* to south-east Asia; *Baeckea* as far north as China; *Osbornia,* a mangrove, occurring in northern Australia and New Guinea. Of special interest are *Metrosideros* and *Leptospermum,* both of which occur in the modern Australian and New Zealand floras; the pollen types of both genera (probably ancient lineages of the modern genera) have been recorded for the Upper Cretaceous in New Zealand (Couper, 1953, 1960). *Metrosideros* also occurs today on South America (the species now being placed under *Tepualia)* and some of the Pacific Islands including Hawaii, where it exists as a polymorphic complex (Corn and Hiesey, 1973). Fossil data and modern distributions suggest that the Leptospermoideae was probably derived from the succulent-fruited Myrtoideae in Gondwanaland and that at least two lineages were established in the Cretaceous-*Metrosideros* and *Leptospermum* types. The occurrence of *Tepualia* in South America could be used to infer that *Metrosideros* was established before the separation of

Table 4.3. Genera of the Myrtaceae, arranged under subfamilies (Bentham, 1866) and according to habitat. Occurrence in more than one habitat indicated by arrows. Number of species and geographic distribution in brackets. W. = confined to the west; E. = confined to the east. W. and E. indicates disjunction. Endemic genera marked #.

Semi-arid and arid (mostly low fertility)	Lowest Fertility (heaths, mallee-heaths and some woodlands)	Lower Fertility (mainly *Eucalyptus* forests and woodlands)	Higher fertility (Rainforests and their margins)	Mangrove
MYRTOIDEAE				
			Cleistocalyx (2, Trop.)	
			← # *Fenzlia* (4, Trop.)	
			Acmena (6, Trop.)	
			↓ *Austromyrtus* (11, E)	
			Decaspermum (2, E)	
			# *Pilidiostigma* (4, E)	
			Rhodamnia (7, E)	
			Rhodomyrtus (7, E)	
			Syzygium (15, Trop., E)	
LEPTOSPERMOIDEAE				
		↓ *Eucalyptus* (450, wide)	→ *Metrosideros* (2, Trop.)	*Osbornia* (1, Trop., E.)
		↓ *Melaleuca* (140, wide)	*Xanthostemon* (5, Trop.)	
	← # *Balaustion* (6, W.)	↓ *Agonis* (12, W. and E.)	← # *Lysicarpus* (1, SE. Q)	
	# *Beaufortia* (16, W.)	*Baeckea* (70, W. and E.)	*Backhousia* (6, E.)	
	# *Calothamnus* (24, W.)	# *Callistemon* (20, W. and E.)	# *Choricarpia* (1, E.)	
	# *Conothamnus* (3, W.)	# *Kunzea* (30, W. and E.)	↓ *Syncarpia* (4, E.)	
	# *Eremaea* (7, W.)	*Leptospermum* (40, W. and E.)	← *Tristania* (14, E.)	
	# *Hypocalymma* (13, W.)	← # *Regelia* (4, W.)		
	# *Lamarchia* (1, W.)			
	# *Phymatocarpus* (2, W.)			
	# *Scholtzia* (13, W.)			
	Sinoga (1, E.)			
CHAMELAUCIOIDEAE				
# *Thryptomene* (50, W. and E.) →	# *Actinodium* (1, W.)			
# *Wehlia* (5, W.) ⇆	# *Calytrix* (50, W. and E.)			
	# *Calythropsis* (1, W.)			
	# *Chamelaucium* (12, W.)			
	← # *Darwinia* (28, W. and E.)			
	# *Homalocalyx* (2, Trop.)			
	# *Homoranthus* (4, E.)			
	# *Lhotskya* (13, W. S.A.)			
	← # *Micromyrtus* (10, W. and E.)			
	← # *Pileanthus* (3, W.)			
	← # *Verticordia* (50, W.)			

South America and Australasia. On the other hand it is equally possible that *Metrosideros (Tepualia)* arrived in South America by distance dispersal, just as *Metrosideros* must have arrived in Hawaii, these islands being of late Tertiary origin.

For Australia, it seems likely that the ancient *Metrosideros* and *Leptospermum* lineages were the ancestors of large groups of modern genera. The *Metrosideros* type produced modern *Metrosideros* (two species in Australia in the north), and possibly the other genera of Bentham's Subtribe Metrosidereae (Andrews, 1913), which includes *Backhousia* regarded by Bentham as a link between the Myrtoideae and Leptospermoideae), #*Choricarpia*, #*Lysicarpus*, *Metrosideros*, *Osbornia*, *Syncarpia*, *Tristania* and *Xanthostemon*. To these #*Angophora* and *Eucalyptus* are added. All of these genera, either wholly or in part, exhibit one or more of the morphological characters exhibited by *Metrosideros,* viz., tree habit, opposite leaves which are usually broad, penniveined and dorsiventral, terminal cymose inflorescences (variously modified in some to produce distinctive Subgenera). These characters occur also in the Myrtoideae. Furthermore, all of the genera occur in or near the rainforests and many are confined to the rainforests. Those species which occur in the rainforests have mesomorphic leaves, whereas those which occur outside the rainforests shows some xeromorphic characters, the degree of xeromorphy increasing with decreasing soil fertility and/or water availability.

The second ancient lineage, the *Leptospermum* type, is assumed to be the forerunner of a large group of genera, all of which occur outside the rainforests, most of them on soils of low fertility and in all cases the leaves are xeromorphic, most being small (leptophylls and nanophylls). The genera are listed in Table 4.3.

The third subfamily, Chamelaucoideae, with 13 genera (Table 4.3), is confined to Australia, no member having migrated outside the continent. The genera are confined to the soils of low fertility and some are well represented in the semi-arid and arid zones. They are all xeromorphic and many are heath-like. The subfamily is probably derived from the Leptospermoideae.

Chromosome numbers for many Myrtaceae have been determined. Divergence from the basic number, $n = 11$, occurs only in some members of the subfamily Chamelaucoideae, and a very few species of *Eucalyptus*. A few polyploids have been reported in all subfamilies, but polyploidy has played a very minor role in the evolution of the family and none in the origin of the genera (Smith-White, 1959).

5.2.2. *Eucalyptus*

Eucalyptus is of special interest in Australia because of its dominance over large areas, and the numerous species are used to name vegetational alliances. The genus, with c. 450 species, is confined to Australia, except seven species found on islands north of Australia, as far as the Philippines; five of these occur in Australia (see Chap. 8). Economically, many species are of great value in Australia for timber, firewood, oils, honey, paper pulp, and several minor products (Penfold and Willis, 1961).

Bentham (1866) in his Flora Australiensis recognized 135 species and 43 varieties, using a classification based on anther type. The classical works of Mueller (1879–1884), Maiden (1903 to 1931) and Blakely (1934) added a large number of species and varieties to Bentham's list. Blakely (1934, reprinted with some additions in 1955, 1965) is still the most readily available and comprehensive work on the subject. It includes descriptions of over 600 species and varieties, some of which were recognized as hybrids by Blakely.

Hybrids occur fairly commonly and hybrid swarms, involving backcrossing of the hybrid with one or other of the parents or the progeny sometimes occur. Pryor (1976), from field observations and experimental crossing, has shown that the hybridization between species of different subgeneric groups does not occur. Species in mixed stands commonly belong to different subgenera and consequently the various species in a stand are genetically isolated. He has shown also that when hybrids do occur they are usually confined to the narrow zone between the associations. When the ecotone between the 2 associations is broad and with a gradation of habitats, there is a greater abundance of hybrids than in ecotones between more sharply defined communities.

The genus is identified by the operculum, composed of fused perianth segments, and the structure varies in its origin. In some cases it is derived entirely from fused petals, the sepals being free and sometimes persistent on the summit of the mature fruit. In some species the operculum is composed of fused sepals, the petals being absent. In most species the operculum is double, consisting of an outer calycine cap covering an inner corolline cap. In such cases the two opercula may be separable, the outer being shed first, or inseparable and often fused together so firmly that the two can be identified only under the microscope.

Recent workers on the genus have pointed out the value of characters other than anther – type (as used by Bentham and Blakely) in indicating taxonomic relationships, especially leaf-venation, the inflorescence and operculum, as well as characters not so readily observable, such as the ovule and seed-coat structure, pollen morphology, bark and wood anatomy, oil glands and essential oils (Baker and Smith, 1920; Carr and Carr, 1959a, b; Pryor and Johnson, 1971; Johnson, 1972; Pryor, 1976). Pryor and Johnson (1971) classify the species, subspecies or varieties into 7 subgenera. They recognize c. 500 species and subspecies, of which c. 60 are still to be described. They regard *Angophora* as another subgenus, the others being: *Blakella, Corymbia, Eudesmia, Gaubaea, Idiogenes, Monocalyptus* and *Symphyomyrtus*.

Eucalyptus is closely related to several other myrtaceous genera (Bentham, 1866). For example, an operculum occurs also in the succulent-fruited *Cleistocalyx,* and the leaf-venation, leaf arrangement and inflorescence type of several eucalypts resemble both Myrtoideae and some Leptospermoideae, especially # *Angophora* and *Metrosideros* (Andrews, 1913).

Leaf-fragments, regarded as eucalyptus, have been recorded for the Eocene (Chapman, 1921, 1926), and several pollen types from Eocene-Oligocene sediments from the Ivanhoe district, N.S.W. (lat. 33°S.), recovered by Martin (1973) are Myrtaceous and could belong to *Cleistocalyx, Metrosideros* or *Eucalyptus tessellaris* (subgenus *Blakella*). The precursor of *Eucalyptus* is unknown, but since the closely related *Metrosideros* is recorded for the Early Tertiary in New Zealand (Chap. 3), this genus might be

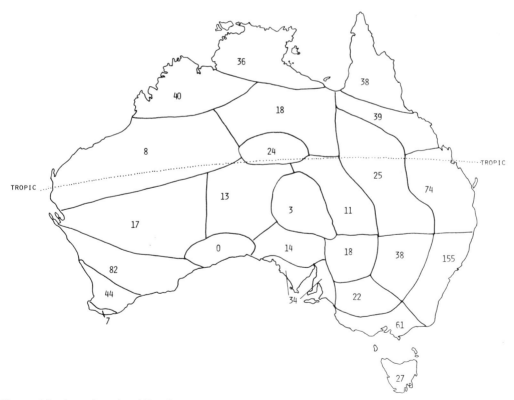

Fig. 4.3. Numbers of species of *Eucalyptus* in various zones across Australia.

regarded as the ancient ancestor. The possibility of *Eucalyptus* being polyphyletic is mentioned by Johnson (1972), who regards the subgenus *Gaubaea* as closely related to *Tristania*, whereas # *Angophora,* and the subgenera *Blakella* and *Corymbia* may well be more closely related to *Arillastrum,* a New Caledonian genus. # *Angophora* is often assumed to be the precursor of *Blakella* and *Corymbia*.

The genus occurs over almost the whole of the continent, with two main centres of species-abundance, one in eastern New South Wales, the other in semi-arid to arid Western Australia (Fig. 4.3).

Variations in size, bark characters and habit are useful for the identification of species and for the classification of communities. The tallest eucalypt, *E. regnans,* found in southern Victoria and Tasmania, reaches a height of 90 m (possibly c. 110 m). At the other extreme, forms of a few of the alpine species in Tasmania are more or less prostrate, though the stems are a few metres long.

Bark characters are variable. Some species are smooth-barked (Gums), the outer bark being shed annually in long ribbons or large scales and this shedding precludes the establishment of epiphytes on the trunk; this bark type is found in all subgenera and is of limited value taxonomically. On the other hand several taxonomically related groups have similar barks (illustrated in Anderson, 1968). Bloodwoods (subgenera *Blakella* and *Corymbia*) have short-fibred, tessellated barks (a few are Gums). Boxes, Ashes and Peppermints have barks which are short-fibred, the aggregates of fibres being a few centimetres long. Stringybarks (including Mahoganies) have long-fibred barks, the aggregates of fibres being removable as "ropes" several centimetres or even metres long. Ironbarks have fibrous barks so heavily impregnated with kino that the fibres are inseparable; the species belong in two taxonomic series and are dealt with collectively in Chap. 12).

Most eucalypts have a single trunk and the length of the bole relative to the depth of the crown is used to distinguish between forest (long boles) and woodland trees. However, many species are naturally many-stemmed and are referred to as mallees. The stems arise from a lignotuber, which is a woody structure originating in the axils of the cotyledons and sometimes in the axils of the first two or three pairs of seedling leaves as well (Kerr, 1925). The growth of the various buds produces a knot of tissue and the buds proliferate in the bark of the lignotuber. In most cases the buds remain dormant, unless the main stem is damaged. Most eucalypts produce a lignotuber (exceptions are *E. camaldulensis, E. diversicolor, E. regnans* and a few others), and the character is genetically controlled (Pryor, 1957; Mullette and Bamber, 1978). Lignotubers become very large in the mallees which mostly are many-stemmed (see Chaps 10 and 14) as a result of the growth of dormant buds located in the lignotuber, which is a storage organ and is fire-resistant, but is not concerned with vegetation reproduction. However, rhizomatous systems which may produce clones are reported for three tropical species by Lacey (1974).

5.3. Proteaceae

The Proteaceae appear to have originated in Gondwanaland, and is now well represented in South Africa, South America and Australia. The presence of four genera in South America and Australia (Table 4.4) suggests a closer relationship between these two continents than between Australia and Africa. Johnson and Briggs (1975) divide the family into 5 subfamilies (formerly 2), all of which are represented in Australia, and this suggests an origin of the family in the Australian region.

Habitat preferences of the early Australian taxa cannot be defined with certainty. However, palynological data, albeit scanty, suggest that some Proteaceae occurred in *Nothofagus* forests. Possibly they occurred also in the subtropical rainforests, as do many of the modern genera (Table 4.4).

The modern genera appear to have been established in the mid-Tertiary. Cookson and Duigan (1950) describe 6 species of *Banksieaephyllum* from Oligocene brown coal in Victoria, in which the internal anatomical and cuticular features can be discerned, the leaves being xeromorphic. However, identification with living *Banksia* and # *Dryandra* is not possible. For the Oligocene-Miocene, Cookson (1954) records 16 species of pollen belonging to *Banksiaeidites* and *Beaupreaidites*. Two of the pollens are identical with living Australian Banksieae, and two are identical with two living New Caledonian species of *Beauprea*. She suggests that some of these fossil species were the lineages of the modern # *Adenanthos,* # *Isopogon,* # *Symphyonema* and # *Xylomelum,* though these relationships with modern genera have been queried by Martin (1973).

Many of the modern Australian genera occur in the rainforests, # *Agastachys* and # *Cennarrhenes* in the *Nothofagus* forests of Tasmania, the remainder in the northern rainforests on the mainland (Table 4.4). However, many of the genera and most of the species occur on the soils of low fertility and these exhibit extreme xeromorphy, most of the taxa being sclerophyllous. Three genera, *Grevillea, Lomatia* and *Orites* occur both in the rainforests and as xeromorphic species on soils of low fertility.

5.4. Rutaceae

There are 20 Australian endemic genera in the Rutaceae. Two of these occur in the subfamily Aurantoideae (fruit succulent), viz., *Microcitrus* with 6 species in Australia (one extending to New Guinea), which occur in the north-eastern rainforests, and # *Eremocitrus,* of rainforest ancestry but occurring in the semi-arid and arid zones (Chap. 5). Six genera in the tribe Zanthoxyleae (subfamily Rutoideae, with a dry fruit) also occur in the rainforests.

The noteworthy development is in the tribe Boronieae (subfamily Rutoideae) to which 12 endemic genera belong. The Boronieae, possibly derived from a rainforest lineage of the Zanthoxyleae, are almost entirely restricted to the soils of low fertility, with most of the species concentrated in the south-west and south-east. The leaves show some xeromorphic characters but are not sclerophyllous.

Cytological studies on the family (Smith-White, 1959) have enabled the following relationships to be established. The basic chromosome number of the family is $x = 9$ and is found in all the tribes. Of the Boronieae, some species of # *Boronia* have $x = 9$, but secondary numbers occur in the genus (7, 8 and 11) as well as polyploids on bases 8 and 9. The remaining genera of the Boronieae are presumably derived from # *Boronia,* and in the formation of new genera both polyploidy and structural changes leading to the loss or gain of centromeres have been involved. In these derived genera $x = 16$ occurs in two genera, $x = 14$ in two genera, $x = 13$ in two genera, and in *Eriostemon* $x = 14$, 17 and 28. Smith-White suggests that generic polyploidy occurred in the pre-Miocene, which is consistent with the east-west disjunction exhibited by the larger genera (see Section 7, below).

The largest genera of the Boronieae are: # *Boronia* (70 species), *Eriostemon* (40, and one in New Caledonia), *Phebalium* (40, and one in New Zealand), # *Zieria* (22), # *Asterolasia* (15), # *Correa* (15).

5.5. Papilionaceae and Mimosaceae (*Acacia*)

The Papilionaceae appear to be very old and to have been widely distributed in the warmer regions before the isolation of the continents (Andrews, 1914). Of the 11 tribes of the family, all are represented in Australia, except Vicieae (Fabeae). However, some tribes are poorly represented, e.g., Abreae (*Abrus* – 1 sp.) and Trifolieae (*Trigonella* – 1 sp.), and at least one other (Dalbergieae – *Derris* and *Lonchocarpus*) appear to have migrated recently into the rainforests from Asia, as have some genera of other tribes, e.g. *Mucuna* (Phaseoleae).

Endemic genera have developed in only five tribes and these tribes must be regarded as long-established residents of the continent and possibly derived from Gondwanaland taxa. In the case of the Podalyrieae, Sands (1975) suggests a Cretaceous origin of the tribe, which occurs in South Africa, Asia and America. Her cytological data indicate a monophyletic origin for the Australian segment of the tribe. The tribes (in order or increasing abundance) and endemic genera are as follows (species numbers in brackets): i. Phaseoleae, # *Kennedya* (16); ii. Sophoreae, # *Barklya* (1), # *Podopetalum* (1); iii. Galegeae, # *Lamprobium* (1), # *Paratephrosia (1),* # *Ptychosema (1), Swainsona* (50, also 1 sp. in New Zealand); iv. Genisteae, # *Bossiaea* (50), # *Goodia* (2), # *Hovea* (6); # *Platylobium* (4); # *Templetonia* (3); v. Podalyrieae, with 21 genera the largest of which are # *Pultenaea* (100), # *Daviesia* (70), # *Jacksonia* (46), # *Gastrolobium* (43), # *Oxylobium* (27), # *Gompholobium* (26), # *Mirbelia* (24), # *Dillwynia* (16), # *Aotus* (15), # *Brachysema* (15).

The host of species belonging to the Genisteae

Table 4.4. Australian Proteaceae genera arranged according to distribution in the continent and elsewhere. Subfamilies indicated by: PE. = Persoonioideae; PR. = Proteoideae; S. = Sphalmioideae; C. = Carnarvonioideae; G. = Grevilleoideae. Haploid chromosome numbers (Johnson and Briggs, 1975) follow the generic names.

Genera confined to the west	Species in xeromorphic communities	Genera occurring east and west	Species in rainforests	Genera confined to the east	Distribution
	0		1	PE. *Placospermum* 7	Endemic
	0		1	*Austromuellera* 14	Endemic
	0		1	*Buckinghamia* 11	Endemic
	0		1	*Cardwellia* 14	Endemic
	0		2	*Darlingia* 14	Endemic
	c. 247		3	*Grevillea* 10	Malays.; N.G.
	0		1	*Gevuina* 14, 13	S. Amer.; N.G.
	0		6	*Helicia* 14 (y)	SE. Asia; N.G.
	0		2	*Hicksbeachia* 14	Endemic
			1	G. *Hollandaea* 14	Endemic
			3	*Lomatia* 11	S. Amer.
	5		5	*Macadamia* 14	Celebes. N.Cal.
	0		1	*Musgravea* 14	Endemic
	0		1	*Opisthiolepis* 11	Endemic
	0		2	*Oreocallis* 11	S. Amer.
	0		1	*Orites* 14, 15	S. Amer.
	7		4	*Stenocarpus* 11	Malays.; N.G.; N.Cal.
	0		1	C. *Carnarvonia* 14	Endemic
	0		1	S. *Sphalmium* 12	Endemic
	50	*Banksia* 14	0		N.G.
	c. 140	*Hakea* 10	0		Endemic
	10	G. *Lambertia* 14	0		Endemic
	3	*Strangea* 11	0		Endemic
	4	*Xylomelum* 14	0		Endemic
	20	*Adenanthos* 13	0		Endemic
	50	PR. *Conospermum* 11	0		Endemic
	35	*Isopogon* 13	0		Endemic
	40	*Petrophile* 13	0		Endemic
	72	PE. *Persoonia* 7 (x)	0		Endemic
G. *Dryandra* 14	50				Endemic
Franklandia 14, 28	2				Endemic
PR. *Stirlingia* 13	6				Endemic
Synaphaea 11	8				Endemic

PR.	*Symphionema* 10	2	Endemic
G.	*Telopea* 11	4	Endemic
	Tasmania only		
PE.	*Bellendena* 5	1	Endemic
PR.	{ *Agastachys* 13	0	Endemic
	{ *Cenarrhenes* 14	0	Endemic
Totals		756	42

(x) includes # *Acidonia* and # *Pycnonia*;
(y) includes # *Athertonia, Floydia* and # *Triunia* (Johnson and Briggs, 1975).

and Podalyrieae are shrubs up to c. 2 m high and they occur almost entirely on the soils of low fertility with two main areas of concentration, the south-west and the south-east (Table 4.6). The species are xeromorphic and occur in xeromorphic forests and heaths. The only other large genus in *Swainsona,* with most of its species occurring in the arid zone (see Chap. 5).

Rhizobial nodules have been recorded on many species of Papilionaceae from all habitats (Bowen, 1956; Lange, 1961; Beadle, 1964). Since many species of Papilionaceae require a specific rhizobium, several species of *Rhizobium* have probably been present in Australia for a long time.

Acacia, with over 700 species in Australia is the largest genus in the continent. It possibly originated in Gondwanaland, as discussed in Chap. 3, and the Australian species belong to four of the sections identified by Bentham (1875). Some criticism has been levelled at Bentham's classification (Vassal, 1972), but the more recent extensive comparative studies of Pettigrew and Watson (1975) support the retention of Bentham's sections.

The genus is represented in almost every habitat. Fewest species occur in the coldest regions, only one, *A. alpina,* being found in the alpine tract on the mainland (Costin, 1954). Hopper and Maslin (1978) show that the area with the largest number of species is the semi-arid region adjacent to the south-west botanical province in Western Australia. Minor centres of richness are associated with mountainous areas in the west and these acted as refugia during the recent arid period and, later, as foci of recent evolutionary divergence, when the climate improved. In arid and semi-arid climates *Acacia* replaces *Eucalyptus* as a community dominant (see Chap. 18).

The sections of the genus are distributed as follows. The Gummiferae are confined to the tropics, except *A. farnesiana* (a possible introduction) which appears to be invading from the north, especially through some grasslands (Chap. 20). The Pulchellae has its greatest development in the wetter south-west with a small development in the east; the species are shrubs occurring in the understory of forests, or in scrubs and heaths. The Botryocephalae are best developed in the wetter south-east and some of the species are trees, a few being well known as cultivated ornamentals, especially *A. baileyana, A. dealbata* and *A. elata* (the last occurs in rainforest margins). The Phyllodineae not only contain the largest number of species but also spans the greatest variety of habitats, from rainforests in both the tropical and temperate regions, through the *Eucalyptus* forests to the heaths on the coast, and throughout the arid interior, while several species are dominants on coastal sand dunes.

5.6. Euphorbiaceae

With 15 endemic genera and a total of 47 genera in the continent, the Euphorbiaceae are well represented in Australia. The endemic genera occur in 4 of the 5 tribes in Australia; 7 of these genera belong to the tribe Stenolobeae and contain a total of 63 species. The largest genera are: #*Bertya* (20 species); #*Beyeria* (15); #*Monotaxis* (9); #*Pseudanthus* (8). All of the genera occur predominantly on the soils of low fertility, some extending into the semi-arid regions. In the tribe Phyllantheae, 3 endemic genera occur, two (#*Dissilaria* and #*Neoroepera*, each with 2 species) are confined to the rainforests, the third, #*Petalostigma*, covers a wide range of habitats, including arid. The other endemics are #*Calycopeplus* (3 species in arid areas) of the tribe Euphorbieae, and #*Adriana* (5 species) of tribe Crotoneae covering a range of habitats. Near-endemic genera of the tribe Stenolobeae are *Poranthera* (9 species in Australia, 2 in New Zealand), and *Ricinocarpus* (15 species in Australia, one in New Caledonia).

The development in the Euphorbiaceae parallels that in the other families discussed above, in that endemism of both genera and species is at a low level in the rainforests in comparison with other habitats. Also, the extensive development of both endemic genera and species occurs in one tribe.

5.7. Epacridaceae and Ericaceae

The Epacridaceae, having its main centre of development in Australia with a lesser segment in New Zealand, are usually thought to have originated in the Australian region. The modern epacrids are confined almost entirely to the wetter south, being more or less equally developed in the south-west and south-east, but are almost absent from the arid zone (see Table 4.6).

The Epacridaceae are closely related to Ericaceae, the former being distinguished mainly by the 1-locular anthers dehiscing by longitudinal slits, the latter having 2-locular anthers dehiscing by terminal pores. It is usually considered that the Epacridaceae were derived from the Ericaceae, which is best developed in Africa and the cooler parts of the Northern Hemisphere. The Ericaceae are poorly represented in Australia, where 5 genera containing a total of 9 species occur. Two species of *Agapetes* and one of *Rhododendron* occur at higher altitudes in tropical Queensland and appear to have been introduced by distance dispersal from Asia where the genera are well represented. The monotypic, #*Wittsteinia* occurs on mountains in Victoria. Three species of *Gaultheria* occur on the eastern highlands and in Tasmania, and two species of *Pernettya* occur in Tasmania. The last two genera occur abundantly elsewhere, on either side of the Pacific Ocean, mainly in the south. It is doubtful if any of these ericaceous genera have any significance with regard to the origin of the Epacridaceae.

The Epacridaceae include three subfamilies, the most primitive being the Prionoteae, distinguished by the 5-locular, succulent fruit with several ovules per loculus. The subfamily is represented in Tasmania by the monotypic #*Prionotes* and in Patagonia and Tierra del Fuego by the monotypic *Lebetanthus*. These occurrences suggest the origin of the family in the cooler parts of Gondwanaland before the separation of South America, Antarctica and Australia. On the other hand, the succulent-fruited ancient lineage might have been transported from Australia to South America in the distant past, where it developed into the modern *Lebetanthus*.

The other subfamilies, the Styphelieae (fruit 5-locular, succulent and with 1 ovule per loculus) and the Epacrideae (fruit dry) have presumably been derived from the Prionoteae and, with the exception of a few species of the succulent-fruited *Leucopogon*, which are found in south-east Asia and some of the Pacific islands, the taxa are restricted to Australia and New Zealand.

Evolution in the Epacridaceae has been studied by Smith-White (1959) through chromosome numbers. Basic numbers in the family vary from 4 to 33 and in some genera the number is variable (apart from polyploidy). *Leucopogon* shows the greatest variation, with different base numbers, which led Smith-White to suggest that the genus is a composite one. There seems to be no correlation between chromosome number and habitat and it is suggested that the chromosome variation indicates antiquity.

The family is best developed today on the soils of low fertility in the warmer south, extending into the coldest areas both in Tasmania and on the mainland. It is usually assumed that the family evolved in low fertility habitats, such as the heaths where the Epacridaceae are always prominent in the south. However, the contrary view is put forward here, namely: i. that the family originated in or at the margins of *Nothofagus* forests; ii. that all subfamilies were established in this habitat; iii. that the earliest members of the Styphelieae and Epacrideae were either tall shrubs or small trees; iv. that the last two subfamilies migrated from the rainforests to the low fertility habitats, with a reduction in size of the plant and leaves.

This view is supported by the occurrence of certain taxa in the modern rainforests. In the *Nothofagus cunninghamii* forests or their margins in Tasmania, the Prionoteae are represented by #*Prionotes*, a small epiphytic shrub on tree trunks and rocks, the Styphelieae are represented by *Cyathodes glauca* and *Trochocarpa gunnii* (both shrubs 1–2 m high) and the Epacrideae are represented by #*Richea pandanifolia*, a shrub 2–3 m high with leaves up to 60 cm long. On the mainland *Trochocarpa laurina*, a tall shrub to 8 m high, occurs in the *Nothofagus moorei* rainforests and extends northward in the subtropical and tropical rainforests to New Guinea and Celebes. Three of these genera, *Cyathodes*, #*Richea* and *Trochocarpa* occur also on soils of low fertility.

5.8. Casuarinaceae

The family is unique among the angiosperms, and, having no close relatives, it is assigned to an order of its own, Casuarinales. The jointed cladodes bearing whorls of scale leaves and resemble *Equisetum*. Two genera, *Casuarina* and *Gymnostoma* are now recognized, though the latter, until recently, was placed under a section of *Casuarina*. The family is distributed throughout Australia (*Gymnostoma* only in the north-east), with extensions to south-east Asia (into India), New Guinea, Mascarene Is., New Caledonia and Fiji. The genus has had a long residency in Australia, fossil pollen being recorded from the Early Tertiary. A megafossil in Miocene rocks in South America is recorded by Frenguelli (1943).

About 45 species of *Casuarina* occur in the continent and some are ecologically noteworthy because they fix atmospheric nitrogen. The phenomenon was first studies by McLuckie (1923) through the discovery of irregularly shaped nodules attaining diameters of several centimetres on the roots of *C. cunninghamiana*. The nitrogen-fixing symbiont proved to be an actinomycete. Several species from the wetter parts of the continent produce similar nodules, but nodules have not been recovered from species growing in semi-arid or arid regions.

Barlow (1959a) has examined chromosome numbers of 37 species of *Casuarina* and found that the tropical species are at the primary diploid level, n being 8 or 9. Another three Australian species, *C. cunninghamiana*, *C. glauca* (eastern species) and *C. cristata* (occurring across the continent, see Chap. 18) have n=9. The remaining Australian species have somatic chromosome numbers ranging from 20 to 28 and Barlow suggests that chromosome evolution possibly involved "an ancient polyploidy followed by stepwise reduction in number". In the *C. distyla* group of 13 species, sexual tetraploid forms have been recorded in nine species and apomictic triploids in four species (Barlow, 1959b). The various species of this group occur mainly in low fertility habitats from New South Wales to southern Western Australia.

The many species of *Casuarina* cover almost the complete range of environments and some of them are community dominants, especially in the littoral zone (Chap. 23), the heaths (Chap. 16) and the arid and semiarid zones in the inland (Chap. 18). Some species occur as understory species in forests.

5.9. Some Monocotyledonous Families

5.9.1. Liliales

The Liliaceae, widely distributed in the world, are well represented in Australia, and 22 of

Fig. 4.4. Plant of *Johnsonia pubescens* (Liliaceae) growing in a heath near Jurien Bay, W.A.

the total of 33 indigenous genera are endemic. Three endemic genera occur in the rainforests (# *Drymophila*, # *Kreysigia* and # *Schelhammera*). A few of the endemic genera occur in tribes which are well represented in Africa, viz., # *Blandfordia*, # *Anguillaria* and # *Burchardia*. These genera are widely distributed in wetter regions of the south, occurring mainly in *Eucalyptus* forests. # *Stypandra* (Dianelleae) occurs in the same general areas as the last three, and it is closely related to *Dianella* which extends into Asia. # *Milligania* (Milliganieae) is endemic to Tasmania; the Tribe is restricted to the Southern Hemisphere (except Africa).

The greatest development of endemic genera belong to the Tribe Johnsonieae, which is confined to Australia. The genera are # *Allania* (1 species), # *Arnocrinum* (1), # *Borya* (2), # *Hensmannia* (1), # *Johnsonia* (3) (Fig. 4.4), # *Laxmannia* (8), # *Sowerbaea* (3), # *Stawellia* (2) and # *Tricoryne* (7). The species belonging to these genera occur mostly in the xeromorphic communities on soils of low fertility in the wetter parts of the continent in the south. The species exhibit xeromorphic characters, many having rigid, filiform leaves.

A most outstanding development in Australia are the Liliales now referred to the Agavaceae (Fig. 4.5) and Xanthorrhoeaceae. The latter family contains 8 genera, 7 of which are endemic in Australia. The largest genus, *Lomandra*, with 30 species in Australia, extends

Fig. 4.5. *Doryanthes excelsa* (Agavaceae) with inflorescences. The trees are *Eucalyptus piperita* and *Angophora costata*. Royal National Park, south of Sydney, N.S.W.

Fig. 4.6. Plants of *Xanthorrhoea australis* in full flower, growing among granite boulders. The inflorescences are spikes consisting of numerous flowers with white corollas. Near Bundarra, N.S.W.

to New Guinea and New Caledonia. All members of the family exhibit extreme xeromorphy or sclerophylly. Some species of # *Xanthorrhoea* (Fig. 4.6) and # *Kingia* (Fig. 4.7) are arborescent. The family has most of its species on the soils of low fertility.

A similar development on the soils of low fertility is seen in the Haemodoraceae. The largest of the endemic genera are # *Conostylis* (30 species) and # *Anigozanthos* (10) both of which are confined to the south-west. *Haemodorum* (20 species) occurs disjunctly in the south-west and south-east but is not endemic, one species occurring in Malaysia.

5.9.2. Cyperaceae

The Cyperaceae is well represented in Australia both as genera (42) and species (530). Unlike most of the other families discussed in this chapter, the large number of species of Cyperaceae in Australia is due to a high proportion of species belonging to more or less cosmopolitan genera rather than to endemic genera. Most of the Australian species occur in the following genera: *Cyperus* (82 species); *Fimbristylis* (80); *Schoenus* (80); *Carex* (45); *Lepidosperma* (40, genus mostly Australian, extending to Asia and New Zealand); *Scirpus* (40); *Eleocharis* (24); *Gahnia* (19, mainly Australia, extending to Asia).

There are 12 endemic or near-endemic genera in Australia, which together contain only 33 species. Eight of these genera belong to the tribe Rhynchosporeae, viz.; *Caustis* (8 species, one in New Guinea): # *Mesomolaena* (4); # *Cyathochaete* (3); # *Arthrostylis* (2); # *Evandra* (2); # *Reedia* (1); # *Remirea* (1); # *Trachystylis* (1). The remaining genera are # *Chorizandra* (4 spp.) and # *Exocarya* (1) belonging to the tribe Hypolytreae, and # *Crosslandia* (1) belonging to the tribe Sclerieae. The species, especially Rhynchosporeae, are xeromorphic and many of them, together with xeromorphic *Lepidosperma* spp., occur in the xeromorphic communities and commonly replace grasses in the herbaceous layer. Of interest also are the three Southern Hemisphere genera *Carpha*, *Oreobolus* and

Fig. 4.7. *Kingia australis* (Xanthorrhoeaceae), some plants with inflorescences, which are often produced before the new crown of leaves after the plants have been burned. In a recently burned woodland of *Eucalyptus calophylla*. Stirling Ranges, south-west, W.A.

Uncinia which link Australasia with South America (see Chap. 3).

5.9.3. Restionaceae

The Restionaceae have its main centre of development in Australia but are represented in Africa and South America (Chap. 3). Thirteen genera are endemic in Australia and 10 of these are confined to the sands of low fertility in the south-west, viz.: #*Loxocarya* (8 species), #*Anarthria* (5) and the monotypic #*Chaetanthus*, #*Dielsia*, #*Ecdeiocolea*, #*Harperia*, #*Hopkinsia*, #*Lyginia*, #*Meeboldina* and #*Onychosepalum*. More widely distributed are #*Lepyrodia* (16, also New Zealand); *Leptocarpus* (12, also south-east Asia and Chile); *Calorophus* (3, also New Zealand). All species are rush-like plants occurring mainly in the wetter south on soils of low fertility, often in swamps.

5.9.4. Orchidaceae (section supplied by B. J. Wallace)

The Australian orchid flora contains 91 genera (18 wholly endemic and 10 largely so) and c. 550 spp., and thus is small in comparison with the rich orchid flora of New Guinea (c. 130 genera and 2000 spp., see Dockrill, 1972) and Malaysia. The flora can be divided roughly into two groups, firstly, the terrestrial genera with their diversity centred in southern Australia, probably the older group originating in the south, and secondly, the epiphytes and tropical terrestrials.

A few species from the first group are at times found in rainforest from the tropics (higher altitudes, e.g. *Corybas abellianus*) to subtropics (e.g. *Acianthus amplexicaulis, Pterostylis baptistii* and *P. pedunculata*) to temperate *Nothofagus* forests (e.g. *Corybas dilatatus* in Victoria and *Townsownia viridis* in Tasmania), but are otherwise restricted to various open communities, particularly those associated with low fertility soils (see Table 4.7). They appear to have had long residency in Australia and possibly the modern taxa are derived from the Gondwanaland flora, though such a postulation conflicts with the view that the Orchidaceae evolved in Malaysia (Garay, in Sandford, 1974). However, the restriction of the tribe Diurideae (in the sense of Lavarack, 1976) to the Southern Hemisphere and of the subtribe Chloraeinae to South America and Australia, suggests a Gondwanaland relationship. Australian genera belonging to this subtribe include *Caladenia* (see Fig. 4.8), *Thelymitra, Calochilus, Acianthus* and *Chiloglottis*. Most of the genera are disjunct within the continent (Table 4.7), the flora being divided into two segments by the arid zone.

The Australian epiphytes occur mostly in the north-east in the rainforests, tapering off in numbers to the south (one lithophytic *Dendrobium* and one epiphytic *Sarcochilus* in Tasmania), and a few extending into drier regions, especially *Cymbidium canaliculatum* which occurs on trees of *Eucalyptus microtheca* (Fig. 5.8) in the semi-arid and a few localities in the arid zone. The epiphytic habit is usually regarded as a derived one and possibly a rela-

Fig. 4.8. Orchids of the Australian centred tribe Diurideae: *Glossodia major* R. Br. (above) and *Caladenia carnea* R. Br.; on right is *Anguillaria dioica* R. Br., an endemic lily. These are typical of communities on low fertility soils.

tively recent development, though the time of evolution cannot be fixed (Sanford, 1974). The habit is apparently polyphyletic as a number of genera have both epiphytic and terrestrial species e.g. *Calanthe* and *Liparis*; the latter has Australian species which are terrestrial with pseudobulbs underground *(L. habenarina)* and above ground *(L. simmondsii)*, lithophytes (*L. reflexa* and *bracteata*) and true epiphytes (*L. coelogynoides, L. nugentae* etc.).

About 37 genera of the Australian orchids are epiphytic, the largest being *Dendrobium* (c. 55 spp.), *Bulbophyllum* (24) and *Sarcochilus* (12).

Dendrobium, with a total of c. 1500 taxa occurs in south-east Asia, extending to Polynesia; *Bulbophyllum* (1500–2000 taxa) occurs throughout the tropics but this genus and *Dendrobium* have their greatest diversity in New Guinea where 500–600 spp. of each genus occur. *Sarcochilus,* in the strict sense, is endemic in Australia and there are 6 other genera of the same subtribe, the Vandinae, also endemic, viz: *Plectorrhiza* (3 spp., one occurs on Lord Howe Is.), *Mobilabium* (1), *Peristeranthus* (1), *Papillilabium* (1), *Schistotylus* (1), *Rhinerrhiza* (1); *Drymoanthus* of the same group, has three spp., one in north-east Queensland, another in New Zealand, and a third in New Caledonia.

Nineteen terrestrial genera with diversity centres in Asia occur mostly in the Australian tropical rainforests and related communities. These are (with world spp. followed by Australian spp. in brackets): *Anoectochilus* (40, 1); *Aphyllorchis* (20, 2); *Apostasia* (7, 1); *Calanthe* (120, 1); *Cheirostylis* (15, 1); *Corymborkis* (6, 1); *Dipodium* (22, 4, one sp. facultatively terrestrial or epiphytic); *Epipogium* (2, 1); *Eulophia* (200, 5); *Galeola* (70, 2); *Gastrodia* (20, 2); *Geodorum* (16, 2); *Goodyera* (40, 2); *Habenaria* (600, 16, mostly open communities); *Liparis* (250, 10); *Malaxis* (300, 2); *Tainia* (25, 1; *Zeuxine* (75, 2).

These data suggest an Asian origin of the Australian epiphytic and tropical terrestrial orchids with some local development in Australia. The taxa were apparently transported by distance dispersal, a topic discussed in general terms for the orchids by Sanford (1974). Furthermore, the occurrence of 19 of the near-endemic Australian terrestrial genera in New Zealand where each genus is represented by one or a few species (some the same as in Australia) adds support to the probability of distance dispersal of orchids (Hatch, 1963). Within Australia, Green (1964) has suggested that orchid seed may be transported from west to east; he based his conclusions on the occurrence in Western Australia and Victoria of certain disjunct species.

6. Xeromorphy and Sclerophylly

6.1. General

The terms xeromorphic, sclerophyllous and xerophytic derive from the earlier botanists Warming (1895) and Schimper (1903); the history of the concept of xeromorphism has been traced by Seddon (1974).

Xeromorphic plants occur in Australia in wetter regions on soils of lower fertility, and in dry regions. In both cases the leaves of the plants exhibit one or more of the following characters: small, usually harsh leaves (sclerophylls), thick cuticles, sunken stomates, the presence of hairs, leaf-rolling and, in some cases, succulence of the leaves and/or stems. The leaf-form of xeromorphs contrasts with the large, broad, soft (mesomorphic) leaves of rainforest plants, though the two types commonly intergrade, sometimes within a genus. Xeromorphism is seen also in alpine plants and consequently it appears that xeromorphism is a morphological response to one or more extremes of the environment, viz., low fertility, reduced water supply or cold[1]. The most xeromorphic leaves develop when two extremes are operative, e.g., low fertility and minimal water supply, as in those species of the grass, # *Triodia,* inhabiting sands in the arid zone. The term xerophyte is used in this book, according to current usage, to describe plants which are drought resistant.

In Australia, xeromorphic taxa occur on a large scale under wet conditions on soils of low fertility. The species are not especially drought resistant and most of these taxa are not found in the arid zone. Nevertheless, in some areas, notably the south-west the xeromorphic plants are subjected to extreme summer drought, possibly

---

[1] Xeromorphism is therefore mostly a peinomorphism (πείνη in Greek = hunger, lack, need) i.e. response to lack of water, phosphorus, nitrogen, oxygen in wet soils etc. (the editor).

with the exception of those with a deep root system (Grieve and Hellmuth, 1970). These authors measured water potentials as low as −39.4 atmospheres in the shallow-rooted *Hibbertia hypericoides* on the Swan Plain (relatively wet) in contrast to −59.2 atmospheres in *Acacia craspedocarpa* at Cue, in the arid zone.

The fact that some of the most extensive stands of xeromorphs occur in the wetter parts of the continent, especially in the Sydney region on low fertility sands, and that these stands often adjoin rainforest on soils of higher fertility, measured in terms of phosphorus content (Beadle, 1954) suggests that soil fertility levels are important in segregating stands of xeromorphic and mesomorphic plants.

Since all the larger families containing xeromorphic taxa occur both in the rainforests on soils of higher fertility and on soils of lower fertility, which was first noted by Diels (1906), an origin of the xeromorphs from mesomorphic lineages might well be assumed. This was first proposed by Andrews (1916) who named nitrogen and calcium as controlling factors in the adaptation to the low fertility environment. Beadle (1966) named phosphorus as the limiting nutrient, which also controls the amount of nitrogen fixed by both symbiotic and non-symbiotic nitrogen – fixing micro-organisms. It follows then, that if the xeromorphs originated as wet climate adaptations to low fertility, any adaptation to a limited water supply, such as the drought in the south-west, is likely to be a secondary adaptation.

The time when xeromorphic taxa commenced to develop is not known. However, Cookson and Duigan (1950) described fossil leaves of Proteaceae from the Early Tertiary, which are xeromorphic and closely resemble the leaves of existing species of *Banksia* and # *Dryandra*. Thus it is possible that the adaptations were initiated on Gondwanaland.

The process of adaptation is a physiological one, with morphological expressions which include chiefly a reduction in the number of cells formed (but without a reduction in cell-size) and a reduction in the length of the internodes. These two processes result in a reduction in both leaf size and plant size, and in some cases the processes can be reversed by the addition of fertilizer (Beadle, 1968).

Adaptations to low fertility probably commenced at the margins of rainforests and proceeded along gradients of fertility from high to low. Few genera maintained their identity on the low fertility media but, rather, new genera evolved. The process of adaptation and the production of new genotypes continued within the new genera (to form many species), and additional new genera evolved. These evolutionary trends are well illustrated in the Myrtaceae (see Table 4.3, in which the genera are arranged to illustrate this point). The same trends in the reduction of leaf-size and plant size can be seen in those few genera which span a range of habitats, especially *Grevillea, Lomatia* and *Orites* (Proteaceae), *Leptospermum* (Myrtaceae) and *Hibbertia* (Dilleniaceae).

6.2. Some Characters of Xeromorphs

Most xeromorphic species have leaves of leptophyll size and are usually referred to as "ericoid". Some of the largest leaves are found in *Eucalyptus* (notophyll size) and many species, especially those from lower fertility soils are isobilateral. The smallest leaves are usually harsh and often pungent-pointed (sclerophylls) but many taxa have relatively soft, even semi-succulent leaves which vary form leptophyll to notophyll size; these are referred to as soft-leaved xeromorphs, and they grow mixed with sclerophylls in the same communities. In addition, many truly mesomorphic species may grow in xeromorphic communities, notably ground orchids, which usually have subterranean storage organs.

Sclerophylls are characterized by a high proportion of sclerenchyma, usually a very thick cuticle and, in a few cases, by the presence of silica in the epidermal walls. The last occurs usually in monocotyledons, e.g., species of *Cladium,* # *Cyathochaete, Gahnia* and *Lepidosperma* (Cyperaceae), rarely in dicotyledons, *Hibbertia acicularis* being a notable exception. The soft-leaved xeromorphs, on the other hand, have a low proportion of the materials mentioned above which add to leaf-rigidity, but they possess thick cuticles; commonly a high proportion of "water-storage" tissue occurs and such tissue may also occur in sclerophylls, e.g., the acicular-leaved species of # *Hakea*.

Carey (1938) in her comparison of rainforest and xeromorphic species growing in adjacent areas in the Sydney region (mean annual rainfall

1120 mm) provides some quantitative data on leaf-characters. The two groups of species showed much the same variation in leaf-thickness. Many of the sclerophylls had sunken stomates (*Eucalyptus* and *#Hakea*), but in some species the stomates lie on the same level as the epidermis, and in some the stomates are raised above the level of the epidermis and are protected by cuticular ridges or hairs (*Angophora cordifolia* and others). A similar variation in the position of the stomates in xeromorphs in the south-west is reported by Grieve and Hellmuth (1970). Carey indicates also that stomatal indices (percentage of epidermal cells which become stomates) are similar in both rainforest plants (indices 7.7 to 19.9) and the xeromorphs (6.6 to 24.8), and the stomates are the same size in both groups of plants. One difference between the groups is that the stomates of the xeromorphs occur on both leaf surfaces, but in rainforest plants on the lower surface only. A second difference concerns the vascular system: For the xeromorphs $37 \pm 2.91\%$ of the leaf-surface is occupied by veins, and for the rainforest plants $21 \pm 6.17\%$.

The leaves of the wet climate xeromorphs are invariably low in ash, unless they contain silica in abundance. Of the ash constituents phosphorus shows by far the lowest values (Beadle, 1962 and unpubl.). The phosphorus contents of sclerophylls range between 104 and 450 ppm dry weight, the soft-leaves xeromorphs 100–760 ppm, and rainforest species 830–2500 ppm.

6.3. Occurrence of Xeromorphy in Families and Genera

Sclerophylly occurs in c. 20 families, chiefly Epacridaceae, Proteaceae and Myrtaceae. Other dicotyledonous families in which many members are sclerophyllous are Dilleniaceae *(Hibbertia)*, Euphorbiaceae (Tribe Stenolobeae), Mimosaceae (phyllodineous *Acacia* spp.), Papilionaceae (Tribe Podalyrieae), and Rhamnaceae (*#Cryptandra* and *#Spyridium*). In some taxa, the leaves are reduced to scales, the stems being photosynthetic, e.g., Casuarinaceae, Euphorbiaceae *(#Amperea),* Papilionaceae (some species of *#Bossiaea, #Jacksonia* and *#Sphaerolobium*), Santalaceae (most species of *#Choretrum, Exocarpos, #Leptomeria* and *#Omphacomeria*) and Tremandraceae (*#Tetratheca juncea)*. Among the monocotyledons, the Xanthorrhoeaceae is almost entirely sclerophyllous, the condition occurring also in some Cyperaceae, Haemodoraceae. Iridaceae and Xyridaceae. In contrast, there are several families which rarely exhibit sclerophylly, even when the species are growing in the same community with highly sclerophyllous species of the families mentioned above. These non-sclerophyllous families include, in particular, Compositae, Goodeniaceae, Pittosporaceae, Rutaceae, Sterculiaceae and Umbelliferae. Members of these families exhibit xeromorphic features in low fertility habitats (or in the arid regions) and in most of these species the leaves or leaflets are of leptophyll or nanophyll size, but are soft and not rigid.

6.4. Geographic Distribution of the Xeromorphic Flora in Wetter Areas

The sclerophylls, and soft-leaved xeromorphs which occur in association, are most numerous on the sandy soils of low fertility in the higher rainfall areas, mainly in the south, with two main centres of maximum expression, the south-west and the Sydney-Blue Mountain region in the east, the two areas being separated by a distance of c. 3,000 km. Some 2,400 xeromorphic species (in 220 genera) occur in the south-west, representing about three-quarters of the total number of species in that area. In the Sydney-Blue Mountain district about 620 species (in 100 genera) occur, representing one-half to one-third of the total flora of that region. The remainder of the floras of both regions is soft-leaved and includes herbaceous perennials (orchids, Compositae and some grasses), a few annuals, several aquatics, many semi-aquatic and littoral species and, in the east, some rainforest species. The last are most significant, since it is this small group of rainforest plants which have provided material for comparisons between mesomorphs and xeromorphs growing in the same climate.

In addition to these two main centres of abundance, small assemblages of species belonging to the same genera as those in the two main centres occur as isolates in various parts of the continent. The modern distribution pattern suggests a one-time more extensive occurrence of this total flora which appears to have been decimated in the centre of the continent by aridity, as discussed in 7, below.

6.5. Habitat Factors

The wet climate xeromorphs have their maximum abundance in areas with a mean annual rainfall of 800–1200 mm. Some extend into drier areas, especially the south-west, and in this area the plants are subjected to summer drought. The species are not subject to severe frosts in most areas, lowest minima being mostly above −5°C. In the coldest areas the numbers of xeromorphic taxa decrease with increasing cold, as indicated by the data in Table 4.5.

Soil conditions are most important for the survival of the xeromorphs. In the areas of maximum development, the soils are invariably sandy, with little or no clay, and acid in reaction (pH mostly 4.5–6.5). They are so low in mineral nutrients that agriculture and even invasion by aggressive weeds are next to impossible. Of the several nutrients which are usually deficient for agricultural species, phosphorus is in lowest supply. In the Sydney area phosphorus values of 30–70 ppm are usual, but much lower values are recorded elsewhere. McArthur and Bettenay (1960) report phosphorus levels as low as 10 ppm in the surface metre of some soils which support heaths in Western Australia, and Hutton (in Specht and Rayson, 1957) reports values as low as 7 ppm in the surface soil layers and 55 ppm in deeper layers below heaths in South Australia. For comparison, tall *Eucalyptus* forests develop on soils with 100–150 ppm phosphorus and rainforests when the phosphorus content lies between 200–1000 (sometimes higher). Detailed analyses for other nutrients are not available and it is relevant that these, with the exception of nitrogen, are not limiting and that in many cases supplies are replenished from the ocean as an intake of cyclic salt, especially near the coast where much of the heath-land occurs. The amount of nitrogen fixed in the xeromorphic communities appears to be closely related to the phosphate content of the soil, some of it being fixed by legumes (Hannon, 1956, 1958). The application of fertilizers to stands of xeromorphic heath in the field (Specht, 1963; Connor and Wilson, 1968; Heddle and Specht, 1975) confirms the "phosphate hypothesis" set out above.

7. History of the Xeromorphic Flora

The low fertility xeromorphs, established in relatively wet climates as new genotypes in the Early Tertiary, gradually expanded their areas during the Tertiary, replacing the rainforests.

Table 4.5. Numbers of xeromorphic species in communities zoned by temperature and wind in the Kosciusko region. Montane and subalpine communities are dominated by eucalypts (adapted from Costin [1954]).

	Montane	Subalpine	Alpine
Possible lowest minimum temperature (°C)	−14	−20	−22
Casuarinaceae *(Casuarina)*	3	0	0
Dilleniaceae *(Hibbertia)*	2	0	0
Epacridaceae	18	11	7
Euphorbiaceae	4	2	2
Mimosaceae *(Acacia)*	14	1	1
Myrtaceae			
Eucalyptus	31	4	0
Other genera	15	5	2
Proteaceae	20	6	2
Rhamnaceae	5	0	0
Rutaceae	9	5	1
Tremandraceae	2	0	0
Totals	123	34	15

Table 4.6. Numbers of species (genera in brackets) of the main wet climate xeromorphic taxa in various floristic zones in Australia, illustrating the effects of aridity in eliminating xeromorphic taxa.

	SW. W.A. Semi-arid to humid	W.A. Arid	S.A. Arid	Centr. Aust. Arid	Arnhem Land, N.T. Humid	N.S.W. Semi-arid to humid	Number of genera common to SW. and SE.
Epacridaceae	161 (14)	19 (5)	1 (1)	1 (1)	0	76 (13)	(6)
Mimosaceae							
Acacia	159 (1)	118 (1)	51 (1)	56 (1)	50 (1)	120 (1)	(1)
Myrtaceae							
Eucalyptus	73 (1)	57 (1)	29 (1)	23 (1)	37 (1)	190 (1)	(1)
Other genera	337 (28)	165 (20)	17 (6)	13 (6)	25 (9)	108 (14)	(9)
Papilionaceae	297 (25)	84 (20)	6 (5)	10 (8)	24 (12)	148 (20)	(20)
Proteaceae	412 (15)	61 (6)	17 (2)	17 (2)	18 (6)	122 (12)	(10)
Rutaceae	74 (12)	22 (7)	2 (1)	1 (1)	4 (1)	86 (8)	(6)
Total genera	(96)	(60)	(17)	(20)	(30)	(69)	(53)
Total species	1513	506	123	121	168	830	

Table 4.7. Genera occurring mainly or entirely on low fertility soils in south-west W. A. and eastern N. S. W.

Western genera	Spp. in S.W.	Genera occurring in both west and east	Spp. in S.E.	Eastern genera
EPACRIDACEAE	100		58	
# Andersonia # Coleanthera # Conostephium # Cosmelia # Lysinema # Needhamia # Oligarrhena # Sphenotoma		# Acrotriche # Astroloma # Brachyloma Leucopogon # Monotoca Styphelia		Dracophyllum Epacris # Lissanthe # Rupicola # Sprengelia # Woollsia
MYRTACEAE	198		84	
# Actinodium Agonis # Astartea # Beaufortia # Calothamnus # Conothamnus # Chamelaucium # Eremaea # Hypocalymma # Lhotzkya # Pileanthus # Regelia # Scholtzia # Verticordia		Baeckea # Callistemon # Calytrix # Darwinia # Kunzea Leptospermum Melaleuca # Micromyrtus # Thryptomene		# Angophora
	33	Eucalyptus	80	
PAPILIONACEAE	206		104	
# Brachysema # Burtonia # Euchylopsis # Eutaxia # Gastrolobium # Isotropis # Jansonia # Latrobea # Nemcia # Ptychosema # Sclerothamnus # Templetonia		# Aotus # Bossiaea # Chorizema # Daviesia # Dillwynia # Gompholobium # Goodia Hardenbergia # Hovea # Jacksonia # Kennedia # Mirbelia # Oxylobium # Phyllota # Pultenaea # Sphaerolobium # Viminaria		# Platylobium
PROTEACEAE	247		70	
# Adenathos # Dryandra # Franklandia # Stirlingia # Synaphea		Banksia # Conospermum Grevillea # Hakea # Isopogon # Lambertia # Persoonia # Petrophile # Strangea # Xylomelum		Lomatia # Symphionema # Telopea
RUTACEAE	50		39	
# Chorilaena # Diplolaena # Muriantha # Nematolepis		# Asterolasia # Boronia # Crowea Eriostemon Phebalium # Philotheca		# Correa # Zieria
ORCHIDACEAE†	120		245	
# Drakaea # Elythranthera # Epiblema # Spiculaea		Acianthus Caladenia # Calochilus Corybas Cryptostylis # Diuris # Eriochilus Gastrodia # Leporella Lyperanthus Microtis # Paracaleana Prasophyllum Pterostylis Thelymitra		Adenochilus Arthrochilus # Burnettia Calanthe Caleana Chiloglottis Dipodium Galeola # Genoplesium # Glossodia Orthoceras # Rimacola Spiranthes

* excluding epiphytes, which do not occur in the South-west.

This expansion occurred on the existing low fertility soils derived largely from sandstones and certain types of granite, or on sands derived from these rocks. The formation of laterite during the Miocene favoured the xeromorphs in some areas, at least, since the mobilization of iron during the process of laterite formation reduced the available phosphate through the precipitation of insoluble iron phosphate.

A further, probably very rapid expansion of the xeromorphs occurred when the laterites were weathered and the sands, deprived of most of their nutrients, were freed from the ferruginous cement and, in some cases, were spread over large areas. The xeromorphs probably reached their maximum extension before the development of the semi-arid and arid zones in the late Miocene, when they occurred over much, if not most, of southern Australia from west to east, with smaller areas in the north.

As the semi-arid and arid zones developed, progressing across the continent from the north-west to the south-east, the wet climate xeromorph was gradually eliminated. Today, the arid zone separates the two main xeromorphic assemblages in the south, one in the south-west, the other in the east (Sydney area). The former assemblage occurs on fossil laterites or on sands derived from these, the latter mainly on Triassic sandstones. Smaller areas occur elsewhere, as mentioned above.

Very few of the wet climate xeromorphs now occur in the arid zone (Table 4.6). Those in Central Australia exist mainly in refuges in the ranges and belong chiefly to the Papilionaceae (genera: # *Brachysema*, # *Burtonia*, # *Daviesia*, # *Gastrolobium*, # *Jacksonia*, # *Kennedia* and # *Ptychosema*). The Epacridaceae (except *Leucopogon*) and Rutaceae (except *Eriostemon*) were almost completely eliminated. The Myrtaceae survived through the genera *Eucalyptus, Baeckea*, # *Calytrix, Melaleuca* and # *Thryptomene*, the Proteaceae by *Grevillea* and # *Hakea*. In the south-west, very many more genera adapted to aridity, which may be due to the continuous sandy habitat from humid to arid in this area. A similar adaptation does not occur in the east probably because the wet area supporting the xeromorphs is separated from the arid zone by a tract of clayey soils laid down by the eastern rivers.

A comparison of the western and eastern xeromorphic assemblages (selected families in Table 4.7) shows that very many genera are common to the two areas, indicating the one-time continuity of this flora across the southern part of the continent. However, in both areas, but mainly in the west, restricted genera occur, indicating independent evolution in each of these two areas.

At the species level, however, there are differences in almost every genus. No more than half-a-dozen species occur in both areas, among them *Acacia myrtifolia* and # *Viminaria denudata*. Vicariiads can be identified in a host of genera, of which *Eucalyptus calophylla* from the west and *E. gummifera* from the east will serve as an example. Each of these species dominates forest or woodland stands, in both cases with an understory of xeromorphic shrubs and herbs, most of which belong to the genera common to the west and east. Consequently, to the casual observer the two stands could appear identical though they are readily distinguished through the identification of the species.

5. The Flora of the Arid Zone

1. Introduction

The arid zone is defined as the area receiving a mean annual rainfall of 250 mm or less, with a possible extension in the north, where temperatures are higher, to the 350 mm isohyet. These boundaries correspond approximately with the drier limit of eucalypt – dominance in the continent, viz., the subtropical and tropical *Eucalyptus* woodlands in the north and north-east (excluding areas of clay which support tussock grasslands), and the mallee in the south.

Ecologically the arid zone is far from uniform. Differences in mean annual rainfall exist, there being two very arid regions, one over the Nullarbor Plain, the other over the Lake Eyre basin and the Simpson Desert. Also, the northern segment receives most of its rain during the summer months, whereas the southern portion has a predominantly winter rainfall.

Topography, soil parent materials and soils likewise vary tremendously. Although the arid tracts are generally of low relief, hills and low mountain ranges occur and many of these are refuges for plants, especially where permanent or semi-permanent water occurs within the hills. The rock types represented vary, and include mainly sedimentary rocks, granites and lateritic (chiefly ferruginous) residuals. At the other extreme, there are depressed areas in the form of watercourses and lakes where water occurs permanently, occasionally or rarely. In some of these lakes, which are usually dry, salt has accumulated in such quantities that plant growth is impossible. The ranges, watercourses and lakes occupy relatively small areas. Most of the arid zone is covered by sand sheets, often in the form of dunes, the sand being derived mainly from the weathering of ancient laterites but partly from sandstones or from alluvial deposits laid down along watercourses or in basins, especially in the east. In contrast to the sands, extensive areas of clay occur, much of which is alluvium associated with watercourses and lake beds. Also, some of the clay soils are derived from Cretaceous shales and mudstones which are subsaline and strewn with stones (gibbers).

Of the various rocks which occur in the arid zone, limestones have the most significance on the distribution of the flora. The most extensive tract of limestone occurs on the Nullarbor Plain, and this area, apart from producing highly alkaline soils *in situ,* has been a source of calcium carbonate which has been blown eastward where it has produced extensive areas of Solonized Brown Soils (Crocker and Wood, 1947).

It follows from the above that the arid zone presents a large number of habitats which vary from fully wet to extremely dry, from sandy to clayey, from extremely saline to non-saline, from acid to alkaline, and each of these conditions is likely to occur in either a summer or a winter rainfall régime. The result of the great diversity of habitats is a relatively rich total flora and the presence in the arid zone of all the recognized life-forms of plants. The plants are adapted to extremes of the environment (except the hydrophytes), and these include water deficiencies, high summer temperatures (mean monthly maxima 33–41°C) and low winter minima (lowest minimum as low as $-7°C$).

2. The Flora

2.1. General

Only the vascular flora is dealt with in detail in this book, but it is mentioned that non-vascular plants, fungi and micro-organisms are all represented in the arid zone, sometimes abundantly. The arid zone biota have been selected from wetter types of vegetation through the adaptation to aridity of certain plastic genotypes, and the various biological processes such as nutrient-cycling (Charley, 1972) and nitrogen-fixation occur in the arid zone, but at slower rates than in wetter communities. Bacteria, actinomycetes and fungi occur in the soils and some nitrogen-fixation through the activity of

Azotobacter has been reported (Tchan and Beadle, 1955). Nitrogen-fixation through rhizobia in association with certain legumes, especially *Acacia, Lotus, Psoralea, Swainsona* and *Trigonella* occurs (Beadle, 1964). Larger fungi include *Psalliota, Coprinus* and several gasteromycetes (Eardley, 1946).

Several genera of green algae, e.g., *Chlamydomonas, Chlorella, Ulothrix* have been recorded in permanent or temporary waterholes (Moewus, 1953) but are of only local significance. More important is the widespread occurrence of soil algae in both saline and non-saline soils (Parberry, 1970). Of these, the blue-green algae, *Anabaena* and *Nostoc*, have some significance as nitrogen-fixers in some areas (Tchan and Beadle, 1955).

Crustaceous lichens occur in several habitats, especially as epiphytes on the trunks of shrubs and trees, on soils which contain some or much clay, and on rock surfaces (Rogers and Lange, 1972; Rogers, 1972). Liverworts are relatively uncommon, but occur in moist situations or after rain, *Riccia* being the most common. Mosses occur in some habitats, especially in moist situations among rocks, often in association with lichens.

2.2. The Vascular Flora

2.2.1. Pteridophytes and Gymnosperms

The pteridophytes are poorly represented in the arid zone. Arid-adapted taxa include *Ophioglossum coriaceum*, found in rocky areas and on sandy soils, and the rock-ferns, occurring in rock ledges, viz., *Gymnogramma reynoldsii*, two species of *Pleurosorus* and four species of *Cheilanthes*. Widely distributed in areas subject to occasional flooding are 4 species of *Marsilea*, the most common being *M. drummondii* (see Chap. 22). The free-floating aquatic, *Azolla* (2 spp.) is found on some permanent or semi-permanent waters, and the aquatic *Isoëtes muelleri* is reported from a rock pool on Ayer's Rock by Chippendale (1959), who records also a small collection of true ferns in a refugium near permanent water in the Macdonnell Ranges, viz., *Adiantum hispidulum, Cyclosorus gongylodes, Histiopteris incisa, Lindsaea ensifolia, Nephrolepis cordifolia* and *Pteris tremula*.

The gymnosperm flora consists of two species of *Callitris*. *C. Preissii* ssp. *verrucosa* occurs mainly in the mallee in the south, and *C. "glauca"* is widespread (Fig. 5.1) on sandy soils or in rocky terrain; it reaches small tree proportions and is sometimes cut for timber for local supplies of fence-posts. *Macrozamia macdonnellii* occurs in a refugium in Central Australia.

2.2.2. Angiosperms

The number of species in the arid zone is 2607 in 503 genera and 102 families (Table 5.1). The families with the highest representation of species are listed in Table 5.2. The Myrtaceae holds first postion for species-abundance in the arid zone, as it does in Australia, and most of the

Fig. 5.1. Pure stand of *Callitris "glauca"*. Grass sward dominated by *Aristida lacunaria*. 40 km west of Cobar, N.S.W.

Table 5.1. Number of families, genera and species of angiosperms in the arid zone.

	Families	Genera	Species
Dicotyledons	83	409	2,302
Monocotyledons	19	94	305
Totals	102	503	2,607

families abundant in Australia are well represented in the arid zone; exceptions are Orchidaceae, Epacridaceae and Rutaceae. Several small families of lesser importance in the total flora are relatively common in the arid zone, especially Frankeniaceae and Zygophyllaceae.

Genera with the highest species-numbers in the arid zone are listed in Table 5.3. *Acacia* is the most abundant as species and many of them dominate the vegetation. # *Eremophila*, endemic in the continent, has most of its species in the arid zone, the other species occurring in semi-arid areas. *Eucalyptus* and *Melaleuca* are the only other woody taxa well represented.

The Australian arid zone flora, distinctive because of its endemics (see 5, below) contains c. 50 genera which are found in all the hot, arid zones of the world (Table 5.4). The explanation of this is obscure, but may be due to the fact that most of the genera are cosmopolitan and many of them are halophytic and readily dispersed around sea-coasts; from the littoral zones of the continents they have migrated into the adjoining arid areas. Notable among such plants is the cosmopolitan *Salsola kali*. A greater similarity between the arid zone floras of Asia-Africa and Australia exists than between America and Australia. Some of the African and Australian communities are strikingly similar both floristically and structurally. In both areas, *Acacia* commonly forms the dominant stratum (bipinnate in Africa, phyllodineous in Australia), with shrubs of *Cassia* below, and a continuous grass sward made up of species of *Aristida* and *Eragrostis*.

2.2.3. The Naturalized Flora

Many exotic species have been introduced into the arid zone. Some are carried into the drier regions by rivers and are most abundant on

Table 5.2. Numbers of genera and species of the families with the highest representation of species in the arid zone, and numbers for the same families in the total Australian flora.

Order of abundance as species in arid zone	Family	Arid Zone		Australia		Order of abundance as species in Australia
		Genera	Species	Genera	Species	
1	Myrtaceae	21	362	47	1300	1
2	Compositae	56	253	104	705	5
3	Mimosaceae	2	c. 200	10	658	6
4	Gramineae	48	190	135	760	4
5	Papilionaceae	40	183	86	846	2
6	Chenopodiaceae	17	136	31	226	13
7	Goodeniaceae	11	123	12	306	11
8	Myoporaceae	2	105	2	130	24
9	Proteaceae	6	70	37	840	3
10	Malvaceae	9	63	16	144	19
11	Amaranthaceae	6	54	9	139	22
12	Solanaceae	7	47	9	127	23
13	Euphorbiaceae	13	45	47	290	12
14	Cyperaceae	8	43	42	530	7
15	Labiatae	9	42	21	183	15
16	Sterculiaceae	11	41	21	152	18
17	Chloanthaceae	10	40	10	66	36
18	Cruciferae	8	39	24	85	31
19	Zygophyllaceae	4	33	4	33	59
20	Frankeniaceae	1	32	1	49	44

Table 5.3. Numbers of species of genera which are represented in the arid zone by 16 or more species. # indicates genus is endemic in Australia.

Genus	Family	Number of spp. in arid zone	Number of spp. in Australia
Acacia	Mimosaceae	c. 190	c. 750
# Eremophila	Myoporaceae	c. 100	c. 110
Eucalyptus	Myrtaceae	86	c. 450
Goodenia	Goodeniaceae	61	120
Helipterum	Compositae	57	60
Melaleuca	Myrtaceae	53	140
Ptilotus	Amaranthaceae	48	100
Swainsona	Papilionaceae	45	50
Maireana (Kochia)	Chenopodiaceae	42	45
Grevillea	Proteaceae	40	250
# Bassia	Chenopodiaceae	39	48
Frankenia	Frankeniaceae	32	49
Atriplex	Chenopodiaceae	31	40
Sida	Malvaceae	25	40
Baeckea	Myrtaceae	24	70
Cyperus	Cyperaceae	24	82
# Dampiera	Goodeniaceae	23	58
Helichrysum	Compositae	23	100
Scaevola	Goodeniaceae	23	70
Eragrostis	Gramineae	22	45
Solanum	Solanaceae	22	80
# Verticordia	Myrtaceae	22	50
Dodonaea	Sapindaceae	21	55
Pimelea	Thymelaceaceae	20	36
# Hakea	Proteaceae	19	140
Stylidium	Stylidiaceae	17	110
# Daviesia	Papilionaceae	17	70
Brachycome	Compositae	16	52
Stipa	Gramineae	16	55
Zygophyllum	Zygophyllaceae	16	16

the clays forming the floodplains to these rivers, or around lakes. Other species have been introduced with domestic stock or into gardens established in towns or near homesteads. However they were originally introduced, they are now being rapidly dispersed by animals, winds and motor vehicles into most parts of the arid zone where domestic stock are grazed.

Few woody species occur, the most common being *Lycium* * *ferocissimum* which is found mainly near habitation, especially around stockyards, and *Nicotiana* * *glauca* which has become naturalized in the east, and is commonly a local dominant along minor watercourses.

The herbaceous species number c. 60, of which the following are most abundant in the east: * *Avena fatua, Cenchrus* * *ciliaris, Eragrostis* * *cilianensis,* * *Hordeum leporinum,* * *Schismus barbatus* (Graminaeae), * *Acetosa rosea,* * *Emex australis* (Polygonaceae), *Amaranthus* * *viridis* (Amaranthaceae), * *Citrullus lanatus* (Cucurbitaceae), * *Carthamus lanatus, Erigeron* * *floribundus, Sonchus* * *oleraceus,* * *Xanthium spinosum* (Compositae), *Erodium* * *cicutarium* (Geraniaceae), * *Sisymbrium orientale,* and the Papilionaceae * *Medicago* and * *Vicia* spp. which usually bear rhizobial nodules.

3. Distribution of the Angiosperm Flora

3.1. General

Many factors, operating either singly or in union, control both species-abundance in the various habitats of the arid zone, and the dis-

Table 5.4. Angiosperm genera which occur in all the arid zones of the world.

Family	Genera
DICOTYLEDONS	
Aizoaceae	*Mollugo, Trianthema*
Amaranthaceae	*Amaranthus*
Boraginaceae	*Cynoglossum, Heliotropium*
Caesalpiniaceae	*Cassia*
Chenopodiaceae	*Atriplex, Chenopodium, Kochia (Maireana), Salicornia, Salsola, Suaeda*
Compositae	*Gnaphalium, Senecio*
Convolvulaceae	*Convolvulus, Cressa*
Crassulaceae	*Crassula*
Cruciferae	*Lepidium*
Euphorbiaceae	*Euphorbia*
Frankeniaceae	*Frankenia*
Geraniaceae	*Erodium*
Lythraceae	*Lythrum*
Malvaceae	*Abutilon, Sida*
Mimosaceae	*Acacia*
Papilionaceae	*Crotalaria, Glycyrrhiza, Indigofera, Lotus, Psoralea*
Plantaginaceae	*Plantago*
Plumbaginaceae	*Limonium*
Portulaceae	*Portulaca*
Sapindaceae	*Dodonaea*
Solanaceae	*Lycium, Solanum*
Umbelliferae	*Daucus, Eryngium*
Zygophyllaceae	*Tribulus, Zygophyllum*
MONOCOTYLEDONS	
Gramineae	*Aristida, Bromus, Chloris, Digitaria, Enneapogon, Eragrostis, Panicum, Sporobolus, Stipa*
Typhaceae	*Typha*
Totals: 25 families	50 genera

tributions of the individual species. The distributions of the various species are determined largely by proximity to the more humid flora adjoining various segments of the arid zone. Other factors include mean annual rainfall, seasonal distribution of the rain and various edaphic properties. Many arid zone species have more or less continuous areas from west to east across the arid zone, either in the north or the south, whereas others have restricted areas and some exhibit disjunction.

3.2. Species – Abundance in Various Habitats

The total arid zone flora of c. 2600 species might be regarded as a small flora if considered on a species per unit area basis (i.e., c. 1 species per 2500 km^2). Such a low density stems from the repetition of habitats over such vast areas.

The only habitat which supports no flora is the highly saline playa. High salinity is far more effective in reducing species-abundance than is low rainfall. For example, in the Lake Eyre – Simpson Desert region, with a mean annual rainfall of c. 120 mm, the floristically poorest community is made up of one species *(Arthrocnemum halocnemoides)* on soils of high salinity. In contrast, the *Zygochloa paradoxa* hummock grasslands growing on dune crests contains at least 16 species (Boyland, 1970). In the two driest areas, each of which presents at least two habitats, the total numbers of species recorded are 76 for the Simpson Desert (Eardley, 1946), and 84 on the western portion of the Nullarbor Plain (Johnson and Baird, 1970). See also Symon (1969) and Wiedemann (1971).

In less arid parts, species-assemblages in the one habitat (community) are mostly of the order of c. 100 species (sampling area c. 10 km²).

On a regional basis, for areas receiving similar rainfall (c. 200–250 mm per annum), the western segment of the arid zone is richer in woody species than the eastern segment. However, the north-western segment (Great Sandy Desert) is much poorer in species than the areas directly to the east (Central Australia) or to the south. The reason for this is possibly that the Great Sandy desert, being almost entirely sandy, contains fewer habitats than other segments of the arid zone.

3.3. Distributions of the More Abundant Taxa

The distributions of the species of the more abundant genera (listed in Table 5.3) indicate the following (for quantitative data, see Beadle, in press):

Most of the genera are well represented as species in all parts of the arid zone, though some are more abundant in the west (*Acacia* and *# Eremophila*) and some are more abundant in the south, especially *Frankenia*, a typically southern genus, and the xeromorphic genera *Melaleuca, Grevillea, Baeckea* and *# Dampiera* which are concentrated in the arid zone in the south-west of the area adjacent to the south-western xeromorphic element of the Australian flora. These xeromorphic genera, and also *Scaevola, # Verticordia, # Hakea* and *# Daviesia* (Table 5.3) which have similar distributions, illustrate the impact of the xeromorphs on the arid zone flora in the south-west only.

Relatively few species of any genus listed in Table 5.3 occur from west to east across the arid zone. The most important of these, because they are community dominants, are *Acacia aneura* (on sandy soils mainly in the north), *Eucalyptus camaldulensis* (along watercourses), *Maireana sedifolia* (on lime-rich soils) and *Atriplex vesicaria* (on subsaline clays).

Two other groups of genera which produce an asymmetry in the arid zone flora are mentioned: In the west, the grass *# Plectrachne* is concentrated in the north-west and several Verbenaceae are confined to the west, occurring in both the arid zone and wetter areas, viz., *# Dicrastylis, # Lachnostachys, # Newcastelia* and *# Pityrodia*. Also in the south-west, several genera from the xeromorphic assemblage, not mentioned above, have adapted to aridity. These belong chiefly to the Proteaceae *(Banksia, # Dryandra, # Persoonia)*, Myrtaceae *(# Balaustion, # Beaufortia, # Calothamnus, # Calytrix, # Chamelaucium, # Darwinia, Leptospermum, # Micromyrtus, # Pileanthus, # Thryptomene* and *# Wehlia)*, Rutaceae *(# Boronia, # Correa, Eriostemon, Phebalium* and *# Philotheca)*, Epacridaceae *(# Astroloma, # Brachyloma* and *Leucopogon)*, and Papilionaceae *(# Brachysema, # Burtonia, # Gastrolobium, # Jacksonia, # Mirbelia* and *# Pultenaea)*. The second group of genera occurs mainly in the north and especially in the north-east. This group is of rainforest origin and further details are provided in 4.2, below.

3.4. Factors Affecting the Distribution of the Flora

3.4.1. Mean Annual Rainfall

The effects of very low rainfall are to reduce the number of species (see 3.2, above) and to eliminate woody species. The two most drought-resistant perennial species in non-saline environments are the grasses, *# Zygochloa paradoxa* and *# Triodia basedowii,* which occur on unconsolidated sands in the Simpson Desert. In this area the most common shrub (though always of rare occurrence) is *Crotalaria cunninghamii* (Fig. 5.2). The most drought resistant perennial on clays is *Atriplex vesicaria.*

3.4.2. Seasonal Rainfall Incidence

That portion of the arid zone which receives a predominantly summer rainfall supports a flora which is markedly different from the southern regions which receive a predominantly winter rainfall. These differences are summarized here and further details are provided in the various chapters dealing with the vegetation.

At the generic level, the northern flora contains a high proportion of species derived from rainforest lineages (see 4.2, below). The summer rainfall belt is characterized by a preponderance of grasses, few of which occur also in the south.

Fig. 5.2. *Crotalaria cunninghamii* growing on red sand. Eastern fringe of the Simpson Desert. 65 km west of Windorah, Qld.

The main genera are: *Aristida*, #*Plectrachne* and #*Triodia* (also 2 spp. in the south) on sands; #*Astrebla*, *Dichanthium* and *Iseilema* on clays. In the south, the Chenopodiaceae becomes increasingly conspicuous, especially *Atriplex* and #*Maireana (Kochia)*. Herbaceous plants are mainly forbs (except *Stipa*) and belong mostly to the Compositae, Cruciferae and Zygophyllaceae. *Acacia* and *Eucalyptus* occur both in the north and south, but are represented by different species, details of which are provided in Chaps 18 and 14. A few other genera are widely distributed from north to south, but usually with specific differences in the two regions.

3.4.3. Edaphic Factors

In all segments of the arid zone, soil properties segregate the flora into assemblages which determine the plant communities. Details of these are provided in Chap. 6 and the factors which operate are listed here as: soil texture; soil moisture; salinity; the presence or absence of calcium carbonate; soil nutrients.

3.5. Disjunctions

Disjunctions have arisen either from changes in the patterning of soil types (mostly clays and sands), or as a result of climatic changes dating back to the time when aridity was imposed on the continent. In both cases large populations of plants were broken into smaller segments separated by barriers (either edaphic or climatic) which prevented the re-uniting of the separated populations. In some cases isolation has resulted in the evolution of new genotypes.

The first type is described as edaphic disjunction, and in most if not all cases isolation is the result of the movement of sand sheets which cover areas of clay, the latter persisting in isolated patches remote from the main clay areas. Thus, former rivers with a flood plain of clay have been filled in by sandy alluvium and/or covered by sand sheets. The effect is most noticeable in the north-west, where isolated patches of clay support *Astrebla* grasslands lying within dune-fields and sand sheets which support *Acacia* scrubs. Similarly in the south, many of the numerous playas, now isolated within dune areas, were probably once connected by watercourses. The playas support a halophytic flora, the dunes *Acacia* scrub or mallee.

The second type of disjunction stems from

climatic changes in the past and appears to date back to the time when the first contraction of the more humid floras commenced. Patches of vegetation, both of the rainforest type and the wet climate xeromorphs, have persisted in refugia in the more favourable areas in the semi-arid and arid zones. The floristically richest of these refugia occur in the Macdonnell Ranges in Central Australia (Chippendale, 1959), and the Hammersley Range in the north-west (Burbidge 1959). The species of rainforest origin are mentioned in 4.2., below. Many isolated lineages have produced new genotypes, e.g., *Macrozamia macdonnellii* and many others of rainforest origin (see 4.2 below). A few species belonging to xeromorphic genera also occur, e.g., species of *Brachysema*, *Burtonia* and *Jacksonia* in Central Australia (Chippendale, 1972).

Several species appear to be disjunct as a result of the fragmentation of the semi-arid woodlands by a reduction of the mean annual rainfall from c. 300 mm to 250–200 mm. The best examples are seen in *Eucalyptus*, e.g., *E. intertexta*, *E. thozetiana*, *E. morrissii* and several others (see Chap. 14).

Many disjunctions occurring within the arid zone appear to be the result of the fragmentation of stands during the Holocene arid period. The patchy distribution of several species of *Eucalyptus* could be accounted for in this way, and also *Acacia peuce*, a tree reaching a height of c. 16 m, occurring in two patches, one on either side of the Simpson Desert separated by c. 300 km (Crocker and Wood, 1947), and several smaller plants including *Solanum ferocissimum* and *Cassia planitiicola* (Randell and Symon, 1977). Of special significance in this respect is *E. socialis* (formerly *E. oleosa*), which has a more or less continuous distribution in the south, but is found also in isolated patches both to the north and the east (Chap. 14). These isolated occurrences possibly serve to fix the one-time fringe of the arid zone flora during the Holocene arid period. The possibility that the isolated stands of the eucalypt were established by distance dispersal cannot be overlooked. However, the occurrence of the eucalypt with species usually associated with it in the south, especially *Triodia scariosa*, lends support to the view that the stands are community fragments rather chance occurrences.

In a few cases the isolated populations show divergence which is only minor but sufficient to identify new species according to taxonomists.

Among these are *Acacia loderi* and *A. sowdenii* (Chap. 18), now separated by the Flinders Range in the south (Crocker and Wood, 1947), and the Central Australian populations of *E. socialis* show some divergence from the more southerly populations.

4. Origins of the Angiosperm Flora

4.1 General

The arid zone was probably initiated in the Miocene when Australia drifted northward into the dry latitudinal belt and, if this be so, the north-west coast was the first strip of semi-arid and, later, arid land in the continent. As the continent continued its northerly drift, at the same time rotating in an anti-clockwise direction, semi-arid and arid conditions would have progressed from north-west to south-east. It is possible that, as the central part of the north and north-west coasts moved into the tropics, the climate became wetter in the extreme north under the influence of the monsoonal rains. There is no positive evidence for this either in the flora or in dryland-relict materials such as salt and gypsum which are readily removed by high rainfall.

If these latitudinal positions of Australia are correct, the oldest segment of the arid zone should be in north-western Western Australia, the youngest in the south-east, and it might be expected that the flora should be richest in the north-western segment. Data to test this possibility are presented by Beadle (in press), but at best the evidence is flimsy. On the contrary, the flora we see today in the arid zone is made up of assemblages selected from wetter areas peripheral to the modern arid zone, or in a few cases to the local expansion of patches of vegetation formerly isolated in refugia during the Holocene arid period.

When semi-arid and arid conditions were imposed upon the continent, three major segments of the flora had developed in the continent, viz., the rainforests, the xeromorphic communities dominated by *Eucalyptus*, and a littoral assemblage of halophytic or semi-halophytic taxa. In addition, two other groups of taxa now occur, which cannot be assigned to these segments: one group has obvious tropical affinities, the species being mainly shrubs and grasses; the other group is an assemblage of herbs belonging mostly to families typical of cool to cold climates. The groups of taxa are discussed below under the five headings indicated.

In adapting to semi-aridity and finally to aridity, very few species (genotypes) of wetter areas persisted and, if so, they are now restricted to refugia. On the contrary, new genotypes were selected; in many cases new species evolved and in some cases new genera developed. As far as vegetative characters are concerned, the dryland taxa are smaller than their wetland relatives; they have smaller leaves and show an accentuation of xeromorphic characters.

Table 5.5. Woody genera which occur in both the rainforests and the arid zone. For additional rainforest taxa which occur in the semi-arid zone, see Chap. 7. For endemic genera in arid zone derived from rainforest lineages, see text. Genera endemic in Australia marked #. C.A. = Central Australia.

Family	Genus	Habit	Distribution in arid zone	Species Arid	Australia
Apocynaceae	*Alstonia*	tree	NE.	1	6
	Carissa	shrub	C.A.; NE.	1	5
	Parsonsia	liana	NE.	1	15
Asclepiadaceae	*Cynanchum*	shrub	wide, N.	1	10
	Marsdenia	liana	wide, N.	1	20
	Pentatropis	lianas	wide, N.	2	4
Bignoniaceae	*Pandorea*	liana	wide, N.	1	3
Caesalpiniaceae	*Bauhinia*	tall shrub	C.A.; NE.	1	2
	Cassia	shrubs	wide, most N.	20	35
Capparidaceae	*Capparis*	shrubs, liana	wide, N.	4	18
Combretaceae	*Terminalia*	tree	NW. W.A.	1	c. 30
Ehretiaceae	*Ehretia*	tree	C.A.; NE.	1	6
Euphorbiaceae	#*Adriana*	shrub	wide; C.A., S.	2	7
	Mallotus	tree	NW. W.A.	1	10
	#*Petalostigma*	shrub	C.A.; NE.	1	7
	Phyllanthus	shrubs	wide, N.	8	60
Gyrostemonaceae	#*Codonocarpus*	tree	wide, NS.	1	2
Malvaceae	*Abutilon*	shrubs	wide, most N.	9	30
	Hibiscus	shrubs	wide, most N.	11	40
Meliaceae	#*Owenia*	trees	wide, N.	2	6
Menispermaceae	*Tinospora*	liana	C.A.; NE.	1	5
Moraceae	*Ficus*	tree	wide, N.	1	40
Myoporaceae	*Myoporum*	shrubs, tree	wide, NS.	5	20
Oleaceae	*Jasminum*	liana	wide, most N.	2	9
Palmae	*Livistona*	tree	C.A.	1	7
Papilionaceae	*Erythrina*	tall shrub	wide, N.	1	4
Pittosporaceae	*Pittosporum*	tall shrub	wide, N.	1	9
Rhamnaceae	*Ventilago*	tree	C.A.; NE.	1	3
Rubiaceae	*Canthium*	tall shrub	wide, N.	4	12
Rutaceae	*Flindersia*	tree	NE.	1	14
	Geijera	trees	NE.; S.	2	4
Santalaceae	*Santalum*	tall shrub	wide, N.	1	4
Sapindaceae	*Atalaya*	tree	wide, N.	1	5
	#*Heterodendrum*	tree	wide, NS.	2	5
Solanaceae	*Duboisia*	tall shrub	wide, NS.	1	3
	Solanum	shrubs	wide	24	75
Sterculiaceae	*Brachychiton*	trees	wide	3	18
Ulmaceae	*Trema*	shrub	C.A.	1	3
Verbenaceae	*Clerodendrum*	shrub	wide, N.	1	3
	#*Spartothamnella*	shrub	wide, N.	2	2

4.2. Taxa Derived from the Rainforests

Of the rainforests which once occurred over the area which is now arid and semi-arid, only taxa from the northern rainforests persisted under drier conditions. The whole of the woody flora of the *Nothofagus* forests was completely eliminated by semi-aridity and aridity, leaving neither relicts nor derivatives.

A few species from the northern rainforests or their margins occur mainly in refuges, viz., *Carissa lanceolata, Clerodendrum floribundum, Erythrina vespertilio, Ficus platypoda, Mallotus nesophilus, Terminalia circumalata, Tinospora smilacina, Trema aspera* and a few ferns (listed in 2.2.1, above).

Far more important, because of their widespread occurrence and abundance, are those arid-adapted species of c. 35 genera listed in Table 5.5 Of these, several are lianas and many are trees, noteworthy among them being *Livistona mariae* (Fig. 5.3) in Palm Valley, Central Australia, *Ficus platypoda* which usually grows over rock surfaces (Fig. 5.4.). #*Codonocarpus cotonifolius* which occurs on deep sands over much of the arid zone, including the mallee-fringe in the south (Chap. 14), *Bauhinia carronii* which occurs in country as dry as the sand-dune troughs in the Simpson Desert.

Most of the species occur in the summer rainfall areas, but some extend to the south, notably #*Heterodendrum oleifolium* which is widespread, extending on to the Nullarbor Plain. Many species are shrubs which are common either as understory species in woodlands or they occur in treeless communities (grasslands or halophytic shrublands), the more abundant being *Cassia* and Malvaceae. A noteworthy plant is the shrub *Duboisia hopwoodii* (Pituri) which, like its rainforest relatives and other members of the Solanaceae, contains an alkaloid, and the leaves of this plant, mixed with ashes, were chewed as a narcotic by the aborigines.

The few endemic genera belonging to the rainforest assemblage are dealt with in 5, below.

4.3. Taxa Derived from the Wet Xeromorphic Assemblages

The xeromorphic taxa, which occur mainly in *Eucalyptus* communities, belong mostly to the Epacridaceae, Myrtaceae, Papilionaceae, Proteaceae and Rutaceae. All of these families are now represented most abundantly in the wetter

Fig. 5.3. *Livistona mariae*. Palm Valley, Macdonnell Ranges, N.T.

Fig. 5.4. A single plant of *Ficus platypoda* at the foot of a granite outcrop. Everard Range, S. A.

parts of the continent in the south, and most of the genera belonging to these families are absent from the arid zone (Tables 5.2, 5.3). Fossil evidence Lange (1978) indicates the presence of a myrtaceous flora in arid South Australia (Woomera area) possibly in the mid-Tertiary. *Acacia* is also well represented in the wetter xeromorphic flora, most species being shrubs and of smaller stature than the arid zone species of *Acacia* which presumably have been derived from another segment of the flora, as discussed in 4.6, below.

The Epacridaceae have been completely eliminated from the arid zone, except in the south-west in the area adjacent to the rich xeromorphic segment of the semi-arid to humid areas. The Rutaceae have behaved similarly, *Eriostemon*, with 8 species in the arid zone, being the noteworthy exception. The Proteaceae are represented only by *Grevillea* and # *Hakea* except for "migrants" in the south-west. Some of the species of these genera, because of their large size, may have been derived from rainforest lineages (see 4.6, below). The Papilionaceae are represented in the arid zone by a few genera relict in the ranges (see 3.5, above).

The Myrtaceae, apart from the south-western adaptations, have been the most successful. *Eucalyptus* is represented by c. 86 species, a few of which are confined to watercourses (Chap. 21), and a few species of *Melaleuca* are restricted to saline habitats. *Thyptomene* is mainly a semi-arid or arid zone genus, and *Calytrix* is arid-adapted, mainly in the south-west.

4.4. Taxa Derived from Temperate Genera

The occurrence in the arid zone of a temperate flora is perhaps unexpected. Five genera occur, which link Australia with New Zealand and cooler South America, viz., *Haloragis, Microseris, Muehlenbeckia, Pratia* and *Vittadinia*. In addition, several cool to cold-climate genera are present belonging to the families Caryophyllaceae, Compositae, Cruciferae and Umbelliferae.

The origins of these taxa in the arid zone is obscure. Two possibilities exist: either they are derivatives of a one-time cold climate flora which may have existed on the higher ranges of the inland, or they are migrants from the south-east.

This segment of the flora has considerable significance because of the large number of endemic genera produced in the Compositae and the Cruciferae and to a less extent the Umbelliferae (see 5, below).

Other herbaceous taxa include the swamp- and water-plants which have probably entered the arid zone along watercourses or have been carried inland by birds. The genera are widely distributed in Australia and many are cosmopolitan.

4.5. Taxa Derived from the Littoral Zone

A very large number of halophytic taxa which fringe the continent in saline and subsaline habitats along the coast occur also in the inland, and it must be concluded that the littoral zone flora invaded the drier parts of the continent. This flora contains a large number of endemic genera (see 5, below).

The development of the halophytic flora from an ancient littoral flora which fringed the shores of the inland Cretaceous sea is discounted on the grounds that this sea was filled with sediments and the dry land so produced once supported forests, as evidenced by the fossils recovered from such areas as the Lake Torrens basin (Chap 3). Again, it is improbable that the littoral zone halophytes invaded the continent along the north-west coast, since this area is almost entirely sandy and free from salt today. If invasion did occur in the west, the taxa have been decimated. On the other hand, the floristically richest assemblages of halophytes are found in the south and it must be assumed that the invasions occurred along the drier parts of the south coast where extensive areas of subsaline clay and limestone (Nullarbor region) adjoin the ocean. If this were so, the halophytic segment of the flora was the last to become established in the arid zone.

The taxa which invaded from the coast belong mainly to the Chenopodiaceae, with a smaller number of genera of the Aizoaceae, Frankeniaceae and Zygophyllaceae. The flora included also some grasses, e.g., *Sporobolus virginicus* and a lineage of *Spinifex* from which *Zygochloa* developed.

4.6. Unclassified Genera

The most abundant woody genus in the arid zone is *Acacia,* and very many of the species, especially in the northern parts of the arid and semi-arid regions, reach tree proportions, some attaining a height of 10–15 m (Chap. 18). On the other hand, many species, especially those in the south in the arid region adjacent to the mallee and heaths, are shrubs with highly xeromorphic phyllodes.

The tree species of *Acacia* which occur in the north, including, for example, *A. aneura, A. cambagei, A. peuce, A. salicina* and *A. stenophylla* have relatively large phyllodes (the last reaching a length of 40 cm, the species being confined to watercourses). Their distributions suggest that they are derived from lineages inhabiting the tall, wet forests of the north-east, perhaps even rainforest margins.

Other woody taxa of tree proportions include *Grevillea striata* (with leaves up to 40 cm long), some species of #*Eremophila* (endemic, see 5, below), *Crotalaria,* some *Santalaceae,* and *Casuarina.* All of these are possibly derived from the tall *Eucalyptus* forest growing on better soils.

Herbaceous genera which cannot be ascribed to an "element" of the flora are mostly tussock grasses. With the exception of *Stipa,* a southern genus, they occur in the area receiving a summer rainfall. A few are endemic (see 5, below) but the remainder are widespread in the northern *Eucalyptus* forests and many of the genera occur also in Asia or on all continents (see Chap. 20). The arid zone occurrences of these genera are a southern extension from the tropics. In the tropics the grasses are subjected to extreme drought in winter, when the culms die, and consequently they are pre-adapted to aridity.

5. The Endemic Genera

The number of endemic genera occurring in the arid and semi-arid zones is 107, belonging to 30 families/ (Table 5.6). Many of these are confined to the arid zone, some having restricted areas, especially monotypic genera. Many occur also in the semi-arid zone and a few of these extend into wetter areas usually as rare occurrences. In many cases the endemic genera probably originated in a semi-arid area and some of their species became adapted to aridity.

Most of the genera endemic in the drier regions are herbaceous or shrubs under 2 m high, the chief exception being the largest genus, #*Eremophila,* of which some of the species are small trees, e.g., *E. bignoniiflora* (Fig. 5.5) and *E. mitchellii.* About one-third of the endemic genera occur in the north, the remainder in the southern half of the arid zone.

In terms of the "elements" of the flora, neither the rainforest taxa nor the Australian xero-

Table 5.6. Numbers of endemic and near-endemic genera in the arid zone arranged under families.

Family	Number of genera
Acanthaceae	2
Aizoaceae	3
Amaranthaceae	2
Amaryllidaceae	1
Apocynaceae	1
Boraginaceae	3
Caesalpiniaceae	1
Capparidaceae	3
Chenopodiaceae	13
Chloanthaceae	4
Compositae	23
Cruciferae	14
Euphorbiaceae	2
Goodeniaceae	1
Gramineae	12
Haloragaceae	1
Labiatae	1
Malvaceae	1
Myoporaceae	1
Myrtaceae	2
Papilionaceae	4
Pedaliaceae	1
Rhamnaceae	1
Rutaceae	1
Santalaceae	1
Solanaceae	2
Stackhousiaceae	1
Sterculiaceae	2
Umbelliferae	1
Violaceae	1

Totals: 30 families; 107 genera; 52 monotypic genera

morphic element produced a significant number of endemic genera. On the contrary, most of the endemic genera evolved from lineages of taxa which probably occurred in the understories of forests or woodlands, or in the littoral zone.

Endemic genera of probable rainforest origin belong in Caesalpiniaceae *Petalostylis*, 3 spp.), Capparidaceae (#*Apophyllum*, 1), Malvaceae (#*Alogyne*, 2 – closely related to *Hibiscus*), Rutaceae (*Eremocitrus*, 1 – Fig. 5.6), and Sterculiaceae (#*Gilesia*, 1; #*Hannafordia*, 4).

It is possible that #*Eremophila* evolved on rainforest margins. The genus is closely related to *Myoporum*, which is represented in rainforests (and also in *Eucalyptus* communities, the arid zone and the littoral zone). Some species of #*Eremophila*, especially *E. mitchellii* and less often *E. bignoniiflora* occur in "dry rainforests"

(see Chap. 7), and *E. polyclada* is a weak liana. This view is opposed to Barlow's (1971) comment on #*Eremophila*, which he describes as "part of a southern temperate Australian element in the arid zone flora, and has had its origins in adjacent, temperate areas". His chromosome studies indicate that most species have the base number $x = 18$, including two of those mentioned above (no count for *C. polyclada*). Polyploidy (tetraploids and hexaploids) also occur and, of the 27 species counted, 13 showed infraspecific polyploidy. He suggests that the formation of polyploid biotypes may lead to geographic expansion into new territory.

A similar small development of endemics from the Australian xeromorphic element is seen only in the families Myrtaceae (#*Balaustion*, 6 spp.; #*Thryptomene*, 50 and #*Wehlia*, 5) (see Chap. 4, Table 4.3), and Papilionaceae (#*Erichsenia*, 1).

The remaining genera cannot be confidently assigned to elements. The two large genera *Ptilotus* (c. 100 spp. and one in Malaysia) and *Swainsona* (50, and one in New Zealand) are widely distributed and both occur in more humid areas, including Tasmania.

Of northern occurrence are most of the endemic grass genera, the largest of these being #*Triodia* (33 spp., 8 of which occur in the arid zone, 2 extending into the southern winter-rainfall belt), #*Plectrachne* (10), #*Neurachne* (5) and #*Astrebla* (4). The monotypic #*Zygochloa* is abundant in the Simpson Desert and is closely related taxonomically to the coastal *Spinifex*.

The largest numbers of endemic genera occur in the Compositae (23), Cruciferae (14) and Chenopodiaceae (13), and most of the endemic genera of these families are related to the temperate, southern flora.

The Compositae, with 104 indigenous and 55 endemic genera are well represented in Australia. The family is now regarded as a very old one, possibly extending back to the Cretaceous (Turner 1977). Of the 12 tribes (Cronquist, 1955) all but the Calenduleae are present in Australia. Endemic genera occur in 6 tribes, mostly in the Astereae, Anthemideae and Inuleae, and the 23 endemic genera restricted to the drier regions belong to these tribes. The origins of the Compositae in the continent are obscure and 2 segments appear to exist, one with tropical affinities, the other related to southern taxa. It is this southern group of genera which contains the largest of the endemic genera in the

Fig. 5.5. *Eremophila bignoniiflora*. 32 km south of Mungindi, N.S.W.

arid zone. In the Inuleae, the endemics include #*Ixiolaena* (6 spp.) and #*Waitzia* (6), and a large group (subtribe Angiantheae with compound heads) related to *Craspedia* which occurs in the arid zone, in temperate Australia and New Zealand. The main genera are #*Angianthus* (30 spp.), #*Calocephalus* (18), #*Gnephosis* (17), #*Myriocephalus* (10), #*Chthonocephalus* (3), #*Eriochlamys* (2), and #*Gnaphalodes* (2). In the tribe Astereae, several endemic genera related to *Olearia* (which occurs in temperate Australia, with 30 species in New Zealand) are restricted to inland Australia, viz., #*Minuria* (7 spp.), #*Cratystylis* (3), #*Bellida* (2), and #*Erodiophyllum* (2).

The Cruciferae, a family of temperate regions, are best developed in Australia in the arid zone, where 14 endemic genera have evolved (the only other Australian endemic genus, #*Drabastrum*, occurs in alpine regions). They are all ephemerals, and the main genera are #*Stenopetalum* (8 spp.), #*Phlegmatospermum* (5), #*Menkea* (4), #*Cuphonotus* (3) and #*Harmsiodoxa* (3).

The Chenopodiaceae have a strong development of 13 endemic genera in the arid zone, though many of the genera occur also in the littoral zone, mainly in the south, and some occur in the semi-arid or even humid areas. The main endemic genera are #*Maireana* (formerly included under *Kochia* – c. 50 spp.), #*Bassia* (48), #*Dysphania* (6), #*Threlkeldia* (5), #*Babbagia* (4) and #*Malococera* (2).

Fig. 5.6. *Eremocitrus glauca*. The central tree is the parent plant; the bushes surrounding the parent have been produced from root suckers. Main herbaceous species in foreground is *Enneapogon polyphyllus*. Fowler's Gap, c. 120 km north of Broken Hill, N.S.W.

6. Some Vegetative Features of Arid Zone Plants

6.1. Life Forms

A classification of the 2607 angiosperm species using convenient life-form categories is presented in Table 5.7. The great diversity of forms is due to the diversity of habitats in the arid zone. The large number of woody species, many of them reaching small tree proportions is, perhaps, surprising. The genera to which these belong are dealt with under the flora. Comment on some of the other life-forms is required.

The lianas, some with stems 6–8 m long, are derived from former rainforest communities. They are completely arid-adapted and usually occur on upland, sandy soils, or in rocky terrain. The genera are named in Table 5.5. Some small twiners (e.g. *Glycine*) and plants with tendrils (Cucurbitaceae) also occur and are included under herbs.

The stem parasites are mistletoes (Chap. 2) and are widely distributed on many hosts. Chippendale (1963) records them even in the driest parts, provided birds are present.

The stem succulents include *Sarcostemma australe* which usually forms a small bush, but may assume a liana habit if a support is available, the stems reaching a length of 3–4 m (Fig. 5.7.). The others are samphires (*Arthrocnemum* and *Pachycornia*).

The sole vascular epiphyte is the orchid, *Cymbidium canaliculatum* which occurs (rarely) on *Eucalyptus microtheca* (Fig. 5.8) eastwards from the fringes of the Simpson Desert, and across the northern half of Australia.

Table 5.7. The arid zone species classified under life-forms.

Life-form	Number of species
Trees > 5 m high	66
Shrubs 2–8 m	552
Shrubs to 2 m	839
Lianas	7
Stem parasites	23
Stem succulents	7
Epiphytes	1
Perennial herbs	701
Geophytes	17
Helophytes	22
Hydrophytes	10
Annuals	362
Total spp. = 2,607	

Fig. 5.7. *Sarcostemma australe* (Asplepiadaceae). The only true stemsucculent in Australia (apart from samphires). In country formerly dominated by *Acacia aneura*. Barrier Range, near Broken Hill, N.S.W.

The geophytes are almost entirely monocotyledons, which form either root or stem tubers or bulbs. They belong mostly to the genera *Bulbine*, *Thysanotus* (Liliaceae), *Calostemma* and *Crinum* (Amaryllidaceae), and the ground orchids *Caladenia*, *Prasophyllum*, *Pterostylis* and *Thelymitra* (Bates, 1976).

The helophytes and hydrophytes (the latter including both submerged and free-floating types) are invariably associated with permanent or semi-permanent water and are dealt with in Chap. 21.

In addition to the stem parasites mentioned above, several other heterotrophic plants occur in the arid zone. They are mainly root parasites belonging to the Santalaceae (14 spp.) in the genera #*Anthobolus*, *Exocarpos*, #*Eucarya*, #*Leptomeria* and *Santalum*. They form haustoria on the roots of almost any plants which their roots contact. For example, #*Eucarya acuminata* (Quandong) growing in the southern

Fig. 5.8. *Cymbidium canaliculatum* growing on *Eucalyptus microtheca*. 45 km west of Moree, N.S.W.

mallee was found to parasitize 11 angiosperm species (both monocotyledons and dicotyledons), a gymnosperm and a fern.

Other heterotrophic plants of rare occurrence are the saprophyte, *Orobanche australiana*, recorded from creeks in the neighbourhood of Lake Torrens, South Australia (Black, 1943) and two species of *Drosera* recorded from wet areas in refugia in Central Australia (Chippendale, 1972).

Annuals are far more abundant in arid areas than in any other climate, and the annual habit appears to be an adaptation to the erratic rainfall. Genera containing annual species are listed in Table 5.8.

6.2. Leaves

The leaves of arid zone plants, excluding aquatics, show almost every possible range of size, shape and texture.

"Leafless" plants include the stem-succulents (*Sarcostemma* and the samphires), and a few with xeromorphic, wiry cladodes, e.g., #*Bossiaea walkeri*, *Casuarina* spp., some members of the Santalaceae (*Exocarpos*, #*Anthobolus*, and #*Leptomeria*) and the rainforest-derived *Spartothamnella*.

The smallest functional leaves (c. 2 × 1 mm) are found in the low fertility xeromorphs (which are rare), as in the genera #*Micromyrtus* and #*Thryptomene*. The largest leaf (megaphyll) is found in the palm, *Livistona mariae*. Leaves of small mesophyll size are found in some of the species of rainforest ancestry, e.g., *Brachychiton* and *Ficus*. Many other rainforest-derived species have leaves of large microphyll size (notophyll) and so also many species of *Eucalyptus;* some species of *Acacia* have phyllodes of similar size, though some are smaller (leptophylls).

The shapes of leaves of shrubs and trees show a similar variation. Linear and terete leaves occur mainly on the shrubs derived from the

Table 5.8. Main families and genera with annual species which occur in the arid zone. (Some species possibly biennial or even perennial but behaving as annuals.) # indicates endemic to Australia.

Family	Genera
DICOTYLEDONS	
Aizoaceae	*Aizoon, Glinus, Tetragonia, Trianthema, Zaleya*
Amaranthaceae	*Amaranthus, Ptilotus*
Boraginaceae	*Cynoglossum, Heliotropium*
Brunoniaceae	#*Brunonia*
Chenopodiaceae	*Atriplex, Chenopodium,* #*Dysphania, Salsola*
Compositae	#*Angianthus, Brachycome,* #*Calocephalus, Calotis, Craspedia, Flaveria,* #*Gnaphalodes,* #*Gnephosis, Helipterum,* #*Millotia,* #*Myriocephalus,* #*Podolepis,* #*Schoenia, Senecio,* #*Waitzia*
Convolvulaceae	*Cressa, Evolvulus*
Cruciferae	#*Blennodia,* #*Geococcus,* #*Harmsiodoxa, Lepidium,* #*Menkea,* #*Phlegmatospermum,* #*Stenopetalum*
Cucurbitaceae	*Cucumis, Melothria*
Euphorbiaceae	*Euphorbia*
Geraniaceae	*Erodium*
Papilionaceae	*Clianthus, Indigofera, Lotus, Psoralea, Swainsona*
Portulacaceae	*Calandrinia, Portulaca*
Scrophulariaceae	*Mimulus, Stemodia*
Solanaceae	*Nicotiana*
Umbelliferae	*Actinotus, Daucus, Trachymene*
Zygophyllaceae	*Tribulus, Zygophyllum*
MONOCOTYLEDONS	
Juncaginaceae	*Triglochin*
Gramineae	*Aristida, Chloris, Dactyloctenium, Echinochloa, Enneapogon, Eragrostis, Iseilema, Panicum,* #*Paractenium, Perotis,* #*Plagiosetum, Sporobolus, Tragus, Tripogon, Triraphis, Setaria*

low-fertility xeromorphic assemblage, found mainly in the south-west. Elsewhere, broad-lanceolate to ovate leaves are more common because of the high proportion of *Eucalyptus* and species of rainforest ancestry. In the case of *Acacia*, phyllodes are mainly narrow-lanceolate, as they are in more humid areas, though acicular phyllodes also occur, as in *A. carnei, A. colletioides* and *A. tetragonophylla*.

Simple leaves are more common than compound leaves and both types occur in both the xeromorphic and "rainforest" segments of the flora. Compound, highly xeromorphic, sclerophyllous leaves occur in *Grevillea* and # *Hakea*, whereas compound leaves which are mesomorphic to leathery occur in *Atalaya, Cassia,* # *Owenia* and *Pandorea*. When compound leaves are a character of a family or genus, the character is retained by the arid zone species. A notable exception is *Flindersia maculosa* of the arid zone, which has simple leaves.

Leaf-texture shows a similar variation to that found in the total flora. Sclerophyllous leaves occur in those taxa derived from the low fertility xeromorphs of wetter areas; and relatively mesomorphic leaves are found in taxa of rainforest ancestry. The last group show some xeromorphic characters, detected mostly through the development of a leathery texture. The most xeromorphic of all leaves are found in the hummock-forming grasses *Triodia* and *Plectrachne* and in the phyllodes of a few species of *Acacia* mentioned above.

Leaf-succulence is more common in the arid zone than elsewhere (except the littoral zone, which contains many of the halophytic genera found in the arid zone). Families in which leaf-succulence is a common or consistent character are Aizoaceae, Chenopodiaceae, Portulacaceae and Zygophyllaceae. Leaf-succulence or semi-succulence, occurs also in a few species of genera belonging to families which are not usually succulent, e.g. *Myoporum* spp., a few species of # *Eremophila, Lycium, Scaevola spinescens*, and a few monocotyledons: *Bulbine, Calostemma* and *Crinum*.

The leaves of all annual species are either mesomorphic or succulent, the latter occurring mainly in Chenopodiaceae, Portulacaceae and Zygophyllaceae. Some of these can accumulate salt if growing on saline or subsaline soils (*Atriplex* and some species of *Zygophyllum*), whereas others do not (especially *Calandrinia*). Leaf-succulence[1] is usually regarded as a mechanism for conserving water. However, observations on both saline and non-saline populations of annuals indicate that during the final stages of drying at the onset of a dry period, the lives of plants with succulent leaves are not significantly prolonged beyond the lives of mesomorphic annuals, irrespective of the salt-content of the leaves, i.e. all annuals in any stand shrivel and die at approximately the same time.

Many leaves of arid zone plants, both annual and perennial, have leaf surfaces which appear to protect the leaf against high light intensity or heat. These take the form of very shiny ("varnished") surfaces, as in *Dodonaea* and some species of # *Eremophila,* or hairs of various types, e.g., simple hairs in many Solanaceae, stellate hairs in Malvaceae, or vesicular hairs in *Atriplex* (see Chap. 19). Leaf-colour therefore varies from bright green to white, according to the epidermal covering, and different species exhibiting the range of variation may occur together in the same stand. The efficacy of the "protective" surfaces has been measured quantitatively for some species by Sinclair and Thomas (1970) who report that "the general absence of significant decreases in absorption of heat in the majority of arid zone species examined is interesting, and somewhat unexpected".

It is apparent that no single morphological or physiological character is responsible for the adaptation of perennials to aridity, though each one, such as leaf succulence, epidermal reflectivity, may assist in conserving water. It seems probable that in many cases the morphological character of an arid zone taxon has been inherited from a more mesic ancestor and that the character has been modified by adjustment to aridity, in many cases by a reduction in leaf size, which is likely to accentuate such a character as hairiness. Thus the Malvaceae usually bear stellate hairs whether the plants are growing in or near the rainforests or in the arid zone. Similarly, of the Solanaceae, *Solanum* spp. are either very hairy or glabrous, irrespective of whether they occur in the rainforests or the arid zone;

---

[1] Leaf-succulence might be formed during evolution of genera and species as a mechanism for water storage (genetically fixed xero-succulence). Genetically evolved halo-succulence must not be a mechanism against drought. In many taxa halo-succulence varies with salinity and is mainly modificative. It must be strongly distinguished between xero- and halosucculence (the editor).

Duboisia on the other hand is strictly glabrous both in the rainforests and the arid zone.

The seasonal deciduous habit is shown by extremely few arid zone species, which stands in sharp contrast to the species in the monsoon rainfall zone in the north (see Chaps 7 and 8). Deciduous species of the arid zone include *Brachychiton gregorii* and *Erythrina vespertilio*, both of which are of rare occurrence. However, many arid zone perennials shed some or most of their leaves or phyllodes in times of greatest water-stress, e.g. *Acacia aneura*, and some species of *Myoporum* and # *Eremophila*. Also some species, notably *Atriplex vesicaria* retain all or most of their leaves in a moribund or even dead condition in times of extreme drought, the leaves being shed by the formation of an abscission layer after a drought-breaking rain. Protoplastic drought-resistance of the leaves of such perennials is extremely high, and osmotic potentials of -40 to -50 atmospheres have been recorded (Helmuth, 1971).

7. Reproduction

7.1. General

The breeding systems of very few arid zone angiosperms have been investigated, and it is merely an assumption that they are mostly normal. For example, the grasses appear to be wind-pollinated and many Papilionaceae appear to be self-pollinated. The occurrence of a diverse insect fauna and of birds which can be observed visiting flowers and inflorescences infers cross-pollination in many cases. Flowering is not necessarily an annual event, but is determined by the occurrence of rain, especially for annuals. Since rain is likely to be seasonal, flowering likewise is a seasonal event for most species. However, there are some species, both perennial and annual, which flower at any time of the year after rain, e.g. *Acacia aneura* (which produces seed only after summer flowering), and the annual *Atriplex spongeosa*.

Some aberrant breeding systems have been recorded. Apomixis occurs in a few plants which have been investigated. For example, in the Compositae, Davis (1967) records gonial apospory (embryo derived from the megaspore mother cell, without meiosis) in *Brachycome ciliaris* and somatic apospory (embryo derived from nucellar tissue) in *Calotis lappulacea* and *Minuria integerrima*. In *Cassia* (Caesalpiniaceae), Randell (1970) reports apomictic reproduction through adventitious embryony, which is almost restricted to polyploid forms occurring mainly on the plains. The diploid populations are restricted to relict occurrences in old mountainous areas.

7.2. Fruits

A great variety of succulent and dry fruits occurs and no one fruit type is most common. The arid zone taxa have been selected from a large number of families and genera, each with its characteristic fruit type, and these fruit types have undergone no apparent morphological change through adaptation to aridity. Both true and false fruits occur.

Several families contain genera with succulent fruits (Table 5.9) but the character occurs consistently only in Loranthaceae. Most of these succulent fruits are edible and were eaten by the aborigines, especially the larger and more abundant ones, e.g., *Capparis* spp., *Eucarya acuminata* and Cucurbitaceae. Of the smaller ones, *Enchylaena tomentosa* is of special interest. The true fuit is a tiny nut enclosed in a succulent perianth which attains a diameter of c. 6 mm. The perianth is rich in vitamin C and the "berries" were eaten by members of Captain Charles Sturt's exploration party which was stranded for 6 months by drought at Preservation Creek, near Milparinka, N.S.W. The vitamin saved the lives of the explorers, except James Poole who died of scurvy. Most succulent fruits are eaten by birds which disperse the seeds (see 7.4, below).

The dry fruits are achenes, capsules, caryopses, cypselas, follicles, legumes, nuts, samaras, schizocarps and siliquas. Dehiscent fruits shed their seeds usually within days or weeks after the maturation of the seed, presumably as a result of rapid drying in air of low humidity. This stands in sharp contrast to the slow dehiscence of species of the same genera growing in humid areas where fruits of such plants as *Eucalyptus*, # *Hakea* and *Casuarina* (the latter with woody "cones" enclosing samaras) are held on the plant for months or even years before dehiscence occurs (Beadle, 1940). Indehiscent fruits remain enclosed in the pericarp and, in some species, other floral or inflorescence parts remain

Table 5.9. Taxa with succulent fruits (or with succulent appendages to the true fruit) occurring in the arid zone. # indicates endemic to Australia.

Family	Genus	Type of fruit
Apocynaceae	*Carissa*	berry
Capparidaceae	*Capparis*	berry
Chenopodiaceae	# *Enchylaena*	perianth succulent
	Rhagodia	1-seeded berry
Cucurbitaceae	*Cucumis, Melothria*	berry
Goodeniaceae	*Scaevola spinescens*	drupaceous
Loranthaceae	All genera	berry or drupaceous
Meliaceae	# *Owenia*	drupaceous
Moraceae	*Ficus*	multiple (syconium)
Myoporaceae	*Myoporum* (most spp.)	drupaceous
	# *Eremophila* (some spp.)	drupaceous
Oleaceae	*Jasminum*	berry
Rubiaceae	*Canthium*	drupaceous
Rutaceae	# *Eremocitrus*	berry
Santalaceae	# *Anthobolus*, # *Eucarya*, # *Leptomeria*,	drupe
	Exocarpos	succulent pedicel
Solanaceae	*Duboisia, Lycium, Solanum*	berry
Verbenaceae	# *Clerodendrum*, # *Spartothamnella*	drupaceous
Zygophyllaceae	*Nitraria*	drupaceous

attached to the pericarp and have some significance in seed-dispersal or in seed germination, as discussed below.

7.3. Seeds; Germination; Longevity

The largest seeds, exceeding 10 mm diameter occur in *Eucarya* and *Owenia*. The smallest seeds occur in *Nicotiana* and the Portulacaceae (especially *Calandrinia*) in which the seeds are of the order of 0.2 mm diameter. Since a similar variation in seed size occurs in both arid zone and humid taxa, size has no apparent significance as a survival mechanism in the arid zone.

The permeability of the testa to water leading to a rough division into hard seeds (impermeable testa) and soft seeds (permeable) shows a similar variation in arid zone plants as elsewhere, the character of the testa being typical of the genus rather than associated with habitat. Hard seeds are more abundant than soft seeds in all species of *Acacia* and most Papilionaceae. Hard seeds in small proportions occur also in several other genera, e.g., *Dodonaea, Atriplex (A. spongeosa, A. inflata).* On the other hand, soft seeds are of regular occurrence in most taxa, e.g., most Chenopodiaceae, Compositae, Cruciferae, Cucurbitaceae, Malvaceae, Portulacaceae and Zygophyllaceae.

Arid zones in general are noted for the rapidity of germination of seed after rain and the rapid flowering of seedlings. In Australia, this applies to annual species, in particular, and to short-lived perennials, especially grasses. At the other extreme, there are some Australian species which are rarely, if ever, observed as seedlings in the field, and the seed of which has never been germinated in the laboratory. Among these are *Eucarya acuminata* and *Owenia acidula*. Some such species are known to perpetuate themselves in the field mainly by vegetative reproduction (see 7.5 below).

Most arid zone species germinate freely, especially those liberated from the pericarp and having a soft testa. Delayed germination is exhibited by many species, the delay being caused either by a hard testa or by some kind of inhibitor which is associated with an enclosing pericarp or lemma.

Temperature requirements for germination are available for only a few species. In some cases species germinate over a wide range of temperature, e.g., several species of *Maireana (Kochia)* (range 9–41°C), whereas some species of *Atriplex* germinate best at temperatures below 25°C (Burbidge, 1946; Beadle, 1948; see Chap. 19 for further discussion on halophytes). Field observations suggest that the optimum temperature requirements are lower for species in the winter rainfall areas than for species in the more northerly summer rainfall belt.

The significance of the hard seed is difficult to assess. Hard seeds have a greater longevity than soft seeds of the same species and the hard seeds may conserve seed supplies and spread germination over a number of years. On the other hand, the testas of some seeds are so water-impermeable that the embryos die before the testa becomes permeable, e.g., *Clianthus formosus* and possibly many species of *Acacia,* including *A. aneura.* Theoretically, hard seeds should accumulate in the soil but in practice this does not occur to any significant extent for *A. aneura.* All attempts to recover seed by sieving from soil below this species have failed and it must be concluded that, in this case, seed crops are rapidly removed by animals.

Mechanisms for inhibiting germination have been recorded in some species. For grasses, Myers (1942) showed that germination of *Astrebla lappacea* increased with age, a dormancy period of c. 2 years being indicated. Other grasses from the summer rainfall area appear to behave similarly (Silcock and Williams, 1975). In *Aristida contorta,* Mott (1974) indicates an after-ripening requirement and also an inhibition by the tightly appressed lemma which prevents the passage of oxygen to the embryo. For several species of *Atriplex* (Beadle, 1948), a high salt concentration in the perianth (bracteoles) prevents germination until the salt is removed by water (see Chap. 19). There are some indications that the environmental conditions during seed ripening affect the degree of seed dormancy of at least some species (e.g. *Stipa variabilis* and *Dactyloctenium radulans,* Bird, 1975).

The longevity of seeds of arid zone plants is surprisingly short. The few data available suggest that some arid-zone species have a shorter life-span than species of the same genus growing in humid climates. For example, seeds of the arid zone acacias live for c. 20 years, in comparison with 50–60 years in humid regions (Ewart, 1908). Soft seeds of most species live for less than 12 years and a few only 3–4 years (Ewart, 1908; Beadle, 1948 and unpubl.). Hard seeds have a greater longevity than soft seeds of the same species, with a maximum longevity of c. 20 years.

7.4. Dispersal of Fruits and Seeds

Species with succulent fruits are possibly all dispersed by birds and many of these species are widely distributed from west to east across the arid zone. Dispersal by birds has been observed for several species, especially the mistletoes and a few shrubs. The semi-digested fruits of mistletoes, consisting of endocarp and enclosed seed, are commonly encountered in birddroppings adhering to the branches of shrubs and trees. In most cases the seeds or seedlings are desiccated before the haustorium penetrates the host. Again, a few shrub species appear to be dispersed locally by birds, especially *Enchylaena tomentosa* and *Rhagodia spinescens.* Plants of these species are commonly aggregated beneath the canopies of shrubs and trees and are probably derived from seed dropped by birds which perch in the branches. Also, the spread of *Nitraria billardieri* and the sudden appearance of this species in areas remote from other *Nitraria* communities is likely to be due to bird-dispersal (Noble and Whalley, 1978).

Dispersal by birds must be assumed for water plants which sometimes appear in remote areas, e.g., *Typha domingensis* in pools of artesian water in the Simpson Desert.

Dispersal by animals possibly occurs and a few species have appendages to the fruits which equip them for dispersal on the coats of animals, especially the spines (modified perianth) of *Bassia* spp. The introduced * *Medicago* is undoubtedly dispersed in this manner.

Dispersal by water is more common than might be thought, especially in hilly areas where streams arise in the uplands and flow on to plains, the water being channelled at first, finally entering a playa, or dissipating in fans of sand. Many species follow such watercourses and their distribution is determined by the location of the watercourse, e.g., *Eucalyptus camaldulensis* and *Cymbopogon exaltatus* (see Chap. 21). In a similar manner, some species, especially exotics, have been introduced into the arid zone along the inland rivers which have their sources in more humid regions.

Wind is an active agent in the dispersal of disseminules, especially in treeless country. A few species have winged disseminules, e.g., the samaras of *Casuarina* spp. and the winged seed of *Hakea* spp. and *Flindersia maculosa.* More effective, however, are appendages to the persistent pericarps found, for example, in Compositae, many Chenopodiaceae, and the persistent lemmas of grasses. Movement of such disseminules can be observed on any windy day, though the distances travelled by the disseminules today, when so much denudation of vegetation has occurred, is probably much greater than it was in the virgin condition. Consequently, the significance of wind-dispersal might well be over-estimated, and this is suggested by the failure of most species of some genera with an apparent wind-dispersal adaptation to span the arid zone from west to east.

7.5. Vegetative Reproduction

Vegetative reproduction occurs in many species, both herbaceous and woody. Short rhizomes, as found in most perennial tussock grasses, are by far the most common among herbaceous plants. The short rhizomes maintain the tussock as an entity and add to the diameter of the tussock; also, the rhizomes act as organs of perennation when the culms dry and fragment during drought periods. In the case of hummock grasses, rhizomes increase the diameter of the hummocks and in some cases the hummocks become so large that the centres die and the hummock becomes a ring (see Chap. 20). Rhizomes occur also in most helophytes, enabling a clone to cover considerable areas (especially *Typha*) when water is available, and acting as perennating organs when the swamp dries out. Stoloniferous grasses are rare, occurring in *Sporobolus virginicus*, *S. mitchellii*, and sometimes *Triodia pungens* in which the stolons may form loops 20–30 cm long, establishing new plants where they touch the ground.

Fig. 5.9. # *Owenia acidula* showing development of a clone by root suckering. Trees of *Casuarina cristata* at rear. Grasses in foreground include *Panicum decompositum*, *Dichanthium sericeum* and *Chloris acicularis*. 20 km south of Moree, N.S.W.

Bulbs and corms are uncommon in the native flora and occur in a few monocotyledons, notably *Crinum flaccidum*, the bulbs of which are buried at depths of 50–80 cm in the soil. The bulb is a well-protected organ of perennation, rather than a reproductive structure.

Several woody species propagate by extensive root-suckering, in some cases forming clones covering at least tens of square metres as thickets or woodlands in monospecific stands. Some of the species regularly produce viable seed, e.g., *Acacia loderi*, *Atalaya hemiglauca*, *Casuarina cristata*, # *Eremocitrus glauca* (Fig. 5.6) and *Hakea leucoptera*. On the other hand, a few rarely produce seed, e.g., *Acacia carnei* (Chap. 18) or they produce seed which may be capable of germinating, though all attempts to germinate the seed under laboratory conditions have failed, e.g., # *Heterodendrum oleifolium* and # *Owenia acidula* (Fig. 5.9).

6. The Plant Communities Classification: Factors Affecting the Patterning of the Communities and the Distribution of the Communities in the Continent

1. Introduction

This chapter is designed to summarize the general aspects of the plant communities, to define the classificatory units employed, and to discuss in general terms the various ecological factors which govern the patterning of the communities in the continent.

The communities have been classified on a floristic basis and a vegetation map is supplied in Fig. 6.1. (see between page 124/125). Unfortunately, many communities dealt with in the text cover such small areas that they cannot be shown on this map. This applies in particular to the communities of the littoral zone, the swamps and the woody communities fringing watercourses.

In addition to the floristic units, structural units are needed to indicate the sizes and densities of the various plant communities. These units, which are sometimes used to designate "formations", such as forest, woodland, grassland etc., are of value for general survey work, and are used in the text where appropriate.

For a better understanding of the local distribution of communities the continent has been divided into regions and the interrelationships of the communities in each region are discussed.

In general, the biomass of the vegetation is correlated with moisture, except in the coldest areas. Consequently, the tallest communities (forests) are found in the climatically wettest areas which lie near the coast. The vegetation is zoned in the continent mainly according to mean annual rainfall, i.e. the zones of vegetation follow the isohyets.

Table 6.1. Life forms of vascular plants. Based on Raunkiaer (1934).

Life form	Positions of highest perennating buds. + above soil-surface − below soil or water-surface	Vernacular terms
Megaphanerophyte	$>+30$ m	tall tree
Mesophanerophyte	$+5$ to $+30$ m	small tree
Microphanerophyte	$+2$ to $+5$ m	tall shrub
Nanophanerophyte	$+25$ cm to $+2$ m	shrub
Liana	$+25$ cm to $>+30$ m	climbing plant; vine
Epiphyte	various +	
Stem parasite	various +	
Stem succulent	Mostly $+10$ cm to $+2$ m	
Chamaephyte	0 to $+25$ cm	
Hemicryptophyte	0 to -2 cm (in soil)	
Geophyte	-2 to -20 cm (in soil)	
Helophyte	0 to -20 cm (in wet soil)	marsh or swamp plant
Hydrophyte	0 to c. -10 m (in water)	water plant
Therophyte		annual

2. Classification of the Communities

2.1. General: Life-forms; Leaf-size Classes; Structural Units; Succession

The life forms and leaf-size classes used throughout are based on Raunkiaer (1934) and they are listed in Tables 6.1 and 6.2. The structural units are those currently in use in Australia and are listed in Table 6.3.

Succession is not featured as a means of classifying stands of vegetation, though many groups of communities, especially those in the littoral zone (Chap. 23) are best arranged in zones which might well be regarded as stages of successions. Secondary succession is used in the accepted sense, i.e. succession following a disturbance of a natural stand of vegetation, and a community maintained in a relatively stable condition through the activities of white man is referred to as "disclimax".

The communities are classified on a floristic basis according to the dominants, giving due consideration to structure. In almost all cases the dominants are regarded as the tallest plants in the community. The exceptions are dealt with individually as occasion arises.

2.2. Floristic Units

The units employed are those defined by Beadle and Costin (1952).

The association is regarded as the basic unit for floristic classification and associations are grouped into alliances. The association is defined as a community in which the dominant stratum exhibits uniform floristic composition, the community usually exhibiting uniform structure. The number of dominant species is often one, sometimes two or three, and, rarely, several. However, in some cases, notably the rainforests in the tropics (Chap. 7), many species, more or less randomly distributed, occur in the tallest stratum and in such cases the identification of associations is impossible. Also, in some areas where several associations are present, possibly as fragments resulting from several changes in the environment, "mixed stands" occur and these greatly complicate classification.

The "mixed stand" and its evolution has been discussed by Goodall (1976) who emphasizes the possibility of an ever-changing floristic composition of a stand resulting from the loss or gain of species either through an ever-changing environment, which includes or excludes species according to their ecological tolerances, or to the evolution of new taxa. Mixed stands which represent ecotonal mixtures or isolated fragments of associations are common in most areas and perhaps most common in the east, where coastal and semi-arid species of *Eucalyptus* are present in the one area. In such cases associations do not cover continuous tracts of country but are so mixed with other associations that they can be represented on the vegetation map as mosaics of dots.

3. Factors Affecting the Patterning of the Communities

3.1. General

Climate exercises primary control over the distribution of the communities. Through the ages climate has led to the "shifting" of the flora

Table 6.2. Leaf-size classes, based on Raunkiaer (1934).

		Area mm^2		Approx. length cm
Class 1	Leptophyll	<25		
Class 2	Nanophyll	$(25 \times 9) =$	25–225	<2.5
Class 3	Microphyll	$(25 \times 9^2) =$	225–2025	2.5–7.6
Class 3a	Notophyll		2025–4500	7.6–12.7
Class 4	Mesophyll	$(25 \times 9^3) =$	4500–18,225	12.7–25.0
Class 5	Macrophyll	$(25 \times 9^4) =$	18,225–164,025	25.0–c. 50
Class 6	Megaphyll	$(25 \times 9^5) = $	$>164,025$	>50

Table 6.3. List of structural forms and subforms used in this book. Based on Beadle and Costin (1952) and Specht, in Leeper, 1973).

Structural form	Main subforms	Usual height of woody component	Projected canopy cover 70–100% = closed 30–70% = open <30% = very open	Alternative terms
Rainforest	Mesophyll vine forest	16–42 m	closed	For expanded classification and alternative terms, see Table 7.1 (Chap. 7)
	Notophyll vine forest	21–45 m	closed	
	Microphyll vine forest	6–20 m	closed	
	Nanophyll moss forest	18–42 m	closed	
Forest (trees with long boles)	Tall forest	25–80 m	closed to open	
	Low forest	15–25 m	open	
Woodland (trees with short boles)	Tall woodland	10–15 m	open to very open	
	Low woodland	5–10 m	open to very open	
	Shrub woodland	5–10 m	open to very open	
	Savannah woodland	5–10 m		
Mallee		3–8 m	closed to very open	
Shrubland	Tall shrubland	3–6 m	closed to open	tall scrub
	Low shrubland	2–3 m	closed to very open	low scrub, thicket, shrubbery
	Heath	1–2 m	closed to very open	
	Halophytic shrubland	60–100 cm	open to very open	shrub steppe, saltbush
Savannah	Tree savannah		closed to very open	
	Shrub savannah		closed to very open	
Grassland	Hummock grassland		open to very open	
	Tussock grassland		closed to very open	
	Sod grassland		closed	
Sedgeland			closed	fen, swamp, marsh
Mossland			closed	bog, *Sphagnum* bog
Forbland	Perennial forbland		closed	perennial herbfield
	Annual forbland		varying	annual herbfield

and vegetation (Herbert, 1935), so that in most areas the regional distribution of the communities can be interpreted in terms of the interaction of the various climatic variables, in particular mean annual rainfall, its seasonal distribution, and temperature. These variables determine also specific growth régimes. Within any climatic region, soil properties are important, producing local patterns of vegetation. Past climates and geological events in the past determined the patterning of the communities in earlier times, and the effects of former patternings can be seen today through disjunctions of segments of the modern vegetation. These are dealt with first under "historic factors".

3.2. Historic Factors

Every plant community has been derived from a pre-existing community, chiefly through the evolution of new taxa, as indicated by the vast floristic changes that have occurred since the first angiosperm communities dominated the continent, as discussed in Chaps 4 and 5. In addition to the evolution of new genera and species, communities have been shifting in the continent, partly as a result of changes in climate and partly because of changes in location of soil parent material brought about by geological erosion, the addition of new parent materials, and the redistribution of soil parent materials by water and wind. The most significant effect of past events on the distributions of plant communities are seen in disjunctions of communities or groups of communities which are now separated by climatic or edaphic barriers to migration.

The effect of the arid zone in separating the south-west corner of the continent from the south-east has been mentioned (Chap. 5) and also some disjunctions in the arid zone (Chap. 5). Other disjunctions of major significance are seen in the tropics where the Carpentaria Plains, an area of fine textured soils in a semi-arid climate supporting grasslands (Fig. 6.1), separates the wetter west from the wetter east, both of these areas supporting woodland or forest. Again, in and around the Tropic in eastern Queensland, the extension of the semi-arid zone to near the coast has effectively divided the formerly wetter eastern region and has led to the elimination of the more mesic communities, notably rainforest, from this drier area (Chap. 7).

The elevation of the eastern Great Divide in the Pliocene produced cooler climates at higher altitudes, probably eliminating rainforest from certain areas, and providing for the northern migration of the cooler climate eucalypts and probably some herbaceous taxa from the south. This range has now eroded in places so that cool climate communities now occur in northern New South Wales and southern Queensland, isolated from the stands in the south. Also, the formation of Bass Strait, isolating Tasmania, caused another major disjunction in the cool-climate assemblages.

3.3. Climate

3.3.1. Rainfall

The vegetation is controlled primarily by mean annual rainfall. The isohyets run more or less parallel with the coast (Chap. 1) so that climates become progressively drier from the coast to the inland.

The arid zone divides the wetter parts of the continent into two segments, except in the east. These segments constitute definite floristic and vegetational regions which are indicated in Fig. 6.2 and are discussed individually in Section 4, below. The northern tropical region is divided into two subregions, separated by the semi-arid Carpentaria Plains. The eastern region, which merges with the tropical region in the north is divided into three subregions: i. the coastal low-

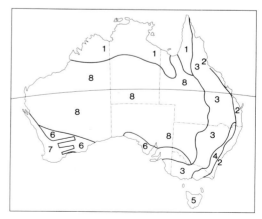

Fig. 6.2. Vegetation regions referred to in this Chapter.

lands; ii. a drier inland region; iii. a cooler (and wet) Tableland region which extends from southern Queensland through New South Wales and Victoria to Tasmania. In the central south the country is largely semi-arid and is divided into two segments by the arid Nullarbor Plain. The south-west corner constitutes a well defined wetter region. The semi-arid and arid zones are treated here as a single region which can be defined as the area in which *Eucalyptus* is absent or subordinate as a community dominant.

The sequence of communities from the wettest to the driest areas, provided soil properties are not controlling vegetation height, is usually: tall forest (either rainforest or *Eucalyptus* forest), tall woodland, open woodland (savannah woodland), low woodland, open low woodland, shrubland, savannah, grassland (arid). However, such a sequence is rarely encountered in a single transect along a straight line. Aberrations are due to edaphic controls which result, for example, in the occurrence of heaths on soils of low fertility in wetter areas, or of tussock grasslands on clays in semi-arid areas.

3.3.2. Temperature

The main effects of temperature are seen on the eastern highlands where low winter temperatures select the dominant eucalypts and in extreme cases prevents the development of trees. At higher altitudes on the Great Divide in the south a tree-line occurs around 2000 m on the mainland and at c. 700–1300 m in Tasmania. Above the treeline the woody communities assume scrub or heath proportions, or herbaceous plants in assemblages of herbfields or grasslands dominate the landscape.

3.4. Edaphic Factors

3.4.1 General

In any climate soil properties control the local distribution of the flora and vegetation, with the result that mosaics of communities are formed, always with floristic differences between the various communities, and in some cases structural differences between contiguous communities may occur. The various soil properties are discussed individually in the following sections.

3.4.2. Soil Depth

Outcrops of rock occur in all the climatic regions and on these Lithosols slowly develop (Chap. 1, Table 1.4, No. 1). Rock surfaces are often colonized by small organisms, possibly firstly by microorganisms, followed by lichens and mosses (see Chap. 16). The mineral soil particles that are produced, accumulate *in situ* or are washed or blown into crevices. Weathering of the rock is accelerated by the rhizoids of lower plants and by organic material, chiefly acidic, derived from the colonizing organisms. Vascular plants become established either in the fragmented rock accumulations or in mats or cushions formed by the non-vascular plants.

Rocky areas in all parts of the continent commonly produce plant communities of considerable height, even in the arid zone, though in these drier regions taller plants are usually scattered. While the habitat is generally drier than deep soils in the same climate, rock crevices present relatively wet habitats, since they receive

Table 6.4. Usual phosphorus levels in soils in relation to the plant communities they support (wetter areas only).

Soil P level ppm (HCl-soluble)	Vernacular description	Plant Communities
5–20	extremely low	heaths; mallee heaths
20–50	very low	heaths; shrublands; mallee; xeromorphic woodlands
50–100	low	woodlands; low forests
100–200	medium	woodlands; tall forests
200–400	high	tall forests; rainforests
>400	very high	rainforests

irrigation water as run-off from the bare rock. Consequently rock-crevice communities, especially in the arid zone, often support unique assemblages of plants, sometimes relicts of former wetter climates (see especially Chaps 5 and 21).

3.4.3. Soil Nutrients

In Chap. 4 the significance of soil nutrients, especially phosphorus and nitrogen, in determining xeromorphy was discussed. In wetter areas soil fertility levels are responsible for the delimitation of many plant communities. The lowest soil fertility levels occur in the sands which support heaths near the coast. In wetter areas, communities with a xeromorphic understory are likewise found on soils of lower fertility, whereas taller woodlands and forests occur on soils of higher fertility, with rainforests occupying the soils of highest fertility (Table 6.4).

In some areas of low fertility in wetter areas the effects of low fertility are superseded by another factor, notably soil waterlogging, salinity, or by frost in very cold climates.

In the semi-arid and arid zones, soil nutrients are of relatively minor importance in determining vegetation patterns, water-availability being the most important factor controlling the communities. Measurements of soil fertility levels by means of pot cultures indicate the soil nitrogen is in lowest supply. However, since many of the arid zone dominants, chiefly *Acacia* spp., are legumes and fix their own nitrogen through a rhizobial symbiosis (Beadle, 1964), nitrogen deficiencies in such communities are unlikely. Similarly, in other communities which contain herbaceous nitrogen – fixing legumes, such as species of *Swainsona, Psoralea* and *Lotus*, a source of nitrogen is apparent and often of measurable significance. In the driest areas, which support grasslands dominated by species of *Triodia* and *Plectrachne*, the sources of nitrogen are obscure.

Soil phosphorus levels in the arid zone are usually relatively high in comparison with coastal soils. Some analyses are provided in Fig. 6.3, and the data show that, in general, clays have a higher phosphorus content than sands. Species with the lowest phosphorus tolerance are *Triodia* spp. and *Eucalyptus socialis*. Most species cover a range of soil phosphorus levels and there is nothing to suggest that phosphorus levels determine vegetation patterns. However, when a woody species occurs over a range of phosphorus levels, plants on soils of lowest fertility are stunted.

3.4.4. Soil Salinity

Salinity, involving high concentrations of sodium chloride, becomes significant along the coast and in the arid interior. Many plant species are adapted to high salt concentrations, some being salt-tolerant, others accumulate salt without being obligate halophytes and others are obligate halophytes. The main halophytic taxa belong in the Chenopodiaceae, Frankeniaceae and Aizoaceae. More detailed accounts of saline environments and halophytes are given in Chap. 23 for the littoral zone and Chaps 19 and 21 for the arid zone.

3.4.5. Soil Texture

In wetter areas soil texture (sandy as opposed to clayey) has relatively little effect on the structure of plant communities, usually both sandy and clayey soils supporting *Eucalyptus* communities. The species on the two different soils may be different, and in most cases the differences are due to soil chemical properties (nutrients, see 3.4.3, above) rather than soil physical properties. If soil texture has any effect on the vegetation in these wetter areas, it usually operates through the development of impervious or slowly pervious pans in the B horizon, which cause underground waterlogging.

In contrast, in the semi-arid and arid zones, soil texture commonly determines the height of the vegetation. Sandy soils produce woodlands, whereas adjacent clayey soils produce grasslands or halophytic shrublands, except in these areas where run-on water, as in watercourses, supplies sufficient additional water to support trees. For example, over much of the semi-arid land in the east sandy duplex soils support woodlands of *Eucalyptus populnea* and *Callitris* "*glauca*", whereas clayey soils support grasslands of *Astrebla* which are traversed by watercourses supporting *E. populnea*. Again, in the arid zone, woodlands of *Acacia aneura* dominate the sandy soils and *Atriplex vesicaria* the clays.

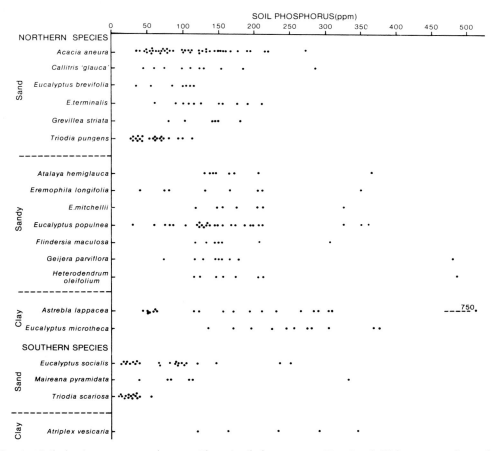

Fig. 6.3. Soil phophorus content of some arid zone soils from western New South Wales, western Queensland and Central Australia, and species occurrences on these soils. Based on c. 200 samples (profiles to 25 cm, or estimates – 15 to 25 cm sample). Sand indicates a deep profile with little or no clay; sandy indicates a sandy A horizon overlying a sandy loam or sandy-clay subsoil (Beadle, unpublished).

The effects of soil texture are not readily explained in simple terms. However, since the sands and clays are contiguous and receive the same amount of rain, the differences in community size might be explained in terms of rate of water – availability from the two contrasting soil types. It is probable that the clays, having greater imbibitional forces, do not release water as rapidly as do the sandy soils. Possibly other factors are concerned, including deeper percolation of water into the sandy soils which would favour larger plants, or possible waterlogging of the clays which might preclude the establishment of trees.

3.4.6. Soil Waterlogging

The occurrence of a water-table above or within a soil profile always has significant effects in selecting the flora. The effects of free water above or at the surface are readily recognized by the occurrence of aquatic or swamp plants (see Chap. 22 for coastal swamps and Chap. 21 for inland swamps). On the other hand, waterlogging below the soil surface is nor always readily observable; it may be either permanent, or temporary and usually seasonal. Some species of *Eucalyptus* which form woodlands are adapted to more or less permanent waterlogging, e.g. *E. gunnii* in Tasmania, and some are apparently dependent upon a water-table or the presence of free water deep in the profile, e.g. *E. camaldulensis* in the arid zone.

Seasonal waterlogging occurs in some regions, particularly those where most of the rain falls in one season, the remainder of the year being dry when the plants in the area are subjected to water stresses. These regions are mainly the wetter tropics, where most of the rain falls during the summer, and the south where most of the rain falls during the winter. In both cases the intensity of the rain is such that the soils cannot absorb all the rain that falls, with the result that some water is shed from the higher to the lower areas – sometimes almost imperceptible depressions – where waterlogging then occurs. In the tropics the large areas of *Melaleuca* woodland (Chap. 22) and some species of *Eucalyptus* (Chap. 8) occupy these seasonally waterlogged tracts. In the south, some species of *Eucalyptus* characterize these wetter areas, e.g. *E. amplifolia* in the east, *E. occidentalis* in the west, and in a few cases small patches of grassland contained within woodlands are controlled by ground-water (Chap. 20). Gilgais (Chap. 1) provide an excellent example of local waterlogging over areas of a few to a few hundred square metres within non-waterlogged areas, and in extreme cases aquatic communities may develop in gilgais.

3.4.7. Soil Acidity/Alkalinity

The pH values of Australian soils lie between c. 4 and 9. Very low values occur in organic soils, especially the Acid Peats, and very high values are recorded mainly in soils from the arid zone, where the alkalinity is due to the accumulation of calcium carbonate. High pH values, mostly around 8.5 (the pH of sea water) occur in some soils (Solonchaks) near the ocean.

In many cases soil acidity or alkalinity selects the flora, and in highly acidic soils the high acidity is induced by the plants themselves by producing acid humus, e.g., Cyperaceae and *Sphagnum* moss. The acidity is increased by microbial activity under anaerobic conditions. It must be assumed that those woody plants which occur in these peaty soils are adapted to high acidity, e.g., many of the alpine heath plants and some species of *Eucalyptus*, especially *E. nitida* and *E. gunnii* in Tasmania. Similarly, heath plants growing on siliceous sands with pH values of c. 4.5 to 5.5 are acidophilous, which has been demonstrated for many species by attempts to cultivate these plants in soils which are only slightly alkaline.

High alkalinity associated with high concentrations of calcium carbonate sometimes selects the flora and leads to patterning of plant communities. This is well illustrated by the arid zone acacias. Thus, *A. aneura* (mulga) prefers acid to neutral conditions and it covers vast areas of red siliceous sandy soils across the arid zone; it sometimes occurs on soils containing calcium carbonate in the lower layers, but the roots do not penetrate deeply into the alkaline layer. When calcium carbonate occurs at or near the surface *A. aneura* is replaced by other species of *Acacia* (see Chap. 18). Also, in the arid zone, *Maireana sedifolia* is an indicator of soils rich in calcium carbonate.

Very many species are relatively indifferent to soil reaction and consequently are found over a wide range of soil types. For example, *Eucalyptus populnea* occurs over a range of pH of c. 4.5 to 8.5; its associated species, however, are possibly more sensitive to soil reaction so that the *E. populnea* alliance is divided into a number of suballiances (see Chap. 13).

3.4.8. Toxic Substances

It is difficult to differentiate between the excess of an essential nutrient or a common non-toxic substance and a poison when dealing with plants. However, in this section substances other than sodium chloride (mentioned above) are dealt with.

3.4.8.1. Gypsum

Gypsum ($CaSO_4 \cdot 2H_2O$) sometimes occurs in large quantities as crystalline deposits in the subsoil in some profiles in the arid zone. It is usually below the level normally explored by roots. In some areas e.g. near Ivanhoe, New South Wales, gypsum is mined for conversion into Plaster of Paris. In playas in the arid zone, gypsum is common below the claypan which forms the floor of the playa. Around some playas, gypsum, by exposure to the wind, has been blown on to the adjacent sand dunes where it has become incorporated with silica sand grains. The gypseous sands, which are acid in reaction, support a very scattered and specialized flora (see Chap. 21).

3.4.8.2. Heavy Metals

In spite of the fact that Australia contains large deposits of heavy metals there is no evidence that these metals are toxic to the surrounding vegetation even when the lodes lie at the surface. Attempts have been made to use plants as indicators of these metal deposits, so far with little success. However, around mining towns, especially Broken Hill, New South Wales and Mt. Isa, Queensland, where lead and zinc sulphides are mined, and around Cobar, New South Wales and Queenstown, Tasmania, where copper, mainly as sulphide is mined, the waste material, piled into huge dumps, is toxic to plants and so also are the leachates which drain from the bases of the dumps and destroy adjoining natural vegetation.

3.4.8.3. Serpentines

Serpentines, ultramafic metamorphic rocks which are often associated with other ultramafic rocks containing heavy metals, occur in small patches in all the States. Serpentines are notoriously infertile the world over. They contain a high proportion of magnesium (as silicate), usually 30–40% (expressed as MgO). They are low in calcium, a few containing none, most having a MgO:CaO ratio as high as c. 100:1 and mostly more than 30:1. (For comparison, basalts contain mostly 5–10% MgO and have MgO:CaO ratios between 1:1 and 1:2). For analyses, see Joplin (1963). The infertility of soils derived from serpentines is usually thought to be due to the high magnesium content or to the suppression of calcium uptake by the concentration of magnesium. In addition, serpentines are commonly low or even lacking in potassium and low in phosphorus.

Serpentine areas can usually be detected by their stunted and sparse vegetation and by the small number of species the soils support. In New South Wales, where most observations on the flora and vegetation have been made, *Xanthorrhoea australis* is possibly the most common species regularly associated with serpentines, together with some or all of the following: *Acacia armata, Bursaria longisepala, Casuarina littoralis, C. stricta, C. torulosa, Exocarpos cupressiformis* (all shrubs), and the trees *Angophora floribunda, Eucalyptus macrohyncha* (drier areas), *E. tereticornis* and *E. propinqua* (wetter areas) (Lyons et al. 1974; W. Hurditch, personal communication). All these species are formed on other rock types, there being no species confined to the serpentines. The possibility of serpentine-adapted ecotypes of the species listed needs investigation.

4. Distribution of the Communities in the Continent

4.1. General

The continent is divided into eight regions as indicated in Fig. 6.2. The regional boundaries bear some relationship to the isohyets (see 3.3.1, above) but seasonal incidence of rain and minimum temperatures are also significant. It is impossible to draw a dividing line between Regions 1 and 2 on the Cape York Peninsula and between Regions 2 and 3 in the tropics, chiefly because the Great Divide in these two areas is so low that it offers no barrier to migration of species as it does in the south.

In the discussion that follows no mention is made of certain vegetation types, especially the littoral zone (Chap. 23) and local swamps (Chap. 22).

The following broad generalizations are made:

i. *Eucalyptus* dominates most of the wetter parts of the continent, the species differing from region to region, with few species occurring in two regions and, then, only when the two regions are contiguous (see especially Chaps 8–15).

ii. *Acacia* replaces *Eucalyptus* as the dominant in the semi-arid and arid zones (Chap. 18), except on some fine textured soils where either tussock grassland (Chap. 20) or shrublands of *Atriplex* (Chap. 19) occur.

iii. In the driest parts of the arid zone, on sandy soils, *Acacia* is replaced by the hummock grasses *Triodia* and/or *Plectrachne* (Chap. 20).

iv. The rainforests (Chap. 7), apparently once extensively developed, have now contracted to small patches on or near the northern and eastern coasts and Tasmania.

v. A small area of highland is covered by treeless alpine communities on the southern highlands and in Tasmania (Chap. 17).

vi. Heaths occur on low fertility, sandy soils mainly on the coastal lowlands from Cape York Peninsula, along the east, south and south-west coasts and in Tasmania (Chap. 16).

4.2. Region 1: Wetter Tropics

The whole area receives a predominantly summer rainfall, with c. 90% of the rain falling between November and March; drought conditions prevail during the winter, which is relatively warm and is referred to as the "dry" in contrast to the summer "wet". Vegetative growth is restricted to the summer wet period. Many woody species are winter-deciduous, including many of the rainforest plants (Chap. 7) and a few eucalypts (Chap. 8). Herbaceous species, which are mainly grasses in upland areas, dry out during the winter, the perennials perennating by rhizomes. Flowering of most species occurs when moisture is adequate. However, some woody species flower either in midwinter when soil moisture is almost depleted, e.g., *Banksia dentata* and *Eucalyptus setosa* (both evergreen), or in the spring when soil moisture is at its lowest, e.g., the deciduous *Brachychiton* and *Cochlospermum* spp.; flowers appear when the stems are leafless and new leaves emerge before the onset of the summer rains. Quantitative data (Specht et al., 1977) on 348 species from Weipa (Qld) show that flowering occurs in both July and December (two times when sampling was done). Somewhat more species (148) were flowering in July than in December (117). Flowering of annual species of fresh water swamps accounted for much of the increase in July. In drier habitats the greatest numbers of species flower in December.

The region is divided into two segments by the Carpentaria Plains, an area of lower rainfall and clay soils which support *Dichanthium* grassland (Chap. 20). The region has a generally low relief with hills in the west (Kimberleys), centre (Arnhem Plateau) and east (Great Divide). In the east the region merges with the eastern coastal lowlands so that a boundary has not been indicated on the map.

Eucalyptus forests and woodlands cover most of the region (Chap. 8), the species being segregated partly by isolation, the western species being different from those in the east, though a few occur across the continent. Soil depth and the proximity of rock to the surface also play a dominating role in the segregation of species in some areas. Also, the temporary waterlogging of the soil in lowland sites accounts for the development of certain woodlands dominated either by *Eucalyptus* (Chap. 8) or *Melaleuca* (Chap. 22) which are adapted to summer waterlogging.

Small patches of rainforest occur, especially in the east, the types ranging from Evergreen Vine Forest in the best watered sites to Deciduous Vine Forest or other "dry" types in drier areas (Chap. 7).

Acacia is poorly represented but is locally dominant on some stony ridges (Chap. 18).

The region is traversed by watercourses, some with a permanent flow. These are often fringed by gallery rainforest, or by tall *Eucalyptus* forest, even in areas which are climatically too dry to support tall types of vegetation. *Casuarina cunninghamiana* is present on some of these watercourses (Chap. 22).

Herbaceous communities in wetter areas are induced by the formation of lagoons or swamps which dry out during the winter. These are dominated by waterlilies, sedges (Chap. 22) or grasses, and they are commonly fringed by *Melaleuca* woodlands.

In the drier south, the *Eucalyptus* woodlands become very open, merging into the savannahs of the semi-arid zone.

The region is unique in the continent because of the rich mangrove flora developed in the littoral zone (Chap. 23).

4.3. Region 2: Eastern Coastal Lowlands

This region receives a high mean annual rainfall, with a distinct summer predominance in the north, the rain being more or less evenly distributed throughout the year in the more southerly districts. Frosts are insignificant in the north but light frosts are experienced in the south, especially inland. Vegetative growth occurs in the warmer months, growth being slowed in the winter by low rainfall in the north and by low temperatures in the south. Flowering is mostly a spring-summer phenomenon, but precise data are available only for the Sydney region where Price (1963) made observations on 180 species. About 100 of these flower mainly during the spring. Thirty-eight species flower only during the summer; thirty flower only during the winter and seven flower mainly of exclusively during the autumn.

The region is well defined in the south where it is flanked by the highlands. A few low sections of the Great Divide have permitted the migration of the inland flora on to the coastal lowlands, notably the Hunter River valley and to a lesser extent the Clarence River valley. North of c. lat. 29°S. where the highlands are relatively low and do not present a migration barrier, the inland and coastal species mix. Similarly, on Cape York Peninsula both the coastal and inland regions merge floristically with the tropical region, as illustrated by the many species of *Eucalyptus* found in this region (Chap. 8).

The region contains a large number of soil parent materials which vary greatly in fertility, this being the main factor segregating the communities. The soils of highest fertility support rainforests which are most complex in the north, least complex in the south (Chap. 7). With decreasing soil fertility *Eucalyptus* becomes dominant, forming tall forests, often with a mesomorphic understory on the better soils (Chap. 9). Xeromorphic forests occupy soils of lower fertility (Chap. 10), and on the least fertile soils, both in the tropics and in the south, trees are absent, the communities being heaths or scrubs (Chap. 16).

Aquatic communities associated with swamps or rivers are commonly encountered (Chap. 22) and the littoral zone supports a rich mangrove flora, especially in the north (Chap. 23).

4.4. Region 3: Eastern Inland Lowlands

This region is best identified as a zone of *Eucalyptus* woodlands, extending from north to south in the continent. Climatically the region is relatively dry (semi-arid) with a summer rainfall predominance and summer growth and flowering in the north, and a winter predominance and growth in the south. Frosts occur but are not severe.

In the north the region is not well defined since it merges with the coastal lowland region, there being no topographic barrier between the two because the Great Divide is low. However, south of c. lat. 29°S. the region is confined by the Great Divide except in the extreme south (Victoria and South Australia) where it reaches the ocean. A few low areas in the Divide enable semi-arid flora to reach the coastal lowlands, especially in the Hunter River Valley. The region merges with the arid zone on the west.

The woodlands of the region consist of two groups. Ironbark woodlands (and forests) which are located mainly in the northern and central segments of the region (Chap. 12), and the Box woodlands which extend from north to south (Chap. 13). For both groups of communities, the species are related and form taxonomic series in latitudinal sequence.

A few other major, unrelated stands of vegetation are found, especially in the summer rainfall zone. Patches of dry rainforest are present and possibly represent remnants of former more extensive stands which have contracted because of reduction in mean annual rainfall. Also, *Acacia harpophylla* and *Casuarina cristata* are locally dominant over relatively small areas (Chap. 18) and appear to be competing for area with both the rainforests and the eucalypts.

Minor communities include small, isolated patches of tussock grassland on clays both in the north and south (Chap. 20), and woodlands associated with watercourses (Chap. 21).

4.5. Region 4: Eastern Highlands

The highlands extend from a lat. 29°S. to Victoria, reaching their greatest height in the south. The range was elevated during the Pliocene-Pleistocene and around the same time a block fault isolated Tasmania, which, however, has been reconnected with the mainland for short periods during the Pleistocene glaciation. The floras of the range on the mainland and of Tasmania are related, many species, including *Eucalyptus,* occurring in both regions. It must be assumed that the flora of the range is derived from a cold-climate assemblage which migrated northward as the range rose. Today, the northern part of the range is isolated from the south by low portions of the range and this has resulted in the disjunction of many taxa, and local endemics have evolved in both the north and the south (and also Tasmania).

The highlands are distinguished from other mainland regions by the lower temperatures, especially during the winter when frosts are severe and inhibit growth. Rainfall is relatively high over the whole region, with a slight summer predominance in the north and winter predominance in the south. However, low temperatures preclude growth in the winter and both vegetative growth and flowering are confined to the warmer months.

Two major groups of communities exist, those dominated by *Eucalyptus* (Chap. 11) and treeless communities at higher altitudes in an alpine or subalpine zone where low winter temperatures, high exposure to wind and/or soil waterlogging control the vegetation (Chap. 17).

Minor communities include small patches of *Nothofagus* rainforest and of the temperate *Ceratopetalum apetalum* assemblage (Chap. 7).

4.6 Region 5: Tasmania

The island is now isolated but supports many mainland taxa, especially in the east. On the other hand, it has retained a strong representation of Gondwanaland taxa in the *Nothofagus* assemblage which is best developed in the west.

Tasmania is well watered and on the basis of mean annual rainfall is divisible into two halves – a wetter west with a mean annual rainfall of 1250–3500 mm and a drier east where the mean annual rainfall is mostly less than 1250 mm. Temperatures are generally lower than on the mainland (except the alpine zone) and frosts occur over the whole island, except the littoral zone. A tree line occurs at c. 700–1300 m. Vegetative growth and flowering are confined to the warmer months.

Three main plant formations are distinguished (Curtis and Somerville, 1949; Jackson, 1965): i. The *Nothofagus* and some Cunoniaceous rainforests (Chap. 7) best developed in the wetter west on soils of higher fertility (Chap. 7); ii. The *Eucalyptus* forests mainly in the drier east but widespread in the island (Chap. 11); The alpine, treeless communities on mountain plateaux, extending to the subalpine zone as "moors" in pockets caused by cold air drainage and/or soil waterlogging (Chap. 17).

Smaller areas are covered by sedgeland (*Gymnoschoenus* – Chap. 22) on organic soils of very low fertility, mainly in the south-west in an area climatically suitable for the development of rainforest. Also, small areas of heath, likewise on soils of low fertility are found close to the coast (Chap. 16).

The communities are zoned with altitude, and in the case of *Eucalyptus,* the species change with altitude (Chap. 11). The *Nothofagus* forests cover an altitudinal range from sea level to c. 1000 m, the trees becoming shorter with altitude. However, simple altitudinal zonations do not always occur, but complex mosaics of different structural forms are common, these resulting from changes in habitat brought about by differences in aspect, soil type and altitude (Jackson, 1965, 1973).

Fire has probably played a major role in altering the pattern of the communities (Gilbert, 1959; Jackson, 1968), leading in particular to the contraction of the rainforests and their replacement by eucalypts or by a mixture of eucalypts and rainforest species. Also, fire may be responsible for the conversion of some *Eucalyptus* stands to more open communities, especially heaths or scrubs in which *Leptospermum* and *Acacia* are prominent (Chap. 16).

4.7. Region 6: The Mallee

The mallee region refers to a tract of country dominated by many-stemmed eucalypts. The term is used also to describe individual plants with the many-stemmed habit, the stems arising from a lignotuber (Chap. 14). The region occurs in the south in areas receiving a mean annual rainfall of c. 375–220 mm and it is broken into two main segments by the arid Nullarbor Plain. The region receives a predominantly winter rainfall and vegetative growth and flowering occur mainly in the cooler months, especially spring and early summer.

The mallee probably covered a larger total area in the past than it does today, having contracted as the arid zone expanded in the late Tertiary, leaving outliers in the arid zone. Also, during the Holocene arid period it probably migrated eastward, as evidenced by the outliers of mallee in the semi-arid zone. The largest of these outliers are shown in Fig. 6.1.

Further details are provided in Chap. 14.

4.8. Region 7: South-western Australia

The area is bounded in the east by the 250 mm isohyet and rainfall increases towards the south-west where the maximum mean annual rainfall is c. 1400 mm. In the wetter areas c. 70% of the total rainfall occurs in the four months, May to August, and in the drier areas (south coast and inland) 55–37% of the rain falls during this period. The vegetation is therefore subjected to summer drought. Frosts are rare but may occur inland, lowest minima being -3 or $-4°C$. Flowering is most abundant during the spring (September–October) at the end of the wet season and continuing to early summer.

The south-west is unique in the continent, for several reasons: There is no rainforest, and only one species of mangrove, restricted to one small area, is represented there (Chap. 23). The area supports the richest assemblage of xeromorphic species, nearly all of which are endemic to the south-west, and many endemic genera are confined to the region (Chap. 4).

The vegetation is mainly forest, woodland, mallee and heath which are delimited partly by mean annual rainfall and partly by soil fertility levels, the last being determined largely by the degree of weathering of the ancient laterites.

Much of the south-west is underlain by granite, some metamorphosed to gneiss. Throughout the granite there are small dykes or intrusions of basic lavas which have been metamorphosed to epidiorite or schists (greenstones), the largest area occurring in the Kalgoorlie-Norseman districts (i.e. drier areas). Along the west coast and in patches along the south coast, limestones and sandstones occur, having been deposited under the ocean following faulting and subsidence of the basal granite. The formation of laterite over most of the area obliterated the granite and metamorphosed rocks, parts of which are now exposed at the surface. Much of the laterite is still present but much has been reweathered and distributed over the modern surface as "strew". This consists of sand sometimes mixed with lateritic nodules. A second source of sand comes from the coastal limestones from which the calcium carbonate has been removed.

The distribution of the plant communities is anomalous in so far as some of the shortest communities (heaths) occur in areas of relatively high rainfall, whereas drier areas support woodlands or even forests. Soil fertility levels account for this anomaly.

The tallest forests (*E. diversicolor* and *E. jacksonii*) have developed on the granite exposures in the wettest areas. On the laterite in the wettest areas either *in situ* or on strew containing much ferruginous material another type of tall forest occurs (*E. marginata* – *E. calophylla*). With decreasing soil fertility (increasing sand content) these forests become stunted or are replaced successively by *Banksia* communities or heath as soil fertility levels decrease (Chap. 16). With decreasing mean annual rainfall on soils of better fertility the tall forests mentioned above are replaced by *E. wandoo* (often an epidiorite) and/or *E. loxophleba,* which usually form woodland stands (Chap. 15). This group of communities is surrounded by a semi-arid zone of mallee on laterite strew (Chap. 14), interspersed with heath occurring mainly on sand (Chap. 16).

In still drier areas there is a second group of woodlands developed mainly on greenstones in the semi-arid zone and extending to the edge of the arid zone. The two alliances are *E. salmonophloia* – *E. salubris* and *E. lesouefii* – *E. dundasii.* The first species is often 20 or 30 m high in this dry climate.

4.9. Region 8: Semi-arid and Arid Areas

Semi-arid and arid areas occupy a large proportion of the continent, reaching the ocean in the north, south and west. The arid and semi-arid climates are relatively recent developments in the continent, as discussed in Chap. 5. The mean annual rainfall varies from c. 125 mm in the Simpson Desert to c. 500 mm in the north. Rainfall is erratic and vegetative growth and flowering occur only after rain. The northern segment of the area receives a predominantly summer rainfall, whereas in the south the rain falls mainly in the winter.

The plant communities are determined partly by mean annual rainfall and its seasonal incidence and partly by soil textures.

In the broadest terms, *Acacia* replaces *Eucalyptus* as the dominant of woodlands, mainly on soils of coarser texture (Chap. 18). The woodlands become progressively shorter with diminishing rainfall and in the drier areas they grade into shrublands which, in the driest areas are replaced by hummock grasslands of *Triodia, Plectrachne* and *Zygochloa* (Chap. 20). The species of *Acacia* vary from district to district, but *A. aneura* is by far the most widely distributed. *Casuarina* is likewise well represented, occurring both on clays and sands, often in association with a species of *Acacia* (Chap. 18).

Eucalyptus is represented by many species, some of which occur in upland, sandy areas often in association with a species of *Acacia* (Chap. 14), whereas other species are restricted to watercourses (Chap. 21).

On soils of fine texture, grasslands or halophytic shrublands occur. Grasslands of the summer rainfall zone are dominated by species of *Dichanthium* or *Astrebla* and others less abundant, and in the winter rainfall zone by *Stipa* spp. (Chap. 20). The halophytic shrublands occupy saline and subsaline soils mainly in drier regions in the south with an extension on to the Riverina Plain in New South Wales (Chap. 19). Other halophytic communities are present on playas (Chap. 21).

7. The Rainforests

1. Introduction

Rainforests are closed communities dominated by trees, the leaves of which are usually mesomorphic, and in which epiphytes are abundant and lianas are often present. The classification of the main rainforest types is given in Table 7.1. The origins of the rainforests and their history in the continent are discussed in Chaps 3 and 4. The communities occur today only in the wetter east from Cape York to Tasmania, and in isolated patches near the north coast. Fragments of rainforest and/or woodland communities derived from ancient rainforest communities occur in the semi-arid and arid zones across the continent, mainly in the north, as discussed below.

2. Factors Governing the Distribution of the Rainforests

2.1. Mean Annual Rainfall and Seasonal Incidence

A mean annual rainfall of about 1400 mm is necessary for the development of the more or less evergreen rainforests in the north and of the evergeen rainforests in the south. Seasonal incidence of the rain rather than the total annual rainfall is important with regard to deciduousness. In the north, those areas with adequate total rainfall (about 1400 mm) but a dry winter, support rainforests containing a high proportion of dry-season deciduous species (Table 7.1).

With diminishing rainfall a sequence of rainforest types may occur. Thus, the larger leaved forests with a complex structure occupy the wettest areas and are replaced in drier areas by "dry" rainforest, with a simpler structure and a different set of dominants. In still drier areas the Vine Thickets occur and in still drier areas (arid zone) the rainforest is represented only by relicts.

2.2. Temperature

Temperature has considerable effect on both structure and floristic composition of the rainforests. In the tropical lowlands and along the east coast, frosts either never occur or are insignificant, and many rainforest species are confined to these more or less frost-free areas. The general effect of lower temperatures during both the summer and winter is a decrease in leaf size and number of tree layers, and the reduction of the numbers of lianas and vascular epiphytes. The effects of decreasing temperature under conditions of equal rainfall can be seen in latitudinal sequence (Table 7.1). Similar sequences occur at any latitude with increasing altitude.

2.3. Edaphic Factors

Soil physical properties have little significance in the distribution of the rainforest species within rainforest stands. Soil waterlogging has the greatest effect on the stands, resulting in both structural and floristic changes (Webb, 1968; Tracey, 1970).

Soil chemical properties are more important. Rainforest species have a high mineral requirement in comparison with species of the *Eucalyptus* forests and heaths. Consequently, rainforests occur on the more fertile soils, notably those derived from basalts, some alluviums, some shales, and some acid rocks (Francis, 1951). Beadle (1954, 1966), Baur (1957) and Webb (1969) indicated that soil phosphate is all-important in determining fertility levels. However, in the tropics, Francis (1951) suggests that soil conditions are not important, and Stocker (see Sect. 7, below) indicates that fires and cyclones play a major role in the distribution of rainforests.

Table 7.1. Classification of rainforest communities (Webb, 1959, 1968 and Baur, 1957, 1964), slightly simplified, the intergrading types omitted. Complex indicates the presence of robust lianas, vascular epiphytes and plank buttresses. Simple indicates that robust lianas and vascular epiphytes are not conspicuous in the upper tree layer, and that spur rather than plank buttresses are prominent. "Evergreen" indicates that the species are mostly evergreen but a few deciduous species may occur.

Leaf or Leaflet Size	Conspicuous Subsidiary Plants	Structural Form	General Characters	Canopy Height m	Emergents and Height m	Number of Tree Layers	Other Names
Mesophyll >4500 mm^2 (leaves exceeding 12.7 cm long)	Vines	Mesophyll Vine Forest	Complex, "evergreen"	21–42	rare	3	Tropical rainforest
			Complex, deciduous	21–42	some	2–3	Monsoon forest
			Simple, evergreen	24–36	some evergreen	2	Tropical rainforest
	Palms and Vines	Mesophyll Palm-Vine Forest	Mostly evergreen	16–23	palms c. 25	2–3	
Notophyll 2025 to 4500 mm^2 (7.6–12.7 cm long)	Vines	Notophyll Vine Forest	Complex, "evergreen"	21–45	some	3	Subtropical rainforest
			Complex, deciduous	21–27	conifers	2	Dry rainforest
			Simple, evergreen	21–33		2	Subtropical rainforest
		Notophyll Vine Thicket	deciduous	5–9	Bottle trees 9–15	1	Bottle Tree scrub
Microphyll 224 to 2025 mm^2 (2.5–7.6 cm long)	Vines	Microphyll Vine Forest	evergreen or deciduous	6–15	conifers 21–36	1–2	Hoop Pine scrub
		Microphyll Vine Thicket	evergreen	5–9	conifers, bottle trees	1	Softwood scrub
	Ferns	Microphyll Fern Forest	evergreen	20		1–2	
		Microphyll Fern Thicket	evergreen	6–10		1	
Nanophyll <225 mm^2	Mosses	Nanophyll Moss Forest	evergreen	18–42		1–2	Temperate rainforest
		Nanophyll Moss Thicket	evergreen or deciduous	6–15		1	

2.4. Margins of Rainforests

Rainforests are sometimes sharply delimited from adjacent communities, especially when soil fertility levels change abruptly from one parent material to another. More often, the margin of the rainforest is not sharply defined but appears to be advancing into *Eucalyptus* forest, some rainforest species forming an understory to tall eucalypts (Fraser and Vickery, 1938; Cromer and Pryor, 1942).

Contraction of the rainforest has not been recorded as a natural phenomenon over a short period of time. However, natural contraction by the erosion of high fertility soil parent materials has occurred in the past leading to the isolation of rainforest stands. Similarly, past climatic changes have led to the fragmentation of the rainforests, possibly with the evolution of new species, and usually to the invasion of the rainforests by eucalypts and acacias, especially *Acacia harpophylla* in Queensland.

Fire is probably the most important factor in controlling the rainforest margins, apart from clearing for agriculture. Rainforest species do not regenerate rapidly after burning, the rainforests being replaced by *Eucalyptus* forests, at least temporarily. Lightning fires have always occurred and man, including the aborigines, has increased the incidence of burning.

3. Characteristics of Rainforest Plants

3.1. Leaf Characters

Leaves of rainforest plants are either simple or compound and both types of leaves occur in some families. Leaf-size (based on Raunkiaer, 1934) range from macrophyll to nanophyll (Webb, 1959, 1968), and are used by this writer as one character which defines structural form (Table 7.1). Predominant leaf-sizes in selected communities are given in Table 7.2.

Macrophylls are relatively uncommon, occurring among the monocotyledons, e.g., some of the palms and the native bananas, *Musa* spp. (Fig. 7.1). **Leaf-shape** varies considerably within communities. However, most species have broad leaves, from broad lanceolate to elliptical or ovate. The leaves are usually held near-horizontal and this broad shape and orientation account for the large amount of shade thrown by the crowns of the trees. **Leaf-margins** in the warmer areas are mostly entire, in the temperate regions mostly serrate or dentate (Table 7.2). **Leaf-texture** varies tremendously from soft (readily crushed to a pulp) to somewhat xeromorphic. The latter condition is found largely in the Proteaceae even in notophylls and mesophylls. **Oil glands** are present in the leaves of many species, being characteristic of certain families, notably Lauraceae, Myrtaceae and Rutaceae. **Drip-tips** are found on most of the broad-leaved species of the tropics and subtropics. **Pulvini** are present on the leaves of most rainforest trees but are not common on the leaves of shrubs. **Domatia** are found on the undersurfaces of the leaves of some plants and are confined to a few families, especially Anacardiaceae and Meliaceae. **Deciduousness** occurs in most of the tree species of the drier tropics, the leaves being shed usually towards the end of the dry winter. In the wetter tropics and subtropics some species shed some or all of their leaves towards the end of the dry winter. Deciduousness is not a character of the southern rainforests, though *Nothofagus gunnii*, confined to Tasmania is winter-deciduous under conditions of high rainfall and low temperatures.

3.2. Life Forms

3.2.1. General

All the recognized life forms (Raunkiaer, 1934), with the exception of annuals and leafless stem succulents, are found in the wetter

Table 7.2. Some leaf-characters of species in selected rainforest communities. The figures are percentages of the number of species in the community (abstracted from Webb, 1958).

Community	Leaf-size Mesophyll	Notophyll	Microphyll	Compound Leaves	Entire Margins
Evergreen Mesophyll Vine Forests (Tropical Rainforest)	50–70	30–50	0–5	30–50	70–90
Deciduous Mesophyll Vine Forests (Monsoon Forest)	c. 30	c. 40	c. 30	c. 20	c. 85
Notophyll Vine Forests (Subtropical Rainforest)	15–30	50–70	10–20	30–40	70–85
Microphyll and Nanophyll Forests (Temperate Rainforest)	0	0–10	90–100	0	0

Fig. 7.1. A wild banana, *Musa banksii,* in disturbed forest by the *roadside,* 16 km east of Atherton. The large pinnate leaves on trees to either side belong to the climber *Raphidophora pinnata.* Photo G.C. Stocker.

rainforests. The annual is apparently excluded from these forests by the low light intensity near ground level.

Geophytes, hemicryptophytes and chamaephytes occur in small numbers, or in relatively large numbers, if pteridophytes are included within these life-forms. Angiosperms belonging to these forms include a few ground orchids, some large-leaved Araceae and a few other monocotyledons. These angiosperms and ferns make up the herbaceous stratum, when it occurs, and their abundance is determined by the amount of light reaching ground level.

The most common life-form in the rainforests are trees (phanerophytes), lianas and epiphytes. A life-form, not usually recognized as a separate class, is the arborescent monocotyledon, typified by the palms, which occur in many of the warmer rainforests and become conspicuous in areas where the soil may be waterlogged.

3.2.2. Woody Plants (Phanerophytes)

Most of the individual trees in any stand occur in the uppermost stratum and collectively make up the canopy of the stand. In addition, many immature trees and some naturally smaller trees and shrubs make up the lower woody strata. In the most complex rainforests three tree strata occur. Two tree layers are characteristic of the rainforests in subtropical and warm temperate areas, whereas in the cool temperate forests only one tree layer may occur. In some cases old trees project above the canopy as emergents.

Other features of some rainforest trees, which distinguish them morphologically from the trees in other communities, is the presence of buttresses described either as plank or spur buttresses (Francis, 1951). These structures are flattened extensions of the stem and roots which project as planks or props at the base of the trunk. Plank buttresses are most commonly found among the species which make up the most complex tropical forests. In cooler climates, both in the tropics and subtropics, smaller and rounder buttresses are found, which are referred to as spur buttresses. There are many species and genera, e.g. *Brachychiton,* which never form either type of buttress, and trees of these species commonly occur in association with buttress-forming trees in both the tropics and subtropics. Buttressing occurs in species in quite unrelated families. The plank buttress reaches its maximum size in *Ficus* (Fig. 7.2).

Cauliflory, the production of flowers either singly or in inflorescences directly on the trunk, occurs in some species, most commonly in the richest of tropical forests. The phenomenon is most common in *Ficus* and *Syzygium*.

3.2.3. Lianas

Lianas are a characteristic feature of the rainforests (Jones and Gray, 1977), being absent or very rare only in the *Nothofagus* forests. The total number of species with this life form is no less than 100. The largest lianas belong to the Apocynaceae, Piperaceae (Fig. 7.2) and Vitaceae, their stems reaching a diameter of up to 20 cm and their lengths exceeding that of the tallest trees whose canopies they cover. Other long lianas, the stems of which are not so robust as the ones mentioned, are *Celastrus* and *Palmeria*. Long monocotyledons include *Calamus*, *Ripogonum* and *Smilax*.

3.2.4. Epiphytes

Vascular epiphytes are most common in the tropics, whereas non-vascular epiphytes, par-

Fig. 7.2. Base of a strangler fig *(Ficus eugenioides)*, with buttresses. Also base of a large liana, *Piper novae-hollandiae*, Bunya Mtn. Qld.

ticularly mosses and less commonly liverworts, are most abundant in the cooler areas and provide the reason for the designation, "Moss Forests" for these communities. Epiphyllic algae and lichens are common in some stands, particularly in the warmer zones. Also epiphytic, saprophytic fungi, especially sooty moulds, are very common in many areas; these organisms are assumed to live on debris or exudations which form films over the leaf-epidermi.

The vascular epiphytes form two main groups: i. woody species, the roots of which ultimately reach the ground to establish the epiphyte as a tree; ii. herbaceous species.

i. Notable among the woody epiphytes are the "strangler figs" (*Ficus* spp.) the seeds of which germinate high on the trunk or branches of trees. The young fig seedling produces a normal shoot and root system, the latter branching and creeping over the bark of the supporting tree. The fig roots ultimately reach the ground and, during the process of growth downward, there is lateral expansion of the fig roots which fuse and ultimately form a mesh over the trunk of the supporting tree, which is ultimately killed. Similar woody epiphytes, which are not stranglers, are *Schlefflera actinophylla* in the tropics, and *Quintinia sieberi* and *Pittosporum bicolor* in the southern rainforests.

ii. The non-woody epiphytes are almost entirely pteridophytes or monocotyledons. Dicotyledonous epiphytes are species of *Dischidia* (Asclepiad.), *Hydnophytum*, *Myrmecodia* (both Rubiac.) and *Peperomia* (Piperac.). The liana, *Fieldia australis* (Fig. 7.3) is sometimes epiphytic. The pteridophytes are mainly ferns, but fern allies are fairly common, among them *Tmesipteris* (all spp.), *Psilotum complanatum*, *P. nudum* and *Lycopodium myrtifolium*. The largest of the ferns are the nest-forming species of *Asplenium*, *Drynaria* and *Platycerium*. Creeping, rhizomatous types belong to many families and are of various sizes, the smallest being the Hymenophyllaceae (Filmy ferns).

Notable among the epiphytes are the orchids of which the best represented genus in Australia is *Dendrobium*. However, in comparison with New Guinea, with a total epiphytic orchid flora approaching 2,000 spp., the Australian rainforests are relatively poor in epiphytic orchids.

The number of epiphytic species decreases with decreasing temperature (either with latitude or altitude within the one geographic zone), and in drier climates.

Fig. 7.3. *Fieldia australis* climbing on the trunk of a tree fern *(Cyathea australis)*. Mt. Wilson area, N.S.W.

There are no records which show that any epiphyte is rigidly restricted to a single tree species, though a few epiphytes appear to have preferences for one tree species. *Dendrobium falcorostrum* usually occurs on *Nothofagus moorei* and *Bulbophyllum weinthalii* is usually epiphytic on *Araucaria cunninghamii*. *Dendrobium aemulum* is restricted to *Tristania conferta* in rainforests but occurs also on ironbark eucalypts in open forest. Fraser and Vickery (1938) report that *Sarcochilus hillii* invariably occurs on *Backhousia myrtifolia* near streams and rivers but it is found on a number of species away from watercourses. Also, *Cymbidium suave* is confined to species of *Eucalyptus* when these occur in the rainforest. This orchid is more common outside the rainforests in the adjacent tall *Eucalyptus* forests. They comment also on the effect of the density of the tree-crown on the abundance of epiphytes on the branches and trunk of the same tree. Tree species with a thin canopy such as *Ackama muelleri* and *Dysoxylum fraseranum* are densely crowded with epiphytes, whereas a species with a denser

Table 7.3. *Approximate numbers of taxa in the rainforests in various districts. Data for New Guinea where available.*

District	Lat. °S	Filicales numbers of genera (spp. in brackets)	Angiosperms Total			Numbers of genera (spp. in brackets) in families				
			Fam.	Gen.	spp.	Laurac.	Myrtac.	Orchid.	Palmae	Proteac.
New Guinea	5–10		148	1300	1900	13	20	150 (2000)	50	6
E. coast Qld	11–29	67 (167)	140	630	540	6 (35)	16 (60)	56 (180)	18 (38)	17 (35)
NE. N.S.W.	29–33	42 (84)	90	260	227	6 (31)	11 (29)	18 (60)	4 (4)	8 (14)
SE. N.S.W.	33–37	35 (70)	72	172	74	5 (6)	8 (8)	6 (16)	1 (2)	3 (3)
Vic.	37–39	25 (58)	47	61		0	4 (4)	3 (5)	1 (1)	1 (2)
"Drier" Tropics W. Qld to W.A.	11–18	24 (37)	70	200	340	2 (2)	5 (10)	5 (7)	7 (7)	2 (2)

crown, such as *Sloanea australis,* supports a less rich epiphytic flora, except near the very top.

4. The Vascular Flora

4.1. Pteridophytes

About 200 species of pteridophytes occur in the rainforests and the more common species are listed under the alliances.

Notable occurrences in the rainforests are: i. The fern allies, *Psilotum, Tmesipteris* and *Lycopodium* (epiphytic); ii. The primitive ferns *Ophioglossum pendulum* (epiphytic, mainly in littoral rainforests), *Marattia* and the giant *Angiopteris erecta* with fronds up to 4 m long (both mainly tropical); iii. The tree ferns, *Cyathea* and *Dicksonia,* which sometimes identify rainforest types (Table 7.1); these are most abundant in the cooler south or at higher altitudes in the tropics; iv. Some large epiphytes form "nests" (see 3.2.4, above). The abundance of pteridophytes is correlated with temperature and moisture levels (Table 7.3), there being a decrease in abundance from north to south (temperature gradient) and from east to west in the tropics (moist to dry).

4.2. Gymnosperms

A few gymnosperms occur in the rainforests, including conifers which are presumably remnants of the coniferous forests which were supplanted by the angiosperms.

Agathis (3 spp.) and *Araucaria* (2) occur north of c. lat. 30°S. in the east. The trees are often emergent above rainforest canopies. *Podocarpus* (3 spp.) occurs north of c. lat. 34°S., with an isolated occurrence in eastern Victoria. The cycads, *Bowenia* in the tropics and *Lepidozamia* (north of lat. 30° S.), occur as understory plants or in rainforest margins. *Callitris macleayana* occurs in rainforest margins north of lat 32½° S. extending to tropical Queensland. In Tasmania, *Athrotaxis, Dacrydium* and *Phyllocladus* occur in pure stands (rarely) or in association with *Nothofagus*.

4.3. Angiosperms

No fewer than 140 families occur in the rainforests, of which 115 are dicotyledons. The low proportion of monocotyledons is probably due to the exclusion of most herbaceous taxa from the understory as a result of low light intensity. The flora becomes progressively poorer in species from north to south (Table 7.3).

Families represented by the largest number of genera are: Orchidaceae (58 genera), Euphorbiaceae (31), Rubiaceae (26), Sapindaceae (20), Proteaceae (19), Meliaceae (16), Rutaceae (16), Palmae (24). The total number of genera is at least 600. Some genera represented by a large number of species are *Dendrobium* (50), *Ficus* (30), *Cryptocarya* (23), *Endiandra* (23) and *Elaeocarpus* (20). The total number of species approximates 1900, and about half of these are trees (Hyland, 1971).

Four endemic families occur, each containing a monotypic genus, viz., Akaniaceae, Austrobaileyaceae, Idiospermaceae and Petermanniaceae. About 80 endemic genera occur, belonging in 35 families and most abundantly in Proteaceae (10), Menispermaceae (5), Rutaceae (5), Monimiaceae (5), Sapindaceae (5), Celastraceae (4) and Escalloniaceae (4).

The monocotyledons are represented most abundantly by the Orchidaceae, with 58 genera and about 184 species. Terrestrial forms are relatively few, with 20 genera, mostly represented by a single species each (total 25 spp.).

Fig. 7.4. *Dracaena angustifolia* in a Deciduous Vine Forest. Near Darwin, N.T.

Four saprophytic, leafless genera occur: *Epipogium* (1 sp.), *Galeola* (2), *Gastrodia* (1) and *Didymoplexus* (1). Epiphytes include four relatively large genera, *Dendrobium* (c. 50 spp. in the rainforests), *Bulbophyllum* (25), *Sarcochilus* (9) and *Liparis* (8).

The only other abundant monocotyledons are those which are either arborescent or lianas. The largest of the arborescent forms belong to the Palmae, the most common of which are *Archontophoenix*, *Licuala*, *Linospadix* and *Livistona*. A noteworthy palm genus is *Calamus* with 6 species, which are lianas.

The Araceae (8 genera), and the Zingiberaceae (7 genera) are the only other monocotyledonous families represented by more than three genera. Other families of note, because of the abundance of some of their species are the Flagellariaceae, Philesiaceae and Smilacaceae. Most of these are lianas. Arborescent monocotyledons occur also in the Liliaceae *(Cordyline)*, Agavaceae (*Dracaena* – Fig. 7.4) and Gramineae (*Bambusa* – Fig. 20.1).

and *Ceratopetalum* and *Schizomeria* on the mainland. The taxa are derived from the southern Gondwanaland flora.

iii. North of lat. 30° S., the southern rainforests are replaced by a much richer flora, dominated by Sterculiaceae (*Argyrodendron* spp.) and containing an abundance of Lauraceae, Rutaceae, Meliaceae and others. These species are derived partly from the ancient northern Gondwanaland flora and partly from a migrant flora from Asia or New Guinea.

iv. In drier areas around and north of lat. 30° S., the *Argyrodendron* alliance is replaced by "dry rainforest" types, referred to as the *Drypetes* alliance, which contains a very high proportion of Euphorbiaceae and Sapindaceae and sometimes *Araucaria*.

v. A small group of rainforests occur on coastal sandy deposits, usually behind the frontal dune, and are referred to as "littoral rainforests".

The main characters of the alliances and sub-alliances based on Baur (1957, 1962), except those dominated by *Nothofagus*, are listed in Table 7.4. A species list provided in Table 7.5.

5. The Alliances

The alliances are distributed mainly according to the climatic tolerances of the flora. They are described under five groups, determined mainly on a climatic basis: i. South of the Tropic in the wetter south-east; ii. North of the Tropic in the wetter north-east; iii. Drier tropics, with a pronounced dry winter; iv. Semi-arid areas and arid-zone relicts; v. Gallery rainforests along watercourses.

6. Alliances South of the Tropic in Wetter Areas

6.1. General

The alliances south of the Tropic are distributed as follows:
i. In the cooler, wetter climates in the south *Nothofagus* spp. are dominant.
ii. In parts of Tasmania and on the mainland in the south-east, the rainforests are dominated by Cunoniaceae, *Anodopetalum* in Tasmania

6.2. The *Nothofagus* Alliances

6.2.1. General

Three species of *Nothofagus* occur in Australia, the ancient forests having contracted to Tasmania and a few isolated occurrences in Victoria and on the eastern highlands (Fig. 7.5) where they occur in cool, humid climates. Each of the species characterizes an alliance. Included with the *Nothofagus* communities are those species which are presumably derived from the original *Nothofagus* assemblage, but which have become dominants in small patches of rainforest either in Tasmania or Victoria (6.2.3, below).

6.2.2. *Nothofagus cunninghamii* Alliance
Myrtle

N. cunninghamii is an evergreen tree dominating Microphyll Mossy Forests and Thickets. It reaches a height of 35–50 m and a diameter of 1–1½ m. It is reduced in size, often

Table 7.4. Main characters of the rainforest alliances south of the Tropic, excluding *Nothofagus* and littoral rainforests.

Alliance	Suballiance	Distribution	Soil Group P Levels ppm	Common Families	Structural Classification
Drypetes australasica – *Araucaria* spp.	*Flindersia* spp.	SE. Qld	Grey Clays	Meliac. Anacard.	Simple Notophyll Vine Forests
Brachychiton discolor – *Flindersia* spp.	*Araucaria* spp.	Mainly SE. Qld Some NE. N.S.W.	Krasnozems Grey Clays	Araucar. Euphorb. Meliac.	Simple Notophyll Vine Forests (Emergents nanophyll)
	Brachychiton discolor	NE. N.S.W., SE. Qld	Grey Clays	Stercul. Euphorb. Sapind.	Simple Notophyll Vine Forests
Argyrodendron spp.	*Dysoxylum fraseranum*	NE. N.S.W., SE. Qld Lower tablelands, 700 to c. 1000 m	Krasnozems 2240–2540	Meliac. Stercul. Laur.	Complex Notophyll Vine Forests
	Argyrodendron spp.	NE. N.S.W., SE. Qld Lowlands to lower tablelands	Krasnozems 650–2060	Stercul. Laur. etc. See text	Complex Notophyll Vine Forests
	Toona australis – *Flindersia* spp.	Mainly lowlands near rivers. Mainly NE. N.S.W.	Alluviums	Meliac. Rutac.	Complex Notophyll Vine Forests
Schizomeria ovata – *Doryphora sassafras* – *Ackama paniculata* – *Cryptocarya*		N.S.W. Central Coast and lower tablelands. Below c. 1150 m	Mainly Krasnozems	Cunon. Monim. Laur.	Simple/Complex Notophyll Vine Forests
Ceratopetalum apetalum	*C. apetalum* – *Argyrodendron* spp.	N. Coast N.S.W. Below 250 m	Alluviums (gullies) 430	Cunon. Stercul. Palmae	Simple/Complex Notophyll Vine Forests
	C. apetalum – *Schizomeria ovata*	Mainly NE. N.S.W. Mostly c. 600 m	Podzolic 170–360	Cunon. Monim. Proteac. Laur.	Simple Notophyll Vine Forests
	C. apetalum – *Diploglottis australis*	S. Coast N.S.W. Below 300 m	Podzolic Krasnozems Alluviums 170–800	Cunon. Sapind. Meliac. Morac.	Simple/Complex Notophyll Vine Forests

Species	Habitat	Soil	Family	Forest type
C. apetalum – *Doryphora sassafras*	Higher altitudes N.S.W., mostly 700–1400 m	Krasnozems 400–1060	Cunon. Monim.	Simple Notophyll Vine Forests
Doryphora sassafras – *Acacia melanoxylon*	S.N.S.W. Robertson. (Degenerate). c. 650 m	Krasnozems	Monim. Mimos.	Simple Notophyll Vine Forests
C. apetalum – *Acmena smithii* – *Tristania laurina*	Mainly coast and lower tablelands. S.N.S.W.; E. Vic. Below 500 m	Alluviums (gullies) 150–300	Cunon. Myrtac.	Notophyll/Microphyll Vine or Fern Forests
Anodopetalum biglandulosum	Mainly lowlands, some subalpine. Tas.	Peat	Cunon.	Microphyll Forests

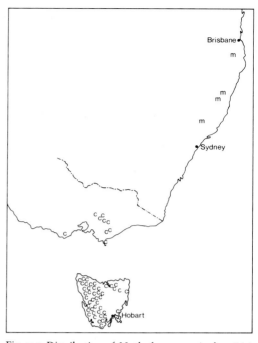

Fig. 7.5. Distribution of *Nothofagus cunninghamii* (c) and *N. moorei* (m).

to shrub proportions, at higher altitudes and on soils of very low fertility. The leaf-blades are triangular-rhomboid to orbicular and 6–18 mm long. The trees are commonly infected by the ascomycete *Cyttaria gunnii*, the ascocarps of which emerge on the branches and are edible. The autecology of the species has been studied by Howard (1973), who has shown that the leaves live for three years and that litter-fall amounts to c. 6.5 tonnes per ha. per annum. Trees in western Tasmania are subject to premature death which appears to be due to attack by a fungus which is carried by the ambrosia beetle. Flowering of the species is annual and seed, with a viability of 45–50% and a longevity of 8–9 months, is prolific (c. 28×10^6 seeds/ha.). Seedlings become established in canopy-gaps on decayed logs or firm soil.

N. cunninghamii, though commonly in pure stands or in association with characteristic species of the alliance, may be a common or even a dominant species of the understory of some *Eucalyptus* forests.

N. cunninghamii occurs in the wetter parts of Tasmania (excluding the treeless plateaux) from sea level in estuaries to c. 1000 m, but in shel-

Table 7.5. List of common species in rainforests in eastern New South Wales and south-eastern Queensland.
T = tree; S = shrub; L = liana.

Communities:
1 Ceratopetalum – Doryphora (Mt. Wilson)
2 Ceratopetalum – Doryphora (N. Tablelands)
3 Ceratopetalum – Diploglottis
4 Ceratopetalum – Schizomeria
5 Schizomeria – Doryphora – Ackama
6 Argyrodendron
7 Drypetes
8 Littoral

Species	Habit	1	2	3	4	5	6	7	8
Coniferales									
ARAUCARIACEAE									
Agathis robusta	T								8
Araucaria bidwillii	T						6	7	
A. cunninghamii	T				4		6	7	
CUPRESSACEAE									
Callitris macleayana	T				4				8
PODOCARPACEAE									
P. elatus	T			3			6		8
Dicotyledons									
AKANIACEAE									
Akania lucens	T				4		6		
ALANGIACEAE									
Alangium villosum	T					5	6	7	
ANACARDIACEAE									
Euroschinus falcatus	T			3		5	6	7	8
Pleiogynium solandri	T						6		
Rhodosphaera rhodantha	T						6	7	
ANNONACEAE									
Polyalthia nitidissima	T						6		8
Rauwenhoffia leichhardtii	L						6	7	
APOCYNACEAE									
Alstonia constricta	T							7	
Alyxia ruscifolia	S		2		4		6	7	8
Chilocarpus australis	S				4	5	6		
Parsonsia brownii	L	1		2	4		6		
P. induplicata	L		2				6		
P. leichhardtii	L		2					7	
P. straminea	L		2	3			6		8
P. velutina	L					5			
ARALIACEAE									
Cephalaralia cephalobotrys	L		2	3		5	6		
Polyscias elegans	T			3		5	6	7	8
P. murrayi	T			3	4	5	6		
P. sambucifolius	S			3		5	6		
ASCLEPIADACEAE									
Tylophora barbata	L	1		3	4				
Hoya australis	L	1					6	7	8
BIGNONIACEAE									
Pandorea pandorana	L	1	2	3	4	5	6	7	8
CAPPARIDACEAE									
Capparis arborea	S			3			6	7	
CELASTRACEAE									
Celastrus australis	S					5	6	7	
Denhamia pittosporoides	S		2		4	5	6	7	8
Elaeodendron australe	S			3	4		6	7	8
Siphonodon australe	T						6	7	
CUNONIACEAE									
Ackama paniculata	T		2		4	5	6		
Aphanopetalum resinosum	L		2		4	5			
Callicoma serratifolia	T	1	2	3	4	5			
Ceratopetalum apetalum	T	1	2	3	4				
Geissois benthamii	T				4		6		
Pseudoweinmannia lachnocarpa	T		2	3	4	5			8
Schizomeria ovata	T					5			
DAVIDSONIACEAE									
Davidsonia pruriens	T						6	7	
EBENACEAE									
Diospyros australis	S			3		5	6	7	8
D. fasciculosa	T						6	7	8
D. pentamera	T			3	4	5	6	7	
LAURACEAE Cont'd									
Litsea leefeana	T				4	5	6		8
L. reticulata	T			3			6		8

147

Species	Habit	1	2	3	4	5	6	7	8
Neolitsea cassia	T		2				6		
N. dealbata	T		2	3		5	6		
MELIACEAE									
Didymocheton rufum	T					5	6		
Dysoxylum fraseranum	T			3		5	6	7	
D. muelleri	T			3			6	7	
Melia azedarach	T						6	7	
Ouenia cepiodora	T			3			6	7	
Pseudocarapa nitidula	T						6	7	
Synoum glandulosum	S			3	4	5	6	7	
Toona australis	T			3		5	6	7	
MENISPERMACEAE									
Legnephora moorei	L			3			6	7	
Sarcopetalum harveyanum	L			3		5	6	7	
Stephania japonica	L			3	4	5	6	7	8
MIMOSACEAE									
Abarema grandiflora	T				4		6	7	8
A. sapindoides	S			3		5	6	7	
Acacia elata	T	1				5	6	7	
A. maidenii	T	1					6	7	
A. melanoxylon	T	1	2			5	6	7	8
MONIMIACEAE									
Atherosperma moschatum	T	1							
Daphnandra micrantha	S		2		4	5	6		
Doryphora sassafras	T	1	2	3	4	5	6		
Hedycarya angustifolia	S	1	2			5	6		
Palmeria scandens	L		2	3	4	5	6		
Wilkiea huegeliana	S			3	4	5	6	7	
MORACEAE									
Cudrania javanensis	T					5	6	7	
Ficus coronata	T			3		5	6	7	
F. henneana	T			3		5	6	7	
F. macrophylla	T			3	4		6	7	
F. obliqua	T			3			6	7	
F. watkinsiana	T				4		6	7	8
Maclura cochinchinensis	L			3			6	7	8
Malaisia scandens	L			3		5	6	7	
Pseudomorus brunonianus	S	1		3			6	7	
MYRSINACEAE									
Embelia australasica	L				4	5	6		8
Rapanea variabilis	S		2	3	4	5	6		8
MYRTACEAE									
Acmena hemilampra	T				4		6		8
A. smithii	T	1	2	3			6	7	8
Austromyrtus acmenoides	S			3			6	7	8
A. bidwillii	T							7	
A. hillii	T						6	7	
Backhousia anisata	T						6		
B. myrtifolia	T			3		5	6		8
B. sciadophora	T					5	6		8
Rhodamnia argentea	T				4		6		8
R. trinervia	S			3	4	5	6	7	8
Rhodomyrtus beckleri	S			3	4	5	6		8
Syncarpia glomulifera	T			3	4		6		8
Syzygium coolminianum	T			3	4		6	7	8
S. francisii	T						6		8
S. luehmannii	T			3		5	6		8
S. paniculatum	T				4	5	6	7	8
Tristania conferta	T						6		8
T. laurina	T		2	3	4	5	6		8
NYCTAGINACEAE									
Pisonia aculeata	L								8
P. brunoniana	T								8
SANTALACEAE									
Exocarpos latifolius	T						6	7	
SAPINDACEAE									
Alectryon subcinereus	S			3	4	5	6		
Arytera foveolata	T				4		6	7	
Atalaya multiflora	T						6		
Castanospora alphandii	T						6		
Cupaniopsis anacardioides	T						6	7	
C. serrata	S								
Diploglottis australis	T			3	4	5	6	7	8
Elattostachys nervosum	T						6	7	8
E. xylocarpum	T						6	7	
Guoia semiglauca	T			3	4		6	7	8
Harpullia pendula	T				4		6	7	8
Jagera pseudorhus	T				4		6	7	8
Mischocarpus pyriformis	S				4		6	7	8
Sarcopteryx stipitata	T								

T = tree; S = shrub; L = liana.

Communities:
1 *Ceratopetalum – Doryphora* (Mt. Wilson)
2 *Ceratopetalum – Doryphora* (N. Tablelands)
3 *Ceratopetalum – Diploglottis*
4 *Ceratopetalum – Schizomeria*
5 *Schizomeria – Doryphora – Ackama*
6 *Argyrodendron*
7 *Drypetes*
8 *Littoral*

Species	Form	1	2	3	4	5	6	7	8
SAPOTACEAE									
Planchonella australis	T			3	4	5	6	7	
P. myrsinoides	T						6	7	
SIMAROUBACEAE									
Ailanthus malibaricum	T						6		
Guilfoylia monostylis	T						6		
SOLANACEAE									
Duboisia myoporoides	S			3	4	5	6		
STERCULIACEAE									
Argyrodendron actinophyllum	T								
A. trifoliolatum	T				4		6	7	
Brachychiton acerifolium	T						6		
B. discolor	T			3	4	5		7	
SYMPLOCACEAE									
Symplocus stawellii	T				3		6		
EHRETIACEAE									
Ehretia acuminata	T				3	5	6	7	
ELAEOCARPACEAE									
Elaeocarpus grandis	T						6		
E. kirtonii	T			3	4		6		
E. obovatus	T				3	5			8
E. reticulatus	T	1		3	4	5	6		8
Sloania australis	T			3	4	5	6		
S. woollsii	T			3	4	5	6		
EPACRIDACEAE									
Trochocarpa laurina	T		2	3	4	5	6		8
ESCALLONIACEAE									
Abrophyllum ornans	S			3		5	6		
Anopterus macleayanus	S		2		4		6		
Cuttsia viburnea	S		2				6		
Polyosma cunninghamii	T		2	3	4	5	6		
Quintinia sieberi	T	1	2	3	4	5	6		
Q. verdonii	S		2	3	4		6		
EUPHORBIACEAE									
Austrobuxus swainii	T				4		6	7	
Baloghia lucida	S			3	4	5	6	7	
Breynia oblongifolia	S			3	4	5	6	7	8
Bridelia exaltata	T						6	7	8
Claoxylon australe	S			3	4	5	6	7	
Cleistanthus cunninghamii	S			3			6	7	
Coelogyne ilicifolia	S			3				7	
Croton insularis	S			3		5	6	7	
C. verreauxii	S			3			6	7	
Drypetes australasica	T							7	
Excoecaria dallachyana	T							7	8
Glochidion ferdinandii	S			3	4	5	6		
Mallotus philippinensis	T				4	5	6	7	
Omalanthus populifolius	S			3	4	5	6		
EUPOMATIACEAE									
Eupomatia laurina	S			3	4	5	6		
FLACOURTIACEAE									
Scolopia brownii	S			3		5	6		8
GESNERIACEAE									
Fieldia australis	L	1	2						
ICACINACEAE									
Citronella moorei	S			3		5	6		
Pennantia cunninghamii	S			3		5	6		
LAURACEAE									
Beilschmiedia obtusifolia	T				4	5	6		
Cinnamomum oliveri	T			3	4	5	6		
C. virens	T				4				
Cryptocarya erythroxylon	T					5	6	7	
C. foveolata	T		2				6		
C. glaucescens	T			3	4	5	6		
C. laevigata	T						6		
C. meissneri	T				4		6	7	

Species	Habit	1	2	3	4	5	6	7	8
C. microneura	T			3	4	5	6	7	
C. obovata	T					5	6		
C. rigida	T			3	4	5			
C. triplinervis	T			3	4	5	6	7	
Endiandra discolor	T			3		5	6		8
E. introrsa	T				4	5	6		8
E. muelleri	T				4	5	6		
E. pubescens	T				4	5	6		
E. sieberi	T			3	4				
E. virens	T				4		6		
OLEACEAE									
Jasminum singuliflorum	L						6	7	
Notelaea longifolia	S		2	3			6	7	8
Olea paniculata	T			3			6	7	
PAPILIONACEAE									
Barklya syringifolia	T						6	7	
Derris scandens	L						6	7	
Erythrina vespertilio	T								
Lonchocarpus blackii	L				4		6	7	
Millettia megasperma	L						6		8
Mucuna gigantea	L						6		
PASSIFLORACEAE									
Passiflora herbertiana	L			3	4		6	7	8
PIPERACEAE									
Piper novae-hollandiae	L			3		5	6		8
PITTOSPORACEAE									
Citriobatus pauciflorus	S	1	2	3	4	5	6	7	
Hymenosporum flavum	S					5	6		
Pittosporum revolutum	S			3	4	5	6		8
P. undulatum	T			3	4		6	7	
PROTEACEAE									
Grevillea billiana	T						6	7	
Helicia youngiana	S			3	4		6		
Hicksbeachia pinnatifolia	T						6		
Lomatia arborescens	S					5	6		
L. fraseri	S		2						
L. myricoides	S	1		3			6		
Macadamia spp.	T				4		6		
Oreocallis pinnata	T				4		6		
Orites excelsa	T		2		4	5	6	7	
Stenocarpus salignus	T		2	3	4	5	6	7	8
S. sinuatus	T			3			6	7	
RANUNCULACEAE									
Clematis spp.	L		2	3	4	5	6	7	
RHAMNACEAE									
Alphitonia excelsa	T			3	4	5	6	7	8
Emmenosperma alphitonioides	T			3		5	6		
Ventilago viminalis	T						6	7	
RUBIACEAE									
Canthium coprosmoides	S						6		8
C. odoratum	T			3				7	
Coprosma quadrifida	S	1	2						
Hodgkinsonia ovatifolia	S						6	7	
Ixora beckleri	S						6	7	
Psychotria daphnoides	S						6	7	
P. loniceroides	S			3	4	5	6	7	
Randia benthamiana	S						6		
RUTACEAE									
Acronychia oblongifolia	T		2	3	4	5	6	7	8
A. simplicifolia	T			3	4		6		
Bosistoa euodiformis	T				4		6		
Evodia micrococca	T		2	3	4	5	6	7	8
Flindersia australis	T						6		
F. bennettiana	T						6		8
F. collina	T						6	7	
F. schottiana	T						6	7	
F. xanthoxyla	T			3			6	7	
Gejiera salicifolia	T						6		
Halfordia kendack	T				4		6		8
Medicosma cunninghamii	T						6	7	
Melicope australasica	T						6	7	
Microcitrus australis	S						6		
Pentaceras australis	T						6	7	
Pleiococca wilcoxiana	T						6	7	
Zanthoxylum brachyacanthum	T						6	7	
TRIMENIACEAE									
Piptocalyx moorei	L		2		4		6	7	
ULMACEAE									
Aphananthe philippinensis	T						6	7	
Celtis paniculata	S			3			6	7	8

T = tree; S = shrub; L = liana.

Communities:
1 *Ceratopetalum – Doryphora* (Mt. Wilson)
2 *Ceratopetalum – Doryphora* (N. Tablelands)
3 *Ceratopetalum – Diploglottis*
4 *Ceratopetalum – Schizomeria*
5 *Schizomeria – Doryphora – Ackama*
6 *Argyrodendron*
7 *Drypetes*
8 Littoral

		1	2	3	4	5	6	7	8
Trema aspera	S			3	4	5	6		
URTICACEAE									
Dendrocnide excelsa	T			3	5	6	7		
D. photinophylla	S			3		6	7		8
VERBENACEAE									
Clerodendrum tomentosum	S			3	5	6	7		8
Gmelina leichhardtii	T			3	4	6	7		8
VITACEAE									
Cayratia clematidea	L			3	5	6	7		
Cissus antarctica	L			3	5	6	7		8
C. hypoglauca	L			3	4	5	6	7	8
Tetrastigma nitens	L			3		5		7	
WINTERACEAE									
Tasmannia insipida	S	1	2	3	4	5	6		
Monocotyledons									
DIOSCOREACEAE									
Dioscorea transversa	L		2	3		5	6	7	8
PALMAE									
Archontophoenix cunninghamiana	T			3			6		8
Calamus spp.	L						6		8
Linospadix monostachys	S		2		4		6		
Livistona australis	T			3					8
PHILESIACEAE									
Eustrephus latifolius	L			3		5	6		
Geitonoplesium cymosum	L			3	4	5	6	7	
SMILACACEAE									
Smilax australis	L	1	2	3	4	5	6	7	8
Ripogonum spp.	L				4	5	6	7	8

tered valleys may extend to 1300 m. In Victoria it occurs in three main areas (Fig. 7.5) from near sea-level to 1300 m. Rainfall is mostly 1250–1500 mm, with a winter predominance, but a minimum monthly summer complement of at least 50 mm. High humidities prevail and fog is of common occurrence. Snow falls during the winter and may be of occasional occurrence even in the summer in Tasmania. Mean minimum temperatures are of the order of -2 to $+2°C$ for the coldest months and lowest minima are probably no lower than -7 or $-8°C$.

The soils vary from Krasnozems derived from dolerites to skeletal soils derived from quartzites. These soils represent the extremes of fertility in Tasmania and soil fertility appears to control the height of the stands, community structure and floristic composition. The tallest and densest stands occur on the better soils, and when soil fertility levels are lower the trees are smaller (even reduced to shrubs) and the stands are open. A feature common to all soils is the accumulation of organic matter on the surface often as a layer of peaty material.

Since the species associated with *N. cunninghamii* in Tasmania are different from those in Victoria the two districts are dealt with separately below. The differences are due partly to the occurrence in Tasmania of some coniferous genera and several angiosperm genera endemic to Tasmania and partly to the fact that in Victoria the *Nothofagus* communities have been infiltrated by species from the warmer rainforests. However, many genera and species are common to the two areas and these are members of the woody temperate flora, some of which extend to New Zealand and/or South America.

6.2.2.1. Communities in Tasmania

These have been described by Gibbs (1920), Sutton (1928), Curtis and Somerville (1949), Gilbert (1959), Mount (1964), and Jackson (1965).

The densest and tallest (c. 50 m) stands of *N. cunninghamii*, with no associates, occur in parts of the west coast and the eastern slopes of the Harz Mts and at Lake St. Clair. The canopy is closed and so dense that the only vascular plants occurring on the forest floor are ferns, such as *Blechnum procera* (Fig. 7.6). Pure stands elsewhere commonly contain a discontinuous understory in which the tree-fern, *Dicksonia antarctica*, is most prominent.

The most common associate of *N. cunninghamii* in lowland areas is *Atherosperma moschatum* (Fig. 7.7). The latter does not attain the same height as the *Nothofagus* but has a conical shape and fills in the gaps between the *Nothofagus* crowns. No litter accumulates below *Atherosperma*. Other associations include *Nothofagus* with the following (Jackson, in Specht et al., 1974): i. *Phyllocladus aspleniifolius* (forest in subalpine area); ii. *Eucryphia millinganii* (forest in subalpine areas); iii. *Dacrydium franklinii* (forest in lowland areas, especially near rivers in the south and west); iv. *Eucryphia lucida* (tall scrub or low forest in lowland areas in the south); v. At higher altitudes bordering the alpine tract *N. cunninghamii* forms thickets, usually in pure stands, or with *Trochocarpa gunnii* as an associate; vi. *Anodopetalum biglandulosum* forms ecotonal associations with *Nothofagus* (see 6.3.1, below).

The total subsidiary flora in closed communities is small. A shrub layer is usually absent. The only angiosperm epiphyte is *Prionotes ceritheroides*, the most primitive member of the Epacridaceae; also *Coprosma billardieri* and *Pittosporum bicolor* start life as epiphytes but finally root in the soil and become shrubs. Epiphytic pteridophytes include species of *Tmesipteris, Microsorium* and several members

Fig. 7.6. *Nothofagus cunninghamii* forest. Only other species visible are mosses. The fern, *Blechnum procera*, occurs on the floor. Cradle Mtn. National Park, Tas.

Fig. 7.7. *Atherosperma moschatum* (Sassafras) forest developed after burning and clearing. Some regenerating *Eucalyptus regnans* on the right and in centre. The ferns are *Histiopteris incisa* and *Hypolepis rugulosa*. Near Maydena, Tas.

of the Hymenophyllaceae. The herbaceous stratum may contain *Dicksonia antarctica, Histiopteris incisa, Hypolepis rugulosa* and *Blechnum* spp. The sedge, *Gahnia psittacorum*, may occur in the wettest places if the tree canopy is open.

In more open stands shrubs or small trees are usually present, including *Anopterus glandulosus, Aristotelia peduncularis, Bauera rubioides, Bedfordia salicina, Olearia argophylla, Pomaderris apetala, Richea pandanifolia, Trochocarpa gunnii* and three Proteaceae, *Agastachys odorata, Cenarrhenes nitida* and *Telopea truncata*. In some areas acacias may occur in the stands, possibly as a result of fire.

The Effects of Fire. *N. cunninghamii* is fire-sensitive and is likely to be killed by fire (Gilbert 1959). The species regenerates readily from seed. Understory species likewise regenerate from seed and a few of them coppice, e.g., *Olearia argophylla* (Mount, 1964). The *Nothofagus* community appears to be capable of regeneration with little or no floristic change, provided species of *Eucalyptus* and *Acacia* do not enter the area. When this occurs a mixed stand of *Eucalyptus* and *Nothofagus* develops. Such a community will revert to *Nothofagus* forest only when the eucalypts die of old age; their life span is estimated at about 350 years. If mixed forests are burned before the eucalypts die of old age, succeeding fires favour the eucalypts and suppress the *Nothofagus*. With increasing fire-frequency species of *Acacia* appear in the stands, mainly *A. melanoxylon* and *A. dealbata*. A very high fire-frequency completely removes *Nothofagus* and its associates, and a forest of eucalypts results.

6.2.2.2. Communities in Victoria

N. cunninghamii occurs in three main regions (Fig. 7.5). The communities occur in gullies within *Eucalyptus regnans* forests under humid conditions, with a mean annual rainfall of 1300 mm or more (Petrie et al., 1929; Patton, 1933; Howard and Ashton, 1973). The communities, described in great detail by Howard and Ashton, are zoned as: i. Tall closed forests up to 42 m high, rich in ferns, at altitudes of 0–650 m; ii. Closed forests, rich in herbs, be-

The Victorian stands are noted for their abundant pteridophyte flora. Howard and Ashton (1973) list 40 spp., none of which are confined to the *Nothofagus* forests. *Dicksonia antarctica, Polystichum proliferum, Belchnum procerum, B. fluviatile* and *Grammitis billardieri* occur in all three zones. A few are confined to the lowlands, e.g., *Asplenium bulbiferum, Blechnum aggregatum, Cyathea australis* and *Tmesipteris billardieri,* and some are confined to the higher country, e.g. *Blechnum penna-marina* and *Lycopodium* spp. The bryophyte flora is even more prolific. Howard and Ashton list 59 spp. of mosses and 47 liverworts.

6.2.3. Some Related Communities

Both in Tasmania and Victoria a few species characteristic of the *Nothofagus* forests occur as local dominants in the same general area as these forests. In some cases the local dominance of these subsidiary species may be due to the removal of *Nothofagus* by fire. The most common are:

i. *Acacia melanoxylon* and sometimes *A. dealbata* may occur in dense stands in lowland situations. These are probably the result of fire.

ii. *Atherosperma moschatum,* mainly in the lowlands, forms forests up to c. 20 m high. In Tasmania the forests are often closed (Fig. 7.7.) and there are no associated shrubs. Light intensity at ground level is so reduced that herbs are virtually absent and the root systems of *Atherosperma* form a reticulum over the surface. In Victoria, Ashton (1969) records shrubs in such stands, viz., *Pittosporum bicolor, Elaeocarpus holopetalus, Telopea oreades* and *Tasmannia lanceolata*. The tree-fern *Dicksonia antarctica* is common and epiphytic mosses are conspicuous.

iii. *Olearia argophylla* occurs in lowland situations both in Tasmania and Victoria as a local dominant, commonly in small pockets in valleys within *Eucalyptus* communities. It is sometimes associated with *Bedfordia salicina,* which may occur in the understory, and *Acacia melanoxylon* frequently occurs in the same area sometimes adjoining the *Olearia.*

iv. *Podocarpus lawrencei* occurs as a local dominant in Gippsland, Victoria, at an altitude of c. 1000 m, as an emergent 10–11 m high in a "rainforest thicket", probably controlled by cloud incidence. The thicket contains *Tasmannia lanceolata, Telopea oreades, Notelaea ligu-

Fig. 7.8. *Nothofagus cunninghamii* dominating a forest in a valley. *Eucalyptus regnans* behind. *Dicksonia antarctica* in foreground. Otway Range, Vic.

tween 650 and 1300 m; iii. Low closed forests 8–11 m high at 1300–1375 m.

The lowland forests resemble the Tasmanian stands in general appearance and structure (Fig. 7.8), but the floristic composition is different. *Atherosperma moschatum,* up to 23 m high, is prominent and persists into the highland stands. *Acacia melanoxylon* occurs up to an altitude of 650 m. At or near its upper limit *Nothofagus* forms an association with either *Leptospermum grandifolium* or *Podocarpus lawrencei* (see 6.2.4, below). The main understory species at lower altitudes are *Hedycarya angustifolia, Coprosma quadrifida* and *Pittosporum bicolor.* The last two of these extend into the middle zone where *Acacia dealbata, Telopea oreades* and *Elaeocarpus holopetalus* are the most frequent shrubs, and the herb *Stellaria flaccida* is common. The upper zone contains a quite different ground flora, including *Uncinia tenella, Libertia pulchella,* and two Ericaceae, *Gaultheria appressa* and *Wittsteinia vacciniacea.*

strina and *Prostanthera lasianthos* (Ashton, 1969).

v. *Eucryphia moorei*. This species (Fig. 7.9) is confined to south-east New South Wales. The genus is represented also in Tasmania by two spp. (see above) which are associated with *Nothofagus cunninghamii*. *E. moorei* is the dominant or co-dominant tree in small patches of rainforest eastward of the Great Divide between 400 and 1000 m. At higher alitudes, e.g., at Mt. Dromedary (near Narooma), *E. moorei* occurs in pure stands, and at lower altitudes it is co-dominant with *Doryphora sassafras*. At altitudes below 400 m these communities are replaced by less cold-tolerant genera, especially *Acmena* and *Ceratopetalum* (see 6.3.2, below).

In the *Eucryphia* communities, associated shrubs are *Pittosporum undulatum*, *Beyeria lasiocarpa* and *Backhousia myrtifolia* (more southerly localities), and *Synoum glandulosum*, *Rapanea howittii*, *Elaeocarpus reticulatus* and *Psychotria loniceroides* (more northerly localities). The tree ferns *Cyathea australis* and *Dicksonia antarctica* form a conspicuous understory. The ground stratum contains the ferns, *Blechnum*, *Hypolepis*, *Lastreopsis* and *Sticherus* spp.

6.2.4. *Nothofagus gunnii* Alliance
Tanglefoot

N. gunnii is usually a shrub with wiry tangled branches, but may reach a height of c. 7 m in more protected situations. The leaves, which are ovate and 1–2 cm long, are winter-deciduous, turning golden brown or red in the autumn before falling (Curtis, 1967). It occurs at elevations above c. 1300 m where it grows under the extremes of cold and exposure to wind. The species occurs also as an understory plant above c. 1000 m in some of the subalpine woodlands or forests, e.g., with *Eucalyptus coccifera*, or *N. cunninghamii* and *Athrotaxis selaginoides*, where it may form a small tree (Sutton, 1928).

As a dominant or co-dominant *N. gunnii* forms thickets varying in height from 1–2½ m (Fig. 7.10). The communities occur mostly on rock-strewn areas (stonefields) or talus slopes near or above the tree-line in shallow depressions, often along streams. When it occurs in pure stands, a few to a few dozen bushes grow together, and as a result of wind-shearing the stands become flat-topped. When associated species occur, some of these may project as emergents above the *Nothofagus*. Associated shrubs include a few which form trees at lower levels, especially *Eucalyptus coccifera* and *Athrotaxis selaginoides*, or some which occur elsewhere on the plateaux, viz.: *Bellendena montana*, *Cyathodes straminea*, *Diselma archeri*, *Tasmannia lanceolata*, *Olearia alpina*, *O. pinifolia*, *Richea pandanifolia* and *Telopea truncata* (Gibbs, 1920).

6.2.5. *Nothofagus moorei* Alliance
Antarctic Beech
or Negrohead Beech

N. moorei is a tree reaching a height of 30 to 50 m and it forms the sole dominant of Microphyll Mossy Forests. The leaf-blades (up to 5 cm long and 2½ cm wide) are larger than the other two species. The trees have long boles without buttresses, but are sometimes knotted near the base (burl) and coppice growth, which contributes to the shrub layer of the forest, sometimes develops from the knots. This growth may be the result of burning. Trees are usually infected with *Cyttaria septentrionalis*, which produces galls on the stem (Wilson, 1937). The fruiting bodies of the fungus are edible. The species rarely produces seed and consequently regeneration is not prolific.

N. moorei occurs in a few isolated localities (Fig. 7.5) in the Barrington Tops area, the Point Lookout – Styx R. – Dorrigo area and on the Macpherson Range. It is confined to soils of

Fig. 7.9. Flowering shoots of *Eucryphia moorei*. The leaves are compound. Near Robertson, N.S.W.

Fig. 7.10. *Nothofagus gunnii* below a stunted *Eucalyptus coccifera*. Above, the treeless slope below the rim of the plateau. Cradle Mt. National Park, Tas.

high fertility, mostly Krasnozems derived from basalt, at altitudes of c. 1000 to 1500 m. It is restricted to cool wet areas of high humidity and high fog incidence, with a mean annual rainfall of 1500–2000 mm with a summer predominance, but with a significant winter rainfall complement which possibly exceeds c. 50 mm per month. Temperature conditions are low during the winter, the mean monthly minimum for the coldest month approaching zero and the lowest minimum c. $-10°$ C. Williams and Hore-Lacy (1963) indicate a greater blanketing effect of the beech forest in contrast to adjacent *Eucalyptus* forest at Point Lookout (altitude 1480 m).

The forests usually have a closed canopy (Fig. 7.11). The trees in the Barrington Tops area occur with a density of 2.4 per 10 m^2 and appear to be evenly distributed (Fraser and Vickery, 1938). A lower tree or tall shrub layer sometimes occurs but is usually discontinuous. The main tree and shrub species, all of which do not occur in the same stand, are: *Acacia melanoxylon, Aristotelia australasica, Atherosperma moschatum* (Barrington only), *Callicoma serratifolia, Doryphora sassafras, Elaeocarpus holopetalus, Lomatia arborescens, Quintinia sieberi, Tasmannia purpurascens, Trochocarpa laurina* and *Weinmannia rubifolia*. The main lianas are *Cissus antarctica, Pandorea pandorana, Parsonsia brownii* and *Streptothamnus beckleri*. Angiosperm epiphytes and *Dendrobium falcorostrum* and *Sarcochilus falcatus; Quintinia sieberi* starts life as an epiphyte.

Fig. 7.11. *Nothofagus moorei*. Forest Point Lookout, N.S.W.

Small shrubs include *Alyxia ruscifolia*, *Rubus moorei* and *R. rosifolius*. Herbaceous plants, sometimes common in light-breaks, include *Acaena anserinifolia*, *Drymophila moorei*, *Elatostema reticulatum*, *Lomandra spicata*, *Stellaria flaccida* and *Uncinia tenella*. Pteridophytes are abundant in light-breaks, especially *Dicksonia antarctica*, and *Histiopteris incisa*; also in these breaks climbing epiphytes occur, especially *Arthropteris tenella*, *Microsorium scandens* and *Pyrrosia rupestris*.

Ecotonal Associations. Where *Nothofagus moorei* adjoins rainforests dominated by *Ceratopetalum apetalum* or *Schizomeria ovata* which occur at lower altitudes (see 6.3, below), the ecotone involving the two alliances is narrow. At Barrington, it extends over a vertical distance of 90 m (Turner, 1976). In this ecotone *Doryphora sassafras* occurs, and the following species from the lower forest are associated with *Nothofagus* (arranged in order of abundance): *Ackama paniculata*, *Daphnandra micrantha*, *Cryptocarya foveolata*, *Orites excelsa* and a few others. Tree ferns are often abundant, e.g., *Cyathea australis* or *C. leichhardtiana* and additional epiphytes occur, e.g., *Sarcochilus falcatus*. In the more northerly stands, *Ceratopetalum apetalum* is a common associate in ecotonal associations with *N. moorei*, and the understory contains species from both the *N. moorei* and *Ceratopetalum* alliances.

The Effects of Fire. Fire usually destroys the beech forest, permitting the entry of eucalypts into the area. Herbert (1963) reports *Xanthorrhoea arborea* in a beech forest in Queensland and regards it as a relict of a former eucalypt community which invaded a burned beech forest, which later regenerated and replaced the eucalypts.

6.3. Alliances Dominated by Cunoniaceae

6.3.1. General

Three species of the Cunoniaceae, *Anodopetalum biglandulosum* in Tasmania and *Ceratopetalum apetalum* and *Schizomeria ovata* on the mainland, dominate the rainforests in the south-east.

These rainforests, presumably derived from the southern Gondwanaland assemblage, are relatively poor floristically and contain a high proportion of species belonging to the Southern Hemisphere families (Table 7.5). The rainforests become progressively richer floristically from south to north, as they mingle with the richer northern rainforest.

6.3.2. *Anodopetalum biglandulosum* Alliance

Horizontal

This monotypic genus is confined to Tasmania. The species has a unique habit in that the stems, when they reach a height of several metres, become top-heavy and, if space permits, curve over towards the ground (Fig. 7.12). Branches then develop from the nodes and grow upwards, while the original stem remains in a more or less horizontal position, becoming thicker with age. The upright stems are therefore

Fig. 7.12. Saplings of *Anodopetalum biglandulosum* which have bent over and have commenced to shoot from axillary buds. Near Hobart, Tas. Photo: Dr. Winifred Curtis.

propped a few metres above the soil surface and an impenetrable woodland-thicket, known as "Horizontal", develops. Light intensity at ground level is so reduced that no associated species can occur. Monospecific stands may reach a height of c. 8 m.

The species occurs mainly in the wetter south in lowland or subalpine situations, often in valleys within stands of *Nothofagus cunninghamii*, which may be an associate, at least during the early history of the communities. *Phyllocladus aspleniifolius* is sometimes an associate (Fig. 7.13) and *Anodopetalum* may occur also below eucalypts.

6.3.3. *Ceratopetalum apetalum* Alliance
Coachwood

6.3.3.1. General

This species characterizes the rainforests of the cooler south-east, between lat. 37° and 28°S. It occupies a variety of habitats from the highlands to the coastal plain. It flourishes on soils of the highest fertility but tolerates soils of the lowest fertility capable of supporting rainforest. In cooler areas there is apparently little competition with rainforest trees which are intolerant of cold, so that *C. apetalum* occupies all soils of sufficiently high fertility to support rainforest. On the other hand, in warmer sites on soils of higher fertility it is partially or completely supplanted by other rainforest trees, chiefly those from the *Argyrodendron* Alliance.

Six suballiances can be recognized (Table 7.4), differentiated by the habitat-preferences of the associated species, which in some cases are migrants from warmer areas.

6.3.3.2. *Ceratopetalum apetalum* – *Doryphora sassafras* Suballiance

These two species are the dominants at higher altitudes (Fig. 7.14). The most southerly stand on the highlands occurs on the basaltic capping at Mt. Wilson (Brough et al., 1924), where *Ceratopetalum* forms pure stands or is co-dominant with *Doryphora*. Similar patches occur on the Northern Tablelands where the stands may adjoin *Nothofagus moorei* forests. The most common woody species in these two stands are listed in Table 7.5.

Floristically the communities are distinguished by the high frequency of the Southern

Fig. 7.13. *Phyllocladus aspleniifolius* with subsidiary *Nothofagus cunninghamii* and *Anodopetalum biglandulosum*. The scale is a man in left-hand corner. Near Maydena, Tas.

Hemisphere families Cunoniaceae, Monimiaceae, Escalloniaceae and Winteraceae. All species are not represented in all stands, which is possibly due to the highly fragmented nature of the alliance. The northern stands are floristically richer than the Mt. Wilson stand, due partly to the proximity of the richer northern rainforests.

The highland communities contain few lianas and angiosperm epiphytes. Noteworthy occurrences are the climbing epiphyte *Fieldia australis* (Fig. 7.3), and the liana, *Piptocalyx moorei*. Angiosperm epiphytes are *Dendrobium pugioniforme*, *D. teretifolium* and *Sarcochilus falcatus*. Epiphytic pteridophytes belong to the genera *Asplenium*, *Dictymia*, *Grammitis*, *Hymenophyllum*, *Microsorium*, *Polyphlebium*, *Pyrrosia* and *Tmesipteris*.

The understory is often dominated by the tree ferns *Dicksonia antarctica*, *Cyathea australis* and *Todea barbara* (rarely the three in the one stand). The herbaceous layer is usually dominated by ferns, the constants for the communities being *Blechnum patersonii*, *Dennstaedtia davallioides*, *Histiopteris incisa*, *Lastreopsis decomposita* and *Pellaea falcata*.

6.3.3.3. *Doryphora sassafras* – *Acacia melanoxylon* Suballiance

This assemblage occurs at Robertson (lat. $34^{1}/_{2}°$ S.) at an altitude of c. 650 m. The communities, now almost entirely cleared for agriculture, occur on Krasnozems derived from basalt. The original composition of the stands is not known. As well as the present dominants, *Acmena smithii* and *Eucryphia moorei* have been recorded (Phillips, 1947), showing a relationship with the flora of the *Ceratopetalum apetalum* alliance on the one hand and with the *Nothofagus* – *Eucryphia* assemblages on the other hand (see 6.2.3, above).

6.3.3.4. *Ceratopetalum apetalum* – *Acmena smithii* – *Tristania laurina* Suballiance

C. apetalum often occurs on soils of relatively low fertility, e.g., in valleys on the Triassic sandstones at Mt. Wilson, where it forms communities more or less identical with those on the basalt. Similar communities occur throughout the sandstone areas from the highlands to the coast in the south, with *Ceratopetalum* as the dominant but *Doryphora* is absent at lower altitudes. Eucalypts, especially *E. pilularis* or *E. saligna*, often form a tall forest above or adjacent to the strip of rainforest which usually follows a creek.

Associated species include: *Acmena smithii*, *Tristania laurina* (each sometimes locally dominant), *Backhousia myrtifolia*, *Clerodendrum tomentosum*, *Elaeocarpus reticulatus*, *Morinda jasminoides*, *Omalanthus populifolius*, *Pittosporum undulatum*, *P. revolutum*, *Rapania variabilis*, *Rhodamnia trivervia*, *Synoum glandulosum* and *Trochocarpa laurina*. The palm, *Livistona australis* (Fig. 7.15) is often present near the coast on wet sands. Vascular epiphytes are uncommon.

The most southerly record of the suballiance occurs in eastern Victoria (Ashton, 1969) at altitudes below 500 m, with a dominant of

Fig. 7.14. Forest of *Ceratopetalum apetalum* with subsidiary *Doryphora sassafras*. Photo from the late R.H. Anderson.

Acmena smithii reaching a height of 20–30 m. *Acacia melanoxylon* is sometimes co-dominant and *Tristania laurina* occurs along creeks. *Livistona australis* occurs sometimes as an emergent to a height of 24 m. Shrubs include *Acronychia laevis*, *Elaeocarpus holopetalus*, *Pittosporum undulatum* and *Rapanea howittiana*. Seventeen species of lianas occur, including *Pandorea pan-*

Fig. 7.15. *Livistona australis* with *Ceratopetalum apetalum* behind. In a valley in Hawkesbury sandstones, Kuring-gai National Park, Sydney.

dorana and *Parsonsia brownii* (which form ropes), and small types including *Cissus hypoglauca, Clematis glycinoides, Eustrephus latifolius, Marsdenia rostrata* and *Smilax australis*.

Many of these rainforest patches have been disturbed by man, with the increase in abundance of *Pittosporum undulatum* and/or the invasion of the rainforest by the exotics, *Cinnamomum * camphora, * Ligustrum lucidum, * Lantana camara, Passiflora * edulis* and *Rubus * vulgaris*.

The suballiance occurs also in the north. Baur

(1962) records the association, *C. apetalum – Tristania laurina*, in exposed areas of high rainfall at Whian Whian (lat. 28½°S.) which he describes as a submontane rainforest with two tree layers and few lianas and vascular epiphytes.

6.3.3.5. *Ceratopetalum apetalum – Diploglottis australis* Suballiance

At altitudes below c. 300 m along the south coast of New South Wales rainforest patches, usually a few hectares in area, occur on high fertility soils, e.g., at Bulli on Triassic chocolate shales and Gosford on recent alluvium. The latter stands were destroyed for agriculture but the few remnants indicate that *Toona australis* occurred in these forests and possibly also in the Bulli stands which are well documented by Davis (1941). A small fragment occurs also near Kiama. As well as high fertility, these rainforests are favoured by higher mean temperatures than the upland stands, being rarely subjected to frosts.

Although *Ceratopetalum* occurs, it is not necessarily dominant and many species typical of the more northerly rainforests occur, some of them consistently in the three localities mentioned above. On the other hand, a few species exhibit disjunction. For example, at Kiama, *Cinnamomum oliveri*, *Diospyros pentamera*, *Gmelina leichhardtii* and *Helicia glabriflora* occur in isolation, being recorded elsewhere only north of lat. 30°S. Species recorded in these stands are listed in Table 7.5. Significant occurrences are *Ficus* spp., *Dendrocnide excelsa*, the large liana, *Piper novae-hollandiae*, and the large epiphytic ferns *Asplenium nidus* and *Platycerium bifurcatum* (the last two extending almost to the Victorian border). Other common epiphytes are *Cymbidium suave* and *Sarcochilus falcatus*.

The herbaceous layer consists mainly of ferns (listed in 6.3.3.2, above) plus *Adiantum aethiopicum*, *A. diaphanum* and *A. hispidulum*.

6.3.3.6. *Ceratopetalum apetalum – Schizomeria ovata* Suballiance

These two species occur in association in north-eastern New South Wales on soils of relatively low fertility, derived from acid igneous rocks and shales. The assemblage often adjoins more complex rainforests (*Argyrodendron* alliance) on more fertile soils, as at Dorrigo and Whian Whian (Baur, 1957).

The main associated trees are *Doryphora sassafras*, *Orites excelsa* and *Endiandra introrsa*. *Araucaria cunninghamii* sometimes occurs as an emergent. Though floristically rich (Table 7.5), few treespecies occur in the canopy, most of the woody species being members of the understories. Lianas are present but are small, and climbing palms are absent. Epiphytes are relatively few, the most common being *Sarcochilus falcatus* and the ferns *Dictymia brownii*, *Grammitis billardieri* and *Pyrrosia confluens*.

Arborescent monocotyledons are sometimes present in depressions (*Cordyline stricta* and *Linospadix monostachya*) and also the tree ferns *Cyathea leichhardtiana* and *Todea barbara*. The ground stratum, if present, is dominated by the ferns, *Asplenium bulbiferum*, *A. faccidum*, *Blechnum patersonii*, *B. procerum*, *Histiopteris incisa*, *Lastreopsis decompositum* and *Pellaea falcata*.

6.3.3.7. *Ceratopetalum apetalum – Argyrodendron* spp. Suballiance

These communities, listed by Baur (1957) and renamed (1962) as the above assemblage, occur on the north coast of New South Wales, especially in the Coff's Harbour area. The suballiance must be regarded as transitional between the *Ceratopetalum apetalum* and the *Argyrodendron* spp. Alliances.

Characteristic of the *Ceratopetalum* alliance are *Callicoma serratifolia*, *Doryphora sassafras*, *Polyosma cunninghamii* and *Trochocarpa laurina*. On the other hand, members of the *Argyrodendron* alliance include an abundance of Lauraceae (chiefly *Beilschmiedia* and *Cryptocarya* spp.). These rainforests are the only ones in New South Wales which contain the three species of palms which occur at these latitudes (excluding the coastal *Livistona*), viz., *Calamus muelleri*, *Linospadix monostachya* and *Archontophoenix cunninghamii*.

6.3.4. *Schizomeria ovata – Doryphora sassafras – Ackama paniculata – Cryptocarya glaucescens* Alliance

This alliance occurs in isolation in the Manning R. Valley and around the tributaries of this river, extending to an altitude of 950–1150 m where it adjoins *Nothofagus moorei* forests. The communities have been described in great detail by Fraser and Vickery (1938) but only a summary of their extensive observations can be included here. The alliance occurs on Tertiary basalts which produce Krasnozems of high fertility. Mean annual rainfall is c. 1500–1800 mm.

The communities have two tree layers, the upper one 20 to c. 30 m tall, the canopy being continuous but with emergents of *Diploglottis australis*, *Dysoxylum fraseranum* and *Ficus henneana*. The last is deciduous and two other deciduous species occur, *Toona australis* and *Ehretia acuminata*. The canopy contains the crowns of 40–50 tree species and no regular associations can be recognized, except in lightbreaks (see below).

The lower tree layer, 5–10 m tall, contains many species, including *Diospyros australis*, *Euodia micrococca*, *Eupomatia laurina*, *Tasmannia insipida* and *Wilkiea macrophylla*, and the tree fern *Cyathea leichhardtiana*. A shrub layer sometimes occurs, containing *Citriobatus multiflorus* at lower altitudes and *Tasmannia insipida* at higher altitudes. Lianas, both large and small, are abundant (Table 7.5), but climbing palms are absent.

Vascular epiphytes are relatively few, their abundance being governed largely by light intensity. Epiphytic mosses, liverworts and lichens are abundant on the trunks of trees, decaying logs and rocks.

The ground stratum consists mainly of ferns and is best developed in light breaks. The main fern species are *Dicksonia antarctica* and *Lastreopsis decomposita* with less abundant *Adiantum formosum*, *A. affine*, *Athyrium japonicum*. A few herbaceous angiosperms are present, especially *Lomandra montana* and a few straggling plants such as species of *Aneilema*, *Galium* and *Pollia*. Mosses and liverworts may be abundant, especially on fallen logs and rocks near watercourses. The large moss, *Dawsonia superba* (Fig. 7.16) is often present on moist banks.

Some variation in the stands occurs in light breaks along watercourses where *Weinmannia rubifolia*, *Alectryon subcinereus*, *Backhousia myrtifolia* and a few others are locally dominant. Lianas are usually more abundant in these situations.

At the margins of the dense forests ecotonal associations occur, containing *Eucalyptus* spp.

Fig. 7.16. *Dawsonia superba*, with sporophytes covered by the calyptra. Scale is a match box c. 5 cm long.

and/or *Syncarpia glomulifera*, *Casuarina cunninghamiana*, *Callistemon salignus* and the rainforest species *Ackama paniculata* and *Trochocarpa laurina*. At lower altitudes the margin contains *Cyathea australis*, *Clerodendrum tomentosum* and *Commersonia bartramia*.

Some altitudinal zonation occurs. The common tree species in higher rainforests are *Dysoxylum fraseranum*, *Acmena smithii*, *Ackama paniculata*, *Doryphora sassafras*, *Orites excelsa*, *Polyscias murrayi*, *Schizomeria ovata* and *Eucalyptus obliqua*. This zonation is similar to the one described in the next alliance.

6.4. *Argyrodendron* spp. Alliance

These communities occur mainly in the valleys of the Tweed and Richmond Rivers, extending to the Macpherson Range and with some isolated occurrences northward to Gympie (Qld). The forests have three tree layers and in this respect resemble the richest rainforests in the tropics. They are perhaps most readily identified by the presence of the climbing palm, *Calamus muelleri*, which is absent from the more southerly rainforests. The forests reach a height of c. 40 m and many of the species are cut for timber. Buttressing of the trunks is a common feature of many tree species, a character absent from the more southerly rainforests.

The communities are developed from near sea-level to altitudes of c. 1000 m, on soils of high fertility, derived mainly from basalt or developed on alluvium (Table 7.4). Mean annual rainfall varies from c. 1400 mm to c. 2400 mm, with a summer predominance (mostly 170–380 mm per month for the 3 wettest months). The winter is relatively dry, with c. 50–100 mm per month for the three driest months. The winter complement, however, is higher than in the tropics, where many stations where rainforests occur record as little as 20 to 50 mm per month for the 5 driest months. Nevertheless, a few winter-deciduous species occur, e.g. *Toona australis.*

Floristically the alliance is very rich, containing a total of c. 350 species. *Argyrodendron actinophyllum* and *A. trifoliolatum,* either individually or together, are consistently in the stands. Other species represented abundantly and in much larger numbers than in the other alliances south of the Tropic belong to the Lauraceae, Simaroubaceae, Rutaceae, Meliaceae, the succulent-fruited Myrtaceae, especially *Syzygium,* and sometimes the palm *Archontophoenix cunninghamii.* The alliance contains several tropical taxa which reach their southern limit in this alliance, e.g. Annonaceae and Papilionaceae (Table 7.5). *Araucaria* is usually absent but occurs in marginal areas close to dry rainforests (see 6.5 below).

The shrub layers are poorly developed and consist largely of young plants of trees; tree ferns are locally abundant *(Cyathea australis* or *C. leichhardtiana). Cordyline stricta,* an arborescent monocotyledon, occurs in some stands.

Epiphytes are numerous and include the pteridophytes, *Asplenium nidus, Arthropteris beckleri, A. tenella, Davallia pyxidata, Dictymia brownii, Microsorium scandens, Platycerium bifurcatum, P. grande, Pyrrosia rupestre* and *Vittaria elongata.* Angiosperm epiphytes are: *Pothos longipes,* and species of *Bulbophyllum, Dendrobium* and *Sarcochilus.*

The ground flora is usually sparse and often lacking over large areas as a result of low light intensity and the accumulation of litter. In light breaks and around margins, or in wetter creek beds and banks, a flora consisting of ferns sometimes mixed with angiosperms usually occurs. The main ferns are *Adiantum formosum, Athyrium australe, Blechnum patersonii, B. procerum, Histiopteris incisa, Pellaea falcata* and *Pteris umbrosa.* Angiosperm herbs include *Alocasia macrorhiza, Alpinia coerula, Calanthe triplicata, Cynoglossum latifolium, Elatostema reticulatum, Helichrysum elatum, Gymnostachys anceps* and *Oplismenus compositus.*

Baur (1962) distinguishes the following main associations:

i. *Argyrodendron* spp., as described above.

ii. *Dysoxylum fraseranum* which becomes the most common dominant at higher altitudes (c. 700 m); this contains most of the species found in i. (Baur, 1957).

iii. *Toona australis – Flindersia* spp. (now mostly removed) which covered the lowlands on the river alluviums.

iv. *Elaeocarpus grandis* which forms narrow bands along major watercourses through the *Argyrodendron* and *Toona – Flindersia* communities.

The alliance is usually delimited by soil parent material (i.e., by soil fertility levels) and when differences in fertility levels are great the alliance is sharply separated from the adjacent *Eucalyptus* communities. A merging of rainforest and tall *Eucalyptus* forest is common, the rainforest species forming an understory to the eucalypts (Fig. 7.17). Again, when an adjacent soil parent material is fertile enough to support rainforest,

Fig. 7.17. A pocket of rainforest in a valley within *E. saligna* forest. Rainforest species are: *Toona australis,* the palm, *Archontophoenix cunninghamiana* and the tree fern, *Cyathea australis.* In foreground, a tangle of the liana, *Cissus hypoglauca.* Ravenbourne National Park, c. 40 km north-east of Toowoomba, Qld.

as at Dorrigo, the soil of lower fertility support the *Ceratopetalum – Schizomeria* rainforest (see 6.3.3.6, above).

6.5. *Drypetes australasica – Araucaria* spp. – *Brachychiton discolor – Flindersia* spp. Alliance

In areas receiving a mean annual rainfall of c. 1100–1400 mm (with a winter component of no less than 50 mm per month), "dry" rainforests replace the more luxuriant *Argyrodendron* alliance. These drier types occur in the Clarence R. valley and northward, being best developed in south-east Queensland, especially in the valley of the Mary R. (Swain, 1928). They occupy fine-textured soils of relatively high fertility. They occur both east (in rain-shadows) and west of the Great Divide, and the inland sites are subjected to relatively low temperatures (lowest min. at Widgee is $-5°C$). Communities in New South Wales have been described by Baur (1957, 1962) and those in Queensland by Swain (1928), Blake (1941) and Cromer and Pryor (1942).

The forests contain two tree layers, the upper one (emergents) discontinuous and reaching a height of c. 50 m *(Araucaria)* (Fig. 7.18). The

Fig. 7.18. *Araucaria cunninghamii* with subsidiary *Alphitonia excelsa, Carissa ovata* and *Cissus hypoglauca*. Grass bald in foreground. Bunya Mt. Qld.

Fig. 7.19. Interior of a dry rainforest. Main species is *Brachychiton discolor* (almost leafless). Also, *Dendrocnide excelsa, Toona australis* and trunk of *Araucaria cunninghamii* (beside man). Liana is *Cissus hypoglauca* and main epiphyte *Dictymia brownii*. Bunya Mt. Qld.

lower tree layer is continuous and mostly c. 20 m high with emergents *(Brachychiton, Flindersia)* to c. 30 m tall (Fig. 7.19). A scattered lower tree/tall shrub layer is commonly present. Buttressing of the trees is rare and vascular epiphytes are uncommon. Lianas are abundant but climbing palms are absent.

The communities, often in undulating or hilly country, commonly merge with the *Argyrodendron* alliance, the latter occupying lower, moister areas, the former the drier upland areas. Also over a gradient in mean annual rainfall, the dry rainforests occupy an intermediate position between the *Argyrodendron* alliance (wet areas) and the Vine Thickets (see Sect. 9, below).

The alliance is floristically rich in spite of the lower rainfall. The Euphorbiaceae and Sapindaceae are the most abundant families; many species are common to the wet and dry rainforests.

The alliance is named from the most constant species, *Drypetes* (formerly *Hemicyclia*) *australasica* and the emergents. The latter identify the suballiances. For the communities in

New South Wales, Baur (1957) identifies two suballiances: i. *Araucaria cunninghamii*, and ii. *Brachychiton discolor*. The latter occurs in the more adverse sites. In Queensland *A. cunninghamii* and *A. bidwillii* both occur as emergents, the latter in moister sites and often in the ecotone with the *Argyrodendron* alliance; these communities represent a northern variant of the suballiance in New South Wales. A third suballiance occurs in Queensland, with *Flindersia* spp. and/or *Euroschinus falcatus* as emergents (Swain, 1928).

Watercourses dissecting the dry rainforest support "gallery" forests identified by *Castanospermum australe* and *Grevillea robusta*. These are not unique to the dry rainforests and are dealt with in more detail below (Sect. 10).

6.6. Littoral Rainforests

6.6.1. General

Small patches of rainforest occur on coastal sand dunes, usually behind the frontal dune (Chap. 23). In many areas the forests occur on deep sands, but in some cases the sands adjoin rock. Analyses of the soils supporting the rainforests indicate extremely low levels of all nutrients, comparable to the levels found in heath soils. For example, Nicholls (1966) reports 57 ppm P in the surface 4 cm of soil and <15 ppm below this level.

Analyses of the leaves and wood of the rainforest species indicate levels of nutrients comparable with those found in inland rainforest stands. The source of the nutrients is not known but it is assumed that material washed in from the ocean, such as marine animals and plants provided the original nutrient supply, with the probable exception of nitrogen. A rough calculation of the weight of a rainforest community and its nutrient content indicates that an original P-content of 400 ppm in the surface metre of soil would provide the necessary phosphorus for the communities. Most of the nutrients are now held in the vegetation.

South of c. lat. 32°S. many species usually associated with rainforests occur in hind dune areas; these do not form rainforests, but occur as an understory to *Eucalyptus* or *Banksia*. The main species are: *Cupaniopsis anacardiodes, Synoum glandulosum, Breynia oblongifolia, Clerodendron tomentosum,* and less commonly *Acmena smithii, Rapanea variabilis* and *Notelaea longifolia*. A few small lianas usually occur, viz., *Stephania japonica, Similax australis, Parsonsia straminea, Tylophora barbata* and *Geitonoplesium cymosum* (Davis, 1941).

North of lat. 32°S these species aggregate with or without eucalypts or banksias to form strips of rainforest, scrub like in more southerly stands but increasing in height from south to north, and dominated by *Cupaniopsis anacardioides*. Also, from south to north the stands become floristically richer, chiefly through the development of more complex rainforests in communities protected from shearing effects of winds by the stands of *Cupaniopsis* behind the frontal dune. The total flora in these forests between lat. 32° and 25°S. is listed in Table 7.5. Two alliances are recognized.

6.6.2. *Cupaniopsis anacardioides* Alliance

This species forms scrubs, sometimes as little as 1 m high, to low forests c. 8 m high behind frontal dunes. The seaward face of the community is usually wind-shorn. *Banksia integrifolia* is a common associate and sometimes *Eucalyptus gummifera*. Associated species belong to the rainforests and include the species mentioned above in the more southerly stands, together with the following in more northerly stands: *Alectryon coriaceus, Elaeodendron australe, Pittosporum revolutum, Rhodomyrtus psidioides, Glochidion ferdinandii* and *Rapanea variabilis*. Additional species occur when the community adjoins richer rainforest, especially *Ficus stenocarpus, Mischocarpus pyriformis, Flindersia* spp., *Syzygium* spp. and *Podocarpus elatus*. Lianas are often present especially *Cissus* spp., *Deeringia amaranthoides* and *Maclura cochinchinensis* (Nicholls, 1966; Baur, unpub.).

6.6.3. *Tristania conferta* Alliance
Brush Box

Baur (1965) describes stands dominated by *Tristania conferta* in northern New South Wales mainly on headlands, rather than on deep sands. *Tristania* occurs with *Cupaniopsis,* with an admixture of *Elaeodendron australe, Acmena* and *Syzygium* spp. and *Crytocarya triplinervis*. The communities, being exposed to strong winds, are invariably wind-shorn and vary from c. 2 m to c. 14 m in height. Taller stands dominated by *Tristania conferta* occur in some localities when the dune sands adjoin rock, and also inland (Fig. 7.20).

7. Rainforests of Tropical North-eastern Australia*

7.1. Distribution – Past and Present

Tropical rainforests are found in patches along the wetter parts (over 1500 mm mean annual rainfall) of the north-eastern coast of Queensland (Fig. 7.21). Although the aboriginal inhabitants did not affect the distribution of rainforests in this region by practising shifting agriculture, their repeated use of fire in adjacent open eucalypt forest communities may have gradually eroded the rainforest edge especially on seasonally dry sites (Stocker and Mott, in press). Another feature of the rainforest environment in this region is the periodic occurrence of intense tropical storms (called cyclones in Australia). Disturbance to the forest canopy by hurricane force winds associated with these storms can result in a characteristic physiognomy locally known as vine or cyclone scrub (Webb, 1958). Debris produced by hurricane force winds may also facilitate the penetration and subsequent destruction of rainforest communities by fire. At the commencement of settlement by Europeans, rainforests covered a total area of about 1.2 million hectares (less than 0.2% of the land area of the continent). About half this area has now been cleared for agricultural development while most of the remainder is within State Forest or National Parks. The current distribution of rainforests in the Cooktown-Townsville portion of the region has been

* Prepared by G.C. Stocker and B.P.M. Hyland of C.S.I.R.O., Division of Forest Research, Atherton, Queensland.

Fig. 7.20. *Tristania conferta* dominating a patch of rainforest in a valley. Main subsidiary species are: *Eucalyptus deanei, Acacia maidenii, Ceratopetalum apetalum, Cissus hypoglauca, Cassinia longifolia* and *Cyathea australis*. 60 km east of Walcha, N.S.W.

mapped by Tracey and Webb (1975) at a scale of 1:100,000.

7.2. Floristic Composition

At the specific level, the flora of the tropical rainforests of north Queensland is poorly documented. The standard reference for Australian rainforest trees (Francis 1970) does not adequately cover the region since it mentions only 247 of the 800 tree species recorded by Hyland (in preparation) in the second edition of his field key to rainforest trees found north of Townsville. Although studies have mainly been confined to tree species, it is apparent that there are still many plants in the rainforests of the region which are either undescribed or in need of revision to correct obvious errors in current nomenclature.

The maximum floristic diversity within the region is found in stands between Cooktown and Ingham. High diversity in this part of the

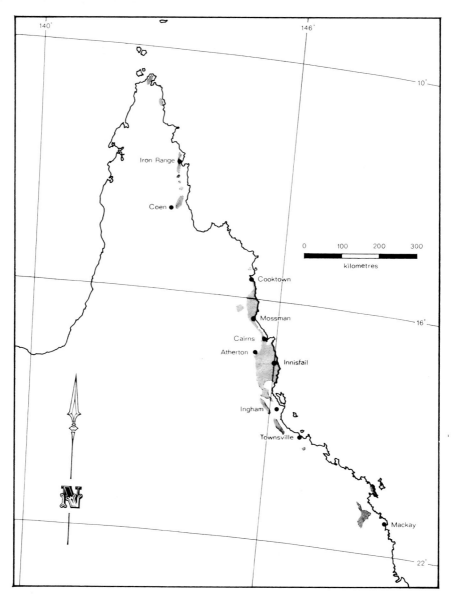

Fig. 7.21. Tropical northeastern Australia. Shading indicates areas where rainforests are the dominant natural vegetation type. Map compiled by the authors with supplementary data from Carnahan (1976), Volck (1968), Lavarack and Stanton (1977) and large scale maps produced by the Queensland Department of Forestry.

region appears to be the result of the mountain ranges providing convenient refuges during periods of climatic stress. The great variety of habitats created by wide variations in the geology, topography and climate of this central area also seems to have played a part. Prominently absent from rainforests to the north and south of this area are many of the Proteaceae and *Flindersia* species which often made up a significant part of the upper canopy in the central area. Although most of the rainforests north of Cooktown are floristically impoverished when compared to those immediately to the south, some contain a few interesting species such as *Tet-*

rameles nudiflora and *Gulubia costata* which may be recent arrivals from Papua New Guinea. The most southern rainforests of the region are those in the Eungella ranges west of Mackay. The forests here appear transitional in that they contain a few species characteristic of forests further north, e.g., *Archontophoenix alexandrae, Eugenia* sp. (Hyland, 1971 #37), *Cryptocarya angulata, Elaeocarpus foveolatus* and *Balanops australiana,* and some from the subtropical rainforests of southern Queensland and northern New South Wales, e.g., *Archontophoenix cunninghamiana, Gmelina leichhardtii* and *Flindersia australis*.

Unfortunately, the notion that the rainforest flora of this region is basically Indo-Malayan is well entrenched in the scientific and popular literature. For example, Richards (1968) in his classic work on the rainforests of the world observed that "in tropical Queensland this

Fig. 7.22. Cyclone damaged lowland rainforest near Innisfail. Vines in the crown of the tree include *Bambusa moreheadiana* (lanceolate leaves) and *Merrima dentata* (heart shaped leaves).

forest closely resembles the rainforests of Indo-Malaya in general aspect and, except for the complete absence of Dipterocarpaceae, in floristic composition". However, similarities with the Indo-Malayan flora should not overshadow the essentially Australian elements which are often a major component of the emergent tree strata. This local element includes species of *Flindersia, Cardwellia, Musgravea, Placospermum, Buckinghamia, Darlingia, Backhousia, Blepharocarya, Castanospermum, Ceratopetalum* and *Doryphora*. The presence of several primitive and restricted angiosperm genera – *Idiospermum, Austrobaileya, Sphenostemon, Bubbia, Ostrearia, Neostrearia, Eupomatia* and *Galbulimima* – add a further distinctive character to the rainforests of the region.

7.3. Characteristics of some Representative Forest Types (See Figs. 7.22 to 7.28)

Although aspects of the ecology of north Queensland rainforests are described by Swain (1929) and Webb and co-workers (Tracey 1969; Tracey and Webb 1975; Webb 1958, 1966, 1968, 1969; Webb *et al.* 1967a, 1967b, 1970; Williams *et al.* 1969) there is still much to be

Fig. 7.23. *Dysoxylum schiffneri* is one of several caulicarpous species in the north Queensland rainforests. Each fruit is about 3 cm diameter. The climbing aroid is *Raphidophora* sp. and the canes with prominent nodes are stems of the climbing bamboo, *Bambusa moreheadiana*.

Table 7.6. Environmental and structural characteristics of some tropical rainforest communities in north east Queensland based on 0.5 ha plots.

Plot No.	Ele-vation (m)	Latitude (S)	Average Annual Rainfall (mm)	Parent material	Soil characteristics (0–30 cm)						Tree component (>10 cm diameter)						Basal area (m^2)	Canopy height (m)
					pH	Organic Carbon (%)	Total Nitrogen (%)	Total Phosphorus (%)	C.E.C. (meq/100 g)	No. of species	No. of indi-viduals	Diameter class distribution (%)					
												10–40 cm	40–80 cm	80+ cm			
1	20	12° 44'	2000	Alluvial	–	–	–	–	–	60	231	85	12	3	21.6	21–41
2	20	16° 07'	3500	Granitic	4.8	4.8	0.48	0.53	19.1	54	394	89	11	–	22.0	25–32
3	20	17° 00'	2200	Metamorphic	5.8	2.6	0.24	0.16	8.2	68	484	93	7	–	17.9	10–29
4	80	17° 31'	4000	Metamorphic	4.8	1.6	0.16	0.05	8.2	43	238	80	17	3	20.5	22–44
5	200	16° 22'	2600	Metamorphic	4.5	1.7	0.17	0.12	9.8	68	489	94	5	1	20.4	17–28
6	380	17° 26'	4000	Alluvial	4.7	2.7	0.32	0.16	15.2	62	296	82	16	2	24.0	29–45
7	400	13° 45'	2000	Metamorphic	4.3	2.4	0.17	0.05	7.5	51	439	96	4	–	13.3	19–30
8	680	18° 30'	2000	Granitic	4.1	2.5	0.18	0.08	13.2	64	399	91	8	1	20.3	22–38
9	720	17° 10'	1200	Granitic	5.3	3.5	0.44	0.02	14.5	35	454	97	3	–	14.2	15–23
10	720	17° 18'	1500	Basaltic	6.5	2.9	0.42	0.38	35.1	44	313	81	17	2	31.3	40
11	800	17° 10'	2200	Granitic	4.7	2.9	0.32	0.03	16.4	87	442	89	10	1	25.1	26–40
12	880	21° 15'	2000	Basaltic	5.1	4.3	0.54	0.26	28.3	32	380	89	9	2	29.9	33–41
13	1000	16° 30'	3000	Granitic	5.1	5.6	0.33	0.05	12.3	82	449	85	14	1	29.7	24–32
14	1000	16° 17'	1500	Granitic	4.5	7.5	0.51	0.03	26.0	68	551	89	10	1	29.4	30–39
15	1100	17° 9'	2000	Granitic	4.4	3.3	0.22	0.01	9.4	63	507	86	13	1	32.8	30–36
16	1100	17° 25'	2000	Rhyolitic	6.1	5.1	0.57	0.08	30.7	62	376	89	10	1	22.6	27–38
17	1200	17° 33'	3000	Rhyolitic	4.3	3.1	0.24	0.07	5.9	49	460	94	6	–	20.0	19–30

Table 7.7. Upper canopy tree species in rainforest plots in north east Queensland.

Species	Plot Nos.
Coniferales	
ARAUCARIACEAE	
Agathis atropurpurea	14
A. robusta	9
PODOCARPACEAE	
Podocarpus ladei	13
P. neriifolius	3
Dicotyledons	
ANACARDIACEAE	
Euroschinus falcata	10
APOCYNACEAE	
Alstonia muellerana	5, 9
A. scholaris	6
Cerbera inflata	9
ARALIACEAE	
Polyscias elegans	3
BALANOPACEAE	
Balanops australiana	14
BLEPHAROCARYACEAE	
Blepharocarya involucrigera	3, 7, 9
BURSERACEAE	
Canarium baileyanum	7
C. muelleri	8
CAESALPINIACEAE	
Storckiella sp.	2
CUNONIACEAE	
Ceratopetalum succirubrum	11, 13, 14
Geissois biagiana	16
Pseudoweinmannia lachnocarpa	9
ELAEOCARPACEAE	
Aceratium doggrellii	14, 17
Elaeocarpus bancroftii	3, 15
E. foveolatus	17
E. ferruginiflorus	17
E. largiflorens	17
Sloanea macbrydei	13, 17
EUPHORBIACEAE	
Aleurites moluccana	10
Croton insularis	9
GUTTIFERAE	
Calophyllum sil	7
IDIOSPERMACEAE	
Idiospermum australiense	2
LAURACEAE	
Beilschmiedia bancroftii	2, 5, 6, 11, 13
B. obtusifolia	1
Cryptocarya angulata	5, 6, 12, 17
C. corrugata	8, 16, 17
C. hypoglauca	17
C. mackinnoniana	8
C. rigida	16
C. sp. aff. C. corrugata	13
C. sp. aff. C. hypospodia	7
Endiandra cowleyana	8, 10
E. palmerstonii	11

	Plot Nos.
E. sp. aff. E. hypotephra	5
Litsea leefeana	2, 5, 16
MELIACEAE	
Dysoxylum decandrum	1
D. pettigrewianum	4, 10
Toona australis	10
MIMOSACEAE	
Acacia aulacocarpa	3, 5, 7, 15
MONIMIACEAE	
Doryphora aromatica	6, 10
MORACEAE	
Antiaris toxicaria	1
Ficus destruens	7
F. pleurocarpa	16
F. watkinsiana	10
MYRTACEAE	
Backhousia bancroftii	4
B. hughesii	5
Eugenia gustavioides	6
E. hemilampra	7
E. kuranda	5
E. sp. (Hyland, 1971 #42)	14, 17
E. sp. (Hyland, in prep. #639)	1
E. sp. (Hyland, 1971 #37)	8, 13, 15, 17
E. sp. (Hyland, 1971 #147)	8, 13
Lindsayomyrtus brachyandrus	2
Xanthostemon chrystanthus	7
X. whitei	3
PAPILIONACEAE	
Castanospermum australe	1
PITTOSPORACEAE	
Pittosporum sp.	17
POLYGALACEAE	
Xanthophyllum octandrum	5, 7, 8
PROTEACEAE	
Austromuellera trinervia	2
Cardwellia sublimis	4, 6, 13, 15
Carnarvonia araliifolia	3, 16
Darlingia darlingiana	15
Grevillea pinnatifida	3, 7
Musgravea heterophylla	15
M. stenostachya	14
Opisthiolepis heterophylla	6, 11
Placospermum coriaceum	13
Stenocarpus reticulatus	3
S. sinuatus	16
RHAMNACEAE	
Alphitonia petriei	12
A. whitei	11
ROSACEAE	
Prunus turnerana	2
RUBIACEAE	
Nauclea orientalis	1
N. undulata	1
RUTACEAE	
Euodia bonwickii	10
E. vitiflora	8

	Plot Nos.
Flindersia acuminata	3 15
F. bourjotiana	3 5 8 11 14 15
F. brayleyana	3 8 10 13 15 16
F. laevicarpa	11
F. pimenteliana	11 15
F. schottiana	9
Halfordia scleroxyla	17
SAPINDACEAE	
Jagera sp.	8
Toechima daemeliana	1
SAPOTACEAE	
Planchonella euphlebia	13
P. macrocarpa	5
P. obovoidea	2 4
P. papyracea	15
STERCULIACEAE	
Argyrodendron actinophyllum ssp. *diversifolium*	8
A. peralatum	2 6 10
A. polyandrum	1
Brachychiton acerifolius	9
Sterculia laurifolia	16
S. shillinglawii	1
SPHENOSTEMONACEAE	
Sphenostemon lobosporus	17
SYMPLOCACEAE	
Symplocus stawellii	7
TETRAMELACEAE	
Tetrameles nudiflora	1
ULMACEAE	
Aphananthe philippensis	1

learned about their dynamics, structure and floristics. Because of the complexity of most tropical rainforests, it is difficult, if not impossible, to intuitively assign broadly recognisable forest types to conventional alliances. Consequently, in this treatment, categorization at this level will not be attempted. However, data will be presented to illustrate the range of floristic, structural and environmental characters encountered in the rainforests of this region. These data were obtained from a number of 0.5 ha plots established and maintained by the staff of the C.S.I.R.O. Division of Forest Research as reference areas for use in studies of rainforest dynamics. These plots have been established in unlogged forest and are thought to be broadly representative of most of the rainforest communities in this region.

Some environmental, structural and floristic features of these plots are summarised in Tables 7.6 and 7.7. Although these plots are numbered in an altitudinal sequence, consistent changes in the data are not obvious. However, the variability of the data between adjacent plots is high and altitudinal trends may be more apparent when additional plots are established.

Most of the rainforest soils examined are infertile by agricultural standards. However, despite their low fertility, they only appear to have marked effect on structural development, in terms of canopy height and basal area, on two sites (9 and 17). A contributing factor on one of these sites (9) is probably low rainfall.

Rainforest soils derived from granites are mostly associated with mountain range systems. They are typically well drained, very variable in depth and range from pink to yellow in colour. Those from metamorphosed shales and sandstones are characteristic of the foothills and tablelands. A feature of the latter soils in this region are the heavy yellow clays usually encountered within 30 cm of the surface. As a

Fig. 7.24. Lowland palm swamp dominated by *Archontophoenix alexandrae* 5 km north of Innisfail. This species is often an emergent in these communities and may be 20–25 m high.

consequence, these soils appear to have poor internal drainage. Basalts and rhyolites are found mainly on the Atherton Tableland and the soils derived from them vary considerably in age, depth of weathering and fertility. Although basaltic soils are usually well drained and indicated fertility levels are higher than most other soils, tree growth data (Volck, 1975) indicate that they are not the most productive in terms of useful wood produced following logging and silvicultural treatment.

The number of tree species (over 10 cm diameter) in each of the half hectare plots varies from 32 to 87. Comparison with data for other tropical rainforests (Whitmore, 1975) shows that the species richness of the forests in this region is greater than tropical rainforests in Africa and the Americas but less than those in some parts of south-east Asia. The paucity of species in plot 9 may be the result of poor soils and low rainfall. However, this explanation is inadequate for plot 12 where low diversity is more probably related to its isolation from the main rainforest belt.

The upper canopy tree species observed in the plots are listed in Table 7.7. Of the 105 species listed, 77 were found only in one plot, 15 occurred in two plots but only 1 in more than five plots as upper canopy species.

Plots 2 and 4 are in wet lowland areas and are unusual in that they each tended to be dominated by a single upper canopy species (*Lindsay-

Fig. 7.25. Part of the interior of plot 9. The epiphytes on the largest tree are *Platycerium superbum* (upper) and *Asplenium australasicum* (lower). The fronds of *A. australasicum* are about 50 cm long.

omyrtus brachyandrus and *Backhousia bancroftii* respectively). These species account for a third of the total number of trees in each of these plots. This tendency for a single species to dominate each of these sites cannot currently be explained. In several other lowland plots (3, 5 and 7) the palm *Licuala ramsayi* is the main component of the mid canopy strata. This palm is seldom found in the upper canopy except on some permanently waterlogged sites. The ground strata in all the plots generally consists of regenerating trees and shrubs. Occasionally a single shrub or small tree species is very conspicuous. For example, in plot 10 the shrub *Hodgkinsonia frutescens* is a major component of the understory.

The number of trees over 10 cm diameter in each plot varies from 231 to 511. 80% or more of the trees in all the plots are less than 40 cm in diameter. The presence of a few large trees in a plot appears to reduce total tree numbers considerably. Plot basal area varies from 13.3 to 32.5 m^2. Local forest management experience suggests that basal area could be expected to increase with increasing elevation but this trend is not particularly obvious in the data. The canopy height data can be taken as an approximate measure of canopy roughness. Site 10 is somewhat unusual in that its canopy is relatively compact and uniform. In general, canopy roughness is greatest in the lowland forests, presumably because of more frequent disturbance by hurricane force winds.

Specialised life forms such as vines, epiphytes

Fig. 7.26. The palm *Licuala ramsayi* is a prominent feature of the small tree strata at several lowland sites. The leaf blades of this species are almost 1 m across.

and pachycauls are present in most plots. Vines are usually numerous and include several species of climbing palm (*Calamus* spp.). Other common vines are *Bambusa moreheadiana*, *Petraeovitex multiflora*, *Faradaya splendida*, *Entada scandens*, *Flagellaria indica*, *Pothos longipes*, *Rhaphidophora australasica*, *R. pinnata*, *Austrobaileya scandens*, *Smilax australis*, *S. glycyphylla*, *Piper* spp., *Tetracera nordtiana* and *Palmeria scandens*. Vines generally become less conspicuous with increasing altitude. Palms, especially *Archontophoenix alexandrae*, *Licuala ramsayi*, *Orania appendiculata* and *Linospadix* spp. and tree ferns, mainly *Cyathea* spp. are often encountered in the mid to lower canopy. Epiphytic ferns and orchids can be conspicuous, especially in the mountain plots. While many species may be encountered, most belong to the following genera: – *Asplenium*, *Platycerium*, *Drynaria*, *Dendrobium* and *Bulbophyllum*. Strangling figs (*Ficus* spp.) and *Schefflera actinophylla* are recorded in most of the plots. This latter group of species begin life as epiphytes but ultimately establish a root system which extends down the trunk of their host and into the soil.

Two interesting communities in which plots have not been established are the lowland swamp rainforests and the low forests and shrublands of the mountain tops. The swamp forests are limited in extent occurring mainly in the wettest parts of the coastal zone between Tully and Cairns. They are rather inaccessible and consequently their structure and floristics

Fig. 7.27. Part of the interior of plot 10. The large tree (about 105 cm diameter above buttress) is *Dysoxylum pettigrewianum*.

have not been examined in detail. However, palms such as *Archontophoenix alexandrae*, *Licuala ramsayi* and *Calamus* spp. are usually numerous and are often the most conspicuous species in the upper canopy. In the lower strata *Pandanus* spp. and sedges are often encountered. Climbing ferns (particularly *Stenochlaena palustris*) and pandans (*Freycinetia* spp.) are also often seen in the lower strata.

The vegetation on most exposed mountain slopes above 1,200 m is usually reduced to low submontane forest or thicket. On the summits of several of the highest mountains (e.g. Bartle Frere 1530 m and Thornton Peak 1375 m), there are open areas of shrubland and bare rock. Species characteristic of the mountain top communities include *Flindersia unifoliolata*, *Cinnamomum propinquum*, *Planchonella singuliflora*, *Acronychia chooreechillum*, *Rockinghamia brevipes*, *Trochocarpa laurina*, *Syzygium erythrodoxum*, *Polyscias bellendenkerensis*, *P. willmottii*, *Orites excelsa*, *Dracophyllum sayeri*, *Leptospermum wooroonooran*, *Ceratopetalum corymbosum*, *Rhododendron lochae*, *Eucryphia* sp. and *Garcinia brassii*. Some of these species are widespread while others appear to be confined to one or two peaks.

Fig. 7.28. The butt of *Agathis atropurpurea*, an important emergent in plot 11. The trunk of this tree is 113 cm diameter and the tree 39 m high.

8. Semi-Evergreen Vine Forests (Monsoon Forests)

8.1. General

This type of rainforest, designated Monsoon Forest by Schimper (1903), is distinguished by a high proportion of deciduous plants, chiefly the dominants. They occur in areas with a mean annual rainfall of 1300–1500 mm and a very dry winter with little or no effective rain (see Chap. 1, climate exemplified by Darwin). They may occur in areas with a lower rainfall where they extend along watercourses, or around lagoons, or they occupy patches of soil fed by springs or runoff water from the uplands. The stands are never subjected to frost. The areas occupied by the rainforests are very small, from c. 1–25 hectares, and they are best developed on the Arnhem Peninsula (Christian and Stewart, 1952, Specht, 1958), with the largest areas on Melville Is. (Stocker, 1968). Smaller areas occur in Western Australia (Miles et al., 1975; Beard, 1976).

The forests occur on a wide range of soils, including Lateritic Red Earths, Lateritic Podsols, and even skeletal soils and dune sands. In Western Australia Beard (1976) reports the communities only on soils derived from basalt. Soil fertility levels do not appear to separate the rainforests from *Eucalyptus* forests on Melville Is. where values for total P are as low as 56 ppm for the rainforest and 48 ppm for *Eucalyptus* forest (Stocker, 1968).

The communities reach a height of c. 15 m, sometimes with emergents projecting c. 5 m above the canopy. Leaf-sizes are mainly mesophyll and notophyll and the degree of deciduousness varies from stand to stand. Some communities in wet areas show little or no leaf-fall, whereas in others leaf-fall is complete. The forests usually have two tree layers in wetter situations, one in the drier, and a few species have buttressed trunks. Lianas are abundant but epiphytes are rare or absent.

The total flora consists of c. 340 angiosperms and 37 pteridophytes (Table 7.3). The flora is closely related to the eastern flora at the generic level and many species are common to the wetter eastern and drier western forests. The monsoon forest flora appears to have been selected from the wetter eastern types and most of the eastern families have persisted in the drier localities where they are represented by fewer genera and species. Most families are represented by only one, two or three genera, exceptions being: Euphorbiaceae (7 genera), Rubiaceae (5), Rutaceae (5), Apocynaceae (4).

Fig. 7.29. Trunk of *Cleistocalyx operculata* in Deciduous Vine Forest. Also trunks of the palm, *Livistona benthamii* and seedling of the palm, *Carpentaria acuminata* (left). Near Darwin, N.T.

8.2. Associations

The communities cannot be regarded as a distinctive alliance comparable to the eastern stands. On the contrary, a former large alliance has apparently been broken into a number of fragments with no species occurring in all the stands.

In the largest stands on Melville Is., Stocker (1968) states that "among the dominant species *Bombax ceiba* is the most common in the drier sites. *Canarium australianum* prefers the intermediate sites and *Gmelina dalrympleana* the wetter". These species occur in other stands but do not form a similar pattern.

The main tree species in stands from Melville Is. and the Darwin area (Christian and Stewart, 1953) are, in addition: *Acacia aulacocarpa, Alstonia actinophylla, Buchanania arborescens, Cheistocalyx operculata* (Fig. 7.29), *Ficus benjamina, F. coronulata, F. lacor, F. racemosa, Nauclea orientalis, Parinari corymbosus, Peltophorum pterocarpum, Sterculia quadrifida, Terminalia sericocarpa, Vitex glabrata* and the palms, *Carpentaria acuminata* and *Ptychosperma bleeseri*.

The lower story trees and tall shrubs include a number of genera, among them: *Canthium, Croton, Diospyros, Exocarpos, Litsea, Livistona, Myristica, Pongamia* and *Trema*.

Lianas are abundant the more common ones being: *Asparagus racemosus, Capparis umbellata, Flagellaria indica, Jasminum didymum, Malaisia scandens, Opilia amentacea, Parsonsia velutina, Pisonia aculeata, Secomone elliptica, Smilax australis* and *Tinospora smilacina*.

Epiphytes are represented by the orchids *Dendrobium dicuphum, D. canaliculatum* and *Luisia teretifolia*, and the ferns *Drynaria quercifolia* and *Stenochlaena palustris*.

Herbs are rare and include: *Amorphophallus glabra, Dracaena angustifolia* (Fig. 7.4), *Geodorum pictum, Tacca leontopetaloides,* the grasses *Oplismenus compositus* and *Panicum trichoides,* and the sedge *Cyperus ramosii*.

Other assemblages which have been recorded are:

i. An association containing *Eugenia subordicularis, Cochlospermum fraseri* and *Buchanania obovata* (Specht, 1958) which form tall scrubs or low woodlands on the Arnhem Peninsula.

ii. An association dominated by *Alstonia actinophylla, Albizia lebbek, Alphitonia excelsa,*

Bombax ceiba and *Myristica insipida,* described as a semi-deciduous microphyll vine thicket and occurring in the Prince Regent River Reserve (Miles et al., 1975).

iii. A closed forest, occurring in a gully in the same area as ii and adjoining a mangrove swamp, containing *Calophyllum* sp., *Melaleuca leucadendron, Carallia brachiata* and *Myristica insipida* (Miles et al., 1975).

iv. A low woodland occurring in the same area as ii, containing *Brachychiton viridiflorus, Acacia delibrata, Canarium australianum* and *Santalum lanceolatum* (Miles et al., 1975).

v. A rainforest of greater floristic complexity (47 spp.) in the Admiralty Gulf is described by Beard (1976). The communities extend from near sea-level to an altitude of c. 250 m. The riverain forests on the banks of streams contain *Calophyllum australianum, Carallia brachiata* (c.f. ii), *Pandanus spiralis, Sterculia quadrifida* and *Tristania grandiflora*. The low level forests are dominated by *Zyzyphus quadriloculalis,* with *Abrus precatorius* the common liana. The high level forests are dominated by *Albizia lebbek, Atalaya variifolia, Cochlospermum fraseri, Pouteria sericea,* with *Flagellaria indica* the common liana.

8.3. The Effects of Fire

In the area occupied by the semi-evergreen rainforest and the tall *Eucalyptus* forests, Stocker (1968) regards regular burning by the aborigines as a potent factor in controlling the boundaries of the communities, with the probable continual contraction of the rainforests.

In their undisturbed condition the rainforests are rarely burned but, if disturbed, for example by the aborigines digging for yams, or by cyclones, or by the introduced buffalo, they become more susceptible to fire. Consequently, when the more open *Eucalyptus* forests are purposefully burned the margins of the rainforests may be damaged by fire. These communities can regenerate, usually by the initial development of stands of *Acacia* which protect the young rainforest plants as they appear in the stand. However, the seral community, if regularly burned, is wiped out and replaced by eucalypts. He provides positive evidence for the replacement of semi-evergreen rainforest by tall open *Eucalyptus* forest by the presence in the latter of mounds of the jungle fowl *(Megapodius freycinet)*. This species is confined to the rainforest and the birds scratch surface litter and soil into mounds where they lay their eggs which are hatched, chiefly through the warmth produced in the decaying leaf-mould. The mounds reach a diameter of several tens of metres and a height of several metres. Mounds in use are scratched over regularly by the birds who tend the eggs, and they support no vegetation. Abandoned mounds now occur in places in the *Eucalyptus* forests; they are considerably flattened and are clothed with grass, and sometimes support a few cycads or *Eucalyptus* saplings.

8.4. Littoral Rainforests

Small patches of littoral rainforest occur, being developed either behind mangroves (see association iii, above) or on sand adjoining the coastal dunes (see association i, above). The stands vary in floristic composition and commonly contain species of *Acacia* (*A. aulacocarpa, A. auriculiformis* and *A. mangium*). Other species of common occurrence are: *Canarium australianum, Flagellaria indica, Gyrocarpus americanus, Hibiscus timorensis, H. tiliaceus, Peltophora ferruginea, Pongamia pinnata, Pouteria sericea, Sterculia quadrifida* and *Wrightia pubescens*.

9. Rainforest-derived Communities in the Semi-Arid and Arid Zones

9.1. General

Evidence for the one-time extensive development of rainforest across the northern part of the continent comes from the relict occurrences of some rainforest species in refuges in the semi-arid and arid zones (Chap. 5), and from the occurrence in these drier regions of arid-adapted species of genera which otherwise occur in the wetter rainforest (exceptions being *Adansonia* and *Cadellia*). These species, most of which are tall shrubs or trees, and a few are lianas, commonly form distinctive communities, some with a single dominant, others with several dominants. Many of the species form the understory below eucalypts and some of them occur as scattered individuals in the *Acacia* woodlands (Chap. 18), or in the savannahs where they are emergents above grasslands dominated by species of *Astrebla*, *Plectrachne* or *Triodia* (Chap. 20).

The number of species exceeds 100. They are most abundant in the area receiving a predominantly summer rainfall, but some extend into the southern winter-rainfall zone, especially *Heterodendrum oleifolium*. The greatest concentration of species occurs in western Queensland close to the wetter rainforests; the communities in these areas are sometimes referred to as "Softwood Scrubs". The species and the communities which the species constitute are in competition with *Eucalyptus* and *Acacia* (especially *A. harpophylla*) and the area of softwood scrub appears to be diminishing, chiefly as a result of fire which favours invasion of the scrubs by either *Eucalyptus* or *Acacia*, or by both.

Fig. 7.30. Dry rainforest dominated by *Cadellia pentastylis*. Understory containing *Heterodendrum diversifolium*, *Geijera parviflora*, *Croton phebalioides*, *Capparis mitchellii*. Near Injune, Qld.

The communities are difficult to classify but are regarded as climaxes. Those which exhibit constancy in floristic composition have been given the rank of "alliance". They are arranged below in order of decreasing affinity with the rainforests, determined by floristic composition and structure.

Of interest is the fact that many of the species appear rarely to reproduce from seed, but monospecific stands are developed by root suckering (see Chap. 5).

9.2. *Cadellia pentastylis* Alliance

Ooline

This species forms closed Microphyll Vine Forests (Fig. 7.30) up to c. 18 m tall in scattered localities (Fig. 7.31) where the annual rainfall is 500–720 mm. The species occurs also in the "softwood scrubs" (next section) and in association with *Acacia harpophylla* (Chap. 18).

The forests occur on sandy or skeletal soils with an apparently high organic content, usually in rocky situations where some additional water is received from runoff. Light frosts are common (lowest minima c. $-5°C$). The communities at their maximum development are more or less closed, though the canopy is not dense, and a well developed shrub layer occurs. The stands are commonly surrounded by *Eucalyptus* woodlands.

Cadellia is usually the sole dominant but *Brachychiton australis* and *B. rupestris* may be present in some stands in Queensland (Blake, 1938; Story, 1967). The associated species are numerous. The main shrubs in the Queensland stands are *Alectryon subdentatus, Alphitonia excelsa, Backhousia angustifolia, Canthium odoratum, Capparis mitchellii, Carissa ovata, Citriobatus spinescens, Croton phebalioides, Ehretia membranifolia, Elaeodendron australe, Eremophila mitchellii, Exocarpos latifolius, Geijera parviflora, Heterodendrum diversofolium, Notelaea microcarpa, Pittosporum phylliraeoides* and *Spartothamnella juncea*. Lianas include *Cissus opaca, Jasminum simplicifolium, Marsdenia* sp. and *Parsonsia eucalyptophylla*. The sole epiphyte is *Cymbidium canaliculatum*. The herbaceous layer contains the grasses *Stipa ramossissima, Ancistrachne uncinulatum* and species of *Aristida*,

Fig. 7.31. Distribution of *Brachychiton rupestre* (r), *B. australe* (a), and *B. grandiflorum* (g), from Guymer (1978). Recorded localities for *Cadellia pentastylis* (c).

Chloris, Eragrostis and the sedges *Carex inversa* and *Cyperus gracilis*.

9.3. *Brachychiton* spp. Alliance

9.3.1. General

Three species of *Brachychiton* occur as dominants of Vine Thickets in the east (Fig. 7.31). The communities are known also as Bottle Tree Scrubs. *Araucaria cunninghamii* is present in some stands and this species, together with many of the smaller trees and tall shrubs, link these Vine Thickets with the Dry Rainforests which occur in wetter country to the east. Three suballiances are recognized and are described in south to north sequence.

Several other species of *Brachychiton* are found in the semi-arid and arid zones. They occur mainly as understory species in *Eucalyptus* communities. The genus has been intensively studied by Guymer (1978).

9.3.2. *Brachychiton rupestris* Suballiance
Bottle Tree

This species (Fig. 7.32) dominates Microphyll Vine Woodlands and Semi-evergreen Vine Thickets in Queensland under a mean annual rainfall of 550–1000 mm. The communities, which occur on deep, fine textured soils, are mostly 6–8 m tall, with emergents of *B. rupestris* up to 20 m high. The stands are sometimes composed entirely of rainforest-derived species but commonly contain sclerophyllous species, especially *Eucalyptus* spp., *Acacia harpophylla* or *Casuarina cristata*. Also, *Eremophila mitchellii* typical of drier areas, is a common constituent, and occasionally *Araucaria cunninghamii* and *Brachychiton australis*.

The main associated rainforest derivatives are *Flindersia australis*, *Bauhinia carronii*, *Macropteranthes leichardtii*, *Alphitonia excelsa*, *Heterodendrum oleifolium* (Speck, 1968). A shrub layer 1–3 m high is present, the common species being *Exocarpos latifolius*, *Carissa ovata*, *Acalypha nemorum* with fewer *Hovea longipes*, *Heterodendrum diversifolium* and *Citriobatus spinescens*. Lianas include species of *Cissus*, *Jasminum*, *Passiflora*, *Marsdenia*, *Sarcostemma*, *Tylophora* and *Malaisia*. The ground stratum is dominated by grasses (*Ancistrachne* and *Chloris* spp.) (Speck, 1968).

9.3.3. *Brachychiton australis* Suballiance
Bottle Tree

This suballiance occurs mainly in tropical Queensland under a mean annual rainfall of 325–625 mm, the main stands being inland,

Fig. 7.32. Semi-evergreen vine thicket dominated by *Brachychiton rupestris* with a lower tree layer of *Excoecaria dallachyana*, *Planchonella cotinifolia*, *Elattostachys xylocarpa* and *Geijera paniculata*. 37 km W. of Biggenden, Qld. Photo: Gordon P. Guymer.

though small patches occur near the drier coast. The communities are deciduous Vine Thickets 8–16 m high, and are developed on a variety of soils, including Krasnozems, fossil laterite and skeletal soils. Soil type affects both the height of the stands and floristic composition, the best developed stands being found on the deeper, finer soils (Stocker et al., 1961).

In stands on deep soils, the emergents are *Brachychiton australis, Gyrocarpus americanus, Ailanthus malabaricum, Eucalyptus tereticornis* and an ironbark eucalypt. The main species in the canopy are *Notelaea longifolia, Strychnos psilosperma, Plumbago zeylanicum* and *Drypetes australasica*. A discontinuous shrub layer of *Carissa ovata, Citriobatus spinescens* occurs. The main lianas are *Ipomoea tuba, Marsdenia microlepis* and *Cordia dichotoma*. Epiphytes are *Dendrobium mortii* and *Platycerium grande*. In addition, some species which occur in the wetter rainforests, are recorded, e.g. *Ficus obliqua, Geijera salicifolia, Melia azedarach* and *Trema aspera*.

9.3.4. *Brachychiton grandiflorus* Suballiance

This species is confined to Cape York Peninsula and is deciduous. The trees reach a height of 10–15 m and occur as emergents in Deciduous Vine Thickets. The communities occur on skeletal soils derived from granodiorite on slopes or in gullies at altitudes of 150–500 m, under a mean annual rainfall of 1000–1250 mm. The main associated tree species are *Bombax ceiba, Cochlospermum gillvaraei* and *Gyrocarpus americanus*. Several smaller trees and small shrubs occur in the understory, belonging to the genera *Alectryon, Alphitonia, Buchanania, Drypetes, Erythrina, Ficus* and others. The common lianas are *Abrus precatorius, Capparis quinquiflora, Jasminum didymum* and *Tinospora smilacina* (Guymer, 1978).

9.4. Mixed Stands

In semi-arid New South Wales and Queensland mixed stands containing up to c. 25 species occur on rocky outcrops (usually basalt) as far south as lat. 31°S. Lowland areas support *Eucalyptus* woodlands, commonly dominated by *E. albens* or *E. melanophloia*.

The communities are tall scrubs or low woodlands and were possibly almost closed in the virgin condition (Fig. 7.33). Eucalypts sometimes

Fig. 7.33. A mixed stand of rainforest relicts containing *Atalaya hemiglauca, Heterodendrum oleifolium, Geijera parviflora, Eremophila mitchellii, Exocarpos aphylla, Ehretia membranifolia* and *Spartothamnella juncea*. Near Gunnedah, N.S.W.

occur as emergents. The most common plants in these stands *Atalaya hemiglauca, Heterodendrum oleifolium, Geijera parviflora, Ehretia membranifolia, Alstonia constricta, Capparis mitchellii, Pittosporum phylliraeoides, Alphitonia excelsa, Ventilago viminalis* and *Canthium oleifolium* in the upper story, with *Citriobatus spinescens, Breynia oblongifolia, Carissa ovata, Notelaea microcarpa* and *Spartothamnella juncea* in the shrub layer. Lianas include *Capparis lasianthos, Jasminum lineare, Pandorea oxleyi* and *Parsonsia eucalyptiphylla*. A common associate, sometimes locally dominant is the tall shrub *Eremophila mitchellii* and the tree *Brachychiton populneus*.

9.5. *Erythrophleum chlorostachys* Alliance

Ironwood

This species (noted for its hard wood, which is used for railway sleepers) reaches a height of c. 14 m and occurs both as an understory species in the tropical *Eucalyptus* forests and in monospecific stands where it displaces eucalypts.

The woodlands dominated by *Erythrophleum* (Fig. 7.34) occur mostly on flat or undulating country but sometimes on rocky hills in the semi-arid tropics. Associated species occur only on hills, mostly *Bauhinia cunninghamii* and *Terminalia platyphylla* (west) and *T. platyptera* (east).

9.6. *Adansonia gregorii* Alliance

Baobab

The origin of this species in Australia is uncertain. The genus is represented in Africa and Madagascar and it is possible that it entered Australia by ocean dispersal in relatively recent times. It is confined to a small area in the northwest (Fig. 7.35) and it appears to be migrating eastward. The species is peculiar because of its swollen trunk but is not related to the eastern Bottle Trees (*Brachychiton* spp.).

A. gregorii reaches a height of c. 12 m and the trunks may be as much as $4^{1}/_{2}$ m diam. It sometimes occurs in pure stands, or in association with rainforest species (wetter areas), or with eucalypts in drier areas, especially along watercourses (Fig. 7.36) where some rainforest species may also occur.

The main associates include *Bauhinia cunninghamii, Buchanania obovata, Celtis philippenensis, Cochlospermum fraseri, Ficus opposita, Gyrocarpus americanus* and *Terminalia canescens*. The ground stratum is always grass-dominated, chiefly species of *Chrysopogon* and *Plectrachne* (Speck and Lazarides, 1964).

Fig. 7.34. Pure stand of *Erythrophleum chlorostachys*. c. 110 km east of Normanton, Qld.

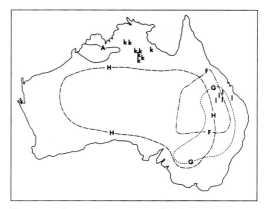

Fig. 7.35. Approximate areas of *Adansonia gregorii* (A), *Geijera parviflora* (G), *Flindersia maculosa* (F) and *Heterodendrum oleifolium* (H). Recorded localities for *Macropteranthes keckwickii* (k) and *M. leichhardtii* (l), from Byrnes, (1977).

Fig. 7.36. *Adansonia gregorii* (leafless – photo taken 22-ix-72) with subsidiary *Bauhinia cunninghamii*. Ground stratum dominated by *Eriachne obtusa*. Hill behind with *Eucalyptus dichromophloia*. c. 16 km east of Wyndham, W.A.

9.7. *Terminalia* spp. *Bauhinia cunninghamii* Alliance

9.7.1. General

Communities dominated by various species of *Terminalia* either as the sole dominant, or with *Bauhinia cunninghamii* and/or other species (defined below) have been described by Gardner (1942), Perry and Christian (1954), Speck (1960), Perry and Lazarides (1964), Story (1970) and Byrnes (1977). They occur in the semi-arid tropics (mean annual rainfall 500–1000 mm with a very dry winter) from Western Australia to the western portion of Cape York Peninsula. Some of the communities occur in even drier country along watercourses. Several of the species associated with the characteristic taxa, especially *Ventilago viminalis* and *Atalaya hemiglauca,* extend far to the south and are used to name other alliances (below). Most of the species occur in the understory of the tropical *Eucalyptus* forests and woodlands (Chap. 8) and many of the species occur also in the *Brachychiton* alliances (9.3 above).

9.7.2. Terminalia spp. Suballiances

Terminalia is almost confined to the tropics and some species occur in the wetter rainforests. The species are deciduous. The following communities occur (arranged from west to east).

i. *T. volucris* in pure stands forms woodlands on grey clays in the north Kimberleys between the 870 and 1000 mm isohyets (Speck, 1960). The species occurs also in Northern Territory. The understory is dominated by *Dichanthium fecundum*. In the wetter parts the woodland is almost closed, and with decreasing rainfall the woodland becomes more open, merging into a grassland of *Dichanthium*.

ii. *T. grandiflora* forms pure stands 5–8 m high in the Barkly region, Northern Territory, on small areas of calcareous loamy soil derived from igneous rocks. There are no associated shrubs. The herbaceous layer is dominated by *Aristida browniana, A. arenaria* and *Enneapogon* spp. (Perry and Christian, 1954).

Other species: These include *T. aridicola* (N.T., W.A.), *T. canescens* (W.A. to Qld.), *T. ferdinandiana* (N.T., Qld.), *T. platyptera* (W.A. to Qld.). These species, either singly or in pairs, usually with another "softwood" species form woodlands or scrubs. The factors segregating the species of *Terminalia* are not clear. In some cases base status of the soil appears to be important. For example, *T. volucris* (see i, above) appears to be more abundant on soils with a high base status, whereas *T. canescens* appears to prefer soils of low fertility. The last species occurs in western Queensland, sometimes with low-growing or stunted eucalypts, e.g. *E. setosa* or *E. terminalis*.

The most constant of the associated "softwood" species is the deciduous *Cochlospermum fraseri*, which sometimes forms pure stands (Fig. 7.37).

The *Terminalia* communities in Western Australia and Northern Territory have few or no associated shrub species. The woodlands reach a height of 5–7 m and in wetter parts (mean annual rainfall c. 850–1000 mm) the understory is dominated by grasses belonging to the *Andropogoneae*, and in drier areas (c. 750–850 mm) by species of *Aristida* and *Enneapogon*.

In Queensland the communities are floristically richer. Perry and Lazarides (1964) record *Terminalia platyptera* (Fig. 7.38) and *T. aridicola* as the most common species in woodlands c. 7 m high on rocky outcrops, especially granites. Associated species are *Cochlospermum* sp., *Ficus* spp., *Gyrocarpus americanus* and

Fig. 7.38. *Bauhinia cunninghamii* (right, with pods showing) and *Terminalia platyptera*. *Carissa ovata* and *Aristida pruinosa* below. 32 km west of Georgetown, Qld.

Petalostigma banksii, with the following less common: *Dolichandrone heterophylla*, *Bauhinia cunninghamii*, *Gardenia* spp., *Diospyros humilis*, *Erythrophleum chlorostachys*, *Erythrina vespertilio* and *Brachychiton* spp. Where the mean annual rainfall exceeds c. 1000 mm, *Schlefflera actiniphylla* and *Tristania suaveolens* occur in some stands.

9.7.3. *Bauhinia cunninghamii* Suballiance

This species, a deciduous tall shrub or small tree to c. 8 m tall, is widespread from west to east in the tropics, sometimes occurring as a dominant but also a common understory species in *Eucalyptus* woodlands (Chap. 8). In a stunted form, 2–3 m high, it extends into the arid scrubs and savannahs (Chaps. 18 and 20).

As a dominant or co-dominant in woodland or tall scrub communities the following main associations are recorded:

i. *B. cunninghamii* – *Tristania suaveolens* woodland 5–10 m high in the west Kimberleys (rainfall 500–750 mm), in watercourses. Associated species are *Adansonia gregorii* and *Terminalia platyphylla*.

ii. *B. cunninghamii* – *Atalaya hemiglauca* – *Grevillea striata* (sometimes with *Ventilago*

Fig. 7.37. *Cochlospermum fraseri*, leafless but in flower (photo taken 8-ix-72). *Eucalyptus grandifolia* behind. Grass layer containing *Eriachne* sp. and *Aristida* sp. Termite mounds. 32 km south of Normanton, Qld.

viminalis). These species form open woodlands from west to east in the drier tropics (see also 9.11, below).

iii. *B. cunninghamii – Gyrocarpus americanus* form woodlands to 7 m high on deep, red, lateritic sands, with a grass understory, in the Barkly region.

iv. *B. cunninghamii* in pure stands c. 8 m high in Northern Territory and Queensland on clays near watercourses (often near woodlands of *E. microtheca*). There are no associated woody species. The understory is grass *(Dichanthium – Eulalia)*.

v. *B. cunninghamii – Terminalia* spp. woodlands in Queensland. *T. platyptera* (Fig. 7.38) is the most common species but *T. canescens* is often present. Other species (which vary from site to site) include *Acacia bidwillii, Petalostigma banksii* and *Tristania suaveolens*.

9.8. *Macropteranthes* spp. Alliance

Two species occur in the semi-arid tropics and each may dominate softwood scrubs (Byrnes, 1977). Both species are deciduous.

M. kekwickii (Bulwaddy) has been recorded in Northern Territory (Fig. 7.35), forming dense thickets 3–3.5 m tall on sandy or gravelly lateritic soils under a mean annual rainfall of 400 mm (Perry, 1970). Associated species are rare, the ground being almost bare.

M. leichhardtii (Bonewood) is confined to Queensland (Fig. 7.35). It may form patches of scrub on clays and it is sometimes the most abundant species in scrubs 8 m tall with emergents of *Acacia harpophylla, Brachychiton rupestris* and *Geijera parviflora* (Pedley, 1967).

9.9. *Excoecaria latifolia* Alliance

Gutta Percha

This species forms small patches of open scrub (Fig. 7.39) on fine-textured, cracking grey or brown soils on flats subject to flooding. It occurs mostly adjacent to woodlands of *Eucalyptus microtheca* in the semi-arid tropics and it is sometimes an understory plant in these woodlands (Christian and Stewart, 1954). It has been recorded also from subsaline, coastal mudflats.

There are usually no associated shrubs, but *Acacia farnesiana* and *Muehlenbeckia cunninghamii* may occur locally. The grasses layer contains *Iseilema* spp., *Paspalidium jubiflorum, Chrysopogon fallax* and *Dichanthium fecundum*.

Fig. 7.39. *Excoecaria latifolia*. Grass layer dominated by *Dichanthium fecundum*. c. 50 km south of Karumba, Qld.

9.10. *Geijera parviflora – Flindersia maculosa – Heterodendrum oleifolium* Alliance

9.10.1. General

In parts of semi-arid New South Wales, chiefly on texture-contrast soils developed on ancient river alluviums these three species form a well defined alliance, which alone justifies grouping them together. The species have completely different total areas (Fig. 7.35) and all three occur commonly as understory plants in *Eucalyptus* woodlands (Chap. 13). The numerous associations are best described under three headings.

9.10.2. Associations Containing *Geijera parviflora*

G. parviflora (Wilga) covers a range of climates with a mean annual rainfall of c. 750 mm near the Tropic (summer predominance) to c. 250 mm in the south (winter predominance). In the more southerly areas the species occurs in the understory and the eucalypts are commonly removed leaving the *Geijera*, which is palatable to stock, as an apparent dominant (Fig. 7.40). Consequently only the northern stands, developed in areas with a summer rainfall predominance need be considered here.

In New South Wales *Geijera* forms closed tall scrubs to low woodlands in association with *Flindersia maculosa*. *Heterodendrum* commonly occurs and less common are *Eremophila mitchellii*, *Capparis mitchellii*, *C. lasiantha*, *Eremocitrus glauca*, *Owenia acidula*, *Apophyllum*

Fig. 7.40. *Geijera parviflora*. Grass layer contains *Stipa aristiglumis* and *S. variabilis*. 60 km south of Moree, N.S.W.

anomalum, *Ventilago viminalis* and *Exocarpos aphylla*. Species of *Acacia* are sometimes locally common (*A. pendula* and *A. homalophylla*) and also *Casuarina cristata*. The shrub layer contains *Atriplex* spp., *Rhagodia spinescens* and *Maireana aphylla* (Beadle, 1948).

In Queensland near the Tropic, the woodlands reach a height of 10 m (Tree Wilga) where *Geijera* is associated with *Eremophila mitchellii* and *Terminalia oblongata* (Pedley, 1967). Further south in Queensland the communities contain *Geijera* with *Bauhinia carronii*, *Atalaya hemiglauca* and those taller species listed for the New South Wales stands. In the more northerly stands *Carissa ovata* is the common shrub and the herbaceous layer is composed of grasses (*Ancistrachne*, *Chloris*, *Cymbopogon* and *Paspalidium* spp.).

9.10.3. Associations Containing *Flindersia maculosa*

F. maculosa (Leopard Tree) occurs in semi-arid and arid Queensland and New South Wales. The young plants, produced either from seed or suckers, are unique in that they develop as a mass of thorny, almost leafless branches, from which the main trunk ultimately emerges. The species occurs most commonly either in pure stands (Fig. 7.41) or in association with *Geijera parviflora*, *Eremophila mitchellii*, or *Atalaya hemiglauca*. In pure stands it forms low, open woodlands 5–9 m tall, usually on elevated areas where a sandy surface soil overlies a compact, clayey subsoil which often contains lime. The soils are susceptible to wind erosion and the removal of the surface sandy layer by wind appears to be responsible for the profuse suckering of the roots.

The understory species are mostly grasses responding to summer rains, especially *Chloris*, *Eragrostis* and *Tripogon* spp., or chenopods, especially *Bassia uniflora*, *B. divaricata* and *Salsola kali*.

9.10.4. Associations Containing *Heterodendrum oleifolium* (Rosewood)

This species occurs over most of inland Australia (Fig. 7.35). The species varies in height from c. 10 m in the semi-arid zone to 3–4 m in the arid zone, and the leaves exhibit a similar decrease in size with increasing aridity.

H. oleifolium occurs in association with *Geijera parviflora* and *Flindersia maculosa* (as above) as well as with *Casuarina cristata* and *Acacia aneura* (Chap. 18). Its most significant development occurs in the south-east of its area between the 200 and 350 mm isohyets where monospecific stands of low woodland or tall scrub have developed as a result of root suckering (Beadle, 1948).

9.11. *Atalaya hemiglauca* – *Grevillea striata* – *Ventilago viminalis* Alliance

These three species, singly or in combination form scrub or woodland communities 4–10 m high mainly between the 300 and 600 mm isohyets, the rain falling mainly in the summer. Minor extensions occur into both wetter and drier country, especially *Atalaya*. The species have no specific soil preferences but are found on more clayey soils in wetter areas and sandy soils in the driest areas.

The species have different areas which overlap considerably (Fig. 7.42). *A. hemiglauca*, the most widely distributed, occurs in the wetter northern parts of its range with *Ventilago vim-*

Fig. 7.41. *Flindersia maculosa*, mature trees and juvenile thicket growth. 16 km north of Blackall, Qld.

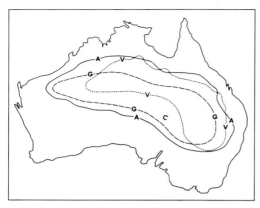

Fig. 7.42. Total areas of *Atalaya hemiglauca* (A), *Ventilago viminalis* (V) and *Grevillea striata* (G).

Fig. 7.44. Pure stand of *Grevillea striata* with grass understory of *Sorghum* and *Eriachne* spp. 80 km north-west of Fitzroy Crossing, W.A.

inalis and subsidiary *Bauhinia cunninghamii*, *Ehretia saligna*, *Erythrina vespertilio*, *Hakea* spp. and *Owenia acidula*. Smaller shrubs are *Carissa lanceolata* and *Capparis umbonata*. The ground stratum consists of grasses (*Triodia*,

Fig. 7.43. *Atalaya hemiglauca* with understory of *Maireana pyramidata*. Scale is a man at right. 40 km west of Thargominda, Qld.

Fig. 7.45. Gallery rainforest containing *Flindersia schottiana, Ficus* sp. and *Glochidion ferdinandii*. Walsh's Pyramid behind with *Eucalyptus polycarpa* and *E. tessellaris*. Grass is *Panicum* maximum. Near Gordonvale, north Qld.

Plectrachne, Aristida spp.) (Christian and Stewart, 1952, Speck and Lazarides, 1964). In the east, Pedley (1967) records *A. hemiglauca* and *Bauhinia carronii* with a grass layer of *Astrebla*, and *Atalaya hemiglauca* is a very common emergent in the *Astrebla* savannahs. Pure stands of *A. hemiglauca* occur in the drier areas, usually with an understory of Chenopodiaceae (Fig. 7.43).

Ventilago viminalis does not usually form pure stands but *Grevillea striata* commonly dominates open scrubs or low woodlands, especially in the drier tropics (Fig. 7.44).

9.12. Other Species

Several other shrubs or small trees occur in the arid zone and are dealt with elsewhere (especially Chap. 5). They are: *Codonocarpus cotinifolius, Duboisia hopwoodii, Eremocitrus glauca, Ficus platypoda, Owenia reticulata, Livistona mariae, Pittosporum phylliraeoides* and *Sarcostemma australe*.

10. Gallery Rainforests

10.1. General

Gallery rainforest is a term used to describe strips of rainforest which edge watercourses. Two types can be identified: i. Distinctive species which edge watercourses within stands of rainforest, and ii. Strips of rainforest which extend along watercourses in areas which are climatically too dry to support rainforests.

Several examples of the first type have already been given. For example, *Dacrydium franklinii* along watercourses in the *Nothofagus cunninghamii* alliance (6.2.2.1, above), *Weinmannia rubifolia* etc. in the *Schizomeria ovata Doryphora* etc. alliance (6.3.4, above), and *Elaeocarpus grandis* in the *Argyrodendron* alliance (6.4, above).

The second type is distinctive chiefly because the rainforest strips contrast sharply both floristically and structurally with the adjacent, usually xeromorphic vegetation (Fig. 7.45). Also the rainforest strip is commonly associated with (sometimes forming an understory to) xeromorphic species characteristic of watercourses in drier areas, viz., certain species of *Eucalyptus* and *Casuarina cunninghamiana*. The communities are described below under three headings.

Fig. 7.47. *Grevillea robusta* (left and smaller tree) and *Casuarina cunninghamiana* fringing a creek. Widgee, Qld.

10.2. Communities between the Clarence R. and the Tropic

Castanospermum australe (Fig. 7.46) dominates gallery rainforests in the most favourable sites (highest moisture, highest nutrients) and

Fig. 7.46. A "wall" of *Castanospermum australe* fringing a creek. Smooth-barked tree is *Eucalyptus tereticornis*. *Callistemon viminalis* in foreground, growing in bed of creek. Widgee, Qld.

forms rainforest stands containing c. 40 species (in N.S.W. – Baur, 1957). Some of the common associates are *Grevillea robusta, Dysoxylum fraseranum, Melia azedarach, Toona australis, Podocarpus elatus* and (Qld only) *Ailanthus malabarica* (Cromer and Pryor, 1942). The communities are sometimes fringed by *Callistemon viminalis* (usually growing on sand in the watercourse) and may be overtopped by odd trees of *Eucalyptus tereticornis*.

In "less favourable localities" Baur (1957) reports that *Syzygium floribundum* dominates the stands. In still more unfavourable sites, usually remote from rainforest stands, *Grevillea robusta*, occurs with or without *Casuarina cunninghamiana* (Fig. 7.47). Associated species are *Acmena smithii* var. *minor, Tristania laurina, Callistemon viminalis, Leptospermum brachyandrum* and *Melaleuca bracteata*.

10.3. Communities North of the Tropic in the East

In wetter areas gallery forest commonly contains most of the species which occur in lowland rainforest, which is probably due to the high rainfall and the lack of distinction between the two habitats; e.g., *Castanospermum australe* is common in both lowland and gallery rainforest. Other species which occur regularly in gallery rainforests are *Tristania laurina, Elaeocarpus grandis*, *Ficus racemosa* and *Nauclea orientalis*. Closest to the water, *Hibiscus tiliaceus*, *Archontophoenix alexandrae* and *Flagellaria indica* are often conspicuous. In some areas the freshwater mangrove, *Barringtonia gracilis* may occur at the water's edge.

In drier areas ribbons of rainforest are commonly dominated by *Nauclea orientalis* and *Tristania grandiflora* and occur in association with *Eucalyptus camaldulensis* and/or *E. raveretiana*, *Casuarina cunninghamiana*, *Melaleuca mimosoides* and *M. saligna*. *Pandanus whitei* may be present in the understory (Perry, 1953).

10.4. Communities in the Drier Tropics

Inland extensions of the Deciduous Vine Forests occur along watercourses which usually support *Eucalyptus camaldulensis* and *Casuarina cunninghamiana*, as well as a rainforest assemblage (Perry and Christian, 1954; Speck, 1961). The tallest rainforest trees, 17–23 m high, are *Nauclea orientalis*, *Terminalia platyphylla* and *T. erythrocarpa*. Other trees are *Abarema monilifera*, *Barringtonia acutangula* (Fig. 22.1), *Brachychiton* spp., *Eugenia suborbicularis*, *Metrosideros eucalyptoides*, *Tristania grandiflora* and *T. lactiflua*. The understory may contain *Livistona* sp., *Pandanus aquaticus* and *Bambusa arnhemica*.

8. Eucalyptus Communities of the Tropics

1. Introduction

Most of tropical Australia, which is characterized by summer rainfall and winter drought, supports *Eucalyptus* communities. Other communities occupy relatively small areas (Chap. 6).

The *Eucalyptus* forests and woodlands are divided into two segments by the Gulf of Carpentaria and the Carpentaria Plains, an area of clay soil, supporting grasslands, which extends southward from the Gulf.

About 80 species of *Eucalyptus* occur in the tropics and are listed in Table 8.1. The assem-

Table 8.1. Distribution of *Eucalyptus* species occurring in the tropics, excluding arid zone species. Species are arranged alphabetically within the groups, except Groups 8, 9 and 10 where the species are placed in the approximate order of appearance in a N.-S. transect, with inland species on the left, coastal species on the right. Arrows indicate an extensive latitudinal range and for species in Group 1 the occurrence of the species on islands north of Australia.

W.A.		N.T.	Qld
GROUP 1. West and east:			
alba ↑, confertiflora ↑, dichromophloia ↓, grandifolia, miniata, nesophila, papuana ↕, phoenicea, polycarpa ↕, tectifica, tetrodonta			
GROUP 2. Wetter W.A. and N.T.: apodophylla, bigalerita, bleeseri, clavigera, ferruginea, fitzgeraldii, foelschiana, herbertiana, jensenii, latifolia, mooreana, oligantha, patellaris, ptychocarpa		GROUP 5. N.T. only: porrecta, jacobsoniana, umbrawarrensis	GROUP 8. Confined to tropics: leptophleba, staegeriana, cullenii, torelliana, abergiana, brassii, microneura, gilbertensis, shirleyi, howittiana, brownii, whitei
GROUP 3. W.A. to Qld, mainly in drier areas: argillacea, brevifolia ↑, pruinosa, setosa, terminalis ↓			GROUP 9. Mainly tropical, with extensions southward: tessellaris, crebra, drepanophylla, melanophloia, phaeotricha, trachyphloia, exserta, citriodora, cambageana, raveretiana, similis, intermedia, peltata
GROUP 4. Drier W.A. collina, cupularis, houseana, lirata, perfoliata, zygophylla	aspera, brachyandra, cliftoniana	GROUP 6. Drier Qld with outliers in N.T.: normantonensis, thozetiana	
		GROUP 7. West to east in watercourses: camaldulensis ↕, microtheca ↕	GROUP 10. Mainly south of the Tropic: populnea, tereticornis, orgadophila, pellita, grandis, acmenoides, tenuipes, fibrosa
		ARID ZONE	

blage of species is unique in the continent in that all 7 subgenera are represented. Also, the assemblage contains a high proportion of subgenera *Blakella*, *Corymbia* (Bloodwoods) and *Eudesmia*, and many of these species are community dominants. Eleven species (Table 8.1, Group 1) occur in both the west and the east on either side of the Gulf of Carpentaria; four of these species have migrated outside Australia either to Timor (*E. alba*) or New Guinea. The Carpentaria Plains separate the western species from the eastern ones, except in drier areas (Table 8.1, Group 3). However, in these drier areas certain species are confined to the west (Group 4) and others to the east (Group 6).

In the east, many species are confined to the tropics (Group 8). However, many extend southward, some being more abundant in the north (Group 9), whereas others have their maximum development south of the Tropic (Group 10). Most of the species belonging to these three groups are dealt with in later chapters.

A few of the tropical species are present in the arid zone where they exist usually in a stunted form, especially *E. brevifolia*, or as relicts, e.g., *E. terminalis*. Conversely, the two watercourse species, *E. camaldulensis* and *E. microtheca* which are characteristically semi-arid or arid zone species, extend into the wetter tropics.

In this chapter species belonging to Groups 1–5 in Table 8.1 are dealt with. They constitute 10 alliances (Table 8.2), found mainly west of the Carpentaria Plains, or on Cape York Peninsula.

The interrelationships of the alliances are not readily explained in terms of single variables.

The following broad generalizations are made:

i. The *E. tetrodonta* – *E. miniata* – *E. polycarpa* alliance occupies the wettest areas in the north, occurring on deep, well drained, usually sandy or gravelly soils.

ii. The *E. tectifica* – *E. confertiflora* – *E. grandifolia* alliance occurs in the same general area as i., but is confined to soils of finer texture which are subject to minor flooding.

iii. The *E. ptychocarpa* alliance is found mainly along watercourses in wetter climates.

iv. The *E. alba* alliance occurs in fragments along river flats or near swamps mainly in the same areas as i. and ii.

v. The *E. phoenicea* – *E. ferruginea* alliance occurs mostly within the *E. tetrodonta* etc. alliance, being confined to shallow, well drained soils on rocky outcrops.

vi. The *E. papuana* alliance is widely distributed in the tropics, mostly on river flats.

vii. The *E. argillacea* – *E. terminalis* alliance replaces the *E. tectifica* etc. alliance on fine textured soils in the drier south.

viii. The *E. dichromophloia* alliance occurs on stony areas or on deep sands in the same general areas as vii., mainly in the west.

ix. The *E. pruinosa* alliance on deep soils, and the *E. brevifolia* alliance on rocky areas occur mainly in the south, the latter adjoining the arid zone.

This sequence from north to south, determined partly by mean annual rainfall and partly by soil texture which controls drainage, is complicated by the northern extension of the "hilltop alliances" (*E. dichromophloia* and *E. brevifolia*) into the higher rainfall zone. Also the rivers and the wetter habitats of the river flats carry the "watercourse alliances" (*E. papuana*, and the *E. camaldulensis* und *E. microtheca* alliances, the last two best developed in the arid zone) through all the alliances from south to north.

The descriptions of the communities are based on: Gardner (1923); Blake (1953); Brass (1953); Christian and Stewart (1953); Perry and Christian (1954); Perry (1956; 1970); Specht (1958); Speck (1960, 1961, 1964); Perry and Lazarides (1964); Stocker (1968, 1972); Story (1969, 1970, 1976); Aldrick and Robinson (1970); Pedley and Isbell (1971); Hall et al. (1970); Hall and Brooker (1972–3); Specht et al. (1977); Byrnes et al. (1977); Lavarack and Stanton (1977).

2. Descriptions of the Alliances

2.1. *Eucalptus tetrodonta* – *E. miniata* – *E. polycarpa* (sens. lat.) Alliance

Darwin Stringybark – Woollybutt – Long-fruited Bloodwood

This is the most northerly of the *Eucalyptus* alliances, occurring in the climatically wettest areas between the 700 and 1500 mm isohyets, mostly on flat to undulating terrain with deep, well drained soils. The three characteristic species, all evergreen, are not consistently present throughout the alliance. *E. tetrodonta*, probably the most abundant of the three species, occurs disjunctly from west to east in the wettest areas. *E. miniata* is absent from the northern portion of Cape York Peninsula (Fig. 8.1), where it is replaced by *E.* aff. *polycarpa* (= *E. nesophila?*). *E. polycarpa* is widely distributed from west to east (Fig. 8.1) forming associations with many species, some with restricted areas, so that associations involving *E. polycarpa* are different from west to east. These variations have necessitated the identification of three suballiances.

2.1.1. *Eucalyptus tetrodonta* – *E. miniata* Suballiance

Darwin Stringybark – Woollybutt

The suballiance occurs mainly west of the Carpentaria Plains. The communities occur on flat to undulating country and are either tall, open to almost closed forests to 30 m high in the wettest areas (Fig. 8.2), or, in the drier areas, mainly woodlands 10–25 m high (Fig. 8.3). The communities occupy deep, well drained sandy or gravelly soils, mainly Red and Yellow Earths, less commonly Lithosols. The soils are derived from alluviums, fossil laterites, quartzites, sandstones or granites. On lithosols, enclosures of the *E. phoenicea* or the *E. dichromophloia* alliances are common, especially in the rugged terrain of the North Kimberleys and Arnhem Land.

The two characteristic species usually occur in equal proportions, though either may be dominant or occur in pure stands. Stocker (1968) states that *E. tetrodonta* has a long taproot and forms pure stands on deep, sandy Red Earths, whereas *E. miniata*, which is shallow-rooted, forms pure stands on extremely sandy soils. *Cal-*

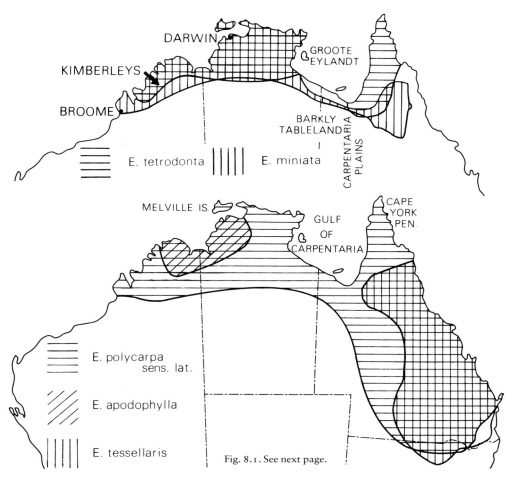

Fig. 8.1. See next page.

litris intratropica, now mostly removed for timber, once formed local associations on deep sand, either in pure stands or in association with *E. miniata*. Of minor importance as associates are *E. polycarpa* in wetter areas, and *E. bleeseri* usually on elevations.

The understory species vary with mean annual rainfall. In wetter areas the tall forests often have a lower tree layer, either of eucalypts from adjoining alliances, e.g., *E. confertiflora* and *E. clavigera*, or of species confined to the suballiance viz., *E. nesophila* (Melville Is.) and *E. porrecta* (Darwin area). A form of *E. setosa*, as a tall shrub, occurs in the Darwin area. Characteristic of the wetter stands is a small tree or tall shrub layer of broad-leaved, mesomorphic species many of which occur in the rainforests, small stands of which are sometimes enclosed by the *Eucalyptus* forests. The tall mesomorphic species include most of those listed for the semi-evergreen vine forests (Chap. 7), and, rarely, the palm *Livistona eastonii* (Hnatiuk, 1977). A few lianas occur, notably *Tinospora smilacina* and *Cayratia trifolia*. One epiphyte, *Cymbidium canaliculatum*, is recorded. *Cycas media* and *Livistona benthamii* are often conspicuous, possibly because they have survived repeated burnings which removed the mesomorphic tall shrubs. The herbaceous layer is dominated by grasses, often 2 m high, which have become increasingly abundant following burning. The main species are *Heteropogon triticeus*, *Sorghum plumosum*, *Coelorhachis rottboellioides*, *Chrysopogon latifolius*, *Themeda australis* and members of the genera *Alloteropsis*, *Aristida*, *Cymbopogon*, *Eragrostis*, *Eriachne*, *Panicum*, *Setaria* and *Thaumastochloa*. Forbs include species of *Borreria*, *Didiscus*, *Fimbristylis*, *Glossogyne*, *Gomphrena*, *Goodenia*, *Haemodorum*, *Pachynema* and *Xyris*.

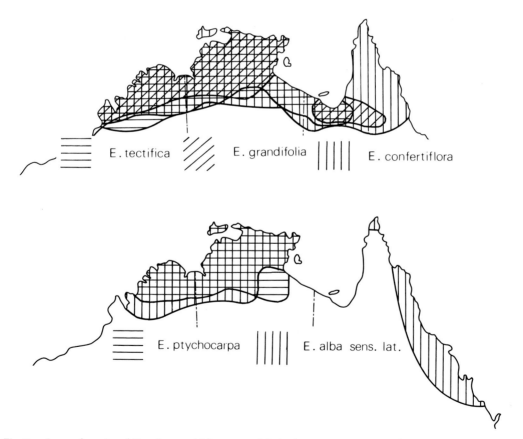

Fig. 8.1. Areas of species of *Eucalyptus* which occur mainly in the wetter tropics. Based on Hall et al. (1970) and Blake (1953).

Fig. 8.2. *Eucalyptus miniata* (with smooth bark on upper trunk) and *E. tetrodonta* in the background. The broad-leaved tree on the left is *Buchanania obovata*. *Cycas media* and *Livistona humilis* present in the herbaceous layer, which is dominated by *Heteropogon triticeus* and *Themeda australis*. Near Darwin, N.T.

In drier sites, including the climatically drier southern segment and ridges in the wetter area, the communities are woodlands and mesomorphic shrubs are rare or absent. Xeromorphic shrubs are sometimes present, especially on ridges in the north where they may form a shrub layer c. 2 m high. They are: *Bossiaea phylloclada, Calytrix microphylla, Jacksonia argentea, J. dilatata, J. thesioides, Verticordia cunninghamii* and *Acacia* spp. The grass layer is usually discontinuous; *Chrysopogon fallax, Sorghum plumosum* and *Heteropogon triticeus* are most abundant on ridges in the north, while the hummock grasses, *Plectrachne pungens* and *Triodia pungens* become increasingly prominent towards the south where they dominate the grass layer (Fig. 8.3).

2.1.2. *Eucalyptus tetrodonta* – *E. polycarpa* Suballiance

Darwin Stringybark – Long-fruited Bloodwood

These species dominate large areas on Cape York Peninsula (Pedley and Isbell, 1971), occurring on flat to undulating country between the

Fig. 8.3. Dry facies of *Eucalyptus tetrodonta* – *E. miniata* association with *E. miniata* locally dominant. Grass layer dominated by *Plectrachne pungens*. 16 km west of Croydon, Qld.

1000 and 1500 mm isohyets. *E. polycarpa* extends far to the south (Fig. 8.1). *E. miniata*, absent from the northern part of the peninsula, occurs south of c. lat 17° S. (Fig. 8.1), where it forms an association with *E. tetrodonta* similar to the community in the drier western portion of the alliance. The soils are Red or Yellow Earths, with smaller areas of Yellow Podsolics. Ironstone or bauxite gravel usually occurs in the subsoil and in some areas the profile overlies massive laterite at a depth of 1–2 m. The communities vary considerably in structure with no change in composition of the dominants; over most of the area they are open forest 12–25 m tall, but towards the drier limits of the suballiance the species constitute woodlands mostly 10–15 m tall.

In the forests, the two tree species may occur in equal proportions, but either is locally dominant. A lower layer occurs and contains combinations of *Alphitonia excelsa, Erythrophleum chlorostachys, Grevillea pteridifolia, Pandanus* sp., *Parani nonda, Planchonia careya* and *Persoonia falcata,* as well as *Acacia rothii* and *A. aulacocarpa* which may be conspicuous or locally dominant in the lower tree or tall shrub layer. *Cycas media* and *Xanthorrhoea johnstonii* sometimes occur locally. The herbaceous layer consists of grasses belonging mainly to *Alloteropsis, Aristida, Chrysopogon, Eriachne, Heteropogon, Panicum, Schizachyrium, Thaumastochloa* and *Themeda*.

In the woodlands, *Melaleuca stenostachya, M. viridflora* and *Grevillea glauca* are the most abundant shrubs, with minor occurrences of the species from the open forests. In the lowest of the woodland, which occur on undulating country, *Petalostigma banksii* becomes a conspicuous shrub locally, in addition to *Melaleuca* spp.

2.1.3. *Eucalyptus polycarpa* – *E. apodophylla* Suballiance
Long-fruited Bloodwood – Whitebark

West of Carpentaria Plains, *E. polycarpa* is usually restricted to flats, commonly along stream lines on deep sandy soils. It extends southward into drier country (mean annual rainfall c. 500 mm) along watercourses. It is commonly associated with *E. apodophylla* (Figs. 8.1, 8.4) and less often with *E. oligantha*, forming woodland 10–15 m high. Mesomorphic shrubs or trees occur in the stands, especially *Erythrophleum chorostachys* or, in drier areas, *Melaleuca viridiflora*. The grass layer is well developed, consisting of various aggregates of *Alloteropsis semialata*, *Aristida hygrometrica*, *Chrysopogon* spp., *Plectrachne pungens*, *Sehima nervosa*, *Sorghum plumosum* and *Themeda australis*.

2.1.4. *Eucalyptus polycarpa* – *E. tessellaris* Suballiance
Long-fruited Bloodwood – Carbeen

In the eastern tropics, *E. polycarpa* is often associated with *E. tessellaris* (Fig. 8.1). The species form open woodlands 15–25 m tall,

Fig. 8.4. *Eucalyptus polycarpa* woodland with a small admixture of *E. apodophyllum* (white trunks, on right). Scattered small trees of *Erythrophleum chlorostachys*, *Bauhinia cunninghamii* and *Terminalia canescens*. Grass layer dominated by *Themeda australis*. South of Katherine, N.T.

Fig. 8.5. Pure stand of *Eucalyptus tessellaris*. Herbaceous layer too fragmented for identification. 8 km south of Rolleston, Qld.

mainly between the 1100 and 1350 mm isohyets on Red or Yellow Earths or Yellow Podsolics, on lowland slopes or plateaux. The alliance is very fragmented and forms mosaics with several others, including small patches of rainforest, forests of *E. crebra* or *E. tereticornis* and woodlands of *E. leptophleba, E. confertiflora, E. alba, E. melanophloia* and, in the south, *E. populnea*. The two characteristic species extend south of the Tropic approximately to the limit of the summer rainfall zone, and in this southern extension the mean annual rainfall is as low as 500 mm.

The two species are often in association but each may form pure stands (Fig. 8.5). A species of restricted distribution, *E. torelliana*, is sometimes an associate around patches of rainforest. The understories contain scattered tall shrubs or small trees, especially *Erythrophleum chlorostachys, Planchonia careya, Pandanus* sp., and less commonly *Acacia flavescens, A. leptocarpa, Melaleuca acerosa, M. viridiflora* and *Petalostigma pubescens*. The grass layer contains species of *Alloteropsis, Bothriochloa, Chrysopogon, Dichanthium, Eulalia, Heteropogon, Imperata, Panicum, Sorghum* and *Themeda*.

2.2. *Eucalyptus tectifica* – *E. confertiflora* – *E. grandifolia* Alliance

Northern Box –
Broad-leaved Carbeen –
Large-leaved Cabbage Gum

This is a complex alliance consisting of the characteristic species (which occur in both the west and the east – Fig. 8.1), together with four others – *E clavigera*, *E. foelschiana*, *E. latifolia* and *E. patellaris* which are confined to the west. The species are deciduous, except the last one. They occur in various combinations mainly in the wetter parts of Northern Territory and Western Australia in the same general area as the preceding alliance. The *E. tectifica* etc. alliance occurs on soils containing some to much clay in the profile, or on areas subject to minor flooding. It is common on flat country but extends on to hills where the underlying rock gives rise to clayey soils (Fig. 8.6).

The alliance is described below under three suballiances which are by no means clear-cut. In addition to mixtures of the seven species named above, ecotonal associations with species from other alliances are common (Table 8.2). The six deciduous species are sometimes found as a discontinuous lower tree layer in the *E. tetrodonta* – *E. miniata* suballiance.

2.2.1. *Eucalyptus tectifica* – *E. confertiflora* – *E. foelschiana* Suballiance

Northern Box –
Broad-leaved Carbeen –
Fan-leaved Bloodwood

The suballiance is found mainly west of the Carpentaria Plains. Only one association is recorded for the east, this being dominated by *E. confertiflora*. The communities occur between the 630–1400 mm isohyets and are mostly savannah woodlands 10–15 m high. The three species are commonly found together, but one is often lacking and pure stands are common. The

Fig. 8.6. Zonation from a flat (subject to some flooding) to a hill. On flat: *Tristania lactiflua* with a few *Livistona humilis* below. On foothills: *E. confertiflora* (still with leaves). *E. tectifica* (leafless) on hill at rear. Termite mound at left. c. 120 km south of Darwin, N.T.

Table 8.2. Associations of *Eucalyptus* forests and woodlands confined to the wetter tropics.

Alliance or Suballiance	Main associations	Species forming minor associations with the dominants	Main species forming ecotonal associations
E. tetrodonta – *E. miniata*	*E. tetrodonta* – *E. miniata*; *E. tetrodonta*; *E. miniata*; *Callitris intratropica*	*E. setosa*; *E. porrecta*; *E. nesophila*; *E. bleeseri*; *E. umbrawarrensis*; *E. clavigera*; *E. jacobsoniana*	*E. polycarpa*; *E. confertiflora*; *E. tectifica*
E. tetrodonta – *E. polycarpa*	*E. tetrodonta*; *E. polycarpa*		*E. alba*; *E. cullenii*; *E. leptophleba*
E. polycarpa – *E. apodophylla*	*E. polycarpa*; *E. polycarpa* – *E. apodophylla*; *E. polycarpa* – *E. oligantha*		*E. confertiflora*; *E. tectifica*
E. polycarpa – *E. tessellaris*	*E. polycarpa* – *E. tessellaris*; *E. tessellaris*	*E. torelliana*	*E. crebra*; *E. alba*; *E. leptophleba*; *E. tereticornis*
E. tectifica – *E. confertiflora* – *E. foelschiana*	*E. tectifica*; *E. tectifica* – *E. confertiflora*; *E. foelschiana*	*E. patellaris*; *E. jensenii*; *E. latifolia*; *E. clavigera*	*E. tetrodonta*; *E. polycarpa*; *E. terminalis*; *E. papuana*
E. grandifolia	*E. grandifolia*		*E. tectifica*; *E. foelschiana*; *E. confertiflora*; *E. apodophylla*; *E. pruinosa*
E. latifolia	*E. latifolia*		*E. polycarpa*; *E. pruinosa*
E. ptychocarpa	*E. ptychocarpa*	*E. houseana*; *E. clavigera*	*E. confertiflora*; *E. camaldulensis*
E. alba	*E. alba*		*E. apodophylla*; *E. papuana*; *E. polycarpa*; *E. tessellaris*
E. phoenicea – *E. ferruginea*	*E. phoenicea* – *E. ferruginea*	*E. setosa*	*E. dichromophloia*; *E. miniata*; *E. collina*; *E. brevifolia*
Eucalyptus spp. – *Calytrix* complex	*E. phoenicea*	*E. bleeseri*	*E. miniata*; *E. dichromophloia*
E. papuana	*E. papuana*		*E. polycarpa*; *E. alba*; *E. camaldulensis*; *E. confertiflora*; *E. tessellaris*
E. argillacea – *E. terminalis*	*E. argillacea*; *E. argillacea* – *E. terminalis*; *E. terminalis*		*E. tectifica*; *E. confertiflora*; *E. microtheca*
E. dichromophloia	*E. dichromophloia*	*E. cliftoniana*; *E. fitzgeraldii*;	*E. cullenii*; *E. crebra* (East);

Continued next page.

Table 8.2 continued.

Alliance or Suballiance	Main associations	Species forming minor associations with the dominants	Main species forming ecotonal associations
		E. jacobsoniana; *E. lirata;* *E. mooreana;* *E. brachyandra* (all West)	*E. miniata;* *E. ferruginea;* *E. phoenicea* (West)
E. dichromophloia – *E. herbertiana –* *E. collina*	as alliance		*E. ferruginea;* *E. terminalis*
E. dichromophloia – *E. zygophylla –* *Acacia* spp.	as alliance	*E. perfoliata*	*Acacia pachycarpa*
E. aspera	*E. aspera*		*E. dichromophloia;* *E. pruinosa*
E. pruinosa	*E. pruinosa*		*E. brevifolia;* *E. terminalis;* *E. grandifolia;* *E. confertiflora;* *E. polycarpa*
E. brevifolia ssp. *brevifolia*	*E. brevifolia* ssp. *brevifolia*	*E. cupularis*	*E. dichromophlia;* *E. pruinosa;* *E. phoenicea;* *E. brachyandra*
E. brevifolia ssp. *confluens*	*E. brevifolia* ssp. *confluens*		*E. dichromophloia*

Fig. 8.7. *Eucalyptus confertiflora* with an admixture of *E. foelschiana*. "Orchard country". Recently burned, note regenerating shoots. Grass layer of *Themeda australis*. c. 100 km south of Darwin, N.T.

woodlands usually contain broad-leaved mesomorphic tall shrubs or trees, especially in the higher rainfall areas. *Erythrophleum chlorostachys* is constant but several others occur, e.g., *Buchanania obovata*, *Cochlospermum fraseri*, *Croton arnhemicus*, *Eugenia bleeseri*, *Gardenia megasperma* and *Xanthostemon paradoxus*, with species of *Acacia* and *Grevillea*. The grass layer, which is usually dense, contains species of *Eriachne*, *Heteropogon*, *Plectrachne*, *Sorghum* and *Themeda australis*, with annual grasses and Cyperaceae. Stunting of this assemblage in the climatically wetter areas is common, producing a type of low woodland referred to as "orchard" country by Christian and Stewart (1952), usually dominated by *E. confertiflora* and/or *E. foelschiana* (Fig. 8.7). The communities appear to be restricted to areas with impeded drainage or shallow soils.

The main associations are:

i. *E. tectifica* (Fig. 8.8) often becomes the dominant on rises, where the clayey soils (Red Earths and Yellow Podsolics) are derived from basalts, andesites, granites or metamorphic rocks.

ii. *E. foelschiana* in pure stands (Fig. 8.9) occurs mainly in drier areas (750–850 mm mean annual rainfall) on Red Earths derived from limestone or igneous rocks.

iii. Of local importance on the northern part of the Arnhem Peninsula and the Kimberleys are pure stands of *E. clavigera* (Fig. 8.10).

iv. *E. patellaris* forms pure stands in the drier south (Fig. 8.11) and it sometimes forms an association with *E. terminalis*. The associated

Fig. 8.8. Pure stand of *Eucalyptus tectifica* with a new growth of leaves. Ground flora too fragmented for identification. 13 km west of Mary R., N.T.

Fig. 8.9. *Eucalyptus foelschiana* (foreground) with *E. tectifica* and a few *Erythrophleum chlorostachys* behind. Grass layer is a mixture of *Heteropogon triticeus* and *Themeda australis*. c. 56 km west of Katherine, N.T.

shrub and small tree species, except *Erythrophleum*, are those of the drier climatic zone.

v. *E. jensenii*, the only ironbark west of the Carpentaria Plains, has been recorded in pure stands (Perry, 1970) or in association with *E. tectifica* (Speck, 1960). Hall and Brooker (1973) indicate also an occurrence with *E. umbrawarrensis*. It forms woodlands c. 7 m high with an understory of *Plectrachne pungens* or *Triodia pungens*. It occurs also on Groote Eylandt at the edge of a *Melaleuca* swamp (Specht, 1958).

East of the Carpentaria Plains the suballiance is poorly represented, only three of the species being recorded for this area: *E. confertiflora*, *E. grandifolia* and *E. tectifica*. Only two communities have been described:

i. An association of *E. confertiflora* and *E. papuana* on Cape York Peninsula; the community is an open, grassy woodland 6–10 m tall, with a grass layer dominated by *Heteropogon contortus* (Pedley and Isbell, 1971).

ii. An association containing *E. confertiflora* and *E. cullenii* at Weipa (Specht et al., 1977).

2.2.2. *Eucalyptus grandifolia* Suballiance

Large-leaved Cabbage Gum

This species occurs in open woodland formation 10–12 m high, usually in pure stands (Fig. 8.12). It is found between the 600 and 1200 mm isohyets on flats or gentle slopes or near watercourses, on Red Earths, Yellow Podsolics or deep sands.

Associated shrub species are few and the plants scattered. The grass layer is well developed, consisting of *Aristida hygrometrica*, *Heteropogon contortus*, *H. triticeus*, *Plectrachne pungens* and annual sorghums.

2.2.3. *Eucalyptus latifolia* Suballiance
Round-leaved Bloodwood

This smooth-barked tree, reaching a height of 12–15 m, dominates open woodlands resembling the *E. grandifolia* woodlands. It is found mostly in pure stands in slight depressions subject to flooding, and is commonly fringed by *Melaleuca* scrubs. It occurs also in association with other species of the alliance and with *E. polycarpa* and *E. pruinosa*. The grass layer is similar to that in the preceding alliance.

2.3. *Eucalyptus ptychocarpa* Alliance
Red or Swamp Bloodwood

E. ptychocarpa, a tree 8–12 m high with conspicuous pink-red buds and flowers, occurs in the wetter tropics in the north-west (Fig. 8.1) and is confined to watercourses, edges of swamps or areas receiving additional water from runoff. The soils are usually sandy or gravelly and are waterlogged during the wet season. The species rarely forms extensive stands, but occurs in elongated belts mainly along watercourses which traverse the tropical *Eucalyptus* forests. Ecotonal associations with adjacent communities are common, especially

Fig. 8.10. Pure stand of *Eucalyptus clavigera*. Lower growth mainly shoots from bases of saplings killed by fire. 16 km west of Mary R., N.T.

Fig. 8.11. *Eucalyptus patellaris* and sapling growth. Tall shrubs are *Bauhinia cunninghamii*, *Atalaya hemiglauca* and *Terminalia canescens*. Grass layer dominated by *Chrysopogon fallax*. c. 100 km north of Dunmarra, N.T.

with *E. confertiflora*, *E. clavigera*, and *E. camaldulensis*. In the west Kimberleys *E. houseana*, a smooth-barked tree to c. 20 m tall, is sometimes associated with it. Scattered rainforest species or species of *Pandanus* or the palm *Livistona benthamii* may occur in the stands.

A woodland community dominated by *E. ptychocarpa* has been described by Perry and Christian (1954) for the Barkly Region, in which *Melaleuca leucadendron*, *Grevillea pteridifolia* and *Erythrophleum chlorostachys* occur in a lower tree layer. The stand contains no lower shrub layer except scattered plants of the herbaceous liana *Passiflora foetida*. The grass layer, to 2 m high, is sparse and dominated by *Coelorachis rottboellioides*. Species of *Drosera* and *Stylidium* occur in a discontinuous lower herbaceous layer on the moist soil surface.

2.4. *Eucalyptus alba* (sens. lat.) Alliance

Poplar Gum

E. alba is used here to include all the poplar-leaved eucalypts described as varieties of *E. alba*, *E. bigalerita* and *E. pastoralis*. Though possibly taxonomically distinct, insufficient has been recorded on their ecology to sort the forms into communities. The species occurs disjunctly in northern Australia (Fig. 8.1) and extends to New Guinea and westwards on some of the northern islands to Timor. In Australia it occurs between the 750 and 1300 mm isohyets, usually on flats along watercourses, on other flat areas receiving runoff water, around swamps and less commonly on elevated areas of clayey soil in the wetter part of its range.

Fig. 8.12. *Eucalyptus grandifolia*. Shrubs are *Grevillea heliosperma* and *G. pteridifolia*. A leafless *Adansonia gregorii* in centre. Grass layer dominated by *Plectrachne pungens*. *Eucalyptus pruinosa* on ridge in background (left). c. 50 km east of Wyndham, W.A.

E. alba reaches a height of c. 20 m and forms woodlands in both its western and eastern segments. Associated species differ in the two segments which are described separately.

i. In the west, *E. alba* is uncommon and is usually associated with one or more eucalypts typical of flats and waterways, especially *E. apodophylla*, *E. papuana* and *E. polycarpa*, and less frequently *E. clavigera* and *E. confertiflora*. It occurs rarely on elevated clayey areas when it is sometimes associated with *E. tectifica*. Shrub species are rare or lacking (possibly the result of fire) but a dense grass sward develops after summer rains, dominated by species of *Chrysopogon*, *Heteropogon* and *Sorghum* spp.

ii. In the east, *E. alba* occurs in pure woodland stands c. 13 m tall (Fig. 8.13) on sandy Solonized Soils. No shrubs are recorded. The grass layer is either tall, dominated by *Heteropogon contortus* and *Chrysopogon fallax*, or short and dominated by *Chloris* spp. and the sedge *Fimbrystylis dichotoma*. Usually, *E. alba* is a co-dominant or subsidiary species in woodlands with *E. polycarpa*, *E. tessellaris* and *E. leptophleba*. Most of these woodlands contain tree or tall shrubs including *Acacia aulacocarpa*, *A. leptocarpa*, *Alphitonia excelsa*, *Erythrophleum chlorostachys*, *Melaleuca viridiflora*, *M. nervosa*, *Pandanus* spp., *Planchonia careya*, *Tristania suaveolens*. The grass layer is well developed and contains *Heteropogon contortus*, *Imperata cylindrica*, *Sorghum plumosum*, *Schizachyrium* spp., and *Themeda australis*.

2.5. *Eucalyptus phoenicea* Alliance

Scarlet Gum

E. phoenicea occurs on either side of the Gulf of Carpentaria north of c. Lat. 16° S. It is confined to rocky outcrops, which are usually silice-

Fig. 8.13. Pure stand of *Eucalyptus alba*. Understory of grasses (too fragmented for identification). 64 km south of Sarina, Qld.

ous and commonly sandstones; it is less common on deep sands. It occurs between the 625 and c. 1200 mm isohyets but in the higher rainfall areas the effective rainfall is greatly reduced as a result of rapid runoff.

In the east, the species is relatively uncommon and is a subsidiary species to *E. dichromophloia* on rocky outcrops (Pedley and Isbell, 1971). In the west, *E. phoenicea* is more abundant and forms a fairly well defined group of communities (Speck, 1960; Speck and Lazarides, 1964) described below as a suballiance. *E. phoenicea* occurs also on the Arnhem Peninsula in rugged rocky country (Fig. 8.14), the vegetation of which is difficult to classify and has been regarded as an "edaphic complex" by Specht (1958).

2.5.1. *Eucalyptus phoenicea* – *E. ferruginea* Suballiance
Scarlet Gum – Rusty Bloodwood

The communities dominated by these species occur in the Kimberleys between the 625 and 700 mm isohyets and are woodlands 10–12 m high in wetter areas, decreasing to c. 5 m high in drier areas, where *E. ferruginea* tends to dominate (Fig. 8.15). The soils are deep Yellow Sands or Lithosols derived from sandstones or quartzites.

E. phoenicea forms pure stands in the wetter areas with *E. ferruginea* as a subsidiary species. Associated tree and tall shrub species include: *Brachychiton* spp., *Callitris intratropica*, *Gardenia* spp., *Grevillea agrifolia*, *G. cunninghamii*, *G. pteridifolia*, *Planchonia careya*, *Petalostigma quadriloculare* and *Ventilago viminalis*. In drier areas *E. ferruginea* becomes more prominent and finally dominant with the following species in the shrub understory: *Acacia sericata*, *A. tumida* with subsidiary *Brachychiton*, *Gardenia* and *Grevillea*, as above. In all areas *Plectrachne pungens* dominates the grass layer, with subsidiary *Aristida hygrometrica*, *Eriachne obtusa* and annual sorghums. Ecotonal associations are common (see Table 8.2).

2.5.2. *Eucalyptus* spp. – *Calytrix* – *Triodia* Complex

In the northern part of the Arnhem Peninsula, where rainfall is sufficient to support forests of *E. tetrodonta* and *E. miniata*, tracts of rugged hills, composed mainly of quartzites and hardened sandstones occur (Fig. 8.14). Extensive areas of bare rock lie at the surface and deep gorges dissect the plateau. Plant communities are confined mainly to the lower slopes and valley floors, while the rocky crags support vegetation only where pockets of sand have collected. Specht (1958) described the communities as a complex. In the rockiest situations the hummock grass, *Triodia microstachya*, dominates a xeromorphic shrub savannah with the following shrubs as emergents: *Acacia humifusa*, *Brachysema bossiaeoides*, *Calytrix laricina*, *Grevillea angulata*, *G. drysandra*, *G. pungens*, *Hibbertia tomentosa*, *Sarcostemma australe* and *Solanum quadriloculatum*. Where deeper sands have accumulated, additional species are found and form a scrub, viz., *Calytrix microphylla* (common) and *Acacia mountfordae*, *Hibiscus zonatus*, *Mackinlaya confusa*, *Petalostigma quadriloculare*, *Pityrodia jamesii* and *Platysace arnhemica*. On deeper sands or in rock crevices *Eucalyptus papuana* and many of the mesomorphic shrub species normal to the *E. tetrodonta* – *E. miniata* forests become established, e.g., *Gardenia*, *Livistona*, *Owenia*, *Xanthostemon* spp.

On gentle slopes on the tops of hills woodlands 8–12 m high of *E. phoenicea* may occur, with an occasional *E. bleeseri*, *E. dichromophloia* or *E. miniata*. On the deepest soils tall open forests of *E. tetrodonta* with *Callitris intratropica* develop.

Fig. 8.14. Rugged sandstone country, Arnhem Land Plateau, N.T. Photo: David Gibson.

Fig. 8.15. *Eucalyptus ferruginea* and *E. setosa* (behind). Shrub layer of *Acacia lysiphloia, A. orthocarpa* and *Grevillea wickhamii*. Grasslayer dominated by *Triodia pungens*. c. 250 km north of Tennant Creek, N.T.

2.6. *Eucalyptus papuana* Alliance
Ghost Gum

E. papuana is widely distributed in northern Australia around and north of the Tropic. It extends over a wide climatic range, from c. the 1500 mm isohyet in the north to the 250 mm isohyet in Central Australia. Over most of its range it receives no frost but in Central Australia subzero temperatures occur in mid-winter with the lowest minima approaching − 7°C. Over most of its range *E. papuana* occupies river flats or areas receiving some flood-water without being waterlogged, and in such situations the communities have been referred to as "frontage woodland". Also, it forms distinctive woodlands on some coastal dunes in the north and east. In drier areas it is found both on broad river flats and on the adjacent hills.

i. Frontage Woodlands. These occur mainly in Northern Territory and Western Australia between the 750 and 1500 mm isohyets on flats along watercourses on deep alluvial sands. Pure stands of *E. papuana* are common (Fig. 8.16) and often they adjoin the *E. camaldulensis* and *E. microtheca* woodlands which line the watercourses. Ecotonal associations with tree species from adjacent communities are common (Table 8.2). *Adansonia gregorii* is sometimes a co-dominant in the Kimberleys. A scattered lower tree/tall shrub layer is sometimes present, usu-

Fig. 8.16. *Eucalyptus papuana* on flat beside a creek. *Atalaya hemiglauca* behind. Grasses are *Heteropogon contortus* and subsidiary *Themeda australis*. 96 km north of Camooweal, Qld.

ally containing *Terminalia* spp. and *Planchonia careya*. A dense grass layer is always present, consisting of *Sorghum plumosum* and subsidiary *Alloteropsis semialata*, *Botriochloa* spp., *Dichanthium* spp., *Heteropogon contortus*, *Sehima nervosum* and *Themeda australis*.

ii. Coastal Dune Woodlands. These woodlands are found on spits of dune sand adjoining narrow coastal beaches on parts of the Gulf of Carpentaria and in the east, but they are not a regular feature of the coast-line. In some areas the dunes separate the coastal beach and an inland saline mudflat. The woodlands reach a height of c. 12 m, the trees being branched close to ground level. *E. polycarpa* is sometimes an associate or even co-dominant. Mesomorphic shrubs or small trees are usually present, e.g., *Erythrophleum chlorostachys*, *Ficus* spp., *Diospyros humilis*, *Myoporum montanum*, *Pandanus* sp. The grass layer is usually dense and consists of *Chrysopogon fallax* (common), *Aristida browniana*, *Sporobolus virginicus* and *Vetiveria pauciflora*.

iii. Semi-arid to Arid Woodlands. In lower rainfall areas *E. papuana* occurs along watercourses or on floodplains and sometimes on the ranges where it grows in pockets of soil which receive additional water as runoff from the bare rocks. The woodlands, 6–10 m high, are unexpectedly tall for such a low rainfall. The communities are always open, the trees being increasingly further apart with decreasing rainfall, and in the most adverse areas the communities are best described as tree savannahs rather than woodlands. Pure stands of *E. papuana* are common. In semi-arid areas *E. pruinosa* and *E. terminalis* form ecotonal associations; in the arid zone *Acacia estrophiolata* and *A. aneura* are common associates, usually as understory species or in the gaps between trees. Four types of understory dominants occur in these drier areas (Perry, 1962): a. *Eragrostis eriopoda* on sandy floodplains; b. *Chrysopogon fallax* on Red Earths on floodplains; c. *Triodia pungens* on the skeletal soils on hillsides; d. *Muehlenbeckia cunninghamii* on billabongs (usually with clayey soils).

2.7. *Eucalyptus argillacea* – *E. terminalis* Alliance

Western Box – Inland Bloodwood

E. argillacea replaces *E. tectifica* on fine textured soils in drier areas (Fig. 8.17) with a mean annual rainfall of 500–750 mm. *E. terminalis* occurs over much the same range of climate but extends also into the arid zone where relict stands occur on the west coast and in the centre of the continent (Fig. 8.17). The soils are fine textured, deep grey clays of alluvial origin or Yellow Earths derived from basic parent materials. They occur in flat areas, sometimes adjacent to stream-lines and in some cases are flooded and waterlogged for a short period.

The species are commonly found together (Fig. 8.18) but pure stands of each occur. The woodlands are 8 m high in drier areas and up to 16 m high in wetter areas. A few tall shrubs are

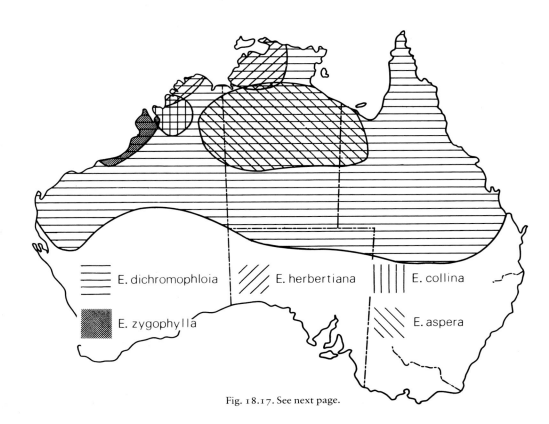

Fig. 18.17. See next page.

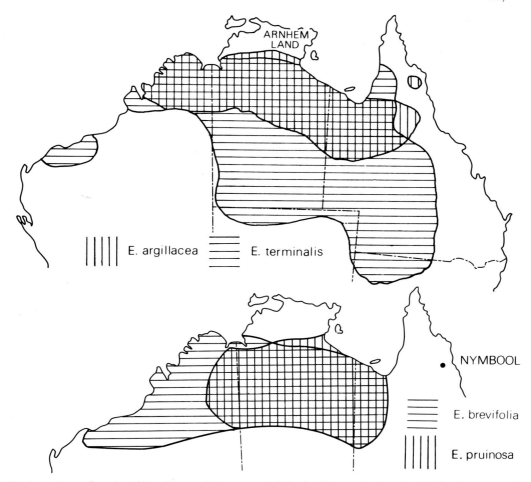

Fig. 8.17. Areas of species of *Eucalyptus* which occur mainly in the drier tropics. Based on Hall et al. (1970) and Blake (1953).

sometimes present, especially *Atalaya hemiglauca* and *Bauhinia cunninghamii*; *Carissa lanceolata*, a shrub 1–2 m high, is almost constant. Other shrubs 1–2 m high become more prominent in drier areas, especially *Cassia oligoclada*, *C. sturtii* and *Eremophila latrobei*. In wetter areas tussock grasses predominate, mostly species of *Chrysopogon*, *Dichanthium*, *Eulalia*, *Heteropogon*, *Sehima* and *Themeda australis*. In drier areas *Aristida* spp. or *Triodia pungens* dominate the grass layer, the latter being replaced by *Plectrachne pungens* in the west.

Pure stands of *E. argillacea* occur mainly on grey alluvial clays in western Queensland (Fig. 8.19), usually adjacent to *E. microtheca* woodlands, with associated shrubs typical of the wetter areas, as above. Pure stands of *E. terminalis* occur mainly in the drier areas, usually on shallow calcareous soils derived from limestone, with the dryland assemblage of understory species, as listed above.

2.8. Eucalyptus dichromophloia Alliance

E. dichromophloia occurs over most of northern Australia between the 250 and 1300 mm isohyets (Fig. 8.17) in well drained situations either on rocky outcrops or on deep sands. It varies in height from c. 20 m in the wettest areas

Fig. 8.18. *Eucalyptus terminalis* (left, in flower) and *E. argillacea* (right and behind). Grass layer dominated by *Themeda australis*. 5 km north of Mt. Isa, Qld.

on deep soils to 4 m in the more arid situations. As well as variations in height, the total population exhibits variation in bark characters (proportion of rough to smooth bark on the trunk and branches), fruit size and leaf size. The species is less abundant in the east than in the centre and west. In the east it occurs usually in association with *E. crebra* and *E. drepanophylla* (Chap. 12). In the central region it commonly forms pure stands. In the west it is often associated with another eucalypt. Four suballiances provide for the central and western communities. In addition to these, *E. dichromophloia* extends southward into the arid zone in the west, where it is a common emergent in some of the *Plectrachne – Triodia* savannahs (Chap. 20).

2.8.1. *Eucalyptus dichromophloia* Suballiance

Variable – Barked Bloodwood

Pure stands of *E. dichromophloia* as open woodlands occur on skeletal soils on rocky terrain throughout the range of the species. In the wetter north, the woodlands may reach a height of 20 m on deeper sands, and in such cases associated small trees of *Erythrophleum*

Fig. 8.19. Pure stand of *E. argillacea*. Shrubs: *Atalaya hemiglauca* and *Carissa lanceolata*. Grass layer: *Dichanthium fecundum*, *Eriachne ciliata* and *Sporobolus australasicus*. 32 km south of Normanton, Qld.

chlorostachys, *Gyrocarpus americanus* and *Hakea arborescens* may occur, with a grass layer of *Plectrachne pungens* and *Chrysopogon* spp. On the other hand, where rock covers most of the land surface, *E. dichromophloia* is stunted and gnarled and there are no associated taller species. The grass layer consists of the hummock grasses *Triodia microstachya* on acid rocks and *T. wiseana* on limestones. In the drier south the woodlands are mostly 6–8 m high (Fig. 8.20). A few xeromorphic shrub species are present, including species of *Acacia*, *Cassia* and *Grevillea*, and *Carissa lanceolata*. The common dominant in the grass layer is *Triodia pungens*.

2.8.2. Eucalyptus dichromophloia – E. herbertiana – E. collina Suballiance
Variable-barked Bloodwood – Kalumburu Gum – Silver-leaved Bloodwood

These three species form woodlands 10–12 m high in the north Kimberleys, mostly on rugged sandstone and quartzite hills on skeletal soils between the 630 and 1000 mm isohyets. *E. collina* occurs mainly in the south (Fig. 8.17). Scattered *Brachychiton* spp., *Gardenia* spp. and *Terminalia* spp. may occur, and, rarely, *Livistona humilis*. The grass layer is dominated by *Plectrachne pungens* and annual sorghums are abundant in summer.

Fig. 8.20. *Eucalyptus dichromophloia* on a rocky outcrop with *E. ferruginea* on deeper soil behind. Scattered *Plectrachne pungens* forms the grass layer. 56 km south-east of Kununurra, W.A.

2.8.3. *Eucalyptus dichromophloia – E. zygophylla – Acacia* spp. Suballiance

Variable-barked Bloodwood – Broome Bloodwood

This suballiance is the southern counterpart of the preceding one, extending between the 400 and 870 mm isohyets in the Kimberleys (Fig. 8.17). The communities occur mainly on deep sands, less commonly on low sandstone hills and slopes. Associated eucalypts include *E. perfoliata* (confined to this alliance) and a form of *A. argillacea*. The main associations are:

i. *E. dichromophloia – E. perfoliata* Association. This community is a "scrubby woodland" 3–7 m high, with an "orchard appearance" and occurs mainly on sandplain (Speck and Lazarides, 1964). Associated tall shrubs include *Bauhinia cunninghamii, Cochlospermum fraseri, Dolichandrone heterophylla, Grevillea* spp. and *Hakea* spp. The grass layer contains mainly *Plectrachne pungens* and *Chrysopogon pallidus*.

ii. *E. dichromophloia – E. zygophylla* Association. These two species form woodlands 5–8 m high and characterize a type of country known as "pindan" (Fig. 8.21). The association covers large tracts of flats to undulating country between the 500 and 700 mm isohyets in the southern and western parts of the Kimberleys. The soils are Red or Yellow Earthy Sands or Red Siliceous Sands and are porous, so that runoff is negligible and watercourses are absent. Small depressions occur infrequently and these support a few trees of *E. grandifolia* or *E. argillacea*. The shrub layer is usually dense and consists mainly of *Acacia holosericea, A. sericata, A. ancistrocarpa, A. tumida* and the species listed for the previous association. Firing of the communities is frequent, resulting in the removal of the shrubs which regenerate from seed after the first heavy rains. The grass layer is

Fig. 8.21. *Eucalyptus dichromophloia* (left) and *E. zygophylla* (right). Shrubs are *Acacia tumida* (dead through burning, and regenerating). Grass layer consists of *Plectrachne pungens, Aristida browniana* and *Eriachne mucronata*. 160 km west of Fitzroy Crossing, W.A.

dominated by *Plectrachne pungens* and *Chrysopogon* spp.

With decreasing rainfall (in the south and south-west), the acacias become more prominent, equalling the eucalypts in height, and at about the 500 mm isohyet (south of Broome) the acacias overtop the eucalypts, the woodland community thus merging into an *Acacia* scrub (*A. ancistrocarpa* alliance, Chap. 18).

2.8.4. *Eucalyptus aspera* Suballiance
Brittle Bloodwood

E. aspera is a small, white-barked tree or shrub, sometimes with a rough tessellated bark below, reaching a maximum height of 8 m, but usually not exceeding 5 m. It has been recorded from widely separated localities on stony hills in Northern Territory and Western Australia southward of c. 650 mm isohyet (Fig. 8.17). It extends into the arid zone where it is found in sandy watercourses in the *Acacia aneura* scrubs of Western Australia or in association with smooth-barked *E. microtheca*. The only community description available is that of Perry and Christian (1954) for Northern Territory where *E. aspera* forms an open scrub 3–5 m high on rock outcrops. A scattered shrub layer 1–2 m high of *Calytrix microphylla* and *C. brachychaeta* occurs throughout. The sparse ground layer consists of *Triodia pungens*, with scattered *Cymbopogon bombycinus*.

2.9. *Eucalyptus pruinosa* Alliance
Silver-leaf Box

E. pruinosa occurs in small stands mainly west of the Carpentaria Plains between the 450 and 750 mm isohyets (Fig. 8.17). It is a small, gnarled tree to 10 m high, and often mallee-like. It is closely related taxonomically to *E. melanophloia* and *E. shirleyi* which occur in the east. *E. pruinosa* is found in the same general area as *E. brevifolia*, the latter occupying more elevated areas, the former occurring mostly on flat or undulating country, where it commonly forms mosaics with *Melaleuca* scrub, particularly *M. viridiflora*, or with "softwood scrub" which contain mixtures of *Atalaya hemiglauca*, *Grevillea striata* and *Terminalia canescens*. The soils are mostly Red and Yellow Earths, but the species may occur on Lithosols especially in ecotonal areas adjoining the *E. brevifolia* alliance.

Open woodlands to 10 m high (Fig. 8.22) and open mallee-like scrubs dominated by *E. pruinosa* occur in patches rarely exceeding a few sq. km in area. Woody understory species are few and rare; they include the tall shrubs: *Bauhinia*

Fig. 8.22. *Eucalyptus pruinosa* low woodland. *Grevillea striata* at left. Grass layer dominated by *Triodia pungens* with subsidiary *Aristida pruinosa*. 110 km north of Camooweal, Qld.

Fig. 8.23. General view of *Eucalyptus brevifolia* woodland. *E. dichromophloia* present in small quantities. The grass is *Triodia pungens*. 112 km north of Camooweal, Qld.

cunninghammi, *Excoecaria parvifolia* (depressions) and *Hakea arborescens;* shrubs 1–2 m high are *Carissa lanceolata* (common) and species of *Acacia, Capparis, Celastrus* and *Phyllanthus*. The herbaceous layer is dominated by tussock grasses on deeper soils, especially *Aristida pruinosa, Chrysopogon pallidus, Cymbopogon bombycinus, Eriachne* spp. and *Sehima nervosa*. On shallow soils *Triodia pungens* is dominant.

2.10. *Eucalyptus brevifolia* Alliance

Snappy Gum or Northern White Gum

E. brevifolia dominates large tracts of country between the 320 and 670 mm isohyets west of the Carpentaria Plains, extending to the Indian Ocean (Fig. 8.17). A small outlier occurs in the east at Nymbool, Queensland. Two subspecies are recognized (see below). The soils are mainly lithosols, but smaller areas of the communities are found on deep, coarse sands and rarely on Yellow or Red Earths. The rock types producing these siliceous, acid, low fertility soils are mainly sandstones, quartzites, or residual laterite profiles eroded to the pallid, mottled or ferruginous zones. Since the alliance extends across the continent it contacts many other alliances and forms mosaics with them. In the wetter areas it sometimes adjoins the *E. tetrodonta* etc. alliance, as in the Kimberleys. In drier areas it forms mosaics with the *E. dichromophloia* alliance (Fig. 8.23), and the *E. pruinosa* alliance. In the drier south it adjoins the arid communities, especially the *Acacia aneura* and the *Triodia – Plectrachne* savannahs. The tropical mallee stands of *E. pachyphylla* and *E. odontocarpa* occur largely within this alliance (Chap. 14).

E. brevifolia ssp. *brevifolia* dominates open woodlands which in the wettest sites are c. 9 m tall (Fig. 8.24); in drier areas the plants are stunted and mallee-like. In the taller woodlands on deep sands, tall shrubs may occur, but never form a well defined stratum, viz., *Atalaya hemiglauca, Cochlospermum fraseri, Grevillea striata, G. pteridifolia, Hakea lorea, Petalostigma banksii* and *Terminalia canescens*. In drier

Fig. 8.24. *Eucalyptus brevifolia* ssp. *brevifolia* low open woodland on lateritic gravel. Grass sward of *Triodia pungens*. Note termite mounds. 100 km north of Cloncurry, Qld.

areas, shrubs 1–2 m high are common in most communities, sometimes forming a shrub layer. Most widespread and abundant are *Acacia chisholmii*, *A. hilliana*, *A. leursenii*, *A. lysiophloia*, *A. orthocarpa* (= *xerophila*), *Calytrix* spp., *Carissa lanceolata*, *Cassia* spp., *Dodonaea oxyptera*, *Grevillea wickhamii*, *Petalostigma quadriloculare* and *Petalostylis labichioides*. The herbaceous stratum consists mainly of grasses and these become increasingly prominent as the communities become more open, except in very rocky situations. Tussock grasses, which occur in most stands, are more conspicuous in the north, whereas hummock grasses are conspicuous in the drier south or in rocky areas. Tussock grasses include *Aristida pruinosa*, *Chrysopogon* spp., *Cymbopogon bombycinus* and *Heteropogon contortus*. Of the hummock grasses, *Triodia pungens* is most frequent, especially in the central southern and eastern portions of the alliance, being replaced locally by: *Plectrachne pungens* (centre and west), *Triodia longiceps* (centre, rare), *T. wiseana* and *T. inutilis* or *T. stenostachya* (west.) Dicotyledonous herbs are usually abundant after rain, especially species of *Gomphrena* and *Ptilotus*.

In the eastern outlier, near Nymbool, *E. brevifolia* is dominant on rocky hills, forming an open, mallee-like woodland. Associated with it are *Callitris*, *Eucalyptus cullenii*, *E. shirleyi*, *E. dichromophloia* and *E. similis;* there is a sparse grass layer and *Xanthorrhoea* is abundant (Story, 1970).

E. brevifolia ssp. *confluens* (Kimberley Gum) occurs in the Ord River area, and in the Kimberleys and characterizes a second suballiance. It occupies rocky terrain and is often the sole dominant, or it is associated with *E. dichromophloia*.

9. The Tall *Eucalyptus* Forests of the Eastern Coastal Lowlands Mostly on Soils of Higher Fertility

1. Introduction

The tall forests of the east occur mainly south of the Tropic, with the greatest development on the north coast of New South Wales, extending to southeastern Queensland. Small stands of some species are found north of the Tropic, where they are possibly relict and are separated from the more southerly stands by the relatively dry area around the Tropic (Fig. 9.1). The forests occupy the soils of higher fertility on the warmer coastal lowlands, with minor extensions on to the cooler tablelands. The soils of lower fertility in the same climatic areas are dominated by different species of *Eucalyptus* (Chap. 10). The tall forests are mostly 30–60 m high but are not the tallest in the continent, these being found in Victoria and Tasmania (Chap. 11).

The tall forests commonly adjoin rainforests and a few of the eucalypts may occur in rainforest. Conversely, rainforest species often dominate the understory in the tall forests, but this is not invariable. On the contrary, several of the tallest dominants tolerate a wide range of habitats so that the understory may be mesomorphic in some stands and xeromorphic in others. In many cases repeated firing of the stands has led to a reduction in the density of the understory, especially when this is mesomorphic, and in extreme cases the elimination of all woody understory species has led to the formation of a grass understory. An understory dominated by grasses or helophytes is natural to only one kind of forest, the swamp forest (*E. robusta* alliance) occurring adjacent to the littoral zone.

About 30 eucalypts characterize the tall forests, and *Syncarpia glomulifera* and *Tristania conferta* (Fig. 9.1) are fairly regular associates. In addition, ecotonal associations are formed with some species from the soils of lowest fertility, notably *E. gummifera* and *E. intermedia*, and also with semi-arid species which have reached the coastal lowlands by migration across the Great Divide, notably *E. crebra*. The communities have been described under 10 alliances which are listed in Table 9.1. South of the Tropic, the alliances are delimited partly by latitude (Fig. 9.1), partly by soil fertility levels and partly by soil waterlogging. On the most fertile soils in the wettest habitats (adjacent to rainforest), *E. grandis* is dominant. With increasing moisture (impeded drainage), *E. grandis* is replaced by *E. tereticornis* and this by *E. robusta* in coastal swampy situations. Again, *E. grandis* is replaced on less fertile soils usually in less moist sites by *E. saligna* which, with decreasing fertility, is replaced successively by *E. resinifera* etc. and by *E. pilularis*. The last species is replaced by *E. maculata* in the least fertile and/or drier areas. With increasing altitude, the coastal alliances are replaced by *E. campanulata* and/or *E. obliqua* (Chap. 11).

2. Descriptions of the Alliances

In the text that follows the alliances are described in north to south sequence as far as possible, and from high to low fertility where 2 or more alliances span the same latitudinal zone. The classification is based on Baur (1962, 1965). Other references include: Blake (1941, 1942); Davis (1941); Pidgeon (1941); Story (1967, 1970); Pedley (1967); Anderson (1968); Speck (1968); Hall et al. (1970); Clifford (1972); Coaldrake et al. (1972); Austin (1975) and others mentioned below. In many cases the understory assemblages are supplied by the writer from field records.

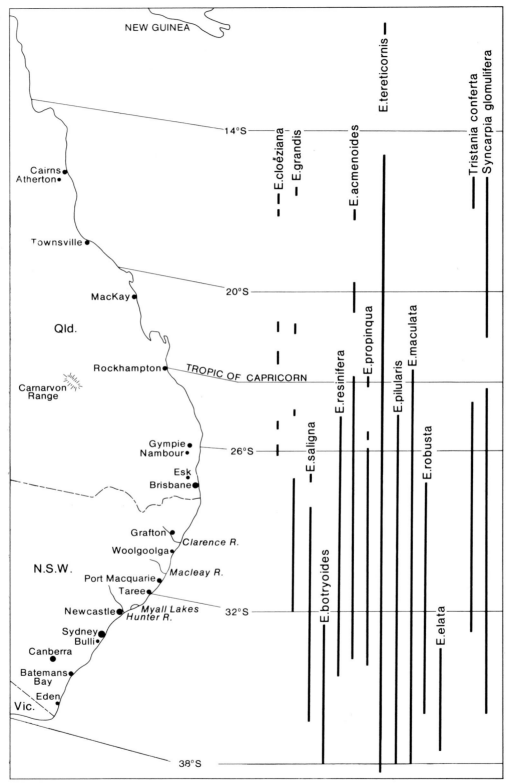

Fig. 9.1. Latitudinal ranges of species occurring in the coastal tall forests.

Table 9.1. Associations in the tall *Eucalyptus* forests found on the eastern coastal lowlands.

Alliance or Suballiance	Main associations	Species forming minor associations with the dominants	Main species forming ecotonal associations
E. cloëziana	*E. cloëziana*		
E. grandis	*E. grandis*		
E. saligna;	*E. saligna;* *E. saligna –* *E. microcorys;* *E. dunnii;* *E. deanei*	*E. largeana;* *E. quadrangulata;* *E. paniculata*	*E. saligna;* *E. pilularis* (Qld) *E. grandis;* *E. pilularis;* *E. resinifera;* *E. acmenoides;* *E. propinqua;* *E. intermedia;* *E. campanulata* (higher altitudes in north)
E. botryoides	*E. botryoides*	*E. paniculata;* *E. punctata;* *E. pellita;* *E. longifolia*	*E. gummifera;* *E. pilularis*
E. resinifera – *E. acmenoides –* *E. propinqua*	*E. resinifera –* *E. acmenoides;* *E. propinqua –* *E. acmenoides;* *E. major –* *E. resinifera;* *E. propinqua –* *E. siderophloia*	*E. eugenioides;* *E. paniculata;* *E. carnea;* *E. deanei*	*E. microcorys;* *E. saligna;* *E. pilularis;* *E. maculata;* *E. tereticornis;* *E. crebra;* *E. intermedia*
E. tereticornis	*E. tereticornis*	*E. exserta;* *E. siderophloia;* *E. glaucina;* *E. eugenioides;* *E. major;* *E. carnea;* *E. globoidea;* *E. benthamii*	Tropics: *E. alba;* *E. polycarpa;* *E. tessellaris* South of Tropic: *E. tessellaris;* *E. grandis;* *E. maculata;* *E. crebra;* *E. intermedia;* *E. propinqua;* *E. moluccana*
E. amplifolia	*E. amplifolia*	*E. bosistoana;* *E. benthamii* *E. bauerana;* *E. parramattensis*	*E. tereticornis*
E. seeana	*E. seeana*		*E. tereticornis*
E. pilularis	*E. pilularis*		See other suballiances below
E. pilularis – *E. microcorys*	*E. pilularis –* *E. microcorys;* *E. pilularis –* *E. acmenoides*		*E. resinifera –* *E. propinqua;* *E. intermiedia;* *E. grandis;* *E. saligna*
E. pilularis – *E. intermedia –* *E. siderophloia*	*E. pilularis –* *E. intermedia –* *E. siderophloia;* *E. pilularis –* *E. intermedia*	*E. acmenoides;* *E. phaeotricha;* *E. major*	*E. resinifera –* *E. propinqua;* *E. tereticornis*
E. pyrocarpa	*E. pyrocarpa*	*E. punctata;*	

(continued next page)

Table 9.1. Continued.

Alliance or Suballiance or	Main associations	Species forming minor associations with the dominants	Main species forming ecotonal associations
E. pilularis – E. saligna – E. paniculata	E. acmenoides E. pilularis – E. saligna – E. paniculata; E. pilularis – E. saligna	E. gummifera E. acmenoides; E. eugenioides	E. resinifera
E. pilularis – E. gummifera	E. pilularis – E. gummifera; E. pilularis – E. sieberi	E. globoidea	E. maculata
E. pilularis – E. piperita E. pilularis – Angophora costata E. maculata	E. pilularis – E. piperita E. pilularis – Angophora costata E. maculata; E. maculata – E. rummeryi	E. punctata; Angophora costata E. signata (North); E. racemosa (South) North: E. tessellaris; E. trachyphloia; E. polycarpa; Angophora costata; E. decorticans; E. exserta; E. phaeotricha; E. carnea; E. tetraphleura; E. fibrosa; E. siderophloia South: E. paniculata; E. agglomerata; E. globoidea; E. muellerana; E. maidenii; E. bosistoana	E. crebra; E. tereticornis; E. moluccana; E. intermedia; E. propinqua; E. pilularis; E. gummifera
E. robusta	E. robusta	E. longifolia; E. pellita	E. tereticornis; E. botryoides; E. resinifera
E. elata	E. elata; E. elata – Casuarina cunninghamiana	E. bauerana; E. angophoroides	E. viminalis; E. tereticornis

2.1. *Eucalyptus cloëziana* Alliance

Gympie Messmate

The main area of this species occurs near Gympie, Queensland, under a mean annual rainfall of 1000 mm, but the species is recorded in six other isolated localities, with a climatic range of 750–1600 mm, as far north as Atherton. In the Gympie area it occurs in pure stands 40–50 m high (Fig. 9.2) on deep Krasnozems and alluvial soils over hilltops and in valleys. In the more favourable sites, especially near watercourses, the understory species are plants of the rainforests and include: *Alectryon subcinereus, Aphananthe philippinensis, Elaeocarpus obovatus, Endiandra sieberi, Ficus obliqua, Glochidion ferdinandii, Jagera pseudo-rhus, Litsea leefeana, Mallotus philippinensis, Rhodomyrtus psidioides, Solanum mauritanium* and *Wickstroemia indica*, and the small liana *Stephania hernandiifolia*. After logging and firing the tall

Fig. 9.2. *Eucalyptus cloëziana* with mesomorphic understory. *Ficus* sp. left as shelter for stock. Near Gympie, Qld.

shrub understory is more or less removed, and if any species remain they are *Alphitonia excelsa, Casuarina littoralis, Trema aspera* and *Tristania suaveolens*. The herbaceous layer is dominated by *Imperata cylindrica*.

On the other sites, *E. cloëziana* occurs as a subsidiary species in association with a variety of eucalypts and a xeromorphic understory. C. L. Bale (personal communication) records it near Herberton (south of Atherton) with a number of eucalypts, e.g., with *E. abergiana, E. brassiana, E. drepanophylla, E. carnea, E. citriodora* and *E. intermedia* (all in the one stand). Species of *Hakea, Melichrus* and *Xanthorrhoea* occur in the shrub layer and grasses dominate the herbaceous layer.

2.2. *Eucalyptus grandis* Alliance

Flooded Gum

E. grandis characterizes tall forests on soils of high fertility, commonly on valley floors along watercourses and in areas subject to minor flooding. It occurs disjunctly (Fig. 9.1), with a lesser development in the tropics. In those areas when *E. saligna* also occurs, the latter occupies slightly less favourable sites with respect to water availability and soil fertility. *E. grandis* forests often adjoin rainforest and the species sometimes occurs within rainforest stands, where it probably established after a fire, later to be invaded by the regenerating rainforests.

Common understory tall associates of the tall forests are *Archontophoenix cunninghamii, Syncarpia glomulifera* (Fig. 9.3) and *Tristania conferta*. Other species are numerous, including *Toona australis, Dendrocnide excelsa* and *Acacia maidenii,* and those listed for the northern segment of the *E. saligna* alliance (next section).

In the tropics, *E. grandis* occurs at the margins of rainforests, or near rainforests, sometimes in association with *E. torelliana* which is confined to this habitat. In the more open stands, other eucalypts occur in association with *E. grandis*, especially *E. acmenoides, E. intermedia* and *E. tereticornis*. A shrub layer is usually absent, possibly as a result of repeated burning. The herbaceous layer is grass, *Themeda australis* and *Imperta cylindrica* being constants (Isbell and Murtha, 1972). An outlier of *E. grandis* occurs near permanent water in the Carnarvon Ranges, Queensland.

2.3. *Eucalyptus saligna* Alliance

Sydney Blue Gum

This alliance overlaps the *E. grandis* alliance in north-eastern New South Wales and southeast Queensland (Fig. 9.1). When the two

Fig. 9.3. *Eucalyptus grandis, Archontophoenix cunninghamii* and *Syncarpia glomulifera*. Near Nambour, Qld.

species occur together, *E. grandis* occupies the most favourable sites and contains the more luxuriant understory (see last section).

E. saligna and *E. microcorys* (the latter as far south as the Hunter R. area) form tall mesomorphic forests on soils of relatively high fertility (Table 9.1) commonly on valley floors, typically between rainforest stands (or *E. grandis*, if it occurs in the area), and the *E. resinifera* – *E. propinqua* alliance which occupies upper slopes. On more siliceous parent materials, the latter is replaced by *E. pilularis*. Both *E. saligna* and *E. microcorys* provide valuable timber and their autecology and regeneration have been studied by Van Loon (1966).

The understory in undisturbed stands is mesomorphic (Fig. 9.4) and is an extension of some rainforest species below the eucalypts.

Some of the common small trees and tall shrubs are: *Ackama paniculata, Archontophoenix cunninghamii, Callicoma serratifolia, Cryptocarya rigida, Elaeocarpus reticulatus, Endiandra sieberi, Euroschinus falcatus, Glochidion ferdinandii, Mallotus philippinensis, Orites excelsa, Schizomeria ovata* and *Synoum glandulosum*. In many stands *Syncarpia glomulifera* and/or *Tristania conferta* are present as large trees. The tree fern, *Cyathea australis*, is usually abundant, and also wiry lianas, especially *Smilax australis*. A few other tall plants, not abundant in or absent from the rainforests are sometimes present, notably *Acacia binervata, Acacia elata* and *Casuarina torulosa*. The herbaceous layer is poorly developed and is often dominated by ferns, especially *Culcita dubia*.

The understory is dramatically changed by

fire which may completely eliminate the rainforest species. *Syncarpia* and *Tristania* commonly survive the fires and species of *Acacia*, especially *A. binervata* and *A. irrorata,* and *Casuarina torulosa* may become more abundant through regeneration from seed. Repeated burning sometimes completely eliminates all woody understory species, so that the forest consists only of tall trees and a grass layer, the latter dominated by *Imperata cylindrica* at lower altitudes and *Themeda australis* at higher altitudes.

In northern New South Wales some extensive variation (apart from ecotonal assemblages) occurs as the result of the local dominance or co-dominance of species which are confined to this alliance. These are *E. dunnii* (Dunn's White Gum) which reaches a height of 30–40 m and is a local dominant, or it occurs as a co-dominant with either *E. saligna* or *E. microcorys*. Also *E. largeana* (Craven Grey Box) occurs as a co-dominant in a restricted area between the Hunter R. and Macleay R.; *E. deanei* occurs in some stands but is more common at higher altitudes (Chap. 11).

South of the Hunter R., *E. saligna* is less abundant and may occur in pure stands on soils of high fertility, which in this area are rare. For example, a pure stand formerly occurred on volcanic breccia near Hornsby (Sydney area) but has now been removed. Elsewhere in the area the soils are of medium fertility, except in narrow valleys, which restricts the habitat required by *E. saligna*. However, on soils of medium fertility it occurs in association with *E. pilularis*, the latter being more abundant in upland areas, partly because of lower fertility levels and partly because of reduced water supply. The southern stands resemble the northern ones superficially in that the understory consists of rainforest species. However, the understory assemblage is completely different, partly because of lower soil fertility and partly because of the latitudinal restriction on the rainforest assemblage. The main species are: *Backhousia myrtifolia, Pittosporum undulatum, Exocarpos cupressiformis, Ceratopetalum apetalum, Trochocarpa laurina, Pittosporum revolutum, Casuarina torulosa, Notelaea longifolia, Rapanea variabilis, Acacia*

Fig. 9.4. *Eucalyptus saligna* in pure stand. Understory of *Tristania suaveolens, Syncarpia glomulifera, Alphitonia excelsa, Rhodomyrtus beckleri* and *Dodonaea triquetra*. Herbaceous stratum (not visible) dominated by *Blechnum cartilagineum*. Near Mountain Creek, south of Gympie, Qld.

binervata, *A. implexa* and *Rhodamnia trinervia*. The compositon varies from site to site and is possibly determined by fire, which, as in the north, reduces the stands to grassy forests with the grass layer of *Imperata* and/or *Themeda*.

E. deanei occasionally replaces *E. saligna*, especially in the lower Blue Mts, where it occurs on valley floors or on shale outliers in sandstone country. It is commonly associated with *E. paniculata*, *E. eugenioides*, *E. resinifera* and *Syncarpia glomulifera*. The understory is composed of rainforest species (above) in more favourable sites, or it is dominated by *Casuarina torulosa* and/or *Acacia* decurrens in drier sites (possibly as a result of fire) (Phillips, 1947).

South of Sydney *E. saligna* intergrades with *E. botryoides* (next section).

2.4. Eucalyptus botryoides Alliance

Bangalay

E. botryoides intergrades with *E. saligna* south of Sydney, the trees of *"saligna"* developing a "sock" of rough bark which becomes progressively longer on the trunk towards the south. *E. botryoides* has a fully rough bark and occurs in this form in Victoria, northward to Bateman's Bay and also close to the ocean as far north as Raymond Terrace (near Newcastle). The species is therefore discussed under two habitats.

i. Remote from the sea and to an altitude of c. 200 m, *E. botryoides,* or forms intergrading with *E. saligna*, replace *E. saligna* in tall forest formation up to 40 m high, sometimes intermixed with *E. saligna*. The stands are rarely pure, but contain admixtures of *E. gummifera*, *E. pellita* (often locally dominant), *E. paniculata* and sometimes *E. punctata*. The stands occupy soils of relatively high fertility, intermediate between those supporting rainforest on more fertile soils and xeromorphic forest or less fertile soils. The understory is mesomorphic, the common species being *Acmena smithii, Callicoma serratifolia, Elaeocarpus reticulatus* and *Tristania laurina* (Austin, 1975). The ground stratum invariably contains a high proportion of *Pteridium esculentum,* indicative of fire-damage.

In the extreme south of its range, *E. botryoides* is confined to river flats, sometimes edging permanent streams or occurring and mingling with *Casuarina glauca* in coastal, subsaline swamps.

ii. *E. botryoides* also occurs as a dominant of hind-dune forest up to 15 m high, along the south coast of New South Wales. The forests have an understory of *Banksia serrata* and *B. integrifolia*, which form the dominants on coastal dunes on the seaward side of the forests. On the landward side and intermixed with *E. botryoides*, *E. longifolia* and *E. pilularis* commonly occur. In the Bulli district, Davis (1941) lists the following common understory species which include a high proportion of mesomorphic plants: *Acacia linearis, Breynia oblongifolia, Clerodendrum tomentosum, Cupaniopsis anacardioides, Pittosporum undulatum, Synoum glandulosum,* the small lianas *Geitonoplesium cymosum, Parsonsia straminea, Stephania japonica, Tylophora barbata, Cassytha pubescens* (parasitic), the small shrubs *Brachyloma daphnoides, Pimelea linifolia,* and the herbs *Haloragis teucrioides, Themeda australis* and *Viola hederacea*.

2.5. Eucalyptus resinifera – E. acmenoides – E. propinqua Alliance

Red Mahogany – White Mahogany – Small-fruited Grey Gum

These species form an alliance of variable composition mainly north of the Hunter R., the alliance separating the more mesomorphic tall forests (e.g., *E. saligna – E. microcorys*) from the more xeromorphic forests (*E. pilularis* or *E. maculata*). In some areas it adjoins dry rainforest, especially in Queensland where *E. major* replaces *E. propinqua* (Fig. 9.5). In some areas *E. eugenioides* is a conspicuous component. The alliance occurs on both flat and hilly country and in most cases a zoning of the species occurs. *E. resinifera* and *E. acmenoides* occupy the most favourable sites, usually near the bottoms of valleys, whereas *E. propinqua* occurs higher on the slopes where it may be associated with *E. siderophloia*.

The understory is mesomorphic, except on upper slopes. Typical species in northern New South Wales are: *Tristania suaveolens* (often forming a tall shrub/small tree layer), *Elaeocarpus reticulatus, Glochidion ferdinandii,*

Fig. 9.5. *Eucalyptus resinifera*, *E. major* and *E. intermedia*. Lower tree layer mainly *Tristania suaveolens* and *Elaeocarpus reticulatus*. Shrubs are *Synoum glandulosum*, *Dodonaea triquetra*, *Persoonia pinifolia* and *Breynia oblongifolia*. Herbaceous layer mainly *Imperata cylindrica* and the fern, *Culcita dubia*. Near Cooroy (south of Gympie), Qld.

Synoum glandulosum, *Exocarpos cupressiformis*, *Stenocarpus salignus*, *Backhousia myrtifolia*, *Persoonia linearis*. Small lianas may occur: *Clematis aristata*, *Kennedia rubicunda* and *Smilax australis*. On the upper slopes *Casuarina torulosa* is often prominent and occasional shrubs of *Breynia oblongifolia* and *Dodonaea triquetra* may occur, and are scattered in the grass layer of *Imperata cylindrica*.

2.6. *Eucalyptus tereticornis* Alliance

Forest Red Gum

E. tereticornis is a member of the Red Gum Series (Tereticornes), a group of closely related and sometimes intergrading species which occur in various habitats across the continent (Chap. 4). It is the most easterly and most widely distributed member of the Series in the east (Fig. 9.1), with occurrences in New Guinea. The species shows some taxonomic variation within itself, chiefly in the sizes of fruits and buds, and field evidence suggests that ecotypes might exist. Several closely related taxa, given specific rank, occur within the geographic area of *E. tereticornis*, including *E. glaucina* and the species which are used to designate suballiances.

2.6.1. *Eucalyptus tereticornis* Suballiance

Forest Red Gum

E. tereticornis dominates forests 30–50 m high thoughout its range (Fig. 9.1). It occurs between the 500 and 300 mm isohyets, but these figures are spurious because the forests almost invariably receive run-off water from the slopes above. The species is dominant on coastal alluvial flats adjacent to stream lines and in narrow or broad valleys; it occurs also on slopes with clayey, poorly drained soils. The stands adjoin rainforest, both the wet and drier types, other tall forests, and occasionally mangroves or *Casuarina glauca* woodlands of the littoral zone.

In the tropics, *E. tereticornis* usually forms pure stands on flats subject to flooding, these being surrounded by other *Eucalyptus* communities on more elevated areas. A shrub layer does not exist except along watercourses, where

Melaleuca lineariifolia and/or *Callistemon viminalis* may occur. The herbaceous layer is dominated by the grasses *Bothriochloa intermedia*, *Heteropogon contortus* and *Themeda australis*.

Between the Tropic and lat. 35°S., *E. tereticornis* and *Angophora subvelutina* usually occur together, either as co-dominants, or the latter forms a tree layer below the eucalypt. In the wettest sites a continuous tall shrub/low tree layer of mesomorphic species occurs, the floristic composition of the assemblages varying with soil moisture and latitude. *Tristania suaveolens* is almost constant and many species typical of the local rainforest assemblages are often present, e.g., *Ficus* spp., *Archontophoenix cunninghamii* and *Castanospermum australe*. The herbaceous layer is sparse and usually contains a few ferns, of which *Culcita dubia* is common. In drier sites tall woody plants form a discontinuous layer below the eucalypts especially *Alphitonia excelsa*, *Casuarina torulosa* and species of *Acacia*. *Angophora floribunda* is sometimes locally common in these areas. The herbaceous layer is dominated by grasses, *Themeda australis* being the most common, often with subsidiary *Bothriochloa macera*, *Eragrostis leptostachya* and *Echinopogon caespitosus*. In many areas the understory has been so severely damaged by fire that *Imperata cylindrica* forms a continuous sward below the trees. In such stands waterlogging of the soil is increased and pools of water persist after rain, supporting a semi-aquatic flora, uncluding Cyperaceae (*Cyperus* and *Fimbristylis* spp.), *Juncus* and *Marsilea*.

South of c. lat. 35°S. *E. tereticornis* forms pure stands or is co-dominant with *E. moluccana*, as on the Penrith Plains west of Sydney (Chap. 13). Near its southern limit, *E. tereticornis* occurs mainly as a forest fringing watercourses, or in ecotonal associations.

2.6.2. *Eucalyptus amplifolia* Suballiance
Cabbage Gum

This species replaces *E. tereticornis* in areas which are extensively waterlogged and stands of the two species are commonly contiguous. *E. amplifolia* is more common above 600 m, but some stands occur on the coastal lowlands and the species is recorded from near Canberra to the Tropic. The stands in both lowland and tableland sites are usually pure but associated species may occur, especially in the south (Table 9.1). *Angophora floribunda* sometimes forms a scattered lower tree layer. Shrubs are usually absent, rare occurrences being *Bursaria spinosa* and *Dodonaea viscosa,* and, in the Sydney are, Phillips (1947) records species of *Melaleuca* (*M. genistifolia, M. lineariifolia* and *M. styphelioides*) which suggest subsalinity. The herbaceous layer is dominated by grasses. Phillips (1947) lists *Aristida vagans, Eragrostis brownii* and *Dichelachne sciurea*.

2.6.3. *Eucalyptus seeana* Suballiance

E. seeana has been recorded from many localities between the Hunter Valley area to south-east Queensland (Blakely, 1955) but apparently is a relatively rare species found in association with other species. It is closely related to *E. tereticornis* from which it differs in having a much longer operculum. It is known to the writer in pure stands on flats in the Clarence R. valley south of Tabulum on deep sandy alluvium subject to some waterlogging. Here it occurs as a woodland c. 25 m high (Fig. 9.6) with a discontinuous understory of *Tristania suaveolens* and an herbaceous layer of *Imperata cylindrica* (possibly fire induced). Rare associated understory species include *Jacksonia scoparia* and *Themeda australis*.

2.7. *Eucalyptus pilularis* Alliance
Blackbutt

E. pilularis, one of the most important hardwoods in the east, is confined to the coastal regions, extending on to the lower tablelands (Fig. 9.1). The best and tallest commercial stands lie between Taree and Woolgoolga, N.S.W., though in this area the largest trees, attaining a diameter of c. 2 m have been felled. The species regenerates readily from seed, which is prolific, germination occurring at the beginning and end of the summer wet season (Floyd, 1962). Seedling establishment is sometimes inhibited in some soils, possibly by "direct microorganism antagonism" (Florence and Crocker, 1962).

The species is dominant or co-dominant in a great variety of habitats within the area receiving a mean annual rainfall of 900–1500 mm. It

Fig. 9.6. *Eucalyptus seeana*. Lower tree layer of *Tristania suaveolens;* grass layer of *Imperata cylindrica*. 40 km south of Tabulum (Clarence R. valley), N.S.W.

occurs on soils of very low to very high fertility, in gullies, on flats and on the drier ridges in rocky situations. On soils of intermediate fertility it usually occurs in pure stands and reaches its maximum height. The uniformity of these stands (referred to below as the *E. pilularis* suballiance) is preserved because the soils are of too low fertility to permit the entry of high nutrient species such as *E. saligna*, and low fertility species such as *E. intermedia* or *E. gummifera* cannot compete with *E. pilularis*. Because of its ecological plasticity, which may possibly be associated with ecotypic variation (though this has not been established) it may be found in combination with almost every tall eucalypt species growing within its area; the most common of these are listed in Table 9.1. However, several distinct combination of species, given the rank of suballiances, can be identified and related to fairly well defined habitats.

2.7.1. *Eucalyptus pilularis* Suballiance
Blackbutt

E. pilularis is often the sole dominant (Fig. 9.7) or is the by far the most common species in the stand, especially in the central portion of its area but also in the north and the south. Associated dominants, if they occur, are of low frequency (Table 9.1). The heights of the tallest stands vary from c. 60 m to as little as 15 m.

Understory species vary with latitude, moisture, and fertility levels. In the tallest stands in the north and especially near watercourses, the understory is mesomorphic, common species being *Acacia binervata*, *Breynia oblongifolia*, *Elaeocarpus reticulatus*, *Glochidion ferdinandii*, *Notelaea longifolia* and *Rhodamnia trinervia*. In less favourable sites *Casuarina torulosa* is a near-constant and some common shrubs are *Acacia longifolia*, *Exocarpos cupressiformis*, *Helichrysum diosmifolium*. *Hibiscus heterophyllus*, *Leucopogon lanceolatus* and *Persoonia linearis*. The more abundant small shrubs and herbs include: *Hibbertia scandens*, *Kennedia rubicunda* (liana), *Platylobium formosum*, *Rubus parviflorus; Entolasia stricta*, *Imperata cylindrica*, *Themeda australis* and the ferns *Culcita dubia*, *Dennstaedtia davallioides* (wetter sites) and *Pteridium esculentum*.

A forest described for the Bulli area by Davis (1936) typifies the southern stands on sandstones. *Eucalyptus paniculata* and *Syncarpia*

Fig. 9.7. *Eucalyptus pilularis*. Pure stand, recently burned. Scattered shrubs are *Jacksonia scoparia*, *Acacia decurrens* and *Casuarina torulosa* (all regenerating from seed). Grass layer contains *Themeda australis* and *Aristida ramosa*. 25 km south-west of Esk, south-east Qld.

glomulifera are present. The understory is mainly xeromorphic, with *Casuarina torulosa* prominent in the small tree layer and the tall shrubs *Acacia binervata*, *Notelaea longifolia* and *Persoonia linearis*. The lower shrub layer contains a mixture of xeromorphic and small-leaved mesomorphic species, e.g., *Acacia myrtifolia*, *A. suaveolens*, *Indigofera australis*, *Pimelea ligustrina*, *Prosthanthera sieberi*, and *Zieria smithii*. The herbaceous layer contains *Imperata* and *Pteridium* with some *Doodia aspera* and *Lomandra longifolia*.

2.7.2. *E. pilularis – E. microcorys* Suballiance

Blackbutt – Tallow-wood

This suballiance occurs north of c. lat 33°S where *E. microcorys* reaches its southern limit. Several other tree species occur, especially *E. grandis* and *E. saligna*, *E. acmenioides*, *E. resinifera*, *E. propinqua*, *E. major*, all of which are found in other alliances, and *Syncarpia glomulifera* and *Tristania conferta*. These species are found on soils of relatively high fertility in the wetter areas. Florence (1963) distinguishes two main groups of associations: i. *E. pilularis – E. microcorys*, and ii. *E. pilularis – E. acmenoides*. *E. pilularis* has a maximum frequency of 90% and *E. microcorys* makes up no more than 50% of the stands. *E. acmenoides* occurs in less favourable sites and in such areas *E. resinifera, E. propinqua, E. umbra* and *E. intermedia* may be present.

In the tallest stands on soils of higher fertility, mesomorphic trees and tall shrubs form a continuous understory, the more abundant species being: *Backhousia myrtifolia, Elaeocarpus cyaneus, Endiandra sieberi, Eupomatia laurina, Glochidion ferdinandii, Rhodamnia trinervia, Synoum glandulosum* and *Trochocarpa laurina*. Wiry-stemmed lianas are usually present, especially *Smilax australis* and *Hibbertia dentata*. The herbaceous stratum, usually poorly developed, usually contains the ferns, *Culcita dubia* and *Pteridium esculentum*. In many cases the mesomorphic understory has been removed by fire and *Casuarina torulosa* occurs as a small tree, sometimes in almost continuous stands, with *Dodonaea viscosa* and *Jacksonia scoparia* below, and *Imperata cylindrica* dominates the herbaceous layer.

In south-eastern Queensland, Swain (1928) describes more luxuriant assemblages in the understory. He distinguishes a "Blackbutt – Hoop Pine" *(Araucaria cunninghamii)* region from which much of the commercially useful timber has been removed. The forests contain many species in the understory not recorded above, viz., *Cupaniopsis* spp., *Diploglottis australis, Duboisia myoporoides*, and on Fraser Is., *Agathis robusta, Podocarpus elata, Findersia schottiana, Gmelina leichhardtiana, Syncarpia hillii* and others.

2.7.3. *Eucalyptus pilularis – E. intermedia – E. siderophloia* Suballiance
Blackbutt – Pink Bloodwood – Grey Ironbark

These species occur north of lat. 31°S., forming an assemblage on soils of low fertility in a variety of habitats from valley bottoms to the tops of ridges. *E. intermedia,* found mostly on soils of low fertility, sometimes occurs as a shaft-like tree in valley bottoms in association with taller eucalypts, including *E. pilularis*. Several other species occur in these "mixed stand", including *E. propinqua, E. resinifera, E. acmenoides, E. phaeotricha, E. major* and *E. tereticornis*. In some cases a zonation of species from the valley floor to a hilltop is apparent, the most common of which is *E. pilularis – E. resinifera* on the lowest parts, *E. pilularis – E. intermedia – E. propinqua* on the lower slopes and *E. intermedia – E. siderophloia* in the higher situations. Associations can always be recognized locally but grouping them into a suballiance, as has been done here, is an indication of the variety of mixtures that can exist. The condition is similar in the south where *E. gummifera* replaces *E. intermedia*.

2.7.4. *Eucalyptus pyrocarpa* Suballiance
Large-fruited Blackbutt

This species (formerly *E. pilularis* var. *pyriformis*) is restricted to northern New South Wales and grows within stands of *E. pilularis*. It occurs on exposed ridges on shallow, skeletal soils and dominates closed forests to 40 m high. Pure stands are rare, there being usually a small proportion of *E. acmenoides, E. gummifera* or *E. punctata* (Baur, 1965). The understory is xeromorphic.

2.7.5. *Eucalyptus pilularis – E. saligna – E. paniculata* Suballiance
Blackbutt – Sydney Blue Gum – Grey Ironbark

These species may occur in the one stand throughout the range of *E. saligna* but the combination of species is most commonly encountered on the central coast, especially the wetter portion of the Wianamatta Shale area of the Sydney region. *E. resinifera* and *E. eugenioides* are common associates and are often locally dominant along the junction of shale and sandstone areas in valleys. *Syncarpia glomulifera* is usually present. The proportions of *E. pilularis* and *E. saligna* varies, the latter becoming dominant in valleys.

The understory is largely mesomorphic and the following species list from St. Ives (Sydney) is typical of the stands (species arranged within the groups in order of abundance: Tall shrubs:

Pittosporum undulatum, Acacia implexa, Casuarina littoralis, Hakea salicifolia, Rapanea variabilis, Tieghemopanax sambucifolius, Dodonaea triquetra, Exocarpos cupressiformis, Notelaea longifolia, Pittosporum revolutum, Trochocarpa laurina. Lianas: *Smilax australis, Eustrephus latifolius, Pandorea pandorana, Kennedia rubicunda, Smilax glyciphylla.* Small shrubs: *Leucopogon juniperinus, Zieria smithii, Persoonia laurina, Platylobium formosum, Leucopogon lanceolatus.* Herbs: *Themeda australis, Lomandra longifolia, Caladenia carnea, Pterostylis nutans.* Ferns: *Culcita dubia, Adianthum aethiopicum, Blechnum cartilagineum.*

2.7.6. *Eucalyptus pilularis* – *E. gummifera* Suballiance
Blackbutt – Red Bloodwood

This combination of species occurs from the central north coast of New South Wales to the south coast and it can be regarded as an ecotonal assemblage of *E. pilularis* with the low fertility alliances on sandy soils. In the south *E. sieberi* and *E. globoidea* occur as co-dominants in some areas. The stands are usually 15–20 m high (Fig. 9.8), except in valleys where they may reach 30 m. The understory is xeromorphic and includes a large number of shrub species, a few of which are listed below Fig. 9.8. In valleys,

Fig. 9.8. *Eucalyptus pilularis* and *E. gummifera*. Main understory species: *Casuarina littoralis, Acacia botrycephala, A. myrtifolia, Persoonia pinifolia.* 12 km north of Eden, N.S.W.

mesomorphic species are more abundant, especially *Dodonaea triquetra* and *Persoonia linearis*.

2.7.7. *Eucalyptus pilularis* – *E. piperita* Suballiance
Blackbutt – Sydney Peppermint

This combination of species is an ecotonal grouping found mainly in valleys, chiefly on the Sydney sandstones. *E. piperita*, a species on low fertility soils (Chap. 10), occurs typically on the sides of sandstone valleys and, where *E. pilularis* forms tall forests on the valley bottom, the two species commonly form a mixed stand. The higher fertility requirements of *E. pilularis* in this low fertility parent material are met either from organic accumulations downslope or by seams of shale interbedded with the sandstones. As well as *E. pilularis*, *Syncarpia glomulifera* and *Pittosporum undulatum*, both usually stunted, may also occur. In such situations *E. pilularis* is commonly only 15 m high and the trees are often rooted among boulders. In contrast, trees on the valley floor are c. 40 m tall.

Additional occasional trees are *E. punctata* and *Angophora costata*, and the tall shrub, *Ceratopetalum gummiferum*. The understory is a mixture of mesomorphic and xeromorphic shrubs of which *Dodonaea triquetra* and *Pultenaea flexilis* are often abundant.

2.7.8. *Eucalyptus pilularis* – *Angophora costata* Suballiance
Blackbutt – Smooth-barked Apple

These species characterize the tall forests, 15–35 m high, which occupy the deep sands behind coastal dunes in New South Wales. The same combination of dominants occurs in sandstone country on alluvial and/or colluvial soils with a water-table occurring within the rooting zone of the trees.

In the coastal forests *Callitris columellaris* is sometimes locally abundant in the north, and *E. signata* (north) and *E. racemosa* (south) are occasional to rare trees in the stands. The understory is commonly dominated by *Banksia serratifolia* and *B. serrata*, and *Leptospermum attenuatum* forms thickets 3–4 m tall. Characteristic shrub species include: *Acacia botrycephala*, *A. suaveolens*, *Calytrix tetragona*, *Conspermum taxifolium*, *Dillwynia ericifolia*, *Epacris pulchella*, *Gompholobium latifolium*, *Hibbertia linearis*, *Platysace linearis* and *Styphelia viridis*. The small lianas *Hardenbergia violacea* and *Kennedia rubicunda* are often abundant. *Macrozamia communis* is sometimes locally dominant in the understory. Repeated burning of many stands has led to the complete removal of all the original understory species in some areas and their replacement by *Imperata cylindrica* and *Pteridium esculentum*.

2.8. *Eucalyptus maculata* Alliance
Spotted Gum

E. maculata occurs as a dominant or co-dominant over a wide range of climates and its area is broken into two segments by the low fertility soils of the Sydney sandstones. Its wide latitudinal range, in addition to its wide climatic tolerance, accounts for the large number of species with which it can be associated (Table 9.1). Furthermore, although the soils are mainly Podzolic, they vary considerably in fertility, from low in wetter areas to relatively high in drier areas. These variations have an effect on the understory.

In the wetter coastal areas mainly in the north, the forests attain a height of c. 50 m (Fig. 9.9) and often adjoin and mix with *E. pilularis* and *E. tereticornis* and rarely with *E. saligna*. In such cases a mesomorphic understory occurs, the species being those in the *E. pilularis* – *E. microcorys* suballiance (2.7.2 above). In some of these stands *E. rummeryi* (Steel Box) is an associate, together with *E. propinqua*.

On less fertile soils and in drier sites *E. maculata*, either in pure stands or in association with one or more other species (Table 9.1), forms open forests with a tall xeromorphic understory, often with some subsidiary mesomorphic species, of which the following are typical for northern New South Wales and south-eastern Queensland: *Acacia cunninghamii*, *A. penninervis*, *Alphitonia excelsa*, *Banksia intergrifolia*, *Casuarina torulosa*, *Exocarpos cupressiformis*, *Helichrysum diosmifolium*, *Jacksonia scoparia*, *Persoonia levis*, *Petalostigma quadriloculare*, *Pittosporum undulatum* and *Tristania suaveolens*. The herbaceous layer usually contains *Themeda australis* and, in wetter areas, *Imperata cylindrica*. Subsidiary species

Fig. 9.9. *Eucalyptus maculata*. Pure stand with understory of *Imperata cylindrica*. Near Grafton, N.S.W.

are *Aristida leichhardtiana* or *A. ramosa*, *Cleistochloa subjuncea*, *Entolasia stricta* and *Paspalidium gracile*.

Stands south of Sydney are either pure or mixed (Table 9.1) and the understory shows a similar variation to the north. The main mesomorphic tall shrub is *Backhousia myrtifolia*. Many of the stands contain *Macrozamia communis* as the dominant in the understory. It is probable that in most areas mesomorphic species have been eliminated by fire, leaving the following as the most abundant: *Acacia longifolia, Casuarina littoralis, Exocarpos cupressiformis, Helichrysum diosmifolium, H. argophyllum* and *Pittosporum undulatum*. The herbaceous layer contains *Themeda australis, Imperata cylindrica* and *Entolasia stricta*, with subsidiary or local *Danthonia pallida, Poa sieberana* and *Pteridium esculentum*.

2.9. *Eucalyptus robusta* Alliance

Swamp Mahogany

This species usually dominates closed forests in swampy areas adjoining the littoral communities dominated by either *Casuarina glauca* or *Melaleuca quinquenervia* (Chap. 23), and both of these species occasionally occur as understory plants below *E. robusta*. Inland occurrences of *E. robusta* are rare, but the species is sometimes found in forests dominated by *E. amplifolia, E. tereticornis* or *Melaleuca* spp.

The forest floor is almost permanently wet and pools of water may be dotted throughout the stand. The soils are subject to very occasional flooding by sea water, though they are acid throughout the profile (pH 4.5–4.0). On the most heavily waterlogged areas few or no associated woody species exist. Hydrophytes and helophytes are present in free water that may occur, e.g., *Villarsia reniformis* and *Baumea juncea*, and patches of *Sphagnum* sometimes occur. Constant herbaceous species are *Blechnum indicum, Gahnia grandis, Restio tetraphyllus* and *Haloragis micrantha*. On better drained areas, usually sand-patches, *Imperata cylindrica* and *Pteridium esculentum* are locally dominant.

Shrubs are not of constant occurrence, and the few that do occur come from the rainforests, e.g., *Acronychia laevis, Backhousia myrtifolia, Cupaniopsis anacardioides, Endiandra sieberi, Ficus coronata* and *Livistona australis*. The last

Fig. 9.10. *Eucalyptus elata* on a river flat. *Casuarina cunninghamiana* fringing the river in the background. Understory species include *Flagellaria indica, Trema aspera, Acacia floribunda* and *Tristania laurina*. Main herbaceous species are *Stipa elegantissima* and *Adiantum aethiopicum*. Colo R., north-west of Sydney.

is often the most abundant and the most conspicuous because of its height. A few lianas typical of the littoral rainforests (Chap. 7) usually occur.

On the landward side, the swamp forests occasionally adjoin patches of rainforest of varying floristic composition, or *Eucalyptus* forest containing one or more of the following species: *E. botryoides* (south), *E. longifolia, E. pellita, E. resinifera* or *E. tereticornis*. Alliances typical of well drained soils may adjoin the swamp forest if the gradient from the swamp to higher country is steep.

2.10. *Eucalyptus elata* Alliance

River Peppermint

E. elata is a southern species (Fig. 9.1) usually associated with watercourses. It occurs mainly on the lowlands but may extend on to the lower tablelands to an altitude of c. 600 m. It forms narrow belts of forest on flats adjacent to streams which are commonly fringed by *Casuarina cunninghamiana*. The soil are mostly Alluvial or Podzolics, less commonly Brown Earths, and are of moderately high fertility as a result of the accumulation of organic matter on the valley floor.

E. elata usually occurs in pure stands mostly c. 30 m tall (Fig. 9.10) but is sometimes associated with other eucalypts which have a high moisture requirement, especially *E. viminalis, E. tereticornis, E. bauerana* and *E. angophoroides*. The understory is mesomorphic and usually contains species which occur in the *Ceratopetalum apetalum* rainforests, the more common ones being *Acmena smithii* and *Tristania laurina*. A few lianas occur, especially *Clematis aristata* and *Stephania japonica*.

10. Eucalyptus Woodlands and Forests on Soils of Low Fertility Chiefly on the Eastern Coastal Lowlands

1. Introduction

These communities occur on soils of low fertility chiefly on the lowlands from the Tropic to Victoria, with extensions to South Australia and Tasmania. The forests and woodlands commonly adjoin heaths and *Banksia* scrubs which occupy soils of even lower fertility (Chap. 16) and they commonly adjoin and mingle with the tall forests which develop on soils of higher fertility (Chap. 9). In contrast to the tall forests which usually have mesomorphic species in the understory, the forests and woodlands on soils of lower fertility have understories of xeromorphic shrubs, and the xeromorphic assemblages become floristically richer from north to south, with maximum development on the sandstones of the Sydney region.

The alliances are described below in latitudinal sequence. The two alliances with the greatest latitudinal range (Fig. 10.1) are characterized by closely related pairs of species, *E. intermedia* and *E. gummifera* being "Bloodwoods", and *E. signata* and *E. racemosa* "Scribbly-gums". Some of the other species have extensive north-south ranges, among them *E. sieberi* which occurs from the Sydney region to Tasmania and links the mainland and Tasmanian alliances. Also, *E. globoidea* occurs over most of the coast of New South Wales, extending to eastern Victoria, and together with *E. sieberi* and many of the xeromorphic understory species, relate the east coast to the south coast alliances. However, in the south, *E. baxteri* has a restricted distribution and so also have the Tasmanian species, except *E. nitida* (Fig. 10.16).

Table 10.1. Main associations in the *Eucalyptus intermedia* – *E. acmenoides* – *E. signata* – *E. nigra* Alliance.

Suballiance	Main Associations	Other species forming minor associations	Species forming ecotonal associations
E. intermedia – *E. acmenoides*	Tropics: *E. intermedia* – *E. acmenoides*	*E. citriodora*; *E. exserta*; *E. carnea*; *E. abergiana*	*E. crebra*; *E. tereticornis*; *E. drepanophylla*; *E. tessellaris*; *E. alba*; *E. cloëziana*
	South of Tropic: *E. intermedia* – *E. acmenoides*; *E. intermedia*	*E. citriodora*; *E. exserta*; *E. siderophloia*; *E. carnea*; *Angophora costata*	*E. tereticornis*; *E. pilularis*; *E. signata*; *E. crebra*; *E. maculata*; *E. nigra*
E. signata – *E. intermedia* – *E. nigra*	*E. signata* – *E. intermedia*; *E. nigra*; *E. signata*; *Callitris columellaris*	*E. trachyphloia*; *E. curtisii*	*E. acmenoides*; *E. tessellaris*; *E. crebra*; *E. paniculata*; *E. planchoniana*; *E. pilularis*

ing a mean annual rainfall of 1000–1500 mm. It is divided into two segments by the dry climate around the Tropic, as indicated by the distribution of *E. intermedia* in Fig. 10.1. The forests and woodlands occupy the sandy soils of low fertility derived mainly from Jurassic and Cretaceous sandstones or from granites or sandy alluviums. Two suballiances are recognized. The more northerly of these is characterized by *E. intermedia*, which occurs throughout the alliance but is less abundant in the south. *E. signata*, which characterizes the second alliance, becomes more conspicuous on deep sands in the south, especially near the ocean. The other two species occur throughout the alliance. *E. acmenoides* covers much the same area as *E. intermedia* but extends southward to Sydney. This species is a common constituent of some tall forests (Chap. 9) and it is possible that ecotypes may exist. *E. nigra* (= *E. phaeotricha*) has a similar range to *E. intermedia*.

2.1.1. *Eucalyptus intermedia* – *E. acmenoides* Suballiance
Pink Bloodwood – White Mahogany

These species occur as forests 15–20 m high in the tropics and disjunctly to the south (Fig. 10.1). In the tropical segment the communities are open forests or woodlands and several other species form mixed stands (Table 10.1). There is a poor development of shrubs, of which only *Tristania suaveolens* is common. The herbaceous layer is dominated by *Imperata cylindrica* and *Themeda australis* with lesser *Heteropogon triticeus* (Isbell and Murtha, 1972).

South of the Tropics, *E. intermedia* and *E. acmenoides* occur either intermixed (Figs. 10.2, 10.3) or in association with a number of eucalypts, especially *E. citriodora*, *E. exserta* (Fig. 10.4), and in south-eastern Queensland where the alliance is best developed, with *E. siderophloia* and *Angophora costata*. Mixed stands with *E. crebra* and *E. maculata* are common, and this assemblage extends into north eastern New South Wales.

The shrub layer in these stands is better developed than in the tropics. *Tristania suaveolens* is often present but the species are mostly xeromorphs and include *Grevillea banksii*, *Casuarina littoralis*, *Xylomelum pyriforme* and *Xanthorrhoea* sp. *Macrozamia communis* is sometimes common near the coast. The herbace-

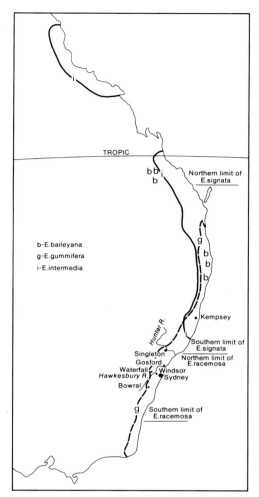

Fig. 10.1. Distribution of species of *Eucalyptus* on soils of low fertility on the east coast. Based on Hall et al. (1970).

2. Descriptions of the Alliances

2.1. *Eucalyptus intermedia* – *E. acmenoides* – *E. signata* – *E. nigra* Alliance
Pink Bloodwood – White Mahogany – Scribby Gum – Stringybark

This alliance extends over flat or undulating country from the tropics to the central coastal lowlands of New South Wales, the area receiv-

Fig. 10.2. General view of xeromorphic forests with a high proportion of *Eucalyptus intermedia*. In the foreground, *Xanthorrhoea* sp. and *Macrozamia communis*. Near Rockhampton, Qld.

ous layer is dominated by grasses, especially *Themeda australis*, *Cymbopogon refractus*, *Bothriochloa macra* and *Aristida ramosa*; small shrubs and herbs are present, e.g., *Pimelea linifolia*, *Stylidium graminifolium*, *Hypoxis glabella*, *Fimbristylis dichotoma* and *Lomandra glauca*.

2.1.2. *Eucalyptus signata* – *E. intermedia* – *E. nigra* Suballiance

Scribbly Gum – Pink Bloodwood – Stringybark

This suballiance, characterized by *E. signata*, is found close to the coast on low fertility sands and gravelly sands in a relatively wet climate (mean annual rainfall mostly 1100–1500 mm). The communities are xeromorphic woodlands varying in height from 8 to 15 m (Fig. 10.5) and they commonly adjoin the *Banksia* scrubs and heaths (Coaldrake, 1961; Westman and Rogers, 1977) in which both *E. signata* and *E. intermedia* may occur in mallee form. *E. acmenoides* is sometimes present as an occasional species. Also *Callitris columellaris* is an occasional associate (sometimes locally dominant) near the coast on deep sands (Fig. 10.6). Other eucalypts of local and usually rare occurrence are *E. curtisii*, *E. trachyphloia* and *E. tessellaris*.

The shrub layers were probably well developed before repeated burning occurred. Today one or two layers may exist. The tall shrub layer when present consists of one or more of *Alphitonia excelsa*, *Banksia integrifolia*, *B. serratifolia*, *Casuarina littoralis*, *Tristania suaveolens* and stunted *Syncarpia glomulifera* and *Tristania conferta*. Small shrubs (in various combinations) include: *Acacia concurrens*, *A. longifolia*, *Banksia aspleniifolia*, *Bossiaea heterophylla*, *Breynia oblongifolia*, *Hakea dactyloides*, *Jacksonia*

Fig. 10.3. *Eucalyptus intermedia, E. acmenoides* and *E. nigra*. Understory: *Casuarina littoralis, Banksia integrifolia, Tristania suaveolens*. The grass layer is dominated by *Themeda australis*.

Fig. 10.4. *Eucalyptus citriodora, E. exserta* and *E. intermedia*. Shrub layer of young eucalypts. Herbaceous layer contains *Cymbopogon refractus, Themeda australis, Paspalidium gracile, Dianella revoluta* and the small liana *Hardenbergia violacea*. Between Gin Gin and Childers, Qld.

Fig. 10.5. *Eucalyptus signata* (smooth bark) and *E. intermedia* (rough bark). Understory: *Pultenaea villosa, Persoonia cornifolia, Hakea dactyloides* and young plants of *Syncarpia glomulifera*. Near Beerwah, Qld.

scoparia, Lomatia salaifolia, Melastoma denticulatum, Monotoca scoparia, Persoonia cornifolia, Pimelea linifolia, Platylobium formosum and *Pultenaea villosa*. The herbaceous layer varies with the frequency of burning as indicated by the presence and abundance of *Imperata cylindrica* and *Pteridium esculentum*. Other species are: *Alloteropsis semialata, Caustis blakei, Dampiera stricta, Fimbristylis dichotoma, Haemadorum planifolia, Hypoxis glabella, Lepidosperma laterale, Stylidium graminifolium, Themeda australis, Xanthorrhoea macronema* and *X. media*.

2.2 *Eucalyptus baileyana* – *E. planchoniana* – *E. bancroftii* Alliance

Bailey's Stringybark –
Bastard Tallow Wood – Orange Gum

These species occur on low fertility skeletal, sandy or gravelly soils in a few isolated localities in north-east New South Wales and south-eastern Queensland (Fig. 10.1) where the mean annual rainfall is c. 1100–1700 mm. In New South Wales the stands lie on the coastal lowlands at an elevation below 170 m, but in

Fig. 10.6. *Callitris columellaris* mature tree and young plants on right. *Eucalyptus intermedia* behind. Broad-leaved shrubs are stunted *Tristania suaveolens*. Bribie Is. Qld.

Queensland they occur on the Blackdown Tableland at an elevation of 500–800 m.

The stands are variable in composition. *E. baileyana* often forms pure stands and in New South Wales all three species commonly occur together, sometimes with subsidiary *E. intermedia* and/or *E. signata* (Fig. 10.7) and *Angophora subvelutina*. The shrubs are mostly xeromorphic and form two layers. Tall shrubs (mainly on deeper soils) include *Acacia cunninghamii, Banksia integrifolia, Casuarina littoralis* and *Jacksonia scoparia*, and the mesomorphic *Alphitonia excelsa, Petalostigma quadriloculare* and *Tristania suaveolens*. Small shrubs, which become more prominent on shallow soils, include: *Banksia aspleniifolia, Bossiaea rhombifolia, Daviesia ulicifolia, Isopogon anemonifolius, Lambertia formosa, Leptospermum attenuatum* and *Persoonia cornifolia*. Herbs and dwarf shrubs form a discontinuous layer composed of: *Alloteropsis semialata, Anisopogon avenaceus, Amperea xiphoclada,* *Cymbopogon refractus, Echinopogon caespitosus, Gahnia clarkei, Goodenia hederacea, Haemodorum planifolium, Laxmannia gracilis, Lobelia dentata, Poranthera microphylla* and *Psilotum triquetrum*.

2.3. *Eucalyptus gummifera* – *E. racemosa* *E. sieberi* Alliance

Red Bloodwood – Scribbly Gum – Black Ash

2.3.1. General: Climate; Soil Parent Materials; Soils; Topography

This alliance is developed on the low fertility, sandy soils derived mainly from sandstones in the central coast of New South Wales, with

Fig. 10.7. *Eucalyptus baileyana, E. intermedia* and *Angophora subvelutina*. Shrubs are *Alphitonia excelsa, Casuarina littoralis* and *Jacksonia scoparia*. Grass in foreground is *Imperata cyclindrica*. Near Copmanhurst, N.S.W.

some extensions north and south on sandy alluviums. The alliance is very complex, partly because of the great diversity of xeromorphic species which it contains, and partly because of the variety of communities of different structural form developed in the many habitats presented within the sandstone areas. The communities have been described for most of the sandstones by Pidgeon (1937, 1938, 1940, 1941) and for restricted areas by Petrie (1925), Davis, (1936, 1941) and Burrough et al. (1977). The mean annual rainfall varies from c. 750 to 1500 mm and is fairly evenly distributed throughout the year. Because of the rugged topography in some areas, run-off is considerable where rock lies at the surface, resulting in dry habitats. Conversely run-on into valleys produces almost permanently wet environments. Close to the coast, temperatures are equable, frosts being rare. In the highlands, frosts occur during the winter, absolute minima being as low as $-7°C$ and this has some effect on the distribution of the communities.

The soil parent materials are almost entirely sandstones, mainly of Triassic age and assigned to the Hawkesbury Series. This series was at one time covered by Triassic shales of the Wianamatta Series. Other Triassic rocks belonging to the Narrabeen Series, laid down below the Hawkesbury Series, outcrop in certain areas and, below these, sandstones of the Permian Upper Coal Measures and Upper Marine Series (oldest) sometimes lie at the surface. As a result of warping in the central coastal region and the elevation of the Blue Mts, the rocks are not always found in successively older layers with depth. On the contrary, the youngest Wianamatta Series has its main area west of Sydney close to sea level and successively older rocks occur in sequence north and south, with the Coal Measures c. 300 m below Sydney. In the Blue Mts the Hawkesbury Sandstones lie at the surface at an elevation of c. 1000 m, where they form a plateau. Furthermore, on the tablelands, remnants of the Tertiary basalt flows still exist, e.g., at Mount Wilson, Bowral and Moss Vale and these support rainforests dominated by *Ceratopetalum apetalum,* which stand in sharp contrast to the xeromorphic flora of the sandstones and illustrate the selecting effect of soil fertility on the flora and vegetation.

The sandstones of various geological ages are commonly interbedded with seams or lenses of shale, and in some areas the effects of vulcanism are seen in the form of dykes of basic rock. The sandstones in general are quartz sandstones, more rarely conglomerates, the quartz being cemented by kaolin or illite, some containing fragments of feldspars or micas. They are very low in most plant nutrients. The shales on the other hand consist mainly of kaolin and illite and are richer in plant nutrients than the sandstones. The sandstones on erosion produce sand which has accumulated as alluvium along some streams as sand or gravel patches, and along the coast as sanddunes or sand flats, on to which the sandstone flora has migrated.

The soils are mainly Lithosols and Yellow Podzolics with small areas of Red and Lateritic Podzolic Soils and Alluvial Soils in valleys. They have in common a general sandiness throughout the profile, and are mostly very acid, with pH values of 4.5–5.0. In rugged terrain rock commonly lies near or at the surface, plants being rooted in cracks along bedding planes or among boulders. The soils are almost useless agriculturally, a condition first discovered by the earliest settlers in 1788. This is due partly to the extensive areas covered by rock and to the shallowness of the soils, and also to the generally low fertility. It is only in some valleys or on some plateaux where the soils are deeper sometimes derived from shale lenses or seams that agriculture is now practised (chiefly orchards, vegetable crops and cut-flowers).

Soil fertility levels are invariably low in the soils derived *in situ* from the sandstones or in alluvium derived from the sandstones, except in valleys where organic matter has accumulated from upper sandstone areas or from shales. Phosphorus levels in upland areas vary between 30 and c. 80 ppm P. In valleys phosphorus levels are about twice as high. In contrast, seams and lenses of shale within the Hawkesbury Sandstones mostly have phosphorus levels of 150 to 250 ppm P and shales of the Narrabeen Series (which support tall forests or rainforests) have levels varying between 150 and 1200 ppm P. Nitrogen levels have been studied by Hannon (1956, 1958) who records c. 180 ppm in Hawkesbury sandstone rock and c. 600 ppm. in the surface soils, and carbon/nitrogen ratios of 18–49. The nitrogen capital of the systems appears to be dependent mainly on symbiotic nitrogen fixation by legumes and species of *Casuarina*, the photosynthetic organs of which contain a much higher nitrogen content (c. 12,000 ppm) than species without a nitrogen-fixing mechanism (7,000 ppm). Other nutrients might well be in low supply, though a source of nutrients from the ocean in cyclic salt must occur in those areas close to the sea.

Over much of the area the country is rugged, both on the coast and in the mountains. Since the sandstones are jointed, they erode in blocks leaving vertical cliffs which, on the coastal plains, are up to 50 m high and in the mountains a few hundred metres. Narrow valleys are therefore common and waterfalls of various sizes occur. Although plateaux are referred to, these are rarely continuously flat areas, but are dissected by narrow gorges. The effect of the rugged topography is the occurrence of a variety of habitats from "arid" on the rocky outcrops on the summits to ever-moist in deep ravines, especially under waterfalls. In between these two extremes, a gradation of moisture régimes exists which is reflected in a gradient in the flora and vegetation. Aspect effects are sometimes observable within valleys and ravines, sometimes in the dominant stratum and sometimes in the understory, the north-facing slope being drier, having a more xeric flora.

2.3.2. The Flora

The flora is of special significance since it represents the richest assemblage of xeromorphic species in the east and is a remnant of the one-time xeromorphic assemblage which once spanned the continent, at least in the south (Chap. 4). This vast tract of xeromorphs is now divided by the arid zone, and in the east isolated segments are to be seen from Tasmania to southern Queensland and as far west as south-eastern South Australia.

The total vascular flora consists of c. 1140 species. These occur over a total area of c. 22,000 square km. However, so many of the species are widely distributed that half of the species usually occur in a small fraction of this area. For example, Rose and Evans (1974) record 37 pteridophytes, 3 gymnosperms and 483 angiosperms (i.e. half the total vascular flora) in an area of 147 square km. in the Ku-ring-gai Chase National Park near Sydney.

The pteridophytes are relatively well represented, especially the more primitive ones – *Psilotum, Lycopodium, Selaginella, Leptopteris* and *Todea, Schizaea* and the *Gleicheniaceae*. These occur mainly in sands, rock-ledges or near waterfalls *(Leptopteris fraseri)*. The more advanced ferns, with a few exceptions, are found in valleys or in shaded situations in forests.

The gymnosperms are represented by *Macrozamia communis* and *M. spiralis* on the coast, and *M. secunda* in the Blue Mts. *Podocarpus spinulosus* is found in valleys, and *Callitris muelleri* and *C. rhomboidea* are widespread but of little importance ecologically.

The angiosperm flora is too extensive to list (see Beadle et al., 1975) but the most abundant families are listed in Table 10.2. The Orchidaceae is most abundant as species; almost all are ground orchids, the plants perennating by underground tubers. Of the woody perennials, the Myrtaceae, Papilionaceae and Proteaceae are most abundant and usually dominate either the upper story or the understory, or both. Other prominent woody xeromorphic families are Epacridaceae Mimosaceae *(Acacia)* and Rutaceae. The Cyperaceae owes its apparent abundance to the occurrences of sedges mainly in swamps, but three genera, *Caustis*,

Table 10.2. Families with the greatest representation of species on the sandstones of the central coast of N.S.W. and adjacent tablelands. Species occurring in gullies in or near rainforests are not included. Data from Beadle et al. (1975).

	Genera	Species
Orchidaceae	24	143
Myrtaceae	12	105
Papilionaceae	19	99
Proteaceae	12	75
Cyperaceae	15	67
Gramineae	22	58
Epacridaceae	13	51
Compositae	12	43
Mimosaceae	1	39
Rutaceae	7	34
Goodeniaceae	5	23
Liliaceae	13	22
Labiatae	6	22
Dilleniaceae	1	19
Xanthorrhoeaceae	2	18
Euphorbiaceae	8	17
Umbelliferae	6	16
Restionaceae	5	14

Cyathochaete and *Lepidosperma* occur in well drained sandy soils. The Gramineae, though well represented as species, is usually subsidiary in most communities and is replaced in the herbaceous layer by other monocotyledons, especially Xanthorrhoeaceae, Liliaceae, some Cyperaceae and, mainly in wetter areas, by Restionaceae and Xyridaceae. The few grasses that become abundant belong mainly to the genera *Anisopogon, Aristida, Entolasia, Paspalidium* and *Stipa*.

Most of the genera occurring in this assemblage are endemic in Australia but few endemic genera are confined to this alliance. Two noteworthy endemics are *Atkinsonia ligustrina*, a rooted parasitic member of the Loranthaceae restricted to a small area in the Blue Mts, and the giant "Gymea Lily", *Doryanthes excelsa*, confined to the coast north and south of Sydney (Chap. 4).

Many of the species are widely distributed over the entire area, though they occupy specific habitats. On the other hand, some species have restricted distributions determined mainly by altitude. This is well illustrated by the distributions of the eucalypts (Table 10.3). The wide

Table 10.3. Tree and mallee species occurring on the sandstones of the central coast of N.S.W. and adjacent tablelands. M = mallee. Arrows indicate occurrence outside the sandstone areas: ↑ indicates extending to the north; ↓ indicates extending to the south; ↕ indicates extending north and south.

Species typical of the sandstones		Species forming ecotonal associations
Higher Altitudes only	High and Low Altitudes	Higher Altitudes
E. apiculata M		E. cypellocarpa ↕
E. baeuerlenii M		E. mannifera ↕
E. blaxlandii		E. radiata ↕
E. ligustrina M ↑		
E. moorei M ↕	E. agglomerata ↓	See Chap. 11
E. oreades ↑	E. eugenioides ↕	
E. rupicola M	E. gummifera ↕	
E. triflora M	E. haemastoma	
	E. oblonga	
Lower Altitudes only	E. piperita ↓	Lower Altitudes
E. camfieldii M	E. racemosa	E. botryoides ↓
E. capitellata	E. sieberi ↓	E. crebra ↑
E. eximia	E. stricta M	E. deanei ↑
E. globoidea ↓	Angophora costata	E. paniculata ↕
E. luehmanniana M		E. pilularis ↕
E. notabilis ↑		E. resinifera ↑
E. obtusiflora M		E. saligna ↕
E. punctata ↑		E. tereticornis ↕
E. squamea		
E. umbra ↑		See Chap. 9

distribution of species inevitably poses a problem in the floristic classification of the communities and this is further complicated by variations in the sizes of some of the dominant species which range from tall trees in some habitats to mallees in others.

Introduced species have had relatively little effect on the total flora, apparently because of the difficulty of establishing in the low fertility soils. Exotics have become abundant only in those areas where nutrients have been added locally to the soils, notably along roads and railway lines, and near habitation.

Fires are of regular occurrence and have altered the composition of the understory species in many areas, those species which regenerate rapidly after a fire gradually becoming locally dominant. The eucalypts are remarkably fire-resistant, though in many cases trunks are killed and replaced by several stems from epicormic shoots to produce a mallee plant. Most species survive fires either through the survival of vegetative parts or seed (Beadle, 1940). The former include buds protected by thick bark in trunks of woody plants, lignotubers or caudices and buds from underground organs including rhizomes, tubers and roots. Many species have woody fruits (Myrtaceae, Proteaceae) or cones *(Casuarina)* which protect the seeds from high temperatures, and many species have hard seeds (Papilionaceae and *Acacia*) which survive fires and are stimulated to germinate after fire by the cracking of the testa by high temperatures or by hydration of the testa by hot water or steam. The high proportion of Myrtaceae, Proteaceae, Papilionaceae, *Casuarina* and *Acacia* is maintained, and probably increased, by repeated burning. Another effect of fire is the temporary exposure of the soil surface to the elements, followed by movement of soil particles downslope, which tends to increase the area of rock at the surface in upland areas and to decrease the area of swampy habitat in more lowland sites. Local change in the patterning of the flora follow such soil movements.

2.3.3. The Plant Communities: General

Eucalypts (Table 10.3) dominate large areas of the sandstones and in successional terms *Eucalyptus* forests may be regarded as the climax vegetation, as described by Pidgeon (1938–1941). However, in many areas, communities other than forests and woodlands occur over large tracts of country, or even within forested areas where some local topographic feature prevents the establishment of trees, such as rocky outcrops, or swamps. Pidgeon arranged the various communities as successions (lithoseres commencing on bare rock surfaces, hydroseres commencing in waterlogged areas), which culminate in forest communities. Whether the various communities are regarded as successional stages in time (projected into the future) or as sequences with the ecological variables is of little consequence as far as the community descriptions are concerned. The one difference lies in the interpretation of the climaxes, which in the account that follows are regarded as the forests which occur on the deep soil profiles derived from the sandstones without accretions of material from external sources. Communities which occur in valleys where soil enrichment has occurred (notably *E. pilularis* tall forests – see Chap. 9) are regarded as fragments for other alliances which may form ecotonal associations with the communities of the sandstones. Similarly, at higher altitudes ecotonal associations occur with the tableland communities (Chap. 11).

The communities are described below under two main headings: i. communities in rocky terrain and in swamps and ii. communities dominated by *Eucalyptus*.

2.3.4. Communities in Rocky Terrain and in Swamps

Pidgeon (1938) described the succession on sandstone rocks as follows (for details, see Chap. 16): The first colonizers of dry rocks are algae, followed in sequence by crustaceous and foliose lichens, xeric mosses and fruticose lichens, and by herbaceous perennial vascular plants. Shrubs become established in these herbaceous micro-communities and with increasing soil depth the variety of shrubs increases and a heath community containing several dozen species develops. Taller shrubs establish as soil depth increases, and these species form the understory in the *Eucalyptus* communities. On rock surfaces which are usually wet, the lichen stage is rapidly succeeded by the moss and herbaceous stages and in moist situations the sequences of communities is somewhat different from those in drier habitats as described in Chapter 16.

Fig. 10.8. Swamp on Hawkesbury sandstone dominated by *Hypolaena fastigiata* and *Xanthorrhoea resinosa*, with *Sowerbaea juncea* and *Blandfordia nobilis*. Shrub is *Banksia aspleniifolia* and *Sprengelia incarnata* also present. *Eucalyptus racemosa* behind. Near Waterfall, N.S.W.

Table 10.4. Main associations in the *Eucalyptus gummifera* – *E. racemosa* – *E. sieberi* alliance, grouped under suballiances. Species forming ecotonal associations in brackets.

South of Hawkesbury R.	North of Hawkesbury R.
E. sieberi – *E. piperita* – *E. racemosa* Suballiance	Drier western parts
Higher altitudes (Blue Mts and Wingecarrabie Tableland)	*E. eximia* – *E. punctata* Suballiance
E. sieberi ± *E. piperita*, *E. racemosa*, *E. oblonga*, *E. gummifera*, *E. stricta*, (*E. oreades*), (*E. radiata*), (*E. mannifera*)	*E. eximia* ± *E. punctata*, *E. gummifera* – *E. oblonga*, *Angophora bakeri*, *E. racemosa*, *A. costata*, *Callitris rhomboidea*, *E. piperita*
E. piperita ± *E. sieberi*, *E. racemosa*	*E. punctata* ± *A. bakeri*, *E. oblonga*, *E. gummifera*, *E. racemosa*, *E. eximia*, *A. costata*, (*E. crebra*)
E. racemosa ± *E. piperita*, *E. stricta*, *E. sieberi*, *E. gummifera*, *E. consideniana*	*E. racemosa* ± *A. bakeri*, *E. oblonga*, *E. gummifera*, *E. punctata*
Lower altitudes	Wetter eastern parts
E. sieberi ± *E. gummifera*, *E. oblonga*, *E. racemosa*, *E. piperita*	*E. gummifera* ± *E. racemosa*, *E. haemastoma*, *E. piperita*, *A. costata*, *E. capitellata*, *E. punctata*, *E. sieberi*, *E. eugenioides*
E. piperita ± *Angophora costata*, *E. punctata*, *E. agglomerata*, (*E. pilularis*)	*E. haemastoma* ± *E. racemosa*, *E. capitellata*, *E. oblonga*
E. racemosa ± *E. gummifera*, *E. sieberi*, *E. oblonga*, *E. consideriana*	*E. racemosa* ± *E. gummifera*, *E. haemastoma*
	A. costata ± *E. gummifera*, *E. punctata*, *E. oblonga*, *E. piperita*, *E. umbra*
	E. piperita ± *A. costata*, *E. punctata*, (*E. pilularis*)

coastal lowlands of New South Wales, extending to eastern Victoria. *E. globoidea* is often the dominant of tall, xeromorphic forests, sometimes in association with *E. sieberi* or other eucalypts (Table 10.5).

The shrub layer is xeromorphic; a typical assemblage on a Podzol is listed under Fig. 10.17. On more sandy deposits closer to the sea *Banksia serratifolia* sometimes dominates the tall shrub layer with subsidiary *Casuarina littoralis, Acacia longifolia* and *Platylobium formosum*.

On deep sands, which are subject to some waterlogging, the stringybark is replaced by *E. cinerea* ssp. *cephalocarpa* (Fig. 10.18) with an understory of xeromorphs of which *Leptospermum obovatum* is a near-constant. Other species include *Banksia serratifolia* (mainly near the sea), *Melaleuca squarrosa* (waterlogged patches), *Acacia longifolia, Epacris impressa, Casuarina pusilla*. The more common herbs are *Lepidosperma longitudinale* and *Schoenus imberberis*.

2.5. *Eucalyptus baxteri* Alliance
Baxter's Stringybark

E. baxteri is confined to South Australia and Victoria (Fig. 10.16) and is restricted to soils of low fertility, chiefly Podzolics with deep, white, sandy A horizons in which lateritic gravel or pans may occur. It is found mainly in areas with a mean annual rainfall of 625–750 mm and the plants vary in size according to soil fertility levels rather than moisture availability.

Pure stands of *E. baxteri,* usually in woodland form up to 15 m high, develop on Podzols with an A horizon c. 2 m deep, often with ironstone concretions. *E. viminalis* or *E. obliqua* occur in some stands and the understory is xeromorphic, containing *Astroloma conostephioides* and *Hibbertia sericea* and sometimes *Xanthorrhoea australis* (Wood, 1937). With increasing soil fertility levels, *E. obliqua* increases in abundance and the two species form a common association throughout the range of *E. baxteri*.

Table 10.5. Main associations in alliances on soils of low fertility in south-eastern New South Wales to South Australia and Tasmania.

Alliance	Main associations	Species forming minor associations with the characteristic species	Species forming ecotonal associations
Eucalyptus globoidea (N.S.W., Vic.)	*E. globoidea*; *E. cinerea* ssp. *cephalocarpa*	*E. polyanthemos*; *E. bridgesiana*; *E. consideniana*	*E. sieberi*; *E. cypellocarpa*; *E. botryoides*; *E. obliqua*; *E. sideroxylon*
E. baxteri (Vic., S.A.)	*E. baxteri*; *E. baxteri* – *E. cosmophylla*	*E. alpina* (Grampians, Vic.)	*E. obliqua*; *E. viminalis*; *E. diversifolia* (Kangaroo Is.); *E. radiata* (Vic.); *E. nitida* (Vic.); *E. fasciculosa* (S.A.)
E. amygdalina (Tas.)	*E. amygdalina*	*E. pulchella*; *E. risdonii*; *E. tenuiramis*; *E. barberi*; *E. cordata*; *E. morrisbyi*	*E. sieberi*; *E. rubida*; *E. dalrympleana*; *E. delegatensis*; *E. viminalis*; *E. obliqua*; *E. ovata*
E. nitida (mainly Tas., rare Vic.)	*E. nitida*		*E. ovata*; *E. baxteri* (Vic.)

With decreasing soil fertility levels *E. baxteri* becomes progressively shorter and in extreme cases is reduced to shrub proportions, the plants being less than 1 m high. Such low woodlands and scrubs (Fig. 10.19) usually contain *E. cosmophylla* (Cup Gum) which is a mallee on soils of lowest fertility but may attain a height of 10 m on less infertile soils. The *E. baxteri* – *E. cosmophylla* association occurs on Mt. Lofty, the Fleurieu Peninsula and Kangaroo Is. The soils are Podzols containing massive ironstone pans. *E. baxteri* is dominant on the deeper sands, and *E. cosmophylla* dominates where ironstone lies near the surface. Where the ironstone is exposed, *Casuarina striata* forms local thickets, and on loose sand within the association *Hypolaena fastigiata* dominates a sedgeland. The understory species below *E. baxteri* are all

◁ Fig. 10.17. *Eucalyptus globoidea* (Stringybark nearest camera) and *E. cypellocarpa* (smooth bark, left). Understory: *Acacia botrycephala, Exocarpos cupressiformis, Melaleuca armillaris* (shrubs), and *Stypandra glauca, Lepidosperma longitudinale, Imperata cylindrica* and *Pteridium esculentum.* Near Orbost, Vic.

Fig. 10.18. *Eucalyptus cinerea* spp. *cephalocarpa* (foreground), with a tall forest stand of *E. globoidea* behind. Shrubs in foreground are *Leptospermum obovatum, Acacia longifolia, Casuarina pusilla* and *Epacris impressa.* 15 km south of Buchan, Vic.

Fig. 10.19. *Eucalyptus baxteri* scrub with an understory of *Leptospermum myrsinoides* and *Casuarina muelleriana*. The grass is *Stipa semibarbata*. 25 km south of Keith, S.A.

xeromorphic (Wood, 1937; Baldwin and Crocker, 1941; Welbourn and Lange, 1968; Specht, 1972).

2.6. *Eucalyptus amygdalina* Alliance

Black Peppermint

E. amygdalina occurs mainly in the eastern half of Tasmania (Fig. 10.16) in contrast to the closely related *E. nitida* which is found mainly in the western half of the island (see next section). *E. amygdalina* is closely related to *E. radiata* and also to *E. pulchella*, *E. risdonii* and *E. tenuiramis*, the last three being included in this alliance.

E. amygdalina, which forms woodlands (Fig. 10.20) or forests mostly 15–20 m tall (rarely to 30 m), develops mainly on Podzolics of low fertility derived from siliceous parent materials, including sandstones, conglomerates, mudstones, quartzites and granites. It extends from near sea level to altitudes of 800 m (rarely 1000) where the mean annual rainfall is 500–900 mm. When it occurs in wetter areas, usually at higher elevations, it occupies ridges where the soils are shallow. Light frosts are experienced over most of the area (absolute minima c. $-4°C$).

E. amygdalina forms pure stands over most of its range. The understory consists of a more or less continuous shrub layer of xeromorphs (which are reduced in abundance by fires) consisting of the following common species: *Acacia dealbata*, *A. mucronata*, *A. myrtifolia*, *A. stricta*, *A. verticillata*, *A. vomeriformis*, *Banksia marginata*, *Beyeria viscosa*, *Casuarina monilifera*, *Daviesia latifolia*, *Dodonaea viscosa*, *Exocarpos cupressiformis*, *Persoonia juniperina*, *Pultenaea juniperina* and *Tetratheca*

Fig. 10.20. *Eucalyptus amygdalina* woodland. Understory is mainly *Acacia dealbata* (suckering), *Exocarpos cupressiformis*, *Tetratheca ericifolia* and *Lomandra longifolia*. Derwent valley, near New Norfolk. Tas.

ericifolia. The herbaceous plants are scattered or form clumps. They are mostly xeromorphic monocotyledons belonging to the genera *Deyeuxia, Dianella, Lepidosperma, Lomandra* and *Stipa*.

The *E. amygdalina* forests sometimes form mosaics with forests occupying soils of higher fertility and ecotonal associations are formed with the various species of these alliances, e.g. *E. delegatensis* and *E. dalrympleana* at higher elevations, and *E. viminalis, E. obliqua, E. ovata* and *E. globulus* at lower altitudes. In these ecotones the understory is partly mesomorphic.

Several species of restricted occurrence in Tasmania form associations with *E. amygdalina* (Martin, 1939; Curtis and Somerville, 1949; Jackson, in Specht et al., 1974), and some of these species occur also on the mainland, while others are Tasmanian endemics. Mainland species are *E. sieberi* which occurs in the north-east, and *E. rubida* which occurs in the south-east and east. The Tasmanian endemics are *E. pulchella* (formerly *E. linearis*) which occurs on dry dolerite ridges either in pure stands or in association with *E. amygdalina, E. barberi* or *E. tenuiramis*. The last species occur mainly on dry mudstone ridges and extends in a stunted form into coastal heaths (Chap. 16). *E. cordata* is an occasional species in the south-east (Fig. 10.16) and is sometimes locally dominant. *E. risdonii* also occurs in the south-east (Fig. 10.16) on skeletal soils or ridges where it forms pure stands, locally replacing *E. amygdalina*. *E. morrisbyi* occurs on poor sandy soils from sea level to an altitude of c. 160 m (Curtis and Morris, 1975).

2.7. *Eucalyptus nitida* Alliance

Smithton Peppermint

E. nitida (formerly *E. simmondsii*) replaces *E. amygdalina* on the western, wetter parts of Tasmania. *E. nitida* also occurs on King and Flinders Islands, and on the mainland in association with *E. ovata* or *E. baxteri*. In Tasmania, *E. nitida* varies in size from a mallee 3–4 m tall to a tree c. 25 m tall. It occurs from near sea leavel to an altitude of c. 500 m, under a mean annual rainfall of 900–2200 mm. In the lower rainfall habitats the species is found in swamps where the plants are often mallee-like. In all sites, the soils which are highly organic and acid and of very low fertility are permanently wet or waterlogged. The soils are derived from siliceous rocks, including schists, conglomerates, quartzites and mudstones which are overlain, with little or no profile differentiation, by a layer of peat built up by plants, especially the sedge, *Gymnoschoenus sphaerocephalus*. The growth of the hummocks of sedge lead to better drainage and the establishment of woody plants, including *E. nitida*.

In the best drained sites *E. nitida*, in pure stands, forms closed forests up to c. 25 m tall, which commonly form mosaics with other types of vegetation, e.g., *Nothofagus cunninghamii*, often in a stunted form (Fig. 10.21), or with *Gymnoschoenus* on badly drained patches. In such stands, the number of understory species is small, the common ones being *Acacia mucronata, Bauera rubioides, Banksia marginata, Cassinia aculeata, Leptospermum lanigerum, L. scoparium, Phebalium squameum* and *Sprengelia incarnata*. The main herbaceous species is *Gymnoschoenus sphaerocephalus*.

With increasing soil fertility, *Nothofagus cunninghamii* becomes increasingly abundant in the understory and finally dominates the tall shrub layer. Such communities commonly merge with the *E. obliqua* forests, *Nothofagus* persisting in the understory. Conversely, with decreasing soil fertility, which is usually accompanied by increased waterlogging, shrubs become progressively less abundant and *Gymnoschoenus* dominates the understory.

In most districts *E. nitida* forms mosaics with scrub or woodland communities dominated by species of *Leptospermum* and/or *Acacia*. It is probable that these stands are successional stages following the destruction of the *E. nitida*

Fig. 10.21. *Eucalyptus nitida* (paler foliage) and *Nothofagus cunninghamii* (darker). Gordon R. Tas.

Fig. 10.22. Recolonization of a denuded area once covered by *Eucalyptus nitida* forest. Left: The primary colonizers, *Restio tetraphyllus* and *Gahnia psittacorum* (taller). Right: Shrub stage in the secondary succession: *Acacia mucronata*, *Phebalium squameum*, *Bauera rubioides* and *Sprengelia incarnata*. Herbs are *Restio tetraphyllus* and *Gahnia psittacorum*. Some of the original forest is visible in the background. Queenstown, Tas.

communities by fires, which sometimes ravage the countryside in spite of the high rainfall. In the Queenstown area the forests have been killed by sulphur dioxide produced during the smelting of copper ore from the local mines, and in this case the communities are successional and can be arranged in sequence according to distance from the town.

In this area disturbance of the *E. nitida* forests initially involved the stripping of the peaty surface soil to a depth of 20–30 cm, followed by further removal of soil by accelerated erosion and the killing of plants by sulphur dioxide. The exposed parent material, largely Cambrian sediments or metamorphic rocks, almost white and extremely low in all nutrients, is colonized by plants at a very slow rate, determined mainly by fertility levels. The first coloniser is *Restio tetraphyllus* (Fig. 10.22). Other early colonizers, particularly in depressions, are *Gahnia psittacorum*, *Centrolepis fascicularis*, *Xyris operculata* and **Aira caryophyllea*. The first shrub to establish in the herbaceous community is *Acacia mucronata* followed by *Bauera rubioides*, *Sprengelia incarnata* and *Phebalium squameum* (Fig. 10.22), and later by *Leptospermum lanigerum*, *L. scoparium* and *Cenharrenes nitida*.

11. Eucalyptus Communities of the Cooler Climates of the Eastern Highlands, Lowland Victoria and Tasmania

1. Introduction

The Great Divide of eastern Australia south of lat. 29° S. (Chap. 1) supports a large number of cold-adapted species of *Eucalyptus* which are dealt with in this chapter. The mountain range has an average altitude of c. 1000 m over much of its length. In the northern portion, north of the Hunter – Goulburn Rivers, referred to as the Northern Tablelands, isolated peaks occur, some reaching an altitude of c. 1700 m. The southern part of the range is referred to as the Southern Tablelands, and the highest country, including the Kosciusko plateau and its extension into Victoria, is referred to as the Southern

Table 11.1. Species of *Eucalyptus* found mainly on the tablelands and/or in lowland Victoria or in Tasmania. Species arranged in groups to show their distributions.

GROUP 4 Northern and Southern Tablelands N.S.W.	GROUP 3 Northern Tablelands to Vic.	GROUP 2 Northern Tablelands only	GROUP 1 Northern Tablelands to Tas.
E. amplifolia	E. bauerana	E. acaciiformis	E. dalrympleana
E. ligustrina	E. bicostata	E. andrewsii	E. obliqua
E. moorei	E. camphora	E. approximans	E. pauciflora
E. oreades	E. cypellocarpa	E. bancroftii	E. radiata
	E. dives	E. banksii	E. viminalis
GROUP 5	E. fastigata	E. caleyi	
Southern Tablelands	E. macrorhyncha	E. caliginosa	GROUP 9
N.S.W. only	E. mannifera	E. cameronii	Tas. only
E. baeuerlenii	E. nitens	E. campanulata	E. amygdalina
E. blaxlandii	E. nortonii	E. laevopinea	E. archeri
E. macarthuri	E. stellulata	E. mckieana	E. coccifera
E. parvifolia		E. michaeliana	E. gunnii
	GROUP 6	E. nicholii	E. johnstonii
GROUP 7	Southern N.S.W.	E. nova-anglica	E. rodwayi
Vic. only	and Vic.	E. scoparia	E. urnigera
E. alpina	E. aggregata	E. youmanii	E. vernicosa
E. crenulata	E. cinerea		
E. mitchelliana	E. fraxinoides	GROUP 8	
E. neglecta	E. glaucescens	Southern N.S.W.	
E. yarraensis	E. kybeanensis	Vic. and Tas. (x);	
	E. maidenii	or Vic. and Tas.	
GROUP 10	E. muellerana	only (y)	
Species of Groups 1–8	E. niphophila	E. delegatensis (x)	
extending to S.A.	E. polyanthemos	E. ovata (x)	
E. macrorhyncha	E. smithii	E. perriniana (x)	
E. obliqua		E. rubida (x)	
E. ovata		E. globulus (y)	
E. pauciflora		E. regnans (y)	
E. rubida			
E. viminalis			

Alps. A further subdivision is required to define the distribution of certain species, viz., the Central Tableland, which refers to the Blue Mts and the Wingecarribee Tableland (Moss Vale area) to the south. The range effectively separates the eastern, wetter flora from the western, drier flora, but low areas on the range, notably at the head of the Hunter R. valley are low enough to permit the migration of the western flora on to the coastal lowlands. The climate on the range is wet and cool to cold. Mean annual rainfall is mostly above 800 mm and frosts occur for several months during the winter. The extremes of temperature are experienced on the Southern Alps where a tree-line occurs at altitude of 1800–2100 m and snow lies for a few months during the winter into the early summer.

About 80 species of *Eucalyptus* are found in these cooler regions, some of which are widely distributed, whereas others are local developments and are restricted either to the north or the south. Thus, the total assemblage on the Northern Tablelands is different from that on the Southern Tablelands (Costin, 1954; Lang, 1970), and in Tasmania several species are confined to that island (Jackson, 1965). The species have been classified into groups on their latitudinal distribution as indicated in Table 11.1.

In general, species with a wide latitudinal range occur at progressively lower altitudes from north to south. However, the factors which determine the local patterning of the same species at any latitude are similar, and these are: frost intensity, mean annual rainfall, soil fertility

Table 11.2. The main *Eucalyptus* alliances found in cooler climates in the south-east and their interrelationships.

I. NORTHERN TABLELANDS

	E. pauciflora (1400–1100 m)			
	E. dalrympleana (lower fertility)	*E. obliqua – E. fastigata* (higher fertility)		
	E. nova-anglica (waterlogging)	*E. viminalis*		
WEST				**EAST**
E. melliodora – E. blakelyi Chap. 13	*E. andrewsii* *E. youmanii* (lower fertility)	*E. caliginosa*	*E. laevopinea*	*E. campanulata* Lowland Forests; Chap. 9

II. SOUTHERN TABLELANDS TO LOWLAND VICTORIA

	Tree-line			
	E. niphophila or	*E. moorei – E. glaucescens* (rocky)		
	E. pauciflora			
	E. delegatensis – E. dalrympleana or (protected)		*E. viminalis – E. rubida* (more exposed)	
	E. obliqua – E. fastigata			
	E. macrorhyncha – E. rossii (drier)		*E. mannifera E. radiata* (lower fertility)	
WEST				**EAST**
E. melliodora – E. blakelyi Chap. 13	*E. bicostata* *E. ovata* *E. regnans*	*E. globulus* *E. radiata* *E. viminalis*	*E. cypellocarpa –* *E. muellerana –* *E. maidenii*	*E. sieberi* Chap. 10

III. TASMANIA

	Plateau *E. vernicosa*			
	E. subscrenulata *E. coccifera* *E. gunnii*			
	E. delegatensis – E. dalrympleana			
	E. rodwayi *E. pauciflora*			
WEST				**EAST**
E. nitida Chap. 10	*E. ovata* *E. obliqua*	*E. globulus*	*E. rubida –* *E. viminalis*	*E. amygdalina* Chap. 10

levels, and soil waterlogging. Many species have wide altitudinal ranges, which is due mainly to the fact that cold air drainage creates frost pockets at lower altitudes and in such pockets species from high altitudes are often dominant. Also, several species show clinal or ecotypic variation with altitude, as indicated below. The relationships between alliances is indicated in Table 11.2.

2. Alliances Occurring from the Northern Tablelands to Tasmania

2.1. *Eucalyptus pauciflora* Alliance

E. pauciflora and a number of related species, all of which have the main veins of the leaves more or less parallel, dominate the woodlands at higher altitudes on the mainland. Four species have been used to identify suballiances and, of these, only *E. pauciflora* extends to Tasmania. See Table 11.3.

2.1.1. *Eucalyptus pauciflora* Suballiance
Snow Gum

E. pauciflora usually dominates woodland stands up to 20 m high, but the species shows considerable variation in size and is often mallee-like (Fig. 11.1). It exhibits variation in other taxonomic characters (Green, 1969a, b). On the Northern Tablelands, this species forms the uppermost zone of *Eucalyptus* communities, but in the Southern Alps a zone of *E. niphophila* stands at higher altitudes and forms the tree line (see 2.1.3, below). *E. pauciflora* commonly extends to much lower altitudes on the mainland (to c. 1000 m in the north and 700 m in the

Table 11.3. Main associations in the *Eucalyptus pauciflora* alliance.

Suballiance	Main associations	Species forming minor associations	Species forming ecotonal associations
E. pauciflora	N. Tablelands *E. pauciflora*	*E. acaciiformis*	*E. stellulata*; *E. viminalis*
	S. Tablelands *E. pauciflora*	*E. debeuzevillei*; *E. kybeanensis*; *E. mitchelliana*	*E. rubida*; *E. viminalis*; *E. dalrympleana*
	Tasmania *E. pauciflora*		*E. dalrympleana*; *E. gunnii*; *E. ovata*; *E. rodwayi*; *E. viminalis*
E. stellulata	N. Tablelands *E. stellulata*		*E. pauciflora*; *E. viminalis*; *E. nova-anglica*
	S. Tablelands *E. stellulata*	*E. aggregata*; *E. camphora*; *E. parvifolia*;	*E. pauciflora*; *E. viminalis*; *E. ovata*;
E. niphophila	*E. niphophila*		*E. pauciflora*; *E. glaucescens*
E. moorei – *E. glaucescens*	*E. moorei*; *E. glaucescens*; *E. moorei* – *E. glaucescens*	*E. perriniana*; *E. kybeanensis*	*E. niphophila*; *E. pauciflora*

Fig. 11.1. A mallee-like stand of *Eucalyptus pauciflora*. Main shrubs are *Bossiaea foliosa* and *Helichrysum hookeri*. Mt. Hotham, Vic. at c. 1900 m.

south) in valleys where it follows lines of cold air drainage into frost pockets. Ecotonal associations with species which dominate at lower altitudes are therefore common (Table 11.2). The *E. pauciflora* suballiance occurs in areas with a mean annual rainfall of c. 800–1250 mm; frosts are of regular occurrence for several weeks on the more southerly areas. The species is indifferent to soil parent material, occurring on a range of rock types from basalt to granites. Costin (1954) lists 11 main soil types, which range from Podzolics to Prairie Soils. The soils have in common a high organic content and are not waterlogged, though they are almost perennially wet. Where drainage is impeded, *E.*

pauciflora is replaced by *E. stellulata* (next section) or by *Poa* grassland (Chap. 20).

On the Northern Tablelands, *E. pauciflora* forms open woodlands 10–20 m tall mainly on basalt at altitudes of 1200–1400 m. The woodlands usually have a poorly developed shrub layer, which may include *Acacia melanoxylon*, *A. dealbata*, *Tasmannia lanceolata*, *T. purpurascens*, *Gaultheria appressa* and some xeromorphic species (chiefly *Leptospermum*, *Lomatia* and *Banksia* spp.). The xeromorphs become more abundant on shallow soils or in exposed situations. A more or less continuous grass layer of *Poa sieberana* occurs.

On the Southern Tablelands, similar com-

Fig. 11.2. *Eucalyptus pauciflora* woodland. Shrubs are *Banksia marginata* and *Cyathodes parvifolia*. Grass layer dominated by *Holcus lanatus* and *Agropyron *repens*. Interlaken (Central Plateau), Tas.

munities occur with a sparse development of shrubs and a strong development of the grass layer (Costin, 1954). Three main grass layers are identified by Costin: i. in the driest areas *Stipa variabilis* complex, with subsidiary *Themeda australis* and *Poa sieberana*; ii. *Themeda – Poa* in cooler areas; iii. *Poa sieberana* in the coldest and moistest sites. Local variations in the tree stratum include the occurrence of shrub or small tree species within stands of Snow Gum, viz., *E. debeuzevillei* on plateaux of the Snowy Mountains, *E. kybeanensis* in similar situations and extending into Victoria, and *E. mitchelliana* which occurs among granite boulders on Mt. Buffalo in Victoria.

In Tasmania, *E. pauciflora* is found below 1000 m and may form pure stands (Fig. 11.2), and it forms ecotonal associations with several species (Table 11.2). A small outlier of the species is recorded in South Australia (Black, 1943–1957).

2.1.2. *Eucalyptus stellulata* Suballiance

Black Sally

E. stellulata reaches a height of c. 15 m and is readily identified by its smooth, greenish-leaden coloured bark. It occurs only on the mainland in the same general area as *E. pauciflora*, with which it is often associated. However, *E. stellulata* occurs on waterlogged soils, commonly at the foothills of slopes in frost-pockets occupied by *Poa* grassland (Fig. 11.3). On the Northern Tablelands, *E. nova-anglica* sometimes occurs as an associate mainly on the valley floor, and *E. viminalis* in more elevated situations. Shrubs are

Fig. 11.3. *Eucalyptus stellulata* woodland on foothills surrounding a grassland dominated by *Poa sieberana* with subsidiary *Amphibromus neesii*. Herbaceous layer in the woodland dominated by *Poa sieberana*, plus *Echinopogon caespitosus, Helichrysum bracteatum, Viola betonicifolia, Glycine tabacina, Asperula conferta* and *Rumex brownii*. A clump of Black-berry *(Rubus *procerus)* is visible in the grassland beyond the fence. Near Glen Innes, N.S.W.

Fig. 11.4. *Eucalyptus niphophila* near the tree-line. Main understory species: *Grevillea australis, Phebalium ovatifolium* and *Celmisia longifolia*. Charlotte Pass, Mt. Kosciusko, N.S.W.

usually absent and the grass layer is continuous (for typical species list, see Fig. 11.3). On the Southern Tablelands, *E. stellulata* forms similar communities in pure stands. Associated eucalypts are a few which tolerate waterlogging (Table 11.2).

2.1.3. *Eucalyptus niphophila* Suballiance
Alpine Snow Gum

E. niphophila, which might be regarded as a glaucous form of *E. pauciflora*, is a shrub or crooked tree attaining a height of 6–10 m. The trunks are disproportionately large, attaining a diameter of c. 1 m (Fig. 11.4). The plants are usually wind-shorn and asymmetrical, with a similar asymmetry in the trunk which is less developed on the windward side.

E. niphophila forms the tree-line on the Southern Alps at altitudes around 2000 m (Costin, 1954). The woodlands receive a mean annual rainfall of 750–2000 mm and are covered by snow for 1–4 months each year. The soils are well drained and are highly organic to a depth of c. 50 cm. The profile is acid throughout (pH 4–5). At higher elevations the alpine communities occur, and at lower elevations *E. niphophila* adjoins woodlands of *E. pauciflora*, or forests of *E. delegatensis* or *E. dalrympleana*. Waterlogged soils within the suballiance support swamps, heath or grassland.

The communities are characterized by a complete ground cover of herbaceous plants, at higher altitudes *Celmisia longifolia* and *Poa hiemata*, and at lower altitudes *Poa* and *Danthonia nudiflora*. Shrubs form a discontinuous layer throughout, especially *Bossiaea foliolosa*, *Orites lancifolia* and *Oxylobium ellipticum*.

2.1.4. *Eucalyptus moorei* – *E. glaucescens* Suballiance
Narrow-leaved Sally – Tingiringi Gum

These species are both mallees or small trees to c. 8 m high. *E. glaucescens* is confined to the Australian Alps. *E. moorei* occurs disjunctly along the Tablelands northward from the Kosciusko area to Gibraltar Range north-east of Glen Innes, occurring on Lithosols derived from sandstones (Blue Mts.) or granite (Gibraltar). It is most abundant in the south.

The 2 species form mallee thickets (Costin, 1954) in rocky terrain, usually in the most exposed situations in the Kosciusko area, mostly at elevations of c. 1500–1900 m. Mean annual rainfall is probably 750–1500 mm but runoff is high, so that the plants are often subjected to dry conditions. Frosts are severe, at least during the winter months and snow is common. The communities are subject to every climatic and edaphic extreme, including shallow (usually skeletal) soil.

As well as the co-dominants, *E. perriniana* and *E. kybeanensis* may be present in the stands. Costin (1954) describes the communities as 3-layered, the uppermost layer being almost continuous. The continuous shrub layer consists of xeromorphs, with species of *Acacia*, *Banksia*, *Grevillea*, *Persoonia*, *Gompholobium* and many others typical of the Australian xeromorphic assemblage. The herbaceous layer is poorly developed, consisting mainly of *Poa* sp., *Danthonia racemosa* and *Agropyron retrofractum*.

2.2. *Eucalyptus delegatensis* – *E. dalrympleana* Alliance
Alpine Ash – Mountain White Gum

On the Northern Tablelands *E. dalrympleana* ssp. *heptantha* is found from the Barrington Tops area to a few km north of the Queensland border, mainly at altitudes of 950–1250 m where the mean annual rainfall is c. 800–1150 mm. It occurs on soils of low fertility, mainly Yellow Earths and Podzolics derived from some granitic rocks. It sometimes forms pure stands but is usually associated with other eucalypts, especially *E. caliginosa*. The species dominates forests 20–30 m high, and a stand developed on leuco-adamellite, described by Williams (MS), contains the following species: There is a low tree layer consisting of *Banksia integrifolia*, *Casuarina littoralis*, *Acacia falciformis*, *A. filicifolia* and *Exocarpos cupressiformis*. The shrub stratum, usually dense, is dominated by xeromorphic species, the most common being: *Jacksonia scoparia*, *Daviesia latifolia*, *Persoonia cornifolius*, *Hakea dactyloides*, *Banksia collina*, *Monotoca scoparia*, *Leucopogon lanceolatus*, *Leptospermum brevipes*, *Bossiaea neo-anglica*

Fig. 11.5. Rugged country dominated by eucalypts. *Eucalyptus delegatensis* and *E. dalrympleana* on east-facing slopes and in valleys with dead trunks (trees killed by fire); *E. pauciflora* on west-facing slopes, as at left. Near Mt. Hotham, Vic.

and *Hibbertia stricta*. The vine, *Hardenbergia violacea* is usually common and regenerates rapidly on road cuttings. The herbaceous layer is dominated by *Poa sieberana*, and *Pteridium esculentum* is usually present.

In the south *E. delegatenis* and *E. dalrympleana* ssp. *dalrympleana* dominate the stands. The forests cover hilly to mountainous country (Fig. 11.5) below the *E. pauciflora* alliance on the mainland and below *E. coccifera* or *E. gunnii* in Tasmania. See Table 11.4.

These tall forests have altitudinal limits of c. 1000–1500 m on the mainland and 350–1000 m in Tasmania. Mean annual rainfall varies between 1000 and 1500 mm; subzero temperatures are experienced for 3–4 months during the

Table 11.4. Associations in the *Eucalyptus delegatensis – E. dalrympleana* Alliance.

District	Main associations	Species forming ecotonal associations
Northern Tablelands	*E. dalrympleana* ssp. *heptantha* – *E. caliginosa*	*E. andrewsii; E. radiata; E. pauciflora; E. acaciiformis*
Southern Tablelands	*E. delegatensis;* *E. delegatensis – E. dalrympleana* ssp. *dalrympleana;*	*E. pauciflora; E. obliqua; E. viminalis, E. radiata; E. cypellocarpa*
Tasmania	As Southern Tablelands	*E. subcrenulata; E. coccifera; E. gunnii; E. pauciflora*

winter and snow is a regular feature. The soils vary from Krasnozems to Podzolics and, rarely, Lithosols. The alliance has been described by Morland (1949, 1951), Costin (1954) and in Tasmania by Martin (1939).

E. *delegatensis* reaches a height of 60–70 m and occupies the most favourable sites with respect to soil fertility and moisture and in such forests the understory is mesomorphic, usually with an abundance of tree ferns. In less favourable situations, *E. dalrympleana*, mostly 35 to 45 m tall becomes co-dominant, and it is dominant in the least favourable sites such as higher altitudes, poorer soils or lower moisture.

Xeromorphs become progressively abundant as *E. dalrympleana* becomes more abundant.

In Tasmania, a similar differentiation of the species occurs. Common mesomorphic understory species in the most favourable sites are: *Atherosperma moschatum, Bedfordia salicina, Tasmannia lanceolata, Pomaderris apetala, Nothofagus cunninghamii,* and *A. melanoxylon.* The xeromorphic species in more open stands, especially below *E. dalrympleana* are *Acacia dealbata, Cyathodes parvifolia* and *Lomatia tinctoria.* In all stands *Gahnia psittacorum* and *Poa* spp. are the more common herbs.

Fig. 11.6. *Eucalyptus obliqua* forest with an understory dominated by *Acacia melanoxylon* and *Olearia argophylla.* Mt. Macedon, Vic.

2.3. *Eucalyptus obliqua – E. fastigata* Alliance

Messmate (Stringybark in the south) – Brown Barrel

These species form tall forests in cooler climates and both provide useful timber. The species have different areas, the former extending to Tasmania with an isolated occurrences in South Australia; the latter is confined to New South Wales with a minor extension into Victoria. *E. obliqua* reaches a height of c. 70 m but is mostly 30–50 m tall. Stunted plants, forming woodlands, occur in some localities in Victoria and Tasmania. Several varieties of the species, distinguished mainly on fruit characters have been recognized (Blakely, 1955). Also, Green (1971), who grew seed from plants growing in different habitats, has shown that genetic variation occurs. *E. obliqua* is found at higher altitudes in the north (900–1400 m) and at progressively lower altitudes towards the south. In Tasmania it occurs only below 600 m, occasionally forming wind-shorn scrubs 3–4 m tall on cliffs overlooking the sea. *E. fastigata* reaches a height of 30–45 m and shows little variation in size. It is found mainly on the highlands between 650 and 1300 m, extending to altitudes of c. 160 m in the south.

The species are commonly found in association in areas with a mean annual rainfall of 750–1200 mm. The best stands occupy fertile soils, especially Krasnozems, Alluvial Soils, and some Podzolics. The understory species vary from mesomorphic to xeromorphic, the former on the more fertile soils, the latter on less fertile ones (Podzolics).

On the northern Tablelands, *E. obliqua*, either in pure stands or with *E. fastigata*, is found mostly on basalt where the understory, now mutilated by fire in many areas, is a mixture of mesomorphic and xeromorphic species. *Acacia melanoxylon* and *Banksia integrifolia* occur as small trees above a discontinuous shrub layer containing *Hakea eriantha, Leucopogon lanceolatus, Senecio linearifolius* and *Olearia nerstii*. Ferns are often abundant, especially *Dicksonia antarctica* and *Polystichum aculeatum*. The herbaceous layer is dominated by *Poa sieberana* and *Lomandra longifolia*, with subsidiary small herbs including *Stellaria flaccida, Plectranthus parviflorus* and *Veronica calycina*. Fire removes the shrubs so that the understory is often reduced to a grass layer.

On the Southern Tablelands and in Victoria the stands (Fig. 11.6) are similar to those in the north but the subsidiary associated eucalypts are different (Table 11.5), and the understory contains *Acacia melanoxylon* as a constant small tree, together with *Olearia argophylla, Bedfordia salicina, Exocarpos cupressiformis* and

Table 11.5. Associations in the *Eucalyptus obliqua – E. fastigata* Alliance.

District	Main associations	Species forming ecotonal associations
Northern Tablelands	*E. obliqua*; *E. obliqua – E. fastigata*; *E. fastigata*; *E. obliqua – E. nitens*; *E. nitens*	*E. pauciflora; E. laevopinea*; *E. viminalis; E. campanulata*; *E. cameronii; E. microcorys*; *E. saligna*
Southern Tablelands	*E. obliqua – E. fastigata*; *E. fastigata*; *E. obliqua – E. nitens*; *E. fraxinoides*	*E. viminalis*; *E. dalrympleana* ssp. *dalrympleana*; *E. delegatensis; E. cypellocarpa*; *E. radiata; E. bicostata*; *E. muellerana*
Lowland Victoria	*E. obliqua*	*E. viminalis; E. regnans*; *E. radiata; E. cypellocarpa*
Tasmania	*E. obliqua*	*E. regnans; E. viminalis*; *E. dalrympleana* ssp. *dalrympleana*; *E. nitida; E. amygdalina*
South Australia	*E. obliqua*	*E. viminalis; E. leucoxylon*; *E. fasciculosa; E. baxteri*

Pomaderris apetala. In the driest sites *E. obliqua* becomes stunted and the understory is xeromorphic.

In Tasmania, *E. obliqua* sometimes occurs in pure stands, but is more commonly associated with *E. viminalis*. The forests on the most fertile soils in the wettest habitats reach a height of c. 70 m and adjoin the *E. regnans* forests. The understory is mesomorphic, consisting mainly of *Eucryphia lucida*, *Olearia argophylla*, *Pomaderris apetala* and *Zieria arborescens*. *Acacia dealbata* and *A. melanoxylon* are sometimes abundant and *Dicksonia antarctica* is usually present. With decreasing soil fertility the forests become shorter and *E. amydalina* is a common codominant, the understory being xeromorphic. With increasing soil waterlogging *E. obliqua* is replaced by *E. nitida,* the two species mixing, and with permanent waterlogging of the soil, eucalypts disappear, the country being clothed with sedgelands of *Gymnoschoenus sphaerocephalus* (Chap. 22), found chiefly in the southwest of the island. *E. obliqua* sometimes forms ribbons of woodland along drainage lines through the sedgelands.

In South Australia, *E. obliqua* occurs in the Mt. Lofty Ranges, where it reaches a height of 25 m and is regarded as an ecotype of the eastern populations (Specht and Perry, 1948; Martin, 1962). In this area, it occurs on a variety of soils, chiefly Podzolics of low fertility, with a mean annual rainfall of 870–1200 mm. In the more favourable sites, especially the moister, south-facing slopes, it forms pure stands. In less favourable sites it occurs with *E. viminalis* on south-facing slopes and with *E. leucoxylon* on north-facing slopes, or with *E. baxteri* and *E. fasciculosa* in the driest areas. It occurs also on Kangaroo Is. mixed with other eucalypts (Baldwin and Crocker, 1941).

2.4. *Eucalyptus viminalis – E. rubida – E. huberana* Alliance

These species are closely related and easily confused with each other, and also with *E. dalrympleana*. *E. viminalis* is widely distributed from southern Queensland to Tasmania, whereas the other two species have restricted distributions within the geographic area of *E. viminalis*. Two suballiances are recognized. See Table 11.6.

2.4.1. *Eucalyptus viminalis – E. rubida* Suballiance
Ribbon or Manna Gum – Candle bark

E. viminalis occurs at altitudes of 700 to 1300 m on the Northern Tablelands and at progressively lower altitudes in the south, being a lowland species in Tasmania at altitudes up to

Table 11.6. Associations in the *Eucalyptus viminalis – E. rubida* Suballiance.

District	Main associations	Species forming ecotonal associations
Northern Tablelands	*E. viminalis*	*E. obliqua; E. pauciflora; E. laevopinea; E. caliginosa; E. dalrympleana* ssp. *heptantha; E. nova-anglica; E. stellulata; E. melliodora*
Central Tablelands	*E. viminalis; E. viminalis – E. blaxlandii*	*E. radiata; E. pauciflora*
Southern Tablelands	*E. viminalis; E. viminalis – E. rubida; E. rubida*	*E. dalrympleana; E. fastigata; E. obliqua; E. pauciflora; E. muellerana; E. radiata; E. melliodora; E. bridgesiana; E. macrorhycha*
Lowland Victoria	*E. viminalis*	*E. obliqua; E. radiata; E. ovata; E. rubida; E. globulus*
Tasmania	*E. viminalis*	*E. obliqua; E. ovata*

Fig. 11.7. *Eucalyptus rubida* tall woodland with subsidiary *E. pauciflora*. Main understory species: *Daviesia ulicinia*, *Cassinia aculeata* and *Poa sieberana*. At c. 1500 m, Mt. Kosciusko area, N.S.W.

c. 500 m. *E. rubida* (Fig. 11.7) is found only on the southern Tablelands and extends to altitudes of 100–200 m above *E. viminalis*.

E. viminalis is variable in most of its characters and many populations may be of hybrid origin. Variation in some Victorian populations has been studied by Ladiges and Ashton (1977). Throughout most of its range, *E. viminalis* occurs as both tall forest and woodland trees. The former are often associated with *E. obliqua*, *E. fastigata* and *E. dalrympleana* (see 2.2 and 2.3, above).

As woodlands 20–30 m tall, *E. viminalis* occupies areas somewhat drier (mean annual rainfall 800–900 mm) than is required for tall forest development. It requires soil of medium to high fertility, mostly grey-brown Podzolics and Chocolate Soils. Where climatic requirements are satisfied, but soils are of low fertility, *E. viminalis* is replaced by stringybarks on the Northern Tablelands (especially *E. laevopinea* or *E. caliginosa*), or by peppermints in the south, especially *E. radiata*. Further, *E. viminalis* appears to be topographically controlled, the stands covering the sides of broad valleys or occurring in pockets in hills or along watercourses. It may therefore extend into drier alliances, especially the *E. melliodora* – *E. blakelyi* alliance or, at higher altitudes in association with *E. pauciflora* which extends downward in frost pockets.

In the wettest areas the understory in *E. viminalis* tall woodlands, which in some cases exhibit forest form, is often partly or entirely mesomorphic. Fraser and Vickery (1939) record the same species as in the *E. fastigata* forests in the Barrington Tops area. In the Blue Mts, Petrie (1925) records a mesomorphic understory of *Pomaderris apetala, Cassinia longifolia* and *Prostanthera lasianthos* and tree ferns. In the south, *Olearia argophylla* commonly forms a dense low tree layer (Fig. 11.8); other species

Fig. 11.8. *Eucalyptus viminalis* forest in a valley with an understory of *Acacia melanoxylon* and *Olearia argophylla*. Near Mt. Macedon, Vic.

less abundant are *Atherosperma moschatum*, *Elaeocarpus holopetalus* and *Telopea oreades*, often with *Dicksonia antarctica*. In drier areas, usually at lower elevations, a discontinuous xeromorphic shrub layer is present and the grass layer is commonly well developed; the latter becomes continuous in the driest areas, possibly as a result of fire.

2.4.2. *Eucalyptus huberana* Suballiance
Rough-barked Manna Gum

This species, formerly included under *E. viminalis*, occurs in western Victoria and south-east South Australia. It forms open to closed woodlands on Grey-Brown Podzolic Soils on level country sometimes pitted with gilgais which form pools after rain. The mean annual rainfall is 600–750 mm with a winter predominance.

E. huberana usually occurs in pure stands as an open woodland 20–25 m tall. A discontinuous shrub layer is present, containing *Acacia pycnantha*, *Banksia marginata*, *Bursaria spinosa*, *Casuarina stricta* and *Dodonaea viscosa*. The shrubs are destroyed by fire, when *Pteridium esculentum* becomes dominant (Fig. 11.9). Ecotonal associations are formed with *E. baxteri* on sandy soils of lower fertility, with *E. ovata* in areas with impeded drainage, with *E.*

Fig. 11.9. *Eucalyptus huberana* woodland. The understory has been removed by fire and replaced by *Pteridium esculentum*. Near Mt. Gambier, S.A.

camaldulensis when a deep water-table occurs, and with *E. obliqua* with increasing soil fertility.

2.5. *Eucalyptus radiata* Alliance
Narrow-leaved Peppermint

This species, which is valued as an oil-producer, extends from the Northern Tablelands to Tasmania, being most abundant on the central and lower parts of the Southern Tablelands. In the south it is sometimes referred to *E. robertsonii*, a subspecies of *E. radiata*. It occurs usually in areas with a mean annual rainfall of c. 1000 mm and forms woodlands and forests, mostly at altitudes of 300–1100 m.

On the Northern Tablelands it is found on soils of low fertility, rarely in pure stands but usually in association with *E. acaciiformis* (especially on soils with impeded drainage) or with *E. dalrympleana* ssp. *heptantha*, or *E. youmanii*. The understory consists mainly of xeromorphic shrubs (*Acacia, Banksia, Hakea, Leucopogon* etc.).

On the Central Tablelands it is a constituent of the xeromorphic forests on sandstones, commonly in association with *E. piperita* and *E. sieberi*, or with *E. mannifera*; the understory is xeromorphic (*Acacia, Banksia* etc.). In this area it occurs also on better soils in pure stands, or in association with *E. viminalis* and *E. cypellocarpa* as tall forests, usually with a mesomorphic understory (e.g., *Acmena smithii, Doryphora sassafras*), in which *Acacia melanoxylon* and/or *A. implexa* are prominent and ferns are abundant in the ground layer.

On the Southern Tablelands it is found mainly on shallow soils below the *E. delegatensis* – *E. dalrympleana* alliance (with which it forms ecotonal associations), extending sometimes in valleys in pure stands, where it forms tall forests (Fig. 11.10). An understory of tall shrubs is common, the main species being *Acacia melanoxylon, A. dealbata* and *Exocarpos cupressiformis*. The grass layer contains *Poa sieberana* and *Themeda australis*. At lower altitudes *E. radiata* forms ecotonal associations with *E. bicostata, E. viminalis* and *E. mannifera*.

E. radiata (robertsonii) is recorded in Tasmania as a subalpine woodland in association with *E. dalrympleana* in a few small areas (Jackson, 1965).

Fig. 11.10. Forest dominated by *Eucalyptus radiata*. Main understory species: *Acacia melanoxylon, A. mearnsii, Cassinia longifolia, Exocarpos cupressiformis, Pomaderris apetala, Asterolasia floccosa, Dicksonia antarctica, Blechnum cartilagineum*. 40 km north-east of Orbost, Vic.

3. Other Alliances Confined to the Mainland

3.1. *Eucalyptus mannifera* Alliance
Brittle Gum or White Gum

E. mannifera, formerly known as *E. maculosa*, is divided into five subspecies (Johnson, 1962) of which two are relatively abundant. All the populations occur on soils of low fertility (Podzolics and Lithosols) derived from sedimentary rocks or metamorphosed sediments. The communities are found under a mean annual rainfall of c. 500 to 1200 mm, at altitudes of c.

1100 m on the Northern Tablelands to 300 m in Victoria.

E. mannifera ssp. *elliptica* (Bendemeer White Gum) occurs on the Northern Tablelands in patches from the Barrington Tops district with an isolated occurrence in the Nandewar Range. The most extensive occurrence is in the Bendemeer – Walcha Road area where the subspecies forms woodlands c. 13 m tall either in pure stands or in association with *E. malacoxylon*, *E. macrorhyncha*, *E. bancroftii* or *Angophora floribunda*. The stands form patches within *E. melliodora* – *E. blakleyi* alliance which occupies the better soils. The understory is xeromorphic, the chief shrubs being *Acacia dealbata*, *Daviesia latifolia* and *Lissanthe strigosa*. The herbaceous layer is almost continuous and is dominated by *Aristida ramosa*, with subsidiary *Agropyron scabrum*, *Cymbopogon refractus*, *Danthonia laevis*, *Dichelachne sciurea*, *Poa sieberana*, *Sorghum leiocladum* and *Themeda australis*. A few forbs are present, especially *Helichrysum apiculatum*, *Ammobium alatum* and *Acaena ovina*.

E. mannifera ssp. *maculosa* occurs on the Central and Southern Tablelands. It sometimes forms pure stands in woodland formation mostly 10–15 m high but is more commonly associated with other eucalypts, especially *E. macrorhyncha*, *E. dives*, *E. cinerea*, *E. rossii*, *E. rubida*, *E. polyanthemos* and *E. melliodora*. A. discontinuous shrub layer and an almost continuous grass layer usually occur (Fig. 11.11).

3.2. *Eucalyptus macrorhyncha* – *E. rossii* Alliance

Red Stringybark – Scribbly Gum

These species form a well defined alliance in New South Wales, *E. macrorhyncha* extending into Victoria. The communities are usually open woodlands 20–25 m high (Fig. 11.12) and occur mainly on the drier parts of the lower tablelands (mean annual rainfall c. 600 mm) on soils of low fertility (mostly Brown Podzolics or Lithosols) derived from siliceous sedimentary rocks or granites. On east-facing slopes the woodlands occur at elevations of 200–400 m and west-facing slopes at 500–1000 m. Other species of *Eucalyptus* are usually present, the species varying from north to south (Table 11.7). *E. rossii*

Fig. 11.11. *Eucalyptus mannifera* (smooth trunk) ssp. *mannifera* with *E. macrorhyncha* behind. Scattered shrubs of *Acacia buxifolia*. Grass layer dominated by *Poa sieberana*. Near Wallerawang, N.S.W.

Fig. 11.12. *Eucalyptus macrorhyncha* (fibrous bark) and *E. rossii* (smooth bark). Main shrubs are *Acacia stricta*, *Grevillea lanigera* and *Exocarpos cupressiformis*. Herbaceous layer dominated by *Poa sieberana*. Canberra, A.C.T.

Table 11.7. Associations in the *Eucalyptus macrorhyncha – E. rossii* Alliance.

District	Main associations	Species forming ecotonal associations
Northern Tablelands and Western Slopes	*E. macrorhyncha*; *E. macrorhyncha – E. rossii*; *E. rossii*; *E. macrorhyncha – Angophora floribunda*	*E. blakelyi; E. dealbata*; *E. crebra; E. bridgesiana*
Southern Tablelands	*E. macrorhyncha*; *E. macrorhyncha – E. rossii*; *E. rossii*	*E. mannifera; E. dives*; *E. polyanthemos; E. rubida*; *E. goniocalyx; E. blakelyi*
Lowland Victoria	*E. macrorhyncha*	*E. polyanthemos; E. dives*; *E. mannifera; E. sideroxylon*; *E. baxteri; E. obliqua; E. goniocalyx*
South Australia	*E. macrorhyncha*	*E. goniocalyx*

intergrades with *E. racemosa* (Chap. 10) on the sandstones in the wetter eastern parts of its range. A species of *Callitris* is often present in the stands, usually *C. endlicheri* but occasionally *C. "glauca"*, or *C. muelleri* in the wetter east.

A dense shrub layer of xeromorphs is always present, the genera being those found on the sandy soils of low fertility on the coast, and some coastal species extend to some of the drier areas, e.g., *Actinotus helianthi, Hibbertia stricta* and *Styphelia triflora*. The xeromorphic assemblage in the north, resembles the "mid-western" assemblage which occurs in some of the ironbark forests (Chap. 12). Some of the more common and constant species in these stands are *Acacia buxifolia, Brachyloma daphnoides, Daviesia latifolia, Dillwynia retorta, Hibbertia linearis, Lissanthe strigosa* and *Stypandra glauca*.

In the north, stands in which *E. macrorhyncha* is dominant and *E. rossii* absent occur mostly on soils of somewhat higher fertility, though sometimes in rocky terrain, and in these woodlands there is a tall shrub layer containing *Acacia doratoxylon, A. neriifolia, A. penninervis, Brachychiton populneum* and the liana *Pandorea pandorana*. In such areas *E. dealbata* or *E. dwyeri* and *Callitris endlicheri* are usually associated with *E. macrorhyncha*. The rare cycad *Macrozamia heteromera* is present in some stands.

On the southern Tablelands of New South Wales, *E. macrorhyncha* is often associated with *E. rossii* (Fig. 11.12) with a xeromorphic understory (Costin, 1954; Pook and Moore, 1966) in which species of *Brachyloma, Daviesia, Dillwynia, Grevillea* and *Monotoca* are prominent. The ground layer is dominated by grasses but other herbaceous species are numerous, especially monocotyledons with rhizomes or tubers. In Victoria (Patton, 1937), the communities are similar but *E. macrorhyncha* is associated with other eucalypts (Table 11.7); the main genera represented in the xeromorphic shrub layer are *Acacia, Acrotriche, Astroloma, Bossiaea, Dillwynia, Epacris* and *Leucopogon* and 14 species of ground orchids are recorded for the herbaceous layer.

3.3. *Eucalyptus cypellocarpa* – *E. muellerana* – *E. maidenii* Alliance

Mountain or Monkey Gum – Yellow Stringybark – Maiden's Gum

These species characterize tall forests in valleys mainly in the south-east from the lower tableland area to the coastal lowlands. The species, singly or pairs, rarely form large tracts of forest, but occupy pockets in large or small valleys separated by drier ridges on which different alliances occur. See Table 11.8.

The species have different areas. *E. cypellocarpa* occurs disjunctly from the Grampians to the Northern Tablelands. At its northern limit

Table 11.8. Associations in the *E. cypellocarpa* – *E. muellerana* – *E. maidenii* Alliance.

District	Main associations	Species forming ecotonal associations
Northern Tablelands	*E. cypellocarpa*	
Central Tablelands	*E. cypellocarpa;* *E. cypellocarpa* – *E. oreades;* *E. cypellocarpa* – *E. blaxlandii;* *E. blaxlandii*	*E. radiata; E. sieberi*
Southern Tablelands	*E. cypellocarpa;* *E. cypellocarpa* – *E. maidenii;* *E. cypellocarpa* – *E. fraxinoides*	*E. viminalis; E. radiata*
Lower elevations and Lowland Victoria	*E. muellerana* – *E. cypellocarpa;* *E. muellerana* – *E. maidenii;* *E. maidenii;* *E. muellerana* – *E. smithii;* *E. muellerana* – *E. bosistoana*	*E. sieberi; E. agglomerata; E. globoidea*
Grampians	*E. cypellocarpa*	*E. obliqua; E. baxteri*

Fig. 11.13. *Eucalyptus maidenii* in a valley, with an understory of *Acacia melanoxylon*, *Pomaderris elliptica* and *Cyathea australis*. 11 km east of Sale, Vic.

(around the gorges east of Armidale) the trees develop a rough bark (in contrast to the smooth bark of southern populations) and are possibly hybrid with *E. goniocalyx*. Isolated stands are found in the Blue Mts where *E. cypellocarpa* occurs in pure stands, or in association with *E. blaxlandii*, or, rarely, *E. viminalis*, *E. radiata* or *E. sieberi*. In this area, *E. blaxlandii* may form pure stands on drier western ridges. The understory is mesomorphic on the more fertile soils, xeromorphic on soils of lower fertility.

The three species occur together in the southeast, *E. maidenii* and *E. cypellocarpa* increasing in abundance towards and in Victoria. All species are tall trees, mostly 30–45 m high,

E. maidenii (Fig. 11.13) reaching a height of c. 75 m.

4. Alliances Confined to the Northern Tablelands

Several species are confined to the Northern Tablelands and form well defined groups of communities in specific habitats, but the species are commonly mixed with species of wider distribution along the tablelands. The stringybarks are grouped together under one alliance, and

Table 11.9. Associations occurring in the alliances confined to the Northern Tablelands.

Alliance or Suballiance	Main associations	Species forming ecotonal associations and mixed stands
E. laevopinea Suballiance	E. laevopinea; E. laevopinea – E. cameronii	Wetter: E. pauciflora; E. viminalis; E. obliqua; E. saligna; E. deanei; E. dalrympleana Drier: E. macrorhyncha; E. bridgesiana; E. caliginosa
E. caliginosa Suballiance	E. caliginosa; E. caliginosa – E. nicholii; E. mckieana	E. melliodora; E. blakelyi; E. bridgesiana; E. laevopinea; E. youmanii; E. dalrympleana ssp. heptantha
E. youmanii Suballiance	E. youmanii; E. youmanii – E. bancroftii; E. youmanii – E. ligustrina; E. youmanii – Callitris endlicheri	E. andrewsii; E. caliginosa
E. campanulata Alliance	E. campanulata; E. campanulata – E. cameronii; E. scoparia; E. benthamii – E. campanulata	Higher: E. obliqua; E. laevopinea; E. dalrympleana; E. viminalis Lower: E. saligna; E. deanei; E. notabilis; E. carnea; E. eugenioides; E. acmenoides
E. andrewsii Alliance	E. andrewsii; E. andrewsii – E. bancroftii; E. andrewsii – E. banksii; E. andrewsii – E. caleyi; E. caleyi; E. andrewsii – Angophora floribunda; E. andrewsii – Callitris endlicheri	E. laevopinea; E. deanei; E. caliginosa; E. blakelyi; E. macrorhyncha; E. youmanii; E. dealbata; E. sideroxylon; E. acaciiformis
E. nova-anglica Alliance	E. nova-anglica	E. stellulata; E. acaciiformis; E. viminalis; E. pauciflora; E. melliodora; E. bridgesiana

three other alliances are recognized. See Table 11.9.

4.1. *Eucalyptus laevopinea* – *E. caliginosa* – *E. youmanii* Alliance

Silver-top Stringybark – Broad-leaved Stringybark – Youman's Stringybark

4.1.1. General

Eight species of true Stringybarks are found on the Northern Tablelands and five of these are restricted to this area. The remaining three are *E. macrorhyncha* which extends south mainly in drier country (see 3.2, above), *E. ligustrina* (Privet – leaved Stringybark) recorded only on Gibraltar Range and to the south on the Central Tablelands, and *E. eugenioides* (White Stringybark) which occurs on the eastern lowlands and tablelands as far north as the Bunya Mt. area. The last species is usually of occasional occurrence on better soils. However, it becomes locally dominant in a few areas in south-eastern Queensland on better soils, including Krasnozems, forming forests up to c. 25 m tall.

The Stringybarks are a difficult group taxonomically, with intergrading populations, and probable hybrids and hybrid-swarms. Identification depends partly on the characters of the juvenile leaves, which are rarely available and consequently the identify of some populations is often vague. Species not already mentioned are *E. mckieana*, which is included under *E. caliginosa* and *E. cameronii* which is of occasional occurrence in several alliances. The three species used to identify the suballiances below are distinct in their "typical" forms, and occupy fairly distinct habitats, but both species and habitats commonly intergrade. The communities occur at altitudes of c. 800–1300 m, mainly on the western side of the Great Divide under a mean annual rainfall of 800–950 mm.

4.1.2. *Eucalyptus laevopinea* Suballiance
Silver-top Stringybark

This species occurs mainly on hill tops or the flanks of gorges, mostly at elevations of 1100–1300 m on Podzolic soils or Lithosols derived largely from granites, or on more clayey soils derived from basalt. On granitic soils the forests reach a height of 20–25 m and on more fertile soils 40 m. Associated eucalypts are numerous (Table 11.9). In moister habitats, especially on the more fertile soils, there is a well developed tall shrub layer containing *Acacia implexa, A. dealbata, Bursaria spinosa* and *Casuarina torulosa;* some lianas may occur, e.g., *Pandorea pandorana* and *Geitonoplesium cymosum*. In ecotonal associations in the east (e.g. with *E. microcorys* etc.) the understory is mesomorphic. In drier sites towards the west, the understory is mostly xeromorphic, containing *Cassinia aculeata, Brachyloma daphnoides, Leptospermum attenuatum, Dillwynia retorta, Lissanthe strigosa, Pultenaea juniperina* and others, with a grass layer containing *Poa sieberana, Themeda australis, Cymbopogon refractus* and *Aristida ramosa*. In areas which have been repeatedly burned the shrubs are absent and the grass layer is more or less continuous.

4.1.3. *Eucalyptus caliginosa* Suballiance
Broad-leaved Stringybark

This is the common stringybark of the lower portions of the Tablelands west of the Great Divide, at elevations of c. 900–1100 m, where the mean annual rainfall is c. 800 mm. It forms pure stands, mostly in elevated situations, usually on rounded hills with shallow soil. Rock type appears to be of little significance, but granites and basalts are the most common and medium to high fertility of the soil is assumed. The stands (Fig. 11.14) are commonly retained by graziers partly as a supply of straight timber and partly as protection for stock; they occur mainly within the *E. melliodora* – *E. blakelyi* alliance. In these pure stands shrubs are either absent or rare, though this may be the result of grazing. The grass layer, potentially continuous but often mutilated by grazing or camping by stock, contains the following assemblage: *Poa sieberana* (dominant) with varying amounts of *Themeda australis, Danthonia laevis, Agropyron scabrum, Dichelachne sciurea, Bothriochloa macra* and *Eragrostis leptostachya*.

E. caliginosa occupies an intermediate position between *E. laevopinea* (better conditions) and *E. youmanii* (worse conditions) with regard to soil fertility, soil depth and water supply.

Fig. 11.14. *Eucalyptus caliginosa* with an understory of introduced grasses and forbs. c. 10 km south of Armidale, N.S.W.

Consequently the three stringybark suballiances intergrade. In drier areas towards the west, *E. caliginosa* is sometimes co-dominant with *Callitris endlicheri*, when there is a xeromorphic understory similar to that under *E. youmanii* (next section).

4.1.4. *Eucalyptus youmanii* Suballiance
Youman's Stringybark

E. youmanii is usually a small tree branching near ground level and rarely exceeding 20 m tall (Fig. 11.15). It occurs in the western portion of the tablelands from Moonbi to Stanthorpe where the mean annual rainfall is 750–900 mm. Typically it forms woodlands in rocky terrain on Lithosols or shallow Podzolic soils derived mainly from quartz porphyry or granite. *E. bancroftii* (Orange Gum) is a common associate and other trees may be present (Table 11.9). *E. youmanii* occurs also in the *E. andrewsii* alliance and the suballiances merges with the *E. caliginosa* suballiance on deeper soils.

A xeromorphic shrub layer is invariably present, sometimes with emergent tall shrubs of which *Acacia neriifolia* is often common. A typical shrub assemblage (from the Bendemeer area) contains: *Brachyloma daphnoides* (common), *Dillwynia sericea* (locally common), *Melichrus urceolaris*, *Correa reflexa*, *Monotoca scoparia*, *Lissanthe strigosa*, *Acacia pruinosa*, *Dodonaea viscosa*, *Olearia microphylla* and others less abundant. The herbaceous layer is poorly developed but relatively rich in species, with a high proportion of grasses for such a poor soil, viz.: *Aristida ramosa*, *Danthonia laevis*, *Dichelachne sciurea*, *Echinopogon caespitosus*, *Sorghum leiocladum* and *Themeda australis*. Forbs are numerous and include *Cheilanthes tenuifolia*, *Drosera peltata*, *Goodenia hederacea*, *Helichrysum apiculatum*, *Hydrocotyle peduncularis*, *Hypoxis glabella*, *Schoenus ericitorum* and *Swainsona oroboides*.

In some areas where large granite tors occur, some unusual species are found, as a result of additional water shed from the bare rock surfaces, e.g., on the Moonbi Range, *Ficus rubiginosa* (Port Jackson Fig), and the lianas, *Clematis aristata* and *Pandorea pandorana*.

Fig. 11.15. *Eucalyptus youmanii* with subsidiary *E. bancroftii*. The shrub layer has been removed by fire. Herbaceous layer contains *Themeda australis*, *Danthonia linkii*, *Echinopogon caespitosus* and *Wahlenbergia gracilis*. Near Bendemeer, N.S.W.

4.2. *Eucalyptus campanulata* Alliance

New England Blackbutt

E. campanulata identifies a tall alliance 30–40 m high which occurs mostly between altitudes of 450 m (on the eastern fall of the range) and 1250 m. The alliance extends from the Barrington Tops area (Fraser and Vickery, 1939) to just over the Queensland border. The communities receive a mean annual rainfall of 900–1500 mm. The soils are Podzolics, derived mainly from granites and are of intermediate fertility. The profiles are often several metres deep, but may be shallow and skeletal, and in many areas, even when the profiles are deep, granite boulders commonly project from the soil surface.

In the *E. campanulata* forests (Fig. 11.16) the understory varies considerably and has probably been strongly influenced by fire. In protected valleys, especially along creeks, species typical of the cooler rainforests form a distinct but usually discontinuous layer, e.g., species of *Ackama*, *Elaeocarpus*, *Endiandra*, *Omalanthus*, *Synoum* and *Trochocarpa* (Williams, 1976); small lianas may also be present, e.g. *Eustrephus latifolius* and *Stephania japonica*. Characteristic of most stands remote from watercourses is a tall discontinuous shrub layer in which *Acacia melanoxylon*, *Casuarina torulosa*, *Exocarpos cupressiformis* and *Banksia integrifolia* are constant. Other common shrubs are *Acacia dealbata*, *Helichrysm diosmifolium*, *Indigofera australis*, *Persoonia linearis*, *Leucopogon lanceolatus*, *Daviesia latifolia* and the small vines *Hardenbergia violacea* and *Kennedia rubicunda*. The herbaceous layer contains *Imperata cylindrica*, *Poa sieberana* and *Lomandra longifolia* as constants, with subsidiary *Themeda australis*, *Echinopogon caespitosus*, *Danthonia pilosa*, *Danthonia pallida* and the small leguminous herbs, *Desmodium varians* and *Glycine clandestina*; the ferns, *Pteridium esculentum*, *Culcita dubia* and *Blechnum* spp. are common in some areas.

4.3. *Eucalyptus andrewsii* Alliance

New England Blackbutt

This species identifies a group of communities growing on low fertility soils, which are either Lithosols or Podzolic, derived from certain types of granite normally low in plant nutrients. The communities occur mainly at altitudes of 800 to 1200 m on the western portion of the Northern Tablelands where the mean annual rainfall is c. 750–880 mm. Structurally the communities vary from low woodlands c. 10 m high in the drier areas on boulder-strewn hills to forests up to c. 40 m high on the deepest soils. The communities always have a well developed xeromorphic understory of shrubs which form one layer in the drier areas on shallow soils, two layers in wetter areas on deep soils. *E. andrewsii* may occur in pure stands, but it is commonly associated with other species (Table 11.9), one of which is a Stringybark. The xeromorphic shrub layer is always well developed and very rich floristically.

In drier areas the shrub layer is relatively poor in species, of which *Dillwynia retorta*, *Grevillea sericea*, *Olearia microphylla*, *Brachyloma daphnoides* are usually most abundant, with a very discontinuous herbaceous layer where *Danth-*

Fig. 11.16. *Eucalyptus campanulata* with subsidiary *E. cameronii*. Tall shrub dominant in understory is *Casuarina torulosa*. Herbaceous layer dominated by the fern *Culcita dubia*. 94 km east of Walcha, N.S.W.

onia paradoxa, Dichelachne sciurea and *Patersonia sericea* are most conspicuous.

In wetter areas, e.g., Torrington, an assemblage of c. 240 shrub species has been listed (Williams and Wissman, MS.). A tall shrub layer is often present and composed of various mixtures of *Acacia neriifolia, A. penninervis, Banksia integrifolia, Casuarina littoralis, Leptospermum attenuatum* and others. The smaller shrubs are too numerous to list here but the following genera of the common families typical of low fertility habitats in Australia are represented: *Hibbertia* (Dilleniaceae); *Brachyloma, Epacris, Leucopogon, Lissanthe, Melichrus, Monotoca, Styphelia* (Epacridaceae); *Amperea, Phyllanthus, Pseudanthus, Poranthera* (Euphorbiaceae); *Acacia* (Mimosaceae); *Baeckea, Callistemon, Calytrix, Kunzea, Leptospermum, Melaleuca* (Myrtaceae); *Aotus, Bossiaea, Daviesia, Dillwynia, Gompholobium, Hardenbergia, Hovea, Indigofera, Jacksonia, Mirbelia, Pultenaea* (Papilionaceae); *Banksia, Conospermum, Grevillea, Hakea, Isopogon, Lomatia, Persoonia, Petrophile* (Proteaceae); *Boronia, Correa, Phebalium, Zieria* (Rutaceae); *Actinotus, Platysace, Xanthosia* (Umbelliferae).

The herbaceous layer is rarely well developed. The most abundant species are *Entolasia stricta, Caustis flexuosa, Stypandra glauca, Xanthorrhoea australis, Poa sieberana, Dichelachne sciurea* and *Imperata cylindrica. Macrozamia heteromera* is recorded for the Howell area.

4.4. *Eucalyptus nova-anglica* Alliance

New England Peppermint

E. nova-anglica occurs mainly at elevations of 900–1300 m, usually in pure stands as savannah woodlands 15 m high or tall woodlands 20 m high on flats along rivers and watercourses, where the soils are occasionally waterlogged (Fig. 11.17). The stands are sometimes broken by patches of grassland *(Poa sieberana)* which are controlled either by waterlogging or frost pockets. The soils are Podsolics derived from granite or alluvium. The woodlands often adjoin woodlands of *E. stellulata* (Fig. 11.3) and other species may occur in association with *E. nova-anglica* (Table 11.9). *E. nova-anglica* is subject to partial or even complete defoliation by beetles during the summer, so that in most areas trees almost invariably carry some juvenile foliage or even a full crown of juvenile leaves.

A shrub layer rarely exists in the woodlands, but *Acacia dealbata* sometimes occurs in patches up to c. 3 m high. *Melichrus urceolaris* is an

Fig. 11.17. *Eucalyptus nova-anglica* in pure stand in a valley. Herbaceous species include *Poa sieberana* (in valley), and *Eragrostis leptostachya, Bothriochloa macra* and *Panicum effusum* in higher ground (foreground). 10 km north of Walcha Road, N.S.W.

occasional small shrub. The grass layer is usually continuous, with *Poa sieberana* dominant and *Themeda australis* as a common (rarely locally dominant) species. Of occasional to rare occurrence are the grasses: *Danthonia laevis, Echinopogon caespitosus* and *Sorghum leiocladum,* and the herbs: *Asperula conferta, Convolvulus erubescens, Epilobium billardierianum, Geranium potentilloides, Hydrocotyle laxiflora, Luzula campestris* and *Leptorhynchos squamatus*.

5. Alliances Occurring Mainly in Victoria and Tasmania

5.1. *Eucalyptus ovata* Alliance

Swamp or Black Gum

E. ovata is confined to the cooler lowlands in the south, extending to altitudes of c. 800 m on the southern Tablelands. It occurs in areas with a mean annual rainfall of 500–1000 mm, on soils with impeded drainage, which are usually of low fertility and are classified as Podzols or Gley Podzols. In the best drained sites it reaches a height of c. 70 m and forms tall woodlands. With increased soil waterlogging the trees become stunted and form scrub thickets. *E. nitida* is sometimes a co-dominant.

The tall woodlands in Victoria have a discontinuous tall shrub layer consisting mainly of *Casuarina littoralis, C. stricta, Banksia marginata, Bursaria spinosa* and *Acacia pycnantha*. Ecotonal associations occur with *E. viminalis* and *E. leucoxylon*. In some areas near the coast of Victoria, other eucalypts of restricted distribution occur mostly as understory species to *E. ovata,* viz. *E. crenulata, E. yarraensis* and *E. neglecta*. In stunted stands on soils with impeded drainage the communities have an understory of xeromorphic shrubs, dominated by *Leptospermum pubescens* (Fig. 11.18).

At higher altitudes, extending into New South Wales, *E. ovata* is usually associated with other swamp-adapted species, *E. aggregata, E. macarthuri, E. camphora* and *E. stellulata*.

Fig. 11.18. Stunted *Eucalyptus ovata* with a shrub layer of *Leptospermum pubescens* and *Melaleuca squarrosa*. Herbaceous layer contains *Gahnia trifida, Poa poiformis, Xanthorrhoea australis* and *Lomandra longifolia*. Near Millicent, S. A.

Fig. 11.19. *Eucalyptus regnans*, c. 80 m tall. Shrubs below are *Atherosperma moschatum* and *Nothofagus cunninghamii*. Near Maydena, Tas.

In Tasmania E. *ovata* is widely distributed around the island on the lowlands. On the west coast, in particular, it is commonly associated with E. *nitida* (see Chap. 10).

5.2. *Eucalyptus regnans* Alliance

Mountain Ash (mainland)
or Swamp Gum (Tasmania)

E. *regnans* is the tallest of the eucalypts, and may possibly reach a height of 130 m. The largest trees have been cut for timber and the greatest authenticated height is 114 m. Existing stands contain trees mostly 70–90 m tall (Fig. 11.19), the trees in each stand being approximately the same size and probably the same age. No saplings or seedling occur in the understory, since establishment of seedlings is prevented by low light intensity at ground level. The autecology of the species have been studied by Ashton (1956, 1958) and Eldridge (1968). The species occurs mostly in valleys in pure stands, but may extend over hill tops within valleys, in areas with a mean annual rainfall of 750–1500 mm. In Victoria it occurs at altitudes of 160–1000 m and in Tasmania from near sea-level to c. 650 m. It occurs on deep, rich, well drained loamy soil and is probably delimited within its climatic range by soil fertility levels, as suggested by its local replacement by other species of *Eucalyp*-

tus, e.g., by *E. amygdalina* on chert ridges near Maydena, Tasmania (Mount, 1964).

At the periphery of pure stands other tall eucalypts occur with *E. regnans*, many attaining a height of 60 m, especially *E. obliqua* and *E. viminalis* and at higher altitudes *E. delegatensis*. Ecotonal association are formed also with *E. ovata, E. nitida* and *E. globulus* in Tasmania.

The understory is mesomorphic. The tall eucalypts dwarf the understory which contains trees up to 30 m tall. The main species in both Victoria (Petrie et al., 1929) and Tasmania (Martin, 1939) are *Olearia argophylla, Pomaderris apetala* and *Bedfordia salicina*. Species of *Acacia* (*A. dealbata* and *A. melanoxylon*) have probably always been present (see effects of fire, below). Shrubs are usually not abundant. *Correa lawreniana, Prostanthera lasianthos* and *Zieria smithii* are constants. In Tasmania, *Aristotelia peduncularis, Pittosporum bicolor* and *Tasmannia lanceolata* occur, and in Victoria, *Lomatia fraseri, Persoonia arborea* and species of *Cassinia* and *Olearia*. Tree ferns are sometimes conspicuous, especially in more open areas along creeks and in litter cones built up at the bases of trees of *E. regnans*. *Dicksonia antarctica* is widespread and *Cyathea australis* is confined to the Victorian stands. Herbs are rare, *Gahnia psittacorum* sometimes occurring in wetter areas, and in Victoria the

Fig. 11.20. *Eucalyptus bicostata*. Note saplings of *E. bicostata* with broad juvenile leaves. Behind: *E. radiata*. Main shrub is *Cassinia longifolia*. 16 km north-east of Khancoban, N.S.W.

Fig. 11.21. *Eucalyptus globulus*. Pure stand with an understory of *Beyeria viscosa, Dodonaea viscosa, Pomaderris apetala, Exocarpos cupressiformis*. Herbs are *Lomandra longifolia, Senecio australis* and **Dactylis glomerata*. Near Hobart, Tas.

ground ferns *Polystichum aculeatum* and *Blechnum* spp.

Nothofagus cunninghamii is sometimes the dominant as an understory species in the forests and in some cases at least it represents a successional stage following burning. The effects of fire have been studied by Jarrett and Petrie (1929), Gilbert (1956), Ashton (1956), Cremer and Mount (1964), Mount (1964), Cunningham and Cremer (1965). In Tasmania, Gilbert has shown that the persistence of *E. regnans* depends upon the burning of the forests. If not burned, *E. regnans*, with a life-span of c. 350 years, is replaced by *Nothofagus cunninghamii*, which prevents the establishment of seedlings of *E. regnans*. Mixed stands of *E. regnans* and *Nothofagus* are produced when the intervals between fires are less than 350 years. A fire frequency of once or twice per century favours *E. regnans* and suppresses *Nothofagus* which is replaced by *Pomaderris apetala, Olearia argophylla* and *Acacia* spp. The effect of more frequent firing (10–20 year intervals) is to remove the fire sensitive *E. regnans* which is replaced by *E. obliqua* and/or *E. delegatensis* which have thick, fibrous bark. Similarly, *E. viminalis* and *E. ovata* may be present in stands of *E. regnans* as a result of fire.

Since *E. regnans* is killed by fire (it lacks a lignotuber) it must regenerate from seed, and in forests which are felled for timber "seed trees" were left in a stand which provide seed for regeneration after the trash on the forest has been removed by a controlled burn. This procedure has now been replaced by the sowing of seed from aircraft. Regeneration of the forest from seed, which falls into an ashbed following firing, is as follows: The seeds require light for germination (Clifford, 1953). Few species survive burning, the only important ones being ferns with rhizomes (*Histiopteris incisa*, *Hypolepis rugulosa* and *Pteridium esculentum*). Bryophytes are the first colonizers of the ash bed, viz., the liverwort, *Marchantia polymorpha*, and the mosses, *Funaria hygrometrica* and *Ceratodon purpureus*. The seedlings of angiosperms establish in the mats of bryophytes, the most abundant of which is the Fireweed, *Senecio minimus*, which reaches its peak (up to 79% cover) after 2 years. After 3 years the bryophytes mentioned above are replaced by another moss, *Polytrichum juniperinum*. The density of the regenerating angiosperms is governed in part by the density of the ferns, which may be so abundant that they suppress the establishment of the angiosperms.

5.3. *Eucalyptus bicostata* – *E. globulus* Alliance

Victorian Blue Gum or Eurabbie – Tasmania Blue Gum

These species are closely related taxonomically. *E. bicostata* (including *E. st.johnii*) produces its flower in umbels of three in the leaf-axils, whereas *E. globulus* produces only one bud in the leaf-axil.

E. bicostata occurs on the mainland with a small occurrence on Flinders Is, where it forms associations with *E. ovata* and *E. viminalis*. On the mainland it is distributed disjunctly as far north as the Northern Tablelands where it is rare and apparently relict. It is characteristically a species of valleys, occurring in pockets or ribbons within other alliances, especially *E. obliqua*, *E. cypellocarpa* and *E. radiata* (Fig. 11.20). The understory is usually mesomorphic, the species being those in the surrounding alliance. *E. globulus* is restricted to Tasmania, except for two small, possibly relict stands in southern Victoria. It likewise occurs in valleys. In Tasmania it forms forests to 50 m tall on the coastal lowland to an altitude of c. 350 m, replacing *E. obliqua* at these lower altitudes (Martin, 1939). Ecotonal associations occur with *E. viminalis*, *E. pulchella* and *E. tenuiramis*. The understory is mesomorphic, *Bedfordia salicina* and *Pomaderris apetala* being common species; other species are listed in Fig. 11.21.

6. Alliances Restricted to Tasmania

6.1. General

Several species are confined to Tasmania and some of these show considerable variation in habit, among them a group of three which form an altitudinal series as follows: *E. vernicosa* is restricted to the mountain plateaux where it may be prostrate, the branches growing over boulders, or upright in mallee form occurring as a dominant or co-dominant in heaths or scrubs (see Chap. 17). In more protected situations it forms a tall shrub to c. 4 m high, and below the edges of the plateaux it becomes taller and intergrades with *E. subcrenulata*, which, with decreasing altitude merges with *E. johnstonii* (see next section). The relationship between the three species is regarded as clinal (W.D. Jackson in Pryor and Johnson, 1971) and the three taxa are sometimes regarded as subspecies of *E. vernicosa*. Curtis and Morris (1975) give the three taxa specific rank. Similarly with *E. gunnii* and *E. archeri* (see 6.3, below).

6.2. *Eucalyptus coccifera* – *E. subcrenulata* Alliance

Tasmanian Snow Gum

This alliance forms the tree-line in Tasmania, except in the north-east (see *E. gunnii*, next section). These species are the most cold-resistant of the eucalypts, except the shrubby *E. vernicosa* which is found on the plateaux, as described in the last section. The alliance occurs between 700 and 1400 m (Fig. 11.22) in areas with a mean annual rainfall of 1000–2000 mm. Snow may lie on the ground for 2–3 months each year. The species reach a height of 40 m at lower altitudes and form closed forests, but become shorter with increasing altitude. *E. coccifera* grows on a variety of soils and forms pure stands on deep, humus-rich soils derived from dolerite. On more acid rocks, especially quartzites, it is partially or wholly replaced by *E. subcrenulata* at higher altitudes, and by *E. johnstonii* (Yellow Gum) at lower altitudes especially on sandstones in the south-east. Also, in the south-east *E. urnigera* (Urn Gum) becomes an important constituent of the forests; this species exhibits clinal variation (Thomas and Barber, 1974). The communities occur in a variety of habitats, including valleys, hilltops within valleys, among boulders at the foot of slopes and surrounding scree on steeper slopes. In some areas trees are commonly rooted among boulders which overly a deposit of boulder clay 1–2 m deep. The soils are usually well drained.

The understory varies considerably and is determined by soil depth, soil waterlogging, soil fertility and exposure to wind and cold. At lower altitudes *Nothofagus cunninghamii* is

Fig. 11.22. General view of the subalpine woodlands dominated by *Eucalyptus coccifera* and *E. subcrenulata*. The treeless plateau above, standing at c. 1300 m. Conifer on this side of lake is *Arthrotaxis cupressoides*. Cradle Mt. National Park, Tas.

Fig. 11.23. *Eucalyptus gunnii* woodland near tree line. Understory of *Cyathodes parvifolia*, *Tasmannia lanceolata*, *Bellendena montana*; *Poa* sp. dominating the grass layer. Ben Lomond, Tas., near the tree-line.

common in the understory, sometimes reaching tree proportions in valleys; this species persists in less favourable sites, even at higher altitudes, in a stunted form, occurring in such sites with a xeromorphic assemblage. In some areas the understory is suppressed, possibly as a result of "rooting space" which is at a minimum where boulders cover the surface. Some typical assemblages are described below.

On Mt. Wellington, near Hobart, the following common shrubs are found in the *E. coccifera* association at higher altitudes: *Helichrysum baccaroides*, *Oxylobium ellipticum*, *Podocarpus lawrencei*, *Lissanthe montana*, *Richea gunnii* and *Tasmannia lanceolata*. In the same area the *E. coccifera* – *E. urnigera* association 15 m tall occurs on dolerite in less exposed areas, with an understory of *Anopterus glandulosus*, *Bauera rubioides*, *Cyathodes parvifolia*, *Gaultheria hispida*, *Leptospermum lanigerum*, *Richea dracophylla* and *Telopea truncata* (Martin, 1939; Curtis and Somerville, 1949; Ratkowsky, 1977). In the Cradle Mt. area, Sutton (1928) describes a mallee-like community in which *E. coccifera* is co-dominant with *Nothofagus cunninghamii* and *Phyllocladus aspleniifolius*. The area is boulder-strewn but supports a wealth of shrubs, including the two conifers, *Microcachrys tetragona* and *Podocarpus lawrencei*, and angiosperm species belonging to the genera *Orites*, *Richea*, *Cyathodes*, *Epacris*, *Boronia*, *Bellendena*, *Hibbertia* and others.

E. *subrenulata* and *E. johnstonii* are either co-dominant with *E. coccifera* or replace it on siliceous parent materials. The former occur at higher altitudes, usually with a xeromorphic understory. The latter occurs at lower altitudes usually with *Nothofagus cunninghamii* (and its associates, see Chap. 7) in the understory in valleys (Sutton, 1928), or with a xeromorphic understory on well drained soils, or with an understory of sedges (*Gahnia* and Restionaceae) on badly drained soils (Martin, 1939).

6.3. *Eucalyptus gunnii* Alliance
Cider Gum

E. gunnii occurs mainly on the Central Plateau and in the north-east; in the former area it may be associated with *E. coccifera*. *E. gunnii* is dominant between altitudes of c. 700 to 1100 m, in areas with a mean annual rainfall of 750–1100 mm. Frost are common during the winter and snow may lie for a few months. *E. gunnii* exhibits clinal variation with increasing altitude. At lower elevations it reaches a height of 20–25 m and with increasing altitude it becomes smaller. Near or on mountain plateaux a shrubby or mallee-like form occurs, which is sometimes referred to as *E. archeri*.

E. gunnii is usually the sole dominant in woodlands 10–20 m tall which occupy a zone between *E. coccifera* and *E. delegatensis* when all three species occur in the same area. When *E. coccifera* is absent, *E. gunnii* forms the tree line on the mountain, as on Ben Lomond. The woodlands occur on a variety of parent materials, the effects of which are masked by the accumulation of deep organic layer in the soil built up largely by grasses and sedges in the herbaceous layer.

The understory species vary with altitude and drainage. In better drained areas, especially at lower altitudes, where the communities are tall woodlands, *Nothofagus cunninghamii* and *Telopea truncata* are usually abundant. At higher altitudes these are replaced by smaller

Fig. 11.24. *Eucalyptus rodwayi*. Pure stand with understory of *Callistemon viridiflorus* and *Leptospermum lanigerum*. East of Steppes, Central Plateau, Tas.

shrubs such as *Tasmannia lanceolata, Bellendena montana* and *Cyathodes parvifolia* (Fig. 11.23). In all habitats, grasses (*Poa* spp., in better drained areas), and sedges are common; the latter include *Gahnia psittacorum, Gymnoschoenus sphaerocephalus* and *Carpha alpina*.

Treeless areas within stands of *E. gunnii* are common and result either from waterlogging or from cold-air drainage and the occurrence of frost-pockets. These are commonly dominated by *Gymnoschoenus* (Chap. 22), or by other sedges, the moors being dotted with shrubs, especially *Hakea lissocarpa, H. microcarpa* and other species (see Chap. 17).

6.4. *Eucalyptus rodwayi* Alliance

Swamp Peppermint

E. rodwayi is a tree up to 20 m high and occurs from near sea level to an altitude of c. 1000 m, being most abundant on the Central Plateau at altitudes of 700–800 m. It is dominant in areas with a mean annual rainfall of 900–1500 mm. Winter frosts are severe, because the species occurs in open situations on or around bleak moors which are temporarily or permanently waterlogged.

The species dominates open to closed woodlands and usually occurs in pure stands. An association with *E. perriniana*, which occurs also on the mainland, is recorded. *E. rodwayi* commonly adjoins woodlands of *E. gunnii* and *E. pauciflora* forms ecotonal associations with these species. The understory species are relatively few and are those which tolerate waterlogging. They are typically watercourse species, e.g., *Leptospermum lanigerum* and *Callistemon viridiflorus* (Fig. 11.24). Away from flowing water the understory is usually herbaceous and is dominated by *Poa* and **Anthoxanthum*, together with species which occur on the moors, especially *Acaena, Calocephalus, Carpha, Hypericum* and *Plantago* spp. (For the moors, see Chap. 17).

12. The Ironbark Forests and Woodlands of the East (Cape York Peninsula to Victoria)

1. Introduction

Ironbarks are identified by their rough, deeply furrowed, dark-coloured, often black barks. They all belong to the subgenus *Symphyomyrtus,* and are placed in 3 of the Series defined by Pryor and Johnson (1971). Most of the species belong to Series *Pruinosae* (Table 12.1) and occur mainly in the northern zone receiving a predominantly summer rainfall. In the south the ironbark communities extend into Victoria, with a predominantly winter rainfall, where they are represented by *Eucalyptus sideroxylon,* a member of series *Melliodorae* (Table 12.1). The remaining ironbarks, *E. beyeri, E. caleyi, E. panda, E. paniculata* and *E. tetrapleura* belong to Series *Paniculatae* and occur mainly near the coast in the southern half of the continent, though a few of the species are sometimes found with members of Series *Pruinosae* or with *E. sideroxylon.*

The Series *Pruinosae* is almost entirely eastern (Table 12.1). *E. jensenii,* a rare species, occurs west of the Carpentaria Plains (Chap. 8), and *E. pruinosa* is found mainly west of the Carpentaria Plains, with restricted occurrences in the east (Chap 8). The remaining species of the Series dominate the ironbark forests and woodlands from Cape York Peninsula to northern New South Wales. The areas of the widely distributed species overlap considerably and in many districts it is difficult to determine the fac-

Table 12.1. The ironbark species of *Eucalyptus* dealt with in this chapter. The species are arranged under Series of subgenus *Symphyomyrtus* (Pryor and Johnson, 1971).

W.A.	N.T.	Eastern States		Latitudinal range °S.	Notes
SERIES *Pruinosae*					
E. jensenii					
	E. pruinosa				Adult leaves opposite
		E. cullenii		11–18	
		E. staigeriana		13–17	Leaves lemon-scented
		E. drepanophylla		17–28	Juvenile leaves oblong-ovate. Adult leaves 5.5–8 times as long as broad (Blake, 1953). Compare with following sp.
		E. crebra		17–35	Juvenile leaves linear to linear – lanceolate. Adult leaves 7–11 times as long as broad
		E. shirleyi		17–21	Adult leaves opposite
		E. melanophloia		18–34	Adult leaves opposite
		E. whitei		19–22	Leaves glaucous
		E. decorticans		24–25	Branches smooth
		E. fibrosa ssp. *fibrosa*		23–36	Formerly *E. siderophloia*
		E. fibrosa ssp. *nubila*		26–33	Formerly *E. nubilis.* Leaves slightly glaucous
		E. siderophloia		25–37	Formerly *E. decepta*
SERIES *Melliodorae*		*E. sideroxylon*		26–38	

tors which delimit the species. Three species, *E. shirleyi,* closely related to *E. melanophloia,* and *E. whitei* and *E. decorticans,* closely related to *E. drepanophylla* and *E. crebra,* have relatively small areas and occur within the area of their nearest relatives. Identifying the species in the field is, in some cases, very difficult, especially with *E. crebra* and *E. drepanophylla,* and ecologists have sometimes abandoned the separation of these two, referring them to "narrow-leaved ironbark".

The ironbark communities usually occupy soils of low fertility and cover large tracts of country in eastern Queensland from the coast across the Great Divide to the semi-arid inland. Further south they occur in drier areas adjoining the tablelands on the west. A few species are found on the eastern lowlands, mostly in rain-shadow areas, notably in the valleys of the Clarence and Hunter Rivers and on the Penrith plains west of Sydney. The communities are classified under three alliances which are subdivided into suballiances where appropriate. Many other species of *Eucalyptus* and *Angophora* are found in association with ironbarks and some species of *Eucalyptus* are restricted to the ironbark communities. The ironbark forests and woodlands contact most of the *Eucalyptus* alliances in the east, excluding those on the higher tablelands, and consequently a large number of ecotonal associations exist, as indicated in Table 12.2.

2. Descriptions of the Alliances

2.1. *Eucalyptus cullenii* Alliance

Cullen's Ironbark

This species is dominant in relatively small patches on Cape York Peninsula, chiefly on the Great Divide (Fig. 12.1). It occurs in frost-free areas with a mean annual rainfall of 900 to 1500 mm, with about 90% of the rain falling in the summer. The soils are mainly Red Podsolics derived from gneiss or granite developed on undulating country or the lower slopes of ridges, or skeletal soils on hill tops.

E. cullenii, which is sometimes deciduous during dry winters, dominates open woodlands 7–12 m high (Story, 1970; Pedley, 1972; Hall and Brooker, 1973). Pure stands are uncommon, the species being associated in most areas with another eucalypt, especially *E. dichromophloia* (throughout), *E. tetrodonta* (north), and less commonly with other species (Table 12.2). *E. staigerana* (Lemon-scented Ironbark) occurs within the alliance on undulating or hilly country, often in skeletal soils among massive granite boulders.

Scattered small trees and tall shrubs occur in the taller woodlands, chiefly *Erythrophleum chlorostachys* and *Melaleuca viridiflora,* less commonly *Alphitonia excelsa, Atalaya variifolia, Persoonia falcata,* and *Planchonia careya. Xanthorrhoea johnsonii* is sometimes locally common, especially among boulders. In the grass layer, *Themeda australis* is usually dominant, with subsidiary *Aristida, Arundinella, Bothriochloa, Chrysopogon, Heteropogon* and *Sorghum* spp. In low woodlands in drier sites the most important shrubs are *Petalostigma banksii, Melaleuca acacioides* and *M. viridiflora;* the grass layer is dominated by *Heteropogon triticeus* and *Sorghum plumosum.*

2.2. *Eucalyptus drepanophylla* – *E. crebra* Alliance

2.2.1. General

These 2 species cover large areas in Queensland, *E. crebra* extending into new South Wales (Fig. 12.1). They are sometimes found in the same district, *E. crebra* occupying the drier sites. The species are very alike, being separable only on minor characters (Table 12.1) and in some cases it is possible that ecologists have not distinguished between the species. Nevertheless each species has been used to characterize a suballiance and, when in doubt, either in the field or when dealing with literature, the writer has assigned narrow-leaved ironbarks in Queensland to *E. drepanophylla.* In New South Wales there is no confusion and the descriptions of the *E. crebra* communities are based mainly on these stands. In addition to the ironbarks, groups of communities occur which are intimately associated or form mosaics with the ironbark communities. These are characterized by *Angophora costata* which is used as the characteristic species of a third suballiance.

Table 12.2. Associations in the alliances dominated by ironbarks.

Alliance or suballiance	Main associations	Species forming minor associations with the dominants	Species forming ecotonal associations
E. cullenii	E. cullenii	E. staigeriana	E. dichromophloia; E. tetrodonta; E. polycarpa; E. melanophloia; E. leptophleba; E. shirleyi
E. drepanophylla	E. drepanophylla	E. exserta; E. decorticans; E. trachyphloia; E. whitei; E. setosa	E. crebra; E. maculata; E. dichromophloia; E. polycarpa; E. melanophloia; E. tessellaris
E. crebra (Tropics)	E. crebra	E. peltata; E. watsoniana; E. trachyphloia; E. bloxsomei; E. exserta; E. citriodora; E. tenuipes	E. drepanophylla; E. dichromophloia; E. melanophloia
E. crebra (south-east Qld)	E. crebra	E. citriodora; E. siderophloia; E. fibrosa ssp. fibrosa; E. exserta; E. trachyphloia	E. drepanophylla; E. maculata; E. intermedia; E. melanophloia; E. moluccana; Angophora costata
E. crebra (western N.S.W.)	E. crebra	E. trachyphloia; E. carnea; E. punctata	E. albens; E. melanophloia; E. sideroxylon; E. moluccana; E. macrorhyncha; E. intermedia; E. dealbata; Angophora costata; E. maculata; E. micrantha
E. crebra (Central coast N.S.W.)	E. crebra; E. crebra – Angophora bakeri	E. sideroxylon; E. fibrosa ssp. fibrosa; E. beyeri; E. eugenioides	
Angophora costata – Eucalyptus spp.	E. costata – E. polycarpa; A. costata – E. chloroclada; E. chloroclada; A. costata; Callitris "glauca"	E. peltata; E. watsoniana; E. citriodora; E. exserta; E. thozetiana; E. trachyphloia; E. panda; E. cloëziana	E. crebra; E. acmenoides; E. intermedia; E. maculata; E. dichromophloia; E. fibrosa ssp. nubila
E. melanophloia (Tropics)	E. melanophloia; Callitris "glauca"; Casuarina luehmannii		E. orgadophila; E. dichromophloia; E. similis; E. crebra; E. papuana
E. melanophloia (south-east Qld)	E. melanophloia; Callitris "glauca"; C. endlicheri	E. trachyphloia; E. peltata	E. crebra; E. populnea; E. dichromophloia; E. tessellaris; E. albens
E. melanophloia (N.S.W.)	E. melanophloia; E. melanophloia – Angophora floribunda		E. albens; E. crebra

2.2.2. Eucalyptus drepanophylla Suballiance

Narrow-leaved Ironbark

E. drepanophylla is the common ironbark in eastern Queensland and dominates vast tracts from north to south (Fig. 12.1). It occurs in areas with a mean annual rainfall of 1000 to 2000 mm, with a summer rainfall incidence and few or no frosts. *E. crebra* may occur in the same district and, if so, it occupies the drier sites, being found usually on ridges. The soils are generally of low fertility and sandy, being classified as Red or Yellow Earths, Red or Yellow Podsolics and, less commonly Lithosols.

E. drepanophylla usually dominates woodlands 12–20 m tall (Fig. 12.2), except in the most favourable sites, as in shallow valleys where it may form forests 30 m tall. In such forests it is usually in association with another eucalypt (Perry, 1953; Perry and Lazarides, 1964; Pedley, 1967; Speck, 1968). The woodlands often surround patches of dry rainforest (often Bottle-tree) or of *Acacia harpophylla* woodlands (Brigalow scrub), which occupy soils of finer texture. Near the coast, mosaics occur with other alliances, especially *E. polycarpa* – *E. tessellaris*, *E. alba* and *E. tereticornis* which occupy slight depressions subject to minor waterlogging. A few other eucalypts are confined to the alliance, forming local variant associations, especially *E. whitei* occurring mainly in pure stands on small patches of Krasnozem derived from basalt, and *E. decorticans* on sandstone outcrops. Many ecotonal associations occur (Table 12.2). *E. setosa*, a shrub or mallee, which is sometimes present as an understory species in the ironbark woodlands, may replace the ironbark locally and dominate mallee or low woodland patches (Fig. 12.3).

In the understory, small tree/tall shrub species are rare, except in depressions where they may form a well defined layer. The main species are: *Erythrophleum chlorostachys, Planchonia careya, Hakea arborescens, Acacia longispicata, H. lorea* (mainly tropics); *Alphitonia excelsa, Lysicarpus angustifolius, Grevillea striata, Petalostigma pubescens* and *Casuarina inophloia* (throughout, but mainly drier areas).

The grass layer is always dense, with *Bothriochloa ewartiana, Heteropogon contortus* and *Themeda australis* as the main constituents. On shallow, coarse textured or gravelly soils other grasses usually dominate, especially *Ari-*

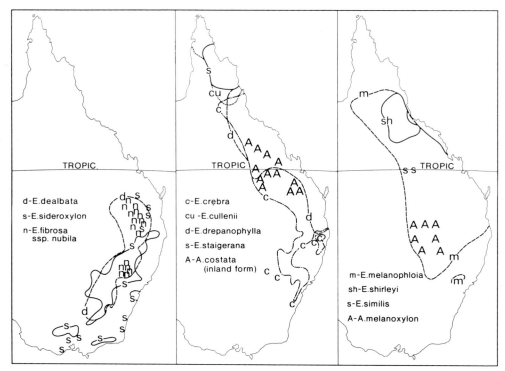

Fig. 12.1. Distributions of main species of *Eucalyptus* and *Angophora* dealt with in Chap. 12. Based on Hall et al. 1970.

stida spp., *Eriachne mucronata, E. obtusa, Triodia mitchellii* or *T. hostilis*.

2.2.3. *Eucalyptus crebra* Suballiance
Narrow-leaved Ironbark

E. crebra occurs over much the same area as *E. drepanophylla* but with an extensive southern extension into New South Wales (Fig. 12.1). It is found between the 500 and 2000 mm isohyets but has its main centre of abundance in the drier regions (500 to c. 800 mm mean annual rainfall). When the two species occur together, *E. crebra* occupies the drier sites, usually on hills. The *E. crebra* suballiance is replaced in some areas by the *Angophora costata* suballiance and towards the south it is divided into two segments by the Great Divide. In the southern drier part of its range it is replaced by the *E. fibrosa* ssp. *nubila* alliance, and on the drier west it adjoins the box woodlands (Chap. 13).

E. crebra dominates tall woodlands or forests 15–25 m high on sandy soils (Red and Yellow Earths, which are often mottled, deep sands), or Lithosols, and in drier areas Solodics. The soils are of low fertility and are unsuitable for cropping, the areas often being reserved as commercial forests, *E. crebra* providing excellent hardwood timber.

In the northern segment of the alliance around or north of the Tropic, *E. crebra* often forms pure stands or occurs with another eucalypt (Table 12.2). *Callitris "glauca"* is sometimes a co-dominant. In some areas a discontinuous shrub layer is present, the main shrubs being *Alphitonia excelsa* and *Lysicarpus angustifolius*. A grass layer containing *Themeda australis, Heteropogon contortus* and *Cympopogon refractus* occurs and in some areas *Xanthorrhoea johnstonii* is present (Speck, 1968).

South of the Tropic *E. crebra* often occurs in pure stands, usually or ridges, and it extends into valleys as a tall tree and a component of the tall forests with a mesomorphic understory (Chap. 9). A very common association with *E. maculata* occurs in south-eastern Queensland, extending on to the coastal lowlands of New South Wales. The communities are forests with a discontinuous tall shrub layer of *Acacia cunninghamii, Alphitonia excelsa, Petalostigma pubescens, Jacksonia scoparia* and *Dodonaea cuneata*. The grass layer is dominated by *Aristida caput-medusae, A. ramosa* and *Cymbopogon refractus* (Cromer and Pryor, 1942; Baur, 1965; Coaldrake et al. 1972). Local variations in the tree layer occur, chiefly through associations of *E. crebra* with *E. fibrosa* spp. *fibrosa* or *E. moluccana*. In the most southerly stands of *E. crebra* which are found west of Sydney on Tertiary gravels and alluvium, Phillips (1947) records *Angophora bakeri* as the common associate; other common trees are *E. eugenioides, E. fibrosa* ssp. *fibrosa* and *E. sideroxylon*. The communities form ecotonal associations with the low fertility coastal assemblages, and the understory layers are completely xeromorphic, containing Proteaceae, Mytaceae etc., as found on the sandstones (Chap. 10). These stands resemble the *E. crebra* woodlands found in the drier parts of New South Wales.

In the woodlands in the drier western parts of New South Wales, *E. crebra* sometimes forms pure stands, but *Callitris "glauca"* (on deep sands) or *C. endlicheri* (on shallow or very coarse soils) (Fig. 12.4) are usually present, the former often co-dominant with *E. crebra*. On Solodic soils *Casuarina luehmannii* often forms a distinctive understory. Many of these stands have no well defined shrub layer, but only a grass understory consisting mainly of *Aristida echinata, Stipa variabilis, Chloris acicularis*. On the other hand, some stands, chiefly those in drier areas on coarse, shallow soils, contain a more or less continuous xeromorphic shrub

Fig. 12.2. *Eucalyptus drepanophylla* woodland. *Hakea lorea* (right) only shrub species. Grass layer with co-dominants of *Heteropogon contortus* and *Themeda australis*, with *Chrysopogon fallax* as a subsidiary species. 72 km north of Hughenden, Qld.

Fig. 12.3. *Eucalyptus setosa* as a local dominant within woodlands dominated by *E. drepanophylla*. Grass layer dominated by *Triodia mitchellii*. Near Charters Towers, Qld.

layer in which the following are common: *Acacia buxifolia, A. deanei, A. hakeoides, A. lineata, A. spectabilis, Calytrix tetragona, Cassinia aculeata, Exocarpos cupressiformis, Lissanthe strigosa, Melaleuca ericifolia, M. pubescens, M. uncinata* (Biddiscombe, 1963).

2.2.4. Angophora costata – Eucalyptus spp. Suballiance

A. costata (Smooth-barked Apple), best known from its occurrences near the coast, is found in drier country to the north-west (mean annual rainfall 600–1000 mm) from Warialda, New South Wales, to central Queensland (Fig. 12.1). In this drier region it is associated with several eucalypts (Table 12.2) in various combinations, forming distinctive communities usually referred to as "sandstone woodlands" (Pedley, 1967; Story, 1967; Speck, 1968). The communities form mosaics and ecotonal associations with narrow-leaved ironbark communities, especially *E. crebra*.

These sandstone communities are regarded as dry-climate derivatives of more extensive tracts of xeromorphic communities which possibly once extended over much of eastern Australia and are now represented more abundantly in the east. They contain a mixture of wetter eastern species and dryland species which are probably local developments from wetter taxa. Further to this, associated with the *Angophora* assemblage are many xeromorphic genera which have their maximum development mainly on the coastal sandstones where *Angophora costata* is abundant.

The tree species occur in various combinations and are floristically related through the constant occurrence of *Angophora costata*, except in some of the driest areas.

In Central Queensland *Angophora costata*, *E. cloëziana* and *E. tenuipes*, often with admixtures of *E. chloroclada*, *E. peltata*, *E. polycarpa* and *E. watsoniana*, form closed forests 20–30 m tall in the wetter habitats, especially in valleys. *Callitris "glauca"* is often present, either as a co-dominant (sometimes locally dominant or even in pure stands), and admixtures (ecotonal groupings) of *E. crebra* (common), *E. dre-*
panophylla, *E. maculata*, *E. tereticornis* and *E. citriodora* (all rare) occur. *Angophora costata* is constant in the stands and often the dominant, but never forms pure stands. The proportion of the various eucalypts varies from site to site and there is no obvious explanation for the patterning.

A lower tree layer of *Casuarina inoploia* and/or *Lysicarpus angustifolius* is often present, or the following species form a tall shrub/small tree layer: *Acacia complanata*, *A. excelsa*, *A. flavescens*, *A. juncifolia*, *A. jucunda*, *Alphitonia excelsa*, *Grevillea longistyla*, *Leptospermum attenuatum*, *Notelaea longifolia* and *Petalostigma glabrescens*. Shrubs are rare and mostly xeromorphic, including *Astrotricha pterifolia*,

Fig. 12.4. *Eucalyptus crebra* forest, with *Callitris endlicheri*, *Casuarina luehmannii* forming a lower tree layer. Main shrubs are *Dodonaea* aff. *attenuata* and *Acacia deanei*. Herbaceous layer dominated by *Aristida jerichoensis* with subsidiary *Chloris acicularis* and *Cheilanthes tenuifolia*. 26 km south of Narrabri, N.S.W.

Fig. 12.5. *Angophora costata*. Pure stand, with the herbaceous layer dominated by *Aristida echinata*. Warialda, N.S.W.

Boronia rosmarinifolia, B. glabra, B. bipinnata, Cryptandra amara, Dodonaea vestita, Hibbertia stricta, Mirbelia pungens, and *Pultenaea cunninghamii.* The herbaceous layer is poorly developed and contains scattered plants of *Triodia mitchellii* with subsidiary *Cymbopogon refractus*.

The stands in southern Queensland and northern New South Wales contain fewer tree species. *Angophora costata* is usually present and sometimes locally dominant (Fig. 12.5). *E. polycarpa* is uncommon, and *E. chloroclada* becomes more abundant and often forms pure stands on gravelly sands near stream lines (Fig. 12.6). Other small trees or tall shrubs are *Callitris "glauca", C. endlicheri, Casuarina inophloia* and *Acacia concurrens*. A shrub layer of xeromorphs is present in some stands, some of the common ones being: *Acacia conferta, A. deanei, Hovea longifolia, Jacksonia scoparia, Melichrus urceolaris, Notelaea microcarpa* and *Persoonia nutans*. The grass layer is well developed and contains the following in various combinations: *Aristida echinata, A. ramosa, Chloris ventricosa* and *Eragrostis lacunaria*.

2.3. *Eucalyptus melanophloia* Alliance

E. melanophloia is widely distributed in the east, and within its area three other groups of communities dominated by *E. shirleyi, E. similis* and *Angophora melanoxylon* occur (Fig. 12.1). Each of these species has been used to designate a suballiance.

2.3.1. *Eucalyptus melanophloia* Suballiance

Silver-leaved Ironbark

E. melanophloia occurs most abundantly west of the Great Divide, mostly between the 370 and 750 mm isohyets. It is found also east of the Divide both in Queensland and New South Wales in areas with as high a rainfall as 1500 mm. The woodlands often form mosaics with woodlands of *E. drepanophylla* in Queensland or of *E. crebra* and especially *E. populnea* in western Queensland and New South Wales. The ecological factors determining the distribution of the species are not clear. In addition to the wide climatic range, it is found on a variety of soils including gravelly and coarse sands, and texture-contrast soils with a light clay B horizon. A large number of ecotonal associations occur (Table 12.2).

E. melanophloia commonly dominates woodlands 10–20 m high, sometimes forming pure stands which rarely exceed a few sq. km in extent. *Callitris "glauca"* is the most common associate. A shrub layer is rarely present though a relatively high number of shrubs are recorded (Blake 1938; Beadle 1948; Holland and Moore, 1962; Biddiscombe, 1963; Baur, 1965; Pedley, 1974). The small tree/tall shrub species recorded are: (a) in the north, *Acacia coriacea, A. rhodoxylon* and *Eremophila mitchellii;* (b) in the west, *Acacia aneura, A. excelsa, Grevillea juncifolia* and *Hakea purpurea;* (c) in the south, *Angophora floribunda, Brachychiton populneum, Callitris endlicheri, Casuarina inophloia* and *C. luehmannii*. Shrubs 1–2 m high are uncommon or absent, and the following have been recorded: *Dodonaea attenuata, Eremophila glabra, Hovea longipes, Micromyrtus minutifora* and *Notelaea microcarpa*.

The grass layer show great variation from north to south. In or near the tropics *Bothriochloa ewartiana, Heteropogon contortus* and *Themeda australis* are the most common species, usually forming a dense sward. In subtropical areas *Triodia mitchellii* becomes the dominant, extending to c. lat. 30°S. (Fig. 12.7) with *Aristida jerichoensis* sometimes conspicuous. In the same area *Eriachne mucronata,* with subsidiary *Aristida jerichoensis* becomes the dominant on stony rises. The latter is common also in the most southerly stands, but *Stipa variabilis* and *Danthonia* spp. are usually common or dominant in the south.

Fig. 12.6. Pure stand of *Eucalyptus chloroclada* on gravelly sand near a watercourse. Herbaceous layer contains *Aristida echinata, Helichrysum apiculatum, Calotis lappulacea* and *Cynoglossum suaveolens*. c. 60 km southeast of Coonabarabran, N.S.W.

Fig. 12.7. *E. melanophloia*. Pure stand with *Triodia mitchellii* dominating the herbaceous layer. Cumborah, N.S.W.

2.3.2. *Eucalyptus shirleyi* Suballiance
Shirley's Silver-leaved Ironbark

E. shirleyi occurs within the area of *E. melanophloia* (Fig. 12.1), with which it is closely related (Table 12.1). It is found between the 650 and 1200 mm isohytes in the summer rainfall zone, on sandy soils derived from acid rocks, including Red Earths and Podsols, or Lithosols on ridges. In the wetter parts of its range it reaches a height of 13 m and forms open woodlands. In drier areas the trees are stunted and shrub-like and in all cases the trunks are usually crooked. Community descriptions are available for only two stands. In the Townsville-Bowen region Perry (1953) describes an open woodland 5–7 m high with an orchard appearance. The grass layer is usually a sward of *Sehima nervosa*, *Heteropogon contortus*, *Aristida browniana* and *Themeda australis* (Fig. 12.8). In places (shallow soils?) *Plectrachne pungens* dominates the lower story.

2.3.3. *Eucalyptus similis* Suballiance
Desert Yellow Jacket

E. similis occurs in a restricted area around the Tropic in the Jericho district (mean annual rainfall c. 500 mm, with a summer incidence). Pedley (1967) records it as a dominant in woodlands 8–15 m tall on Red and Yellow Earths, with some inclusions of *E. melanophloia*. *E. similis* occurs in groves across slopes, the groves being separated by grasslands. Associated eucalypts may occur, *E. trachyphloia*, *E. setosa* or *E. drepanophylla*. The shrub layer 1½ to c. 3 m high, is dominated by *Acacia leptostachya*. The dense grass layer contains *Bothriochloa ewartiana*, *Heteropogon contortus* and *Themeda australis*, with some *Aristida glumaris* and *Erichne* spp. The grasslands separating the groves of eucalypts contain the same species and *Triodia mitchellii*.

2.3.4. Angophora melanoxylon Suballiance
Coolabah Apple

This species occurs mainly within the *E. melanophloia* Alliance in south-eastern Queensland (Holland and Moore, 1962), with smaller occurrences in the *E. populnea* woodlands in the same district and in north-western New South Wales. The species reaches a height of c. 10 m and forms woodlands between the 320 and 430 mm isohyets on deep, coarse sands or gravelly sands. *Callitris "glauca"* occurs in some stands.

The woodlands contain distinctive and rich (in comparison with the adjacent *Eucalyptus* woodlands) assemblages of shrub and herbaceous species, including the tall shrubs *Grevillea albiflora*, *G. juncifolia*, *G. pterosperma*, with subsidiary (sometimes locally dominant) *Acacia calamifolia*, *Eremophila longifolia*, *Leucopogon mitchellii*, *Myoporum acuminatum* and *Petalostigma pubescens*. Shrub thickets of *Eremophila glabra* sometimes occur. The herbaceous layer is usually dominated by *Triodia mitchellii* (Fig. 12.9), with subsidiary *Aristida echinata* and *Eragrostis lacunaria*. Annual herbs are common between the grass hummocks after rain, noteworthy among them being *Actinotus paddisonii* and *Brunonia australis*.

Fig. 12.8. *Eucalyptus shirleyi* woodland. Scattered shrubs of *Bursaria multisepala* and a grass layer of *Heteropogon contortus* and *Aristida browniana*. c. 250 km north of Hughenden, Qld.

Fig. 12.9. *Angophora melanoxylon* with a shrub layer of *Grevillea albiflora*. *Triodia mitchellii* dominates the herbaceous layer. 48 km east of Cunnamulla, Qld.

2.4. *Eucalyptus sideroxylon* Alliance

Mugga

2.4.1. General

E. sideroxylon characterizes this alliance, which is the most westerly of the forests in the eastern part of the continent, and occurs between the 350 and 650 mm isohyets, with a few outliers to the east in wetter country. The alliance is restricted to ridge-tops or plateaux, where the soils are of very low fertility, and is unique in the semi-arid zone since it contains a rich assemblage of xeromorphic species which relate it to the xeromorphic forests of the eastern sandstones (Chap. 10). However, these two related xeromorphic assemblages are no longer connected. The *E. sideroxylon* alliance occurs on an isolated, elongated tract (Fig. 12.1) of sandy soil surrounded by more fertile, mostly texture-contrast soils which support box woodlands (Chap 13), except in the north where it adjoins segments of the *E. crebra* and *E. macrorhyncha* – *E. rossii* alliances. The *E. sideroxylon* alliance is sharply delimited from the box woodlands, neither the eucalypts nor the understory species mixing to any great extent.

This isolated segment of the eastern sandstone flora deserves further comment. About 280 spp. are listed for the alliance and c. 120 of these are found also on the coastal sandstones where the mean annual rainfall is about double that recorded for the *E. sideroxylon* alliance. The remaining species are mainly local developments in genera which are widespread in the southeast, including 75 species of *Acacia*. A few western species (c. 30) are recorded but the influence of the adjoining semi-arid flora on the alliance is small.

E. sideroxylon is often associated with *E. dealbata* and, in the northern part of the alliance, *E. fibrosa* ssp. *nubila* is locally domin-

Fig. 12.10. *Eucalyptus sideroxylon* and *Callitris endlicheri*. Understory dominated by *Helichrysum diosmifolium* with subsidiary *Acacia armata* and *Hibbertia obtusifolia*. Bunbury Range, west of Manildra, N.S.W.

Fig. 12.11. *Eucalyptus dealbata* and *Callitris endlicheri* co-dominant on rocky ridges. *E. crebra* and *E. macrorhyncha* also present. Shrubs include *Casuarina inophloia, Dillwynia retorta, Melichrus urceolaris, Leucopogon mutica* and *Hibbertia obtusifolia*. Main herbs are *Cymbopogon refractus, Xanthorrhoea australis* and *Echinopogon caespitosus*. 30 km south of Emmaville, N.S.W.

ant. Two suballiances are therefore recognized, which are related through the xeromorphic understory rather than the dominant eucalypts.

2.4.2. *Eucalyptus sideroxylon – E. dealbata* Suballiance
Mugga-Tumbledown Red Gum

These species dominate low forests or woodlands on hills composed mainly of sandstones, conglomerates and, rarely, granites mainly between the 400 and 600mm isohyets (Beadle, 1948; Moore, 1953; Biddescombe, 1963). The area of *E. sideroxylon* is very fragmented (Fig. 12.1). It is found mainly west of the Great Divide, with small outliers east of the Divide and in Victoria. It occurs also as an occasional to rare species in some of the mallee outliers in New South Wales, usually close to hills on which *E. sideroxylon* is dominant. *E. dealbata* occupies an area similar to that of *E. sideroxylon* (Fig. 12.1) and extends eastward on to the Tablelands (see below). The soils are Lithosols, rock often lying at the surface, or deep sands or gravelly sands with no profile differentiation or, rarely, with a weakly cemented B horizon. They are acid throughout the profile (pH 4.5–5.5) and sometimes contain an accumulation of dark, finely divided organic matter in the A horizon which greatly retards wetting.

E. sideroxylon commonly occurs in pure stands (Fig. 12.10) on deep soils, forming closed forests to 20 m high (rarely to 30 m). With decreasing soils depth *E. dealbata* becomes a subsidiary species or is co-dominant. On rocky

Fig. 12.12. *Eucalyptus fibrosa* ssp. *nubila* (background) and *E. trachyphloia* (centre, foreground). Main shrubs are *Casuarina rigida, Cassinia aculeata, Dodonaea viscosa, Boronia bipinnata, Brachyloma daphnoides, Acacia pilligaensis* and *Dampiera stricta*. 64 km north of Coonabarabran, N.S.W.

outcrops *E. dealbata* often forms gnarled woodlands and in these situation it may be replaced by *E. dwyeri*. *Callitris endlicheri* and *C. glauca* are usually present in the stands, sometimes becoming local co-dominants. The understory sometimes contains small trees or tall shrubs, including *Angophora floribunda*, *Brachychiton populeum*, *Acacia doratoxylon* (often locally dominant in rocky areas), *A. baileyana* (confined to the Cootamundra – Temora area, *Casuarina stricta*, *Exocarpos cupressiformis*, *Eucarya acuminata* and *Pittosporum phyllirаeoides*. Two lianas, *Pandorea pandorana* and *Parsonsia eucalyptiphylla*, are sometimes present in rocky terrain, the plants usually rooted in wet depressions near the bases of small cliffs or boulders.

Shrubs up to 2 m high are usually abundant, often forming a continuous layer. The flora is richest in the north decreasing in species abundance towards the south. Common species in northern and central New South Wales, some of which form local thickets, are *Acacia buxifolia*, *Brachyloma daphnoides*, *Cassinia laevis* (thickets), *Dillwynia floribunda*, *Grevillea floribunda*, *Hibbertia linearis*, *H. obtusifolia*, *Leptospermum parvifolium* (thickets), and *Melaleuca erubescens*. The herbaceous stratum is poorly developed, though floristically rich, and the more common species include: *Amphipogon strictus*, *Aristida ramosa*, *Danthonia pallida*, *Eragrostis lacunaria*, *Paspalidium gracile* and *Stipa verticillata*. Interesting occurences are *Macrozamia spiralis* and several delicate herbs (Droseraceae and Orchidaceae) which occur in small soaks, often at or near the bases of rocky outcrops.

In southern New South Wales and Victoria (Patton, 1944; Moore, 1953), *E. sideroxylon* is usually associated with *E. woollsiana* with an understory dominated mainly by semi-arid species of *Acacia* (*A. buxifolia*, *A. difformis*, *A. hakeoides* and others), but xeromorphic shrubs are present as well, especially *Brachyloma daphnoides*, *Cassinia aculeata*, *Dillwynia floribunda*, *Grevillea lanigera*, *Phyllanthus thymoides* and *Pultenaea foliolosa*. The herbaceous layer is dominated by *Stipa falcata* and *Danthonia* spp.

E. dealbata extends on to the lower tablelands in northern New South Wales and is usually co-dominant with *Callitris endlicheri* or it occurs in mallee-like stands on exposed rock surfaces (Fig. 12.11). *E. caleyi*, sometimes locally dominant, and *E. macrorhyncha* are often present in the stands. The shrub layer in this communities is very rich floristically, consisting entirely of xeromorphs and the assemblages resemble the understory found in the *E. andrewsii* woodlands (Chap. 11).

2.4.3. *Eucalyptus fibrosa* ssp. *nubila* Suballiance
Blue-leaved Ironbark

This subspecies dominates woodlands 13–20 m high, between the 470 and 600 mm isohyets, with the centre of abundance in the drier areas. The woodlands occur on coarse and gravelly sands and Lithosols derived from sandstones and conglomerates and they often adjoin woodlands and forests of *E. crebra* or *E. sideroxylon*. When the three species occur in the same district, as in New South Wales, *E. fibrosa* ssp. *nubila* appears to select the coarsest soils, probably of lowest fertility.

Pure stands of the subspecies occur but *E. trachyphloia* is usually present (Fig. 12.12) and is commonly the dominant on rocky outcrops, where it forms gnarled woodlands. Ecotonal associations occur with the *E. sideroxylon* – *E. dealbata* suballiance with which it is closely related through its understory species. *Callitris endlicheri* is a common species, usually occurring in the understory but sometimes as a co-dominant.

A dense understory of shrubs is always present, often with local thickets, especially on deep sands. Some of the more common species are *Acacia doratoxylon*, *A. gladiiformis*, *Brachyloma daphnoides*, *Calytrix tetragona*, *Cassinia aculeata*, *Casuarina distyla*, *C. littoralis*, *Dodonaea attenuata*, *Grevillea floribunda*, *Melaleuca uncinata* (see below) and *Phyllanthus thymoides*. The herbaceous layer is discontinuous. *Aristida jerichoensis* is a constant species; *Macrozamia spiralis* is sometimes locally common and also *Lomandra* spp.

Variations in the assemblages are brought about by the presence of rock at the surface, or by impeded drainage in shallow depressions. On rocky outcrops *E. trachyphloia* is usually dominant or associated with *E. dealbata* or *Callitris endlicheri*. Shrub are abundant but usually form a discontinuous layer in rocky terrain. *Acacia doratoxylon*, 3–4 m tall, is common and a few mesomorphic species, including *Alphitonia excelsa* and *Maytenus cunninghamii*

Fig. 12.13. *Melaleuca uncinata* thicket with subsidiary *Casuarina rigida,* and smaller plants of *Dampiera stricta, Pultenaea boormanii* and *Westringia cheelii.* 80 km north of Coonabarabran, N.S.W.

are often present. Common xeromorphic shrubs are *Daviesia* spp., *Grevillea triternata, Leucopogon* spp. and *Pultenaea foliolosa.* The herbaceous layer is scattered and usually contains *Entolasia stricta* and *Stipa densiflora* with a variety of soft-leaved herbs, including the typically rock-ledge species *Cheilanthes distans* and *Isotoma axillaris.* In slightly depressed areas, commonly a few hectares in extent, trees are absent and thickets develop. The absence of trees appears to be due to the shallowness of the soils and possibly to the underlying water-impervious substratum (ferruginous laterite) which induces temporary waterlogging. *Melaleuca uncinata* (Broombush) usually dominates the thickets (Fig. 12.13).

13. The Box Woodlands of the East and South-East

1. Introduction

Eucalyptus woodlands, dominated chiefly by species referred to as "Boxes", extend in an area from Cape York Peninsula to South Australia. They occupy climatically drier areas and constitute a zone which separates the forests of the wetter coastal areas from the arid interior. Some of the woodlands extend to the coast in the north, east and south, where they occur in drier areas which are often rainshadows, especially in the south-east.

The communities have considerable economic importance. They all provide grazing for domestic stock. South of c. lat. 26° S. large tracts of woodland have been cleared of timber for the cultivation of grain crops, the area constituting the main wheat belt in the east. The eucalypts themselves have little commercial value as timber, mainly because of the short boles of the trees. However, they provide unsawn logs for fencing, excellent fuel and some of the best honey produced in the continent. *Callitris glauca*, a common associate in some of the woodlands, is cut for milled timber which is used mainly for flooring. Many of the understory species, particularly in the north and east, are eaten by stock and are lopped during drought for emergency feed, especially *Brachychiton populneum* (Kurrajong) and *Geijera parviflora* (Wilga).

The eucalypts which dominate the various communities all belong to subgenus *Symphyomyrtus*. The species are closely related taxonomically and occur in latitudinal sequence, with

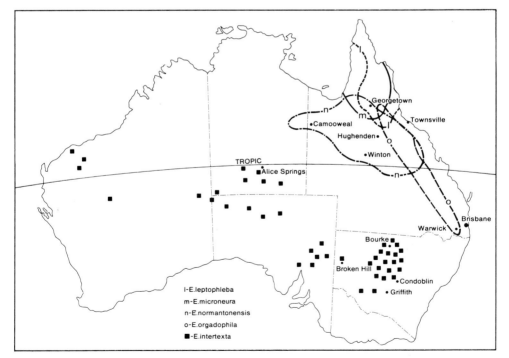

Fig. 13.1. Distributions of species of *Eucalyptus* forming woodlands in drier regions mainly in the north. Based on Hall et al. (1970).

Table 13.1. Associations in the woodland alliances and suballiances of eastern and south-eastern Australia.

Alliance or suballiance	Main associations	Species forming minor associations with the dominants	Species forming ecotonal associations
E. leptophleba	E. leptophleba		E. polycarpa; E. argillacea; E. papuana; E. alba; E. confertiflora; E. tessellaris; E. microtheca
E. microneura	E. microneura ± Erythrophleum chlorostachys		E. microtheca
E. normantonensis	E. normantonensis		E. terminalis; E. melanophloia; E. brevifolia
E. orgadophila	E. orgadophila		E. populnea; E. melanophloia; E. crebra; Acacia harpophylla
E. intertexta – Acacia spp.	E. intertexta – A. excelsa or A. aneura		E. populnea
E. intertexta – Callitris "glauca"	E. intertexta – Callitris "glauca"		E. populnea
E. brownii	E. brownii		E. crebra; E. drepanophylla
E. populnea	E. populnea		E. albens; Acacia harpophylla; A. pendula
E. populnea – Casuarina cristata	E. populnea – Casuarina cristata		Acacia harpophylla; A. pendula
E. populnea – Eremophila mitchellii	E. populnea – Eremophila mitchellii		E. melanophloia; E. crebra
E. populnea – Callitris "glauca"	E. populnea – C. "glauca"; E. populnea – Geijera parviflora		E. melanophloia; E. intertexta; E. woollsiana E. crebra; E. pilligaensis
E. populnea – Geijera – Flindersia	E. populnea – Geijera – Flindersia		Casuarina cristata; Acacia harpophylla
E. populnea – Acacia spp.	E. populnea – Acacia excelsa or A. aneura		E. intertexta
E. melliodora – E. blakelyi	E. melliodora – E. blakelyi; E. melliodora; E. blakelyi	E. bridgesiana; E. conica; E. dives; E. polyanthemos	E. albens; E. caliginosa; E. viminalis; E. sideroxylon; E. leucoxylon
E. moluccana	E. moluccana	E. fibrosa ssp. fibrosa; E. exserta; E. siderophloia; E. beyeri	E. tereticornis; E. maculata; E. crebra; E. sideroxylon
E. albens	E. albens; E. albens – Callitris "glauca"	Angophora floribunda	E. melliodora; E. melanophloia; E. crebra; E. woollsiana
E. woollsiana	E. woollsiana ssp. microcarpa; E. woollsiana ssp. woollsiana		E. melanophloia; E. populnea; E. melliodora; E. sideroxylon; E. odorata; E. populnea;

some overlapping which leads to the intergrading of species and the development of many ecotonal associations (Table 13.1).

2. Descriptions of the Alliances

2.1. *Eucalyptus leptophleba* – *E. microneura* – *E. normantonensis* Alliance

These species constitute the Box woodlands of the eastern tropics. Taxonomically the species are related to *E. argillacea* and *E. tectifica* which form the Box woodlands mainly west of the Carpentaria Plains (Chap. 8).

The three species are sometimes found together (Story, 1970) and are sometimes mixed with *E. argillacea* and *E. microtheca*. They are restricted to fine textured soils and the woodlands are usually sharply delimited by soil texture from adjacent communities of bloodwoods or ironbarks on sandy soils. The areas of the species overlap but do not coincide (Fig. 13.1). Each species is used to identify a suballiance.

2.1.1. *Eucalyptus leptophleba* Suballiance
Gum-leaved Box

E. leptophleba occurs between the 1000 and 1200 mm isohyets on alluvial flood-plains along rivers on Yellow or Red Earths. In drier areas it occurs on Solodic Soils or small areas of Krasnozem derived by basalt.

The communities are woodlands (Fig. 13.2) reaching a height of 25 m in the wetter areas and 10–15 m high in the drier south-west. *E. leptophleba* rarely grows in pure stands, *E. polycarpa* being the most regular associate. Small tree or tall shrub species are usually present, especially *Erythrophleum chlorostachys* and species of *Acacia, Melaleuca, Petalostigma* and *Planchonia*. The herbaceous stratum, usually well developed, consists of species of *Alloteropsis, Bothriochloa, Dichanthium, Eulalia, Heteropogon, Sorghum* and *Themeda*. In general appearance the community resembles the *E. microneura* woodlands (next Section).

Fig. 13.2. *Eucalyptus leptophleba* woodland. Grass layer with *Bothriochloa ewartiana* and *Themeda australis*. Near Einsleigh, Qld. Photo: Dr. Betsy Jackes.

Fig. 13.3. *Eucalyptus microneura* with *Erythrophleum chlorostachys* (tree nearest camera) and *Terminalia ferdinandiana* (at rear and below eucalypts). Grasses are *Chrysopogon fallax, Aristida pruinosa* and *Themeda australis*. 40 km west of Georgetown, Qld.

2.1.2. *Eucalyptus microneura* Suballiance
Georgetown Box

E. microneura occurs between the 700 and 1100 mm isohyets on flat to undulating country on Red and Yellow Earths, Red and Yellow Podzolics, Solodized and Solonetz Soils. Its distribution is shown in Fig. 13.1. The species, usually in pure stands, forms open woodlands 7 to 15 m high. *Erythrophleum chlorostachys* is a regular associate and may equal the eucalypts in height (Fig. 13.3). Tall shrubs often from a discontinuous layer, the most important being *Alphitonia excelsa, Bauhinia cunninghamii, Dolichandrone heterophylla, Melaleuca viridiflora, Terminalia* spp. and *Ventilago viminalis. Carissa lanceolata* is often a common shrub c. 1 m high, sometimes forming local thickets. The grass layer consists mainly of species of *Aristida, Chrysopogon, Heteropogon,* and *Themeda australis*.

2.1.3. *Eucalyptus normantonensis* Suballiance
Normanton Box

E. normantonensis occurs between the 600 and 700 mm isohyets on alluvial flats where the soils are mostly fine-textured. The species is variable in habit but is mostly a tree to 13 m high when it forms open woodlands. At its drier limit it sometimes exhibits a mallee habit, e.g., in the outlier occurrences in the driest regions (Fig. 13.1).

In woodland formation it usually has no associated tall species, *Casuarina luehmannii* being the only one recorded (Pedley, 1967). *Carissa lanceolata*, a shrub c. 1 m high is often locally abundant. The grass layer is scanty and consists either of species of *Aristida, Bothriochloa, Chloris, Cymbopogon, Eriachne, Heteropogon* and *Themeda*, or of *Triodia mitchellii*.

2.2. Eucalyptus orgadophila Alliance

Gum-top Box

E. orgadophila occurs between the 630 and 1250 mm isohyet, with a summer rainfall maximum (Fig. 13.1). It forms woodlands 12–16 m high (Perry, 1964; Pedley, 1967). Pure stands are restricted either to basalt outcrops or to dark, cracking clays (often Black Earths) derived mainly from basalt or basaltic alluvium. In the tropics the communities are grassy woodlands, the main grasses being *Eulalia fulva, Dichanthium fecundum* and *Astrebla* spp.

South of the Tropic, the woodlands are similar (Fig. 13.4) to those further north, but tall shrub or tree species are sometimes common, especially in ecotonal areas (usually with the *Eucalyptus populnea* woodlands). The trees and tall shrubs are *Alphitonia excelsa, Atalaya hemiglauca, Brachychiton populneum, Capparis lasiantha, Casuarina cristata, Eremophila mitchellii, Geijera salicifolia* and *Ventilago vinalis*. The grass layer is usually continuous and consists of mixtures of *Astrebla* ssp., *Bothriochloa ewartiana, Heteropogon contortus, Sporobolus elongatus, Stipa aristiglumis* and *Themeda australis*.

2.3. Eucalyptus intertexta Alliance

Gum Coolabah

E. intertexta, a tree to 25 m high, occurs disjunctly across the drier parts of the continent (Fig. 13.1). Its main centre of abundance lies around the 350 mm isohyet in New South Wales, where it is often associated with *E. populnea*. The occurrences in the arid zone are regarded as relicts of more extensive woodlands that existed prior to the development of the arid zone. In its main centre of abundance two main

Fig. 13.4. *Eucalyptus orgadophila* woodland. Shrubs are *Capparis lasiantha, Geijera salicifolia* and *Ventilago viminalis*. Grass layer contains *Stipa aristiglumis, Panicum queenslandicum* and *Sporobolus elongatus*. 3 km west of Pittworth, Qld.

Fig. 13.5. *Eucalyptus intertexta* woodland with a tall shrub layer of *Acacia aneura*. Tree on extreme right (foreground) is *E. populnea*. Near Byrock, N.S.W.

suballiances can be recognized defined according to the associated tall shrub and tree species.

2.3.1. *E. intertexta – Acacia* spp. Suballiance

E. intertexta with *Acacia excelsa* and/or *A. aneura* forming a small tree/tall shrub layer (Fig. 13.5) forms shrub woodlands on Non-calcic Brown Soils mainly in the Bourke – Cobar area. A large number of tall shrubs may be present, some of them becoming locally prominent, especially *Eremophila mitchellii* and less commonly, *Heterodendrum oleifolium*, *Flindersia maculosa* and *Acacia homalophylla*. Occasional tall shrub/small tree species include *Brachychiton populneum* (at its western limit at this latitude), *Codonocarpus cotonifolius* (on sand patches) *Geijera parviflora*, *Grevillea striata* and *Atalaya hemiglauca*. Shrubs 1–2 m high are usually common, sometimes forming thickets, especially *Eremophila sturtii* and *Dodonaea attenuata*; Less common are *Cassia artemisioides*, *Myoporum deserti* and *Scaevola spinescens*.

The density of the grass sward is determined by the density of the upper stories and is usually discontinuous, or even absent if the shrubs form thickets. The main species include: *Aristida jerichoensis*, *Danthonia bipartita*, *Enneapogon avenaceus*, *Eragrostis eriopoda*, *E. lacunaria*, *Neurachne mitchelliana*, *Paspalidium gracile* and *Stipa variabilis*. Forbs are abundant after rain, especially the winter-growing *Erodium cynorum* and Compositae.

2.3.2. Eucalyptus intertexta – Callitris glauca Suballiance

On deep red sands, *E. intertexta* is associated with *Callitris "glauca"*, the latter often being very dense and suppressing the entry of other woody species into the community. The herbaceous species listed above occur also in this suballiance but their densities are low, except in open patches, some of which have possibly been produced by the removal of *Callitris* for timber. *Aristida jerichoensis* is usually the dominant species, with subsidiary *Eragrostis lacunaria* and *Enneapogon avenaceus* (Fig. 13.6).

The arid outliers of *E. intertexta* probably belong to this suballiance. They occur mostly in or near watercourses, often in rocky terrain. The trees are usually scattered and often up to 20 m tall, but rarely form woodland stands. At Yudnapinna, S.A., Crocker and Skewes (1941) record *Callitris "glauca"* as a co-dominant, with *Acacia burkittii* as an understory shrub, and *Cymbopogon exaltatus* (a typical "dry" watercourse species) in a discontinuous grass layer. In still drier areas in Northern Territory and Western Australia, associated species have not been mentioned, but the localities suggest an understory of *Triodia* spp.

2.4. *Eucalyptus populnea* Alliance

Bimble Box or Poplar Box

2.4.1. General

E. populnea, a tree 12–23 m high, is dominant in woodlands over a vast area between the 300 and 500 mm isohyets (Fig. 13.7). In the northern part of this area there is a summer rainfall maximum, and in the south a slight winter predominance. Minor frosts occur (lowest minima to $-6°C$). It is a vigorous species with reliable seed-production and germination, and a rapid growth rate. It has a remarkable range of edaphic tolerance, occurring on soils of any texture, from gravelly sands to clays, covering the pH range c. 5 to 8.5. The species appears to be increasing in area, partly by migration into the *Acacia* shrubs in the drier west and partly by establishment in other communities, especially the outlier mallee stands in the south (Chap. 14) and the "softwood scrubs" in the north-east (Chap. 7).

In the east, the *E. populnea* woodlands adjoin and often form mosaics with either other woodlands (see below) or the ironbark forests (Chap.

Fig. 13.6. Open woodland of *Eucalyptus intertexta* and *Callitris "glauca"*. Grass layer dominated by *Aristida jerichoensis*, with subsidiary *Danthonia bipartita* and *Eragrostis lacunaria*. 64 km west of Cobar, N.S.W.

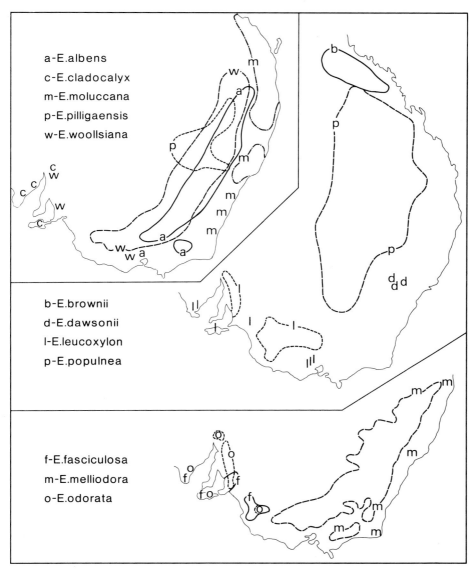

Fig. 13.7. Approximate areas of species of *Eucalyptus* dealt with in Chap. 13. Based on Hall et. al. 1970.

12). In the drier west they are deliminated by soils of fine texture in the north and south, these supporting either grassland (mainly north) or halophytic shrublands (south); in the central position *E. populnea* forms a broad ecotone with the *Acacia* scrubs. Enclosures of all the communities mentioned above occur within the *E. populnea* woodlands.

Since *E. populnea* is dominant in such a variety of habitats, it is associated with a very large number of species which have narrower ecological tolerances, many being restricted to specific soil types. This leads to the occurrence of fairly well defined assemblages of species which have been given the rank of suballiances. In the various habitats *E. populnea* shows some variation in leaf size and shape, which is correlated with soil texture. A narrow-leaved form (subspecies) occurring in the tropics is sometimes referred to *E. brownii*. Some narrow-leaved forms are hybrids, usually with *E. crebra* or *E. microtheca*.

The descriptions that follow are based on

Blake (1938); Beadle (1948); Clark (1948); Holland and Moore (1962); Biddiscombe (1963); Pedley (1967); Chinnick and Key (1971), and others quoted below.

2.4.2. *Eucalyptus brownii* Suballiance
Reid River Box

This species is sometimes regarded as a narrow-leaved form of *E. populnea* and specimens with a leaf-width intermediate between those typical of the two species occur (Pedley, 1969). *E. brownii* is found northward of *E. populnea* but is rarely mixed with it. The suballiance occurs in the tropics (Fig. 13.7) in areas with a mean annual rainfall of 600–800 mm. It occupies lower slopes, often adjacent to watercourses, or depressions on fine-textured Solodized-Solonetz soils (Perry, 1953; Perry and Lazarides, 1964). The communities are mostly open woodlands 10–16 m high, with *E. brownii* the sole dominant. Ecotonal associations are rare but sometimes occur with *E. crebra* or *E. drepanophylla*. A tall shrub layer, usually very open, sometimes occurs, the most constant species being *Eremophila mitchellii* with subsidiary *Erythrophleum australe* and *Petalostigma* sp., and less commonly *Alphitonia excelsa* and *Grevillea parallela*. The low shrub *Carissa lanceolata*, c. 1 m high, is sometimes common. The grass sward varies within the one stand, sometimes being interrupted by bare ground. Andropogoneae dominate some patches, especially *Dichanthium fecundum* and *Eulalia fulva* (in more favourable sites), *Bot-*

Fig. 13.8. Pure stand of *Eucalyptus populnea* on grey clay soil. Grass layer contains *Eragrostis parviflora*, *Chloris ventricosa* and *Sporobolus caroli*. 24 km north of Warren, N.S.W.

hriochloa intermedia, Chrysopogon fallax and *Heteropogon contortus*. On sandier soils presenting a less favourable water régime, species of *Aristida* dominate the sward, with subsidiary *Eragrostis* and *Eriachne* spp.

2.4.3. *Eucalyptus populnea* Suballiance
Bimble Box

E. populnea forms open grassy woodlands 10–20 m high throughout the area of the species on light clay soils on higher floodplains or along minor flattish watercourses. Associated tall shrub or small tree species are absent or rare (Fig. 13.8), the most frequent being *Heterodendrum oleifolium, Apophyllum anomalum* and *Capparis mitchellii*.

The grass layer is well developed. In the summer rainfall zone *Bothriochloa ewartiana, Dichanthium sericeum, Eulalia fulva, Heteropogon contortus* and *Themeda australis*, either as local dominants or in mixtures, are the main species. In areas with a predominantly winter rainfall *Stipa aristiglumis* is the main dominant. Grasses of lesser importance, sometimes prominent as a result of selective removal of the original dominants by domestic stock, are *Chloris truncata, Eragrostis parviflora* and *Sporobolus actinocladus*. In depressions *Paspalidium jubiflorum* and *Stipa verticillata* are often locally common.

2.4.4. *Eucalyptus populnea – Casuarina cristata* Suballiance
Bimble Box – Belah

E. populnea forms woodlands in association with *Casuarina cristata* mainly between the 350 and 500 mm isohyets. The communities occur usually on fine textured grey clays which are commonly gilgaied, or in depressions within large areas of Red Brown Earths. When gilgais occur, *Casuarina* occupies the rims of the depressions, whereas the shelves between depressions support the eucalypt. The gilgais commonly contain species adapted to waterlogging, e.g., *Muehlenbeckia cunninghamii, Marsilea drummondii, Eleocharis pallens* and *Juncus polyanthemos*.

The suballiance extends from north to south and the associated species vary according to the seasonal incidence of rain. In the northern summer rainfall zone *Heterodendrum oleifolium* is the most frequent tall shrub; rare species are *Atalaya hemiglauca, Eremocitrus glauca* and *Acacia harpophylla*. *Geijera parviflora* and *Acacia oswaldii* (both rare) occur more or less from north to south, with a greater abundance in the north.

The herbaceous stratum shows a similar zonation. In the summer rainfall zone *Astrebla lappacea, Dichanthium sericeum, Eulalia fulva* and *Sporobolus actinocladus* are the most abundant. In winter rainfall areas *Stipa aristiglumis* is often the dominant, or *Danthonia semiannularis* with *Chloris* spp. The latter combination is possibly induced by the removal of *Stipa* from the sward.

In some areas the density of *Casuarina* is so high that the herbaceous sward is almost completely suppressed, the ground being strewn with *Casuarina* litter in which scattered Chenopodiaceae occur, mainly *Rhagodia hastata* and *Bassia quinquecuspis*.

2.4.5. *Eucalyptus populnea – Eremophila mitchellii* Suballiance
Bimble Box – Budda

These two species occurs in association in the summer rainfall zone, mainly north of lat 28°S. The soils are mostly loams to clay loams above with clay subsoils, slightly acid in the surface, alkaline below. The communities are shrub woodlands, the spaces between the tree-crowns being occupied by tall shrubs, which may occur also as an understory below the trees (Fig. 13.9). Other tall shrubs include *Atalaya hemiglauca, Eremocitrus glauca* and *Heterodendrum oleifolium*. Shrub 1–2 m high are sometimes locally abundant, e.g., *Carissa lanceolata* and *Eremophila maculata*.

The grass layer is usually very discontinuous and variable, possibly as a result of grazing. Genera represented commonly are *Aristida, Bothriochloa, Chloris, Chrysopogon, Eragrostis, Enneapogon* and *Heteropogon*.

2.4.6. *Eucalyptus populnea – Callitris glauca* Suballiance
Bimble Box – White Pine

These two species dominate tall woodlands or shrub woodlands 12–20 m high on texture con-

Fig. 13.9. *Eucalyptus populnea* with a tall shrub layer of *Eremophila mitchellii*. South of Rolleston, Qld.

trast soils, commonly Red Brown Earths. The species are mostly co-dominants (Fig. 13.10), but on the more clayey soils the eucalypt tends to dominate, whereas on soils with deep sandy A horizons, or on deep sandy profiles, *Callitris* becomes more abundant, sometimes forming pure stands. The communities occur mainly on gently undulating or flat country, but sometimes on hills where *Brachychiton populneum* forms associations with one or other of the dominants, or both.

Floristically the communities are rich, containing 30–40 common shrub or small tree species and at least 130 herbaceous species. Several of the larger species become local co-dominants in shrub woodlands, giving distinctive associations, usually with a decrease in abundance of *Callitris*, especially *Casuarina luehmannii* (usually on Solodized Soils), and *Geijera parviflora*. Another consistent local variant on deep sands is an understory of *Alstonia constricta* (with a distinctive grass layer of *Perotis rara, Eragrostis lacunaria* and *Tragus australianus*). Three lianas occur: *Capparis lasiantha, Jasminum linearis* and *Parsonsia eucalyptiphylla*.

Additional common tall shrubs are : *Acacia homalophylla, Apophyllum anomalum, Eremophila longifolia, Hakea leucoptera, Heterodendrum oleifolium*. Some common smaller shrubs are: *Acacia deanei, A. rigens, Dodonaea viscosa* and *Olearia pimelioides*.

The herbaceous layer in the virgin condition was probably dominated by grasses. Ground cover is determined largely by the density of the tree and tall shrub canopies. *Callitris*, in particular, suppresses the growth of herbs, and in undisturbed woodlands grass density is considerably reduced under the crown of *E. populnea*. The following species are listed for New South Wales. Species of *Stipa* and *Danthonia* dominate more compact soils (*S. falcata, S. scabra, S. setacea* and *S. variabilis, D. caespitosa* and *D. semiannularis*), with *Dichanthium sericeum, Eragrostis parviflora* and *Themeda australis* in depressions. *Chloris truncata* occurs throughout. On sandier soil-phases *Aristida* spp., particularly *A. jerichoensis* is dominant. A common winter-growing herb, *Erodium cygnorum* (crowfoot) is mentioned because of its value as a pasture plant.

2.4.7. *Eucalyptus populnea* – *Geijera parviflora* – *Flindersia maculosa* Suballiance

Bimble Box – Wilga – Leopard Tree

Geijera parviflora identifies this suballiance and occurs from north to south in association with *E. populnea*. *Flindersia maculosa* occurs north of c. lat. 28°S. The communities intergrade with the two preceding suballiances.

2.4.8. *Eucalyptus populnea* – *Acacia* spp. Suballiance

At its drier western margin *E. populnea* forms shrub woodlands in a broad ecotonal zone with *Acacia excelsa* and *A. aneura* between the 250 and 350 mm isohyets (Fig. 13.11). The soils are mainly Non-calcic Brown Soils, with sandy surface horizons varying from a few cm to c. 1 m deep. The depth of the surface horizon is largely responsible for the patterning of species. The acacias occur either in an understory to the eucalypt, or as dominants between scattered eucalypts in the driest areas. The woodlands are tall for such a low rainfall, *E. populnea* reaching a height of c. 20 m while the acacias are commonly 10–12 m high in the wettest areas. The woodlands often enclose or are intermixed with woodlands dominated by *E. intertexta* (see 2.3 above) and *E. terminalis,* at its southern limit, occurs occasionally. Outliers of *Callitris glauca,*

Fig. 13.10. *Eucalyptus populnea* and *Callitris "glauca"*. Grass layer contains *Stipa variabilis* and *Chloris ventricosa.* 16 km north of Goondiwindi, Qld.

Fig. 13.11. *Eucalyptus populnea* with tall shrub layer of *Acacia aneura*. Grass layer dominated by *Eragrostis eriopoda*. Near Cunnamulla, Qld.

often in dense stands occur on some deep sandy soils.

Floristically the suballiance is rich in both woody and herbaceous species. The main tall shrub/small tree species are: *Acacia homalophylla, Atalaya hemiglauca, Brachychiton populneum, Casuarina cristata, Codonocarpus cotonifolius* (deep sands), *Eremophila mitchellii, Flindersia maculosa, Geijera parviflora* (east), *Grevillea striata, Hakea leucoptera, Heterodendrum oleifolium* and *Myoporum platycarpum*. Shrubs 1–3 m high are: *Acacia colletioides* (sands), *Canthium oleifolium, Capparis mitchellii, Cassia artemisioides, Dodonaea attenuata* (sands), *Eremophila longifolia* and *E. sturtii*.

The herbaceous layer is dominated by grasses. Summer-growing species on sands include: *Aristida jerichoensis, Danthonia bipartita, Digitaria brownii, Eragrostis eriopoda* and *Neurachne mitchelliana*, plus the annuals, *Dactyloctenium radulans* and *Tragus australianus*. In depressions, chiefly below *E. populnea*, *Dichanthium sericeum, Eragrostis parviflora* and *Themeda australis* occur. Winter rains produce a sward of *Stipa variabilis* and a wealth of annuals, mainly Compositae (*Helipterum* spp.), Cruciferae (*Blennodia* spp.) and *Erodium cygnorum*.

2.5. *Eucalyptus melliodora* – *E. blakelyi* Alliance

Yellow Box – Blakely's Red Gum

These two species, the areas of which almost correspond (Fig. 13.7) form a well defined alliance between the 400 and 800 mm isohyets, extending from the western slopes at an altitude of c. 170 to c. 1200 m on the Northern Tablelands. The area receives rainfall at all times of the year, with a slight summer maximum in the north and a slight winter maximum in the south. At higher elevations frosts are severe in the winter, lowest minima of −10°C being recorded in several districts. The communities are intensively grazed by domestic stock and much of the area has been pasture-improved through the introduction of herbaceous legumes, *Trifolium* spp. in the wetter parts and *Medicago* spp. in the drier areas, and of grasses, especially species of *Lolium*. Large areas have been cleared for the growing of wheat and oats.

E. melliodora reaches a height of 30 m and is renowned for its high production of nectar and high quality of the honey. The species extends outside the main body of the alliance. In the east a few patches occur near the coast, and the species extends into the semi-arid west along the main rivers beyond its normal climate range. *E. blakelyi* reaches a height of c. 23 m and at its western limits it intergrades with *E. dealbata*.

The alliance forms mosaics with the *E. albens* Alliance, the former occurring mainly on river flats, the latter on the adjacent undulating country. The soils are mainly Grey-brown and Brown Podsols, Non-calcic Brown Soils and Solodics, or sandy or sandy-clay alluviums with little or no profile differentiation.

The species usually occur in association in savannah woodland formation (Fig. 13.12). When differences in topography or soil parent material exist, the species sometimes segregate into pure stands. *E. melliodora* tends to dominate on flats, especially near watercourses, where it is often associated with *E. conica* in the central part of the alliance. *E. blakelyi* tends to dominate on more elevated areas which are sometimes stony or very sandy. Associated smaller trees, other than eucalypts, occasionally lend a distinctive appearance to the communities, though they are usually not co-dominants, e.g., *Callitris "glauca"* (on sandy soils,

Fig. 13.12. *Eucalyptus melliodora* (fibrous bark) and *E. blakelyi* (smooth, blotched trunk). Grass layer contains *Stipa variabilis, Bothriochloa macra, Dichanthium sericeum* and *Danthonia linkii*. Near Scone, N.S.W.

mainly west), *C. endlicheri* and *Brachychiton populneum* (stony areas) and *Casuarina stricta* (stony areas, south-east).

A shrub stratum does not occur and shrubs are usually rare, possibly as a result of grazing and burning. Species which are constant in the alliance are: *Acacia implexa* (tall shrub or small tree at higher elevations) and *Acacia armata, Bursaria spinosa, Exocarpos cupressiformis, Hibbertia linearis, Jacksonia scoparia, Lissanthe strigosa* and *Melichrus urceolaris*. The herbaceous layer is dominated by grasses, of which only *Themeda australis* is constant throughout the alliance. In the area with a predominantly winter rainfall, *Stipa scabra* and *S. variabilis* are common dominants with subsidiary *Danthonia* spp. and a smaller representation of Angropogoneae (*Bothriochloa macera, Dichanthium sericeum* and *Themeda australis*). On soils of fine texture, especially adjacent to watercourses or in ecotones with *E. albens, Stipa aristiglumis* is common or even locally dominant, especially in the central part of the area. On the Northern Tablelands where summer rainfall predominates and winter minimal temperatures are often below zero, *Stipa* is absent and the swards are dominated by Andropogoneae, especially *Themeda australis, Dichanthium sericeum, Sorghum leiocladum, Eulalia fulva, Cympopogon refractus* and *Bothriochloa macera*. At higher altitudes both in the north and the south, species of *Dichelachne* and *Echinopogon* are sometimes locally common; at the uppermost limits of the alliance, adjacent to the *E. pauciflora* woodlands, *Poa sieberana* is usually dominant in the grass layer.

The herbaceous flora contains over 200 native species and c. 130 naturalized species (Costin, 1954, for the southern area). The latter are mainly Gramineae, Papilionaceae and Compositae. Several of these are purposeful introductions for pasture improvement, especially species of *Lolium, *Medicago and *Trifolium. Weed species are abundant in some areas, notably *Hordeum murinum, Cruciferae (*Cardaria draba, *Sisymbrium spp.) and Compositae (*Carduus tenuiflorus, *Carthamus lanatus, *Chondrilla juncea and *Silybum marianum).

Fig. 13.13. *Eucalyptus moluccana* (bark fibrous below, smooth above) and *E. fibrosa* ssp. *fibrosa*. Shrubs are *Acacia fimbriata* and *A. falcata*. Grass layer dominated by *Themeda australis*. 7 km west of Childers, Qld.

Fig. 13.14. *Eucalyptus moluccana* woodland with a discontinuous shrub layer containing *Eremophila mitchellii, Capparis mitchellii, Eremocitrus glauca, Heterodendrum floribundum* and *Carissa ovata*. 42 km south of Rockhampton, Qld.

2.6. *Eucalyptus moluccana* Alliance

Grey Box

E. moluccana, one of the Boxes (Series Moluccanae), is closely related to the other three species in the Series (see the following three alliances) and intergrades with some of them. The species has a very fragmented area, with centres of development in rainshadows near the coast, e.g., the drier segment of the Triassic Wianamatta shales south-west of Sydney and the floors of the Hunter and Clarence River valleys. It occurs patchily in the intervening areas, with outliers in the south, and a northern extension in small isolated stands as far as the Tropic (Fig. 13.7). Small outliers are found also on the Northern Tablelands. The species reaches a height of 24–30 m in wetter areas, forming tall woodlands between the 650 and 1100 mm isohyets, the rain being fairly evenly distributed throughout the year. It is found on soils classified as Yellow, Brown and Grey Podzols, rarely Solodic Soils, all of which are fine grained and derived mainly from shales and are relatively fertile.

E. moluccana sometimes forms pure stands but is commonly associated with *E. fibrosa* ssp. *fibrosa* (Fig. 13.13) or *E. tereticornis.* The latter becomes more abundant on lower slopes and on flats subject to minor flooding, where it may be replaced by *E. amplifolia* or *Angophora floribunda*. Ecotonal associations occur with many species (Table 13.1).

The Understories vary with latitude and three areas are mentioned:

i. On the Wianamatta shales, south-west of Sydney, shrub species are rare but *Bursaria spinosa* occurs throughout, and *Melaleuca linariifolia* and *M. styphelioides* are often abundant in depressions and along watercourses. The herbaceous stratum was probably dominated by grasses and the following possibly occurred in the natural assemblage: *Agrostis avenaceus, Amphibromus neesii, Bothriochloa macera, Chloris truncata, Danthonia racemosa, Dichanthium sericeum, Dichelachne sciurea, Echinopogon ovatus, Eragrostis leptostachya, Eriochloa pseudoacrotricha, Imperata cylindrica, Microlaena stipoides, Sporobolus elongatus, Stipa scabra, Themeda australis*.

ii. In the Hunter Valley, where *E. moluccana* occurs in association with *E. crebra*, a typical assemblage contains: *Aristida warburgii, Bothriochloa macera*, and *Themeda australis,* with subsidiary *Chloris ventricosa, Cymbopogon refractus, Danthonia purpurascens, Panicum effusum*, and the forbs and subshrubs *Calotis lappulacea, Convolvulus erubescens, Glycine tabacina, Helichrysum semipapposum, Myoporum debile* and *Sida corrugata*. On eroded soil, denuded of its A horizon, *Atriplex semibaccata* and/or *Maireana tamarascina* are the first colonisers.

iii. At its northern limit at the Tropic, *E. moluccana* (sometimes associated with *E. tereticornis* or *E. populnea*) occurs in open stands (Fig. 13.14), with an understory of shrubs typical of the semi-arid zone, viz., *Acacia salicina, Atalaya hemiglauca, Capparis mitchellii, Casuarina luehmanii, Eremocitrus glauca, Eremophila mitchellii, Flindersia dissosperma, Grevillea striata, Heterodendrum floribundum, Petalostigma pubescens* and *Pittosporum phyllireaeoides*. Small shrubs include *Carissa ovata* and *Kochia tamarascina*. A small liana, *Cissus opaca* and the epiphytic orchid *Cymbidium canaliculatum* are rare. The herbaceous layer is dominated by grasses, *Heteropogon contortus* with subsidiary *Themeda australis* and species of *Aristida, Arundinella, Bothriochloa* and *Chrysopogon*.

2.7. *Eucalyptus albens* Alliance
White Box

E. albens reaches a height of 20–25 m and dominates woodlands. Glaucous and green-leaved forms of the species exist, the latter sometimes resembling *E. moluccana*. The species occurs mainly between the 500 and 700 mm isohyets (Fig. 13.7), with the rainfall distributed more or less evenly throughout the year, or with a slight summer maximum in the north and a slight winter maximum in the south. It is subjected to minor frost and intolerance to frost appears to prevent its migration on to the more elevated areas of the Tablelands. The species occurs mostly on gently undulating country or on hills and the alliance often forms mosaics with the *E. melliodora* – *E. blakelyi* Alliance, which occupies flats, often along watercourses, whereas *E. albens* dominates the more elevated sites. It grows on soils belonging to several Groups, including Red, Yellow and Grey-Brown Podzolics, Brown Earths, Red Loams, Terra Rossas and Brown Calcareous Clays. They have in common a high base status, particularly calcium, and are generally of relatively high fertility. The most fertile soils are derived from basalt or basaltic alluvium, the less fertile ones from siliceous rocks, especially granites. Soil type has some effect in determining associations. Large tracts of soil are used for wheat-growing (Fig. 13.15).

Pure stands of *E. albens* occur on soils with a high clay content, the communities being closed to somewhat open woodlands (Fig. 13.16). Associated tall species are rare or absent, *Brachychiton populneum* being the most abundant. This species, a valuable fodder tree, is usually preserved when the land is cleared for cultivation and is lopped during droughts which accounts for the globular crowns induced in the trees (Fig. 13.15). Other species of rare occurrence, but sometimes locally common, especially on hills in the summer rainfall zone, are *Atalaya hemiglauca, Capparis mitchellii, Ehretia membranifolia, Eremophila mitchellii, Geijera parviflora, Heterodendrum oleifolium, Notolaea microcarpa* and the lianas *Jasminum lineare, Pandorea pandorana* and *Parsonia eucalyptiphylla*. Shrubs of constant occurrence, but rare, are *Acacia buxifolia* and *Lissanthe strigosa*.

The herbaceous layer is dominated by *Stipa*

Fig. 13.15. *Eucalyptus albens* country which has been cleared for cropping. The trees left in the paddocks are *Brachychiton populneum* the leaves of which provide good stock feed. *E. albens* on the ridge in the background. 4 km north of Parkes, N.S.W.

aristiglumis, which grows during the cooler seasons, reaching a height of 1–2 m and constituting at least 90% of the sward. Subsidiary species are *Dichanthium sericeum* * *Hyparrhenia hirta* which become increasingly important after summer rains. Less common are: *Agropyron scabrum*, *Bothriochloa macera*, *Chloris ventricosa* and *Danthonia* spp., and the forbs *Calotis lappulacea*, *Craspedia chrysantha*, *Cymbonotus lawsonianus*, *Dichondra repens* and *Swainsona galegifolia*. In small depressions subject to waterlogging, *Marsilea drummondii* is sometimes common.

Treeless areas of grassland dominated by

Fig. 13.16. *Eucalyptus albens* woodland, the trees probably spaced as they were before white settlement. The lower layers have been altered by burning. Grass layer dominated by *Themeda australis*. Near Bundarra, N.S.W.

Stipa aristiglumis sometimes break the continuity of the *E. albens* woodlands. These areas are subject to waterlogging and in some cases are the beds of former lakes. The grasslands are often fringed with *Acacia pendula* and/or *Casuarina cristata*.

E. albens is associated with *Callitris glauca* when a sandy surface horizon occurs (Moore, 1953; Costin, 1954; Biddiscombe, 1963). The *Callitris* rarely attains the height of the eucalypt and may occupy the spaces between the eucalypts. Shrubs are often locally conspicuous in these stands, especially *Acacia armata, A. buxifolia, Brachyloma daphnoides, Bursaria spinosa, Cassinia longifolia* and *Dodonaea cuneata*. On hills in the south *Acacia implexa, A. pycnantha* and *Casuarina stricta* are also recorded. The herbaceous stratum usually contains some *Stipa aristiglumis* but with increasing sandiness this is replaced by *S. falcata* and *S. scabra* with subsidiary *Agropyron scabrum, Chloris truncata, Dichelachne sciurea* and *Themeda australis*. Forbs are abundant but subsidiary to the grasses, especially *Bulbine bulbosa, Cheilanthes tenuifolia, Glycine tabacina, Oxalis corniculata, Rumex brownii* and *Sida corrugata*.

2.8. Eucalyptus woollsiana Alliance

Grey Box

E. woollsiana extends from Queensland to South Australia (Fig. 13.7) with its main centre of abundance in New South Wales. Two intergrading subspecies occur. A broad-leaved form (ssp. *microcarpa*) occupies wetter areas and is found over the range of the species, intergrading with *E. moluccana*. A narrow-leaved form (ssp. *woollsiana*) is confined to New South Wales and intergrades with *E. pilligaensis*. *E. woollsiana* dominates woodlands between the 380 and 600 mm isohyets, the rain being fairly evenly distributed throughout the year or with a winter predominance in the south. The soils are mostly Red Brown Earths and Grey Clays, with smaller occurrences on deep undifferentiated sands in some drier areas, and Yellow Podsolics in wetter areas. Two main groups of associations can be identified, according to the subspecies present.

In the north, *E. woollsiana* ssp. *microcarpa* forms woodlands 20–25 m high on Grey Clays or Red Brown Earths, usually with no or few shrubs in the understory, *Casuarina luehmannii* being the most common (Fig. 13.17). Dwarf shrubs are sometimes conspicuous, especially *Maireana tamarascina* and *Myoporum debile*.

Fig. 13.17. *Eucalyptus woollsiana* ssp. *microcarpa*. Pure stand with understory of *Casuarina luehmannii* (scarcely visible because it has been eaten by stock). Grass layer dominated by *Chloris ventricosa* with subsidiary *Aristida ramosa* and *Rumex brownii*. 25 km west of Warwick, Qld.

Fig. 13.18. *Eucalyptus woollsiana* ssp. *woollsiana* and *Callitris* "*glauca*". Tall shrub is *Casuarina luehmannii*. The shrubs c. 2 m high are *Dodonaea viscosa* and *Cassia artemisioides*. Grass layer contains *Stipa variabilis*, *Chloris truncata* and *Agropyron scabrum*. 56 km north of Narrandera, N.S.W.

The herbaceous stratum is suppressed through the accumulation of litter. *Chloris ventricosa*, *Aristida ramosa*, *Rumex brownii* and *Lomandra brownii* are the most common herbs.

In the south *E. woollsiana* ssp. *microcarpa* sometimes occurs in pure stands but is usually associated with *E. melliodora*, *E. sideroxylon*, *E. leucoxylon* (Victoria) or *E. odorata* (South Australia). In these stands shrubs are more abundant than in the north. *Casuarina luehmannii* is usually present and species of *Acacia* are often abundant, the assemblages varying from district to district. Some common species are *A. buxifolia*, *A. hakeoides*, *A. rigens*, *A. pycnantha* (extreme south) and also *Cassia artemisioides*, *Cassinia laevis*, *Dodonaea cuneata* and *D. viscosa*. The common grasses are *Stipa* aff. *variabilis*, *Agropyron scabrum* and *Chloris truncata*.

E. woollsiana ssp. *woollsiana* usually occurs in association with *Callitris glauca* (Fig. 13.18) in central and southern New South Wales on soils with a sandy surface horizon, including some Red Brown Earths, or on deep sands. *Casuarina luehmannii* often occurs in the lower tree layer and, rarely, *Brachychiton populneum*. Scattered tall shrubs are usually present, especially *Eremophila longifolia*, *Geijera parviflora*, *Heterodendrum oleifolium* and *Pittosporum phylliraeoides*. The herbaceous layer is dominated by grasses.

2.9. *Eucalyptus pilligaensis* Alliance

Pilliga Box

E. pilligaensis replaces *E. woollsiana* ssp. *woollsiana* in the north-west and the two species intergrade taxonomically. It dominates woodlands 15–25 m high mainly between the 400 and 500 mm isohyets, the rain being more or less evenly distributed throughout the year. The woodlands occur on flat or undulating country, rarely in hilly terrain. The soils are Red Brown Earths, Solodized-Solonetz or, less commonly deep sandy deposits along watercourses. *E. pilligaensis* rarely forms pure stands but is usually associated with *Callitris glauca* forming a lower tree layer (Fig. 13.19); this association occurs on Red Brown Earths and in many areas the *Callitris* has been cut out for timber. A second association, found on Solodized-Solonetz Soils, involves *E. pilligaensis* with a well defined lower tree layer of *Casuarina luehmannii*.

Shrubs are infrequent and distinct shrub lay-

Fig. 13.19. *Eucalyptus pilligaensis* with young plants of *Callitris "glauca"*. Grass layer contains *Chloris ventricosa*, *Eragrostis parviflora* and *Digitaria brownii*. 16 km south of Gilgandra, N.S.W.

ers rarely occur, except locally when some species form thickets. Tall shrub species include *Acacia homalophylla*, *Callitris endlicheri*, *Eremophila mitchellii*, *Geijera parviflora* and *Heterodendrum oleifolium*. Shrubs mostly 1 to 2 m high include *Acacia buxifolia*, *A. deanei*, *A. spectabilis*, *A. triptera*, *Cassia artemisioides*, *Cassinia aculeata*, *Dodonaea viscosa*, *Helichrysum diosmifolium*, *Melaleuca pubescens* and *Myoporum deserti*. The herbaceous layer is usually discontinuous, except in open areas which may be natural but are possibly induced by the felling of trees. Grasses are dominant: *Aristida jerichoensis* and *Eragrostis lacunaria* (sands), *Bothriochloa macera*, *Danthonia caespitosa*, *Chloris ventricosa* (usually in slight depressions), *Digitaria brownii*, *Eragrostis parviflora*, *Stipa variabilis* and *S. verticillata* (usually in depressions). Common forbs are *Calotis lappulacea* and *Vittadinia australis*.

2.10. *Eucalyptus dawsonii* Alliance

Slaty Gum

E. *dawsonii* is an uncommon species taxonomically related to E. *fasciculosa* (see below, 2.11). It occurs disjunctly in the Upper Hunter Valley and westward across the Great Divide to Mudgee and Dunedoo (Fig. 13.7). This area receives a mean annual rainfall of c. 600 mm. E. *dawsonii* forms pure woodland stands 20–25 m high on lower slopes and small ridges in broad valleys. The soils are Grey and Brown Podzols derived mainly from shales. The upper slopes and ridge-tops are commonly sandstones and support communities of ironbarks, which do not mix with E. *dawsonii*.

Shrubs are abundant, sometimes forming a continuous layer (Fig. 13.20), or occurring in clumps and increasing in density downslope near watercourses. The understory assemblages vary from site to site. *Acacia cultriformis*, *A. decora*, *A. melanoxylon*, *A. salicina*, *A. verniciflua*, *Bursaria longisepala*, *Cassia nemophila*, *Geijera parviflora*, *Hovea longipes*, *Myoporum montanum* and *Rhagodia hastata* are recorded (Williams, MS.).

2.11. *Eucalyptus leucoxylon* – E. *fasciculosa* Alliance

Yellow Gum – Pink Gum

E. *leucoxylon* occurs disjunctly in Victoria and South Australia (Fig. 13.7). In Victoria it is often associated with E. *sideroxylon*. E. *fascicul-*

Fig. 13.20. *Eucalyptus dawsonii*. Pure stand with a shrub layer dominated by *Acacia verniciflua*. Baerami (upper Hunter Valley) N.S.W.

osa is confined to South Australia (Fig. 13.7). The species cover approximately the same climatic range, *E. leucoxylon* extending between the 750 and 450 mm isohyets, *E. fasciculosa* between the 750 and 370 mm isohyets, the rain falling mainly in the winter. The communities they dominate vary in structure from low open forests to low woodlands. The species are recorded on a variety of soils, including Yellow, Red, Grey-Brown and Meadow Podzolics, Red Brown Earths and Terra Rossas (Specht, 1972) and Solodic soils and unconsolidated sands on lunettes (Litchfield, 1956; Connor, 1966). On Podzols, the species are often in association. *E. leucoxylon* is more abundant on Red Brown Earths and *E. fasciculosa* is more abundant on soils with a deep, sandy surface horizon, or on skeletal soils. Two distinctive associations, which intergrade are described below.

E. leucoxylon Association, described by Wood (1937) for South Australia, is a woodland 10–30 m tall with a discontinuous tall shrub layer of *Acacia pycnantha* and local patches of *Casuarina stricta* on shallower soils. In more humid areas (south-facing slopes) the shrub layer contains *Hibbertia sericea*, *H. acicularis* with subsidiary *Tetratheca ericifolia*, *Astroloma humifusa*, *Dillwynia hispida* and *Leptospermum myrsinoides*. The herbaceous layer is characterized by geophytes (Liliaceae and Orchidaceae), *Helichrysum scorpioides*, *Vittadinia triloba* and *Leptorrhynchus squamatus*. In drier sites (east-facing slopes) the woody species are absent, except scattered *Acacia pycnantha*. The herbaceous layer is dominated by *Danthonia penicillata*, *D. carphoides*, *Stipa variabilis*, *S. eremophila* and *Vittadinia triloba*.

In the same areas as the preceding association, *E. fasciculosa* forms woodland stands 8–10 m tall on shallow podzols or skeletal soils, with a xeromorphic shrub layer, in which *Xanthorrhoea semiplana* is often the most conspicuous species in the herbaceous layer. Scattered shrub species include *Banksia marginata, B. ornata, Leptospermum myrsinoides, Melaleuca gibbosa* and other xeromorphs (Crocker, 1944).

2.12. *Eucalyptus odorata* – *E. porosa* Alliance

Peppermint – Black Mallee Box

E. odorata, normally a tree to c. 16 m high, occurs disjunctly in South Australia, with a minor extension into Victoria (Fig. 13.7). The populations show considerable variation, with the result that the taxonomy of the "*odorata* group" is confusing. Both tree and mallee forms occur, the latter in drier areas. Narrow-leaved forms are referred to *E. odorata* var. *angustifolia* and have been recorded both along the coast (referred to as "Seaside Mallee") and inland on the Flinders Range. *E. calcicultris,* once included as a variety of *E. odorata,* is now referred to *E. porosa.* The species, *E. odorata* and *E. porosa,* cover the climatic range 700–360 mm mean annual rainfall, with a winter maximum. They form woodlands in the wetter areas, merging into low woodlands and mallee in drier climates. The species occur on Grey Brown and Yellow Podzolics soils in wetter areas, on Red Brown Earths in the intermediate zone, and on Terra Rossas in the driest areas. Occurrences on skeletal soils, at least some of them on fossil lateritic residuals, are also recorded.

Woodland communities of *E. odorata* once covered the area adjacent to Adelaide but have now been removed to provide land for cultivation. Wood (1937) describes the woodlands as pure stands, with a sparse tall shrub layer dominated by *Acacia pycnantha.* The shrub, *A. armata,* and the small liana, *Hardenbergia violacea,* appear to have been constant species. The grass layer was probably dominated by *Themeda australis* and *Danthonia penicillata,* with a strong representation of the geophytic herbs, *Anguillaria dioica, Burchardia umbellata, Bulbine bulbosa* and *Dichopogon strictus.* He describes two variant societies in the understory, one dominated by *Schoenus apogon* forming mats on silt deposits in winter – waterlogged areas, with some emergent shrubs, especially *Hibbertia stricta* and *H. sericea.* The second occurs on lateritic residuals where thickets of *Leptospermum juniperinum* and *Lissanthe strigosa* are dominant.

In the wettest parts of the Mt. Lofty Ranges

Fig. 13.21. *Eucalyptus odorata* low open woodland with a discontinuous shrub layer of *Melaleuca lanceolata.* Eyre Peninsula, S.A.

and the southern parts of the Flinders Range, a much richer flora occurs (listed by Specht, 1972), which includes *E. woollsiana* as a co-dominant, the small tree, *Acacia melanoxylon*, and tall shrubs *Callitris preissii*, *Casuarina stricta* and others. Several small shrubs are recorded, especially species of *Acacia* and many herbaceous plants.

In drier areas the communities are stunted (Crocker, 1946; Specht, 1951). Crocker describes a mallee stand on Terra Rossas developed on travertine on the Eyre Peninsula, with an understory of *Melaleuca lanceolata* (Fig. 13.21), *Callitris glauca* and subsidiary *Acacia ligulata* and *Geijera linearifolia*. Admixtures of the mallee *E. diversifolia* occur ecotonally. The grass layer is dominated by *Stipa eremophila* and *Danthonia semiannularis*. Specht describes a stand close to the Victorian border as a savannah woodland on Red Brown Earths with a grass layer dominated by *Danthonia caespitosa*.

2.13. *Eucalyptus cladocalyx* Alliance

Sugar Gum

E. cladocalyx occurs in several small disconnected areas in South Australia (Fig. 13.7). The tree reaches a maximum height of c. 30 m and is better known outside its natural area, since it is commonly planted as a windbreak. The young foliage contains a cyanogenetic glycoside and is poisonous. The natural stands occur between the 380 and 630 mm isohyets in the winter rainfall region, and the tallest stands are forests on Yellow Podzolic soils which often contain lateritic gravel. Woodland stands are found in drier areas on Skeletal Soils, Solonized Brown Soils and deep sands.

The tallest forests, found on Kangaroo Is., are c. 30 m tall with an almost closed canopy and an understory up to 5 m high, consisting of *Beyeria leschenaultii*, *Lasiopetalum discolor*, *L. schulzenii*, *Adriana quadripartita* and *Correa rubra*. The parasitic *Cassytha pubescens* commonly forms tangled masses with the crowns of the shrubs. Similar pure stands of open forest are recorded in the Flinders Range near watercourses and on flooded areas with an understory of *Bursaria spinosa*, *Banksia marginata* and *Dodonaea viscosa* (Wood, 1937).

Woodland stands occur on soils which are shallow or sandy, and in the drier climates. Crocker (1946) records pure stands of xeromorphic woodland dominated by *E. cladocalyx* on lateritic Podzolics on the Eyre Peninsula. The understory is dominated by *Xanthorrhoea tateana*, *Melaleuca uncinata*, *Hakea cycloptera* and *H. rugosa*. On skeletal soils in the same area the eucalypt dominates savannah woodlands with xeromorphic shrubs forming a tall heath between the eucalypts.

14. The Mallee and Marlock Communities

1. Introduction

1.1. General

The term "mallee", an aboriginal word, describes a eucalypt of shrub proportions with many stems which arise from a large lignotuber. The term is also used to identify communities dominated by eucalypts with the mallee habit. Marlock is used in Western Australia to describe single-stemmed, usually compact eucalypts with a shrub or small tree habit, mostly 3–7 m high. Many marlock species do not produce lignotubers. Heights of mallee plants vary from c. 1 to c. 8 m and the numbers of stems per plant vary from about a dozen to one. Mallees with several to many thin stems (2–5 cm diam.) are commonly designated "whip-stick" mallee, and the term "bull" mallee is applied to single-stemmed plant with trunks reaching a diameter of c. 25 cm. Lignotubers of mallees vary from c. 20 cm to c. 1 m across and in the case of bull mallees they sometimes scarcely exceed the diameter of the base of the trunk. Many species exhibit a tree and mallee form, especially in Western Australia, e.g., *E. loxophleba, E. megacarpa, E. patens* and *E. wandoo;* the mallee form of the last species is referred to as *E. redunca*. Many mallee species exhibit both whip-stick and bull forms but marlocks do not exhibit the same variation in form as do the mallees. Stunting of marlocks is caused by adverse habitat conditions but the plants remain single-stemmed at the base, though stunted plants usually branch closer to ground level than do well grown specimens.

"The mallee" is used to identify a vegetation region (Chap. 6) referred to here as the "southern mallee region", which is characterized by the dominance of mallee eucalypts, the absence of watercourses and to some extent by its soils. However, all mallee communities do not belong to this regions, but some occur in the arid zone, as far north as the Tropic and these are referred to here as the "arid zone mallee communities".

In the southern mallee region large areas of the communities have been cleared for farming, chiefly for the cultivation of grain-crops and for grazing. Adjacent to the major rivers, chiefly the Murray, cleared mallee communities provide much of the land irrigated for fruit crops. A subsidiary industry of the region is the production of eucalyptus oil, the leaves and branches being harvested either by hand or mechanically and the oil extracted on the spot by steam distillation (Penfold and Willis, 1961). For a bibliography of literature on all aspects of the southern mallee region, see Ives (1973).

1.2. Distribution; Climate, Soils

Mallee and marlock communities occur in the south between 375 and 220 mm isohyets, with outliers in slightly wetter areas. In the arid zone the communities occur as scattered patches. In the southern mallee region the communities dominate the landscape and are divided into two main segments separated by the arid, calcareous Nullarbor Plain (Parsons, 1970). The outliers, which consist of community fragments, suggest that the mallee communities have expanded and contracted in the past with changes in climate. Mallee outliers in the wetter areas are interpreted as relicts of an extensive mallee formation which extended to the north-east of its present location during the Holocene arid period. On the other hand, fragments of the same communities, could indicate a wetter climate in the central and western parts of the continent in the more distant past.

In the south the rain falls mainly in the winter, but precise seasonal incidence of rain appears to be a subsidiary factor in determining distribution, since many of the component species occur in areas where the rainfall is equally distributed throughout the year and some are found as outliers in areas where summer rainfall predominates.

Soils play a significant role and, in general terms, the mallee communities occur on soils of coarse texture which are well drained. Barrow and Pearson (1970) define three main Soil Groups: i. Solonized Brown Soils (Mallee soils) in the drier areas (west and east); ii. Solodized-Solonetz in wetter areas (mainly east); iii. Deep Siliceous Sands (mainly west). Mallee and marlock communities occur also on outcrops of a variety of rocks including granites, greenstones, sandstones, limestones or ferruginous lateritic residuals, and on redistributed lateritic material containing ironstone gravel and sand. Mallee communities rarely occur on fine-textured soils, and when these occur within the mallee region, as along rivers in the east or on the greenstones in the west, they support woodlands. A few mallee and marlock species have a preference for fine-textured soils, e.g., *E. behriana* in the east and *E. platypus* in the west.

1.3. Community Structure

The tallest of the communities, dominated by single-stemmed bull mallees, might well be classed as low woodlands. Similarly, marlocks sometimes reach small tree proportions and form low woodlands. A common condition is the occurrence of many-stemmed mallee plants 3–6 m high with multiple, flat-topped or domed crowns spaced at fairly regular intervals. In wetter areas the crowns almost touch, in drier areas they are separated by distances of a few to many times the diameter of the crown. Stunted mallee communities 1–3 m high (whip-stick forms) also exist, being found in the most unfavourable sites, especially shallow soils.

Understory layers vary chiefly with the soil, five main types being discernible: i. A halophytic shrub layer *(Maireana, Atriplex, Rhagodia)* on lime-rich soils; ii. A dense thicket-like understory *(Melaleuca* and/or *Casuarina)*; iii. A shrub layer of xeromorphs (Proteaceae-Myrtaceae); iv. An herbaceous layer of hummock grasses *(Triodia);* v. Annuals only, usually with scattered shrubs. The last condition is brought about partly by the suppression of the understory by the accumulation of litter under the crown of the mallee plant and partly by the exclusion of smaller plants between the crowns, possibly resulting from competition for water.

The mallee communities with a high proportion of xeromorphic shrubs, particularly Proteaceae, are of special significance. These communities are intermediate in structure between the mallee shrublands and the heaths and are referred to as mallee-heaths (see below, 2.3).

1.4. The Species of *Eucalyptus*

About 100 species of *Eucalyptus* constitute the dominants in the mallee, marlock and mallee-heath communities (Blakely, 1955; Chippendale, 1973) and they belong to three of Pryor and Johnson's (1971) subgenera, 7 to *Eudesmia*, 9 to *Monocalyptus* and the remainder to *Symphyomyrtus*. The species have been classified into groups on the basis of their geographic occurrence and are listed in their approximate relative positions in Table 14.1. The following additional comments are made:

i. Species of Group 1 occur on both sides of the Nullarbor Plain and six of these species (*E. dumosa, E. foecunda, E. gracilis, E. incrassata, E. oleosa* and *E. socialis*) are more widely distributed than the others. Several of the species have extensive north-south ranges, as indicated by the arrows.

ii. Species of Group 2 are confined to the east, occurring in New South Wales, Victoria and South Australia. They have either restricted (Group 2a) or disjunct distributions.

iii. Species of Group 3 are confined to Western Australia and those on the left have extensive ranges from east to west in the State.

iv. Species of Group 4, 4a, 4b, and 4c are confined to wetter, southern Western Australia, some having extensive ranges from east to west (Group 4), the remainder being more restricted, as indicated. Many of these species are marlocks and a few are found only in mallee-heaths.

v. Species of Group 5 occur mainly along the northern fringe of the southern mallee region in the west. A few are found as outliers in the arid zone (as indicated by the arrows), a few occur also in South Australia, as indicated. *E. tetragona* is disjunct, occurring also in southern Western Australia.

vi. Species of Group 6 are confined to arid regions. The species occur patchily, rather than in a distinct region, though some are widely distributed.

Those mallee species which occur on either side of the Nullarbor Plain (Group 1, Table 14.1) are probably the oldest of the southern mallee species. They are related to taller species occurring in south-western Western Australia,

Table 14.1. Species of *Eucalyptus* found in the mallee, arranged in groups according to their occurrence in the continent. Arrows indicate extensions into areas of adjoining groups.

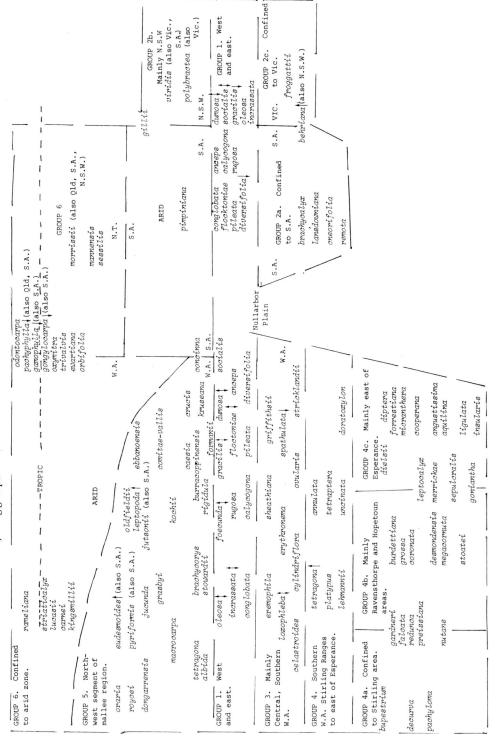

none being closely related to eastern species. Consequently it is probable that the mallee species originated in the south-west (Burbidge, 1947) and that the eastern mallee communities were established by the eastern migration of these western taxa. Local additons of mallee species have occurred over the ages, particularly in the west. Interchange of species between the west and the east and vice versa has been prevented by the arid environment of the calcareous Nullabor Plain. Thus, those western species restricted to siliceous sands in wetter areas cannot survive on the limestone substrate, while migration of these species across the siliceous sands north of the Nullabor Plain is prevented by aridity. However, several of the western species (Group 3) occurring on the northern fringe of the mallee region have migrated eastward into South Australia north of the Nullabor Plain, e.g., *E. pyriformis* and *E. jutsonii*. The few mallee species confined to the east have apparently been derived from related tree species which are confined to the south-eastern part of the continent.

2. Descriptions of the Alliances

2.1. General

The alliances are described under four headings: i. Mallee and marlock alliances occurring on deep soil profiles; ii. The mallee-heaths; iii. Alliances on rocky outcrops in the south-west; iv. The arid zone species.
While the identification of associations is relatively easy, even when stands contain 2, 3, 4 or even 5 species of *Eucalyptus*, grouping of associations into alliances is often very difficult. The reason is that many of the widespread species (Group 2, Table 14.1) occur in combination not only with species of the same group but also with many species of other groups. Consequently many species are listed under two or even several alliances. Within the four groups of communities the alliances are described as far as is practicable from wet to dry. This helps to indicate continuum relationships, when they occur. The eastern alliances are dealt with first, mainly because of their relative simplicity.

2.2. Mallee and marlock alliances on deep soil profiles

2.2.1. *Eucalyptus diversifolia* Alliance
White Mallee

E. diversifolia dominates the wetter mallee communities c. 3 m high, mainly in South Australia between the 650 and 350 mm isohyets (Wood, 1929, 1937; Crocker, 1944, 1947; Smith, 1963; Parsons, 1969). The soils are mainly Terra Rossas or sands containing calcium carbonate which, in drier areas, may lie within a few centimetres of the surface.
E. diversifolia sometimes forms pure stands (Fig. 14.1) but it usually has one or more associated species. On the mainland, *E. rugosa* is possibly the most common associate, but other species also occur, viz., *E. anceps, E. conglobata, E. dumosa, E. gracilis* and *E. oleosa*. In wetter areas the alliance adjoins woodlands and open forests of *E. obliqua, E. cladocalyx, E. baxteri* and *E. porosa*. Under these wetter climatic conditions three mallee species with restricted distributions occur in *E. diversifolia* stands, viz., *E. cneorifolia* (mainly on Kangaroo Is.), *E. remota* (confined to Kangaroo Is.) and *E. lansdowniana* (southern part of Eyre Peninsula). *E. cneorifolia* forms pure stands on leached sands with an understory of xeromorphic shrubs, mainly *Melaleuca lanceolata, Acacia pycnantha, A. calamifolia* and *Daviesia genistifolia*.
The understories are composed of xeromorphic species, many of which occur in the south-east of the continent. Some are endemic to South Australia, especially on Kangaroo Is. A list of plants occurring in the alliance has been compiled by Specht (1972) who records a total of 260 species. The floristic assemblages differ from one area to another, partly as a result of geographic isolation and partly because of differences in mean annual rainfall. The wettest areas in the south contain both the richest flora and the tallest and densest shrub strata. Only three species are present in all stands, viz., *Calytrix tetragona, Melaleuca lanceolata* and *Pultenaea acerosa*. On Kangaroo Is., Wood (1937) lists *Olearia teretifolia* and *Hakea ulicina* as the abundant species, with Epacridaceae common, e.g., *Astroloma conostephioides, A. humifusum, Epacris impressa* and *Acrotriche fasciculiflora*. In addition, this assemblage is unique since it

Fig. 14.1. *Eucalyptus diversifolia* mallee stand. Understory species include *Hibbertia sericea, Exocarpos aphyllus, Lepidosperma congestum* and *Stipa semibarbata*. 16 km west of Meningie, S.A.

contains a number of species endemic to the island, e.g. *Adenanthos sericea, A. terminalis* and *Petrophile multisecta*. On the mainland, the assemblages contain *Banksia marginata* and *Xanthorrhoea australis* as conspicuous members, with species of most of the Australian xeromorphic genera, such as *Acacia, Daviesia, Grevillea, Hakea, Hibbertia*, etc. The herbaceous layer is composed of Cyperaceae (especially *Gahnia, Lepidosperma* and *Schoenus*), Gramineae *(Danthonia, Stipa)*, Goodeniaceae *(Dampiera, Goodenia)*. Orchidaceae are represented by 10 spp. of *Acianthus, Caladenia, Lyperanthus, Pterostylis* and *Thelymitra*. A few small lianas (to 2 m) are recorded, viz., *Billardiera cymosa, Clematis microphylla, Comesperma volubilis* and *Muehlenbeckia adpressa*.

2.2.2. *Eucalyptus incrassata* – *E. foecunda* Alliance

Lerp Mallee –
Narrow-leaved Red Mallee

E. incrassata (including *E. angulosa* and its varieties) and *E. foecunda* (formerly *E. leptophylla*) occur almost invariably on deep sands or sandy Solodic soils between the 240 and 450 mm isohyets. They are found either in association or in pure stands, or in association with other species of *Eucalyptus*. Both species occur also in the mallee heaths. The eastern and western associations are dealt with separately.

In the east, *E. incrassata* and *E. foecunda* are fairly regularly associated, sometimes with admixtures of other eucalypts, especially *E.*

anceps, *E. calycogona, E. diversifolia, E. floctoniae* and *E. pileata* (Crocker, 1946; Jessup, 1946; Specht, 1951; Lichtfield, 1956). In all associations *Melaleuca uncinata* is a common understory species, sometimes forming thickets 1–2 m high. In drier areas *Callitris preissii* is an occasional to common tall shrub. Smaller xeromorphic shrubs occur and these become increasingly abundant in wetter areas. Some of the common and widely distributed species are: *Acacia spinescens, Baeckea behrii, Boronia coerulescens, Calytrix tetragona, Casuarina muelleriana, Hakea cycloptera, Hibbertia stricta, Lasiopetalum behrii* and *Leptospermum laevigatum.* The herbaceous layer is usually dominated by *Triodia scariosa.* Other herbs of constant occurrence are *Dianella revoluta* and *Lepidosperma laterale.* Annuals are rare.

In the west, the two species rarely occur together in a distinctive association (Beard, 1972, 1973). *E foecunda*, either in pure stands or in association with *E. gracilis* or *E. oleosa* forms mallee stands, rarely low woodlands, on low dunes, often around salt lakes. *Melaleuca uncinata* sometimes forms a sense shrub understory in these communities. *E. incrassata* (sometimes with a co-dominant of *E. goniantha*) occurs on deep white sands adjacent to the coast, where it forms dense to open shrubberies with *Acacia cyclops* as the principal understory species. *Melaleuca* spp. sometimes occur in the understory with an occasional *Banksia occidentalis.* Inland, as far north as the 240 mm isohyet, it is found on red or yellow-brown sandhills (Fig. 14.2), sometimes in small pure stands, but usually in association with *E. burracoppinensis, E. leptopoda* and *Brachychiton gregorii* as an occasional tall shrub associate.

2.2.3. *Eucalyptus socialis – E. dumosa – E. gracilis – E. oleosa* Alliance
Giant Mallee – Congoo Mallee – Yorell – Redwood

These species characterize the main mallee alliance in the east which occurs in patches be-

Fig. 14.2. *Eucalyptus incrassata, Callitris preissii* (left) and *Acacia linophylla* on a brown-orange sandhill. The grass is *Plectrachne desertorum* and shows ring development. c. 30 km south of Menzies, W.A.

tween the 400 and 180 mm isohyets. In the north it forms mosaics with arid zone communities, especially tall scrubs and low woodlands of *Acacia* spp., *Callitris "glauca"*, *Heterodendrum oleifolium*, and halophytic shrublands dominated by species of *Maireana* and *Atriplex*. In the wetter south it adjoins other mallee alliances *(E. diversifolia, E. incrassata)* or *Eucalyptus* woodlands. The alliance shows considerable fragmentation, outliers occurring in the east and north and a detached segment occurs west of the Nullarbor Plain. The soils are predominantly Solonized Brown Soils, but the depth of the surface sandy horizon varies so much that soil conditions are not uniform throughout the alliance. In the wetter areas (mean annual rainfall 400–300 mm) the surface sandy horizon is commonly 50–80 cm deep above the nodular limestone platform. This depth decreases to a few centimetres in the drier areas. In the Nullarbor region, where limestone lies at the surface, pockets of skeletal soil with additons of windblown sand and powdered limestone provide the rooting medium for plants. In many parts of the drier segment of the alliance, sand movement by the wind, probably during the recent arid period, has resulted in the redistribution of the surface sandy horizon, with the formation of low sand-ridges 1–2 m high. These dunes are separated by troughs, from which the surface sand has been removed, thus exposing the lime layer. The dunes and troughs support different types of plants.

In addition to the four characteristic species, *E. foecunda* is a fairly regular associate throughout the alliance. In the east, *E. behriana*, *E. viridis* and *E. polybractea* occur occasionally and locally, and *E. froggattii* is confined to Victoria. In South Australia, *E. brachycalyx* is confined to this alliance, and *E. anceps*, *E. calycogona*, *E. pileata* and *E. rugosa* also occur abundantly and are found west of the Nullarbor Plain as well. Also, west of the Nullarbor Plain *E. merrickae* and a mallee form of *E. salubris* are found in the alliance.

In South Australia and Victoria the patterning of species in stands is controlled largely by soil texture (Wood, 1929; Willis, 1943; Jessup, 1946; Patton, 1951; Litchfield, 1956; Connor, 1966; Parsons and Rowan, 1967). *E. behriana* occurs on soils of fine texture, *E. calycogona* on soils of intermediate texture, *E. anceps* on soils liable to temporary waterlogging. *E. oleosa* and *E. gracilis* occur on flats with some clay content. *E. socialis* occurs on sands overlying a more clayey horizon and *E. foecunda* occurs on deeper sands. *E. dumosa* occurs over a range of soil types but is absent from the coarsest sands.

A large number of associations occur (Wood, 1937; Zimmer, 1937; Crocker, 1946; Beadle, 1948; Jessup, 1948). These are determined partly by the regional distribution of the various

Fig. 14.3. *Callitris preissii* and *Eucalyptus socialis*. 30 km south of Mildura, Vic.

Fig. 14.4. *Codonocarpus cotonifolius* (centre) with *Eucalyptus socialis* and *E. oleosa* behind. The hummock grass is *Triodia scariosa*. 100 km south of Broken Hill, N.S.W.

species of *Eucalyptus* and partly (locally) by the response of the species to soil properties. Some of the associations occur throughout the alliance. In many cases both bull and whipstick forms of the same species of mallee are found in the one area, the former usually on more sandy soils, especially coarse sands of low fertility, the latter on soils containing lime near or at the surface, especially troughs between dune ridges. The proportions of co-dominants vary considerably especially with *E. socialis* and *E. dumosa*.

The latter is usually more abundant in the climatically wetter part of the alliance.

The understory layers are of two main types which are determined by mean annual rainfall. In wetter areas (between 400 and 280 mm isohyets) xeromorphic shrubs are abundant and tussock grasses occur in the herbaceous layer. The characteristic tall shrubs are *Melaleuca lanceolata* (often common and locally dominant), *Acacia oswaldii*, *Callitris preissii* (Fig. 14.3), *Exocarpos aphyllus*, *Eucarya acuminata*, *Geij*-

era linearifolia and *Pittosporum phylliraeoides*. Shrubs to 2 m high of regular or constant occurrence are *Acacia calamifolia, A. microcarpa, A. rigens, Bossiaea walkeri, Cassia nemophila, C. sturtii, Duboisia hopwoodii, Grevillea huegelii, Hibiscus farragei, Lycium australe, Melaleuca uncinata* (often abundant, forming thickets) and *Olearia pimelioides*. The herbaceous layer is very discontinuous and contains herbaceous grasses, chiefly *Danthonia setacea* and *Stipa* spp. *Triodia scariosa*, a rare species in the wetter segment, increases in abundance with increasing aridity. Annuals are not abundant, *Zygophyllum* spp. being the most common.

In drier areas the shrub species mentioned above still occur but are rare. The common species are the tall shrubs or small trees *Myoporum platycarpum* and *Heterodendrum oleifolium* and, rarely, *Codonocarpus cotonifolius* (Fig. 14.4). Shrubs to 2 m high include *Cassia* (above) and species of *Eremophila*, especially *E. scoparia*. The lower strata can be divided into two groups, according to the proximity of lime to the surface:

i. On lime-rich soils a low shrub layer 50 cm to 1 m high of chenopods, chiefly *Maireana sedifolia, M. georgei, M. tomentosa, M. triptera* and *Atriplex vesicaria, A. stipitata* and *Bassia* spp. (Fig. 14.5). In western South Australia and Western Australia the saltbush-like composite, *Cratystylis conocephalus* becomes conspicuous and sometimes locally dominant. These species commonly dominate open patches within mallee stands.

ii. On sands, *Triodia scariosa* dominates the understory (Fig. 14.6) and is often the only perennial species of any significance in the community apart from the eucalypts, though emergent shrubs sometimes occur, e.g., *Atriplex stipitata, Acacia ixiophylla* and *A. rigens*. In areas where sand dunes and limy troughs alternate, *Triodia* occupies the dune sands while chenopods occupy the troughs.

Fragments of the alliance occur in New South Wales as far east as the 450 mm isohyet, the largest occurring in the Roto district. The largest of these contain the four characteristic species and *E. foecunda*. Some contain, as well, admixtures of *E. behriana, E. polybractea* and *E. viridis*. The mallee patches are usually surrounded by woodlands of *E. woollsiana* or *E. populnea*.

Fig. 14.5. Mixed stand of *Eucalyptus socialis, E. dumosa* and *E. gracilis*. Understory of *Maireana triptera, Atriplex stipitata, Enchylaena tomentosa* and subsidiary *Atriplex companulata, Bassia paradoxa, B. obliquicuspis, Zygophyllum apiculatum* and *Z. fruticulosum*. 16 km east of Balranald, N.S.W.

Fig. 14.6. *Eucalyptus socialis* and *Triodia scariosa*. Near Pooncarie, N.S.W.

Superficially the stands closely resemble those in the main body of the alliance, but the understory species are eastern, rather than southern. For example, a small outlier in the Macquarie region (near Eumungerie), covering c. 2 sq. km and containing *E. dumosa* only is described by Biddiscombe (1965). The understory shrubs are *Baeckea densiflora, Casuarina distyla, Calytrix tetragona* (widespread in Aust.), *Leptospermum parvifolium* and *Micromyrtus ciliata*.

Outliers in the arid zone contain only *E. socialis* and the populations are possibly not taxonomically identical with the southern populations. The stands usually take the form of bull mallee or low woodlands to c. 7 high and occur mainly in rocky terrain or along creeks in hills on Lithosols containing lime. The species associated with *E. socialis* are usually the arid assemblage, mostly *Myoporum platycarpum* and a Chenopodiaceae understory (Wood, 1937, Carrodus et al, 1965). The most northerly stands recorded are in the Macdonell Ranges, close to the Tropic (Perry and Lazarides, 1962) on the lower slopes of limestone hills. The understory is dominated by hummock grasses, *Triodia longiceps* in one of the stands, *T. clelandii* in the other. A third stand south of the range occurs on limestone plains with a grass layer of *Aristida arenaria* and *Enneapogon* spp.

2.2.4. *Eucalyptus viridis* Alliance
Green Mallee Box

E. viridis, usually a mallee to 6 m high, but sometimes a small tree, occurs disjunctly in the east between the 240 and c. 500 mm isohyets.

The scattered distribution appears to be the result of the fragmentation of more extensive stands. The species occurs typically on ridges or slopes, usually on soils with a sandy, stony or gravelly surface horizon overlying clay and usually derived from shales or from denuded profiles with an ironstone horizon. The mallee or low woodland patches usually cover no more than a few hectares. The stands form mosaics with woodlands of *E. pilligaensis, E. populnea* and *E. woollsiana*, or, in the east, with ironbark forests (*E. crebra* or *E. sideroxylon*). The rare *E. bakeri* is an occasional associate of *E. viridis* in some stands in New South Wales and Queensland (e.g. at Inglewood).

The understory layers are usually poorly developed (Fig. 14.7). In wetter areas, especially adjacent to ironbark forests, xeromorphic shrubs 1–3 m high are sometimes present. *Melaleuca uncinata* occasionally forms an understory thicket 2–3 cm high (Biddescombe, 1963). Other shrubs, usually scattered or in small clumps are: *Acacia buxifolia, A. hakeoides, A. lineata, Calytrix tetragona, Cassinia laevis, Dillwynia ericifolia, Dodonaea viscosa, Melaleuca pubescens* and *Prostanthera chlorantha*. The herbaceous stratum is discontinuous and sometimes absent over areas of many square metres. *Cheilanthes tenuifolia* is the most constant species; others include *Atriplex spinibractea, Danthonia* spp., *Dianella revoluta, Helichrysum semipapposum, Minuria leptophylla, Poa fordeana*, and in small depressions liable to waterlogging after rain *Carex inversa* and *Juncus* sp.

In drier areas in New South Wales, e.g. near Cobar, where *E. viridis* occurs in open mallee stands within woodlands of *E. intertexta*, the understories are well developed and contain species typical of the arid zone, viz., *Cassia artemisioides, Eremophila longifolia, Scaevola spinescens, Sida virgata* (shrubs 1–2 m high), and the subshrubs and herbs *Bassia birchii, B. uniflora, Chloris truncata, Digitaria divaricatissima, Enchylaena tomentosa, Paspalidium gracile, Rhagodia nutans, Sida corrugata* and *Vittadinia australis*.

2.2.5. Eucalyptus redunca – E. uncinata Alliance
Black Marlock – Hook-leaved Mallee

These species occur in association near the south coast of Western Australia between the

Fig. 14.7. *Eucalyptus viridis*. Herbaceous layer of *Danthonia* sp. with scattered *Helichrysum semipapposum, Minuria leptophylla* and *Dianella revoluta*. Near West Wyalong, N.S.W.

Fig. 14.8. Pure stand of *Eucalyptus platypus*. *Melaleuca parviflora* in foreground. c. 30 km north of Hopetoun, W.A.

375 and 500 mm isohyets (Beard, 1972, 1973). The communities occupy the floors and slopes of valleys or flat areas towards the south, and they sometimes adjoin the *E. eremophila* alliance in the drier north. The soils are deep alluvial sands or sands with a clayey subsoil. Mosaics commonly occur with communities on elevated sand dunes (*E. incrassata* and *E. goniantha*) or with low woodland or marlock or mallee communities which occur on clays in depressions (*E. occidentalis, E. platypus, E. dielsii* or *E. annulata*). The stands are mostly 2–3 m high. Associated eucalypts are *E. leptocalyx* (mainly south), *E. forrestiana* (south east), *E. flocktoniae* and *E. conglobata* (throughout).

The understory layers are xeromorphic and appear to vary with the fire-history. Thickets of *Melaleuca uncinata* sometimes occur and other species of *Melaleuca* (*M. substrigosa* and *M. thymoides*) are often abundant. A mixed understory of the genera prominent in the mallee-heaths (see 2.3, below) is common, viz., species of *Acacia, Adenanthos, Banksia, Grevillea, Hakea* etc.

2.2.6. *Eucalyptus platypus* Alliance
Moort

E. platypus is a marlock 4–5 m high, branched near ground level and with a very dense crown. It does not form a lignotuber and is killed by fire. It occurs mostly between the 400 and 550 mm isohyets near the south coast of Western Australia and is confined to depressions where the soils have a high clay content and are sometimes subsaline (Beard, 1972, 1973). The species is usually found in pure stands, the dense crowns of the plants touching or interlacing to form a thicket, which commonly excludes all other species. Shrubs of *Melaleuca* spp. sometimes occur within the thicket or form a fringe at the periphery of the thicket (Fig. 14.8). Occasionally, mallee species which tolerate subsaline conditions are found intermingled with *E. platypus* at the margins of the thickets, especially *E. annulata* and *E. spathulata*.

2.2.7. *Eucalyptus nutans* – *E. gardneri* Alliance
Red-flowered Moort – Blue Mallet

This is a small alliance confined to the Ravensthorpe district (Beard, 1972), a region with a mean annual rainfall of c. 400 mm. *E. nutans*, a mallee or marlock to 4 m high and closely related to *E. platypus*, is confined to this alliance. *E. gardneri*, a mallee or small tree, is more widely distributed, extending westward to the 550 mm isohyet. The alliance occurs on greenstone outcrops and the stands commonly contain other eucalypts, including *E. floctoniae*, *E. leptocalyx*, *E. loxophleba*, *E. spathulata*, *E. tetragona* and *E. uncinata*. The shrub layer is dense and consists either of thickets of *Melaleuca uncinata* or *M. striata* or of mixed xeromorphs belonging to the genera *Banksia*, *Boronia* and *Casuarina*.

2.2.8. *Eucalyptus cooperana* Alliance
Many-flowered Mallee

This mallee species, closely related to *E. floctoniae*, reaches a height of 5 m and is conspicuous because of its snow-white stems. It is restricted to acid sands along the south coast of Western Australia, the largest stands occurring inland from Israelite Bay (Beard, 1973). In this area, the sands are derived from limestones which have been severely leached to leave an infertile red, siliceous profile which overlies the limestone. Associated species sometimes occur, *E.* aff. *salmonophloia*, *E. micranthera* and *E. uncinata*. The mallee stands adjoin mallee communities or low woodlands of *E. oleosa* and *E. socialis* in the north. In the south, the stands are separated from the ocean by *Banksia* scrubs. The understory species are xeromorphic shrubs, including *Acacia sorophylla*, *Callitris roei*, *Calytrix tetragona*, *Dryandra longifolia*, *Templetonia retusa* and species of *Andersonia*, *Beaufortia*, *Boronia*, *Eremophila*, *Eriostemon*, *Grevillea*, *Melaleuca*, *Pultenaea* and *Verticordia*.

Smaller stands of *E. cooperana* occur to the east, close to the ocean, on white sands. Other species which may occur in association are *E. diversifolia* (at its western limit, south of Cocklebiddy), where *Callitris roei* is a common associate or locally dominant. *E. micranthera*, in association with a form of *E. gracilis*, forms dense mallee stands close to the ocean on white dune sand (Fig. 14.9). *E. angustissima* recorded on white sands in the same general area may also belong in this alliance.

2.2.9. *Eucalyptus eremophila* Alliance
Tall Sand Mallee

This species is confined to Western Australia, occurring mainly between the 300 and 450 mm isohyets with minor extensions into drier areas in the east and north-east where it is a subsidiary species. It has been used to designate an extensive and floristically complex alliance because of its constancy in many mallee stands. It is found over a wide range of soil types including Solonized Brown Soils, red brown sands with or without lateritic gravel, and, uncommonly, on sandy clays. These soils are developed on flat or undulating country in broad, shallow valleys.

The mallee communities (Beard, 1972, 1973) contain several other species as well as *E. eremophila* which are patterned to some extent by soil variations, though fire possibly plays some part in determining the patterns. Since *E. eremophila* occupies a central position in the mallee belt in the west, the alliance contacts both northern and southern mallee alliances and ecotonal associations between species from adjacent alliances are common. Further to this, belts of woodland (chiefly *E. loxophleba*, itself sometimes a mallee, and *E. salmonophloia*) and patches of heath traverse the alliance and provide species which enter the mallee stands, especially in the ecotonal zones. Ecotonal associations occur also at the bases of rock outcrops which support their own characteristic species (see 2.4, below).

Fig. 14.9. *Eucalyptus micranthera* and *E. gracilis* on white dune sand with *Acacia rostellifera* (right foreground). *Cakile maritima* on sand dune. Photo taken from the top of coastal dune, looking inland. Great Australian Bight, south of Cocklebiddy, W.A.

E. eremophila sometimes forms pure stands but is found usually in association with up to a dozen other eucalypts, mostly two to four species occurring in any stand. The following assemblages are the more common ones:

i. *E. eremophila* – *E. oleosa*. This association occurs from east to west, the latter often being more prominent; the understory is usually composed of xeromorphic Proteaceae and Myrtaceae, but in drier areas species of *Eremophila* become conspicuous or dominant.

ii. Mixed with the two species above, and in various combinations (except where indicated) are: *E. anceps* (east), *E. calycogona, E. celastroides, E. erythronema* (west), *E. floctoniae, E. leptocalyx* (south), *E. loxophleba, E. merrickae* (east), *E. ovularis* and *E. pileata*. In these communities, emergent small trees of *E. falcata* and *E. gardneri* occur above the mallee-canopy.

iii. In the south-east, *E. eremophila* occurs in association with the distinctive *E. forrestiana* (Fuchsia Mallee, with 4-angled orange-scarlet buds 3–4 cm long). The understory is dominated by *Melaleuca* spp.

iv. In depressions in the south and south-eastern parts of the alliance, *E. annulata, E. spathulata* and/or *E. gracilis* form mixed stands with *E. eremophila*, the last being absent from stands on subsaline soils, the other species forming distinctive assemblages.

Other variations, usually attributable to differences in soil texture, are: On deep sandy soils *E. foecunda* and *E. incrassata* occur, sometimes as small trees, and in the east *E. goniantha;* the sandy areas are sometimes dunes associated with shallow dry lakes. In depressions, usually on fine-textured soils, local dominants of *E. dielsii* and *E. diptera* occur (east and south-east).

Enclosures of the *E. platypus* Alliance in depressions and stands of *E. annulata – E. spathulata* also occur in the south.

The understory in wetter areas consists of a more or less continuous layer 1–2 m high of xeromorphic species. These are common or dominant species from the heaths (Chap. 16) and mallee-heaths (see below, 2.3) which occur in the same area as the mallee community. *Callitris* is represented by *C. preissii* and *C. roei* which may form local thickets. In drier areas these taxa become less abundant but the same genera are found, especially *Grevillea* and *Melaleuca*. In these drier areas species of *Eremophila* and *Olearia* become increasingly important and finally dominant.

2.2.10. *Eucalyptus sheathiana – E. loxophleba – E. oleosa* Alliance

Ribbon-barked Mallee –
York gum – Redwood

These species occupy the northern segment of the mallee region in Western Australia mainly between the 350 and 250 mm isohyets. The alliance is found in relatively small, irregular patches on Solonized Brown Soils, on compacted red-brown sands or at the bases of rocky outcrops which are either granite or sandstone (Beard, 1972). The mallee patches form mosaics either with woodlands of *E. salmonophloia* and *E. salubris* which occupy the more fertile soils often in depressions or on flats, or with mallee heaths and heaths developed on yellow sand. Several other mallee species form associations with one or other of the characteristic species, viz., *E. calycogona, E. erythronema, E. flocktoniae, E. gardneri, E. gracilis, E. griffithsii, E. merrickiae, E. redunca* and *E. rigidula*.

The understory varies with the type of soil. Xeromophic shrubs form a very discontinuous layer on sandy soils, especially *Acacia, Casuarina, Grevillea* and *Melaleuca* and these become increasingly prominent in transitional areas towards heaths. Grasses are sometimes prominent but never form swards. *Triodia scariosa* is locally common; less common species include *Neurachne alopecuroides* and *Stipa elegantissima*. Annuals are abundant after rain, especially *Schoenia cassinioides* and *Waitzia acuminata*.

In rocky situations and in drier areas, genera typical of the arid zone become prominent, especially species of *Cassia, Eremophila* and *Euca-*

Fig. 14.10. *Eucalyptus burracoppinensis* with *Casuarina acutivalvis, Grevillea exelsior* and *Eucarya acuminata*. 40 km east of Merriden, W. A.

rya (shrubs), and herbs and subshrubs of the genera *Aizoon, Bassia, Convolvulus, Danthonia* and *Maireana*.

2.2.11. *Eucalyptus burracoppinensis – Casuarina acutivalvis* Alliance
Burracoppin Mallee

This community occurs on patches of deep, yellow-brown sand around the 300 mm isohyet in the Southern Cross-Burracoppin district. The sand patches are surrounded mainly by Solonized Brown Soils supporting mallee stands dominated by *E. eremophila* or *E. sheathiana* and *E. loxophleba*. *E. burracoppinensis* is usually the only eucalypt present (Fig. 14.10), but *E. leptopoda* is an occasional species on loose sand. The community is more or less closed, 3 to 4 m high, possibly reaching 6 m if unburned (Beard, 1972). Associated shrubs are *Casuarina acutivalvis, Acacia crassiuscula, A. signata* and less abundant *Grevillea* spp., *Callitris preissii, Eucarya acuminata* and *Banksia elderiana*. A lower xeromorphic shrub layer occurs, in which *Thryptomene kochii* is the most common, with *Melaleuca cordata* sometimes locally abundant. Herbs are rare, *Plectrachne rigidissima* forming hummocks locally.

2.2.12. *Eucalyptus pyriformis – E. oldfieldii – E. leptopoda* Alliance
Large-fruited Mallee – Oldfield's Mallee – Tammin Mallee

Several species of *Eucalyptus*, sometimes referred to as sandplain mallees, occur on loose yellow-brown to red-brown sands across the northern fringe of the western segment of the mallee region, almost from the Indian Ocean to

Fig. 14.11. *Eucalyptus leptopoda*. Shrubs are *Duboisia hopwoodii* and *Phebalium microphyllum*. Hummock grass is *Plectrachne desertorum*. On loose yellow-brown sand. 100 km north of Kalgoorlie, W.A.

Fig. 14.12. *Eucalyptus oldfieldii* and *Actinostrobus arenarius*. c. 155 km north of Geraldton, W.A.

arid central South Australia, in areas receiving a mean annual rainfall of 300–200 mm. Some of the species are widely distributed, others have restricted areas. A few of them extend northward into the arid zone, where they occur in isolated stands on stabilised or partially stabilised sand dunes (see below 2.5).

Few of the communities have been described and consequently the species are grouped together into one alliance which includes the three species mentioned above and *E. dongarrensis, E. ebbanoensis, E. eudesmoides, E. juncunda, E. jutsonii, E. pimpiniana* and *E. roycei* (see Table 14.1). In combination with one or more of these species, *E. incrassata* and narrow-leaved forms of *E. oleosa* (*E. grasbyi* and *E. kochii*) may be present. Some of the stands are described below.

i. *E. pyriformis – E. oleosa – E. leptopoda.*

This assemblage is described by Speck (1958) for the Mullewa areas (east of Geraldton) on brown sands on gentle slopes and undulating country. The communities are 3–5 m high and contain *Casuarina acutivalvis, Acacia* spp., *Grevillea eriostachya, Banksia prionotes* and *B. spectrum,* together with a large number of smaller shrubs which occur also in the heaths in the same district. Annual species, especially Compositae are abundant after rain. Associations dominated by the three eucalypts, either singly or in pairs could be recognized. Both *E. pyriformis* and *E. oleosa* are found extensively to the east, the latter occurring in association with many eucalypts. *E. pyriformis* occurs throughout the southern fringe of the arid zone on sand as far east as Ooldea and Tarcoola in South Australia (Adamson and Osborn, 1922) where it forms pure stands on sand. *E. leptopoda,* confined to Western Australia, is a dominant on loose sand-patches, often at the periphery of sandhills. It ranges between the 300 and 220 mm isohyets and in wetter areas may be a subsidiary species in the *E. burracoppinensis* alliance. At its drier extreme it dominates an open whipstick mallee community 2–3 m high with a set of arid zone associates (Fig. 14.11).

ii. *E. oldfieldii* occurs as a mallee 5–6 m high in association with several species of *Eucalyptus*, especially *E. eudesmoides* and *E. jucunda* on yellow or yellow-brown sands in the west and red-brown sands towards the east (e.g. north of Kalgoorlie). In the west the mallee stands adjoin heaths, and the eucalypts in a stunted form extend into the heaths to form mallee heaths. In this area *Actinostrobus arenarius* occurs in the heaths and ecotonal associations with this species and the eucalypts sometimes occur (Fig. 14.12). The understory species are those which occur in the heaths, e.g., *Acacia longispinea*, *Anthrotroche myoporoides*, *Calytrix* spp., *Hakea buculenta*, *Melaleuca* spp., *Micromyrtus peltigera*, *Verticordia ethleliana* and many others. On red brown sandhills, e.g., north of Kalgoorlie in a somewhat drier area, *E. oldfieldii* occurs with *E. incrassata* and *E. ebbanoensis* (Fig. 14.13) the latter on loose sand at the fringes of dunes. *Brachychiton gregorii* as a small tree is an occasional associate. The communities are surrounded by other mallee species (*E. concinna* and *E. oleosa*) on more compact soils. Understory species include *Callitris pressii* and *Acacia linophylla* and the hummock grass *Plectrachne desertorum*.

iii. *Eucalyptus justonii* is recorded for both Western and South Australia. The species dominates mallee stands 3–5 m high on red-brown sands. There is a dense understory of *Casuarina acutivalvis* and *Lamarckia hakeifolia* or of *Acacia colletioides* and *A. ramulosa*. The herbaceous layer is composed of annuals: *Brunonia australis*, *Calandrinia* sp., *Ptilotus alopecuroides*, *Podolepis aristata* and *Waitzia acuminata*. An association containing *E. jutsonii*, *E. roycei*, *E. dongarrensis* and *E. eudesmioides* on red calcareous sand close to Shark's Bay is indicated by Carr et al. (1970).

Fig. 14.13. *Eucalyptus ebbanoensis, Casuarina cristata, Acacia colletioides, Scaevola spinescens* and *Plectrachne desertorum* on red-brown sand. 96 km north of Kalgoorlie, W.A.

2.2.13. *Eucalyptus concinna* Alliance
Victoria Desert Mallee

E. concinna is found as a small tree or bull mallee on the greenstones in the Kalgoorlie area at the fringe of the arid zone. Here it occurs both in the *E. salmonophloia* – *E. salubris* and *E. lesouefii* Alliances (Chap. 15) mostly as an understory or subsidiary associate, particularly with *E. lesouefii*. With increasing aridity it forms pure mallee stands on slightly elevated areas or on ridge tops, the taller species mentioned above occupying the lower areas. These mallee stands have a discontinuous shrub layer composed mainly of *Eremophila scoparia*, *Cassia nemophila*, and *Westringia rigida*. *Triodia scariosa* forms scattered hummocks and *Ptilotus exaltatus* is a common annual herb.

In the more arid areas, possibly with a mean annual rainfall of c. 240 mm, *E. concinna* occurs on red sand deposits, sometimes with associated *E. comitae-vallis*, with which it intergrades, *E. foecunda* and *E. ovularis*. *Callitris preissii* is a common tall shrub associate and *E. incrassata*, which occupies the crests of dunes, forms ecotonal associations with *E. concinna*. The latter extends northward, without associated eucalypts, where the alliance meets the *Acacia aneura* alliance. In the transitional area between the two alliances, *E. concinna* occupies the dune troughs, *Acacia aneura* the dune crests. *Acacia linophylla* occurs as the dominant in the understory in both communities.

E. concinna is reported to occur in the Victoria Desert in small patches and it extends eastward into South Australia as far as Ooldea.

2.2.14. *Eucalyptus dongarrensis* – *E. oraria* Alliance

These species are restricted to the coastal limestones of Western Australia from Dongara (south of Geraldton) to Bernier Is. The species are confined to the coastal limestones, mostly 1–5 km from the ocean and they dominate mallee communities in depressions as enclosures within stands of *Acacia rostellifera* which

Fig. 14.14. General view of mallee-heath. *Eucalyptus tetragona* c. 2 m high in the foreground and scattered through the heath. Patch of mallee (*E. occidentalis*) in centre. Near Hopetoun, W.A., looking toward the ocean.

Fig. 14.15. *Eucalyptus tetragona* with a heath understory dominated by *Casuarina humilis, Hakea crassifolia, H. corymbosa* and *Petrophile fastigiata*. c. 30 km north of Hopetoun, W.A.

characterizes littoral communities in this district (Chap. 23). The communities are described by Beard (1976) as thickets and, on Dorre and Bernier Islands, the stands are only 30 cm to 1½ m tall. Shrub associates are *Thryptomene baekeacea, Melaleuca cardiophylla, Beyeria canescens, Stylobasium spathulatum* and *Brachysema macrocarpum*. Between the shrubs, *Triodia plurinervata* forms hummocks 40–60 cm high.

2.3. Mallee-heaths

2.3.1. General

Mallee-heaths are heath communities with emergent mallee eucalypts. The aggregation, density and height of the latter vary considerably. The mallee plants may occur as scattered individuals or in clumps (Fig. 14.14). In many cases the mallee-heaths are ecotonal, the mallee eucalypts being the same species as those in the mallee communities in the same district. In other cases the mallee eucalypts are stunted species (mallee "forms") of tree species which occur in the same district, and in some cases the mallee form has been produced by fire, e.g., *Eucalyptus gummifera* in coastal New South Wales and *E. calophylla* in the south-west.

The eastern mallee heaths and the mallee eucalypts in these wetter areas are dealt with elsewhere (Chaps 10 and 16). In South Australia, mallee heaths develop adjacent to the mallee communities and are ecotonal, the main eucalypts involved being *E. diversifolia, E. foecunda* and *E. incrassata* (Jessup, 1946; Coaldrake, 1951).

In Western Australia, some of the mallee-heaths are ecotonal but others are unique in so far as the eucalypts occur only in mallee-heaths and many of the species are confined to limited areas and to specific habitats. The communities develop mostly near the coast on soils of low fertility, usually deep sands containing ironstone gravel, under a mean annual rainfall of 320–500 mm. Two main regions exist, one in the wetter south (*Eucalyptus tetragona* – *E. incrassata* alliance), the other near the west coast and inland where several eucalypts are involved. Of the various species, only *E. tetragona* occurs in both regions.

2.3.2. *Eucalyptus tetragona* – *E. incrassata* Alliance
Tallerach – Lerp Mallee

The alliance occurs on the southern coastal plain of Western Australia between the 320 and 420 mm isohyets (Beard, 1972, 1973). *E. tetragona* is found also in small quantities in the mallee-heaths in the west, while *E. incrassata* extends into South Australia. The communities show a considerable variation in pattern brought about by the local presence or absence of eucalypts which are commonly clumped to form open shrubberies with patches of heath between. Depressions commonly support either mallee communities, or woodlands often dominated by *E. occidentalis* (Fig. 14.14), or by scrubs of *Nuytsia floribunda*.

The soils are invariably sandy and often overlie ironstone nodules. When the surface sandy horizon is less than c. 90 cm deep, *E. tetragona* is dominant. Deeper sandy horizons support either heath or scrub-heath. In the western segment of the alliance the glaucous form of *E. tetragona* occurs (Fig. 14.15); towards the east the soils are more consistently sandy and *E. incrassata* is the dominant, with a green form of *E. tetragona* as a subsidiary species. In some areas where rock lies at the surface, the mallee-heath forms mosaics with other mallee alliances and ecotonal associations occur, involving *E. tetragona* and the species of the rocky outcrops (see 2.4, below).

In addition to the two characteristic species, a few other species are confined to this alliance, *E.*

Fig. 14.16. *Eucalyptus tetraptera* with an understory dominated by *Dryandra cirsioides, Hakea crassifolia* and *Melaleuca scabra*. c. 40 km north of Hopetoun, W.A.

bupestrium, *E. decurva* and *E. pachyloma* (east, Stirling Range area), and *E. tetraptera* (mainly south of Ravensthorpe, Fig. 14.16). Several others, which are usually components of the mallee communities, are often present, viz., *E. foecunda*, *E. goniantha*, *E. leptocalyx*, *E. redunca*, *E. spathulata* and *E. uncinata*.

The heath component consists mainly of species c. 1 m high or less, but some taller plants, which characterize scrub-heaths, as defined by Beard, also occur, some of them attaining the height of the eucalypts. The most common of these are *Banksia baueri*, *Calothamnus quadrifidus*, *Dryandra longifolia*, *Grevillea hookeriana*, *Hakea cinerea*, *Isopogon buxifolius* and *Lambertia inermis*. Shrub species, mostly c. 1 m high or less, belong chiefly to the genera *Acacia*, *Adenanthos*, *Andersonia*, *Banksia*, *Calytrix*, *Casuarina*, *Dampiera*, *Darwinia*, *Hakea*, *Isopogon*, *Labichea*, *Melaleuca*, *Petrophile*, *Regelia* and *Verticordia*.

2.3.3. Species on the West Coast and Inland

Several species of *Eucalyptus* are found in the heaths on the western coastal plain, some extending eastward into drier areas. The region receives a mean annual rainfall of c. 500 mm on the coast, decreasing to 300 mm inland. The mallee-heaths are developed on white sands in wetter areas, yellow sand in the drier inland.

Some of the eucalypts are restricted to the mallee-heaths, others extend from the mallee communities into the mallee-heaths. The latter include *E. eudesmoides*, *E. leptopoda*, *E. oldfieldii* and *E. redunca*. Also, *E. todtiana* which often forms a small tree and is common in the *Banksia* scrubs (Chap. 16) sometimes occurs in stunted form in the mallee-heaths. Likewise, *E. calophylla*, normally a tall tree (Chap. 15) becomes mallee-like in this region and is a component of mallee-heaths.

Species restricted to the mallee heaths are: *E. tetragona* (also in the south, see last section); *E. macrocarpa* (Mottlecah – the species with the largest fruit, reaching a diameter of 6–7 cm), found in sandplain heath as a gnarled shrub 2–5 m high, around the 500 mm isohyet; *E. albida* (white-leaved Mallee), which reaches a height of c. 3 m.

The heath component of the mallee-heaths in this region contains the same genera as those in the south but most of the genera are represented by different species, as indicated in Chap. 16.

2.4. Communities on Rocky Outcrops in South-western Western Australia

2.4.1. General

Mountain ranges and hills with bare rock at their summits are scattered throughout the south-western portion of Western Australia from the coast with a mean annual rainfall of c. 850 mm to the arid interior. The largest of these are the Stirling Ranges and the Porongorups where peaks rise to c. 1200 m above sea level. Hills of smaller dimensions occur throughout the area, reaching a height of c. 200 m above the lowlands, with a maximum height of c. 600 m. The most interesting of these botanically are the Barren Ranges in the south.

The rock comprising the hills is mainly granite, residuals of the vast granitic platform which underlies most of the south-western part of the continent. The Stirling Ranges, on the contrary, are composed of Upper Proterozoic metamorphosed sediments, quartzites and sandstones. Elsewhere, ironstones of lateritic origin occur and also small areas of quartzites and basalt.

Most of the hills are domed at the summit but some are flat-topped. The latter are sometimes fissured or traversed by dykes and present a rectilineal pattern of plant communities, the cracks or dykes where soil accumulates supporting taller communities (Beard, 1967). Bare rock surfaces are sometimes pitted with depressions which form pools when rain falls. Hills composed of granite are weathered physically, the granite cracking or exfoliating to produce crevices in which mineral matter and debris accumulate and provide a rooting medium for plants. Downslope, the soils are skeletal and scree has accumulated near the bases of some hills.

Runoff water from bare rock surfaces irrigates crevices, lower slopes and any enclosed valleys that occur. This irrigation water creates wet environments, at least seasonally, and accounts for the occurrence of some outliers of wet-climate species in dry climates, e.g., *E. diversicolor* forests in the Porongorups (Chap.

15), and *Hakea prostata* and *Nuytsia floribunda* in the semi-arid zone.

The hills support distinctive floras including many restricted endemics (Gardner, 1942). In the Stirling Ranges, for example, c. 100 endemic species are confined to this small areas, mainly of *Banksia, Darwinia, Hypocalymma* and *Isopogon*. Also, many species are confined to these hills and identify distinctive communities. The primary colonizers of the rocks and heath communities are described in Chap. 16; the *Eucalyptus* stands are described below.

2.4.2. *Eucalyptus preissiana* Alliance
Bell-fruited Mallee

E. preissiana, a mallee 1–5 m high, is the most common species on the hills which rise out of the plain along the south coast. The plants grow in rugged terrain, often in rock crevices, either in pure stands or in association with another eucalypt. The soils are invariably Lithosols and are derived from quartzites, siliceous conglomerates or from schists. On lower slopes the thickets occur on coarse colluvium derived from mixed rock materials and in this zone other species of *Eucalyptus* are found, some confined to the alliance, others occurring in the adjacent communities, which are commonly *E. tetragona* mallee-heath (Beard 1972, 1973). *Dryandra quercifolia* is a constant associate of *E. preissiana*, sometimes reaching the same height as the eucalypt. Other shrub species of high frequency are *Banksia lehmanniana, Calothamnus pinifolius, Casuarina humilis, Dryandra armata* and *Hakea crassifolia*.

The main species of *Eucalyptus* associated with *E. preissiana*, is *E. lehmannii*, occurring on lower slopes, where it may form pure stands, usually with a dense shrub layer containing *Acacia lasiocalyx, Dryandra* sp. and *Hakea laurina*. A species of very restricted distribution found sometimes in association with *E. preissiana*, or as a local dominant, is *E. sepulcralis* (Weeping Gum). Beard (1972) records it on sandstone in thin, yellow sandy soils. It has a peculiar habit in that, though it reaches a height of 6 m, its stems never exceed a diameter of 5 cm, the tree being willow-like. Its main shrub associates are *Adenanthos sericeus, Banksia baxteri, B. prostata* and *Eucalyptus decurva*.

Other rare eucalypts of restricted distribution recorded by Beard for the hills and lower slopes are:

i. *E. coronata* (Crowned Mallee) on massive siliceous conglomerate in the Barren Ranges in association with *E. preissiana, E. lehmannii* and *Dryandra quercifolia*.

ii. *E. desmondensis* (Desmond Mallee), a shrub to 5 m high, found only in the Ravensthorpe district at the foothills of ranges on stony soils but sometimes extending on to clays.

iii. *E. doratoxylon* (Spearwood Mallee, so called because the aborigines used the stems for spears), a mallee up to 5 m high with a smooth, white bark; it occurs usually in protected situations on granite, sandstone or quartzite outcrops.

iv. *E. grossa,* usually c. 2 m high occurring on lower slopes usually in *Melaleuca* thickets (Gardner, 1960).

2.4.3. Other *Eucalyptus* Species

Several species of very restricted distribution (Table 14.1) are found in rocky situations in the wetter south. They are possibly dominants in mallee or marlock stands but no community descriptions are available. These are:

i. *E. megacornuta* (Warted Yate), a marlock to 8 m high with a smooth ash-brown bark, is confined to the Ravensthorpe Range where it grows in small, pure stands in gravelly declivities (Gardner, 1959). It is peculiar in having an operculum c. 5 m long which is densely warted and about twice as long as the hypanthium.

ii. *E. burdettiana* is a mallee to 2 m high, with erect, smooth branches. It resembles the preceding species, but is smaller in all its parts and the operculum is not warted. It forms mallee thickets in quartzite declivities on East Barren Mountain and the spurs which radiate in a northerly direction from the mountain (Gardner, 1963).

iii. *E. stoatei*, a marlock to 8 m high, is confined to an area of 2–3 sq. km. east of Kundip between Ravensthorpe and Hopetoun "in open scrub country on the tributaries of the Jerdacuttup River" (Gardner, 1952).

iv. *E. aquilina,* a mallee 2–5 m high and *E. ligulata,* a mallee 2–3 m high, in association, are recorded only for the Mt. le Grand area (east of Esperance) on shallow soils among gneissic rocks on the sides of creeks. The stands contain *E.* aff. *goniantha* and/or *E. doratoxylon* (Brooker, 1974).

v. *E. insularis* is recorded in mallee stands 2 m high on Mt. le Grand in association with *E.*

lehmannii. It occurs also on gneissic hills on the islands of the Recherche Archipelego in stands 8 m tall (Brooker, 1974).

vi. In drier areas within the southern mallee region in Western Australia several other species have been recorded on or at the bases of rocky outcrops, chiefly granite. Of these, three are mentioned here; those extending to Central Australia are dealt with in the next section.

vii. *E. caesia* (Gungunnu), a highly ornamental marlock with drooping branches, reaches a height of 6–7 m and occurs in very open stands at the bases of granite outcrops. It is found mainly around the 320 mm isohyet, but has been recorded also in the Victoria Desert.

viii. *E. crucis,* a mallee 3–8 m high, also occurs around the bases of granite outcrops within the area of the last species.

ix. *E. kruseana,* a mallee 3–5 m high, with sessile, glaucous, almost orbicular leaves and bright yellow stamens, grows on shallow loams overlying granite. It is recorded from a few widely separated localities.

2.5. *Eucalyptus* Species in the Arid Zone

Seventeen species of *Eucalyptus* are either confined to or occur mainly in the arid zone and another 30 species extend from wetter areas into the arid zone. Some of these 30 species are restricted to watercourses (Chap. 21), but most are found in upland areas and are probably relicts from the adjacent wetter communities which contracted when the arid zone developed. Most of these relict species belong to the southern mallee region.

Of the 17 species confined to the arid zone none forms a well defined alliance which can be recognized as a constant assemblage of species, including the understory. On the contrary, the various species occur disjunctly across part of the arid zone, the stands of mallee or woodland of any one species being separated by distances of a few to hundreds of kilometres; the understory species in these isolated stands are com-

Fig. 14.17. *Eucalyptus odontocarpa*. Grass layer of *Triodia pungens*. Community is in N.T., 125 km west of Camooweal, Qld.

monly quite different. Alternatively, the species are widely scattered in hummock grasslands to form mallee or tree savannahs. It is possible that the now isolated stands of one species were once more extensive and even contiguous and if this were so, the modern stands represent fragments of a one-time extensive alliance. On the other hand, the isolated stands could well have been established by distance dispersal from a small area which could have been the centre of origin of the species. The more abundant of the species for which community descriptions are available are described below. The rarer species are listed in Table 14.1.

i. *Eucalyptus odontocarpa* (Sturt Creek Mallee) (Fig. 14.17) and *E. pachyphylla* (Thick-leaved Mallee) sometimes occur in association in the tropics on what appears to be most unfavourable sites with respect to water availability and soil nutrients (total phosphorus levels 20 and 30 ppm – the lowest recorded for the tropics). The mallee stands occur within the *E. brevifolia* alliance (Chap. 8) and either species may be locally dominant. Rare associated tall shrubs are *Grevillea striata*, *Atalaya hemiglauca* and *Ventilago viminalis*. A shrub layer may occur in patches, species of *Acacia* (especially *A. lysiphloia*) and *Cassia* being most common (Perry and Christian, 1954). The grass layer is dominated by *Triodia pungens* and a large number of herbaceous species appear between the *Triodia* hummocks after rain, especially *Aristida* spp., *Enneapogon* spp., *Ptilotus exaltatus*, *P. leucocoma* and *Scaevola aemula*. The two mallee species occupy different areas and therefore dominate communities individually. *E. odontocarpa*, confined to the tropics, often grows in rock ledges, when the plants are stunted (1–1½ m high). *E. pachyphylla* extends south of the Tropic and is sometimes found in stunted form on rocky slopes (Fig. 14.18). Alternately it occurs on the sides of dunes associated with *Grevillea juncifolia*, or with *Acacia kempeana* and *Casuarina decaisneana* (Chippendale, 1963).

ii. *Eucalyptus gamophylla* (Blue Mallee) is found in isolated stands rarely exceeding a few hectares in extent in the western half of the arid zone, the stands occurring in two areas separated by the Great Sandy Desert. The species reaches a height of c. 6 m and has opposite adult foliage. It forms pure stands on red sand, commonly on the crests of dunes, or on rocky outcrops. On sands it is co-dominant mostly with *Acacia aneura* (Chippendale, 1958, 1963), but on rocky ridges it is associated with other

Fig. 14.18. Stunted *Eucalyptus pachyphylla* and *Triodia pungens*. 110 km north of Alice Springs, N.T.

Fig. 14.19. *Eucalyptus gamophylla* and *Triodia wiseana*. Hammersley Range, W. A.

Fig. 14.20. *Eucalyptus gongylocarpa*. Near Mt. Magnet, W. A. Photo: M. I. T. Brooker.

eucalypts, e.g. *E. brevifolia* and/or *E. setosa*. The understory layers vary from site to site and appear to be determined by the species available from the surrounding communities. In most cases there is an understory of hummock grasses, sometimes with scattered shrubs. In Central Australia *Triodia pungens* forms the understory north of the Tropic and *T. basedowii* south of the Tropic. In the Hammersley Ranges the species is *T. wiseana* (Fig. 14.19). Chippendale records the following shrubs in a stand near Lake Amadeus: *Cassia nemophila, Chenopodium nitrariaceum, Dodonaea attenuata* and *Scaevola spinescens*.

iii. *Eucalyptus kingsmillii* (Kingsmill's Mallee) is confined to arid Western Australia and occurs mainly on sands, less commonly on rocky outcrops. It is a mallee up to 6 m high and is found mostly as scattered individuals or clumps within the *Acacia aneura* alliance (Speck, 1958) with different grass associates, especially *Triodia concinna, T. basedowii* and *Plectrachne melvillei*, with subsidiary *Danthonia bipartita* and *Eragrostis eriopoda*.

iv. *Eucalyptus gongylocarpa* (Marble Gum) is remarkable in the desert (Fig. 14.20) because of its size, being a tree reaching a height of 8–13 m (rarely 20 m). It occurs either as isolated trees or in pure stands as savannah woodlands on red sandplain, or on the slopes and the crests of dunes (Chippendale, 1963; Speck, 1963). The understory is always dominated by hummock grasses, *Triodia pungens* in the north and *T. basedowii* in the south. Scattered tall shrubs sometimes occur; Speck records *Grevillea juncifolia, Hakea lorea, H. multilineata, H. rhombalis* and the shrubs *Eremophila leucophylla* and *Grevillea pinifolia*.

v. *Eucalyptus striaticalyx* (Kopi Gum) is a mallee, mostly c. 5 m high, or a small tree to 13 m and follows drainage lines in the arid regions of Western Australia. No distinctive set of understory species is associated with the eucalypt. In mulga country *E. striaticalyx* may form small elongated areas of low woodland c. 8 m tall with a sparse tall shrub layer of *Acacia aneura* and a sparse shrub layer of *Cratystylis subspinescens, Plagianthus helmsii* and *Cassia* spp. (Speck, 1963). In watercourses which traverse saltbush country, the understory consists of species of *Atriplex* and *Maireana. E. striaticalyx* has been recorded also on whitish kopi (gypsum) dunes with *Plagianthus helmsii* and saltbush below (Beard, n.d.).

vi. *Eucalyptus lucasii* (Barlee Box) is confined to arid Western Australia and grows to a small

Fig. 14.21. *Eucalyptus gillii* with an understory of *Maireana pyramidata, M. sedifolia, M. triptera, Atriplex vesicaria, Babbagia acroptera* and annual *Atriplex angulata* and *A. inflata*. c. 115 km north of Broken Hill, N.S.W.

tree 13 m high, but in drier sites develops a mallee habit. It is found as scattered individuals or small clumps within the *Acacia aneura* alliance and occupies slight depressions, whereas the *Acacia* occurs on more elevated sites (Speck, 1963). The mallee clumps sometimes contain *E. oleosa. E. lucasii* occurs also at the northern fringe of the main southern mallee region and may be associated with *E. loxophleba, E. oldfieldii* and *E. trivalvis* (Hall and Brooker, 1973).

vii. *E. gillii* (Curly Mallee) is recorded mainly in the Flinders and Barrier Ranges. It is a mallee or small tree to c. 10 m high, with opposite, very glaucous adult foliage. It is found mainly in rocky terrain, on alkaline soils, either Lithosols derived from limestone, or on travertine which lies close to the surface. The mallee stands are mostly 3–4 m high but low woodlands to 10 m tall occur near creeks. Individual stands occupy only a few hectares and are surrounded by scrubs of *Acacia aneura* (usually on slightly acid soils). In New South Wales the understory is composed almost entirely of Chenopodiaceae (Fig. 14.21).

viii. *E. carnei,* a small tree to 9 m high, is restricted to a small area in Western Australia. It occurs on steep slopes among rocks on hill-sides (ironstone or sandstone) usually in clumps in the *Acacia aneura* alliance. Speck (1963) records *Dodonaea* sp., *Eremophila latrobei* and *Cassia nemophila* as shrub associates, with a sparse grass flora dominated by *Eriachne mucronata.*

ix. *E. trivalvis,* a mallee to c. 7 m, is dotted across the western part of the arid zone. It occurs mainly in the *Acacia aneura* alliance, sometimes in association with other eucalypts, especially *E. oldfieldii, E. oleosa* and *E. lucasii* (Hall and Brooker, 1973). It is recorded on limestone rises south-west of Alice Springs with an understory of *Triodia longipes.*

15. *Eucalyptus* Forests and Woodlands in the South-West

1. Introduction

Eucalyptus forests and woodlands occur in the south-west in the triangular segment of the continent bounded in the east by the arid zone. The wettest country lies in the extreme south-west where the mean annual rainfall exceeds c. 1200 mm in a few small areas. However, forests and woodlands do not form a continuous cover over this triangular segment, but low soil fertility levels over considerable areas restrict the heights of communities which are reduced to mallee or heath proportions, as discussed in Chap. 6. Consequently the *Eucalyptus* forests and woodlands are located in two main areas, the wetter south-west and the drier east, the two areas being separated by mallee and heath communities, sometimes with a sprinkling of woodland stands where soil fertility levels are high enough to support taller communities.

In the wettest areas soil parent material determines the patterning of the forest stands. On the more fertile soils derived from granite two tall forests occur over relatively small areas (*E. diversicolor* and *E. jacksonii*). On laterite and lateritic strew *E. marginata* and *E. calophylla* are dominant and on the coastal limestones, *E. gomphocephala*. With diminishing mean annual rainfall these taller forests are replaced by other forests or woodlands. On the coastal limestones, *E. erythrocorys* may be regarded as the dryland replacement of *E. gomphocephala*. Elsewhere, *E. wandoo* and its associates and *E. loxophleba* replace the taller, wetter forests and in still drier areas *E. salmonophloia* – *E. salubris* and *E. lesouefii* – *E. dundasii*. Two alliances are controlled by soil waterlogging – *E. cornuta* and *E. occidentalis*. A third species, *E. rudis*, associated with watercourses or waterlogged soils occurs in the area and is mentioned in this chapter only in ecotonal associations; it is dealt with in more detail in Chap. 21.

A general account of the forests has been published by Gardner (1942) and Speck (1958) discusses interrelationships in some detail. The distributions of the various dominants are shown in Fig. 15.1. The alliances and main associations are listed in Table 15.1.

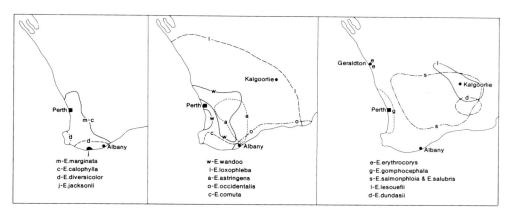

Fig. 15.1. Approximate areas of species of *Eucalyptus* which dominate forests or woodlands in south-western Western Australia.

Table 15.1. Associations of eucalypts in the forests and woodlands in south-western Western Australia.

Alliance or Suballiance	Main associations	Species forming minor associations with the dominants	Species forming ecotonal associations
Eucalyptus diversicolor	E. diversicolor	E. guilfoylei	E. marginata; E. calophylla; E. jacksonii; E. megacarpa; E. patens
E. jacksonii	E. jacksonii	E. guilfoylei	E. calophylla
E. marginata – E. calophylla	E. marginata – E. calophylla; E. marginata; E. calophylla; E. megacarpa; E. patens	E. haematoxylon; E. decipiens; E. staeri; E. lane-poolei; E. laeliae	E. diversicolor; E. wandoo; E. gomphocephala; E. cornuta
E. wandoo	E. wandoo	E. decipiens	E. marginata; E. calophylla; E. loxophleba; E. salmonophloia
E. accedens	E. accedens	E. laeliae; E. gardneri	E. calophylla; E. astringens
E. astringens	E. astringens	E. falcata	E. accedens; E. wandoo
E. gomphocephala	E. gomphocephala	E. decipiens	E. marginata
E. erythrocorys	E. erythrocorys		E. oleosa
E. loxophleba	E. loxophleba ± Acacia acuminata		E. wandoo; E. redunca; E. oleosa; E. celastroides; E. corrugata
E. cornuta	E. cornuta	E. gomphocephala; E. megacarpa; E. rudis; E. lehmannii	E. wandoo; E. loxophleba
E. occidentalis	E. occidentalis	E. decipiens	E. wandoo; E. loxophleba
E. salmonophloia – E. salubris	E. salmonophloia – E. salubris; E. salmonophloia; E. salubris	E. longicornis; E. transcontinentalis	E. lesouefii; E. dundasii; E. kondininesis; E. sargentii
E. lesouefii	E. lesouefii	E. torquata; E. campaspe; E. clelandii; E. corrugata; E. floctoniae; E. griffithsii; E. longicornis; E. transcontinentalis; E. oleosa; E. stricklandii	E. salmonophloia; E. salubris
E. dundasii	E. dundasii	E. brockwayi and the 10 species listed for the previous suballiance	E. lesouefii; E. salmonophloia; E. salubris

2. Descriptions of the Alliances

2.1. *Eucalyptus diversicolor* Alliance

Karri

Karri is one of the tallest eucalypts in Australia, reaching a height of c. 85 m (usually 40–70 m) and a diameter of c. 7 m; it is not buttressed. The tree grows rapidly, annual diameter-increases of 90–130 mm having been recorded. It provides valuable timber, millable into long beams, but is not resistant to termites. It is closely related to *E. deanei, E. grandis* and *E. saligna* which occur in the east (Chap. 9). Karri is restricted to a narrow belt in the south (Fig. 15.1). The mean annual rainfall is c. 1000–1500 mm, and in the lower rainfall portions of its area it occurs only in valleys, and is then usually associated with another eucalypt. It is found almost from sea level to an altitude of c. 300 m. The soils are acid, fine sands to sandy loams derived from granite and are of low nutrient status by agricultural standards.

The forests are two-layered – a towering tree layer 40–70 m high with a more or less closed canopy, and a tall shrub layer 3–7 m high of mesomorphic or semi-xeromorphic plants (Fig. 15.2). In undisturbed areas the shrub layer is so dense that it prevents the regeneration of the eucalypt. The herbaceous layer is poorly developed. Lianas are few, small and thin-stemmed and epiphytes are absent.

E. diversicolor forms pure stands in the wetter areas, except in the extreme south where *E. guilfoylei* (Yellow Tingle), which reaches a height of 40–50 m, is sometimes co-dominant in low-lying situations along watercourses. Ecotonal associations occur mainly at the margins of the

Fig. 15.2. *Eucalyptus diversicolor* forest. Shrub layer containing *Trymalium spathulatum, Acacia pentadenia* and *Casuarina decussata*. c. 48 km south-east of Manjimup, W.A.

forests with species from the adjoining alliances, viz.; *E. calophylla, E. marginata, E. megacarpa* and *E. patens.*

In the shrub layer *Trymalium spathulatum*, if unburned, forms thickets over large areas. Less common shrubs, sometimes locally abundant (depending on the fire-history, see below) are: *Acacia myrtifolia, A. pentadenia, A. pulchella, A. strigosa, A. urophylla, Agonis juniperina* (along creeks as a small tree), *A. parviceps, A. lineariifolia, Albizia distachya, Banksia grandis, Bossiaea aquifolia, Casuarina decussata* (tall shrub), *Chorilaena quercifolia, Chorizema ilicifolia, Crowea dentata, Hibbertia tetrandra, H. montana, H. amplexicaulis, Hovea elliptica, H. chorizemifolia, Leucopogon verticillatus, Logania vaginalis, Persoonia longifolia, Podocarpus drouyniana.* Lianas, with stems 2 to 4 m long are *Clematis pubescens, Hardenbergia comptoniana* and *Kennedia macrophylla.*

Outlier communities, as in the Porongurup Range (Smith, 1962), consist of mixed stands of *E. diversicolor* with *E. calophylla, E. cornuta* and *E. megacarpa*. The stands occur in an area receiving a mean annual rainfall of 850 to 1000 mm, but run-off from the bare granite ranges provides additional water to the valley floor where the communities are located. Most of the shrubs listed above occur also in this area which has, however, a significant addition of xeromorphic shrubs, such as *Mirbelia dilatata* and *Oxylobium lanceolata* and some subshrubs and herbs, e.g., *Scaevola platyphylla* and *Veronica calycina.*

Fig. 15.3. *Eucalyptus jacksonii.* Understory species are *Casuarina fraserana* and *Acacia pentadenia.* Near Nornalup, W.A.

Natural fires probably always passed through the forests, continually changing the pattern of the understory species. Controlled burning is now practised by foresters, partly to protect the trees and partly to ensure regeneration of the eucalypt. *E. diversicolor* does not regenerate from proventitious buds and it produces no lignotuber. Consequently, stands may be destroyed after a severe fire which may eventuate if trash were allowed to accumulate. Controlled burning removes the understory either partially or completely, as well as the trash, but does not harm the trees. Fires are lit when the capsules of Karri are most abundant and ripe, whereby reseeding can be effected. The secondary succession following fires is as follows (J. Havel, pers. comm.): During the first year, the first plants to regenerate are those with rhizomes, which have survived the fire, especially *Pteridium esculentum* and, locally, *Anarthria* sp. and *Lindsaea linearis*. Soon after these have emerged, mats of the moss, *Funaria hygrometrica,* develop on ash-beds, particularly around logs which have been burned. In these moss mats the seeds of karri germinate, producing seedlings several to many cm high in their first year of growth. Seeds of shrubs, many of which have hard testas, germinate also during the first year and become established, but they remain inconspicuous. During the second year seedlings of shrub species grow rapidly and assume dominance, especially *Hovea ovalifolia* and/or *Bossiaea aquifolia*. During the third year, species which were subsidiary during the second year overtop the *Hovea-Bossiaea* shrub canopy and produce a taller shrubbery exceeding 1 m in height, especially *Acacia pentadenia* and/or *Agonis parviceps*. During the fourth year *Trymalium spathulatum* may assume dominance (sometimes only locally) and continues its growth with the ultimate suppression of the other shrub species.

2.2. *Eucalyptus jacksonii* Alliance

Red Tingle

E. jacksonii reaches a height of 70 m (usually 40–60 m) and has a girth of up to 21 m, the trees being buttressed at the base (Fig. 15.3). It has a magnificent timber but supplies are limited, partly by the rarity of the tree and partly because the best stands are preserved in a national park. It dominates tall forests on low hills in valleys along the lower reaches of some of the southern rivers (Fig. 15.1) in frost-free areas with a mean annual rainfall of 1300–1500 mm on deep loams.

The stands are usually pure and with a closed canopy. The understory species are similar to those in the preceding alliance. A tall shrub layer with dominants of *Trymalium spathulatum* and *Acacia pentadenia* occurs when the understory layers are not damaged by fire. Associated tree species are rare. *E. guilfoylei* occurs near streams and *E. calophylla* is a common associate in ecotones.

2.3. *Eucalyptus marginata – E. calophylla* Alliance

Jarrah-Marri

These species form forests and woodlands over most of the south-west between the 650 and 1300 mm isohyets, with a few outliers in drier areas (Fig. 15.1). The alliance occurs on fossil laterite or on the redistributed materials from the weathered laterite profiles, which are either a mixture of lateritic gravel and sand, or sand (Gardner, 1942; Williams, 1932, 48; Speck, 1958).

The alliance is a complex one, in so far as one or other of the characteristic species, or both, are the dominants of communities which range from tall forests to shrublands. Consequently the identification of associations on the basis of dominants alone is impracticable and structure must be used to classify the communities. Furthermore, since the species extend over such a range of climates, communities with the same dominants and similar structure exhibit variations in floristic composition of the understory layers, i.e., wet and dry facies of the same type of forest exist.

The following generalizations are made:

i. *E. marginata* is dominant on laterite *in situ* (Fig. 15.4), with *E. calophylla* as a subsidiary species. The latter becomes more prominent on the lateritic strew containing a mixture of ironstone gravel and sand, and it is dominant on redistributed sand (Fig. 15.5) except when the sands are of such low fertility that they support *Banksia* scrubs or heath (Chap. 16).

Fig. 15.4. Pure stand of *Eucalyptus marginata* (dry facies). Understory species: *Xanthorrhoea preissei, Macrozamia reidlei, Bossiaea linophylla* and *Persoonia longifolia*. c. 30 km north-east of Manjimup, W. A.

ii. The tallest forests occur on the fossil laterite profiles, whereas the communities on sands are usually woodlands. The heights of both forests and woodlands decrease with decreasing mean annual rainfall.

iii. The understory species are governed by the same factors. Thus, assemblages on soils containing ironstone are different from those on sand (some species may be common to the two types), and assemblages in wetter areas are different from those in drier areas.

The alliance contains a surprisingly small number of species of *Eucalyptus* in comparison with the eastern xeromorphic forests (Chap. 10). Only seven other species occur within the alliance. All are relatively uncommon and have restricted distributions, and only two are used to identify associations (below). The others are mentioned here: *E. haematoxylon* (Mountain Gum), a crooked tree to 9 m high, resembling *E. calophylla,* occurs usually as an understory species in the forests in a restricted area in the south (Fig. 15.6). *E. decipiens* (Redheart), a tree to 8 m high, occurs mainly on the eastern fringe of the alliance, often in the ecotone with *E. wandoo*; it occurs on lower slopes in wet situations and is more abundant in the *E. gomphocephala* forests (next section). *E. staeri* (Albany Blackbutt), a tree to c. 15 m high, resembling *E. marginata,* is recorded from the alliance on swampy flats in the Albany-Denmark district. *E. lanepoolei* (Salmonbark Wandoo), a tree to c. 13 m

Fig. 15.5. *Eucalyptus calophylla* forest with an understory dominated by *Bossiaea aquifolia* and *Logania vaginalis*. c. 16 km south-west of Manjimup, W. A.

high, occurs on sands or gravelly sands in open forest formation fringing forests of *E. marginata*, *E. calophylla* or *E. wandoo* on the western coastal plain. *E. laeliae* (Darling Range Ghost Gum) is confined to the Darling Ra., extending to the west coast (Hall and Brooker, 1973).

E. marginata attains a height of c. 40 m and a diameter of 2 m on laterite in the wettest part of its range. Elsewhere it is smaller and it may even assume a mallee form in the least favourable sites. The timber is renowned for its variety of uses, from the heaviest construction work to housebuilding and cabinet making; it is resistant to termites (Willis, 1972). The largest trees have nearly all been felled and replaced by natural regeneration by smaller trees of the same species or by *E. calophylla*. Infection of the roots by the fungus *Phytophthora infestans*, resulting in morbidity in some stands is causing concern among foresters (Ecos, 1978). *E. calophylla* reaches a height of c. 40 m and a diameter of 1.6 m, the tallest trees, with long boles occurring in the wetter south in the ecotone with *E. diversicolor*. Smaller trees of forest form, woodland trees, as well as shrub and mallee forms are found in progressively less favourable sites. The timber is of little commercial value mainly because of the frequency of gum pockets (kino). Consequently it is not felled for timber and mixed stands of jarrah-marri are gradually being converted to pure stands of marri by the removal of jarrah.

2.3.1. Forest Stands of *E. marginata* and *E. calophylla*

Forests up to 40 m high, with an almost closed canopy, occur in the wetter areas on fossil laterite profiles. *E. marginata* usually occupies the less eroded ridges with *E. calophylla* as a subsidiary species. The latter becomes more abundant downslope. In drier areas the forests reach a height of 12–24 m and are more open. Three types of understory occur:

i. In the wettest areas of tall forest, especially in valleys, a lower tree layer up to 13 m high of *Banksia grandis*, *Casuarina fraserana* and *Persoonia elliptica* occurs, and, rarely, *Agonis flexuosa* in the extreme south (Smith, 1972, 73). Below this is a shrub layer 2–3 m high of mesomorphic or semi-xeromorphic species, many of which are present also in the *E. diversicolor* forests, especially *Acacia pentadenia*, *Bossiaea aquifolia*, *Hovea elliptica*, *Leucopogon verticillatus*, *Logania vaginalis*, *Podocarpus drouyniana* and the liana, *Clematis pubescens*.

ii. Xeromorphic shrubs dominate in most of the upland areas in the wetter south-west (listed by Speck, 1958), the species mentioned above occurring in wetter pockets. Microcommunities occur, being delimited by topography, soil depth, the proportion of sand to lateritic gravel, and fire. The main genera are *Banksia*, *Dryandra*, *Grevillea*, *Hakea*, *Isopogon*, *Petrophile*, *Acacia*, *Daviesia*, *Hovea*, *Eriostemon*, *Bossiaea*, *Gompholobium*, *Styphelia*, *Xanthorrhoea* and many others.

iii. In the drier segment of the association xeromorphic shrubs form an open layer. The same genera occur as in wetter areas but the species are different, though a few occur throughout. *Xanthorrhoea preissei* is commonly conspicuous as a tall shrub (Fig. 15.4), and some common shrub species are: *Acacia extensa*, *A. pulchella*, *Bossiaea liniphylla*, *Burtonia scabra*, *Hakea ruscifolia*, *Hypocalymma angustifolia*, *Leptomeria cunninghamii*, *Leucopogon capitellatus*, *Melaleuca thymoides*, *Persoonia longifolia*, *Petrophile serruriae*, *Phyllanthus caly-*

Fig. 15.6. *Eucalyptus haematoxylon* forming an understory in *E. marginata* forest. 32 km west of Nannup, W.A.

cinus and the herbs: *Conostylis setigera* and *Lepidosperma angustata*. In this dry facies *E. wandoo* is found on flats subject to minor waterlogging.

2.3.2. Woodland Stands of *E. marginata* and *E. calophylla*

These woodlands, 10–15 m high, are found mainly on the western coastal on stabilized dune sands which present an undulating landscape with dune crests and troughs. The two species are usually segregated into zones, the former on the high ground, the latter dominating the lower slopes. The troughs are usually swampy, commonly supporting a scrub of *Melaleuca raphiophylla* surrounded in some cases by *Eucalyptus rudis*, the latter forming ecotonal associations with *E. calophylla*.

The woodlands of *E. calophylla*, now mostly cleared for cropping and grazing, contain several tall shrub or small tree species, especially *Nuytsia floribunda* (a root parasite), *Banksia attenuata* (common), *B. grandis* (occasional) and, near the coast, *Agonis flexosa*.

2.3.3. Stunted Communities

In drier areas, in ecotones with mallee heaths and near the sea, both *E. calophylla* and *E. marginata* become stunted.

E. calophylla c. 2 m high and mallee-like (possibly reduced to this condition in some areas by repeated burning) forms the dominant over small areas in mallee-heath, sometimes in association with *E. todtiana* (Chap. 14). A similar formation is seen on shallow skeletal soils on granite or sandstones (e.g., Stirling Ranges), *E. redunca* and *E. decurva* sometimes being associated species. Also, near the sea, as on the south coast. *E. calophylla* forms wind-shorn mallee-like thickets.

E. marginata is reduced to tall shrub proportions on shallow soils on lateritic plateaux, or on granite or sandstones and near the sea. On a lateritic plateau, Smith (1972) describes a shrubland of *E. marginata* associated with *Kingia australis, Banksia grandis, Dasypogon hookeri* and *Casuarina humilis*. An extreme example of stunting is provided by the mallee-heath described by Beard (1972) near Cape Riche on the south coast. Here, *E. marginata* 2–3 m high dominates a mallee-heath in which *E. preissiana* and *E. tetragona* occur as emergents. The heath species belong mainly to *Banksia, Casuarina, Daviesia, Dryandra, Grevillea, Kingia* and *Xanthorrhoea*. The community occurs on bleached sand overlying an ironstone layer in an area with mean annual rainfall of c. 635 mm. On deeper sand *E. marginata* forms a low woodland with an understory of *Banksia* spp.

2.3.4. The Effect of Fire

The forests are periodically swept by fire. Fossil pollen data, correlated with radiocarbon dating of charcoal fragments indicate that the forests have been periodically burned over the past 7000 years (Churchill, 1968). All species appear to have some mechanism for regenerating after fire-damage (Christensen and Kimber, 1975), similar to those exhibited by the species comprising the xeromorphic forests on sandstones in the east (Chap. 10), viz.: epicormic shoot from trunks and lignotubers, underground perennating organs (mainly monocotyledons), caudices (*Macrozamia, Kingia,* and *Xanthorrhoea*), woody fruits (Myrtaceae and Proteaceae), hard seeds (*Acacia,* Papilionaceae) etc.

Many herbaceous species conspicuous after a fire sometimes form local societies before the taller species regenerate, e.g., ground orchids, *Burchardia umbellata* and *Eryngium paniculatum*. These plants regenerate rapidly and assume temporary dominance in the absence of taller competitors. Their growth-rate is enhanced by the increased light intensity and the ready supply of nutrients provided from the ash of the burned vegetation. The societies are supplanted by the taller species as they regenerate.

E. marginata is fire-sensitive only at one stage of its growth. Seedlings form a lignotuber which produces a large number of shoots, the young plant being a bush, which is fire resistant. When the lignotuber attains a diameter of c. 10 cm it supports 6–7 stems each c. 80 cm long. The plant remains in this condition for c. 15 years (up to 50 years in the poorest sites). However, one shoot ultimately gains ascendency and develops into a lank leader, which over a period of 5–10 years reaches a height of 6–10 m. This leader has green fleshy bark which is fire sensitive, and when it develops its thick fibrous bark the trunk becomes fire-resistant (Forest Focus, 1971).

2.3.5. Other Associations

2.3.5.1. *Eucalyptus megacarpa* Association
Bullich

E. megacarpa is a smooth-barked tree reaching a height of 25–30 m. A mallee form occurs in the Stirling Ranges and eastward to the Barren Hills. The tree form, which has a limited distribution in the wetter south, forms tall forests, usually in pure stands, but sometimes with admixtures of *E. patens, E. decipiens* or *E. cornuta*. *E. megacarpa* is restricted to swampy soils (Fig. 15.7) and sometimes fringes swamps supporting *Melaleuca preissei* and/or *Banksia littoralis* which adjoin streams. The understory species in these forests are those found in the moist *E. marginata* forests.

2.3.5.2. *Eucalyptus patens* Association
Swan River Blackbutt

E. patens is a rough barked tree usually 20 to 30 m high (up to 50 m) forming forests on pockets of more fertile soil in the moister habitats of valleys under a mean annual rainfall of 500 to 1250 mm. A shrubby form of the species is recorded in the south from Albany to Esperance. Pure stands are uncommon, the species occurring usually in association with *E. marginata, E. megacarpa* or *E. diversicolor*. It has similar edaphic requirements to the last species, but a lower water requirement, so that it extends northward into country too dry to support *E. diversicolor*. In the drier parts of its range *E. patens* is restricted to valleys which receive additional water from runoff, where the forests sup-

Fig. 15.7. *Eucalyptus megacarpa* and *E. calophylla* behind. Understory dominated by *Agonis parviceps* and *Acacia decipiens*. c. 15 km south-west of Manjimup, W.A.

port understory layers similar to those of the moist *E. marginata* forests.

2.4. *Eucalyptus wandoo – E. accedens – E. astringens* Alliance

These species occur mostly between the 900 and 500 mm isohyets, forming forests and woodlands which adjoin or form mosaics with the *E. marginata – E. calophylla* Alliance (Gardner 1942; Speck, 1958; Smith 1972). Three suballiances are recognized.

2.4.1. *Eucalyptus wandoo* Suballiance
Wandoo

E. wandoo is a tree reaching a height of 30 m. A mallee form also exists which is referred to *E. redunca*. In the wetter part of its range *E. wandoo* forms forests mostly 20–25 m high and in drier areas it forms woodlands 10–15 m high. The timber is valued for heavy and light construction work; it contains a high percentage of tannin, which is extracted commercially, the wood being chipped, the tannin extracted with hot water and recovered by evaporation of the liquors.

The forests occur under a mean annual rainfall of 600–700 mm mainly on clayey soils

Fig. 15.8. *Eucalyptus wandoo* forest. Understory species are *Xanthorrhoea preissii*, *Hibbertia hypericoides* and *Dryandra nivea*. 48 km east of Midland, W. A.

Fig. 15.9. *Eucalyptus accedens* woodland with the understory dominated by *Gastrolobium calycinum*. Near Narrogin, W.A.

derived from dykes of epidiorite, which traverse the basal granite of the Darling scarp, or on clay alluvium (Fig. 15.1). *E. wandoo* forms pure stands (Fig. 15.8) or occurs in ecotonal associations with *E. marginata* or *E. calophylla*. The shrub layer(s) are similar to the dry facies of the jarrah-marri forests, xeromorphic shrubs dominating, with *Xanthorrhoea preissei* and *Macrozamia reidlei* often prominent. A typical understory assemblage east of Manjimup contains: *Astartea fascicularis, Dryandra nivea, Hakea prostrata, Hypocalymma angustifolia, Leucopogon parviflorum* and *Melaleuca preissei* (local, in depressions). The ground layer consists of Liliaceae, Amaryllidaceae and Orchidaceae. Often there are bare pathes with hard, glazed surfaces. Associations with other eucalypts, apart from ecotones are rare. *E. decipiens* occurs occasionally on "wandoo flats" where water accumulates, sometimes with an understory of *Banksia littoralis*.

Woodlands of Wandoo occur both in the drier east and on the coastal plain, usually forming mosaics with other communities, e.g. with *E. calophylla* and less commonly with *E. marginata, E. loxophleba* and *E. salmonophloia*. On the drier coastal plain supporting heath (Chap. 16), *E. wandoo* forms elongated stretches of woodland on clayey soils which are flanked on either side by sands supporting heaths. The understories of the woodlands are always xeromorphic, varying with mean annual rainfall and usually containing species of the genera *Acacia, Baeckea, Casuarina, Dryandra, Grevil-*

lea, *Hakea, Hibbertia, Hypocalymma, Melaleuca* and *Pimelea* (shrubs) and the herbs and subshrubs *Borya, Burchardia, Caladenia, Dampiera, Diurus, Schoenus, Stylidium, Tribonanthes* and *Waitzia*. Outliers of *E. wandoo* woodland sometimes occur in drier country to the east, within the mallee-belt, either in shallow watercourses or at the bases of granite outcrops.

2.4.2. *Eucalyptus accedens* Suballiance
Powderbark Wandoo

E. accedens occurs between the 500 and 850 mm isohyets usually on gravelly (ironstone) clays, often on hills among boulders of laterite, where it forms woodlands 15–20 m high on shallower soils and forests to 30 m high on deeper soils. The species occurs mostly in pure stands (Fig. 15.9), but ecotonal associations are formed with *E. calophylla*. The understories are xeromorphic and often in 3 layers. The tallest layer is discontinuous, consisting of *Casuarina fraserana* and *Dryandra nobilis*, with *Xanthorrhoea preissii* often prominent. A shrub layer 1–2 m high and more or less continuous is present and its composition appears to be controlled by fire. The poisonous *Gastrolobium calycinum* is sometimes locally dominant with subsidiary *Dryandra, Hakea, Isopogon, Synaphaea* spp. The herb and subshrub layer contains *Macrozamia reidlei* as a conspicuous member, and species of *Conostylis, Dampiera, Leschenaultia, Patersonia, Pimelea, Stylidium* and Restionaceae.

Other eucalypts which belong to this subassociation are *E. laeliae,* which is closely related to *E. accedens,* occurring in small pure stands on laterite-free soils in the Darling Range. It reaches a height of c. 20 m and is conspicuous because of its vivid white, powdery trunk. *E. gardneri,* a slender, smooth-barked tree to 12 m high is an occasional associate.

2.4.3. *Eucalyptus astringens* Suballiance
Brown Mallet

E. astringens occurs between the 550 and 350 mm isohyets and is a tree to 18 m high in wetter areas and a mallee in the drier east. It grows either on shallow lateritic soils on hills in the north, or on clays with or without ironstone gravel on flat areas in the south. The tree is of value commercially because of the high tannin content of its smooth bark (40–57%) which is stripped off during the winter when it has a high moisture content. Trees are being cultivated in plantations for tannin production (Forest Focus, 1972).

E. astringens dominates open woodlands with a xeromorphic shrub layer. *Gastrolobium* spp., poisonous to stock, are common in the understory in some of the northern stands, which are notorious "poison country".

2.5. *Eucalyptus gomphocephala* Alliance
Tuart

E. gomphocephala, a tree to 40 m high, is confined to soils derived from limestone or calcareous sands in a narrow belt on the west coast (Fig. 15.1), where the mean annual rainfall is 750–1000 mm. It occurs from sea-level to an altitude of c. 30 m, the areas being frost-free. The species produces no lignotuber. The forests which it dominates are flanked on the seaward side by the hind-dune coastal scrubs (Chap. 23) and inland by the *E. marginata – E. calophylla* Alliance.

E. gomphocephala usually occurs in pure stands but occasionally is found with *E. decipiens* and/or *E. cornuta* which rarely reach the same height as *E. gomphocephala*. The alliance is divisible into two segments on the basis of associated understory species:

i. In the south, and best developed around Wonerup and Ludlow, pure stands 40 m high occur, with a small tree layer of *Agonis flexuosa* often with a continuous canopy (Fig. 15.10). Other species are rare and make up an insignificant fraction of the total biomass of the community. The most common are *Hibbertia hypericoides, Acacia saligna, Phyllanthus calycinus,* and *Stipa compressa* dominates the discontinuous herbaceous stratum. Close to the sea, species from the hind-dune scrubs sometimes occur in the understories, e.g., *Acacia rostellifera, Myoporum serratum, Olearia axillaris* and *Rhagodia baccata*.

ii. In the northern segment, *Agonis* is absent (it extends to c. 5 km S. of Perth) and is replaced by small trees of *Banksia attenuata, B. men-*

Fig. 15.10. *Eucalyptus gomphocephala* with *Agonis flexuosa* forming a tall shrub layer. Also present are *Hibbertia hypericoides* and *Stipa compressa*. Introduced grasses in foreground. 6 km north of Busselton, W. A.

ziesii, *B. grandis* and *Casuarina fraserana*. Xeromorphic shrubs dominate the understory. Most of these shrubs are confined to the limestone area but some are present also in the *E. marginata* forests which adjoin the *E. gomphocephala* forests. The species are as follows (those extending to the *E. marginata* forest are marked (x)): Tall shrubs, *Dryandra floribunda* (x), *Grevillea vestita*, *Hakea prostrata*, *Jacksonia furcellata*, *J. sternbergiana*, *Xanthorrhoea preissii* (x); shrubs to 1 m, *Acacia pulchella* (x), *Casuarina humilis* (x), *Dryandra nivea* (x), *Grevillea thelemanniana*, *G. vestita*, *Hibbertia hypericoides* (x), *H. racemosa*, *Logania vaginalis* (x), *Melaleuca acerosa*, *Oxylobium capitatum* etc., and the small vines *Clematis microphylla* (x) and *Hardenbergia comptoniana* (x). The discontinuous herbaceous layer is composed of the genera *Daucus*, *Eryngium*, *Lobelia*, *Helichrysum*, *Waitzia* and others, with a poor representation of monocotyledons, e.g., species of *Schoenus*, *Scirpus* and *Mesemolaena* (Cyperaceae), and the grasses, *Amphipogon spp.* and *Stipa compressa*.

The forests sometimes adjoin stony areas where limestone comes to the surface and on these rocky outcrops heaths develop, and these contain many of the species comprising the understory of the *E. gomphocephala* forests (Chap. 16).

2.6. Eucalyptus erythrocorys Alliance

Illyarrie

E. erythrocorys is a smooth-flaky-barked tree and is unique among the eucalypts because of its orange-red operculum and yellow stamens connate in 4 bundles. It is restricted to a narrow strip of coastal limestone 5–15 km wide near the coast from south of Geraldton to lat. c. 31°S. (mainly Dongara district) (Fig. 15.1), the area receiving a mean annual rainfall of 300 to c. 450 mm.

Beard (1976) records that it grows where there is no continuous sand cover, the trees being rooted in deep crevices in the limestone. The species is fire sensitive and this rocky habitat may protect it from fire. According to Beard, it "forms isolated groves, growing as a small, misshapen tree up to 8–10 m. The groves have an open shrub understory containing *Acacia blakelyi, A. pulchella, A. spathulata, Dryandra sessilis, Hakea costata, Hibbertia hypericoides, Hibiscus huegelii, Hybanthus calycinus, Scholtzia.* sp. and Restionaceae".

The surrounding vegetation (on deeper sands) is either thickets of *Acacia, Melaleuca,* and *Eucalyptus oleosa* (north of the Arrowsmith R.), or, in the south, in the south scrub-heath with dominants of *Beaufortia squarrosa* and *Banksia sphaerocarpa.*

2.7. *Eucalyptus loxophleba* Alliance

York Gum

E. loxophleba is a rough-barked tree reaching a height of 15 m, when it forms open woodlands. It occurs also in mallee form (mostly 3 to 4 m high) and is often a co-dominant in the mallee shrublands (Chap. 14). The species spans the climatic range 600 to 250 mm mean annual rainfall (Fig. 15.1). In woodland form, however, it is found mainly between the 480 and 300 mm isohyets, between the *E. wandoo* woodlands in the west and the *E. salmonophloia – E. salubris* woodlands in the east. The woodlands vary in height from 15 m in the wetter areas to 8 m in the drier, extending over level and undulating country. The soils are mainly Solodized – Solonetz derived largely from granite or alluvium. The surface horizon, usually a sandy loam, slightly acid and grey to grey-brown, overlies clay subsoils. The latter are acid in wetter areas but slightly alkaline in the drier segments of the woodlands.

E. loxophleba usually occurs in pure stands and in the wetter parts *Acacia acuminata* (Jam) is a constant associate but rarely attains the same height as the eucalypt (Fig. 15.11). *Casuarina fraserana* and *Acacia saligna* are rare tall shrubs. Smaller shrubs are uncommon and include species of *Grevillea, Hakea, Hibbertia* and *Hypocalymma.* A discontinuous herbaceous layer of perennial monocotyledons occurs, including a few grasses (*Amphipogon* and *Poa*) and Liliaceae, Cyperaceae and Orchidaceae. Annual herbs are mainly Compositae. Most stands are now strongly influenced by introduced plants, including the common grasses (*Avena, Vulpia*) and common weeds (*Romulea, Taraxacum*).

With decreasing rainfall the tall shrub layer becomes very discontinuous and the herbaceous layer is composed of species typical of the semi-arid regions, e.g., *Bassia diacantha, Salsola kali* and species of *Atriplex* and *Zygophyllum.* At the fringe of the arid zone, *E. loxophleba* as a small tree, forms a discontinuous upper story in shrub woodlands in which *Melaleuca uncinata* and *Casuarina campestris* dominate. *E. loxophleba* occurs also in the mallee and in some of the mallee-stands it forms a small tree overtopping the other eucalypts to form a type of mallee-woodland with *E. redunca, E. celastroides, E. oleosa* and *E. corrugata* in the lower layer. *Acacia acuminata* persists in these stands and a dense shrub layer is present, consisting of species of *Acacia, Alyxia, Dodonaea* and *Eremophila,* with an ephemeral herbaceous layer of Compositae.

2.8. *Eucalyptus cornuta* Alliance

Yate

This species is relatively uncommon, occurring only in scattered stands in the south (Fig. 15.1) between the 700 and 1370 mm isohyets on the mainland (westward of Esperance) with occurrences on the Recherche Islands (Willis 1953, 1959). The species, valued for its extremly strong timber formerly used mainly for wheel spokes and shafts for carts, may reach a height of 22 m but is usually much less, especially in drier areas where it is described as a marlock (Chap. 16). It occur mostly on soils derived from granite (Gardner, 1961) but sometimes on sandy soils overlying other siliceous rocks, or on alluvium. In most, if not all cases, the soils are subject to a short period of winter waterlogging.

Fig. 15.15. *Eucalyptus lesouefii* woodland with an understory of *Eremophila scoparia* and *Lycium australe*. 70 km south of Coolgardie, W.A.

oleosa, *E. redunca*, *E. sheathiana*, and in slightly saline areas *E. annulata*, *E. gracilis* and *E. spathulata*.

iv. In the drier northern and eastern portion of the alliance, or in saline areas, saltbushes and other species typical of the arid zone dominate the shrub layer. The saltbushes usually dominate on flats (Fig. 15.14), the main species being *Atriplex hymenotheca* and *A. nummularia*, with the saltbush-like Compositae, *Cratystylis conocephala*. Scattered taller shrubs 2–4 m high sometimes occur, especially *Eremophila scoparia*, *Pittosporum phylliraeoides* and *Eucarya acuminata*. The herbaceous flora is typically arid and consists mainly of annual or short-lived perennial species of *Artiplex*, *Bassia*, *Helipterum*, *Ptilotus* and *Zygophyllum*.

2.11. *Eucalyptus lesouefii* – *E. dundasii* Alliance

These two species occur as woodlands mostly 15–20 m high between the 250 and 300 mm isohyets in the Kalgoorlie and Norseman districts (Fig. 15.1) (Beard, 1975). Each species has been taken as the characteristic species of a sub-alliance. The soils are Calcareous Red Earths, Solonized Brown Soils, Solodized-Solonetz, Solonchaks or Lithosols, derived mainly from greenstone or alluvium washed from the greenstone ridges. They have in common a high base-status which is associated with a moderate to high clay-content, and some of the soils are moderately saline.

Fig. 15.16. *Eucalyptus torquata* woodland with an understory of *Eremophila scoparia* and *Atriplex nummularia*. 50 km south of Coolgardie, W.A.

2.11.1. *Eucalyptus lesouefii* Suballiance
Goldfields Blackbutt

This species occurs in the northern segment of the alliance either in pure stands (Fig. 15.15) or in association with several other eucalypts, viz., *E. torquata* (Fig. 15.16), *E. campaspe,* and others (see Table 15.1). These species do not always reach the same height as *E. lesouefii* but may occur as a tall shrub layer, sometimes locally associated with sand patches. The woodlands occupy low ridges and flats, or depressions following minor watercourses which empty into saline lakes. Ecotones with *E. salmonophloia, E. salubris* are common.

There are few understory species. *Melaleuca pauperiflora* sometimes occurs in clumps and other species, mostly scattered but sometimes in clumps, are *Atriplex nummularia, Eremophila scoparia* and *Lycium australe,* and in the most saline areas *Arthrocnemum arbuscula* and *Frankenia pauciflora. Ptilotus exaltatus* is the most abundant herbaceous species.

2.11.2. *Eucalyptus dundasii* Suballiance
Dundas Blackbutt

E. dundasii is the dominant tree (Fig. 15.17) around salt lakes in the southern segment of the alliance (Beard, 1975) and at its northern limit it

Fig. 15.17. *Eucalyptus dundasii* woodland. Undershrubs are *Eremophila scoparia* and *Atriplex nummularia*. Near Lefroy, c. 150 km south-west of Coolgardie, W.A.

occurs with *E. lesouefii*. Throughout its area it is often associated with *E. salubris* and in less saline areas it occurs with *E. salmonophloia* and *E. longicornis*. At the bases of greenstone ridges it may mix with *E. brockwayi*, a species restricted to greenstone ridges within the limits of this suballiance. Other associated trees are those listed above for the *E. lesouefii* suballiance.

The woodlands are often dense and approach forest dimensions, when an understory of tall mallee species, especially *E. flocktoniae*, sometimes occurs. The shrub assemblage varies with salinity. On the least saline sites species of *Acacia* are abundant and with increasing salinity these are replaced by *Eremophila scoparia*, *Cratystylis conocephala*, *Atriplex nummularia* and *A. hymenotheca*.

16. Heaths, *Banksia* Scrubs and Related Communities on the Lowlands

1. Introduction

1.1. Definitions

The term "heath" was originally applied to certain genera of the Ericaceae occurring in Europe and, later, was used to describe communities dominated by these plants. The European heaths, mostly less than 1 m tall, dominated by plants with leptophyllous (ericoid) leaves, bear only a superficial resemblance to the analogous communities in Australia where species of the Epacridaceae, closely related to the Ericaceae (Chap. 4) are common or dominant in some heaths.

In Australia, the ericoid low shrublands are more commonly composed of species of other ericoid or aphyllous genera (e.g. *Leptospermum* and *Casuarina*) and they commonly contain species with very much larger leaves, even notophylls, which are usually species of *Banksia*. The heights of the communities vary considerably. Stands less than 1 m tall are common,

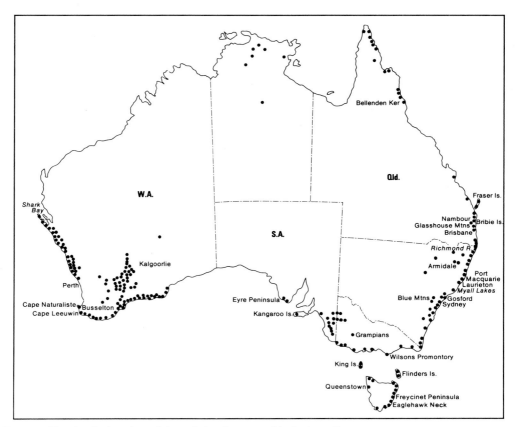

Fig. 16.1. Showing the locations of the main heath communities in Australia.

but in many cases the communities average 1–2 m tall and many, especially those dominated by *Banksia,* are woodlands. Most of the heaths and scrubs occur in areas which are climatically wet enough to support *Eucalyptus,* the development of trees being prevented by a limiting soil property. However, stunted eucalypts are commonly present, the plants usually being mallees or mallee-like and these give the distinctive name "mallee-heath" to the stands (Chap. 14). Nearly all the eucalypts also occur in the *Eucalyptus* forests, woodland or mallee communities, one exception being *E. conglomerata* (Swamp Stringbark) which is restricted to a small area near on coastal flats 60–150 km north of Brisbane where it occurs as a small crooked tree in scrubs with species of *Banksia, Leptospermum* and *Melaleuca* (Hall and Brooker, 1974).

1.2. Location and Habitats

The heaths and scrubs occur mainly in wetter lowland regions near the coast (excluding those on the highlands which are dealt with in Chap. 17). However, some taxa have adapted to semi-aridity or even aridity, so that inland extensions are to be found remote from the coast, especially in Western Australia, with smaller occurrences in South Australia and Victoria. They are best

Table 16.1. Species of *Banksia, Casuarina* and *Leptospermum* which occur commonly in heaths, scrubs or thickets in eastern Australia. Species listed alphabetically in groups according to district and/or habitat.

Species on sands	Species on rock exposures		Species on sands near coast
		5. Glasshouse Mts., Qld (trachyte) *B. absent* *C.* sp. *L. brachyandrum*	1. Tropical Qld *B. dentata* *C. littoralis* *L. fabricia*
	9. Warrumbungle Mts. Trachyte *B.* absent *C.* aff. *distyla* *L.* absent	6. Granites, north-east N.S.W. *B. collina* *B. marginata* *C. rigida* *L.* sp.	2. Southern Qld and Northern N.S.W. *B. aspleniifolia* *B. ericifolia* *B. integrifolia* *B. robur* *B. serratifolia* *B. spinulosa* *C. distyla* *C. littoralis* *C. paludosa* *L. attenuatum* *L. flavescens* *L. liversidgei* *L. semibaccatum*
		7. Sandstones, Central N.S.W. *B. ericifolia* *B. marginata* *B. serrata* *B. spinulosa* *C. distyla* *C. nana* *C. paludosa* *L. arachnoides* *L. parviflorum* *L. squarrosum*	3. Southern N.S.W. and eastern Vic. *B. marginata* *C. paludosa* *C. pusilla* *L. juniperinum* *L. myrsinoides*
Species on sands 10. S.A. *B. marginata* *B. ornata* *C. muellerana* *C. paludosa* *C. pusilla* *L. coriaceum* *L. juniperinum* *L. myrsinoides*		8. Granites, Wilson's Promontory *B. marginata* *C. paludosa* *L. juniperinum* *L. myrsinoides*	4. Tas. *B. marginata* *C. monilifera* *L. glaucescens* *L. grandiflorum* *L. lanigerum* *L. nitidum* *L. scoparium*

Table 16.2. Common species occurring in various heaths and scrub-heaths in south-western Western Australia.
Sources of data: Speck (1958); Seddon (1972); Beard (1976).

	On Calcareous Sands. West Coast (Hill R. to Fremantle)	*Banksia ashbyi* Assemblage	*Actinostrobus arenarius* Assemblage	Heaths on siliceous sands. Western coastal plain	Heaths on fossil laterite. Western Coastal plain	On white siliceous sands. South Coast	On yellow siliceous sands and earthy sands. Inland
CASUARINACEAE							
CASUARINA		*acutivalvis*	*acutivalvis, humilis*	*humilis*	*humilis, microstachya*	*dielsiana, humilis, microstachya, thuyoides*	*acutivalvis, campestris, corniculata, dielsiana, helmsii, humilis, microstachya, pinaster*
DILLENIACEAE							
HIBBERTIA	*aurea, hypericoides, polystachya, racemosa*		*acerosa, conspicua, spicata*	*huegelii, hypericoides*	*aurea, hypericoides, polystachya*	*racemosa, stricta, teretifolia*	*stricta, uncinata*
EPACRIDACEAE							
ANDERSONIA					*aristata*	*brevifolia, caerulea, parviflora, simplex*	
ASTROLOMA	*ciliatum, pallidum, richei, pulchellus, propinquus*				*candolliana*	*serratifolium*	
LEUCOPOGON					*plumuliflorus*	*crassifolius, cucullatus, gibbosus, oppositifolius, reflexus*	*crassifolius, dielsianus, fimbriatus, flavescens*
GOODENIACEAE							
DAMPIERA		*spicigera*		*alata*	*lavandulacea*	*oligophylla*	*juncea, lavandulacea, sacculata, wellsiana,*

Taxon							
(continued)					*concinna, pterigosperma, strophiolata*	*formosa, stenosepala, tubiflora* *thesioides*	*pinifolia*
GOODENIA							
LESCHENAULTIA	*linarioides*	*floribunda, juncoides*	*linarioides*	*biloba*			*expansa*
SCAEVOLA	*canescens, thesioides*			*anchusifolia, glandulifera, globulifera*	*globulifera*		
MIMOSACEAE							
ACACIA	*cuneata, diptera, lasiocarpa, pulchella, xanthina*	*ligulata, longispinea, murrayana, ramulosa, stereophylla*	*scirpifolia, signata, spathulata*	*pulchella*	*acuaria, dilatata, multilineata, pulchella, quadrisulcata*	*baxteri, biflora, cupularis, gonophylla, pilosa, pulchella, salicina, strigosa*	*beauwerdiana, fragilis, jutsoni, lasiocalyx, multispicata, neurophylla, resinomarginea, rossei, stereophylla*
MYRTACEAE							
BAECKEA		*grandiflora, pentagonantha, robusta*		*camphorosmae*		*dimorphandra, uncinella*	*crispiflora, fumana, grandibracteata, leptospermoides, preissiana*
BEAUFORTIA							*heterophylla*
CALOTHAMNUS	*sanguineus*	*dampieri, chrysantherus*	*squarrosa, blepharospermus, chrysantherus*		*pinifolius, sanguineus, villosus*	*micrantha, gracilis*	*chrysantherus, quadrifidus*
CALYTRIX	*brevifolia*		*brevifolia, tenuifolia*		*brevifolia*	*decandra, flavescens*	*brachyphylla, breviseta, decandra, megalopetalum, pauciflorum, virgatum*
CHAMAELAUCIUM						*axillare, megalopetalum*	
DARWINIA			*virescens*		*helichrysoides, neildiana*	*diosmoides, vestita*	
EREMAEA	*beaufortioides*	*pauciflora*				*pauciflora*	
HYPOCALYMMA					*angustifolia, tetrapterum*		*pauciflora*

continued next page

Table 16.2 continued

	On Calcareous Sands. West Coast (Hill R. to Fremantle)	*Banksia ashbyi* Assemblage	*Actinostrobus arenarius* Assemblage	Heaths on siliceous sands. Western coastal plain	Heaths on fossil laterite. Western Coastal plain	On white siliceous sands. South Coast	On yellow siliceous sands and earthy sands. Inland
LEPTOSPERMUM					*spinescens*	*spinescens*	*erubescens, roei*
MELALEUCA	*acerosa, huegelii*	*huegelii, nesophila*	*cordata, leiopyxis, megacephala, scabra*	*cordata, platycalyx*	*cordata, platycalyx, radula, scabra*	*nesophila, platycalyx, pulchella, striata, suberosa, thymoides*	*cordata, holosericea, pungens, scabra, subrigona*
MICROMYRTUS		*imbricata*	*peltigera, rosea*			*elobata*	*imbricata, racemosa*
THRYPTOMENE			*denticulata, urceolaris*				*kochii*
VERTICORDIA			*grandis, picta, grandiflora, lepidophylla, stelluligera*		*chrysantha, grandis, picta*	*chrysantha, densiflora, habrantha, insignis, plumosa, roei*	*acerosa, brownii, chrysantha, grandiflora, humilis, picta, prizelii, roei, serrata*
PAPILIONACEAE							
BOSSIAEA	*eriocarpa*						*spinosa*
BRACHYSEMA			*tomentosum*	*preissei*	*biloba*		*chambersii, daviesioides*
CHORIZEMA						*aciculare, cytisoides*	*aciculare*
DAVIESIA			sp.	*incrassata, pectinata*	*horrida*	*brevifolia, colletioides, juncea, obtusifolia, pectinata, rhombifolia, teretifolia*	*croniniana*

GASTROLOBIUM							*spinosum*
GOMPHOLOBIUM	*tomentosum*						
JACKSONIA	*hakeoides*			*floribunda*			*hakeoides, racemosa*
MIRBELIA	*spinosa*			*spinosa*			*floribunda*
PULTENAEA			*bidens, oxylobioides*	*calycinum, spinosum, pycnostachyum, baxteri, polymorphum*	*marginatum*	*capitata, decumbens*	*spinosa, capitata, dasyphylla, georgei*
						capitata, umbellata	
PROTEACEAE							
ADENANTHOS	*sericea*	*acanthocarpus*					*flavidiflora*
BANKSIA	*ashbyi, sceptrum*	*attenuata, menziesii, prionotes, sceptrum, sphaerocarpa*	*attenuata, candolleana, menziesii*	*tricuspis*	*cuneata, dobsonii, sericea*	*baxteri, dryandroides, media, nutans, repens, petiolaris, prostrata, pulchella, speciosa*	*audax, baueri, elderana, laevigata, lullfitzii, media, prostrata, sphaerocarpa*
CONOSPERMUM	*triplinervium*	*stoechadis*	*stoechadis, triplinervium*	*acerosum, brachyphyllum, incurvum, stoechadis, triplinervium*	*brownii, caeruleum, crassinervium, densiflorum, glumaceum, incurvum, nervosum*	*amoenum, caeruleum, distichum, leianthum, teretifolium*	*brownii, stoechadis, teretifolium*
DRYANDRA	*floribunda, nivea, sessilis*	*fraseri, sessilis, shuttleworthiana*	*nivea, carlinoides, floribunda, tridentifera*	*arctotidis, nobilis, bipinnatifida, nivea, carlinoides, fraseri, formosa, kippistiana,*	*armata, cuneata, fraseri, mucronulata, preissii, pteridifolia, quercifolia, runcinata,*		*cirsioides, erythrocephala, pteridifolia, runcinata*

continued next page

Table 16.2 continued

	On Calcareous Sands. West Coast (Hill R. to Fremantle)	Banksia ashbyi Assemblage	Actinostrobus arenarius Assemblage	Heaths on siliceous sands. Western coastal plain	Heaths on fossil laterite. Western Coastal plain	On white siliceous sands. South Coast	On yellow siliceous sands and earthy sands. Inland
GREVILLEA	crithmifolia, thelemanniana, vestita	eriostachya, gordoniana, obliquistigma, rogersiana, stenobotrya	annulifera, biformis, dielsiana, eriostachya, integrifolia, leucopteris, pterosperma	vestita	patens, polycephala, sclerophylla, shuttleworthiana	seneciifolia	
					acerosa, eriostachya, eryngioides, rudis, saccata, tenuiflora, thyrsoides	fasciculata, hookerana, nudiflora, pauciflora, pulchella	asparagoides, alpiciloba, didymobotrya, biformis, eryngioides, excelsior, hookerana, haplantha, integrifolia, pterosperma, rufa, shuttleworthiana, teretifolia, tridentifera
HAKEA	bipinnatifida, prostrata, trifurcata	stenophylla	adnata, falcata	baxteri, candolleana, conchifolia, sulcata	auriculata, candolleana, cristata, erinacea, conchifolia, incrassata, ruscifolia, sulcata, smilacifolia, undulata, trifurcata	adnata, ambigua, baxteri, ceratophylla, cinerea, corymbosa, crassifolia, cucullata, ferruginea, laurina, nitida, obliqua, pandanicarpa, prostrata, trifurcata, varia	crassifolia, falcata, lehmanniana, lissocarpa, multilineata, platysperma, roei, scoparia, strumosa, subsulcata, trifurcata

ISOPOGON		divergens	linearis, tridens	asper, divergens, dubius, linearis, sphaerocephalus, teretifolius	attenuatus, axillaris, baxteri, buxifolius, polycephalus, tridens, trilobus, uncinatus	attenuatus, buxifolius, scabriusculus, teretifolius, villosus
LAMBERTIA			multiflora	multiflora	ericifolia, inermis	
PERSOONIA	angustifolia		saccata		striata	coriacea, saundersiana, teretifolia, tortifolia
PETROPHILE	serruriae	brevifolia, conifera, divaricata, ericifolia, media	circinnata, linearis, microstachya, shuttleworthiana, latifolia	chrysantha, drummondii, ericifolia, media, megalostegia	ericifolia, fastigiata, heterophylla, linearis, phylicoides, squamata, teretifolia	circinnata, ericifolia, phylicoides, semifurcata, seminuda, striata, trifida
STIRLINGIA			simplex	latifolia, simplex	latifolia, tenuifolia	
SYNAPHEA		sp.	polymorpha	petiolaris, polymorpha	favosa, polymorpha	favosa, petiolaris
RUTACEAE						
BORONIA		caerulescens			spathulata	caerulescens, ternata
ERIOSTEMON			spicatum	spicatum		brucei, coccineus

developed in temperate regions, though some occur in the tropics (Fig. 16.1). They have developed in three main habitats: i. on coastal sands, sometimes with extensions into the drier inland; ii. on coastal headlands; iii. on Lithosols on rock exposures, the rocks being siliceous, especially granites, trachytes, quartzites and related igneous rocks, and sandstones or conglomerates.

The heaths and scrubs dealt with in this chapter are delimited either by low soil fertility (on sands) or by shallow soil, both factors preventing the growth of trees or mallees. Patterning within the assemblages results from slight differences in the limiting factors. Also, the depth of the water-table plays an important role in flat sandy areas. With increasing height of the water-table the heaths are infiltrated by helophytes which dominate sedgelands in waterlogged areas, or by species of *Melaleuca* which tolerate waterlogging (Chap. 22). Consequently, ecotones between the heaths and scrubs and other communities occur along gradients of the controlling factors. This necessitates the identification of transitional structural formations, of which Mallee-heath (Chap. 14) and Sedge-heath (see below and Chap. 20) are the most commonly encountered.

1.3. Flora

Floristically the heaths and scrubs are very rich, especially in the south-west. The main taxa belong to the Australian xeromorphic segment of the flora, mainly Myrtaceae, Proteaceae, Papilionaceae, and Rutaceae with a smaller component from other families, especially Casuarinaceae, *Acacia* and Euphorbiaceae. The Epacridaceae is usually present but rarely dominant and the Ericaceae is represented in only one tropical community (2.2.2, below). Of the herbaceous families, the Cyperaceae and Restionaceae are the most important; Gramineae is poorly represented and often absent. Ground orchids are usually present and often abundant.

Genera which characterize most of communities are *Banksia*, *Casuarina* and *Leptospermum*. The species of each genus vary from region to region (Tables 16.1 and 16.2). An outstanding difference between the eastern and western assemblages is the prominence of *Leptospermum* in the east and its virtual absence in the west.

The communities are described below on a regional and/or habitat basis from north to south and east to west, and within a region from wet to dry.

2. Communities in the Tropics

2.1. General

In the tropics, heaths occur on coastal sands adjoining the littoral dunes, the communities being analogous to those fringing the coast in the south. In addition, several communities on shallow soils on the tops of hills or mountains have been described. Each of these is unique and one (High Mountain Scrub, see 2.3, below) contains two Ericaceae (*Rhododendron* and *Agapetes*) which relate the Australian assemblage to the New Guinean and Asian floras; it contains also *Dracophyllum* which occurs in south-eastern Australia and is abundant in New Zealand. The communities in Northern Territory occur on shallow sandy soils and are possibly remnants of a more extensive xeromorphic flora which existed when the continent lay further south. They contain species of widespread xeromorphic genera which relate the tropical assemblages to those in the south-west, as suggested by the occurrence of *Verticordia*. In all cases, the floristic assemblages are affected by infiltration of local tropical species. Of the other genera, a comment on *Banksia* is made. Only one species, *B. dentata*, occurs in the tropics and it extends into New Guinea. It is present in the coastal heaths of Queensland; elsewhere it is usually an understory species in *Eucalyptus* forest, except for a local occurrence in pure stands or in association with *Tristania lactiflua* or *Grevillea pteridifolia* in wetter Northern Territory (see Chap. 22).

2.2. *Fenzlia – Melaleuca – Leptospermum – Sinoga* Alliance

This alliance occurs on Cape York Peninsula (Fig. 16.1), and has been described as "wet desert" – referring to the low stature of the vegetation in a climate which might well support rainforest. The mean annual rainfall is 1500–1700 mm, falling almost entirely between November and April. However, soil analyses (C.L. Bale, personal communication) indicate that nutrients in the heaths are in low supply, phosphorus levels lying between 6.5 and 25 ppm. The communities are either closed or open, the former being on better drained soils (sandy, mottled Yellow Earths or deep sands), the latter occurring on waterlogged soils (Podzols and Gleyed Podzols). The two types, described by Brass (1953) and Pedley and Isbell (1971) form mosaics determined by drainage. Patches of scrub, dominated by *Banksia dentata*, or *Thryptomene oligandra* and *Melaleuca saligna*, or *Pandanus* sp., and of rainforest or *Eucalyptus* forest sometimes break the continuity of the heath.

In the closed heaths which reach a height of c. 2 m, *Fenzlia obtusa* and *Leptospermum fabricia* are co-dominant and these, together with *Melaleuca* aff. *symphyocarpa* and *Neoroepera banksii* form the canopy. Constant shrubs include *Choriceras tricorne*, *Jacksonia thesioides* and *Sinoga lysicephala*. Less abundant shrubs are *Acacia calyculata*, *Boronia bowmanii*, *Casuarina littoralis*, *Morinda reticulata*, *Grevillea pteridifolia* and stunted *Eucalyptus tetrodonta*. Herbaceous species are *Xanthorrhoea johnstonii* and *Schoenus sparteus*, the latter being the main constituent of the herbaceous layer.

Sinoga lysicephala dominates the open heaths ½–1½ m high. Most of the species listed for the first suballiance occur in this heath but in a stunted form. *Xanthorrhoea johnstonii* and *Schoenus sparteus* are most common in the herbaceous layer, but other sedges, Restionaceae and *Eriocaulon* spp. also occur. Noteworthy occurrences are the three insectivorous plants, indicative of swampy conditions: *Byblis liniflora*, *Nepenthes mirabilis* and *Utricularia chrysantha*.

The alliance appears to be represented on hills to the west at elevations of 300–600 m. Brass (1953) describes "Mountain Turkey Scrub" c. 3 m tall developed under "misty and apparently fairly wet climatic conditions" with a mixture of *Melaleuca* aff. *symphyocarpa*, *Casuarina* sp., *Leptospermum fabricia* and *Agonis lysicephala*.

2.3. "High Mountain Scrub"

At altitudes of c. 1600 m, the rainforests on mountains are replaced by stunted vegetation. Brass (1953) writes:

"Following rainforest in altitudinal sequence and covering the tops of the highest mountains is a densely packed, mossy, low forest or scrub which seems to merit recognition as a major community. We encountered it on Mt. Finnegan and Mt. Bellenden-Ker, the former mountain being now its northernmost known limit. Characteristic elements include species of *Austromyrtus*, *Drimys*, *Bubbia*, *Balanops*, *Orites*, *Quintinia*, *Rhododendron lochae*, *Agapetes meiniana*; and on Bellenden-Ker, *Dracophyllum sayeri* and *Leptospermum Wooroonooran*".

2.4. Communities in Northern Territory

There are several records of xeromorphic thickets and shrubberies from Northern Territory, including Specht's (1958) heath-like stand, previously referred to as a shrub savannah, occurring on quartzite and sandstone exposures within the *Eucalyptus phoenicea – E. ferruginea* alliance (Chap. 8, 2.5.1). The assemblage contains species of *Acacia*, *Brachysema*, *Calytrix*, *Grevillea*, *Hibbertia* and others.

Assemblages containing *Calytrix achaeta*, *C. microphylla* and *Verticordia cunninghamii* are reported by Speck (1965) and Story (1969). The scrubs, which occur under a mean annual rainfall of c. 1000 mm, are up to 3 m tall and contain *Acacia* ssp. as subsidiary components and an herbaceous layer of grasses, chiefly *Eriachne* spp.

An assemblage in a still drier area (mean annual rainfall 450–500 mm) is described from near the Tropic by Perry and Christian (1954). The community, almost 1½–3 m tall, occurs on deep sand and is dominated by *Jacksonia odontoclada* and several species of *Acacia* (*A. lysiphloia*, *A. stipuligera* and others). *Grevillea*, *Oxylobium* and *Bossiaea* are represented in the stands, together with some arid-zone genera. The herbaceous layer is dominated by grasses, mainly *Triodia pungens*, *Aristida pruinosa* and *Cymbopogon bombycinus*. This xeromorphic assemblage extends into some *Eucalyptus* communities, e.g. *E. odontocarpa* – *E. pachyphylla* alliance (Chap. 14).

3. Communities South of the Tropic in the East

3.1. General

This group of communities extends almost from the Tropic to Tasmania. They occur mainly on sands close to the coast, on headlands overlooking the ocean and on rocky outcrops near the coast and inland (Fig. 16.1). The communities are related through the dominants though the species differ from north to south (Table 16.1). The associated species which form the understories are common to many of the stands and some species occur in all the communities from north to south. They are arranged below according to habitat and classified into alliances when possible.

3.2. Communities on Coastal Sands

3.2.1. General

Heaths and scrubs occur on sands adjoining the dunes of the littoral zone, and landwards they usually adjoin woodlands or forests of *Eucalyptus*. The heath-scrub communities cover undulating or flat land, the former representing stabilized dunes, the latter lake beds which have been filled with sand. Soils in the more elevated areas are well drained, whereas those in flat areas or depressions usually support a water-table. The sands are mostly podzolized, the A horizon on the crests of dunes commonly exceeding 1 m. Gleyed Podzolics often occur where a water-table is present and in such soils organic accumulations are prominent, these occurring throughout the profile in swampy areas. Soil-water relationships usually determine the patterning of the communities. On the better drained soils species of *Banksia* dominate and sedges are rare, the latter becoming increasingly abundant with increasing waterlogging. Also, with increased waterlogging species of *Lepto-* *spermum* replace *Banksia* spp. and in heavily waterlogged soils *Melaleuca* spp. or sedges alone replace *Leptospermum*. Consequently, in any area mosaics of communities exist with a gradient in height and structure from woodland to scrub, heath and sedge-heath. In addition, heath-scrub communities are often interspersed with *Eucalyptus* forest or woodland or even rainforest, which occur on soils of relatively high fertility (Coaldrake, 1961).

Many of the heath-scrub communities have been destroyed to recover heavy minerals which have accumulated in the sands. These are rutile, zircon, monazite and ilmenite. Mining companies are obliged to revegetate the mined areas and in most cases regeneration of the native plants is fairly effective, though the topography of the land is somewhat altered (see Chap. 23).

The communities are described below under alliances, in north to south sequence. The dominant species change with latitude, as indicated in Table 16.1. An additional note on *Leptospermum* is inserted here. The genus has had a long history in Australia (Chap. 3); the shrub species which occur in heaths and xeromorphic forests are best known to ecologists. However, several species reach tree proportions, often forming pure stands adjacent to rainforests or along watercourses (Chap. 22 and 2.3.10 below). It is probable that the Early Tertiary species of *Leptospermum* were trees and that the shrub habit is a derived condition, being a response to low soil fertility. This is supported by leaf-texture and the morphology of the fruits, items which need systemic taxonomic-ecological investigation.

3.2.2. *Banksia serratifolia* Alliance

This species is the common dominant on deep sands on Podzols with an A horizon 1–2 m deep. It occurs from Fraser Island to central New South Wales. The scrubs are 3–5 m tall and often contain *Banksia integrifolia* and *Leptospermum laevigatum* from the coastal dunes, and less commonly *Casuarina distyla* and/or *Melaleuca quinquenervia*. Eucalypts are often present on the landward fringe (Fig. 16.2). The understory is xeromorphic, the main species in northern New South Wales and Queensland being *Acacia longifolia, Austromyrtus dulcis, Dodonaea triquetra, Monotoca scoparia, Lep-*

Fig. 16.2. *Banksia serratifolia* tall scrub with emergent *Eucalyptus intermedia*. Other shrubs are *Phebalium squameum, Dodonaea triquetra, Ricinocarpus pinifolius, Leptospermum attenuatum, Monotoca scoparia* and *Tristania conferta*. Bribie Is., Qld.

tospermum attenuatum, *Phebalium squameum, Pimelea linifolia, Platysace billardieri* and *Ricinocarpus pinifolius*. The herbaceous layer contains *Hypolaena fastigiata* as a common species, and *Cymbopogon refractus* and *Imperata cylindrica*.

3.2.3. Banksia aspleniifolia Alliance

This species is found in heaths from Southern Queensland to the sandstones of the Sydney region. In the north it characterizes heaths on sands in which a water-table is present (but not near the surface) or on headlands where some clay occurs in the B horizon (see 3.3, below).

B. aspleniifolia characterizes the assemblages but dominates only in patches which are separated by stretches of sedge-heath (Fig. 16.3). Other species which form local clumps of shrubs are *Strangea linearis* (north only), *Leptospermum flavescens, L. semibaccatum* and *Banksia serratifolia* in a stunted form. Stunted plants of *Tristania conferta* and/or *Eucalyptus intermedia* occur mainly in Queensland. The clumps of shrubs have an understory of slender shrubs, chiefly *Boronia falcifolia, Epacris microphylla,*

Eriostemon lanceolatus and *Sprengelia sprengelioides*. Herbaceous species include *Xanthorrhoea media, Leptocarpus tenax* and *Lepyrodia scariosa*. In the sedge-heath which separates the clumps of shrubs, the sedges dominate and the slender shrubs mentioned above are emergents.

3.2.4. Banksia robur Alliance

This species extends by southern Queensland to central New South Wales and is locally dominant where the water-table occurs close to the surface in sandy soils without organic accumulations. The main associated species are *Calorophus minor* and *Hypolaena fastigiata*. Slender shrubs are usually present, especially *Sprengelia sprengelioides* (north), *S. incarnata* (south), *Epacris obtusifolia* and *Conospermum taxifolium* (Fig. 16.4).

3.2.5. Banksia ericifolia Alliance

B. ericifolia is confined to New South Wales and characterizes heaths and scrubs from the

Fig. 16.3. Mixed heath. Main species in foreground are *Banksia aspleniifolia, Xanthorrhoea media, Eriostemon lanceolatus, Grevillea sericea* and *Epacris microphylla*. Centre: *Strangea linearis*, stunted *Eucalyptus intermedia* and *Gahnia sieberana*. Trees in distance are *Melaleuca quinquenervia*. Coast, east of Nambour, Qld.

Richmond R. region to the sandstones of the central coast. It occurs both on sands and in rocky terrain. On sands it occupies areas with a water-table and sometimes forms thickets c. 2 to 3 m tall, often in pure stands, with subsidiary species only at the margins (Fig. 16.5). Alternatively it occurs in association with a few other species of similar size. In the Gosford area, Siddiqi et al. distinguish two groups of associations for the alliance: i. *B. ericifolia – Casuarina distyla – Hakea teretifolia – Cyathochaete diandra*, and ii. *Casuarina distyla – Themeda australis – Eragrostis* sp. The latter occurs on soils of higher fertility. Other shrub species which are sometimes dominant in patches are *Phebalium squameum* (north), *Lambertia formosa, Banksia spinulosa, Acacia myrtifolia, Melaleuca nodosa* and *Callistemon citrinus*. The heaths have been studied intensively by Siddiqui et al. (1976), who present data on reactions to fire, competition, growth rates and proteoid roots.

3.2.6. *Banksia marginata* Alliance

B. marginata is the most widely distributed species of *Banksia* in eastern Australia and is an understory species over much of its range. It is a co-dominant with *Leptospermum myrsinoides* in some heaths in eastern Victoria (see 3.2.9, below) and it extends into heaths in South Australia (see 4, below).

It is the only species of *Banksia* in Tasmania where it is commonly associated with *Casuarina monilifera* (Fig. 16.6). The heaths occur on patches of low fertility sand around the coast but mainly in the north-east and east. The heaths often adjoin the coastal dunes and *Leptospermum laevigatum* and *Leucopogon parviflorus* from the dunes sometimes occur in the heaths; in such stands *Acrotriche serrulata, Astroloma humifusum* and *Lomandra longifolia* are common understory species (Kirkpatrick, 1977). On the landward side the heaths are bordered by *Eucalyptus* communities, chiefly *E.*

Fig. 16.4. Sedge heath with a clump of *Banksia robur*. *Leptospermum semibaccatum* in foreground. Herbaceous plants are mostly *Calorophus minor*. Near Caloundra, Qld.

Fig. 16.5. Thicket of *Banksia ericifolia*. Subsidiary species are *Persoonia lanceolata, Melaleuca nodosa, Acacia sophorae, Epacris obtusifolia* and *E. microphylla. Calorophorus minor* in foreground and *Melaleuca quinquenervia* behind, with *Eucalyptus pilularis* forest in distance. Near Laurieton, N.S.W.

Fig. 16.6. Patch of heath occurring within a stand of *Eucalyptus tenuiramis*. Main heath species are *Casuarina monilifera, Banksia marginata, Hakea teretifolia, Leptospermum scoparium, Epacris impressa, Leucopogon collinus, Hibbertia stricta, Boronia pinnata, Bauera rubioides*. Herbs: *Tetrarrhena juncea, Calorophus minor, Lepidosperma lineare, Lindsaea linearis* and *Selaginella uliginosa*. On sandstone. Eaglehawk Neck, Tas.

amygdalina in the north-east and east, *E. tenuiramis* in the south-east and *E. nitida* on the west coast. The heath assemblage usually extends below the eucalypts as an understory in woodlands. Associated species are numerous and are variously distributed according to local conditions of soil depth, soil fertility and waterlogging. Consequently, many associations can be identified and these have listed by Jackson (in Specht et al., 1974; Kirkpatrick, 1977).

Either *B. marginata* or *Casuarina monilifera* may be locally dominant or they occur in association forming heaths or scrubs $\frac{1}{2}$–$1\frac{1}{2}$ m high. The main subsidiary species, some of which are local dominants are: *Acacia myrtifolia, A. ericoides, A. villosa, Bossiaea cinerea, Brachyloma ciliatum, Calytrix tetragona, Epacris impressa, E. lanuginosa Gompholobium huegelii, Hakea teretifolia, Haloragis tetragyna, Hibbertia acicularis, Kunzea ambigua, Leucopogon collinus, L. ericoides, Pultenaea daphnoides, P. juniperina* and *Sprengelia incarnata*. The herbaceous layer contains the following, with some variation in composition: *Lepidosperma concava, Lindsaea linearis, Selaginella uliginosa, Xanthorrhoea australis* and *X. minor*.

Species of *Leptospermum* (mainly *L.*

scoparium and *L. glaucescens*) and of *Melaleuca* occur in some heaths and indicate transitions to the *Leptospermum* thickets described in 3.2.10, below.

3.2.7. *Leptospermum flavescens* – *L. attenuatum* Alliance

These species are found mainly in Queensland and central and northern New South Wales on soils subject to waterlogging. Both species occur in the understory of xeromorphic forests and woodlands and each one may be the dominant of heaths or thickets. The factors delimiting the species are not clear. The communities are mostly c. 2 m tall and the associated species vary from district to district. *Banksia serratifolia* is often present at the drier margins and sometimes mallee eucalypts. *Xanthorrhoea media* is usually present and the herbaceous layer contains Cyperaceae, especially *Gahnia sieberana* and *Caustis flexuosa*. On the wetter margins the *Leptospermum* becomes sparser and is replaced by *Baeckea imbricata* and/or *Melaleuca nodosa* with subsidiary *Banksia aspleniifolia*, *Epacris pulchella* with an herbaceous layer of *Xanthorrhoea*, *Gahnia* and *Selaginella uliginosa*.

3.2.8. *Leptospermum liversidgei* Alliance

This species occurs mainly along the north coast of New South Wales, extending into Queensland. It is found in swampy situations where it forms thickets up to 2 m high. The community described by Osborn and Robertson (1939) for the Myall Lake area is possibly typical. The soil has a strong development of peat built up by the herbaceous sward which consists largely of *Blechnum indicum*, *Calorophus minor*, *Restio tetraphyllus* and *Blandfordia* sp. As waterlogging and the depth of peat increase, the community merges into a sedge heath in which Epacridaceae are abundant (see 3.2.12, below).

3.2.9. *Leptospermum myrsinoides* Alliance

This species characterizes the heaths in eastern Victoria. The heaths, 1–1½ m high, now largely destroyed, once extended over extensive sandy Podzols areas east of Port Phillip and have been described by Patton (1933, 1935). The most common species, all of which may be local dominants, are *L. myrsinoides*, *Acacia oxycedrus*, *Olearia ramulosa*, *Casuarina pusilla*, *Ricinocarpos pinifolius* and *Banksia marginata*. The understory shrubs are mainly Papilionaceae (e.g. *Aotus*, *Bassiaea*, *Daviesia*, *Dillwynia* etc.) and the Epacridaceae is well represented, especially *Epacris impressa*. The herbaceous layer contains a mixture of grasses (especially *Themeda australis* and *Stipa semibarbata*) and sedges (*Lepidosperma concava* and *Hypolaena fastigiata*). Liliaceae and Orchidaceae are abundant in the spring. Biomass studies, water use and growth of these heaths have been studied by Jones (1968), Jones et al. (1969) and Specht and Jones (1971).

A stunted, wind-shorn form of this floristic assemblage occurs on Wilson's Promontory (Fig. 16.7) where *Casuarina pusilla* is prominent and the main associated species include *Correa reflexa*, *Hibbertia procumbens*, *Leucopogon virgata* and *Kennedia prostrata*.

3.2.10. *Leptospermum* Communities in Tasmania

Eight species of *Leptospermum* occur in Tasmania, four being endemic to the island (Curtis and Morris, 1975). Two species are exceptional for the genus because they reach tree proportions (c. 18 m tall) and dominate closed forests, either in climatically wet areas or on small patches of soil where drainage waters accumulate. The same species occur in shrub form mostly near the coast. Jackson (in Specht et al., 1974) recognizes three groups of communities:

i. *Leptospermum glaucescens* (formerly *L. sericeum*) is endemic to Tasmania occurring mainly in the west and wetter south-east at lower elevations on peaty soils. Near the coast, especially in the south-east it forms heaths or scrubs up to 4 m high, sometimes with co-dominants of *Leptospermum scoparium*, *L. nitidum* or *Acacia mucronata*. It is found also in the *Banksia marginata* heaths (Kirkpatrick, 1977). Associated species are *Bauera rubioides* and *Agastachys odorata*. In the wettest sites it forms closed forest to c. 18 m tall with a floristically poor understory. Associated species are *Cenarrhenes nitida* and *Bauera rubioides*.

Fig. 16.7. Heath on granite. Main species are *Casuarina pusilla* and *Leptospermum myrsinoides,* with subsidiary *Epacris impressa, Isopogon ceratophyllus* and *Dillwynia ericifolia.* Behind, a single plant of *Banksia marginata* and a thicket of *Leptospermum laevigatum.* Wilson's Promontory, Vic.

Fig. 16.8. Thicket of *Leptospermum lanigerum* with *Gleichenia alpina* in foreground. Hartz Mts Tas. at c. 1100 m.

ii. *Leptospermum lanigerum* shows a similar range in height to *L. glaucescens* and occurs mainly in the north and east and on the Central Plateau from sea level to the subalpine zone. It is confined to wet soils and often forms fringes of scrub or forest along watercourses, the tallest stands occurring in the subalpine zone. There are usually few or no associated woody species in these dense stands. Jackson records *Callistemon viridiflora* on the Central Plateau. *Bauera rubioides* and *Gleichenia alpina* often form closed stands as the margins of the communities (Fig. 16.8).

iii. *Leptospermum scoparium*, which occurs on the mainland in the south-east and in New Zealand, is widespread in Tasmania and is dominant in heaths or thickets mainly in the east and south-east. It is often associated with *Acacia verticillata* to form thickets 2–4 m tall on soils which are periodically waterlogged. The latter is often dominant, especially in estuaries or behind coastal dunes along the north coast. *L. scoparium* is often associated with *Melaleuca gibbosa* on badly drained soils in the same districts as the *Banksia marginata* heaths. It is also a common emergent in the sedge-heaths (Kirkpatrick, 1977) as described in 3.2.12, below.

iv. *Leptospermum grandiflorum*, endemic to Tasmania, sometimes in association with *Hakea rostrata* forms heath-like communities in the south-east, especially on Freycinet Peninsula (on granite outcrops) and some islands.

3.2.11. *Acacia mucronata – Phebalium squameum* Alliance

These two species occur in wet lowland sites in Tasmania forming heaths or thickets mostly 1–3 m tall near the coast. On the west coast, *Baeckea leptocaulis* forms an association with *Acacia mucronata*. In the Queenstown area the alliance covers extensive tracts as a regenerating community following the destruction of woodlands by sulphur dioxide produced in smelting of copper ore (Fig. 16.9). Common associated species are *Leptospermum scoparium, Bauera rubioides, Sprengelia incarnata, Gahnia psittacorum, Restio tetraphyllus, Xyris operculata* and *Centrolepis* sp.

3.2.12. Sedge-Heaths

In areas where the water-table lies close to the surface for long periods, sedges dominate and

Fig. 16.9. Heath regenerating on an area denuded by sulphur dioxide from smelting. Main species are *Acacia mucronata, Leptospermum scoparium, Phebalium squameum, Bauera rubioides* and *Sprengelia incarnata*. Main herbaceous species are *Restio tetraphyllus* and *Gahnia psittacorum*. Near Queenstown, Tas.

constitute sedgelands which are dealt with in more detail in Chap. 22. The sedgelands are often dotted with shrubs which are either randomly distributed or occur in patches. Such communities, which are transitional between the sedgelands and heaths, are referred to as sedge-heaths. Two main groups can be recognized:

i. Those in which little organic matter occurs, the soils being siliceous through the profile, though subsoil pans which lead to impeded drainage may occur.

ii. Those in which extensive organic accumulations are present usually built up by the sedges themselves or, rarely, by *Sphagnum* moss which is more abundant at higher altitudes (Chap. 17). Many of the swamps support woody communities dominated by various species of *Melaleuca* (Chap. 22), whereas in others the shrub species are commonly Epacridaceae. The sedge heaths occur both on coastal sands and on sandstones.

Most common of the Epacridaceae are *Sprengelia sprengelioides* (north), *S. incarnata*, *Epacris microphylla*, *E. obtusifolia* and *E. pulchella* (mainland), and *E. impressa* (Vic. and Tas.), *E. lanuginosa*, *Leucopogon australis* and *L. collinus* (Tas.). The herbaceous species, which form a dense sward are mainly *Calorophus minor*, *Gymnocephalus sphaerocephalus*, *Hypolaena fastigiata*, *Leptocarpus tenax*, *Lepyrodia scariosa*, *Xanthorrhoea* spp. and *Xyris* spp. Other small herbs are numerous, including *Selaginella uliginosa*, *Drosera* spp. and *Utricularia* spp.

Local variations occur from north to south, both in the herbaceous dominants and the woody components, e.g. in New South Wales, *Symphionema paludosa* is restricted to the sedge-heaths and in Tasmania *Leptospermum scoparium* is a common emergent. (Some additional information is provided in Chap. 22).

3.3. Headland Heaths

3.3.1. General

Heaths occur close to the ocean on many headlands composed mainly of sedimentary rocks. The heaths adjoin the littoral assemblages on rock (e.g. *Pandanus* and *Westringia* communities dealt in Chap. 23). The communities exhibit various degrees of wind-shearing and, in extreme cases of exposure, the heaths are 30 to 40 cm high. Where the rock is entirely siliceous, the communities are similar floristically to the heaths and scrubs further from the ocean, as on the Hawkesbury sandstones of the Sydney region. On the north coast of New South Wales the headland rocks, mostly sedimentary, contain some clay and produce podzols with a clayey B horizon. The heaths developed on these soils differ from other heaths in having a grass layer of *Themeda australis*. With increasing clay-content of the soil the xeromorphic shrubs become increasingly scattered and *Themeda* dominates patches of grassland (Chap. 20). At least two alliances are present in south-east Queensland and northern New South Wales, as described below. Further south, analogous communities exist. These contain species which relate them to the littoral flora, with which they are included (Chap. 23).

3.3.2. *Casuarina distyla* – *Jacksonia stackhousii* Alliance

These species form thickets c. 2 m high on deep profiles, the sandy A horizon being c. 30 cm deep. The thickets occur in broad valleys with some protection from the wind. *Casuarina* is usually dominant and the associated species include *Dodonaea triquetra* and *Banksia integrifolia*. A discontinuous lower shrub layer occurs, containing *Lomatia silaifolia*, *Banksia aspleniifolia*, *Hakea teretifolia*, *Epacris pulchella* and *Cryptandra amara*. The herbaceous layer is almost continuous, *Themeda australis* being dominant, with rare occurrences of *Dianella longifolia* and *Thysanotus tuberosus*.

3.3.3. *Casuarina littoralis* – *Banksia aspleniifolia* Alliance

These species are the most common in assemblages in the most exposed situations on headlands (Fig. 16.10), but are not necessarily consistently dominant throughout the heath. The communities occur on shallow podsolized soils with a sandy surface horizon which usually contains quartz pebbles and stones.

The heaths average 30–50 cm in height with emergent patches 1–2 m high, the latter occurring in depressions where the plants are protected from the wind. As well as the characteris-

Fig. 16.10. Headland heath. Taller plants near ocean are *Pandanus spiralis* (on point) and *Banksia integrifolia*. Darker patches are *Acacia sophorae* and **Chrysanthemoides monilifera*. Common heath species are *Casuarina littoralis*, *Banksia aspleniifolia* and *Hakea teretifolia*. Herbaceous species, as in foreground, is mainly *Themeda australis*. South of Port Macquarie, N.S.W.

tic species, the following are commonly local dominants: *Hakea teretifolia*, *Lambertia formosa*, *Kunzea capitata*, *Melaleuca nodosa*, *Ricinocarpus pinifolius* and *Dodonaea triquetra* (the last often in association with mallees of *Eucalyptus intermedia*). Subsidiary shrubs include: *Acacia longifolia*, *A. suaveolens*, *Epacris microphylla*, *Isopogon anemonifolius*, *Persoonia lanceolata*, *P. prostrata*, *Philotheca salsolifolia*, *Phyllanthus thymoides*, *Pimelea linifolia*, *Xanthosia pilosa* and many others. The herbaceous layer always contains *Themeda australis*. Less common are the grasses *Paspalidium gracile*, *Aristida warburgii* and the herbs *Thysanotus tuberosus* and *Helichrysum apiculatum* (often locally common). The twine-like parasite *Cassytha glabella* is often abundant and the fern *Lindsaea linearis* sometimes forms small societies in the protection of bushes.

3.4. Heaths, Scrubs and Thickets on Rocky Outcrops

3.4.1. General

Heaths, scrubs and thickets develop in many rocky situations where the soils are too shallow to support trees. The rocks are mostly trachytes, granites and related igneous rocks, and sandstones. The rock surfaces are colonized initially by microorganisms and non-vascular plants, and some typical successions are described below. Four areas from the east are selected to present a representative account of these communities.

Of interest is the fact that some of the rocky outcrops supporting heaths harbour disjunct taxa, notable among these being *Borya* from the high mountains of Northern Queensland (Bellenden Ker and neighbouring peaks), Glasshouse Mts. (see 3.4.2 below) where the species is *B. septentrionalis*, and in the south-west (*B. nitida*, see 5.5.6 below).

3.4.2. On Glasshouse Mountains, Queensland

The Glasshouse Mountains (Figs. 1.10 and 16.11), composed of trachyte, reach altitudes of 500–600 m and receive a mean annual rainfall of c. 1500–1800 mm. On steep slopes, several herbaceous rock-ledge species form small communities which establish vegetation (Fig. 16.11). The more important of these are the ferns *Cheilanthes tenuifolia* and *Culcita dubia*, two rare species, *Borya septentrionalis* and the grass *Micraira subulifolia*, and several others, especially *Paspalidium gracile*, *Plectranthus parviflorus* and *Xanthorrhoea australis*.

Fig. 16.11. *Leptospermum brachyandrum* thicket with stunted, mallee-like *Eucalyptus tereticornis* as an emergent (taller plant). Other species include *Agonis scortechiniana, Calytrix tetragona, Baeckea* sp., *Jacksonia scoparia, Pimelea linifolia, Arundinella nepalensis* (grass on left) and *Micraira subulifolia* (clumps on rock in foreground). Glasshouse Mts, south-east Qld.

Shrubs become established in these herbaceous clumps, notably *Calytrix tetragona* (sometimes forming extensive heaths) and *Leptospermum brachyandrum* which dominates extensive scrubs or thickets. Other common species include *Acacia pravissima, Jacksonia scoparia* and *Keraudrenia lanceolata*. On deeper soils, mallee forms of local trees become established, especially *Eucalyptus trachyphloia, E. tereticornis* and *Syncarpia glomulifera*.

3.4.3. On Trachyte Exposures Warrumbungle Mountains

In this area, the larger heaths occur on trachytes at elevations of 500–1200 m under a mean annual rainfall of c. 1000 mm. Mrs. G. Harden (person. commun.) provides the following data for four isolated areas of heath where bare rock covers 30–90% of the ground on slopes of 2–20°.

The first vascular plants to establish in rock crevices or on shallow soil accumulations are the pteridophytes *Cheilanthes tenuifolia, C. distans, Pleurosorus rutifolius* and *Ophioglossum lusitanicum,* and *Stypandra glauca*. Herbaceous composites and grasses are sometimes locally dominant on shallow soils. The shrubs comprising the heaths are usually rooted in crevices. Most species are common to the four sites and four species are constant, each being dominant or co-dominant in one or other of the sites, viz., *Micromyrtus ciliata, Calytrix tetragona, Kunzea ambigua* and *Acacia cultriformis*. Other common species are *Cryptandra amara, Phebalium* spp., *Acacia triptera, Dodonaea boroniifolia, Prostanthera nivea, Leucopogon attenuatus* and *Melichrus erubescens*.

The heaths are usually c. 1 m high, but emergents 2 m tall are present in some, *Kunzea ambigua* being the most common; *Casuarina* aff. *distyla* is recorded in one area on steeper slopes. With increasing soil depth, taller plants, overtop the heath assemblage, usually stunted *Eucalyptus dealbata, Acacia cheelii* and *A. doratoxylon* (see Chap. 18).

3.4.4. On Triassic Sandstones, Central New South Wales

3.4.4.1. General

The sandstones of the Sydney area and their rich xeromorphic flora are described in Chap. 10. The area receives a mean annual rainfall of 1100–1400 mm and extends from near sea level to an altitude of c. 1000 m in the Blue Mountains. Colonization and succession on these rocks has been described by Pidgeon (1938) as follows: The first colonisers of dry rock are the algae *Stigonema* and *Gloeocapsa* spp. which give the rocks a red coloration. These are followed in sequence by crustaceous and foliose lichens, xeric mosses and fruticose lichens, and by herbaceous perennial vascular plants. The latter include *Lepyrodia scariosa, Ptilanthelium deustum, Lomandra longifolia, Dianella coerulea* and *Lepidosperma laterale*. Shrubs become established in these herbaceous micro-communities, usually *Epacris pulchella, E. microphylla, Leptospermum squarrosum, Leucopogon microphylla* and *Darwinia fascicularis*. With increasing soil depth the variety of shrubs increases and a heath community containing several dozen species develops. Taller shrubs, notably species of *Banksia, Grevillea, Casuarina, Hakea* and many others establish as soil depth increases, and these species form the understory in the *Eucalyptus* communities.

On rock surfaces which are usually wet, the lichen stage is rapidly succeeded by the moss and herbaceous stages. On rock surfaces which are permanently wet some species which are usually epiphytic may become established, including the fern *Pyrrosia rupestris* and the orchids, *Dendrobium ligniforme, D. speciosum* and *D. striolatum*. Also lycopods are sometimes abundant in some places, especially *Lycopodium cernuum* and/or *L. laterale* and the ferns *Gleichenia microphylla* and *Culcita dubia*.

In moist situations the sequences of communities is somewhat different from those in drier habitats. Additional rock-ledge species include the herbs *Drosera* spp., usually with an abundance of liverworts, *Mitrasacme polymorpha, Actinotus minor,* a variety of ferns, and the shrubs *Bauera rubioides, Dracophyllum secundum, Callistemon linearis, Leptospermum parviflorum* and *L. arachnoides*.

On these shallow soils in both wet and dry habitats heath and scrub communities occur and many dozens of associations could be identified, some occupying small areas, others occurring extensively over many hectares or even many square kilometres. The two most extensive of these are dominated by species of *Casuarina* and are described below. However, many other species are locally dominant over areas of a few square metres to one or a few hectares in rocky terrain, as on rock exposures or in clefts between boulders or seams in the sandstones. The following species dominate heaths less than 1 m tall: *Baeckea brevifolia* (especially in the Blue Mountains in exposed situations), *Calytrix tetragona, Darwinia fascicularis* and *Petrophile fucifolia*. Scrub communities 1–3 m tall are commonly dominated by one or more of the following: *Angophora cordifolia, Banksia ericifolia, B. serrata, Hakea teretifolia* (wetter sites) and *Leptospermum squarrosum*. In all communities the understory of xeromorphic smaller shrubs occurs.

3.4.4.2. *Casuarina distyla* Alliance

C. distyla forms thickets mostly 2–3 m tall (Fig. 16.12) on shallow soils, usually among rocks, on the sandstones mainly at lower elevations. It is found less abundantly on the sandstone tablelands to the west (see next section). Usually it forms pure stands, but may be co-dominant with *Banksia ericifolia*. Stunted *Eucalyptus gummifera* is commonly present in the stands which sometimes form mosaics with mallee eucalypts, especially *E. camfieldi* and *E. luehmanniana* (Chap. 10). A large number of understory species occur, such as *Acacia suaveolens, Bossiaea scolopendria, Calytrix tetragona, Darwinia fascicularis* and many Papilionaceae. The herbaceous layer usually contains *Xanthorrhoea media, Lepidosperma laterale* and Restionaceae.

3.4.4.3. *Casuarina nana* Alliance

C. nana is restricted to the higher parts of the sandstone areas forming extensive heaths in the Blue Mountains and on the Wingecarrabie Tablelands on exposed plateaus or rock outcrops (Petrie, 1925; Burrough et al., 1977).

The heaths are closed and rarely exceed a height of 1 m (Fig. 16.13). On deeper soils within the heaths taller species may occur, notably *Banksia ericifolia, B. marginata, Eucalyptus stricta, Casuarina distyla, B. spinulosa, Lep-*

Fig. 16.12. Thicket dominated by *Casuarina distyla*. Understory species are *Banksia ericifolia, Darwinia fascicularis, Hakea teretifolia* and *Epacris microphylla*. Narrabeen, near Sydney, N.S.W.

Fig. 16.13. Female plant of *Casuarina nana* with mature "cones" containing the fruits (samaras) enclosed by woody bracts and bracteoles. The scale is a match box 50×35 mm. Near Bell (Blue Mts), N.S.W.

tospermum attenuatum, Hakea dactyloides and *Petrophile fucifolia*. In badly drained areas or along watercourses *Hakea teretifolia* is sometimes locally dominant. The subsidiary species are common xeromorphs which exhibit a stunted form in the heaths. Some common species are *Isopogon anemonifolius, Leptospermum arachnoides, Cryptandra amara, Dampiera stricta, Phyllanthus thymoides, Hibbertia serpyllifolia* and *Epacris microphylla*. Common herbs are *Goodenia bellidifolia, Sowerbaea juncea, Patersonia sericea, Lepidosperma laterale* and *Lepyrodia scariosa*.

3.4.5. On Granite Exposures on Wilson's Promontory

Granites at Wilson's Promontory lie at the surface between sea level and an altitude of c. 800 m and receive a mean annual rainfall of 1000–1500 mm. The successions are described by Ashton and Webb (1977). Below 400–500 m the primary colonizers are the crustose lichens *Verrucaria* sp. and *Lecidia* sp., followed by the foliose lichen *Parmelia conspersa* and the moss, *Campylopus bicolor*. In the gravelly shallow soil, succulent annuals become established, *Crassula sieberana* and *Calandrinia calyptrata*, and geophytes of the genera *Drosera, Chamaescilla, Bulbine* and ground orchids. These are followed by shrubs, the dominant being *Kunzea ambigua* with an understory of *Epacris impressa* and *Pultenaea daphnoides*. On the landward side this community adjoins the *Leptospermum – Casuarina* heath described in 3.2.9, above or by a mallee or woodland of *Eucalyptus* spp.

At altitudes above 400–500 m, the rock is colonized by the algae *Ulothrix* and *Gloeocapsa* spp., followed by the mosses *Grimmia laevigata*, and then by larger mosses, *Campylopus bicolor* and *Rhacomitrium* spp. The main herb to follow is *Helichrysum baxteri*. The rock-heath dominating at these altitudes is composed of *Callistemon pallidus* associated with *Leptospermum juniperinum, Epacris impressa* and *Olearia stellulata*. The heaths adjoin *Eucalyptus* woodlands.

A noteworthy occurrence in the same area are heaths c. 1 m high of *Nothofagus cunninghamii*. The communities are windshorn and the plants exist in the protection of rocks. Associated species are *Monotoca elliptica, Epacris impressa, Callistemon pallidus, Tasmannia lanceolata* and *Olearia stellulata*.

4. Communities in South Australia and Western Victoria

4.1. General

The heaths and scrubs in South Australia occur in the south-east of the State and extend into western Victoria (Fig. 16.1). The mean annual rainfall decreases from c. 600 mm near the coast to 470 mm inland. The inland stands are regarded as "deserts" which refers to the low fertility of the soil rather than the climate. On more fertile soils in the same climates *Eucalyptus* woodlands develop.

Unlike the eastern heaths, those in South Australia do not adjoin the coastal dune systems, which are edged on the landward side by mallee *(Eucalyptus diversifolia)* or *Melaleuca lanceolata*, or swamp in which species of *Gahnia* dominate. These swampy areas extend inland for many kilometres and in the better drained parts *Melaleuca squarrosa* and *M. ericifolia* commonly form thickets (Chap. 22).

The heath and scrub communities, dominated by species of *Banksia, Casuarina, Hakea* and *Xanthorrhoea* are structurally similar to those in the east. They have been divided by Specht (1972) into two assemblages, one in the lower south-east (wetter climate), and one in the upper south-east (drier climate).

4.2. *Xanthorrhoea australis – Banksia ornata – Hakea rostrata – Casuarina paludosa* Alliance

This is described by Specht (1972) as an open "wet heath". It reaches a height of about 2 m and occurs on Gleyed Podzols developed on flat land subject to waterlogging in the more southerly parts of south-eastern South Australia. In saline areas (Solodized Solonetz soils), *Melaleuca gibbosa* and *Hakea rugosa* are locally dominant. The total assemblage contains a few shrubs, including *Acacia verticillata, Banksia marginata, Calytrix tetragona, Casuarina pusilla, Hibbertia stricta, Leptospermum juniperinum, L. myrsinoides* and *Leucopogon australis.* Herbaceous species include *Lepidosperma, Schoenus* spp. and *Leptocarpus brownii.*

4.3. Xanthorrhoea australis – Banksia ornata – Casuarina pusilla Alliance

The ecology of this heath, including such items as biomass, water relations and the effects of fire and the applications of fertilizers, has been studied by Coaldrake (1951), Litchfield (1956), Specht and Rayson (1957) and their co-workers (1957–1975), and Lange (1971). An account of the "Little Desert", extending to western Victoria has been compiled by Thiele (1977).

The heaths grow on stabilized sand dunes mainly of the seif (longitudinal) type with minor occurrences of the barchan (crescentric) type, the latter being attached to the longitudinal dunes. There are no surface streams. The sands overlie clay, and where clay lies near or at the surface more fertile soils occur, producing mallee communities, mallee heaths or thickets of *Melaleuca uncinata.* Where the depth of surface sand is c. 40 cm, mallee heaths occur and when surface sand exceeds c. 120 cm the heath species alone can grow. The roots of the heath species occur almost entirely in the topmost 25 cm of sand. Fertility levels of the sands are extremely low. Coaldrake indicates phosphorus levels of c. 20 ppm in surface soils and as little as 4 ppm in subsurface layers. (For comparison, mallee communities in the same area have phosphorus levels at least twice as high.)

The flora is a relatively small one. Specht (1972) records only 76 species. The flora contains many local endemics but many of the species occur to the east, especially in the Grampians or in south-eastern Victoria, and a few species are widespread, but disjunct, across southern Australia, e.g., *Calytrix tetragona* and *Hibbertia stricta.* Affinities with the south-west are slender, apart from the occurrence of the widespread genera, only two western genera, *Adenanthos* and *Lhotzky* being represented in the South Australian assemblage (each by one species, *A. terminalis* and *L. alpestris*).

The heath varies in height from $1/2$ to c. $1 1/2$ m and many associations could be recognized. The characteristic species are not equally distributed throughout the heath. *Xanthorrhoea australis* is continuous throughout, but is absent from lee faces of the dunes. *Banksia ornata,* varying in height from $1/2$ to 1 m, is continuous throughout but more abundant on deep sand ridges. *Casuarina pusilla* is almost continuous. Other local dominants are mallee plants of *Eucalyptus baxteri* on crests and eastern faces of dunes, *Leptospermum myrsinoides* occurring throughout, and *L. coriaceum* restricted to deeper sands of larger dunes, *Banksia marginata* in wetter areas, and *Casuarina muellerana.* Other common woody species belong mainly to the genera *Baeckea, Calytrix, Correa, Cryptandra, Hibbertia, Isopogon, Leucopogon,* and *Phyllota,* and many species show habitat preferences. In addition to *Xanthorrhoea,* the herbaceous layer contain species of *Lepidosperma, Hypolaena, Lepidobolus, Lomandra* and five species of ground orchids.

The heath occurs also on the Eyre Peninsula and northern part of the Flinders Range where it is represented by dominants of *Casuarina muellerana* and *Leptospermum coriaceum.*

5. Communities in South-Western Western Australia

5.1. General

5.1.1. Location and Habitats

In Western Australia the xeromorphic flora reaches its greatest diversity in the heaths, scrub heaths and related communities which occupy deep sands of low fertility, or soils developed on fossil laterite, with a lesser development on limestones near the coast. The total flora contains c. 2000 species and many endemic genera are confined to these communities (Chap. 4). The sandy deposits are referred to as "sandplain" and they are found in two main regions, one on the western coastal plain, the other along the south coast with an extension inland to the arid zone. These two regions are separated by a belt of fossil laterite supporting *Eucalyptus* forests running in a general north-south direction, and the two areas support different species of the same genera (Table 16.2). The narrow strip of coastal lowland which occurs around

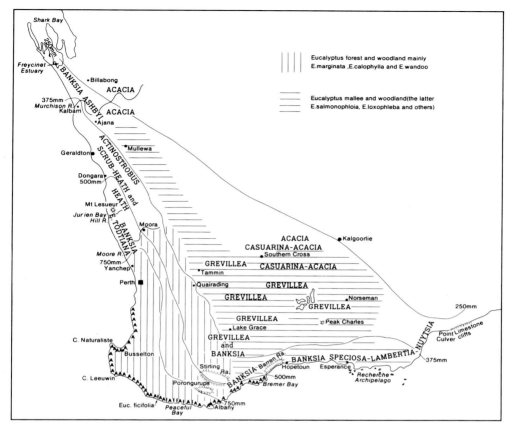

Fig. 16.14. Showing the locations of the main heath communities in south-western Western Australia. The 250, 375, 500, 750 mm isohyets are shown. Abbreviations: A = *Agonis flexuosa*; f = *Eucalyptus ficifolia*.

the south-west "corner" does not connect the western and southern sandplains. On the contrary, this strip of coastal sand forms a third region which is identified by *Agonis flexuosa*. The locations of these regions is shown in Fig. 16.1, and the communities are described below under these regions. In addition, heaths and scrubs occur on rocky outcrops along the coast and inland and the locations of the larger of these are indicated in Fig. 16.14.

The following addition explanatory notes are provided:

i. On the western coastal plain the communities lie between the coastal sand dunes (Chap. 23) and *Eucalyptus* forests, woodlands or mallee, or *Acacia* woodlands in the drier north. The *Banksia* communities and some of the heaths are developed on sandplain, and with decreasing mean annual rainfall the communities become progressively shorter, but often without a change in the dominant species. Thus, in the wetter south, e.g. near Perth, woodlands of *Banksia* occur, and with decreasing rainfall towards the north, the same species dominate scrubs or heaths. The reduction in height of the communities is apparently due to decreasing soil fertility levels which appear to reach a minimum around the 500 mm isohyet, where white sands dominate the landscape and supports heaths. This is deduced from the fact that with decreasing rainfall (in a northerly or easterly direction) the communities become taller and occur on yellow (less leached) sand. At the fringes of the arid zone, where the sands are red and scarcely leached the communities are scrubs 3–5 m tall (see Fig. 16.14).

ii. On the southern coastal plains the heaths and scrub-heaths adjoin the coastal dunes and extend inland on sand to a belt of mallee. North of this mallee belt the heaths and scrub heaths occur in patches northward to the arid zone, occupying yellow sands which form mosaics with sandy lateritic material which supports mallee or woodland. Ecotones with mallee are common (mallee heaths) and other ecotones identified by mixtures of heath species and components of the mallee communities, chiefly *Melaleuca* and *Casuarina* spp. are common. In the drier areas *Acacia* become prominent and near the arid zone *Acacia* scrubs replace the heaths on sand (Fig. 16.14).

5.1.2. Soil Parent Materials and Soils

The soils are derived mainly from fossil laterites formed long ago on granite or gneiss or from coastal limestones.

Reweathering of the fossil laterite has produced vast amounts of siliceous sand and various sandy deposits exhibit different degrees of leaching which has reduced fertility levels. The white sands are presumably most leached. Ferruginous lateritic concretions and rock outcrops (granite or limestone) occur in some areas and usually support a flora different from the sandplain flora. Soils on laterite residuals are possibly of higher fertility than the sands; this is suggested by the development of a taller type of vegetation on sands in which some laterite gravel is incorporated.

The coastal limestones and calcareous dunes have provided much of the sand which now lies close to the west coast. McArthur and Bettenay (1960) recognize three types of sand dunes on the Swan coastal plain:

i. The Bassenden dunes (oldest, possibly Pleistocene) occurring furthest from the coast at the foot of the Darling scarp (fossil laterite); these dunes, now leached of their calcium carbonate, support *Banksia* woodlands and scrub heaths.

ii. The Spearwood system (late Pleistocene) which lies near the coast and consists of a core of aeolianite with a hard capping of secondary calcite, overlain by various depths of yellow or brown sand. The material was originally calcareous but leaching has removed the calcium carbonate from the upper portions. In wetter areas these dunes support *Eucalyptus gomphocephala* forests (Chap. 15) and in the drier north some heaths (below).

iii. The Quindalup dune system (post-glacial) which borders the present coastal line and supports the littoral dune vegetation (Chap. 23).

Phosphorus levels in the Bassenden dunes are measured at 10 ppm P in the surface metre of soil and 30–40 ppm below 1 m; these levels are similar to those in the eastern heaths.

5.1.3. Flora

Of the 2000 species that constitute the total assemblage of xeromorphs, very few are widely distributed, though most of the genera occur throughout the total area, and most genera occur also in the east (Chap. 4). For most genera, different species have evolved in different regions or in specific habitats. This is illustrated for some genera in Table 16.2. These data indicate that the eastern assemblage of species is quite different from the western assemblage and that the inland eastern assemblage is related to the south coast assemblage. Also, within the one region, assemblages on different soils are different as shown by the limestone, sandplain and laterite heaths, all of which occur in the same area, sometimes forming mosaics.

The monocotyledons are not represented in this table and the following generalizations are made: The greatest diversity of species, genera and families occurs in the wetter areas (west and south). Most of the local endemic genera are confined to wetter areas, e.g. *Anigozanthos, Blancoa, Dasypogon, Johnsonia* and *Macropidia. Xanthorrhoea* is most abundant in wetter areas and may be locally dominant, and the same distribution pattern is seen in the Orchidaceae, Liliaceae and Restionaceae. The Cyperaceae is more conspicuous in the drier areas, especially *Lepidosperma* and *Schoenus*. The Gramineae is either poorly represented or is absent, only two genera being recorded commonly, *Amphipogon* and *Neurachne*.

Herbaceous dicotyledons show a decrease in numbers from wet to dry, especially those which are characteristic of swamps, e.g. the carnivorous plants *Byblis, Drosera, Polypompholyx* and *Utricularia*.

5.2. Communities on the Western Coastal Plain

5.2.1. General

This group of communities extends from the Perth district northward to Shark Bay under a diminishing mean annual rainfall of c. 900 mm in the south to c. 250 mm in the north (Fig. 16.14). In the wetter south *Banksia menziesii* and *B. attenuata* form woodlands and these two species extend into drier areas to the north. *Eucalyptus todtiana* and *Casuarina fraserana* occur also in the wetter south and, together with the banksias form distinctive communities. The two species of *Banksia* continue northward, beyond the limit of *E. todtiana,* where they dominate or are important members of scrub heaths c. 2 m high. In still drier areas, around the 500 mm isohyet, where the soils of lowest fertility occur, the same two species of *Banksia* are found in or dominate "sand heaths" above or less than 1 m high. In the same area, the heaths on laterite occur, often forming mosaics with the sand heaths. With decreasing rainfall, the communities are taller and the sands yellow. *Actinostrobus arenarius* characterizes a complex of communities which extend in a general north-south direction between the 500 and 350 mm isohyets. In still drier areas and adjoining the arid zone *Banksia ashbyi* characterizes another group of communities 3–5 m tall which adjoin and are influenced by the arid zone flora.

In addition to these communities there are numerous patches of vegetation referred to as "thicket" in which species of *Acacia, Casuarina* and *Melaleuca* dominate. These are associated with some variation in the environment, such as soil alkalinity, rocky outcrops, salinity or some other factor which cannot be defined. These communities are given separate treatment below.

Fig. 16.15. *Banksia menziesii – B. attenuata* community with *Adenanthos cygnorum* dominant in the tall shrub layer and a lower shrub layer containing *Stirlingia latifolia, Eremaea fimbriata* and *Hibbertia hypericoides.* Herbs: *Burchardia umbellata, Thsanotus multiflorus.* c. 16 km south of Perth, W. A.

The aboriginal name "Kwongan" has been suggested by Beard (1976) to describe the sandplain and its treeless vegetation.

The classification and descriptions are based mainly on Speck (1958) and Beard (1976), and so also the species lists in Table 16.2. The assemblages have been designated "alliances", chiefly to conform with other chapters in this book. Each of the "alliances" could possibly be divided into two or more and doubtless a few hundred associations could be identified if a detailed analysis of the assemblages were done.

5.2.2. *Banksia menziesii* – *B. attenuata* Alliance

These species occur in association (Fig. 16.15) mainly in the wetter southern portion of the sandplain area (mean annual rainfall 750 to 900 mm). *B. prionotes* is sometimes an associate or may form pure stands. *B. ilicifolia* sometimes forms an association with *B. attenuata*, mostly on deep sands near swamps or on sand spits which traverse swamps. *Casuarina fraserana* is sometimes a tree associate, especially in the wetter southern portion. The communities are closed, or almost so, and reach a height of c. 12 m in the wettest areas. Well defined shrub and herbaceous layers are developed, the species being those which occur in the *Eucalyptus calophylla* woodlands which commonly surround the *Banksia* communities (Chap. 15). Depressions within the communities contain patches of heathland sometimes dominated by *Viminaria denudata*, the wetter areas with an herbaceous assemblage in which carnivorous plants are abundant.

North of the 750 mm isohyet *Eucalyptus todtiana* becomes a conspicuous member of the assemblage and is used to identify another group of communities (next section).

5.2.3. *Banksia menziesii* – *B. attenuata* – *Casuarina fraserana* – *Eucalyptus todtiana* Alliance

This group of communities stretches northward over the sandplain between the 750 and 500 mm isohyets (Fig. 16.14). *B. attenuata* and *B. menziesii* are usually the dominants, with subsidiary *Casuarina* (wetter south) and *E. todtiana*. The last becomes prominent in the drier areas where *Casuarina* drops out and *Nuytsia floribunda* is often locally dominant (Fig. 16.16). In the wetter south the communities reach a height of c. 10 m but diminish in size to c. 3 m tall with decreasing rainfall, when the stands become more open, so that the understory shrub layer becomes increasingly important from wet to dry.

Shrubs 2–3 m high are present in wetter areas, mostly species of *Acacia*, *Adenanthos*, *Banksia*, *Calothamnus*, *Casuarina*, *Dryandra*, *Grevillea* and *Hakea*, and *Xanthorrhoea preisii* c. 2 m high is sometimes conspicuous. A layer of shrubs 1 m high is present, the species being those which occur commonly in the sandplain heath (5.2.4, below).

Towards the drier limit of the alliance *Xylomelum angustifolium* occurs in association mainly with *Eucalyptus todtiana*. *Xylomelum*, together with species of *Banksia* continue into drier areas as scrub heaths (next section).

5.2.4. *Banksia* Scrub-heaths

Species of *Banksia*, of which *B. attenuata* is constant, form scrub heaths (Beard, 1976) around the 400 mm isohyet on bleached sand which sometimes overlies laterite, and rarely on yellow sand near the coast. The communities are a northern extension of those containing *Eucalyptus todtiana* which persists in depressions in the southern stands of scrub-heath. Near the coast, *Acacia rostellifera* is sometimes a component of the scrub-heaths and on the drier fringe *Actinostrobus* (5.2.5, below) and its associated shrubs are sometimes present.

The scrub-heath consists of scattered shrubs, mostly 1–2 m high, with a denser understory c. 1 m high. Taller stands may occur, usually near the coast, sometimes reaching a height of 5 m. *Banksia menziesii*, *B. prionotes* and *Nuytsia floribunda* are common in the upper stratum. The composition of the scrub varies from stand to stand. *Banksia prionotes* and *B. menziesii* are consistently present and *B. hookeriana* in association with *Xylomelum angustifolium* is locally common south of Dongara (Speck, 1958). The communities sometimes contain scattered mallees plants of *Eucalyptus dongarrensis*, *E. eudesmoides* or *E. tetragona*.

Fig. 16.16. *Eucalyptus todtiana* and *Nuytsia floribunda* (small trees at rear) and heath (foreground) containing *Banksia attenuata, Casuarina humilis, Synaphea petiolaris, Conospermum stoechadis, Stirlingia latifolia* and *Xanthorrhoea preissii.* c. 64 km north of Perth, W. A.

5.2.5. The Heaths

5.2.5.1. General

The heaths on the western coastal plain occur between the 750 and 300 mm isohyets. They can be divided into three groups, which are edaphically controlled: i. heaths on limestone; ii. heaths on deep sands which do not contain ferruginous lateritic gravel; iii. heaths on laterite. The three groups of communities are floristically different but a few species are common to all (Table 16.2).

The limestone heaths are confined to the coastal limestones and cover relatively small areas in comparison with the other two types. The heaths on sand are a northern extension into drier areas of the *Banksia attenuata – B. menziesii – Casuarina fraserana – Eucalyptus todtiana* alliance, the banksias persisting in the heaths as conspicuous species. The heaths on laterite are the northern extension of the understory layers of the *E. marginata – E. calophylla* alliance which dominates the laterite in the wetter south (Chap. 15). These heaths also contain many characteristic, restricted species. The heaths on sand and those on laterite sometimes form mosaics, and some mixing of the dominant species occurs.

The flora of the sandplain and lateritic heaths is extremely rich in xeromorphic taxa (Diels, 1906) and some species are endemic to restricted areas. For example, *Asterolasia phebalioides, Banksia tricuspis, Darwinia helichrysoides, Hakea megalosperma* and *H. neurophylla* occur only in a small area of hilly country on Jurassaic sediments near Mt. Lesueur (Univ. W. A., 1971).

The heaths are commonly burned by wildfires, possibly once every 5 years and regeneration of most if not all species occurs after a period of 3–5 years (Speck, 1958). If not burned for periods of up to 10 years, growth becomes very retarded, the plants flower sparsely and the bushes consist largely of almost leafless twigs, with the understory greatly reduced. When burned, such heaths are severely damaged.

Speck records the following sequence of stages following burning. The first plants to regenerate are those with underground perennating organs, notably the Amaryllidaceae, Haemodoraceae, Liliaceae, Restionaceae and Xanthorrhoeaceae. These, with some annuals, dominate the heath during the first year; shoots from shrubs, the bases of which survived the fire, are inconspicuous. The shrubs make rapid growth during the second year and assume dominance. He notes that some species are killed by fire, e.g., *Adenanthos sericea* and *Dryandra multiflora*, but these regenerate rapidly from seed and colonise denuded areas or bare sand.

5.2.5.2. Heaths on Limestone

Patches of closed heath 50–80 cm high, elongated parallel to the ocean and rarely exceeding a few hectares in extent, are found on limestone, usually on ridges from which the surface sand has been removed into depressions. They occur mainly under a mean annual rainfall of c. 500 mm (Seddon, 1972), with an isolated occurrence north of the Murchison River, where the mean annual rainfall is 350–300 mm. The heaths usually adjoin the *Acacia* communities on the littoral dunes and they usually contain species which belong to the littoral zone. In the south, they sometimes occupy small areas within the

Fig. 16.17. Heath on limestone. Main constituents: *Dryandra sessilis, Grevillea thelemanniana, Acacia cuneata, Conospermum triplinervium, Templetonia retusa, Senecio lautus* and species of *Eriostemon, Scaevola, Stackhousia* and *Conostylis*. Background: *Banksia attenuata* and *Xanthorrhoea preissii*. Near Yanchep, W. A.

Eucalyptus gomphocephala forests, where the soils are too shallow to support trees (Fig. 16.17).

The most abundant plants in the wetter areas are species of *Acacia* and *Melaleuca*. The main woody species are listed in Table 16.2. The common herbaceous species are *Lepidosperma gladiatum, Anigozanthos humilis, Conostylis aculeata* and *Xanthorrhoea* sp. (local). The assemblage is related to the understory of the *E. gomphocephala* forests rather than to the sandplain heaths which may adjoin the limestone heaths in the drier north. Also, some of the species grow in crevices in limestone cliffs, viz., *Acacia xanthina, Dryandra sessilis* and *Templetonia retusa*. Several of the species relate these heaths to those which are found near the ocean further south (see 5.3.2, below).

In the drier area north of the Murchinson R., Beard (1976) describes heaths on grey shallow sand on exposures of limestone. The heath is dominated by *Melaleuca leiopyxis* with the following associates: *Grevillea stenomera, Conospermum stoechadis, Casuarina humilis, Calothamnus chrysantherus* and *Hakea trifurcata*. It contains also species from adjacent communities, *Acacia ligulata* from the littoral scrubs and occasional plants of *Actinostrobus arenarius* and other species from the scrub heaths on the landward side. An isolated stand of *Eucalyptus erythrocos* (Chap. 15) is recorded in the heath.

5.2.5.3. Heaths on the Sandplain

These heaths (Speck, 1958) are characterized by a wealth of Proteaceae, including *Banksia attenuata* and *B. menziesii* (Table 16.2). They are mostly 60–120 cm high, sometimes with emergents. They occur on deep sands, varying in colour from white to pale yellow (drier areas) and derived mainly from leached siltstones, shales and sandstones from Jurassic sediments or coastal dunes and limestones. The heaths occur mainly north of the Moore R. (Fig. 16.14) between the c. 750 and 300 mm isohyets. They occur on flat or undulating country and are sometimes traversed by watercourses along which some fine-textured alluvium has been deposited on which *Eucalyptus wandoo* forms woodlands. The heaths contain a very large number of xeromorphic shrubs, the more common of which are listed in Table 16.2. The understory to the heaths contains mainly monocotyledons, especially species of *Anigozanthos, Blancoa, Burchardia, Chamaescilla, Conostylis, Diurus, Haemodorum, Patersonia* and *Sowerbaea*.

Associations have not been defined but three groups of communities are identified by: i. *Banksia attenuata – B. menziesii* in stunted form (most prominent in the south); ii. *Conospermum stoechadis*; iii. *Grevillea* spp. Speck (1958).

C. stoechadis, widespread in all heath communities, becomes locally abundant or dominant on the lower parts of gentle slopes (Fig. 16.18). Apart from the dominance of *Conospermum*, the floristic composition of the assemblage is virtually unchanged. *C. stoechadis* extends to the northern extremity of this assemblage and dominates some depressions in the *Actinostrobus* and *Banksia ashbyi* assemblages in drier areas. *Grevillea eriostachya* and *G. leucoptera* become prominent in the north in the Murchinson R. sandplain, often forming the local dominants in heaths 70–140 cm high.

Emergents in the heath occur, apparently haphazardly in upland areas, the emergents reaching a height of up to 2 m. The main emergents are *Adenanthos sericea* (Fig. 16.18) and *Nuytsia floribunda*. Other taller patches are brought about by additional moisture in depressions, often near watercourses where Myrtaceae become prominent, mostly species of *Agonis, Beaufortia, Calothamnus, Chamaelaucium, Leptospermum* and *Melaleuca*.

5.2.5.4. Heaths on Laterite

Heaths occur on laterite, which caps the old plateau, or on redistributed lateritic material, between 630 and 300 mm isohyets. The soils are shallow sands overlying ferruginous laterite, or skeletal and consisting of fragmented ferruginous material. These heaths may be regarded as the extension into drier areas on the understory of the *E. marginata – E. calophylla* forests, with additional species which are confined to the heaths. The communities occur in the same general area as the sandplain heaths and the two types sometimes form mosaics.

Floristically these heaths are the richest of all. The common shrub species are listed in Table 16.2. Although these heaths lie in close proximity or even adjacent to the sandplain heaths, the two floristic assemblages have relatively few species in common. On the laterite the conspicu-

Fig. 16.18. Heath on sandplain. *Conospermum stoechadis* (white flowers) throughout but locally dominant on lower slopes. At left, emergent *Adenanthos sericea* and *Xanthorrhoea reflexa*. Other species: *Calothamnus sanguineus, Dryandra tridentifera, Hakea baxteri, Stirlingia latifolia, Strangea stenocarpoides*. On hill behind, *Xanthorrhoea – Kingia* heath on laterite. Near Jurien Bay, W.A.

ous species are the emergent *Xanthorrhoea reflexa* and *Kingia australis* (Fig. 16.19) and the most common genera in the dominant shrub layer are *Dryandra, Hakea, Grevillea* and *Casuarina. Banksia* is usually absent.

The heaths vary in height from c. 60–130 cm, and are two layered. The upper layer contains the Proteaceae-*Casuarina* dominants, with emergent *Xanthorrhoea* and *Kingia*, and sometimes with emergents of *Nuytsia floribunda* (to 3 m) and rarely of *Banksia tricuspis*, which is confined to laterite. The lower layer is composed of subshrubs, dicotyledonous herbs or tufted monocotyledons. The smaller dicotyledons include species of *Comesperma, Dampiera, Drosera* and *Stylidium*. The monocotyledons include species of *Anigozanthos, Borya, Burchardia, Caustis, Dasypogon, Haemodorum, Johnsonia, Patersonia, Sowerbaea* and *Thysanotus*. The black Kangaroo Paw, *Macrophidia fuliginosa* (Fig. 16.19) is confined to this community. One grass is recorded, *Neurachne alopecuroides*.

5.2.6. *Actinostrobus arenarius* Alliance

This conifer, because of its distinctive shape (Fig. 16.20) readily identifies a group of communities referred to by Speck (1958) as "coniferous scrub heath". They occur between the 500 and 350 mm isohyets on sandplain, the sand being mostly yellow, or white in the wetter areas. The land is flat or undulating and often traversed by sand ridges which rise to a height of a few metres above the plain. Rocky areas within the alliance are rare and some ferruginous lateritic residuals occur, on which *Casuarina* is dominant, usually with acacias (see 5.2.8, below).

The alliance is a mosaic of tall scrub communities which occupy the elevated sand ridges (Fig. 16.20), whereas the lower areas support sandplain heaths. This arrangement applies particularly to the drier areas; in the wetter western segment the taller plants sometimes forms thickets or tall scrubs on the flats.

Tall plants, often 5 m tall are *Actinostrobus arenarius, Adenanthos cygnorum, Banksia attenuata, B. menziesii, B. prionotes, B. sceptrum, Grevillea annulifera* (drier), *G. biformis, G. eriostachya, G. leucoptera, Nuytsia floribunda* (wetter) and *Xylomelum*. Also *Acacia rostellifera* may be found near the coast. Eucalypts occur in some stands, especially *E. oldfieldii*. The assemblages are not uniform throughout. Of the species mentioned, *Adenanthos, Banksia attenuata* and *B. menziesii* are

Fig. 16.19. Heath on laterite with emergent *Xanthorrhoea reflexa* and *Kingia australis*. Main shrubs are *Lambertia multiflora, Hibbertia hypericoides, Casuarina humilis, Calothamnus sanguineus, Hakea conchifolius* and species of *Dryandra, Petrophile, Haemodorum* and others. Note inflorescence of *Macropidia fuliginosa* (Black Kangaroo Paw) against man wearing white pullover. Near Jurien Bay, W. A.

Fig. 16.20. General view of heath-scrub mosaic. On sandhill behind, *Actinostrobus arenarius* and *Banksia* spp.; centre shows heath dominated by *Conospermum stoechadis;* foreground, shrubs of *Grevillea dielsiana*. c. 50 km north of Ajana, W. A.

more abundant in wetter areas on white sand. *Actinostrobus, B. prionotes, B. sceptrum* and *Xylomelum* flourish best on yellow sand (Beard, 1976).

The understory to these scrubs is a layer up to 1 m high of xeromorphic plants of which c. 100 are listed by Beard. These species occur on the flats and constitute heaths, usually without emergents. One of the most common species is *Conospermum stoechadis* (Fig. 16.20) and other more common ones are listed in Table 16.2. The herbaceous layer contains *Conostylis aculeata, Lomandra hastilis, Patersonia occidentalis* and others.

5.2.7. *Banksia ashbyi* Alliance

This species characterizes an alliance on red or red-brown sand reaching the sea at Freycinet Estuary in the north and extending southward around the 280–250 mm isohyets. The assemblage is the most northerly of the series of *Banksia* communities and is remarkable because of its height (4½–6 m). It adjoins the arid zone communties of drier climates and sometimes contains arid zone tall shrubs, especially *Acacia ramulosa, A. ligulata, A. tetragonophylla, Cassia nemophila* and *Duboisia hopwoodii*.

The alliance extends over large areas of stabilized sand dunes, sometimes with limestone at the surface, and the continuity of the sand sheets is sometimes broken by salty patches with support *Melaleuca* spp.

Banksia ashbyi occurs throughout the alliance (Fig. 16.21) and is often co-dominant with *Grevillea gordoniana*, except towards the west where the latter is replaced by *Hakea stenophylla, Melaleuca huegelii* and *Acacia ligulata*. Other tall components are scattered eucalypts (Chap. 14), *Beaufortia dampieri, Bursaria spinosa, Calothamnus chrysantherus* and *Lamarchea hakeifolia*. A large number of shrubs are recorded by Speck (1958) and Beard (1976) which are listed in Table 16.2. Beard records an herbaceous layer of *Plectrachne danthonioides, Ptilotus exaltatus* and *Trichodesma zeylanica*.

Fig. 16.21. *Banksia ashbyi* scrub with an understory of *Acacia longispinea, Micromyrtus imbricata* and *Calytrix brevifolia*. 48 km south of Billabong, W.A.

5.2.8. Thickets Dominated by *Acacia*, *Melaleuca* and *Casuarina*

Several species of these genera constitute thickets within heaths and *Banksia* shrubs. The thickets occur mainly on calcareous, subsaline or stony-gravelly soils (often lateritic), less commonly on sands. Some of the species occur to the east of the semi-arid heath areas (see 5.3.3, below). The following communities are described by Beard (1976).

5.2.8.1. *Acacia rostellifera* Thickets

This species is common on coastal sand dunes (Chap. 23). It occurs inland and close to the coast on soils derived from limestone, mainly north of Dongara. It forms thickets on red soils with some clay, some being alluvial. On deeper soils it is dominant, reaching a height of c. 6 m. Associates are *A. ligulata*, *A. scirpifolia*, *A. xanthina*, *Eucalyptus eudesmoides*, *E. oleosa* and *Melaleuca cardiophylla*. The last becomes increasingly prominent as soils become shallower and limestone approaches the surface (see next section).

5.2.8.2. *Melaleuca cardiophylla* Thickets

This species dominates dense scrubs or thickets up to 4½ m tall on coastal limestones mainly in rocky situations. An inland occurrence is reported for the Murchison R. gorge on "white chalk". The tallest communities are 3–4½ m high, with *Melaleuca* in pure stands or with *Acacia ligulata* as an occasional species. Understory species are few, mainly *Rhagodia obovata* and *Ptilotus obovatus*. Variations include an association of *M. cardiophylla* with *Thryptomene baeckeacea* to form an open heath 30–40 cm high with grass associates (*Plectrachne* sp., *Poa* sp. and *Stipa crinata*). As a possible result of fire, *M. cardiophylla* is replaced by *M. leiopyxis* in some areas (see 5.2.5.4, above).

5.2.8.3. Other Species of *Melaleuca*

M. uncinata, widespread across southern Australia (Chaps. 12 and 14) forms thickets 2–3 m high in depressions subject to waterlogging. *M. thyoides* forms similar thickets on subsaline soils inland, commonly around communities of Samphire.

5.2.8.4. *Melaleuca megacephala* – *Hakea pycnoneura* Thickets

These two species are recorded in association east and south-east of Geraldton on stony slopes. The thickets reach a height of c. 2 m; emergents to 3 m of *Nuytsia floribunda* and *Santalum acuminatum* sometimes occur and, rarely, *Callitris "glauca"* on deeper sand. The thicket is a mixture of several species, the more common associates being *Acacia ericifolia*, *A. saligna*, *A. ulicina*, *Calothamnus quadrifidus* and others.

5.2.8.5. *Acacia – Casuarina – Melaleuca* Thickets

Species of these genera, in various combinations, form thickets 2–3 m tall in the drier areas around the 300–250 mm isohyets. They occur on red sand or on sand over laterite. Those on sand are related to the *Banksia ashbyi* shrubs (see 5.2.7, above). Those containing *Casuarina* spp. occur on more compact soils and the species of *Casuarina* extend eastward (see 5.3.3, below). The species composition varies. The main component species are *Acacia aciphylla*, *acuminata*, *A. blakelyi*, *A. longispinea*, *A. stereophylla*, *Casuarina acutivalvis*, *C. campestris*, *Melaleuca eleutherostachya*, *M. nematophylla*, *M. uncinata*, together with *Dryandra sessilis*, *Eremaea beaufortioides*, *Grevillea eriostachya*, *G. stenobotrya*, *Hakea auriculata*, *H. multilineata*, *Lamarchea hakeifolia* and *Xylomelum angustifolium* (Fig. 16.22). Eucalypts are sometimes present (*E. oleosa*, *E. eudesmoides* and *E. oldfieldii*). A shrub layer c. 1 m high is present, composed of xeromorphic species.

5.3. Communities on the Southern Coastal Plain and Inland

5.3.1. General

These communities occur along the southern coastal plain from the vicinity of Albany, extending eastward to the limestone cliffs at the western end of the Great Australian Bight. The communities close to the ocean are confined in the south behind the low coastal dunes and in the north by a belt of mallee, which separates

Fig. 16.22. *Casuarina campestris* and *Xylomelum angustifolium* (foreground, with fruits). Understory contains *Verticordia acerosa, Hakea platyspermum, Eremaea fimbriata* and *Calectasia cyanea*. Near Quairading, W. A.

the coastal assemblages from the more inland ones which extend northward in patches almost to the boundary of the arid zone (Fig. 16.14). The belt of mallee which separates the two groups of heaths does not appear to prevent the interchange of species between the assemblages since many species which are characteristic of the heaths occur in the mallee and some species occur in both the coastal and inland heaths. It is probable that the inland assemblages were derived from the coastal assemblages when the modern rainfall gradient was established.

The communities are described under two headings and divisions into smaller units are indicated when this is possible.

5.3.2. The Coastal Heaths and Scrub-Heaths

These communities occur close to the coast behind the coastal dunes in an area receiving a mean annual rainfall mostly between 600 and 400 mm, the rain falling mainly in winter; they are subject to summer drought.

The communities occur on white to yellowish sand, mostly 1 m deep or more (Beard, 1972a, 1973b). When lateritic gravel or small amounts of clay are incorporated in the profile eucalypts are present. On lateritic gravel *E. tetragona* or *E. incrassata* occur to produce mallee-heaths. On fine-textured soils, which are rare but sometimes edge watercourses, *E. occidentalis* forms mallee stands or even woodlands (Chap. 14). Also, in many stands stunted mallees, c. 1 m high occur intermixed with the heath species and of the same height, especially *E. conglobata, E. leptocalyx* and *E. lucasii*.

The most conspicuous species in the stands are *Banksia speciosa, Lambertia inermis* and *Nuytsia floribunda,* but these species are not consistently present; some of the variations that exist can be attributed to local waterlogging induced by an impervious layer in the subsoil.

Generally, the heaths vary in height between 40 and 160 cm, sometimes with local emergents 2–3 m high. Beard lists c. 20 species exceeding 120 cm high, which commonly form the upper story, viz., the three mentioned above and *Adenanthos cuneatus, Agonis obtusissima, Baeckea uncinella, Banksia media, Casuarina humilis, Dryandra quercifolia, Grevillea hookerana, Hakea adnata, Isopogon axillaris, Melaleuca nesophila, Petrophile linearis, Pimelea lehmanniana* and a few others. The common smaller shrubs are included in Table 16.2. Plants listed for the herbaceous layer are *Anigozanthos rufa, Conostylis* spp., *Diuris filifolia, Stylidium pseudohirtum,* and Cyperaceae and Restionaceae. Also, *Mac-*

rozamia reidlei and *Xanthorrhoea preissii* are often present and their leaves may project above the general level of the heath.

In the climatically wetter areas, mostly with a mean annual rainfall exceeding 600 mm, *Nuytsia floribunda*, up to 3 m tall is locally dominant in patches where some waterlogging occurs; the heath, c. 1 m high, contains *Banksia speciosa, B. petiolaris, B. media, Hakea adnata, Isopogon alcicornis, Lambertia inermis, Macrozamia reidlei* and *Xanthorrhoea preissii*. More heavily waterlogged areas in depressions in such heaths support scrubs of *Melaleuca preissii* surrounded by a zone of *Banksia littoralis* (Chap. 22).

5.3.3. The Inland Heaths and Scrub-Heaths

The inland heaths and scrub-heaths occur between 350 and 280 mm isohyets, the rain falling mainly during the cooler months. Mean monthly values indicate that about two-thirds of the annual rainfall is more or less equally distributed among the 6 months March to July. The remaining one-third is more or less equally distributed among the remaining months. The climate is therefore semi-arid, bordering on arid in the drier north-east and the remarkable assemblage of plant species is unique in the continent.

The communities are confined to deep yellow sands which form mosaics with soils containing lateritic material, usually in the form of ironstone gravel, and these soils support either thickets of *Casuarina, Acacia* and/or *Melaleuca* which form ecotonal communities with the heaths. Also, the heaths sometimes contain emergent mallee eucalypts which produce mallee-heaths (listed in Table 16.2).

The total flora is relatively large, differing from the assemblage on the south coast in several respects, notably in the abundance of *Grevillea* spp. and their prominence in the communities. The assemblages change from wet to dry, the species in the wetter south being more closely related to the coastal assemblage than are the more northerly assemblages where the effects of aridity are indicated by the occurrence and local dominance of species typical of the semiarid to arid tract in Western Australia. Since a climatic gradient exists from south to north the assemblage has been divided into two segments, a wetter southern and a drier northern one. Beard's (1969, 1972a, b, c) species lists for the common dictyledons are presented in Table 16.2. Lists of monocotyledons have not been published. The main genera recorded in the inland heaths are *Conostylis, Calectasia, Dianella, Lomandra, Lepidosperma, Schoenus* and the grass *Neurachne alopecuroides*.

The two groups of communities have in common an upper shrub layer of mixed composition, mostly c. 60 cm to 1 m tall and a lower layer of shrubs 30–40 cm high. The herbaceous layer is very scattered and the yellow sand is often visible at the surface.

Above the general canopy level emergent shrubs of *Grevillea excelsior* (plus *G. pteridosperma* in the north) rise to a height of 3–4 m. They often have crooked stems which bend over to form an arch (Fig. 16.23). These are not always the most abundant species, but they are conspicuous because of their height. Beard suggests that *G. excelsior* is a short-lived pioneer species, which appears rapidly after fire, later giving way to slower-growing, slower-regenerating shrubs.

The main features of the northern and southern floristic assemblages are summarized as follows (see also Table 16.2):

i. The Southern Communities

Grevillea excelsior is the common emergent, but is not consistently present. Upper stratum dominates commonly forming a discontinuous canopy are: (those marked (b) occur also in the northern assemblage) *Banksia elderana* (b), *B. laevigata, Calothamnus quadrifidus, Casuarina acutivalvis* (b), *Grevillea concinna, G. didymobotrya* (b), *G. integrifolia, G. rufa, Hakea falcata* (b), *H. multilineata* (b), *Isopogon axillaris, I. scabriusculus, Petrophile seminuda, P. semifurcata* (b). Definite associations have not been recognized. The understory contains a large number of small shrubs, listed in Table 16.2.

ii. Northern Assemblages

The emergents include both *Grevillea excelsior* and *G. pterosperma*. The upper shrub layer includes the species marked (b), above, plus *Banksia audax, Casuarina corniculata, C. dielsiana, Grevillea biformis, Petrophile conifera* and *P. ericifolia*. In addition, the assemblages commonly contain taller species which are typical of the more arid regions, e.g., *Callitris preissii* ssp. *verrucosa* and *Santalum acuminatum*, which may reach a height of c. 3 m. Also, species

Fig. 16.23. *Grevillea excelsior* as an emergent in heath containing *G. hookerana, Petrophila semifurcata, P. circinnata, Verticordia chrysantha* and *Dampiera lavandulacea* as the most common species. c. 70 km east of Southern Cross, W.A.

of *Acacia* are more prominent, especially towards the north where the assemblage merges into *Acacia* thickets.

Other noteworthy features are: (a) the occurrence of several genera which are confined to semi-arid regions, e.g. *Balaustion* (confined to this area), some Myrtaceae, especially *Micromyrtus, Thrytomene* (*T. roei* is sometimes dominant in the understory) and *Wehlia,* and some Verbenaceae (*Lachnostachys* and *Pityrodia*). The Myrtaceae are of special interest because they occur in the few arid-zone heaths which have persisted in the drier inland (see 5.6, below); (b) The hummock-forming grasses *Plectrachne rigidissima* and *Triodia scariosa*, common in arid mallee communities, are sometimes minor constituents of the drier heaths.

Two ecotonal communities of regular occurrence are:

i. With *Casuarina* spp. The yellow sands which support the heaths often contain ironstone gravel at the periphery of the deposit and on these gravel/sand mixtures *Casuarina acutivalvis* and *C. corniculata* become dominant and suppress the growth of heath species. However, many of the heath species persist in the *Casuarina* thickets, especially *Banksia audax, B. elderana, Grevillea excelsior, G. hookerana, G. didymobotrya, Hakea multilineata* and many smaller species. In more northerly drier stands species of *Acacia* (Table 16.2) are often common or co-dominant with *Casuarina*.

ii. With *Acacia* spp. At the drier northern extremity, the heaths are gradually replaced by thickets of *Acacia* which become dominant around the 240 mm isohyet. The deep sands in this rainfall region are mostly yellow-brown or brown. The main species in the *Acacia* thickets on sands comparable in depth (>90 cm) to that supporting the heaths is *A. resinomarginea* which may form pure stands 3–6 m high. Some heath species occur in these stands, *Grevillea excelsior, Hakea multilineata, Banksia elderana* and a few others. *Casuarina* spp. (as above) and *Callitris preissii* may also occur.

5.4. Communities on the Coastal Sands along the Wettest Part of the Coast

5.4.1. General

Along the wettest part of the south-western coastline, where the mean annual rainfall is mostly c. 600 mm and up to 1500 mm, the sandy coastal plain is narrow, rarely exceeding 5 km wide, and it is sometimes absent where rocky headlands adjoin the ocean. The stretch of the sand corresponds approximately with the area of *Agonis flexuosa* (Fig. 16.14). On the seaward side the sandy coastal plain adjoins the coastal dunes. On the landward side on the south coast and part of the west coast (to the region of Busselton) the sands adjoin and mix with the lateritic material washed downslope from the fossil profiles. This lateritic material supports stunted communities of *Eucalyptus marginata* and *E. calophylla*. On the west coast, a belt of limestone skirts the ocean northward from the Busselton region and supports forests of *E. gomphocephala*. Small patches of limestone occur in places above the basal gneiss, which is seen at the surface only in a few places along the immediate coast-line, as at Cape Leeuwin.

The sandy coastal plain supports a mosaic of communities, the distribution of which are governed mainly by topography. In the most elevated areas on headlands, heaths occur which, with increasing protection from the wind, are replaced by low scrubs and then by woodlands. On the sandy coastal plain the degree of protection from the southern and westerly winds and waterlogging determine the vegetation patterns.

In waterlogged areas sedgelands occur. These sometimes support scattered or clumped shrubs, mainly ***Leptospermum firmum***, to produce open heaths, or they are traversed by low sand ridges supporting *Banksia ilicifolia*. Also, relatively

Fig. 16.24. *Agonis flexuosa* woodland with an understory of *Phyllanthus calycinus* and introduced grasses. Near Busselton, W.A.

large areas are covered by swamps supporting communities of *Melaleuca raphiophylla* and *M. preissiana*. These waterlogged communities are dealt with in more detail in Chap. 22.

On the better drained sands two sequences of communities from the ocean inland can be recognized:

i. When the coastal dune is high (usually c. 3 m) the steep lee-slope of the dune supports scrubs of *Agonis flexuosa*, and immediately behind the dune and in its shelter, woodlands of this species occur (Fig. 16.24).

ii. Where the coastal dune is low (mostly 1–2 m high) a heath or scrub of *Acacia decipiens* ± *Jacksonia horrida* adjoins the coastal dune and the coastal dune species (e.g. *Acacia rostellifera*, *Olearia axillaris* and others, see Chap. 23) mix with *Acacia decipiens*. *Agonis flexuosa* commonly occurs in a stunted form with *Acacia decipiens* and with increasing distance from the ocean it becomes larger and dominant, forming a woodland.

5.4.2. Heaths on Headlands

Small areas of heath occur on limestones on the west coast on headlands which are elevated above the ocean so that they are not affected by sand-deposits and not significantly influenced by salt spray (Fig. 16.25). In the most exposed situations facing the ocean the heaths are c. 20 cm high but increase in height inland.

The main heath species include *Acacia cochlearis, A. cuneata, A. rostellifera, Baeckea ambigua, Calothamnus sanguineus, Casuarina humilis, Daviesia horrida, Dryandra nivea, Hakea prostrata, Hybanthus calycinus, Hypocalymma angustifolia, Pimelea rosea, Templetonia retusa* and *Thryptomene saxicola*. The herbaceous layer contains *Conostylis seoriflora, Patersonio occidentalis, Senecia lautus* and *Stypandra glauca*. The thread-like *Cassytha micrantha* parasites many hosts. The assemblage has some species in common with the coastal heaths developed on limestone in the north (Fig. 16.25).

Fig. 16.25. Heath on a limestone headland. Main species are *Acacia rostellifera, Olearia axillaris, Thryptomene saxicola, Calothamnus sanguineus* and *Daviesia divaricata*. Cape Naturaliste, W.A.

Agonis flexuosa in a stunted form occurs in these heaths and sometimes stunted *Eucalyptus calophylla*. With increasing protection downslope these species gradually increase in height and form a scrub or low woodland with the following understory associates in various combinations: *Banksia attenuata, B. grandis, Daviesia horrida, Hibbertia tetrandra, Macrozamia reidlei* and *Xanthorrhoea preissii*. Alternatively, on shallow soil overlying limestone *Melaleuca huegelii* forms a close scrub.

5.4.3. *Agonis flexuosa* Alliance
Peppermint

A. flexuosa, a small tree reaching a height of c. 16 m occurs along the south-west coast from the mouth of the Gardner R. to c. 50 km south of Perth (Fig. 16.14). It reaches its maximum size on deep sands in sheltered situations and forms woodlands or low forests between high coastal dunes and the *Eucalyptus* forests on the landward side (Smith, 1972/73). It is often the main or only understory tree in some of these forests, especially below *E. gomphocephala* on the west coast and to a lesser extent the *E. diversicolor* and *E. marginata – E. calophylla* forests in the south-west and south (Chap. 15). It occurs also, usually in a stunted form, on sands in exposed situations and on limestone when it is usually associated with another or other tall species.

As a tall woodland or forest *A. flexuosa* occurs in pure stands, usually with a closed

Fig. 16.26. *Eucalyptus ficifolia, Casuarina fraserana* and *Agonis flexuosa*. Shrub layer mainly *Kunzea preissiana*. Herbaceous layer: *Lepidosperma gladiatum* (foreground), *Anarthria scabra, Patersonia occidentalis, Glossodia emarginata* and *Caladenia flava*. Near Peaceful Bay, W.A.

Fig. 16.27. Low woodland of *Banksia grandis* and *Agonis flexuosa*. Shrubs are *Hakea oleifolia*, *Daviesia* sp. and *Pimelea rosea* (in flower). On sand, c. 3 km from the coast. Near Peaceful Bay, W. A.

canopy (Fig. 16.24). Understory species are usually scanty (which may be the result of burning), the most common being *Acacia decipiens*, *Anthocercis littoralis*, *Leucopogon australis*, *Phyllanthus calycinus* and *Pimelea argentea*, and the herbs *Lepidosperma gladiatum* and *Anthocarpus preissii*.

In more open stands, which are scrubs, or mallee-like as a result of fire, the shrub associates are more abundant and include *Acacia cochlearis*, *A. cyclops*, *Daviesia horrida*, *Dryandra sessilis* (sometimes co-dominant or shallow sands on limestone), *Spyridium globosum*, and locally on limestone *Melaleuca huegelii*. *Macrozamia reidlei* and *Xanthorrhoea preissii* are sometimes present in hollows.

Floristically richer assemblages occur near its eastern limit where the scrubs occupy the sheltered sides of dunes, and in these localities the understory is strongly influenced by the richer assemblages of the district. For example, in the Bremer Bay area, Beard (1972a) records a mallee-like assemblage of *Agonis flexuosa* with an understory of the species mentioned above, plus many others, including *Acrotriche cordata*, *Adriana quadripartita*, *Boronia crenulata*, *Gyrostemon sheathii*, *Hibbertia cuneiformis*, *Phebalium rude*, and the herbs *Conostylis seoriflora*, *Pelargonium littorale*, *Waitzia citrina* and *Zygophyllum apiculatum*.

A noteworthy woodland association contained within the *Agonis flexuosa* alliance is one containing *Eucalyptus ficifolia*, a bloodwood with pink to crimson stamens which is planted extensively in Australia as an ornamental. This species, which reaches a height of c. 10 m, is restricted to an area of c. 50 sq.km near the coast (Fig. 16.14). It rarely occurs in pure stands but is usually associated with *Casuarina fraserana* and/or *Agonis flexuosa* (Fig. 16.26).

5.4.4. Communities Dominated by *Banksia* spp.

In the ecotonal zone between the *Agonis flexuosa* communities and the *Eucalyptus marginata – E. calophylla* stunted woodlands, some species of *Banksia* are locally dominant over small areas. These are *B. ilicifolia* and *B. grandis*.

B. ilicifolia, which occurs extensively to the north on sand, sometimes in association with *B. attenuata*, is usually associated with the margins of sandy swamps (Chap. 22). It is found on the southern coastal plain as a low open woodland on low sand-ridges which rise above and stretch across low swampy flats. The ground has an open cover of shrubs and sedges (Smith, 1972).

B. grandis commonly occurs as a small understory tree in the *E. marginata – E. calophylla* forests. In this southern area in the zone of stunted eucalypts, close to or fringing the *Agonis flexuosa* zone, *B. grandis* forms woodlands either in pure stands over small areas or in association with *Agonis flexuosa* (Fig. 16.27).

With increasing exposure close to or on the ocean front on sands which sometimes overlie rock above rocky bays, *B. grandis* forms dense scrubs 2–3 m high. It occurs either in clumps or in association with other species of the same or similar height, including *Lambertia inermis*, *Nuytsia floribunda*, *Banksia baueri* and *B. baxteri* with an understory of *Beaufortia*, *Boronia*, *Casuarina*, *Dryandra* spp. and others. *Xanthorrhoea preissii* is often present and is sometimes dominant (Fig. 16.28).

5.4.5. *Acacia decipiens* Alliance

A. decipiens occurs as the dominant on exposed sandy flats or undulating areas adjoining low coastal dunes which afford little or no protection from the wind. The communities are open to closed heath-like scrubs 1–2 m high. The associated species are *Jacksonia horrida* which may be locally dominant, with occasional occurrences of *Casuarina humilis*, *Spyridium globosum* and *Leucopogon parviflorus*. *Dryandra sessilis* and *Melaleuca huegelii* occur on shallow soils; *M. lanceolata* occurs in somewhat saline areas and *Agonis juniperina* near swamps (Smith, 1972).

On the seaward side, the stands intermingle with the dune species, especially *Olearia axillaris* and *Scaevola crassiflora*. *Agonis flexuosa* in a stunted form is a rare to occasional constituent and it becomes increasingly abundant and taller with increasing distance from the sea. In these ecotonal zones *Banksia attenuata* and *B. grandis* may occur as tall shrubs, sometimes with *Xanthorrhoea preissii* in the understory.

Fig. 16.28. Windshorn scrub of *Banksia grandis* with *Xanthorrhoea preissii* behind. Other species include *Agonis flexuosa*, *Dryandra sessilis*, *Eucalyptus calophylla* (in mallee form), *Olearia axillaris*, *Kennedia coccinea* and *Hardenbergia comptoniana*. Sandy coast overlooking ocean, near Cape Leeuwin, W. A.

5.5. Heaths and Related Communities on Ranges and Hills

5.5.1. General

Mountain ranges and hills with bare rock at their summits are scattered throughout the south-western portion of Western Australia. The largest of these are the Stirling Ranges (Fig. 16.14) and the Porongorups where peaks rise to a height of c. 1200 m above sea level. Hills of smaller dimensions occur throughout the area, rising to a height of c. 200 m above the lowlands, with a maximum height above sea-level of c. 780 m. The most interesting of these botanically are the Barren Ranges in the south.

The rock comprising the hills is mainly granite, part of the vast granitic platform which underlies most of the south-western part of the continent. The Stirling Ranges, on the contrary, are composed of Upper Proterozoic metamorphosed sediments, quartzites and sandstones. Also, ironstone, presumed remnants of the duricrust, occur and small areas of quartzites and basalt.

The hills support distinctive floras, sometimes unique because of the endemics (Gardner, 1942). In the Stirling Ranges, for example, c. 100 endemic species are confined to this small area, mainly of *Banksia, Darwinia, Hypocalymma* and *Isopogon*. Many species are confined to the various hills in both the wetter and/or drier areas, including several species of *Eucalyptus* (Chap. 14) and many shrubs or small trees of which a few only are mentioned: *Acacia restiacea, A. lasiocalyx, Casuarina huegeliana* (wetter areas) and *Kunzea sericea, Leptospermum erubescens, Plagianthus helmsii* (drier areas).

Most of the hills are domed at the summit but some are flat-topped. The latter are sometimes fissured or traversed by dykes and present a rectilineal pattern of plant communities, the cracks or dykes, where soil accumulates and supports taller communities (Beard, 1967). Bare rock surfaces are sometimes pitted with depressions which form pools when rain falls.

Runoff from bare rock surfaces irrigates crevices, the lower slopes and any enclosed valleys that occur. This irrigation water creates wet environments, at least seasonally, and accounts for the occurrence of some outliers in macroclimates which are otherwise too dry for the species, e.g., *E. diversicolor* forests in the Porongorups (Chap. 15), and the extension of *Hakea prostrata* and *Nuytsia floribunda* into the semi-arid zone.

The hills extend from the coast with a mean annual rainfall of c. 850 mm to the arid zone (250 mm) but it is impossible to estimate the effectiveness of these rainfall figures because of the water runoff factor. There are, however, distinct changes in floristics from wet to dry, especially with the woody dominants, as indicated below.

5.5.2. Colonization of the Rock Surfaces

In the wetter parts of the area, Smith (1962) describes the first colonizers as algae (e.g. *Calothrix*), followed by crustose or foliose lichens (species of *Caloplaca, Cladonia, Parmelia* and *Siphula*). The mosses *Campylopus bicolor* and *Breutelia affinis* then form mats, together with other mosses and liverworts. In these mats and in crevices with accumulated organic material small herbs occur, including the pteridophytes *Cheilanthes tenuifolia, Ophioglossum lusitanicum, Phylloglossum drummondii, Pleurosorus rutifolius* and occasionally *Isoëtes* spp.

In shallow soils a rich herbaceous flora develops in winter, containing *Pelargonium drummondii* as a common species and bulbous Liliaceae, Orchidaceae, *Drosera, Polypompholyx, Stylidium,* Compositae and Centrolepidaceae.

A noteworthy Liliaceae is *Borya nitida* which forms cushions. Its leaves tolerate desiccation (Gaff and Churchill, 1976) and the species occurs in the wet areas described above and extends at least into the semi-arid zone as the first colonizer of the hilltops. The genus occurs disjunctly in the continent (see 3.4.1, above).

With an increase in soil depth heaths develop, commencing in crevices among rocks. The species composition varies from area to area, as described below. Taller species, chiefly of *Casuarina* or *Eucalyptus* (Chap. 14) sometimes establish in the heath patches. *Casuarina huegeliana* which reaches a height of c. 6 m, is one of the characteristic plants confined to granite exposures from the south inland to the 300 mm isohyet. It is usually accompanied by *Acacia lasiocalyx* (also restricted to the rocky outcrops) as an understory dominant and *Borya nitida* forms hummocks below.

The heath communities vary greatly in floristic composition and in the sizes of patches. The smallest patches are composed of a few or even a single plant established in a rock crevice with only mosses or lichens as associated species; the largest may cover a few hectares though the continuity of the canopy is commonly broken by exposed boulders or sheets of rock (Fig. 16.29).

The communities are divisible into three groups: i. Those near the ocean on headlands or in exposed situations which are subjected to the shearing effects of winds and/or to salt spray; ii. Those in the wetter south near the coast but not obviously affected by strong winds or salt spray; iii. Those in the drier inland. The communities are described under these headings.

5.5.3. Heaths on Headlands or on Exposed Hilltops Close to the Ocean

The lists available indicate that floristic assemblages vary from hilltop to hilltop and that no species is constant in all assemblages. *Pimelea ferruginea* appears to be the most commonly represented species in the various stands. The communities have structural characteristics in common, namely the dominance of domed or windshorn shrubs, a condition presumably induced by high wind velocities. Two groups of communities are described:

i. In the Bremer Bay area on a granite pavement near the ocean at Point Gordon, Beard (1973b) records a domed heath consisting of *Dryandra pteridifolia* c. 1 m high, with subsidiary *Banksia dryandroides, Isopogon buxifolius* and *Melaleuca violacea*, with the spaces between the bushes filled in by a dense heath 15–25 cm high of *Andersonia parvifolia* and *Hakea marginata*. On deeper soils in more sheltered places this community is replaced by a scrub of domed *Banksia media, Darwinia diosmoides, Melaleuca pentagona* and *Acacia decipiens*. In the same area but at a higher level above the ocean the bare granite summits are dotted with domed shrubs of *Pimelea ferruginea, Isopogon formosus, Hibbertia cuneiformis, Acacia cyclops, Hakea prostrata* and others.

ii. On the Recherche Archipelago, Willis (1953) records a dwarf cushion shrubbery on granite summits containing *Andersonia sprengelioides* with herbaceous *Borya nitida* and *Stylidium pubigerum* below. On steep slopes on

Fig. 16.29. Granite outcrop with shrubs of *Calothamnus quadrifidus* (common) *Leptospermum* sp. and *Callitris preissei* ssp. *verrucosa. Restio* sp. in foreground. Peak Charles, W.A. Photo: Dr. J.S. Beard.

granite on these islands he records a mixed shrub-thicket of dwarfed plants of *Astartea fascicularis, Dodonaea oblongifolia, Eutaxia obovata, Hakea clavata, Kunzea sericea, Leucopogon obovatus, Pimelea ferruginea* and *Spyridium globulosum.*

5.5.4. Heaths on Rocky Outcrops in Wetter Areas

These outcrops, like those near the ocean, individually support floristically different communities, though many species are common to several or many peaks. These species are: *Astartea ambigua, Borya nitida, Calothamnus quadrifidus, Casuarina campestris, Hakea laurina, Leptospermum erubescens, Melaleuca fulgens* and *Thryptomene australis* (Beard, 1972a, 1973b).

Borya nitida is usually the first vascular coloniser and forms cushions up to c. 25 cm diameter and 5 cm high. This species usually persists in the understory of the heaths, and *Acacia lasiocalyx* and *Casuarina huegeliana* are usually present in the area forming a scrub or low woodland. Eucalypts sometimes occur, *E. preissiana* and others, many of them with very restricted areas (Chap. 14). Rock type appears to play some part in delimiting species and three assemblages are commented on:

i. On granite, the species mentioned above are the main components of the heaths. Additional species belong mainly to the genera *Andersonia, Anthocercis, Boronia, Darwinia, Leptospermum, Platysace, Prostanthera* and *Stylidium.*

ii. On quartzite, Beard records two different assemblages, one from the Barren Ranges, the other from Peak Charles. The former contains, among others, *Regelia velutina, Calothamnus validus, Melaleuca citrina, Banksia quercifolia* and *Baeckea ovalifolia.* The latter supports a quite different assemblage dominated by *Leptospermum* sp. growing in crevices, and below on slopes, thickets 1.5–2 m high of *Calothamnus quadrifidus, C. gilesii, Melaleuca fulgens* and others (Fig. 16.29).

iii. On the metamorphosed sediments of the Stirling Ranges, Diels (1906) records a heath community on the rock plateaux dominated by *Leucopogon unilateralis* with subsidiary *Darwinia meissneri, Hypocalymma myrtifolia, Oxylobium retusum,* and species of *Gastrolobium* and *Lasiopetalum.* In more exposed situations rooted in crevices among rocks are *Leucopogon gnaphalioides* and *Monotoca tamariscina,* with *Sphenotoma drummondii* in more protected pockets.

5.5.5. Heaths on Rocky Outcrops in Drier Areas

In drier areas some of the species mentioned above persist, especially *Borya nitida* and the tall shrubs *Acacia lasiocalyx* and *Casuarina huegeliana.* However, the smaller shrub assemblages, which form local patches of heath or an understory to the taller *Casuarina* and/or *Acacia,* become floristically poorer. Other southern species are *Calothamnus quadrifidus* and *Thryptomene australis.* However, in the driest areas (mean annual rainfall 300–260 mm) new sets of species occur. Some of these are common to several isolated hills, e.g., *Kunzea pulchella* and *Stypandra imbricata,* whereas others have more restricted distributions, e.g., *Thryptomene kochii* and *Wehlia thryptomenoides.* Furthermore, in these areas several species typical of more arid regions make their appearance, notably the tall shrub *Pittosporum phylliraeoides,* the rock-ledge herb *Isotoma petraea, Nicotiana* spp., and some Compositae. Certain species of *Eucalyptus* also are restricted to these hills – *E. caesia, E. crucis* and *E. kruzeana* (Chap. 14).

At the bases of rocky outcrops on sandy soils or sands overlying a pebbly or gravelly/sandy subsoil *Casuarina campestris* commonly forms thickets c. 1 m tall, sometimes in pure stands but usually in association with *Melaleuca uncinata, M. hamulosa* and *Thryptomene australis.* In deeper soils woodlands of *Eucalyptus,* especially *E. loxophleba* and *E. occidentalis,* commonly surround the rocky hills.

5.6. Heaths in the Arid Zone

The Australian xeromorphic flora is poorly represented in the arid zone and presumably has been eliminated by aridity (see Chap. 4, Table 4.6). However, there are several genera, apart from *Eucalyptus,* which are well represented in semi-arid regions and a few of these occur in the arid zone in the south-west, especially *Baeckea, Calytrix, Micromyrtus, Thryptomene, Wehlia*

(Myrtaceae), *Brachysema, Burtonia* and *Jacksonia* (Papilionaceae), as well as the more widespread *Grevillea, Acacia* and *Casuarina*. Most of the shrub species are associated with eucalypts, especially *E. gongylocarpa* (Chap. 14). However, small areas of heath have been recorded by Beard (1974) in the Great Victoria Desert around lat. 28°S. (mean annual rainfall c. 220 mm).

The heaths, which are c. 30 cm high, are dominated by *Thryptomene maisonneuvii* and cover the flanks and lower slopes of some dunes, replacing hummock grasses (*Triodia* and *Plectrachne* spp.) which are of rare occurrence. Associated species are *Micromyrtus flaviflora* and *Calytrix longiflora*. On the summits of the dunes the heath is sparse and taller species occur in very open stands, the main ones being *Grevillea stenobotrya, Acacia salicina, Gyrostemon ramulosus, Crotalaria cunninghamii* and, locally, *Callitris preissii* ssp. *verrucosa* and *Eucalyptus gongylocarpa*.

17. The Alpine Communities

1. Introduction

1.1. General

The alpine zone is defined as the region above the tree-line. In Tasmania this stands at c. 700 m in the south and 1350 m in the north; in Victoria and New South Wales the tree-line stands at about 2000–2200 metres. In both areas the limit of the "trees" depends on exposure as well as altitude, the line being lower on the western and south-western than on the eastern slopes.

Some of the species and communities characteristic of the alpine regions extend into the subalpine or even montane zones below the alpine tracts and some communities occur disjunctly along the Great Divide to the northern Tablelands (lat. 29°S.). The communities are mainly those which are produced under conditions of soil waterlogging and the occurrences at lower altitudes represent downward extensions of the alpine communities into valleys which are either badly drained or subject to cold air drainage, or to both.

1.2. Climate

The alpine communities are delimited by cold and exposure to strong, cold winds. Mean annual rainfall in alpine areas both on the mainland and in Tasmania probably lies between 1200 and 3000 mm, possibly half of this falling as snow. There is greater precipitation during the winter than the summer, but rain (or snow) occurs regularly during the summer months with probable minimum monthly mean values of 70 mm. Consequently, during growing periods plants are usually not subjected to a water stress resultant upon a lack of liquid water in the substrate. However, during droughts, water may be in limited supply and the plants are under a water stress. For 4–8 months of the year the water remains as a snow blanket and patches of snow commonly persist into the summer, especially in shaded situations. In all districts mean minimum temperatures are below freezing for about six months of the year and lowest minima lie between $-13°$ and $-20°C$. The frost – free period is mostly only 2–3 weeks. Winds of gale to hurricane force are common, blowing from the north-west, west and south-west. The effects of winds are most pronounced in rocky areas where there is little soil to support vegetation. In such areas wind-shearing is common and many shrub species assume a prostrate form, appressed to boulders.

1.3. Topography

Although commonly referred to as "mountain plateaux", the summits of the mountain ranges are rarely flat but rather sloping or rounded (Fig. 17.1), with the relatively even surfaces dissected by streams. In Tasmania such surfaces lie adjacent to or are surrounded by hills rising some 200–300 m above the plateaux; these hills form the summits of the mountains. The sides of the hills are commonly precipitous, being composed mostly of columnar dolerite which weathers mainly vertically to produce "organ pipes" (Fig. 17.2). On the mainland, where either granite or basalt lies at the surface, weathering has produced a relatively rounded topography. It is only at the edges of the plateaux that steep gradients or cliffs occur, the more so when the rock is columnar basalt; in such cases, as in Victoria, "organ pipe" columns flank the plateaux.

The mountain plateaux and also the shoulders of the mountains in the subalpine zone, particularly in Tasmania, are sometimes strewn with boulders which may form several layers one upon the other, covering the soil so deeply that vascular plants cannot establish in these areas. These are sometimes referred to as "potato-fields", though the boulders commonly average 1 m diameter (Fig. 17.3). Such areas shed water rapidly and this water, together with water from higher elevations, is channelled into streams which flow below the boulders.

Fig. 17.1. General view of alpine communities on Mt. Kosciusko. *Eucalyptus niphophila* in the foreground. Photo taken at Charlotte Pass, c 1840 m.

Fig. 17.2. Scrub dominated by *Tasmannia lanceolata* and *Bellendena montana* at c. 1000 m. Note "organ-pipe" dolerite above, and the edge of the plateau. Ben Lomond, Tas.

Fig. 17.3. *Podocarpus lawrencei* on dolerite boulders at c. 1200 m. Mt. Wellington, Tas.

Small lakes and tarns (Fig. 17.4 and 17.5) are common on some of the plateaux, especially Tasmania. They have been formed either by glacial action or by the impounding of water into pools by mat plants or cushion plants. Valleys are usually wide in the alpine and subalpine zones (Fig. 17.6) the streams flowing slowly over low gradients until the edges of the plateaux are reached, or until they empty into alpine lakes.

Glacial action, more noticeable in Tasmania than on the mainland, takes the form of cirques, moraines and alpine lakes. As far as the vegetation and soil development are concerned, the chief effects of glaciation are the changes brought about in the drainage patterns. Apart from lakes, large areas of swamp are formed which support characteristic floras and produce soils of distinctive types.

1.4. Soil Parent Materials and Soils

The most common soil parent materials are granites, basalts, dolerites and quartzites, though smaller areas of sedimentary rocks and metamorphosed sediments occur. On the Kosciusko plateau the most extensive parent material is granitic. Basalt occurs at Kiandra in New South Wales and also at Bogong in Victoria. Costin (1975) concluded that there is no correlation between the soil parent material and the plant communities in the Kosciusko area. He suggests that the presence of eutrophic and oligotrophic communities (i.e. the minerally rich fens, and the minerally poor bogs respectively) as separate entities might be a reflection of differences in mineral status. However, the occurr-

Fig. 17.4. Dove Lake. Foreground, *Leptospermum lanigerum* and *Gymnoschoenus sphaerocephalus*. Cradle Mt. behind with *Eucalyptus coccifera* and *E. subcrenulata* on foothills. Cradle Mt. National Park, Tas.

ence of both types of community on granitic rocks points to a redistribution of nutrients as the cause rather than the direct action of parent material *in situ*.

In Tasmania, large areas of alpine plateaux are covered by the remnants of dolerite sills, and quartzites occur as well, sometimes in close proximity. The quartzites are of extremely low fertility, whereas the dolerites contain nutrients in relatively high supply. However, the latter are very much lower in nutrients, especially phosphate (Joplin, 1963), than the mainland basalts.

The soils have formed in three distinct habitats:

i. Exposed, rocky areas where vegetation is scant. In such situations physical weathering of the rock dominates chemical weathering. High wind velocities and the paucity of vegetation preclude the accumulation of what little organic matter is shed from the plants on to the soil surface, and Lithosols develop. These differ chemically, and to a less extend physically, in so far as the original primary minerals in the rock fragments produce differences in nutrient status and in pH.

ii. Less exposed or unexposed areas where a vegetative cover is developed; the soil profiles are well drained, and the "climax" soil results (Alpine Humus Soil).

iii. Areas where waterlogging of the profile occurs for most or all of the year. Waterlogging can result in two ways: Either, below snow-patches from which water continually trickles during most or all of the warmer months, or, in swamps and bogs, under conditions of impeded drainage. The soils produced in these waterlogged areas are mainly Acid Peats.

2. Flora

The flora is listed in Table 17.1, and an illustrated guide to the common species is available (Harris, 1970). The flora is composed of four

Fig. 17.5. Plateau communities at c. 1300 m. Cushion plants are *Ewartia meridithae* (white) and *Donatia novae-hollandiae* (darker). Foreground, *Astelia alpina* and *Richea scoparia*. *Boronia citriodora* common. Note tarn. Cradle Mt. National Park.

segments, the first two being distinctive in Australia at the generic level.

i. Subantarctica genera (see Chap. 3, Table 3.7). Almost all of the herbaceous genera included in this list occur in the alpine zone. Eight of them are confined to the colder areas of Tasmania, viz., *Donatia*, *Drapetes*, *Forstera*, *Gaimardia*, *Gunnera*, *Ourisia*, *Pernettya* and *Phyllachne*. Only one, *Schizeilema*, is confined to the mainland, the remainder occurring both on the mainland and in Tasmania. Several genera, shared between Australia and New Zealand, are confined to the colder climates, e.g. *Celmisia* and *Ewartia* (Compositae), *Cheesemannia* (Cruciferae) and *Pentachondra* (Epacridaceae).

ii. Endemic genera. Eighteen endemic angiosperm genera are confined to the cold climates of the south-east or to Tasmania (Chap. 4). Several of these occur in or are confined to the alpine zone; they are marked #in Table 17.1. Of these genera, *Bellendena*, *Milligania* and *Pterygeropappus* are confined to Tasmania; *Parantennaria* and *Wittsteinia* are confined to the mainland; the remainder occur both on the mainland and in Tasmania.

iii. Australian genera. Relatively few of the Australian xeromorphic genera (Chap. 4) have

Fig. 17.6. General view of a valley in the subalpine zone. Shrubs in foreground are mainly *Grevillea australis* and *Phebalium ovatifolium*. Floor of valley supports *Sphagnum* bog. The telegraph pole is c. 13 m tall. Below Charlotte Pass at c. 1600 m. Mt. Kosciusko, N.S.W.

Table 17.1. The Australian alpine flora. When only one species of a genus occurs, the genus only is listed; numbers after the genera indicate the number of species. Compiled from Rodway (1903), Costin (1954), Wakefield (1955), Curtis (1956, 1963, 1967), Vickery (1970). † indicates angiosperm genera occurring also in southern South America and confined to the Southern Hemisphere (see Chap. 3). # indicates that the genus is endemic to Australia.

PTERIDOPHYTES

Lycopodiaceae	*Lycopodium* 5	Athyriaceae	*Cystopteris*
Selaginellaceae	*Selaginella*	Blechnaceae	*Blechnum*
Isoëtaceae	*Isoëtes*	Gleicheniaceae	*Gleichenia*
Ophioglossaceae	*Ophioglossum*	Grammitidaceae	*Grammitis*
Aspidiaceae	*Polystichum*	Gymnogrammaceae	*Anogramma*
Aspleniaceae	*Asplenium, Pleurosorus*	Hymenophyllaceae	*Hymenophyllum*

GYMNOSPERMS

Cupressaceae	# *Diselma*
Podocarpaceae	# *Microcachrys*, # *Pherosphaera* (= *Microstrobos*), *Podocarpus*
Taxodiaceae	# *Athrotaxis* 2

ANGIOSPERMS

Dicotyledons

Baueraceae	*Bauera*
Boraginaceae	*Myosotis*
Campanulaceae	*Wahlenbergia* 2
Caryophyllaceae	† *Colobanthus* 2, *Spergularia*, *Stellaria* 2
Compositae	† *Abrotanella* 3, *Brachycome* 9, *Cassinia*, *Celmisia* 2, *Cotula*, *Craspedia* 3 *Erigeron*, *Ewartia* 4, *Gnaphalium* 7, *Helichrysum* 14, *Helipterum* 2, † *Lagenophora*, *Leptorrhynchus*, † *Microseris*, *Olearia* 7, # *Parantennaria*, *Podolepis* 3, # *Pterygopappus*, *Senecio*
Cruciferae	*Cardamine*, *Cheesemannia*
Donatiaceae	† *Donatia*
Droseraceae	*Drosera* 2

Continued next page

Table 17.1 continued

Family	Genera
Ericaceae	† Gaultheria, † Pernettya 2, # Wittsteinia
Epacridaceae	Archeria 2, Cyathodes 4, Dracophyllum, Epacris, Leucopogon 2, Lissanthe, Monotoca, Pentachondra 2, # Richea 6, Sprengelia, Trochocarpa
Euphorbiaceae	Poranthera 2
Gentianaceae	Centaurium, Gentianella
Geraniaceae	Geranium 2, Pelargonium
Goodeniaceae	Goodenia, Scaevola
Haloragaceae	† Gunnera, † Haloragis 2, Myriophyllum 3
Hypericaceae	Hypericum
Illecebraceae	Scleranthus 2
Labiatae	Prostanthera
Lentibulariaceae	Utricularia
Lobeliaceae	† Pratia 2
Loganiaceae	Mitrasacme 3
Mimosaceae	Acacia 2
Myrtaceae	Baeckea, Eucalyptus 3, Kunzea, Leptospermum, Melaleuca
Onagraceae	Epilobium 4
Oxalidaceae	Oxalis
Papilionaceae	Bassiaea 3, Hovea, Oxylobium 2, Pultenaea
Plantaginaceae	Plantago 5
Polygalaceae	Comesperma
Portulacaceae	Montia
Proteaceae	# Bellendena, Grevillea, Lomatia, † Orites 3
Ranunculaceae	Anemone, Caltha 2, Ranunculus 8
Rhamnaceae	Cryptandra
Rosaceae	†Acaena 2, Geum, Rubus
Rubiaceae	Asperula 4, † Coprosma 2, Galium, † Nertera
Rutaceae	Boronia 2, Phebalium 3
Santalaceae	Exocarpos
Scrophulariaceae	Euphrasia 3, † Ourisia, Veronica 4
Stackhousiaceae	Stackhousia
Stylidiaceae	† Forstera, † Phyllachne, Stylidium
Thymelaeaceae	† Drapetes, Pimelea 4
Umbelliferae	Aciphylla 3, Actinotus 2, # Dichosciadium, # Diplaspis, Hydrocotyle, Oreomyrrhis 4, # Oschatzia, † Schizeilema, Sesile 2, Trachymene
Violaceae	Hymenanthera, Viola 3
Winteraceae	Tasmannia

Monocotyledons

Family	Genera
Centrolepidaceae	† Gaimardia
Cyperaceae	Carex 7, † Carpha, † Oreobolus 2, Schoenus 6, Scirpus 2, † Uncinia 2
Gramineae	Agropyron 2, Agrostis 3, Danthonia 4, Deyeuxia, Hierochloe 2, Microlaena, Poa 11
Juncaceae	Juncus 2, Luzula 2
Liliaceae	† Astelia, Herpolirion, # Milligania 2
Orchidaceae	Acianthus, Caladenia, Prasophyllum 3, Pterostylis 2, Thelymitra
Restionaceae	Calorophus, Hypolaena, Restio

produced species adapted to the coldest climates, the Epacridaceae alone being well represented. Only two species of *Acacia* and three of *Eucalyptus* occur above the tree-line. The Cyperaceae and Restionaceae, well represented in Australia (Chap. 4), are also well represented in alpine areas, becoming prominent or dominant in swamps, but the endemic genera do not occur in the alpine zone.

iv. Cosmopolitan genera. Many genera of cosmopolitan distribution occur, some of which, common elsewhere in Australia and familiar to Northern Hemisphere botanists, are *Anemone, Caltha, Cardamine, Drosera, Epilobium, Viola* and the monocotyledons *Agropyron, Agrostis, Carex, Juncus, Luzula, Poa*, etc. The pteridophytes are not numerous but several of these, especially *Blechnum, Gleichenia,*

Table 17.2. Specific differences in four genera represented in alpine areas on the mainland and in Tasmania.

Family	Genus	Species Mainland	Tasmania
Compositae	*Abrotanella*	*nivigena*	*fosterioides*
			scapigera
Epacridaceae	*Richea*	*continentis*	*acerosa*
			gunnii
			milliganii
			sprengelioides
Proteaceae	*Orites*	*lancifolia*	*acicularis*
			milliganii
			revoluta
Rutaceae	*Phebalium*	*ovatifolium*	*montanum*
			oldfieldii

Lycopodium and *Selaginella* become abundant in places and are even community dominants.

Cosmopolitan non-vascular genera are invariably present and dominate certain areas, notably the mosses, *Sphagnum*, *Blindia* and *Bryum* which produce bogs and peat in much the same fashion as in analogous areas of the cooler parts of the Northern Hemisphere.

A comparison between the Tasmanian and mainland alpine floras is made. While most of the genera are common to Tasmania and the mainland, there are important floristic differences which lead to the occurrence of some different associations in the two areas. The main floristic differences are: i. In Tasmania there is a concentration of New Zealand – South American genera which are not found on the mainland; ii. Both regions contain endemic genera restricted either to Tasmania or the mainland, Tasmania having more restricted endemics than the mainland, including the endemic conifers.

At the specific level, different species of the same genus often occur on the mainland and in Tasmania; a few examples are given in Table 17.2.

3. Vegetation

3.1. Life-forms

The life forms, with two exceptions, conform entirely with the classes nominated by Raunkiaer (1934). Two somewhat aberrant chaemaephytic forms are the more or less prostrate woody shrub, and the "cushion" or "bolster" plant. The former is represented by several species which, under more favourable climatic conditions, reach shrub or tree proportions. The plants in question produce branches 1–2 metres long which lie appressed against bare rock surfaces or between boulders, and may even form mats. A few species of *Eucalyptus* are included in the group, and other angiosperms such as *Baeckea gunniana* (Fig. 17.7), *Exocarpos humifusus*, *Leptospermum humifusum*, and the gymnosperm, *Podocarpus lawrencei*.

A curious chamaephytic growth form is the cushion plant found in Tasmania (also in New Zealand and South America), occurring mainly on the mountain plateaux but sometimes found at slightly lower elevations. The form has been described by Rodway (1903), Gibbs (1920) and Sutton (1928). The cushions commonly reach a diameter of c. 1 metre. They are composed of vast numbers of moss-like stems which become crowded and appressed, with a terminal rosette of leaves forming the surface of the cushion. Dead leaves remain inside the cushion and do not decompose readily. The surface of the cushion is either flat or domed, the latter condition prevailing when lateral expansion is restricted by neighbouring plants, in which case the cushion grows vertically, becoming domed and attaining a height of 50–60 cm. The age of cushions is not known with certainty but Sutton suggests great age, possibly rivalling many forest trees.

The cushion habit is found in six species belonging to four families: *Abrotanella fosterioides* (Fig. 17.8), *Ewartia meredithae*, *Ptery-*

Fig. 17.7. *Baeckea gunniana* growing over a boulder. *Pimelea alpina* in flower at right. Summit of Ben Lomond, Tas.

geropappus lawrencii (Compositae), *Donatia novae-hollandiae* (Donatiaceae) (Fig. 17.9), *Dracophyllum minimum* (Epacridaceae), and *Phyllachne colensoi* (Stylidiaceae). Also, similar but smaller cushions are produced by *Centrolepis monogyna* and *Pimelea pygmaeum* (Fig. 17.10) and by some mosses, e.g. *Bryum hypnoides*. All of these species may be found together forming a single alliance (see 4.1.4, below).

3.2. Structural Forms of Vegetation and their Interrelationships

Five distinct types of vegetation can be recognized, which are related to habitat. In order of increasing exposure and soil depth, these are:

i. The fjellfields (scattered shrubs in rocky terrain); ii. The woody communities, described by different authors as "dwarf mountain forest", "thickets", "scrub", "shrubberies", or "heath"; iii. Herbfields which are dominated by herbs other than grasses; iv. Tussock grasslands; v. Swamps, including sedge-dominated communities (fens or sedgelands) and moss-dominated communities (bogs or mosslands).

The fjellfields occupy the most adverse climatic conditions with respect to cold and exposure to strong winds. The woody communities occupy less rocky areas in less exposed situations. In still less exposed situations and in less rugged terrain the Herbfields occur on soils which are relatively deep and in which a soil profile has developed (Alpine Humus Soils); this is regarded as the climax community-soil system of the alpine zone (Costin, 1954). The grasslands occupy areas where snow lies longer (in comparison with the herbfields) or areas where some waterlogging occurs. Lastly, the swamps occupy areas which are more or less permanently waterlogged. Mosaics of all or most of these structurally different communities commonly occur in a repetitive pattern (see Figs. 17.1, 17.6, 17.11).

4. Descriptions of the Alliances

4.1. The Fjellfields

4.1.1. General

The term "Fjell" was used originally to describe a hill or mountain, a rocky or barren hill, or a moor. The term was first used in Australia by botanists to describe the rock-strewn areas of the Tasmanian plateaux. It is now used to describe a type of vegetation, defined by Costin under Feldmark (or Fjaeldmark) as an "open subglacial community of dwarf flowering plants, mosses and lichens, usually dominated physiognomically by chamaephytes".

The communities occupy the rockiest habitats. Precipitation in areas is as high as or higher than that in any part of the alpine tract, most of the water being received as snow. However, the plants and communities do not reflect high rainfall responses since much of the snow is blown off these exposed areas, while the water from melting snow and rain is shed as surface runoff or percolates rapidly through the fragmented substratum. It is only at the bases of

Fig. 17.8. *Abrotanella forsterioides*, with *Danthonia pauciflora*. Plants in the surrounding sward include *Carpha alpina*, *Lagenophora stipitata*, *Scirpus* sp., *Plantago* sp. Steppes, Central Plateau, Tas.

Fig. 17.9. Cushion of *Donatia novae-hollandiae* many of the plants with a single terminal flower. Surrounding species are *Astelia alpina, Richea scoparia* and *Boronia citriodora*. Cradle Mt. National Park, Tas. at c. 1300 m.

snowpatches which persist into the summer that more or less permanently wet soils occur. In summer, when temperatures are high during the daytime and winds are strong, desiccating conditions may prevail and it is not unusual for the soil to be dried out to such low levels that dust can be raised when the plant roots are disturbed.

The open nature of the communities is due partly to the presence of rock at the surface and partly to soil movements. The latter stem either from the freezing of water in the soil (frost-heaving), or through the movement of soil or rock downhill under the influence of gravity (Barrow et al., 1968; Costin et al., 1969).

The communities form series which can be arranged in order of decreasing exposure. In the most extreme climatic situations rocks are commonly covered with non-vascular plants, chiefly lichens. The least complex of the communities of vascular plants are those which occur in rock clefts or ledges. The simplicity of these communities is determined by spatial restriction, the community being a single plant (often a clone) growing in a vertical rock cleft a few centimetres wide. Such communities are not necessarily in the coldest environment, since proximity to the rock may protect the plant against the cold and, depending on aspect, against high wind velocities. Areas which are less precipitous, though still dominated by rock, harbour a greater number of species and in such cases associations of a few to several species occur. These are described below under "open communities". A third assemblage assigned to the Fjellfields by Tasmanian workers are the cushion mosaics.

4.1.2. Communities of Rock Ledges and Clefts

Costin (1954) described the *Brachycome nivalis – Danthonia alpicola* Alliance for the

Kosciusko plateau and for Victoria. The alliance, classed as a tall alpine herbfield consists of six associations containing a total of 14 species. The communities occur in rock crevices or under rock ledges which afford some protection from the extremes of temperature and wind. Species which dominate the associations are the two listed above and *Cardamine hirsuta, Alchemilla novae-hollandiae* and the two ferns, *Blechnum penna-marina* and *Polystichum aculeatum.*

Some of the species mentioned above have been listed for the Tasmanian plateaux, e.g. societies of *Cardamine hirsuta* (Gibbs, 1920). Sutton (1928) lists another set of rock-ledge plants with the comment that they are typical dolerite plants, which suggests that different species may be associated with different rock types. The plants are tufted perennials with stout tap-roots, and include chiefly members of the Umbelliferae (species of *Aciphylla, Dichosciadium* and *Oreomyrrhis*), and *Ourisia*, together with woody species, e.g. of *Hibbertia* and *Rubus*. He lists also, in wetter areas, (possibly linking these communities with the cushion mosaics) species of *Claytonia, Cotula, Diplaspis, Erigeron, Mitrasacme* and *Oreobolus*. He records also the endemic Liliaceae, *Milligania*, growing "widely and in close and exclusive association in certain secluded places", without specifying the habitat.

4.1.3. Open Communities

On the Kosciusko plateau, Costin (1954) has defined two alliances as follows:

i. *Epacris petrophila – Veronica densiflora.* This alliance which is confined to the most windswept areas, contains 35 species. He defines six associations, involving the two genera mentioned, and *Helipterum* and *Kelleria*.

Fig. 17.10. *Pimelea pygmaea* forming cushions over rocks in a small pool. Steppes, Central Plateau, Tas.

Fig. 17.11. Mosaic of herbfield-grassland (*Celmisia longifolia* and *Poa hiemata*) and heath (*Orites lancifolia* and *Grevillea australis* with plants of *Helichrysum bracteatum* to left of travel bag). Note snow-patch. Mt. Hotham, Vic. (21-v-71).

In this exposed environment the community exhibits the pattern of coseral development, the community being destroyed on the windward side, replacing itself on the leeward side.

ii. *Coprosma pumila – Colobanthus benthamianus*. This is a small alliance containing only six species, the dominants plus *Colobanthus apetalus*, *Ranunculus anemoneus*, *R. muelleri* and *Epilobium confertifolium*. It is confined to the upper portions of snow-patch areas. Individual plants or groups of plants are well separated by bare stony ground. Costin comments on the rapid flowering of the species after the melting of the snow and suggests that extensive pre-development occurs during the winter months.

4.1.4. Cushion Mosaics

The four cushion plants *Dracophyllum minimum*, *Donatia novae-hollandiae* (Fig. 17.9), *Ewartia meredithae* (Fig. 17.5) and *Pterigeropappus lawrencii* are usually found in association, and with them other cushion plants may also occur, namely *Abrotanella forsterioides* (Fig. 17.8), *Centrolepis monogyna* and *Phyllachne colensoi* (the last only in the south-west, Curtis, 1963). Mosses, which also form cushions are commonly found with the angiosperms, e.g. *Dicranium billardieri* and *Rhacomitrium pruinosum*.

The Alliance is confined to Tasmania and is described by Gibbs (1920) and Sutton (1928). It is restricted to rocky areas on the plateaux but may occur also on talus or rocky moraines in valleys or round alpine lakes. The plants are covered by snow during the winter, being exposed only for a few months during the summer. The substrate supporting the plants is usually wet and mostly waterlogged, though in elevated areas cushions growing on rocks may be subjected to a dry external environment. The cushions, however, with the accumulated mass of internal debris containing the living stems and some roots are unlikely to become desiccated. Individual cushions attain a diameter of c. 1 m and neighbouring cushions coalesce. Plants of other species which occurred between cushions persist in the mosaic after coalition. The cushions themselves act as seed beds for other species, and are the usual habitat for some species, especially *Plantago gunnii, Prasophyllum alpinum* and one variety of *Sprengelia incarnata* (Curtis and Somerville, 1949).

4.2. The Herbfields

4.2.1. General

Alpine Herbfields are closed communities dominated by perennial herbs, mostly forbs and some grasses. On the Kosciusko plateau, Costin (1954) distinguishes Tall and Short Alpine Herbfields. In the former the tallest of the herbs are about 25 cm high and subordinate strata are developed. The Short Alpine Herbfields are composed of carpet-forming herbs, usually so closely appressed to the ground that subordinate strata do not develop. Two alliances are described.

4.2.2. *Plantago muelleri* – *Montia australasica* Alliance

The alliances is a Short Herbfield and occurs at the bases of snow-patches and in similar habitats saturated by water from melting snow. The communities occupy the leeward aspects, occurring on relatively even terrain, both on the mainland (Costin, 1954) and in Tasmania (Gibbs, 1920). The communities are usually one-layered, forming mats. The soils are classified by Costin as snow patch meadow soils, the mineral material being fine-textured colluvium, with occasional shallow variants of poor fen peat. The alliance on the mainland contains 23 species of which the most important are the two mentioned above, *Caltha introloba, Brachycome stolonifera, Ranunculus inundatus* and *Oreobolus pumilio*. Costin comments on the rapid flowering of the species in the communities, which suggests pre-development during the winter months.

In Tasmania, the herbfields contain *Plantago muelleri,* as on the mainland, and the following herbs are also present, either in association with *Plantago* or as local dominants: *Caltha phylloptera, Euphrasia striata, Gentianella diamensis, Millingania densiflora, Mitrasacme archeri, Plantago antarctica, P. paradoxa, Ranunculus nanus, Rubus gunnianus* and *Velliaea montana* (Jackson, in Specht et al., 1974).

4.2.3. *Celmisia longifolia* – *Poa* Alliance

Silver Snow Daisy – Snow Grass

On the Kosciusko plateau the *C. longifolia* – *Poa hiemata* alliance, described by McLuckie and Petrie (1927) and Costin (1954) is regarded as the climatic climax of the alpine tract, developing where the soil profiles are deep and differentiated into horizons (Alpine Humus Soils). The same alliance occurs in Victoria but covers relatively small areas, most of the higher country being occupied by *Poa* grasslands (Carr and Turner, 1959, see Chap. 20). In Tasmania, the alliance contains *Celmisia longifolia* and *Poa gunnii.*

Though *Celmisia* and the grass may occur more or less randomly distributed in the same sward, they are more often segregated into patches, with *Celmisia* dominating some areas, *Poa* other, the two forming mosaics. According to Costin, *Celmisia* dominates areas where snow lies for longer periods. Carr und Turner suggest that in Victoria the balance between the two may be changed by disturbing the communities by grazing or by fire. In Tasmania, Curtis (1969) ascribed the dominance of *Celmisia* to the better drainage of the soils.

The communities occur on a variety of rock types, without apparent change in floristic composition, but where rock lies near or at the surface and the soils are shallow, shrubs appear in the herbfield and become more and more prominent with increasing rockiness. On the other hand, soil waterlogging excludes the communities, there being a gradual transition from herbfield to fen or bog with an increase in height of the water-table.

Celmisia and *Poa* (Fig. 17.11) make up by far the greatest bulk of the communities. However, a very large number of associated species occur. For Kosciusko, Costin lists 130 species, 90% of which are perennials. Most are dicotyledonous herbs, represented most abundantly by Compositae (16 genera, 34 species), Umbelliferae (9, 12), Scrophulariaceae (2, 6) and Caryophyllaceae (3, 4). Most of the monocotyledons are grasses (6 genera 14 species) and Cyperaceae (5, 12). Four pteridophytes are recorded. Many of the species occur in Tasmania, where, however, there are notable differences, e.g. the prominence of *Erigeron pappochroma* and *Craspedia alpina* (Curtis, (1969). Dominants of smaller local associations are *Helipterum incanum* var. *alpinum* and *Chionochloa frigida*.

The growing season is short and flowering of the various species occurs over the same period with a dramatic display of colour. Costin comments on the summer societies of the several species, e.g. *Prasophyllum alpinum, P. suttonii, Ranunculus* spp., *Cardamine hirsuta* etc.

4.3. The Grasslands

These are dealt with in Chap. 20. The main alliances are dominated by *Poa hiemata* (mainland [Fig. 17.11]), *P. labillardieri* (mainland and Tasmania) and *P. gunnii* (Tasmania).

4.4. Heaths and Shrublands

4.4.1. General

Large areas of the plateaux are covered or dotted by small shrubs most of which are angiosperms, but in Tasmania gymnosperms are often prominent and form distinctive associations. These shrub-dominated communities commonly form mosaics with fjellfields, herbfields (Fig. 17.11) or bogs, according to exposure to wind and the drainage pattern. Some of the communities extend downward into the subalpine tract and in some cases different associations are found at these lower altitudes. The alliances can be grouped into two series, one linked with the fellfields in well drained habitats, the other linked with the bogs in wetter habitats; the communities are described under these two headings. Furthermore, the associations on the mainland are different from those in Tasmania, as outlined below.

4.4.2. Communities in Well-Drained Habitats on Plateaux

Two types of communities can be recognized, one in which the shrubs become prostrate and form mats which grow up the sides of boulders and cover or partly cover their tops. Such species identify "boulder communities" (Fig. 17.7). The second type of community, which occurs in better protected and less rocky habitats, consists of erect shrubs up to c. 1 m tall, either closed or open, and with a continuous or discontinuous herbaceous layer (Fig. 17.12).

4.4.2.1. Communities on the Mainland

Costin (1954) describes two main alliances:

i. On rock-strewn plateaux, the *Oxylobium ellipticum* – *Popocarpus lawrencei (alpina)* Alliance which is composed of 14 associations, each with a single dominant belonging to the genera: *Acacia, Baeckea, Bossiaea, Callistemon, Drimys, Hovea, Kunzea, Leucopogon, Lissanthe, Orites, Oxylobium, Phebalium, Podocarpus* and *Prostanthera*.

ii. *Casuarina nana* – *Leptospermum lanigerum* alliance on windswept peaks and plateaux in the subalpine and montane tracts (see also Chap. 16). Other prominent species are *Callistemon pallidus* and *Calytrix tetragona*.

4.4.2.2. Communities in Tasmania

Communities dominated by angiosperms are numerous and the dominants belong mainly to genera of Myrtaceae, Compositae, Epacridaceae and Proteaceae. The main "boulder" communities are formed from *Eucalyptus vernicosa, Leptospermum humifusum, Baeckea gunniana* (Fig. 17.7), *Pimelea alpina, Exocarpos humifusus, Pentachondra pumila* and *Podocarpus lawrencei* (Fig. 17.3).

Heath communities, which are either closed or open are composed of various combinations of the following species (Jackson, in Specht et al.

Fig. 17.12. Rock-strewn plateau with open heath. Main species are *Orites acicularis* (right), *Pimelea alpina* (in flower), *Richea scoparia* (left), *Pentachondra pumila* and *Baeckea gunniana* (mats). Ben Lomond, Tas.

1974): *Bellendena montana, Cyathodes petiolaris, Epacris petrophila, Grevillea australis, Helichrysum backhousei, H. hookeri, Lissanthe montana Monotoca empetrifolia, Olearia ledifolia, Orites acicularis, O. revoluta, Persoonia gunnii, Richea scoparia* and *Tasmannia lanceolata* (Fig. 17.2 and 17.12).

The gymnosperm – dominated communities (except *Podocarpus lawrencei*) are unique to Tasmania. Six of the eight genera in the island are represented on mountain tops. They sometimes form associations with angiosperms.

Podocarpus lawrencei commonly forms monospecific communities (sometimes one large plants) among boulders (Fig. 17.3). It may be associated with *Olearia pinifolia*, a Compositae with gymnosperm-like leaves, or with *Leptospermum humifusum*.

Microcachrys tetragona, a decumbent or semi-erect species, forms tangled heaths, sometimes carpeting the ground in pure stands, or it occurs with *Pentachondra pumila* and *Cyathodes dealbata* as common associates.

Microstrobus niphophilus and *Diselma archeri*, either individually as local dominants or in association, form very dense, tangled heath-like communities or thickets in rocky situations, usually protected by low cliffs or large boulders. In more sheltered habitats, often in the heads of shallow valleys, the association is taller, assuming scrub proportions, the plants reaching a height of 2–3 metres.

Pharerosphaera hookeriana, usually a shrub about one metre high, is sometimes a dominant in alpine shrublands. It occurs most often along small watercourses in some of the herbfields, or around tarns and small lakes. In these wet habitats swamp herbs such as *Drosera* and *Mitrasacme* are common. Shrub associates are *Baeckea gunniana* and *Helichrysum baccharoides* (Gibbs, 1920). The two common species if *Athrotaxis* occur in a stunted form about the treeline (Fig. 17.13). Both these species reach tree proportions at lower elevations.

In the subalpine tract shrublands and heaths

Fig. 17.13. Thicket of *Athrotaxis cupressiformis*. Foreground: *Astelia alpina* and some *Richea scoparia*. Cradle Mt. National Park at c. 1300 m.

sometimes occur within woodlands on talus slopes or in pockets affected by cold air drainage. Species involved in these include most of those from the plateau, plus the following listed by Jackson (in Specht et al. 1974): *Bauera rubioides, Cyathodes straminea, Lomatia polymorpha, Orites densiflora, O. lanceolata, Richea sprengelioides, Telopea truncata, Trochocarpa thymifolia* and *Westringia rubrifolia.*

Unique among the subalpine scrubs are those containing *Richea pandanifolia,* the tallest of the Australian epacrids, which attains a height of 6–8 metres. The leaves, which reach a length of 150 mm and a breadth of 35 mm, are arranged in a terminal crown. This species is found in the subalpine woodlands and forests. It occurs also in the montane forests at lower altitudes and sometimes even on the mountain plateaux in a stunted form. The clumps of plants occur in very moist situations in monospecific stands (also as an understory plant in the *Eucalyptus* woodlands) sometimes under bog conditions with sphagnum moss forming a continuous sheet over the peaty surface. *Phyllocladus aspleniifolius* and the endemic Proteaceae, *Cenharrenes nitida,* are common associates.

4.4.3. Wet Heaths

Wet heaths develop in semi-waterlogged soils and are transitional between the *Sphagnum* bogs and the better drained soils on which other heaths and shrubberies occur. There is some overlap in species in what must be regarded as ecotones. The soils are sometimes deep (Alpine Humus Soils) but in Tasmania, particularly on quartzites, the soil is the peat built up by mosses or cushion plants, and may be as little as 15 cm deep.

Epacris serpillifolia is a common dominant of such heaths both in Tasmania and on the mainland. In the latter area, other species of *Epacris, Kunzea* and *Leptospermum* identify the associations. In Tasmania *Richea scoparia* and *Boronia citrodora* are two of the most common species. Plants of lesser frequency include Cyperaceae and Restionaceae, and species of *Acaena, Epilobium, Euphrasia, Plantago* and *Ranunculus,* among others.

4.5. Communities in Free Water and in Waterlogged Areas

4.5.1. General

Three groups of communities are dealt with, those in free water, the fens (types of sedgeland) and the bogs (mossland). The last two are used by Costin (1954) as follows: Fen is "an hygrophilous community dominated by helophytes and hydrophytes" (usually monocotyledons), from which hummock-forming mosses are absent. Bog is "an hygrophilous community dominated by hummock-forming mosses and acidophilous shrubs or helophytes and hydrophytes"; the mosses control the habitat by maintaining waterlogged conditions.

4.5.2. Aquatic Communities

Several aquatic plants form distinctive associations in lakes and streams, mainly in Tasmania (Sutton, 1928), and the following are mentioned

Fig. 17.14. Subalpine lake edged with *Carex gaudichaudiana* and *Juncus* sp. (taller). *Myriophyllum propinquum* in water; *Eucalyptus pauciflora* behind. Near Kiandra, N.S.W.

as common species or community – dominants: the pteridophyte *Isoëtes gunnii* and the angiosperms, *Claytonia australasica*, *Forstera bellidifolia*, *Gaimardia fitzgeraldii*, *Gunnera cordifolia*, *Liparophyllum gunnii*, *Myriophyllum elatinoides*, *Scirpus fluitans*, *S. inundatus* and *Trithuria filamentosa*.

4.5.3. Fens (Sedgelands)

Three alliances can be identified, one occurring extensively on the mainland, the other two in Tasmania. The dominants are found both on the mainland and in Tasmania, but the species are present at different densities so that the alliances are not equally well expressed in the two areas.

4.5.3.1. *Carex gaudichaudiana* Alliance

The alliance is identified by Costin (1954) who distinguishes 5 alpine and subalpine associations of this sedge with the grasses *Danthonia nudiflora*, *Festuca muelleri* (which sometimes forms pure swards) and *Poa costiniana*. The alliance extends into the montane zone where several other species are added to the alliance, including species of *Eleocharis*, *Phragmites*, *Scirpus* and *Typha* (see Chap. 22). The communities are normally 3-layered, except in acid areas where only 2 layers occur. In such areas small cushions of *Bryum* mosses may be present.

The alliance occurs at higher altitudes on level to gently undulating country along permanent watercourses and in flats within valleys (Fig. 17.6). He identifies two kinds of soil parent materials, transported mineral matter and sedentary peat built up by plants of past generations. He recognizes acid and alkaline fens, as determined by the acidity and base-status of the subsoil and surface waters. *Carex gaudichaudiana* (Fig. 17.14) is a consistent member of all communities. On drier, acid, imperfectly aerated soils in the alpine tracts this species occurs with *Danthonia nudiflora;* under similar soil conditions at lower altitudes, *Carex* occurs with *Festuca muelleri* which may form pure swards. With improved lateral drainage and aeration on slightly acid soils, *Carex* is found in associations with *Poa*.

4.5.3.2. *Astelia alpina* Alliance

Astelia alpina (Liliaceae) occurs in Tasmania, and on the mainland where it is listed by Costin (1954) as the dominant of an association in the *Epacris* – *Sphagnum* raised Bog Alliance. In Tasmania it covers extensive areas as the dominant or co-dominant in waterlogged areas, commonly adjacent to water, but usually without *Sphagnum*. The alliance extends into the subalpine zone where it usually covers swamps adjacent to watercourses. The plant produces stiff erect leaves 20–25 cm long which are silvery-white below. It reproduces by short rhizomes, the plants ultimately becoming so closely aggregated that they form a complete mat over the swamp (Curtis, 1969). Fibrous peat is produced below as the stems grow upward. The mat is flexible and vibrates when walked upon. Pure swards of *Astelia* are common. Equally common is an association of *Astelia* with the fern *Gleichenia alpina* (Fig. 17.15).

4.5.3.3. *Carpha* – *Uncinnia* – *Oreobolis* Alliance

Carpha alpina, *Uncinnia flaccida* and *Oreobolus pumilio* (all Cyperaceae) form an alpine alliance on all the higher mountains in Tasmania. The community is a true sedgeland, forming an open to almost closed sward. *Carpha* is tufted and associations dominated by this species resemble grasslands. *Oreobolus pumilio* on the other hand forms small porcupine-like cushions up to 10 cm diameter which are commonly surrounded by a sward of *Carpha*, sometimes with *Uncinnia flaccida* as a subordinate species. The alliance occurs on soils which are occasionally waterlogged, but may dry out during dry summers particularly where *Carpha* is dominant.

Carpha alpina occurs also as the dominant of subalpine sedgelands in valleys which are sometimes dotted with shrubs 1–1^1/$_2$ m high of *Hakea lissocarpa*, *H. microcarpa*, *Olearia algida* and *O. lepidophylla*, with subsidiary *Lissanthe montana*, *Pernettya tasmanica* (Ericaceae, endemic to Tasmania, see Fig. 17.16). Cushion plants also are sometimes present, especially *Abrotanella fosterioides* (Fig. 17.8) and *Pimelea pygmea* (Fig. 17.10).

The sedgelands are composed mainly of *Carpha*, with a large number of subsidiary herbaceous species belonging mainly to the genera

Fig. 17.15. *Astelia alpina* and *Gleichenia alpina*. Harz Mts. Tas. at c. 1100 m.

Acaena, Hypericum, Lagenophora, Oreomyrrhis, Plantago and *Scirpus*. They are surrounded by low, open woodlands, usually of *Eucalyptus gunnii* and on the Central Plateau by *E. rodwayi* and *E. dalrympleana* as well. The moors are heavily grazed by marsupials, which possibly control their botanical composition, and they are referred to by Jackson (personal communication) as "Marsupial Moors".

4.5.4. Bogs (Mosslands)

4.5.4.1. General

Bogs are identified by the presence of hummock forming mosses which control the substrate conditions. The mosses are species of *Sphagnum* (as in the Northern Hemisphere), or of *Blindia* or *Bryum*. Costin (1954) recognizes two subforms, Valley Bog and Raised Bog. The former, developed under very wet conditions contains only mosses and herbs, whereas Raised Bog supports shrubs as well. Bogs occur from the Northern Tableland to the alpine and subalpine tracts of the Australian Alps (Fig. 17.6). There are some small occurrences on the lowlands (Chap. 22). In Tasmania, extensive bogs are much less frequent than on the mainland. Sutton (1928) and Martin (1939) make reference to hummocks of *Sphagnum* in swamps, often with associated shrubs, e.g. *Richea scoparia*, indicating that Raised Bog and possibly Valley Bog occur there. On the mainland three main alliances have been described.

4.5.4.2. *Carex gaudichaudiana* – *Sphagnum cristatum* Alliance
Valley Bog

Costin (1954) describes the alliance in the Kosciusko area. It is made up of two associations, one dominated by *Carex*, the other by *Sphagnum* and *Carex*. The communities are

Fig. 17.16. *Pernettya tasmanica* (white fruits), with *Olearia lepidophylla* and *Lagenophora stipitata* on a moor. Steppes, Central Plateau, Tas.

confined to level or slightly sloping sites in the alpine and subalpine tracts, where they develop only under acid and base-deficient conditions. *Carex* and *Sphagnum* occur together in the strongly acid areas, and with increasing acidity this association is replaced by raised bog (next Section). The *Carex* association dominates moderately acid areas, and with decreasing acidity it is replaced by fen dominated by the same species (see 4.5.3.1, above).

The general surface of the bog is flat or concave. Slightly elevated banks of moss protrude above the general surface and form a continuous network. The moss banks enclose areas of fen lying at a slightly lower level. The moss in the hummocks forms a turf-like surface and above this stands a stratum of *Carex* which is rooted in the hummock. The intervening fen communities consists of three strata of herbs, the uppermost being a discontinuous stratum of *Carex*. The intermediate stratum is discontinuous and consists of smaller helophytes and hydrophytes. Below this is a continuous ground stratum of dwarf helophytes, herbaceous chamaephytes and non-hummock-forming mosses (*Polytrichum* spp.).

Costin records a total of 31 species of which 20 are helophytes and hydrophytes. Genera which are represented include *Carex, Carpha, Scirpus, Agrostis, Poa, Calorophus, Restio, Juncus, Eriochilus* and a few herbaceous dicotyledons.

4.5.4.3. *Epacris paludosa* – *Sphagnum cristatum* Alliance
Raised Bog

The alliance occurs mainly in the alpine and subalpine tracts extending to the montane zone. It is recorded from the Bogong high plains in Victoria (Carr and Turner, 1959), in the Kosciusko area (Costin, 1954) and at Barrington Tops on the Northern Tablelands (Fraser and Vickery, 1939).

Costin identifies 10 associations for the Kosciusko area; four of them contain shrub species above the moss, the remainder consist entirely of herbs, without *Sphagnum*. These associations are the result of the composite structure of the community which is an aggregation of microcommunities as described below. The total flora consists of 76 species in 32 families.

The bog is composed of two parts: (a) *Sphagnum* areas which are elevated and form hummocks in which hollows occur. This is referred to as the regeneration complex. The outer edges of these hummocks fall away sharply at angles of about 45° and grade into: (b) Flat, fen areas lying below the *Sphagnum* hummocks. These are covered by a sedgeland in which *Carex gaudichaudiana* is the most abundant species, especially in the wettest areas. In less wet areas *Restio australis* and *Carpha nivicola* become prominent or dominant, especially in transitional areas towards the hummocks. In alpine tracts where snow lies for longer periods *Astelia alpina* may become locally dominant. These areas are only slightly acid (pH about 6.5) which contrasts sharply with conditions in the hummock.

The regeneration complex consists of hummocks of *Sphagnum* which support shrubs, and hollows which do not. The hollows are only slightly acid, as are the fen areas, and support the same type of fen vegetation. In contrast, the *Sphagnum* hummocks are highly acid (pH about 4.5). Rooted in the *Sphagnum* are dwarf xeromorphic shrubs of which *Epacris serpyllifolia* is the most common and the most widespread. Other shrubs include *Callistemon sieberi*, *Epacris paludosa* and *Richea continentis*. These shrubs form a type of heath.

The *Sphagnum* hummocks and the hollows are not spatially permanent features in the regeneration complex. On the contrary, *Sphagnum* mounds degenerate and ultimately become hollows, and it is assumed that *Sphagnum* becomes established in hollows within the regeneration complex, ultimately to form mounds. The degeneration of the *Sphagnum* hummocks commences at the top of the hummock, death of the *Sphagnum* probably being due to insufficient moisture at or near the highest patches. A small hollow forms and the shrubs commence to degenerate. The oxidation of some of the organic matter and the liberation of the bases possibly account for an ever decreasing acidity which permits the regeneration of fen species in the hollows. Non-hummock forming mosses of the genera *Polytrichum* and *Breutelia* appear in the hollows together with the sedges. The establishment of *Sphagnum* in the hollows leading to the establishment of a new hummock has not been satisfactorily explained. The pattern of the regeneration complex is therefore continually changing. The time for the conversion of hummock to hollow and *vice versa* is not known.

4.5.4.4. *Epacris breviflora* – *Blindia robusta* Alliance
Raised Bog

This alliance does not occur in the alpine tract but is best developed in the montane regions of the Kosciusko area, the Australian Capital Territory, in Victoria, and on the Northern Tablelands. Since it occurs at lower altitudes it develops under warmer conditions and is less resistant to cold than the previous alliance. It is found only in sloping, broken, base- and mineral-deficient situations where permanent wetness prevails. It covers a variety of rock types, and its associated soils are peats. The alliance may adjoin any of the following communities: wet tussock grassland, fen, heath, savannah woodland or forest.

For the Kosciusko area, Costin (1954) lists 31 species. About half of these are helophytes. Six dwarf shrubs are present; the most abundant of these are *Callistemon sieberi*, *Epacris breviflora*, *E. serpyllifolia* and *Leptospermum arachnoides*. The main herbs are *Carex gaudichaudiana*, *Deyeuxia gunniana* and *Restio australis*. The most important mosses are *Blindia robusta* and *Bryum* spp.

The bog exhibits a structure similar to that of the *Sphagnum* raised bogs. The hummocks, however, are much smaller and the hollow-hummock development is less obvious in the *Epacris* – *Blindia* bogs than in the *Sphagnum* bogs.

On the Northern Tablelands at altitudes above 1067 m, Millington (1954) describes acid bogs developed in valleys under a mean annual rainfall of 1000–1300 mm, with additional water from run-off and seepage. The communities have much in common with the southern communities.

The bogs develop in water along streams and around springs, the colonizers being *Sphagnum, Calorophus minor, Carex gaudichaudiana, Restio gracilis, Isachne australis* and *Myriophyllum* spp. These species develop a "quaking" bog which is composed of rafts of moss floating on semi-liquid organic matter, the rafts being held together by the roots of the vascular plants. The establishment of shrubs in the rafts produces the "regeneration complex", which is a hummock-hollow system. The hummocks provide a better drained habitat for the shrubs but, ultimately, the weight of the shrubs and the compaction of peat below cause the shrubs to sink, when they are immersed and die. The main shrubs which dominate hummocks are *Epacris breviflora, Baeckea gunniana* and *Leptospermum myrtifolium*. Where the peats are drier and denser the "still-stand" complex develops, this being an almost closed heath containing the shrubs mentioned above plus *Callistemon sieberi, Grevillea acanthifolia, Hakea microcarpa, Lomatia ilicifolia* and *L. salaifolia*.

18. *Acacia* and *Casuarina* Communities of the Arid and Semi-Arid Zones

1. Introduction

In the arid zone species of *Acacia* and *Casuarina* are dominant mainly on soils containing a high proportion of sand or gravel in the profile, as on dunes, sandplain or rocky ridges, which they cover as woodlands or scrubs, except in the most arid regions where hummock grasslands replace the woody communities (Chap. 20). On more clayey soils halophytic shrublands of *Atriplex* and *Maireana* (Chap. 19) or tussock grasslands (Chap. 20) replace the taller communities. In the semi-arid zone species of *Acacia* and *Casuarina* become dominant under certain ecological conditions and replace eucalypts; these conditions are either shallow soils on hilltops, or soils of fine texture. The *Acacia* and *Casuarina* communities are influenced by adjacent communities, partly through the development of ecotonal associations and partly through the extension of grasses or chenopods from the grasslands or halophytic shrublands into the

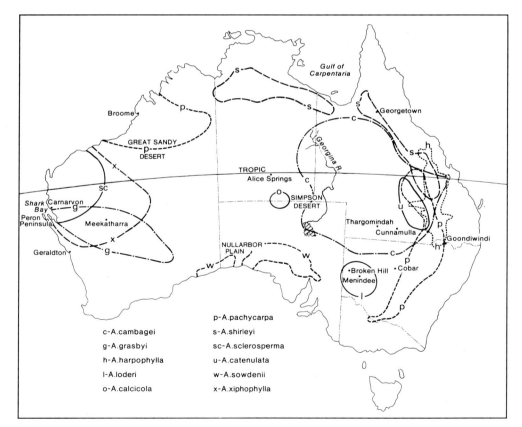

Fig. 18.1. Approximate areas of the more abundant species of *Acacia* found in the semi-arid and arid zones. Data from various sources, as in text.

understory of the *Acacia* or *Casuarina* communities.

2. Communities Dominated by *Acacia*

2.1. General

About half the total number of species of *Acacia* occur in the arid and semi-arid zones (Chap. 4). Many of these are dominant in scrubs and woodlands and a few form forests in the semi-arid zone in the east. The success of the genus in drier regions is due partly to the capacity of the species to fix atmospheric nitrogen through a symbiotic association with rhizobia in root nodules (Beadle, 1964). The highly xeromorphic phyllodes doubtless assist in water conservation, but this characteristic cannot be regarded as the sole key to success since many broad-leaved plants such as *Eucalyptus, Canthium* and *Codonocarpus* spp. (none of which has a nitrogen-fixing mechanism) grow in association with acacias. A few of the species sucker freely from the roots which appears to be an ecological advantage to the species (e.g. *A. carnei, A. harpophylla, A. loderi,* see below).

In addition to the species which constitute the alliances described in this chapter, there are several more which are often locally dominant in *Eucalyptus* communities, usually on hills where the soils are too shallow to support eucalypts or where eucalypts are scattered and stunted. The main species are trees and occur chiefly in the east (see also Chap. 16 for Western Australian species) and include the following: *Acacia doratoxylon, A. cheelii, A. concurrens* (formerly *A. cunninghamii*), mainly in western New South Wales and Queensland; *A. microsperma* (Pedley, 1974) and *A. rhodoxylon* (Speck, 1968) in Queensland.

The distributions of the species which identify the alliances, all of which are phyllodineous, are shown in Fig. 18.1, except *A. aneura* (see vegetation map) and a few of wide and scattered occurrence (see text).

The distributions of the alliances are

Table 18.1. Showing the relative locations of the *Acacia* alliances and their main soil characteristics.

			NORTH			
		A. ancistrocarpa (= *A. pachycarpa*) (deep sands)		*A. shirleyi* (lithosols)		
	A. translucens (subsaline sands)	*A. aneura* (sandy, acid to neutral soils or lithosols)	*A. catenulata* (lithosols)	*A. harpophylla* (clays)		
		A. xiphophylla (clays)		*A. cambagei* (clays)		
WEST	*A. sclerosperma* (sands over lime)	*A. calcicola* (calcareous)	*A. excelsa* (sands)	*A. pendula* (clays)	EAST
	A. grasbyi (calcareous)	*A. ramulosa – A. linophylla* (deep sands)			
		A. acuminata (lithosols)			
	A. ligulata (neutral to alkaline sands)		*A. sowdenii – A. loderi* (calcareous)		
			SOUTH			

Fig. 18.2. Woodland of *Acacia shirleyi* with a grass layer containing species of *Aristida* and *Eriachne*. 67 km north of Elliott, N.T.

explained in the following broad terms. In the tropics and subtropics, *A. shirleyi* occurs on rocky outcrops in *Eucalyptus* woodlands. Three alliances occur on clay soils in the east (*A. harpophylla*, *A. pendula* and *A. cambagei*). The *A. aneura* alliance covers most of the sandy soils and rocky areas of the arid zone, and is replaced by other species either on deep siliceous sands or highly calcareous soils, as indicated in the text and Table 18.1.

The alliances are described below, as far as is possible, in north to south sequence and from east to west, using the *A. aneura* alliance for reference, where applicable.

2.2. *Acacia shirleyi* Alliance

Lancewood

A. shirleyi, a tree to 18 m tall, dominates small patches of woodland or forest on rocky outcrops or steep slopes mainly in the tropics (Fig. 18.1) between the 500 and 1000 mm isohyets. The stands occur within woodlands of *Eucalyptus* on Lithosols (Fig. 18.2) extending occasionally on to sandy or gravelly soils (Yellow or Red Earths) on lower slopes, where *A. shirleyi* mingles with the local eucalypts, especially *E. dichromophloia*, *E. confertiflora*, *E.*

normantonensis and *E. howittiana* (tropics) and *E. citriodora, E. exserta* and *E. crebra* in subtropical Queensland.

The communities usually contain trees of *A. shirleyi* only (Fig. 18.2) and shrub species are usually absent, though a few are recorded in some stands, viz., *Erythropleum chlorostachys* in the tropics, and *Alphitonia excelsa, Petalostigma pubescens* and *Callitris "glauca"* in the subtropics (Perry and Christian, 1954; Speck, 1965, 1968; Story, 1970; Pedley, 1967, 1974). The herbaceous stratum is usually composed of grasses: in the tropics, species of *Eriachne, Heteropogon, Plectrachne, Schizachyrium* and *Triodia*; in the subtropics *Aristida caput-medusae, Cympopogon refractus* and *Cleistochloa subjuncea*.

2.3. *Acacia harpophylla* Alliance

Brigalow

A. harpophylla is a tree reaching a height of 20–25 m; it suckers freely from the roots. It occurs between the 470 and 750 mm isohyets, mainly in Queensland with small outliers in New South Wales (Fig. 18.1). It is found mainly on clay soils on flat or undulating country, rarely on slopes when another community replaces it on the uplands. The country is ecologically suitable for agriculture, chiefly the grazing of cattle and consequently has been studied in some detail.

Brigalow occurs on five types of soil (Isbell, 1962):

i. Deep gilgaied clays, 2–3 m deep, developed on flat country; the surface is alkaline, becoming acid with depth.

ii. Sedentary clays c. 1 m deep, without gilgais, developed on undulating areas; the soils are mostly alkaline throughout.

iii. Alluvial clays without gilgais but subject to flooding; the profile is alkaline throughout.

iv. Miscellaneous deep clays with few or no gilgais, developed on gently to undulating country; the soil surface is alkaline, becoming acid with depth.

v. Light-textured red soils without gilgais occurring as outliers in the south-west (Cunnamulla and north-west of Brewarrina); the surface sandy horizon is acid, the more clayey subsoil is alkaline.

The soils vary in fertility; Russell et al. record total phosphorus levels of 255–350 ppm in the surface and 102–114 ppm at a depth of 60–90 cm. The alluvial clays are the most fertile.

Three growth forms are defined by Johnson (1966):

i. Tall or Virgin Brigalow ranging from 10–25 m tall (usually 40–60; these communities always contain a high proportion of dead, standing trees (Moore et al. 1966) and a dense litter layer. In stands where the soils are gilgaied the largest trees are confined to higher portions around the rims of the gilgais where there is also the highest concentration of roots and the highest organic content (Russell et al. 1967); these authors suggest that the trees may play some part in the formation of gilgais.

ii. Whipstick Brigalow 4–8 m tall, possibly developed from the copious germination of seed following damage to a virgin stand; the communities probably exceed 30 years in age.

iii. Sucker Brigalow to c. 5 m tall with a low branching habit, developed from root-suckering following burning.

The high density of the brigalow stands and the capacity of the plants to sucker freely from roots have necessitated special precise treatments for the removal of the stands. The trees are "pulled" to the ground by a chain or chain and cable dragged by two high-powered bulldozers. "Pulling" is followed by burning of the dried trash at least nine months after "pulling" (Johnson, 1976). Also, aerial spraying with organic poisons has proved successful in killing brigalow, if spraying is done under still air conditions when soil moisture is high (Everist, 1966).

The autecology of the species has received some attention. Coaldrake (1971) from field and glasshouse experiments on seedlings has shown that the species is largely cross-pollinated and that variations occur in the number of flowers in the head, seed size and longevity, seed-coat hardness and germination rates. Much of the variation is clinal. He concludes that variation in germination behaviour, growth and ontogeny suggest why brigalow is able to colonize habitats that vary widely in terms of water relations and soil fertility.

Most of the brigalow communities in Queensland occur in regions which support *Eucalyptus* woodlands and types of dry rainforest, both of which form ecotonal associations with brigalow, the former dominating soils of lower fertility, the latter on more fertile soils especially in wetter sites. Field observations suggest that the brigalow is invading the other two types of communities and that *A. harpophylla*, because of its capacity to regenerate rapidly from seed and to sucker from the roots is expanding its area.

When brigalow occurs in pure stands (Fig. 18.3) it has few understory species. The shrub *Carissa ovata* is possibly most common and others of occasional to rare occurrence are *Heterodendrum diversifolium* and *Eremophila mitchellii*. The herbaceous layer consists mainly of grasses, e.g., *Bothriochloa intermedia, Dichanthium sericeum, Sporobolus caroli* and others. When gilgais occur, they invariably contain *Marsilea* in the bottoms, often with sedges and a few grasses (*Panicum buncei, Leptochloa debilis* and *Sporobolus mitchellii*). Around the rims of the gilgais taller plants may occur, especially *Muehlenbeckia cunninghamii* and *Eremophila polyclada* (Story, 1967). On the western (drier) fringes of the communities *Casuarina cristata* is commonly co-dominant with brigalow, the stands probably representing ecotones (see 3.3, below).

Fig. 18.3. *Acacia harpophylla* woodland. Natural stand on right. Dead grass in foreground is *Dichanthium sericeum* and *Sporobolus caroli*. At left burned and cleared brigalow with much regeneration from root-suckers. Note high fence to control movement of kangaroos. Near Moonie, Qld.

Where brigalow scrubs form mosaics with *Eucalyptus* woodlands or softwood scrubs distinctive ecotonal associations exist, the following being the most common (Isbell, 1962; Johnson, 1966).

i. *Acacia harpophylla* – *Eucalyptus cambageana*. This association occurs mainly on slopes north of the Tropic. A tall shrub layer occurs, containing species of the "softwood scrub" assemblage (below).

ii. *A. harpophylla* – *E. thozetiana*. This association occurs in the same areas as i., on lower slopes.

iii. *A. harpophylla* – *E. populnea*. These two species are found together mainly south of the Tropic. The eucalypt is commonly an emergent, becoming more abundant on upper slopes, or along watercourses. Associated species vary from few to numerous. *Casuarina cristata* is often present and numerous "softwood scrub" species are often present.

iv. *A. harpophylla* – *E. microtheca*. The latter is a species of watercourses or flooded areas and the association is found on flats near rivers.

v. *A. harpophylla* – *Casuarina cristata*. This combination is common throughout the brigalow area, the latter becoming more abundant in waterlogged sites, especially around gilgais. The outliers in New South Wales, except the one on sandy soil, contain these two species as co-dominants.

vi. *A. harpophylla* – Softwood scrub assemblages. Mosaics of brigalow with softwood scrubs (dry rainforest assemblages – Chap. 7) are common in Queensland and numerous tall species occur which are sometimes locally dominant with brigalow or they occur as understory species in the brigalow forests. The more abundant of these are: *Terminalia oblongata* as a co-dominant north of the Tropic; *Macropteranthes leichhardtii* (Bonewood) as an understory species north of the Tropic; *Brachychiton rupestre* (mainly north); *Cadellia pentastylis* (south); *Geijera parviflora* (throughout). The more common tall understory species include *Eremophila mitchellii*, *Heterodendrum diversifolium* and many others. The shrub *Carissa ovata* is a near-constant. The ground layer is usually dominated by grasses, species of *Chloris*, *Paspalidium* and *Sporobolus* being most common. Also, *Santalum lanceolatum* is co-dominant with brigalow in a few areas, and in climatically drier areas *Atalaya hemiglauca* and *Heterodendrum oleifolium*.

vii. *A. harpophylla* – *Bauhinia* spp. This association which includes either *B. carronii* or *B. hookeri* occurs on coarse textured alluviums on some highlands.

viii. *A. harpophylla* – *Callitris "glauca"*. This is a rare combination found only on sandy soils in New South Wales. The understory shrubs are *Eremophila sturtii*, *E. mitchellii* with few *Myoporum deserti*, *Canthium oleifolium* and *Apophyllum anomalum*.

ix. *A. harpophylla* – *Acacia* spp. Several tall species of *Acacia* form ecotonal associations with *A. harpophylla*, mainly in climatically drier areas. The species are *A. cambagei*, *A. excelsa* and *A. pendula*, which are discussed under separate headings in this chapter.

2.4. *Acacia pendula* Alliance

Myall or Boree

This species is a small tree 8–12 m high and forms open woodlands or tree savannahs (Beadle, 1948; Moore, 1953; Biddiscombe 1963). It is confined to the east (Fig. 18.1) between the 400–500 mm isohyets in the north and the 350–450 mm isohyets in the south. It occurs on Grey and Brown Clays, rarely on Red Brown Earths in the south-east. The more clayey soils usually have a self-mulching surface and crack deeply on drying.

Unlike most acacias, *A. pendula* fruits and sets seeds sparingly and seed germination (in the laboratory) is poor. Regeneration in the field appears to be sporadic and, since the foliage is palatable to stock, young plants are often damaged by grazing. Heavy infestations of the mistletoe, *Amyema quondong*, are usual, but the parasite rarely kills the host.

A. pendula, usually in pure stands, dominates communities adjacent to streams or halophytic shrublands (Chap. 19) or in depressions in grasslands. Occasional to rare associates occur mainly in the north and are: *Acacia salicina*, *A. stenophylla* (near watercourses), *A. oswaldii*, *Atalaya hemiglauca*, *Capparis lasianthos*, *Cassia circinnata*, *Casuarina cristata*, *Eremophila bignoniflora* and *Flindersia maculosa*.

The composition of the understory varies. In the north, a continuous sward of tussock grasses occurs, common species being *Astrebla lappacea* (Fig. 18.4), *Dichanthium sericeum* and *Panicum prolutum*. Gilgai depressions sometimes break

Fig. 18.4. Open woodland of *Acacia pendula*. Grass layer dominated by *Astrebla lappacea*. In N.S.W., 48 km south of Goondiwindi, Qld.

the continuity of the grass layer and in the depressions *Eremophila maculata* sometimes forms shrubberies, with *Marsilea drummondii* and sedges in the wettest parts. In the southern half of the alliance, the perennial saltbushes (*Atriplex nummularia* and/or *vesicaria* with subsidiary *Maireana aphylla*) may form a discontinuous shrub layer $1/2$–3 m tall. However, in most areas the saltbushes have been removed by domestic stock and replaced by species which were subsidiary in the virgin stands, viz., *Stipa aristiglumis* and *Eragrostis setifolia*. These grasses also have been reduced in density by overgrazing which leads to the dominance of more fragile species, especially *Chloris truncata*, *Sporobolus caroli* and annuals belonging to the Compositae and Cruciferae. In some areas other native species have become abundant and are regarded as weeds by graziers, especially *Hordeum leporinum* and *Bassia quinquecuspis*. Species of * *Medicago* are naturalized in most stands.

2.5. *Acacia cambagei* Alliance

Gidgee

This species occurs over a wide area (Fig. 18.1) bounded in wetter districts by the 500 mm isohyet. It occupies areas subject to irregular flooding and in drier areas it follows the drainage lines, even into the Simpson Desert (Crocker 1946; Boyland, 1970). The species is notorious for the sewer-like smell it emits when the phyllodes are damp. The plants reach a height of 5–10 m and form scrubs in drier areas and woodlands in the wetter parts of its range, the density of the stands varying from very open to closed. The soils are mostly Grey and Brown Clays, less commonly duplex soils; those in drier areas are sometimes strewn with gibbers. Gilgais are common in the clay soils.

Gidgee usually occurs in pure stands. In wet-

Fig. 18.5. Pure stand of *Acacia cambagei*. Note copious litter and fallen branches and trunks. 27 km north of Tambo, Qld.

ter areas several other species of similar size are sometimes present (Blake, 1939; Beadle, 1948; Davidson, 1954; Perry and Christian, 1954; James, 1960; Perry and Lazarides, 1962; Pedley, 1967, 1974). These are *Atalaya hemiglauca, Heterodendrum oleifolium, Flindersia maculosa, Grevillea striata, Ventilago viminalis* and *Casuarina cristata*. Shrubs are rare but may sometimes form a definite understory, viz., *Apophyllum anomalum, Capparis lasianthos, C. mitchellii, Eremophila mitchellii, Santalum lanceolatum* and, in the south, *Atriplex nummularia*. Around gilgais *Muehlenbeckia cunninghamii* and *Eremophila bignoniflora* may form thickets. Ecotonal associations may occur with *Eucalyptus microtheca, E. largiflorens* (south), and *Acacia harpophylla* (east).

The low shrub and herbaceous layers vary tremendously. In dense stands the ground is commonly strewn with litter including fallen branches and logs (Fig. 18.5), and there is virtually no herbaceous layer. A grass layer is usual in the north and possibly *Astrebla* spp. and/or *Dichanthium* spp. were once dominant. Since the communities have been heavily grazed and used for shelter by stock, especially in drier areas, the grass sward has been considerably modified. Common grasses include *Sporobolus actinocladus* and *Chloris* spp. with a high proportion or even the dominance of *Bassia* spp. In very dry areas of Central Australia, *Triodia pungens* and *Aristida* spp. are dominants. In the south-west *Atriplex nummularia* and/or *A. vesicaria* dominated the understory, often with *Astrebla pectinata* in gilgais.

Two other species are mentioned here, though they possibly characterize distinct alliances.

i. *Acacia georginae*, closely related to *A. cambagei* forms woodlands c. 6 m tall in the Georgina R. district on calcareous or alluvial soils. Associated trees are *Grevillea striata* and *Hakea lorea*. A shrub layer of *Cassia oligophylla, C. sturtii* and *Eremophila latrobei* is present (Perry and Lazarides, 1962).

ii. *Acacia cana* occurs in the semi-arid east. Davidson (1954) reports that in the Longreach district it replaces *A. cambagei* on higher ground as a woodland c. 6 m tall. The species extends into western New South Wales to the Menindee district and sometimes replaces *A. cambagei* along shallow watercourses in the *A. aneura* alliance.

2.6. *Acacia aneura* Alliance
Mulga

2.6.1. General: Autecology of *A. aneura*

A. aneura is the dominant species over vast tracts of country in the arid and semi-arid zone (see vegetation map) and it occurs also as an understory species in some *Eucalyptus* woodlands to the east (chiefly *E. populnea* and *E. intertexta*, Chap. 13). Outliers are found as far as the upper Hunter Valley (near Scone). The main body of the alliance in which *A. aneura* dominates the upper stratum occupies most of the area bounded approximately by the 250 mm isohyets in the west and south and by the 450 mm isohyet in the north east. At its drier limit it disappears at about the 170 mm isohyet and is replaced by hummock grasslands. In these drier areas mulga often persists in more favourable sites as an emergent in the grasslands (Chap. 20). In the north the alliance merges with the tropical *Eucalyptus* woodlands and ecotones involving *A. aneura* and *E. dichromophloia, E. brevifolia* and *E. terminalis* (Chap. 8) are common. In the south the alliance merges with the southern mallee (Chap. 14) or *Casuarina cristata* (3.3, below) and in the east with the *Eucalyptus* woodlands, as mentioned above. The continuity of the alliance across the continent is broken by patches of tussock grassland or saltbush (chiefly on clays) and by watercourses, all of which have some effect on the composition of the alliance, as described below.

A. aneura is a polymorphic species and several ecotypes probably exist, which is suggested by variations in the sizes of the phyllodes (Pedley, 1973) and the habits of plants, especially with respect to branching. Some of these variants appear to be associated with special habitats but these have not been systematically studied. Polyploidy is reported in some populations (De Lacy and Vincent, in Pedley, 1973) the diploids having the wider phyllodes.

A. aneura, being dominant in the upper story, is subjected to great water stress (Slatyer, 1967). Its distribution in relation to rainfall incidence (Nix and Austin, 1973) suggests that both summer and winter rain are necessary for its survival. Vegetative growth occurs after rain at any time of the year and stem-flow after a small shower of rain can benefit the trees (Slatyer,

1965). Flowering also occurs several times in the year after rain but the principal flowering times are spring and autumn. However, Preece (1971) found that only late summer rain leads to seed formation, and he estimates that successful establishment could possibly occur once in every six years. The seeds are mostly hard (up to 98%) and the seeds germinate equally well in the light and dark, the optimum temperatures for germination being 20–30°C. Germination is more rapid in an atmosphere rich in CO_2, which could be provided in organic accumulations under trees (Preece, 1971). However, Burrows (1973) indicates that in Queensland significantly fewer seedlings regenerate under trees occurring at 640 trees per hectare. In the field seedling populations usually appear to be of the one age which suggests simultaneous germination at infrequent intervals (Lange, 1966 and Fig. 18.6).

Mulga produces a long taproot, extending a few metres into deep soils. Seedlings grown in pots with stem lengths of 8–10 cm produced

Fig. 18.6. *Acacia aneura* woodland with an understory of young plants from natural seeding. Note infestations of mistletoe (*Amyema quondong*). Near Thargominda, Qld.

Fig. 18.7. *Acacia aneura* woodland. There are no associated woody species. Grass layer dominated by *Eragrostis browniana*. 17 km north of Wittenoom, W.A.

taproots 2–3 m long. An extensive lateral root system develops in the field with a concentration 5–20 cm below the surface, the roots being so matted that digging with a spade is difficult. The roots produce rhizobial nodules but nodules are usually of rare occurrence, possibly because of the paucity of root hairs. Wilkins (1967) has demonstrated that the arid strain of *Rhizobium* is more heat-resistant than a strain infecting *A. rubida* on the eastern tablelands.

A. aneura possibly has a life-span of a few hundred years. Hall et al. (1964), using growth rates of young plants (c. 1 m every 10 years) estimate the time to reach maximum size is about 100 years. Growth rings, which are not necessarily annual, number 140–200 in trunks 25 cm diameter.

Soil properties determine the distribution of the alliance in areas where the climate is suitable. *A. aneura* occurs mainly on Red Earths and Lithosols. The former cover flat or undulating country which is generally sandy but usually with an accumulation of finer material (iron hydroxide or clay) in the subsurface layers. The Lithosols occupy hills where rock occurs at or near the surface, including sandstones, quartzites, schists, slates, granites and, rarely, basalts, and in many areas, especially in the west, ferruginous fossil laterite.

The Red Earth show considerable variation in the depth of the surface sandy horizons which relates to their mode of origin (various authors, summarized by Dawson and Ahern, 1973). The deepest surface sandy horizons occur in aolian

Fig. 18.8. *Acacia aneura* (left and behind) and *Grevillea striata*. Herbaceous layer mainly *Bassia* spp. Near Thargominda, Qld.

deposits where mulga is sparsest. The soils have in common a neutral to slightly acid reaction, which is significant in so far as the presence of calcium carbonate reduces the abundance of mulga which is replaced by other species (see below). Fertility levels are usually regarded as low, though Dawson and Ahern (1973) indicate adequate calcium, potassium and magnesium levels in the soils they analysed. Phosphate values vary considerably (23 to over 300 ppm total P – see Chap. 6) and at lower phosphate levels the mulga is stunted. Nitrogen levels are generally low but since mulga fixes nitrogen symbiotically, it is likely never to be deficient in this element, though associated species without a symbiont may be, especially under modern grazing conditions which tend to remove the surface organic matter (see 2.6.3, below).

2.6.2. Community Structure, Associations and Subassociations

The various communities dominated by *A. aneura* show some variation in structure. Mulga itself varies in height from c. 3–10 m. The tallest stands are almost closed woodlands with scattered or no shrubs below and a discontinuous grass layer. With diminishing mean annual rainfall or/and with increasing elevation on to hilltops, the trees become shorter, the communities are more open and the shrubs and grass layer increase in density. Open scrub communities 2–3 m tall are found in the drier limits of the range of the species on hilltops capped by ferruginous lateritic materials. At the drier limit on sand mulga often occurs in groves following contours, separated by treeless areas of hummock grassland (Perry and Lazarides, 1962), the mulga occurring in moister sites.

The alliance has been studied for half a century (Collins, 1923, 1924; Wood, 1937; Blake, 1938; Beadle, 1948; Jessup, 1951; Speck, 1963 and others below). Throughout its range *A. aneura* is nearly always the sole dominant (Fig. 18.7) but a few other tall species occur locally as co-dominants, including a few eucalypts, especially in Western Australia (*E. gamophylla, E. kingsmillii* and *E. striaticalyx*). *Callitris "glauca"* occurs as an associate from east to west both on sandy soils and on rocky outcrops, though it usually forms small monospecific patches of scrub or woodland. *Grevillea striata* (Fig. 18.8) and *Canthium oleifolium* are rare associates from east to west and *Acacia pruinocarpa* in the west. In the east *Atalaya hemiglauca, Hakea leucoptera* and *Heterodendrum oleifolium* occur and may be locally common since they sucker from the roots. A few others are less common, e.g. *Hakea lorea, H. intermedia* and *Pittosporum phylliraeoides*. In the south *Casuarina cristata* forms ecotonal associations with mulga. On hills, a few other species of *Acacia* are sometimes locally dominant, the more abundant of these being *A. kempeana* (Central Australia and western Queensland), *A. brachystachya* (throughout), *A. carnei* (South Australia and New South Wales), and *A. tetragonophylla*. Each of these may be locally dominant or occur in association with *A. aneura*. Of special interest is *A. carnei* which rarely sets seeds but reproduces vegetatively by root-suckers to form local patches of impenetrable scrub. It occurs also on lunettes around some lakes with or without mulga (Fig. 18.9).

In contrast to the uniformity of the upper stratum, the understory species, both woody and herbaceous vary from east to west and from north to south. The total understory flora is a very large one, containing several hundred species, of which about one third are shrubs and subshrubs and about half are annual forbs, the remainder being grasses. Noteworthy among the shrubs are the many species of *Eremophila* (Table 18.2). Some shrub species occur from east to west and some are localized. As a rule, the northern assemblages (containing many species of *Eremophila* and *Cassia*) are different from the southern ones, where Chenopodiaceae are more common (on soils containing calcium carbonate or salt). The abundance of many shrub species has been changed by overgrazing and/or a reduction in density of mulga. Notable among these shrubs are species of *Eremophila*, especially in Queensland, and *Cassia* and *Dodonaea* in New South Wales and South Australia (see 2.6.3, below).

The herbaceous stratum is usually dominated

Fig. 18.9. Pure stand of *Acacia carnei* produced by root suckering. Understory plants are *Bassia paradoxa*, *Phyllanthus filicaulis* and *Tetragonia expansa*. Sandhill in Kinchega National Park, near Menindee, N.S.W.

Table 18.2. Main understory shrubs in the *Acacia aneura* alliance in various districts. Genera and species listed alphabetically. Data from various authors in text.

NORTHERN W.A.	CENTRAL AUSTRALIA	QUEENSLAND
Acacia adsurgens	*Acacia adsurgens*	*Acacia brachystachya*
A. burkittii	*A. brachystachya*	*A. kempeana*
A. craspedocarpa	*A. chisholmii*	*A. tetragonophylla*
A. dictyophleba	*A. kempeana*	*Capparis lasiantha*
A. inophloia	*A. lysiphloia*	*Cassia artemisioides*
A. tetragonophylla	*Cassia artemisioides*	*C. nemophila*
Eremophila compacta	*C. desolata*	*C. oligophylla*
E. cuneifolia	*C. nemophila*	*C. phyllodinea*
E. foliosissima	*C. oligophylla*	*C. sturtii*
E. forrestii	*C. sturtii*	*Dodonaea adenophora*
E. fraseri	*Dodonaea attenuata*	*D. attenuata*
E. gilesii	*Eremophila alternifolia*	*D. petiolaris*
E. latrobei	*E. duttonii*	*Eremophila bowmanii*
E. leucophylla	*E. freelingii*	*E. gilesii*
E. macmillaniana	*E. gilesii*	*E. glabra*
E. margarethae	*E. glabra*	*E. goodwinnii*
E. miniata	*E. latrobei*	*E. latrobei*
E. platycalyx	*E. longifolia*	*E. longifolia*
E. spathulata	*E. paisleyi*	*E. mitchellii*
Grevillea juncifolia	*E. sturtii*	*E. sturtii*
	Grevillea juncifolia	*Grevillea juncifolia*
SOUTHERN W.A.	*Pittosporum phylliraeoides*	*Hakea ivoryi*
Acacia brachystachya		*Myoporum deserti*
A. resinomarginea		*Phebalium glandulosum*
A. tetragonophylla		*Pittosporum phylliraeoides*
Atriplex hymenotheca		*Prostanthera suborbicularis*
Cassia artemisioides		*Santalum acuminatum*
C. nemophila		
Eremophila abietina	SOUTH AUSTRALIA and BARRIER RA., N.S.W.	
E. alternifolia	*Acacia brachystachya*	*Eremophila glabra*
E. dielsiana	*A. burkittii*	*E. latrobei*
E. fraseri	*A. carnei*	*E. longifolia*
E. granitica	*A. kempeana*	*E. paisleyi*
E. leucophylla	*A. tetragonophylla*	*E. scoparia*
E. margarethae	*Atriplex vesicaria*	*Lycium australe*
E. scoparia	*Cassia artemisioides*	*Maireana georgei*
Lycium australe	*C. nemophila*	*M. pyramidata*
Maireana pyramidata	*C. phyllodinea*	*M. sedifolia*
M. sedifolia	*C. sturtii*	*Myoporum deserti*
Pittosporum phylliraeoides	*Codonocarpus cotonifolius*	*Pittosporum phylliraeoides*
Santalum acuminatum	*Dodonaea attenuata*	*Prostanthera striatiflora*
S. spicatum	*Eremophila alternifolia*	*Santalum acuminatum*
	E. brownii	*S. spicatum*
	E. duttonii	*Scaevola spinescens*

by grasses, except in very dense stands under trees when the accumulated litter and probable root-competition suppress the growth of any herbaceous species. Annuals, both grasses and forbs are abundant after rain and their potential density in many mulga stands has been increased through the reduction in density of the mulga.

Grass assemblages on sandy soils and on hills with rock exposures are usually different.

On sandy soils several grass layers occur:

i. *Eragrostis eriopoda* dominates from east to west in the north (summer rainfall predominating) on soils containing some finer material in the profile. *Danthonia bipartita* and *Neurachne mitchelliana* are occasional to rare associates. The continuity of the *Eragrostis* sward is sometimes broken be the occurrence of depressions containing more clay and in these other grasses are conspicuous or dominant, especially *Themeda australis, Eragrostis setifolia, Eulalia fulva, Bothriochloa ewartiana* and, rarely, *Astrebla* spp.

ii. In Central and Western Australia *Plectrachne schinzii* (Feathertop Spinifex) dominates in the north-west on sandplains, particularly north of lat. 23°S (Perry and Lazarides, 1962).

iii. *Triodia pungens* (Soft Spinifex) dominates on sandplains and dune fields, especially north of lat. 23°S., and *T. basedowii* dominates south of lat. 23°S.

On hills with rock exposures, *Enneapogon* spp. and *Paspalidium gracile* were probably the most abundant in the east, with *Stipa variabilis* and *S. eremophila* becoming conspicuous in the south (winter rainfall predominating). In the centre and west, in the north, *Triodia pungens* is the most common species. South of lat. 23°S., *T. clelandii, T. spicata* and *T. irritans* are of local occurrence.

The most abundant of the summer-growing annuals are *Aristida arenaria, A. browniana, Enneapogon avenaceus, E. polyphyllus, Paractenium novae-hollandiae* and *Plagiosetum refractum*. Forbs comprise the bulk of the herbaceous winter-growing annuals which are very numerous and belong largely to the Compositae.

In the south where *Maireana* or *Atriplex* occur in the shrub layer, the herbaceous species are mainly species of *Bassia* and Compositae.

Fig. 18.10. Degenerating mulga country. The dead trees in the foreground have died following heavy grazing and drought. Some living plants behind with an understory of *Triodia pungens*. Composites in foreground are the winter-growing annuals *Myriocephalus stuartii* and *Senecio gregorii*. c. 200 km south of Alice Springs, N.T.

2.6.3. The Effects of Grazing; Degeneration and Secondary Successions

The introduction of domestic stock into the mulga country towards the end of the nineteenth century led to the degeneration of the stands through the death of mulga trees. This premature death is apparently initiated by the disturbance of the community through the removal of some trees for stock feed (the phyllodes of mulga are palatable to stock), fuel or fence posts. Further to this, the removal of litter from the soils surface and the eating of mulga seedlings by stock have retarded processes of regeneration which could occur under conditions of minimal or no grazing. Following the removal of part or the whole of the vegetal cover, the soil surface is exposed to erosion by wind or water. Soils with a deep sandy profile are likely to be blown by the wind into dunes. Soils with hardpan in the subsoil are stripped of their surface horizon by wind or water to produce bare water-impervious surfaces (scalds) which are rarely colonised by plants under the existing conditions. Premature death of the mulga has been commented on by many authors. The death of the trees is sometimes a gradual process but more usually occurs over large tracts of country during very dry periods.

Reduction in the density of mulga by felling or premature death is usually followed by an increase in density of the understory species, except when the surface soil is removed and scalds develop. Consequently many of the mulga communities that exist today are not climax but

Fig. 18.11. *Cassia nemophila* and *C. artemisioides* dominant in a regenerating community on windblown sand. Understory species: *Bassia lanicuspis*, *Atriplex stipitata*, *Enneapogon avenaceus* and *Brachycome ciliaris*. Regeneration area, 34 years after enclosure. Broken Hill, N.S.W.

rather undergoing secondary succession. Notable amount the shrubs which are regenerating in gaps in mulga stands or over large tracts of country formerly covered by mulga are species of *Eremophila* in Queensland and species of *Cassia* or *Dodonaea attenuata* in western New South Wales and South Australia. In degenerate communities in which shrubs have not become established the former woodland has been replaced by annuals or short-lived perennials (Fig. 18.10).

The organic matter contained within a mulga community in the virgin condition is surprisingly high for such an arid climate. In a dense stand there are 400–600 trees per hectare, each with a mass of c. 100–150 kg. The surface litter may accumulate to a depth of 5 cm and the surface soil (0–6 cm) contains 2–2.5% organic carbon which drops to c. 1% at a depth of 10 cm, and mostly 0.2% below this level. In severely degenerate and eroded country the organic matter levels are reduced to c. 0.2% and this material provides the substrate for the regenerating communities.

An opportunity to study rates of accumulation were provided by the regeneration areas at Broken Hill. This area, formerly mulga woodland, was denuded mainly by clearing to provide fuel and mine props for the city and by heavy grazing. Windblown sandy material with a small clay content comprised the bare area which surrounded the city. Part of the area was enclosed against stock in 1937 (Morris, 1939) and regeneration of annuals, followed by some perennial herbs and shrubs, (*Cassia nemophila* and *C. artemisiodes*/(Fig. 18.11), was rapid (Pidgeon and Ashby, 1940).

The soils under three different types of vegetation were sampled on four occasions, commencing 1954 and were analysed for organic carbon; the results are summarized in Table 18.3 and show that the organic profile does not approach the possible original condition after a period of 30 years. Although the amount of litter under *Cassia* plants reached a depth of c. 2 cm, the rate of decomposition appeared to be slow, because of the frequent drying out.

2.7. *Acacia catenulata* Alliance
Bendee

This species, previously confused with *A. aneura* and distinguished from the latter by its fluted trunk, occurs in south-western Queensland (Fig. 18.1), overlapping *A. aneura* in the eastern part of its range (Pedley, 1974). *A. catenulata* occurs on shallow soils in rocky situations, often dominating on hills above a zone of *A. aneura* which occurs on deeper soils. In the northern part of its range, *A. catenulata* sometimes adjoins communities of *A. shirleyi*, but without mixing.

A. catenulata dominates scrubs 8–12 m tall (Speck, 1968; Pedley, 1974) usually as pure

Table 18.3. Accumulation of organic carbon, expressed as a percentage of air dried soil in 3 habitats in a stock-free enclosure at Broken Hill. The area was enclosed in 1937–1938 and sampled as indicated. Each figure is the mean of 5 estimates.

Bare areas	% O.C. in the years indicated			
	1954	1958	1967	1971
0– 3 cm	0.17	0.19	0.26	0.22
5– 8 cm	0.23	0.20	0.19	0.16
8–22 cm	0.24	0.22	0.22	0.21
Under *Enneapogon*				
0– 3 cm	0.53	0.50	0.54	0.64
5– 8 cm	0.29	0.22	0.18	0.23
8–22 cm	0.29	0.23	0.23	0.28
Under *Cassia*				
0– 3 cm	0.33	0.41	0.77	0.89
5– 8 cm	0.30	0.21	0.23	0.33
8–22 cm	0.28	0.20	0.28	0.30

stands. Associated shrubs are rare but may include *Lysicarpus angustifolia, Geijera parviflora* and *Eremophila mitchellii*. A sparse ground layer of grasses is present, containing one or more of *Aristida caput-medusae, Ancistrachne uncinulatum, Neurachne mitchelliana* and *Paspalidium caespitosum*.

The scrubs often occur within woodlands of *Eucalyptus*, especially *E. crebra, E. melanophloia* and *E. populnea,* and trees of these species may be present as emergents in the *Acacia* scrub.

2.8. *Acacia excelsa* Alliance

Ironwood

A. excelsa reaches a height of c. 12 m and is one of the largest acacias of the arid and semi-arid zones (Fig. 18.12). Its main centre of abundance is the Cobar – Bourke area, where the mean annual rainfall is c. 280–350 mm; it extends eastward as an understory species in the *Eucalyptus intertexta* and *E. populnea* woodlands (Holland and Moore, 1962) and it is recorded in eastern Queensland in association with *A. harpophylla* or, rarely in pure stands (Gunn and Nix, 1977).

In the Cobar – Bourke districts it occupies deep sandy soils sometimes with pans, with or without lateritic gravel. On the deeper sands it is commonly associated with *Callitris "glauca"* or, rarely, with *Codonocarpus cotonifolius*. On more compact soils the main associates, which often forms the understory, are *Acacia homalophylla, A. aneura, Eremophila mitchellii, Grevillea striata* and *Hakea leucoptera*. A lower shrub layer, often very dense, is composed of *Eremophila sturtii, E. longifolia, Cassia nemophila, C. artemisioides* and *Dodonaea*

Fig. 18.12. Pure stand of *Acacia excelsa*. Understory shrub is *Eremophila sturtii*. Grass layer (heavily grazed) is dominated by *Eragrostis eriopoda*. 56 km north of Bourke, N.S.W.

Fig. 18.13. General view of the *Acacia pachycarpa* shrublands, containing some *Eucalyptus dichromophloia*, *E. setosa* and *Bauhinia cunninghamii*. The photo illustrates also the flat topography typical of the north-west coastal plain. c. 60 km south-west of Broome, W. A.

attenuata. *Eucalyptus intertexta* and/or *E. populnea* are present in some areas as emergents. The ground stratum is dominated by *Eragrostis eriopoda* with *Aristida jerichoensis* as a subsidiary species, the latter becoming more abundant following heavy grazing. In areas with a dense shrub layer the herbaceous layer is almost absent. The total herbaceous flora is very large (Beadle, 1948).

A closely related species of similar size, *A. estrophiolata* occurs in Central Australia north of lat. 24°S., separated from *A. excelsa* by the Simpson Desert. It is often associated with *Atalaya hemiglauca*, *Ventilago viminalis* and *Hakea intermedia* (in various combinations) on shallow soils overlying granite, schist or gneiss (Perry and Lazarides, 1962). The understory is dominated by the grasses *Eragrostis eriopoda* or *Aristida browniana* on sandier areas and *Chrysopogon fallax*, *Themeda avenacea*, *Bothriochloa ewartiana* and *Eulalia fulva* in depressions containing more clay.

2.9. *Acacia pachycarpa* (== *A. ancistrocarpa*) Alliance

Pindan Wattle

This species replaces *A. aneura* in the north-west extending as a common or dominant species through the Great Sandy Desert to the north-west coast (Fig. 18.1). Its mean annual rainfall limits are c. 350–500 mm. In the Great Sandy Desert it forms thickets in or overtopping hummock grasses and, with increasing rainfall it becomes taller and denser. In the wetter parts of its range it is associated with eucalypts (mainly *E. dichromophloia* or *E. setosa*) and in the wettest parts the eucalypts overtop the *Acacia*, mainly north of Broome (Chap. 8), where the communities are woodlands.

A. pachycarpa occurs on deep red Siliceous Sands and may attain a height of c. 8 m in the wettest climates, being reduced to c. 3 m in the driest areas. In the north-west, south of Broome, the shrublands occur on flat or slightly undulating country with no watercourses (Fig. 18.13). The communities are dense and *A. pachycarpa* is the most abundant species, though often with an admixture of *A. impressa*. Other tall species are *Bauhinia cunnighamii*, *Dolichandrone heterophylla*, *Erythrophleum chlorostachys* and *Gyrocarpus americanus*. The shrub layer contains *Acacia translucens*, *Grevillea* spp. and *Jacksonia aculeata*. The main grasses are *Plectrachne schinzii*, *Triodia pungens* and *Eragrostis eriopoda* (Beard and Webb, 1974).

The communities are regularly burned so that the composition of stands varies from year to year. Burning does not kill the eucalypts (Fig. 18.14) but the acacias are almost completely removed and regenerate from seed.

Fig. 18.14. Burned *Acacia pachycarpa* shrubland showing the regeneration of *Eucalyptus dichromophloia* and *E. setosa*. Dead stems are *Acacia pachycarpa*. 150 km south-west of Broome, W.A.

Fig. 18.15. Heath-like scrub of *Acacia translucens* forming a pure stand. No associated species visible but hummocks of *Plectrachne schinzii* are present. 250 km south-west of Broome, W.A.

2.10. *Acacia translucens* Alliance

This species is usually a shrub $1/2-1 1/2$ m tall, with a preference for subsaline conditions. It occurs in the tropics usually close to the sea from the islands of the Gulf of Carpentaria (Bentham 1864) into Western Australia to c. lat. 21°S. It is recorded also in the inland as an occasional species in the hummock grasslands of Western Australia and arid Northern Territory (Chap. 20). In Western Australia, between the 370 mm and 300 mm isohyets, it is dominant on flat country close to the sea, extending inland for some 20? km. In places it extends on to the coastal dunes (Chap 23) and it is common on sandy deposits surrounding saline mudflats. In the north and east the alliance adjoins the *Acacia pachycarpa* alliance and in the drier south it merges with the hummock grasslands.

A. translucens dominates heath-like scrubs mostly 1 m high and on flat subsaline country there are no associated shrubs (Fig. 18.15). Hummocks of *Plectrachne schinzii* grow under some acacia bushes and annuals probably appear after rain. On elevated sandy patches *P. schinzii* become more abundant and emergent shrubs, mostly 2–3 m tall may grow. The main shrubs are *Acacia bivenosa* (a coastal dune species), *A. holosericea*, *A. monticola*, *A. tumida*, *Bauhinia cunninghamii*, *Erythrophleum chlorostachys*, *Eucalyptus setosa*, and *Grevillea eriostachya* (south only).

2.11. *Acacia sclerosperma* – *A. tetragonophylla* – *Eremophila pterocarpa* – *Atriplex* spp. Edaphic Complex

A. sclerosperma, a shrub $1 1/2-2$ m high, dominates red sands or sandy loams in the 200–240 mm mean annual rainfall zone near the west coast (Fig. 18.1). The species is recorded in some inland sites, as a subsidiary species, e.g., at Meekatharra (Speck, 1963). *Eremophila*

Fig. 18.16. Three zones of the *Acacia sclerosperma* complex. Foreground, *Atriplex hymenotheca*; central (near scale), *Eremophila pterocarpa*; background, *Acacia tetragonophylla* (darker) and *Eremophila leucophylla* (paler and smaller). Tops only of *Acacia sclerosperma* visible behind. 70 km south of Carnarvon, W. A.

pterocarpa is confined to the same area. On the other hand, *A. tetragonophylla* is widespread across the arid zone. The complex occurs on Red Brown Earths which have been subjected to wind erosion in the past, resulting in alternating patches of red dune sand and flats with calcium carbonate at or close to the surface; the flats are subject to flooding after rain. The complex of communities can be divided into four zones (Fig. 18.16).

i. Where the surface sandy layer on low dunes is c. 10 cm to 2 m deep. *A. sclerosperma* is dominant, forming open scrubs usually in pure stands. *A. coriacea* is occasional to rare and associated shrubs include *Cassia desertorum, C. sturtii* and *Eremophila freelingii*. Smaller shrubs and forbs include species of *Atriplex, Calandrinia, Chenopodium, Lotus, Myriocephalus, Schoenia* and *Swainsona*. The grass layer consists of *Plectrachne schinzii, Eriachne* sp., *Paractenium novae-hollandiae* and *Triraphis mollis*.

ii. *A. tetragonophylla* forms a narrow zone on soils with a surface sandy layer c. 10–20 cm deep, below the *A. sclerosperma* zone. This zone is 1–3 plants deep and may contain associates of *A. victoriae* and *Heterodendrum oleifolium*.

iii. *Eremophila pterocarpa* forms a third zone 1–1½ m high below the *A. tetragonophylla* zone and a few centimetres above the flat.

iv. *Atriplex hymenotheca* dominates the flat and is accompanied by a variety of small shrubs and annuals (species of *Angianthus, Gnephosis, Maireana* etc. – Chap. 19).

2.12. *Acacia ramulosa – A. linophylla* Alliance

Bowgada

These two species are closely related and the species complex occurs mainly in Western and South Australia and in Northern Territory in areas with a mean annual rainfall of 260–200 mm. The species are most abundant on deep unconsolidated sands, often on or near the crests of dunes, on sandplain or in rocky situations.

Fig. 18.17. *Acacia ramulosa* scrub and a single plant of *Brachychiton gregorii* (right). Herbaceous layer contains *Brunonia australis, Podolepis canescens, Myriocephalus guerinae* and *Waitzia acuminata*. 16 km south of Billabong, W. A.

The communities dominated by the species rarely cover continuous large tracts of country, except in Western Australia, but they occur in isolated patches within the hummock grasslands and savannahs, the *Acacia aneura* alliance or along the fringe of the mallee zone in the south.

In Western Australia, *A. ramulosa* covers large areas on Red siliceous Sands around the Shark Bay area and northward (Beard, 1977). Pure stands are common, but major associations with each of the following occur: *Acacia coriacea, Callitris "glauca", Grevillea stenobotrya, Eucalyptus oleosa* and *E. oldfieldii*. Ecotonal associations with *Acacia aneura, A. grasbyi, A. ligulata* and *A. sclerosperma* also occur. Other species present in these stands vary from district to district and include *Acacia murrayana, Brachychiton gregorii* (Fig. 18.17), *Thryptomene johnstonii* and *Grevillea eriostachya* and the shrubs *Dodonaea inaequifolia, Eremophila platycalyx* and *Labichea cassioides*. In drier areas to the north where *Grevillea stenobotrya* and *Acacia coriacea* are abundant, a different set of understory shrubs is found, viz., *Adriana tomentosa, Calytrix muricata, Crotalaria cunninghamii, Hibiscus pinonianus, Pityrodia loxocarpa, Stylobasium spathulatum* and others, with a grass layer dominated by *Plectrachne schinzii* and the annual *Aristida browniana*.

In South Australia, the communities occur mainly on Siliceous Sands or Solonized Brown Soils (Crocker and Skewes, 1941; Jessup, 1951; Specht, 1972), with a completely different set of associates, viz., *Cassia nemophila,* and occasional to rare occurrences of *Myoporum deserti, Pimelea microcephala* and *Lycium australe*. The grass layer is dominated by *Aristida* and *Enneapogon* spp. and a wealth of annuals appear after rains. The communities are sometimes surrounded by loose sand covered by annuals after rain, and with scattered plants of *Crotalaria dissitiflora* and *Acacia ligulata* which are the first perennials to establish on such sands. In some areas ecotonal associations with *Callitris "glauca"* and *Casuarina cristata* occur in the south and with *Acacia aneura* in the north.

In New South Wales *A. linophylla* is recorded on Lithosols on quartzite on the Dolo Hills west of Wilcannia with an understory of *Atriplex vesicaria, A. stipitata* and *Bassia* spp. and a scattered shrub layer of *Cassia artemisioides* and *Sida virgata*.

2.13. *Acacia acuminata* Alliance

Jam

A. acuminata is the most usual associate of *Eucalyptus loxophleba* (Chap. 15) in woodlands in Western Australia, and with decreasing rainfall the woodlands become stunted. *A. acuminata* is becoming more prominent and finally dominant (often with scattered mallee plants of the eucalypt). The *A. acuminata* scrubs usually occur in rugged country on soils containing some to much clay and sometimes with concretions of calcium carbonate.

The scrubs are mostly c. 4–6 m tall and contain several associates, drawn from neighbouring alliances, including *Acacia rostellifera* (near coast), *A. quadrimarginea, A. ramulosa, A. tetragonophylla, Casuarina huegeliana* and sometimes species of *Grevillea* and *Hakea*. The understory varies: *Melaleuca uncinata* dominates the shrub layer in some stands; in others a mixed understory of southern xeromorphs occurs, containing species of *Calothamnus, Dryandra, Grevillea, Scholtzia* and *Xanthorrhoea* (Beard, 1976).

2.14. *Acacia xiphophylla* Alliance

Gidgee

This species is confined to Western Australia (Fig. 18.1) and occurs in small patches usually in pure stands 3–5 m high. It is sometimes found within the mulga woodlands but is more common on clayey flats in the hummock grasslands or surrounding subsaline areas dominated by species of *Melaleuca*, or in the *Eragrostis xerophila* grasslands. Associated shrubs are rare and include *A. tetragonophylla, A. victoriae* and *Cassia desolata* or, rarely, *A. translucens*. The understory consists most commonly of Chenopodiaceae (Fig. 18.18), especially species of *Bassia, Maireana triptera, M. georgei,* or grasses – *Eragrostis eriopoda* on clayey soils and *Triodia* spp. in sandier sites.

2.15. *Acacia grasbyi* Alliance

A. grasbyi is closely related to *A. aneura* and the two species may occur together in hilly terrain, as far south as the fringes of the Nullarbor

Fig. 18.18. Pure stand of *Acacia xiphophila*. Understory dominant is *Bassia quinquecuspis* var. *lasiocarpa*. 16 km north of Wittenoom, W.A.

limestones (Beard, 1975). *A. grasbyi* is usually found on soils containing calcium carbonate and it is the dominant on a limestone plateau (Beard, n.d.) near the west coast where the mean annual rainfall is c. 210–230 mm (Fig. 18.1). The species reaches a height of c. 3 m and forms almost closed stands on Solonized Brown Soils. The surface sandy layer, which overlies clay contain calcium carbonate, varies in depth and on the deeper sands other tall species may occur.

A. grasbyi occupies the shallowest soils particularly among limestone blocks (Fig. 18.19). Undershrubs include *Eremophila clarkei*, *E. maitlandii* and *Solanum lasiphyllum*. Subsaline depressions sometimes occur within the scrubs and these are dominated by *Atriplex num-*

Fig. 18.19. *Acacia grasbyi* on a limestone ridge, with *A. tetragonophylla* at the foot. Foreground: *Atriplex hymenotheca*, *A. bunburyana*, *Ptilotus obovatus*, *Bassia diacantha* and *Solanum lasiophyllum*. 160 km south of Carnarvon, W.A.

mularia, *A. hymenotheca*, *A. bunburyana*, *Ptilotus obovatus* and occasionally *Arthrocnemum* sp.

On deeper sandy soils *A. tetragonophylla* appears and in slight depressions *A. victoriae* and *A. xiphophylla*. On the deepest sands *A. aneura*, *A. ligulata* and *A. ramulosa* are locally dominant and represent fragments of other alliances in the general area. *Codonocarpus cotonifolius* is sometimes a local dominant, possibly as a result of fire.

2.16. *Acacia calcicola* Alliance

This species has a restricted area in Central Australia (Fig. 18.1) and occurs within the *A. aneura* alliance, replacing this species on calcareous soils. The communities are described by Perry and Lazarides (1962) as woodlands 3–5 m tall occurring under a mean annual rainfall of 180 mm, about two-thirds of which falls in the warmer months. The understory consists of *Atriplex vesicaria*, *Maireana astrotricha* and *Bassia* spp., with short grasses and forbs which are dominant in most areas (possibly as a result of grazing).

2.17. *Acacia sowdenii* – *A. loderi* Alliance

These two species are closely related taxonomically and have possibly developed from a common ancestor following the division of an ancestral population into two segments separated today by the Flinders Ra. and alluvial flats of the Darling – Murray R. systems (Fig. 18.1). The species occur on similar soils (mainly Solonized Brown Soils) in similar climates. Two suballiances are described.

2.17.1. *Acacia sowdenii* Suballiance
Western Myall

This species reaches a height of c. 8 m and forms almost closed to open low woodlands in the wetter part of its range (means annual rainfall c. 230 mm). In lower rainfall areas (to c. 160

Fig. 18.20. Open woodland of *Acacia sowdenii* with an understory of *Atriplex vesicaria*, *Maireana sedifolia* and *Arthrocnemum halocnemoides*. Near Madurah, W. A.

mm) the stands become more open and savannah-like, the understory being chenopodiaceous shrubs rather than grass (Fig. 18.20). The area of the species is broken into two segments by the very arid Nullarbor Plain (Fig. 18.1).

Associated tall species are rare. *Heterodendrum oleifolium* occurs as a co-dominant in wetter areas and *Myoporum platycarpum* in the drier parts. Ecotonal associations occur with *Acacia aneura* on deeper sands. The understory species vary with the depth of the surface sandy horizon and mean annual rainfall. In wetter areas on deep sandy soils a conspicuous or even continuous shrub layer develops, containing *Acacia burkittii, Cassia phyllodinea, C. sturtii, Eremophila scoparia, Lycium australe* and *Templetonia egena*. A discontinuous lower shrub layer 50–70 cm high of *Atriplex vesicaria* and *A. stipitata* occurs in some areas. The ground stratum contains a mixture of species of *Enneapogon* and *Bassia* (Crocker and Skewes, 1941; Jessup, 1951). Annuals are abundant after rain.

In drier areas where the soils have a shallow surface sandy horizon with calcium carbonate at or near the surface, *A. sowdenii* forms open stands with a shrub layer of *Maireana sedifolia* 60–80 cm high, sometimes mixed with the silver-leaved composite *Cratystylis conocephala*, especially in the west. Annuals are abundant after rain.

The communities in the wetter areas are sometimes destroyed by fire, the woody species being killed. Regeneration takes place mainly from seed, and *Cassia* spp. are the first plants to regenerate (Crocker and Skewes, 1941). The communities are grazed by domestic stock which eat both the understory plants and the phyllodes of *Acacia*. Grazing leads to an abundance of species of *Bassia*, especially *B. paradoxa* and *B. obliquicuspis* and the disappearance of *Maireana sedifolia*. Premature death of *A. sowdenii* is reported (see *A. aneura* alliance for a discussion of this topic), and Lange and Purdie (1976) warn that since *A. sowdenii* does not reproduce vegetatively but only from seed, the stands are likely to disappear.

2.17.2. *Acacia loderi* Suballiance
Nelia

This species occurs in south-western New South Wales (Fig. 18.1) on Solonized Brown Soils in an area receiving a mean annual rainfall of 240–280 mm. The species suckers freely from the roots, especially when the roots lie close to or at the surface, with the result that some stands are often dense, the crowns touching. These dense clumps are usually separated by more open areas. The stands are usually interspersed with woodlands of *Casuarina cristata* but the factors responsible for segregating the species are not clear.

A. loderi usually occurs in pure stands (Fig. 18.21), rarely with *Myoporum platycarpum* as a

Fig. 18.21. Pure stand of *Acacia loderi*. Understory contains *Rhagodia spinescens, Bassia paradoxa, B. obliquicuspis, Enneapogon avenaceus* and *Eragrostis lacunaria*. 80 km south of Broken Hill, N.S.W.

co-dominant. The shrubs reach a height of 3–5 m, the shorter stands occurring when plants are clumped. Shrubs 1–2 m high sometimes form a lower stratum, viz., *Grevillea huegelii*, *Lycium australe* and *Templetonia egena* (Beadle, 1948). *Maireana sedifolia* is present in most stands, having its highest density in open stands. It is commonly removed by domestic stock, when other Chenopodiaceae become prominent (see caption to Fig. 18.21). Grasses are sometimes abundant, *Stipa variabilis* being the most common. Under conditions of heavy grazing all perennial understory species are removed and annuals form the herbaceous layer. In this degenerate condition the soils are subject to severe water erosion, firstly to sheet erosion which exposes the limy layer and finally to gully erosion.

2.18. *Acacia ligulata* Alliance

A. ligulata is widely distributed in the arid zone and it reaches the coast where it forms scrubs on coastal dunes in parts of South and Western Australia. It usually occurs on sands or sandy soils which are neutral or slightly alkaline. Although widely distributed, it forms extensive stands of scrub in only a few areas; elsewhere it is an occasional species on sand ridges in the mallee or other *Acacia* scrubs where it appears to be a colonizer of loose sand.

In the east it is possibly most abundant as a tall shrub to 3 m high on lunettes edging lakes, where it is sometimes associated with *Pittosporum phylliraeoides* and/or *Dodonaea attenuata*. It occurs as the only tall shrub on dune fields on the western fringe of the Simpson Desert in ephemeral grasslands of *Aristida browniana*.

In Western Australia it is found also on lunettes, at least in the south. On the west coast, it replaces *A. rostellifera* on coastal dunes (Chap. 23) north of Geraldton to north of the Shark Bay area, where it possibly has its greatest development as a community dominant. In this area it forms open to almost closed scrubs 1–2 m high on sand overlying limestone. The communities occur along the coast, being prominent on most of the off-shore islands and extending inland on the mainland for a couple of kilometers, being replaced on the landward side by *A. ramulosa* or heaths containing *Banksia* (Beard. 1976).

The main associated shrubs in the communities are *Diplolaena dampieri*, *Exocarpos sparteus*, *Heterodendrum oleifolium*, *Alogyne cuneiformis*, *Acacia coriacea* and *Pittosporum phylliraeoides*. Understory shrubs are species of *Anthocarpus*, *Atriplex*, *Cassia*, *Frankenia*, *Melaleuca* and *Myoporum*. The grass layer varies but *Pletrachne plurinervata* is a near constant. In subsaline areas *A. ligulata* is mixed with *Melaleuca cardiophylla* and, on the western side of the Peron Peninsula, *A. ligulata* is associated with *Lamarchea hakeifolia*.

3. Communities Dominated by *Casuarina*

3.1. General

Three species of *Casuarina* dominate forests or woodlands in the drier parts of the continent (Fig. 18.22), and they are surprising tall for such a dry environment. Unlike the species in wetter climates the dryland species do not produce nitrogen-fixing nodules (Chap. 4).

3.2. *Casuarina luehmannii* Alliance

Bull Oak or Buloke

This species, which reaches a height of c. 12 m, occurs in an area from the Tropic in the east (mean annual rainfall 1000–1500 mm, with a summer predominance) to South Australia (mean annual rainfall 400–500 mm). It is commonly an understory species in *Eucalyptus* woodlands, but becomes locally dominant on patches of soil which are periodically waterlogged, this condition being induced by a pan in the subsoil, as in Solodized Solonetz Soils; in such soils the A horizon is sandy. It is also dominant on some Grey Clays and, rarely, on Red Brown Earths. In most areas the soils are gilgaied.

In general *C. luehmanni* is the sole dominant in the stand and no shrub layer occurs, though scattered shrubs may be present, especially in the south; the grass layer is usually discontinuous. In the tropics, Perry (1953) describes a stand with a scattered grass layer of *Aristida browniana*, *Alloteropsis semialata* and *Chry*-

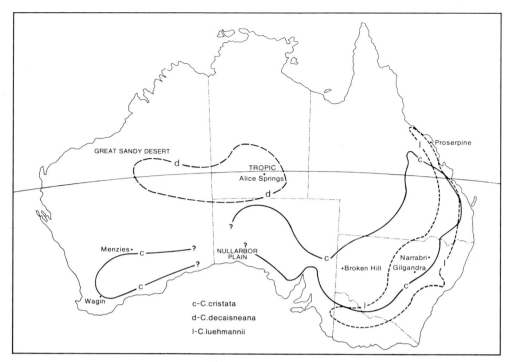

Fig. 18.22. Approximate areas of three species of *Casuarina* found mainly in the semi-arid and arid zones.

Fig. 18.23. Pure stand of *Casuarina luehmannii* on Solodized Soil within a woodland of *Eucalyptus woollsiana*. The grass layer is dominated by *Eragrostis lacunaria*. 16 km south of Narrabri, N.S.W.

sopogon fallax. Other stands contain co-dominants of *Melaleuca viridiflora* and lesser *Grevillea striata*, *Canthium oleifolium*, *Acacia salicina* and *Petalostigma pubescens*. A typical stand in New South Wales, within a woodland of *Eucalyptus woollsiana* is shown in Fig. 18.23. In Victoria and South Australia the stands occur mainly within woodlands of *E. woollsiana* (*microcarpa*) and have been described by Connor (1966) and Perry (1951), the occurrences being on Red Brown Earths in Victoria and mainly on gilgaied Grey clays in South Australia. A much larger flora is reported possibly because the stands are open, including a few shrubs – *Eremophila longifolia*, *Eutaxia microphylla* and *Myoporum deserti*. The herbaceous layer is well developed in open stands, *Stipa eremophila* dominating the shelves of gilgais and *Danthonia* spp. the puffs and crab-holes. Other species are *Panicum prolutum*, *Swainsona procumbens* and others, with an assemblage of Cyperaceae and other small herbs in the wetter crab-holes.

3.3. *Casuarina cristata* Alliance

Belah or Black Oak

C. cristata extends from central western Queensland with a mean annual rainfall of over 1000 mm (summer predominance) across the

Fig. 18.24. Woodland of *Casuarina cristata* on grey clay. Herbaceous layer contains *Panicum queenslandicum*, *Danthonia caespitosa*, *Chloris truncata*, *Paspalidium gracile* and *Sporobolus caroli*. 40 km south-west of Gilgandra, N.S.W.

continent in the south (Fig. 18.22). In the more southerly districts the mean annual rainfall is as low as 250 mm. Over this range of climates it shows some variation in size (tallest in the north), colour of cladodes (changing from dark green in the wetter north to grey-green in the drier south), and size of the cones which are smaller in drier climates. It occurs on different types of soil and is associated with different species from north to south. The species commonly suckers from roots.

In the north the species occurs on Grey and Brown Clays which are usually alkaline, crack deeply on drying and are gilgaied. In this habitat the trees reach a height of c. 13 m and occur in pure stands or in association with *Acacia harpophylla*. Tall shrubs are rarely present, the main ones being *Eremophila mitchellii, Geijera parviflora* and *Bauhinia carronii* (Pedley, 1974). The herbaceous layer is dominated by *Astrebla* spp. and their usual associates (Chap. 20); gilgai depressions usually support *Marsilea drummondii* and Cyperaceae.

In the central part of its range, as in central-northern New South Wales (Beadle, 1948), the stands occur on similar soils and reach the same height (Fig. 18.24). Associated shrubs include *Atalaya hemiglauca, Heterodendrum oleifolium* and rarely *Acacia pendula. A. homalophylla, A. oswaldii, Geijera parviflora* and *Flindersia maculosa*. Grasses dominate the herbaceous layer, including *Astrebla lappacea, Panicum* spp., *Dichanthium sericeum* (summer-growing) and several winter-growing species, notably *Stipa aristiglumis* which becomes more prominent towards the south; in these southerly areas the associated shrubs are less prominent, except *Acacia pendula* and *A. homalophylla*. Gilgai depressions support *Marsilea drummondii, Centipeda cunnighamii, Juncus polyanthemos* and Cyperaceae.

From south-western New South Wales *C. cristata* extends across the continent on Solonized Brown Soils. The depth of the surface sandy horizon, which was blown into dunes during the Holocene arid period, determines the patterning of the communities.

In south-western New South Wales, *C. cristata* forms an alliance with *Heterodendrum oleifolium* (Beadle, 1948; Stannard, 1963). Both these species sucker from the roots, so that monospecific groves of each species commonly

Fig. 18.25. Pure stand of *Heterodendrum oleifolium* with a few scattered plants of *Apophyllum anomalum* below. Herbaceous layer completely removed by stock. South of Wilcannia, N.S.W.

Fig. 18.26. Two trees of *Casuarina decaisneana* c. 9 m tall near the base of a sand ridge. The hummock grass is *Triodia basedowii*. c. 80 km south of Alice Springs, N.T.

occur (Fig. 18.25). The *Acacia loderi* suballiance (see 2.16.2, above) is confined to these stands and on deep sands, usually in the form of local dunes, *Callitris "glauca"* is commonly dominant. A rare associate is *Myoporum platycarpum*. Tall shrubs in the *Casuarina-Heterodendrum* woodlands are rare, the following being recorded: *Acacia homalophylla, A. victoriae, Eremophila longifolia, Eucarya acuminata, Exocarpus aphyllus* and *Geijera parviflora*. On deeper sands *Acacia burkittii, A. ligulata, Cassia nemophila* and *Eremophila sturtii* are sometimes prominent. Along its northern margin ecotonal associations are formed with *Acacia aneura*. A lower shrub layer is sometimes present, dominated by *Maireana sedifolia* with subsidiary *Lycium australe* and *Eremophila scoparia* when calcium carbonate lies close to the surface, and *M. pyramidata* on deeper sands. The grass layer is usually dominated by *Stipa variabilis* and this species may form local patches of grassland between groves; a large number of annual species are recorded. On flats which are subject to minor waterlogging, usually during the winter, an herbaceous layer is almost lacking, the soil surface being strewn with *Casuarina* cladodes, and in such areas *Maireana triptera* occurs on patches of sand which are slightly elevated above the surface.

In South Australia a similar patterning of the communities is seen (Wood, 1937; Crocker, 1946). *Myoporum platycarpum* is more abundant and *Heterodendrum oleifolium* less abundant. On deep sand *Acacia linophylla* is recorded and the main grass is *Stipa nitida*.

In Western Australia the stands are possibly isolated by the arid Nullarbor region, though this needs confirmation. *Casuarina cristata* usually dominates the stands, and *Eremophila scoparia* often dominates the shrub layer. The main associated species are those present in the more easterly stands, especially *Myoporum platycarpum* which is often locally dominant and *Acacia aneura* on the northern fringe. Species of *Maireana* occur in the understory in most areas and may be locally dominant. Shrubs which are confined to the west are *Eremophila miniata* and *Maireana carnosa* (Beard, 1974, 1975). In the west *C. cristata* occurs also in somewhat wetter country, often around subsaline flats, sometimes in association with *Melaleuca acuminata* (e.g. near Wagin, mean annual rainfall 440 mm).

In New South Wales and South Australia, *C. cristata* extends on to the undulating stony downs of the arid zone which support the halophytic shrublands (Chap. 19). The trees occur in small patches, usually in gilgais or near minor watercourses.

3.4. *Casuarina decaisneana* Alliance

Desert Oak

This species is one of the tallest casuarinas and possibly the tallest species inhabiting the arid zone. It reaches a height of 20 m and is renowned for its extremely hard wood. It occurs over a wide area of the arid zone (Fig. 18.22), surviving on as little as 200 mm mean annual rainfall, with a maximum of c. 300 mm. However, since the trees usually occur in small groups at the bases of sandy dunes, it is possible that additional water is supplied to the plants by drainage waters. Groves of the trees, sometimes covering several hectares, are reported by Chippendale (1963) in the Tanami Desert and Beard and Webb (1974) in the Great Sandy Desert.

Where groves occur, the communities are stratified. In the Tanami Desert, *Eremophila longifolia* is a common tall shrub, with subsidiary *Canthium latifolium*, *Cassia nemophila* and *Solanum orbiculatum*. Hummock grasses form the herbaceous layer – *Triodia basedowii* and *T. pungens*, with *Plectrachne schinzii* common in patches. In the Great Sandy Desert, shrubs include *Acacia salicina*, *Carissa lanceolata*, *Eucalyptus pachyphylla*, *Gyrostemon tepperi* and *Melaleuca* spp. The hummock grasses *Plectrachne schinzii* and *Triodia pungens* dominate the herbaceous layer. Annuals are abundant after rain in all communities, especially *Aristida browniana*.

The groves of *Casuarina* are sometimes interspersed with open stands of shrubs, sometimes containing a few trees of *Eucalyptus gongylocarpa* and/or *Brachychiton gregorii*; the shrubs are *Acacia kempeana*, *A. patens*, *Dodonaea attenuata*, *Exocarpus sparteus* and *Grevillea juncifolia* (Chippendale, 1963).

When only a few trees of *Casuarina* occur together, shrubs are absent and the herbaceous layer is dominanted by hummock grasses (Fig. 18.26).

19. The Halophytic Shrublands

1. Introduction

1.1. Definitions

The halophytic shrublands, also known as "shrub steppe" (Wood, 1937), "saltbush formation" (Beadle, 1948), "low shrubland" (Specht, 1972) and "unwooded succulent steppe" (Beard, 1975), are dominated by Chenopodiaceae, chiefly species of *Atriplex* and *Maireana* (formerly *Kochia* – see Wilson, 1975). The shrublands contain few or no trees, but the dominants of the shrublands extend into woodlands as understory dominants, as in some *Acacia* and *Casuarina* communities (Chap. 18), some mallee communities (Chap. 14) and some *Eucalyptus* woodlands (Chaps 15, 21). The halophytic shrublands occur mainly in the southern half of the continent (see vegetation map), mostly on soils with a high clay content which are subsaline or saline, and some are highly calcareous. Similar shrublands occur in the littoral zone and are described in Chap. 23.

As well as *Atriplex* and *Maireana* many other genera of the Chenopodiaceae are represented in the halophytic shrublands, some being confined to these shrublands. An account of the family in the continent is provided in the following section.

1.2. The Chenopodiaceae in Australia

The Chenopodiaceae are well represented in Australia by c. 21 genera (c. 30, if "splits" are recognized), and of these 11 are endemic. Most, if not all, Chenopodiaceae are halophytic, and in Australia the family is represented both in the littoral zone (Chap. 23) and inland, usually in saline and subsaline habitats. The genera are listed in Table 19.1 in which the number of species of each genus is shown, and the endemic genera are indicated. The largest genera in Australia are *Atriplex*, *Bassia* and *Maireana*. Most Chenopodiaceae are found in the arid and semiarid zones, the greatest concentration of species occurring in the south-east where the arid and the subsaline clayey soils adjoin the littoral zone along the Nullarbor plateau and eastward, especially at Spencers Gulf (Fig. 19.1). Several genera are represented in non-saline habitats in the arid and semi-arid zones and a few extend into humid climates e.g. species of

Table 19.1. Australian genera of Chenopodiaceae arranged according to habitat. Number of species of each genus shown in brackets. # indicates endemic to Australia. S = south; W. = west; N. = north.

Genera confined to the littoral zone	Genera in littoral, arid and/or semi-arid zones	Genera confined to arid and semi-arid zones
# *Didymanthus* (1) N.W.; *Suaeda* (1) all coasts, rarely inland; *Tetricornia* (1) Trop; *Theleophyton* (1) S.	*Arthrocnemum* (9); *Atriplex* (c. 50); *Bassia* (c. 48); *Chenopodium* (c. 20); # *Enchylaena* (1); *Rhagodia* (c. 15)[†]; # *Roycea* (2); *Salicornia* (2); *Salsola* (1); # *Threlkeldia* (5)	# *Babbagia* (4); # *Dysphania* (6); # *Maireana* (c. 50); # *Malococera* (2); # *Pachycornia* (2); *Scleroblitum* (1); # *Duriala* (1) semi-arid

[†] endemic, apart from one species in New Zealand.

Fig. 19.1. Junction of the littoral zone (mudflat dominated by *Arthrocnemum halocnemoides*, background) and the arid zone halophytic shrubland dominated by *Atriplex vesicaria* (centre). In foreground, plants of *Disphania australe* (Round-leaved Pigface). Spencer Gulf, S. A.

Bassia, Chenopodium, Duriala, Dysphania, Maireana, Salsola and *Scleroblitum;* in a few cases the same species may be found in both saline and non-saline environments. The most salt-tolerant of all chenopods is *Arthrocnemum* which is sometimes inundated by sea water (Chap. 23) and it occurs also on salt-encrusted playas (Chap. 21). Some species of *Atriplex* also tolerate high salt concentrations and the physiology of some of these is discussed in more detail below.

1.3. Economic Importance

Most Chenopodiaceae are palatable to stock and most of the halophytic shrublands have been grazed by domestic stock for well over a century. Although the various chenopods are not the most palatable species in any assemblage (grasses are usually eaten first), they are accepted by stock and the more palatable ones, especially *Maireana* and perennial species of *Atriplex* have been removed over large areas and replaced by annual plants or by short-lived perennials, chiefly *Bassia*. This genus, with about 48 species, is widespread in the arid interior and is identified by the presence of 2–12 spines which develop on the perianth as the fruit matures. The spines, which are usually pungent-pointed and commonly 1–2 cm long, protect the plant against the grazing animal. Consequently several species of *Bassia* often appear to be codominants with perennial species of *Atriplex* and in highly degenerate communities species of *Bassia* are completely dominant and are regarded as weeds. Notable among these are *B. quinquecuspis* (Electric Burr) and *B. birchii* (Galvanized Burr). The latter is a noxious weed in certain districts in western New South Wales and Queensland (Auld and Martin, 1975) though it rarely occurs in the halophytic shrublands.

In many degenerate shrublands the soil has been eroded resulting in the development of bare patches or "scalds" (Beadle, 1948) some of which are saline, or in undulating country of gullies produced by water erosion. The reclamation of these areas in wetter climates where the land is valued highly has been studied by several authors and cannot be dealt with here (see Knowles, 1954; Condon, 1960; Jones, 1969; Malcolm, 1972, who discuss the problems associated with the establishment of shrubs in saline environments).

1.4. Leaf Anatomy and Physiology of some Chenopodiaceae

The anatomy of the leaves of Chenopodiaceae has been given considerable attention both overseas and in Australia, partly to clarify taxonomic affinities of the genera and partly in relation to physiological studies. Carolin et al. (1975) in a survey of the world genera indicate that seven anatomical types exist, divisible into two groups, non-Kranz and Kranz types. The first includes arrangements of the palisade and spongy mesophyll cells according to the normal dicotyledonous dorsiventral or isobilateral leaves, in all cases the photosynthetic tissue lying adjacent or close to the epidermi; some of the leaves are succulent. These genera have the normal C_3 photosynthetic pathway and many Australian genera have this type of leaf anatomy, e.g., *Babbagia*, *Bassia*, *Chenopodium*, *Enchylaena*, *Maireana*, *Rhagodia* and *Suaeda*. On the other hand, a few Australian genera have the Kranz type, notably *Atriplex*, *Salsola* and *Theleophyton*. The phenomenon was first noted in Australian *Atriplex* by Wood (1925) and was studied in some detail for *A. nummularia* and *A. vesicaria* by Black (1954). In this anatomical type, the chloroplasts are concentrated in the cells of the bundle sheath, with fewer and smaller chloroplasts in the mesophyll cells. Further to this, mitochondria are associated with the chloroplasts of the bundle sheath but not with the chloroplasts of the mesophyll. The Kranz-structure is possibly associated with the C_4-dicarboxylic acid pathway of carbon dioxide fixation which endows the plant with an increased photosynthetic rate (Osmond, 1970). A survey of the Australian species of *Atriplex* (Osmond, 1974) shows that of the 44 species examined all but one exhibited the Kranz arrangement. The exception, *A. australasica*, confirmed chemically as a C_3 plant, is usually regarded as indigenous. It occurs mainly on the wetter parts of the littoral zone. Of the remaining Australian species most occur in the arid zone, with a few in the littoral zone, e.g. *A. cinerea* in the wetter south-west and *A. isatidea* in the drier south-west.

It is usually considered that the Kranz anatomy (and the C_4-pathway) are an ecological advantage to plants in the arid zone. However, it cannot be regarded as the sole mechanism leading to the success of these plants in an arid environment, since species of *Atriplex* are commonly associated with other perennial woody species of similar size, which do not posses the Kranz anatomy, especially *Maireana*, *Rhagodia* and *Bassia*. Also several species of *Atriplex* are annuals, and they do not survive for significantly longer periods than do non-Kranz types, e.g., Compositae, when subjected to desiccation in the field.

Many Chenopodiaceae accumulate salt which is associated with succulence, stem-succulence in *Arthrocnemun* and *Salicornia* and leaf succulence in most other genera. Salt absorption in *Atriplex* has been studied in some detail. All species investigated have the capacity to accumulate salt from even the lowest concentrations of sodium and chloride (Wood, 1925; Beadle et al., 1957). In all cases examined the amount of monovalent ions (sodium and potassium) in the plant greatly exceeds the amount of divalent (magnesium and calcium). Sodium is regarded by Brownell and Crossland (1972) as an essential micronutrient, in addition to the role which it plays in developing high osmotic pressures. Several authors have demonstrated the stimulating effect of sodium chloride on the growth of *Atriplex,* and Ashby and Beadle (1957) have shown that sodium is preferred to potassium by plants grown in solution culture.

In *Atriplex* the high salt concentrations in the leaves are due to the accumulation of chloride, balanced by monovalent cations, in hairs derived from epidermal cells. The hairs are 2–4-celled, and the terminal cell becomes more or less globular (vesicular) and the numerous vesicles cover the leaf-surfaces. These vesicular tips accumulate the salts and finally collapse with age, when the water on the cell sap evaporates leaving salt crystals among the dead hairs. The latter finally become matted to form a "vesicular layer" which is felt-like and apparently protects the leaf against water loss. A second or even a third crop of hairs is usually produced inside the first-formed vesicle layer. Furthermore, since the vesicular layer contains salt, it is hygroscopic and capable of absorbing water from the atmosphere when the humidity is high and also of collecting dew. However, whether the water absorbed in the vesicular layer passes into the living cells is open to doubt, but is unlikely to occur when the osmotic pressure in the vesicle layer is at its highest.

2. Descriptions of the Alliances

Only six species are used to distinguish the communities. Of the species, some show ecotypic variation, especially *A. vesicaria* (see below), and since most of the six characteristic species have overlapping ranges of tolerance, the alliances and suballiances are not always sharply delimited, but ecotonal mixtures of species occur. They are confined to the southern half of the continent, south of lat. 21°S. (Perry and Lazarides, 1962). The alliances and suballiances are listed in Table 19.2 and the main factors delimiting the communities are indicated.

2.1. Alliances Dominated by *Atriplex* spp.

2.1.1. General

Atriplex is identified by the presence in the female flower of two bracteoles (sometimes regarded as perianth-segments) which enlarge after fertilization to enclose the true fruit, which is a nut with a membranous pericarp containing one seed. The bracteoles are usually leathery, and with or without appendages, the latter sometimes inflated (Fig. 19.2). They have a high salt-content (though usually lower than the leaves) and the salt appears to assist in the protection of the seed against germination when soil moisture is inadequate for the establishment of the seedlings through the development of a high osmotic pressure in the bracteoles under conditions of low moisture. When salt is removed from the bracteoles by rain, the seeds germinate (Beadle, 1952).

The autecology of the two common perennial species has been studied in some details. It has been shown that *A. vesicaria* has a shallow root system, rarely penetrating the soil below 25 cm and that after rain it produces a crop of fine roots below the soil surface and that these are shed when the soil dries out (Wood, 1937). On the other hand, *A. nummularia* roots more deeply (up to c. 2m), and this habit may be responsible for the segregation of species in some areas. Sharma (1976) has shown that in plantations *A. vesicaria* reduces soil water to a much lower level than does *A. nummularia* and that *A. vesicaria* has a higher tolerance to desiccation than *A. nummularia*.

The high salt concentrations in the leaves and bracteoles of the perennial species of *Atriplex* lead to a cycling of salt as a result of concentrations in the leaves, the shedding of leaves and bracteoles surrounding the fruits and the washing of this salt into the soil. Charley (1959), Charley and Cowling (1968) and Sharma and Tongway (1973) present quantitative data. A greater accumulation of salt below bushes occurs with *A. nummularia* which contains similar concentrations of salt in the leaves but sheds a greater mass of leaves on to the soil surface.

2.1.2. *Atriplex nummularia* Alliance
Oldman Saltbush

A. nummularia commonly reaches a height of 2 m and forms tall shrublands over relatively small areas, usually on deep clay alluvium. On the Nullarbor Plain, it thrives on apparently shallow (depth of red-brown clay is c. 25–40 cm) but the roots of the saltbush probably extend into cracks in the limestone. The species is widely distributed in the arid and semi-arid zones, from the fringes of the Simpson Desert (mean annual rainfall c. 160 mm, but

Table 19.2. Main alliances and suballiance in the halophytic shrublands. Distribution, main soil characters and approximate mean annual rainfall limits are indicated. Rainfall mainly in the winter, or more or less evenly distributed throughout the year in the north.

Alliance/suballiance (distribution)	Main soil characters	Mean annual rainfall (mm)
A. nummularia (east and west)	deep grey or brown clay alluviums; red-brown clay over limestone	150–420; drier stands mostly in creeks or swamps or among rock
A. vesicaria (east)	deep grey or grey-brown clays; some gilgais; subsaline	250–410
A. vesicaria-Bassia (central)	deep red-brown clays; usually gilgaied; subsaline to saline	200–250
A. vesicaria-Ixiolaena (central)	as previous but soils sometimes shallow and without gilgais	150–240
A. rhagodioides (central)	as previous	140–220
A. hymenotheca (west)	shallow clays over limestone or deep subsaline clay	150–230
Maireana sedifolia (east to west)	lime abundant at or near surface	150–350
M. astrotricha (central)	shallow, slowly pervious soil with lime	180–250
M. pyramidata (east to west)	sandy soils, usually deep, especially lunettes; some subsaline	160–350

with additional moisture in creek beds) to the Deniliquin area (mean annual rainfall 410mm) in the east, and in the west across the Nullarbor Plain. It is commonly an understory species in *Eucalyptus* woodlands both east (below *E. largiflorens* and *E. microtheca*, Chap. 21) and west (below *E. salmonophloia* and *E. salubris* (Chap. 15), and, in the north-east it is sometimes associated with *Acacia cambagei*, and in the south-east with *A. pendula* (Chap. 18). It is probable that the species was once far more abundant then it is today. It is readily removed by continual grazing by domestic stock, especially cattle, and in some cases stock trim young bushes to a height of c. 1 m, preventing them from attaining maximum size. Regeneration from seed occurs under conditions of light or no grazing, even on the Nullarbor Plain (see Fig. 19.12) where the saltbush may be associated with *Maireana sedifolia*. The species is often used as a hedge plant around homesteads and stockyard and these plants provide a source of seed for establishing plants in communities in which it formerly did not occur, e.g., in the Broken Hill regeneration reserves.

When it occurs as a dominant, *A. nummularia* forms pure stands. A lower shrub layer of *A. vesicaria* with or without *Rhagodia spinescens* may occur, and on the Nullarbor Plain *Maireana sedifolia*. The annual plants and sub-shrubs are those which occur with *A. vesicaria* and are usually annual species of *Atriplex* and *Bassia* spp. (Fig. 19.3).

2.1.3. *Atriplex vesicaria* Alliance
Bladder Saltbush

Several forms (ecotypes) of *A. vesicaria* exist which differ morphologically in habit, leaf-shape and the size of the inflated bladders on the bracteoles. The latter vary from very large (Fig. 19.2) to minute or even absent. This morphological variation has not yet been correlated with habitat, except for two forms recognized by Osmond (1961). A broad-leaved form (usually associated with *Maireana astrotricha* – Fig. 19.11) has an upright habit and rudimentary or no bladders on the bracteoles; it occurs in soils in which the surface layer contains calcium car-

Fig. 19.2. Male (right) and female plants of *Atriplex vesicaria*. The female plant bears a crop of false fruits which consist of two bracteoles, with inflated bladders, enclosing a nut with a membranous pericarp enclosing a single seed. Near Woomera, S.A.

bonate which increases in abundance down the profile. The soil surface lacks gibbers but siliceous rock fragments may occur, especially on foothills. A narrow-leaved form with a more spreading habit and conspicuous bladders on the bracteoles is found on saline soils with low levels of calcium carbonate in the surface which is strewn with gibbers and is gilgaied. The two forms may occur in adjoining communities and the forms may intergrade across a few metres in the ecotonal area.

In Western Australia, *A. vesicaria* is referred to *A. hymenotheca*. Forms of one or other of these two "species" are found from the Indian Ocean across the southern half of the continent to about the 420 mm isohyet, in the south-east. In the south-east, *A. vesicaria* is dominant mainly on grey clays (see 2.1.2.1, below); in the central part of its area it occurs mainly on clays strewn with wind-polished gibbers (Stony Downs, discussed in more detail below); on the Nullarbor Plain and westward it is found on clays overlying limestones.

The "stony downs" or "gibber plains" have developed mainly on saline Cretaceous shales and sometimes on alluviums which extend from west of the Darling R. into South Australia. On more sandy or gravelly soils in the same climate, woodlands of *Acacia aneura* replace the saltbush, or minor accumulations of sand, mostly c.

1 m deep, support scrubs of *Cassia* spp. up to 2 m tall or shrublands of *Maireana pyramidata* (2.3.4, below). Also, where lime occurs abundantly *M. sedifolia* is locally dominant.

The gibber-strewn soils vary in depth from a few centimetres to a few metres, the deeper soils developing in lowland sites. Soil depth, salinity and the clay content of the soil determine the amount of gilgai development. Gilgais do not form on soils less than 50 cm deep (Jessup, 1951) though the surface of these is irregular and the depressions show surface salt accumulation which affects the vegetation pattern. Minimum gilgai development occurs on alluviums on flats, especially between ranges, where that alluvium is of mixed origin and contains a high proportion of sand. Maximum gilgai development occurs on deep soils derived from the Cretaceous shales. In these areas the gilgai depressions are usually circular or elliptical in outline and 5–12 m across; sometimes they are elongated along the contours and 50–100 m long. The elevated shelves are continuous around the depressions.

The soils throughout are cracking clays and the mobile soil constituents are distributed as follows: In the depressions, which accumulate water and which usually lack gibbers, lime occurs in the subsurface and gypsum below this; salt concentrations are relatively low (0.1% at the surface and up to 0.5% in the subsurface). In contrast, the shelves, which shed most of their water, contain lime near the surface and salt concentrations are relatively high (mostly 1% in the surface but up to 8.5%) (Jessup, 1951; Charley and McGarity, 1964). An extraordinary finding of the latter authors is the presence of high nitrate levels in the shelves (130 ppm at a depth of 3–5 cm) which is in sharp contrast with nitrate levels in the depressions which approach zero for the whole profile. Since nitrate and chloride levels are correlated, Charley and McGarity postulate that the nitrate originates in the wetter depression and is transported laterally and vertically by diffusion to accumulate near the shelf surface.

Several distinct assemblages can be recognized, which are described below as suballiances.

2.1.3.1. *Atriplex vesicaria* Suballiance (on deep clays in the east)
Bladder Saltbush

In the wettest part of its range on the Riverine Plain in New South Wales (eastern limit around Deniliquin, with a mean annual rainfall of c. 410 mm fairly evenly distributed throughout the year), *A. vesicaria* occurs in almost pure stands c. 60 cm high or in association with *A. num-*

Fig. 19.3. A young stand of *Atriplex nummularia* on deep grey clay alluvium near the Darling R. Trees in background are *Eucalyptus microtheca* and *E. largiflorens*. Ephemeral plants in foreground are *Bassia divaricata* and *Atriplex inflata* (pale grey). Wilcannia, N.S.W.

mularia. The plain is virtually level and built from alluvial deposits consisting mostly of grey clays which are mildly alkaline and contain calcium carbonate in small quantities, especially in the lower layers. Some sandy deposits also occur and *A. vesicaria* is sometimes found on these, but the deeper sands, especially in the west and north support *Maireana pyramidata* and, in the same areas, lime-rich deposits support *M. sedifolia* (see 2.2.2, below).

The clays show some gilgai development and *A. vesicaria* is absent from deeper depressions which commonly support thickets of *Muehlenbeckia cunninghamii* and/or *Chenopodium nitrariaceum*, *Eragrostis australasica* or small hydrophytic communities in which *Marsilea drummondii* is usually conspicuous (Chap. 21). The more common herbaceous species are *Panicum decompositum* and *Eragrostis setifolia*.

A. vesicaria is randomly distributed on the level areas, except young plants which tend to aggregate around mature bushes (Anderson et al. 1969). *Rhagodia spinescens* and *Maireana aphylla,* bushes of similar size, were minor constituents in the virgin condition. Between the bushes a very large number of subshrubs and herbaceous species appear after rain, the composition varying with the season. The various species are described and illustrated by Leigh and Mulham (1965). The more abundant species are members of Chenopodiaceae, Compositae, Gramineae and some of these are mentioned below.

Overgrazing of the *A. vesicaria* stands has led to great changes in the dominants (Beadle, 1948). *A. vesicaria,* though less palatable than many of the associated herbs, especially grasses, has been completely removed from some areas, and so also *Rhagodia spinescens*. These woody species are replaced by *Maireana aphylla,* various species of *Bassia,* especially *B. divaricata* and *B. quinquecuspis,* each of which may be locally dominant. Many stands have been converted to grassland in which *Danthonia* and *Stipa* are prominent (Chap. 20). Again, many areas have degenerated to annual assemblages dominated by *Portulaca oleracea* and/or *Tribulus terrestris* (summer-growing) or *Calocephalus sonderi* and the introduced **Malva parviflora* and crucifers (winter-growing).

Notable among the invaders of the communities, either degenerate or near virgin, is *Nitraria billardieri* (formerly *N. schoberi*). This species, which has an edible succulent fruit, is widely distributed across the southern half of the continent. It has increased in abundance in many districts and is probably distributed by

Fig. 19.4. A hummock of *Nitraria billardieri* extending over an accumulation of sand blown from the surrounding sandy country formerly dominated by *Maireana pyramidata*. The area is saline. Smaller shrubs in foreground are *Rhagodia spinescens*. 24 km north of Broken Hill, N.S.W.

birds (especially emus – Noble and Whalley, 1978). The species is salt-tolerant and forms hummocks several metres in diameter (Fig. 19.4). The plants accumulate wind-blown soil particles and the shoots continue their growth through the accumulated sand thus increasing the height of the hummock.

2.1.3.2. *Atriplex vesicaria* – *Bassia* ssp. Suballiance
Bladder Saltbush – Copper Burr

This complex of communities is developed around the 250 mm isohyet and merges into drier types in which *Ixiolaena* or *A. rhagodioides* become prominent.

A. vesicaria is consistently dominant in gilgai depressions and species of *Bassia* are usually present, especially *B. divaricata* and *B. ventricosa*. Perennial grasses usually occur, especially *Astrebla pectinata*, which becomes increasingly abundant or even dominant in the north, *Stipa variabilis* in the south-east and *S. nitida* in the south-west. *Eragrostis setifolia* and *Chloris truncata* are constants and a wealth of annuals appear after rain. With decreasing mean annual rainfall other species become more abundant, notably *Ixiolaena* and *Minuria* spp., as described in the next section. This assemblage of species occurs also on deeper alluviums which show little or no gilgai development and are commonly free of surface stone.

Shelves in gilgaied soils are usually strewn with gibbers (Fig. 19.5). Their high salt-content alone is likely to retard colonization by plants. In addition the soil expands on wetting and seals the surface, resulting in rapid run-off into the depression. Consequently shelves are usually devoid of vegetation. The first plants to establish after rain are *Bassia brachyptera* and *B. ventri-*

Fig. 19.5. *Atriplex vesicaria* dominant in gilgaied clay soil with subsidiary *Enneapogon avenaceus, E. polyphyllus, Salsola kali, Erodium cygnorum, Bassia limbata* and *B. brachyptera*. The bare area in the foreground is raised above the vegetated area and is more saline and strewn with quartz gibbers. 80 km north of Broken Hill, N.S.W.

Fig. 19.6. View of stony downs (with white quartz gibbers) after a number of effective rains. The shelves (with gibbers) are almost covered by vegetation, especially *Bassia brachyptera* and *B. obliquicuspis*. Depressions with abundant *Atriplex vesicaria*. Darker bushes in the centre background are *Rhagodia spinescens*. c. 100 km north of Broken Hill, N.S.W.

cosa; other less abundant species are *Babbagia acroptera, Bassia lanicuspis,* and, rarely, *Frankenia pauciflora* and *Pachycornia tenuis*. Annual species of *Atriplex* (*A. spongiosa* and *A. inflata*) and Compositae may also occur. The floristic composition and the extent of the ground cover are determined by the amount of rain which percolates into the soil. Successive falls of rain in the one season wash salt deeper into the profile with an increase in the number of species and in the percentage ground cover. Under the most favourable conditions (about twice the average rainfall over a growing season) the shelves may be almost completely clothed with vegetation (Fig. 19.6).

The communities have been grazed by domes-

Fig. 19.7. *Bassia ventricosa* dominant on an eroded soil formerly supporting *Atriplex vesicaria*. A few plants of *Tetragonia expansa* visible. Mundi Mundi Plain, west of Broken Hill, N.S.W.

tic stock for several decades and the chief changes are the removal of *A. vesicaria* and its replacement by species of *Bassia* (Fig. 19.7). Charley and Cowling (1964) provide quantitative data on the amounts of organic matter, nitrogen and phosphorus in healthy stands of saltbush and they indicate the possible losses of nutrients which may occur if the bush is removed and the soils are eroded.

Accelerated soil erosion by wind is uncommon since the surface gibbers protect the soil. Water erosion leading to gullies is not uncommon in undulating country. The accumulation of wind-blown particles in saltbush stands sometimes occurs, resulting in changes in the natural patterning. The wind-blown material is derived either from saltbush stands where the soil is not protected by gibbers or from other communities. The windblown material accumulates within or around saltbush plants (dead or alive) or among gibbers, and the material forms mounds which are colonized by plants, commonly *A. vesicaria* seedlings around mature plants of the same species (resulting in clumping of plants) or of annual species or short-lived perennials on gibber-strewn surfaces. The most significant in the latter case is *Enneapogon avenaceus* which may develop a grass sward.

2.1.3.3. *Atriplex vesicaria – Ixiolaena leptolepis* Suballiance

This assemblage occurs in South Australia under a mean annual rainfall of 240–150 mm. The description is based on the communities north-west of Lake Torrens (Murray, 1931; Jessup, 1951). They cover a stony tableland which is devoid of trees, except along creeks, the topography being relatively flat (Fig. 19.8) so that water may accumulate as sheets after rain. Larger depressions may support thickets of *Muehlenbeckia cunninghamii* or *Chenopodium*

Fig. 19.8. Stony downs (with red-brown gibbers) north west of Lake Torrens showing gilgai patterning. Depressions are vegetated with *Atriplex vesicaria, Ixiolaena leptolepis, Minuria cunninghamii, Frankenia serpyllifolia, Bassia ventricosa* and *Bromus arenarius*. Near Woomera, S.A.

nitrariaceum, or "swamps" dominated by *Eragrostis australasica*.

The soils are gilgaied in most areas. *Atriplex* and *Ixiolaena* dominate in the depressions; other species include *Minuria leptophylla, Abutilon halophilum, Sida trichopoda, Bassia ventricosa*, the grasses *Astrebla pectinata, Eragrostis setifolia, Eulalia fulva, Panicum decompositum*, and a variety of annuals (*Atriplex, Erodium, Goodenia, Helipterum, Lotus* and others). The samphires *Arthrocnemum leiostachyum* and *Pachycornia tenuis* are sometimes present. The shelves are bare except after rain when they support *Bassia ventricosa, B. brachyptera, B. divaricata* and, rarely, the samphires.

2.1.3.4. *Atriplex rhagodioides* Suballiance
Silver Saltbush

A. rhagodioides, a shrub to c. 1m tall, replaces *A. vesicaria* on the gilgaied stony tableland soils in the driest part of the arid zone, mainly in South Australia (Jessup, 1951). The mean annual rainfall is c. 220–140mm. The communities overlap the preceding alliance.

The same patterning of species as in the *A. vesicaria* shrublands occurs. *A. rhagodioides* occupies the gilgai depressions, usually as the sole dominant but sometimes with shrubs of similar size, especially *Maireana aphylla* or *Sarcostemma australe*. Other common subsidiary species are *Abutilon halophilum, Ixiolaena leptolepis, Minuria leptophylla* and *Sida trichopoda*. Grasses are usually present, especially *Astrebla pectinata, Eragrostis setifolia, Eulalia fulva* and the annuals *Aristida anthoxanthoides, Iseilema membranacea* and *Sporobolus actinocladus*. Species of *Bassia* and annual saltbushes are abundant after rain and a few legumes may appear, especially *Swainsona stipularis*.

On the more saline shelves, which shed water into the depressions, plants are sparse or absent but partial coverage of the surface may occur after rain, the main species being *Bassia brachyptera, B. ventricosa* and annual *Atriplex*, especially *A. inflata, A. spongiosa*.

2.1.3.5. *Atriplex hymenotheca (vesicaria)* Suballiance
Bladder Saltbush

A. hymenotheca occurs as a local dominant on the Nullarbor Plain on alkaline red-brown clay overlying limestone (Fig. 19.9). The stands often contain small plants of *A. nummularia* and/or *Maireana sedifolia* and its associates (see 2.2.2, below). Also, it is sometimes dominant around saltlakes in Western Australia (Burbidge, 1941), the lakes being remnants of former watercourses. The saltiest parts of the lake beds support *Arthrocnemum* sp. and less saline areas are dominated by *A. hymenotheca* associated with *Maireana pyramidata, Cratystylis conocephala* and *Lycium australe*. Subordinate species include *Bassia divaricata, Atriplex inflata, A. spongiosa, Rhagodia* and *Frankenia* spp. A similar assemblage occurs on the Gascoyne River floodplan where sandridges supporting *Acacia sclerosperma* are separated by low limey flats dominated by *A. hymenotheca* (Fig. 19.10).

2.1.3.6. *Atriplex vesicaria* – *Maireana astrotricha* Suballiance
Bladder Saltbush – Bluebush

This suballiance is characterized by the broad-leaved ecotype of *A. vesicaria* which is often a co-dominant with *M. astrotricha* (see 2.2.3, below). The community is widespread in New South Wales and South Australia occurring under a mean annual rainfall of 250–180mm, but covers relatively small areas on slopes (sometimes at the foot of a range), around lakes or in watercourses. The stands often adjoin the stony downs on which the more compact and narrow-leaved ecotype dominates. When adjacent to ranges the two species form an understory to the *Acacia aneura* woodland. The soil is distinctive in that the surface horizon is shallow and contains lime; it overlies a clayey lime-rich subsoil, the profile rarely exceeding 1m in depth. Rounded gibbers or rock fragments from the range may occur on the surface (Wood, 1937; Crocker and Skewes 1941; Beadle, 1948; Jessup, 1951).

The characteristic species are usually co-dominants (Fig. 19.11), but either may be locally dominant and other shrubs of equal size are often present, especially *M. georgei, M. pyramidata, M. villosa, M. aphylla* and *Rhagodia spinescens*, and sometimes the shrubs (1–1½m tall) *Sida virgata* and *Eremophila scoparia*. The understory always contains *Bassia* spp. and the herbaceous layer is dominated by *Enneapogon avenaceus* and/or *E. polyphyllus*, and contains a large number of forbs after rain.

Fig. 19.9. *Atriplex hymenotheca* dominant in a shrubland on Nullarbor Plain which extends to the horizon. Near Nullarbor Sta., S.A.

Fig. 19.10. *Atriplex hymenotheca* (paler) in association with *Maireana enchylaenoides* with subsidiary *Gnephosis brevifolia*, *Angianthus plumigera*, *Frankenia pauciflora* and *Calandrinia balonensis*. Sand ridge behind dominated by *Acacia sclerosperma*. 56 km south of Carnarvon, W.A.

Fig. 19.11. Mixed community of *Atriplex vesicaria* (broad-leaved ecotype), more abundant in the background, *Maireana astrotricha* (main bush in foreground) and a few plants of *M. pyramidata*. Smaller plants mostly *Bassia ventricosa* and *Tetragonia expansa*. The peg in the foreground marks a permanent quadrat established in 1955 and is c. 60 cm high. Fowlers Gap, c. 110 km north of Broken Hill, N.S.W.

2.2. Alliances Dominated by *Maireana* spp.

2.2.1. General

Maireana is identified by the presence of five horizontal lobes or wings, which may be connate, on the top of the mature perianth which persists around the fruit (a one-seed nut). Several species are community dominants (below) and they have in common deep root systems, the roots usually following both vertical and horizontal cracks in the soil. The germination of the seeds of *M. sedifolia*, *M. pyramidata* and *M. georgei* has been investigated by Burbidge (1946). She showed that removal of the perianth increased the rate of germination, and that burial below c. 1 cm in the soil reduced the number of seedlings. For all species 20–50% of the seeds germinated within 3 days from wetting and temperature optima were 18–30°C for the first two species and 20–25°C for *M. georgei*. Also the seeds have an after-ripening period, maximum germination being reached after 12 months storage. Under field conditions Carrodus and Specht (1965) indicate that *M. sedifolia* does not re-establish as rapidly as does *Atriplex vesicaria*.

Three alliances are recognized but they are not rigidly delimited by definable habitat factors. Two or three species are often found together and sometimes also with perennial species of *Atriplex*. Species of *Maireana* do not accumulate salt to the same extent as do *Atriplex* spp., though field observations suggest that they are salt-tolerant, especially *M. pyramidata*.

2.2.2. *Maireana sedifolia* Alliance
Bluebush

This species, which forms a compact bush c. 1 m high, is an excellent indicator of calcareous

Fig. 19.12. An almost pure stand of *Maireana sedifolia*, with a few young plants of *Atriplex nummularia*. Behind, a mallee stand of *Eucalyptus oleosa* with some *Myoporum platycarpum*. c. 50 km east of Cocklebiddy, W. A.

soils, the lime often occurring in the surface horizon. It extends across the arid zone in the south as the dominant in shrublands and it sometimes forms the understory in some *Eucalyptus* communities, especially some mallee stands (Chap. 14) and woodlands of *Casuarina cristata* and some species of *Acacia* (Chap. 18).

M. sedifolia is the most common dominant on the limestones of the Nullarbor Plain (Fig. 19.12) where lime occurs close to or at the surface. Mean annual rainfall is c. 150–180 mm. It is replaced on deeper soil by *Atriplex* spp. and is sparse or absent in depressions (dongas), or on limestone outcrops which support only a lichen flora (chiefly *Buellia subalbula*, Johnstone and Baird, 1970). These authors recorded the fol-

Fig. 19.13. *Maireana sedifolia* shrubland with a small admixture of *M. pyramidata* (darker) on undulating country in Barrier Range west of Broken Hill, N. S. W. Note nodular lime on eroded soil in foreground.

Fig. 19.14. *Maireana pyramidata* dominant on a lunette on the eastern side of Lake Cawndilla. Trees in background edging the lake are *Eucalyptus largiflorens*. Near Menindee, N.S.W.

lowing associates of *M. sedifolia* on the western segment of the plain: *Bassia patenticuspis, B. uniflora, Angianthus brachypappus, Gnephosis skirrophora* and *Stipa nitida*, together with a large number of annuals. *M. sedifolia* is rare in the dongas where tall shrubs or small trees replace it, especially *Acacia oswaldii, Eremophila longifolia* and *Pittosporum phyliraeioides*, and less often *Eremophila maculata, Lycium australe* and *Grevillea nematophylla*. On the eastern portion of the plain the communities are similarly arranged. Ising (1922) mentions *Goodenia pinnatifida* and *Podolepis* as common herbs associated with *M. sedifolia*. The dongas contain a similar set of species as in the west, with some additions, especially *Casuarina cristata*.

In the east, *M. sedifolia* forms pure stands on lime-rich soils. Occasional associates include *Eremophila scoparia* and *Maireana triptera. Bassia obliquicuspis* and *B. sclerolaenoides* are common subshrubs and *Enneapogon avenaceus* and *E. polyphyllus* are common in the herbaceous layer. A large number of annual species appears after rains (Jessup, 1951; Hall et al., 1964). *M. sedifolia* is commonly mixed with *M. pyramidata* (Fig. 19.13) and the latter increases in abundance with increasing depth of the surface sandy layer (Beadle, 1948).

2.2.3. *Maireana astrotricha* Alliance
Bluebush

This species resembles *M. sedifolia* in general appearance and is distinguished by the presence of a distinct petiole in the leaf. It replaces *M. sedifolia* on lime-rich, shallow soils, often with surface gibbers, which have a low rate of percolation of water, rendering them more arid than the climate might indicate (mean annual rainfall c. 250–180mm). *Ptilotus obovatus* is a common associate and species of *Bassia* are invariably present. Shrub emergents sometimes occur, e.g. *Eremophila scoparia* and *Pittosporum phyliraeoides* (Jessup, 1951) and the community sometimes extends as an understory to *Acacia aneura*. *M. astrotricha* forms an ecotonal association with *A. vesicaria* (see 2.1.3.6, above and Fig. 19.11).

2.2.4. *Maireana pyramidata* Alliance
Bluebush

M. pyramidata is dominant on soils with a deep sandy horizon, commonly Solonized Brown Soils, and on dunes (lunettes) around lakes (Fig. 19.14). The sands usually contain lit-

Fig. 20.2. Ring formation in *Triodia scariosa*. From a stand of *E. socialis* – *E. dumusa* mallee, near Balranald, N.S.W.

rhizome system. In a few cases "wandering" rhizomes proliferate underground and initiate new tussocks remote from the parent tussock, e.g. *Imperata cylindrica,* the rhizomes of which form a interlacing network in many forests.

A special type of tussock (referred to as a "hummock") occurs in *Plectrachne* and *Triodia* (Burbidge, 1945b). A branching rhizome initiates the culms which produce distichous, xeromorphic, pungent-pointed leaves separated by short internodes. The culms are massed together and the tussock becomes more or less circular at the base, and finally the mass of culms becomes hemispherical, resembling a porcupine. New culms are added peripherally and a hemispherical hummock, flat-topped in the largest hummocks, develops. The inflorescences are produced terminally. In some species stolons are produced which touch the ground remote from the parent tussock and initiate a new tussock. In some species the centre of the hummock dies and an annulus of living rhizomes continues to produce new culms (Fig. 20.2); the annulus frequently becomes broken into a number of smaller hummocks. These grasses constitute the Hummock Grasslands which occur mainly in the arid zone.

Stoloniferous grasses occur but are not abundant. Examples are *Spinifex hirsutus* on coastal dunes, *Sporobolus virginicus* on littoral clays and *Pseudoraphis spinescens,* usually an aquatic with stolons floating on water. Such grasses may form mats over the soil surface and characterize Mat Grasslands.

The annual habit is common and all species of some genera are annual, e.g., *Dactyloctenium* and *Iseilema*. Many genera contain both annual and perennial species, e.g., *Poa* and *Sporobolus*. Many species which are potentially perennial behave as annuals in the arid zone, especially the inland species of *Stipa* and *Aristida*. Annual and short-lived perennial species charcterize the Ephemeral Grasslands of the arid zone.

4. Distribution of the Grasses and Grasslands in Australia

The distributions of the grasses and of grasslands in the world have received much attention (Clayton, 1975) and several Australian authors have made significant contributions to these subjects (Hartley, 1958–73; Moore, 1964). The distributions of many grass genera appear to be climatically controlled but, as Hartley points out, many of the larger genera cover such a diversity of habitats that a precise statement concerning the ecological limits for the genus cannot be made. Indeed, since new species commonly arise through adaptation to a new environment, it is likely that members of larger genera should occupy diverse habitats as a result of genetic selection by the various habitat factors. For example, of the 34 Australian species of *Poa* (Vickery, 1970) a large number are confined to the wetter, cooler south-east where the species are delimited mainly by soil properties. However, several species occur in warmer, drier environments, e.g., *P. fax* is an annual in the arid and semi-arid zones and *P. poiformis* is confined to subsaline soils along the south-eastern and southern coastline. Nevertheless the genera, especially those which are grassland dominants, can be classified into groups according to their habitats, provided several environmental factors are considered. In some cases a soil property, especially salinity or waterlogging, completely overrides the climatic factors in determining the distribution of a taxon. For example, *Sporobolus virginicus* occurs on saline muds around the entire Australian coastline.

The genera can be divided into two main segments, as suggested by the arrangement in Table 20.1. The larger segment consists of the northern tropical and subtropical genera which germinate and make their vegetative growth during the summer months. They occur in areas with a predominantly summer rainfall. Most of these genera belong to the Andropogoneae or Paniceae. Many of the genera are restricted to the wetter tropics, e.g. *Bambusa, Chionachne, Ophiuros, Oryza* and *Thaumastochloa*. Many of the northern genera extend well south of the Tropic, some of them occurring in areas with a predominantly winter rainfall. This is due to the occurrence in these southern areas of a significant summer rainfall which is sufficient to complete the life cycle during the warmer months. The main genera are *Bothriochloa, Cymbopogon, Dichanthium, Digitaria, Imperata, Panicum* and *Themeda*. The last needs a special comment.

Themeda is represented in Australia by 4 species, of which 3 are found in the northern half of the continent. *T. australis*, however, occurs from New Guinea to Tasmania and westward across the continent into the arid zone, where it is found in depressions. It is recorded in most plant communities, including the *Poa* grasslands of the eastern highlands. Although most abundant in woodland communities, it occurs in places as a dominant or co-dominant of natural grasslands in a variety of habitats, from the tropics to the highlands, on coastal headlands and in depressions in the arid zone. Hayman (1960) has shown that the plasticity of the species is due in part, at least, to the polyploid nature of some of the populations. Diploid populations occur exclusively in the cooler south-east, whereas tetraploid populations occur westward to Western Australia; triploid, pentaploid and hexaploid plants are found as individuals in populations of another chromosome number.

The second large segment of genera occurs in the south and the grasses are usually known collectively as the "temperate grasses". Many of them make their growth during the cooler months, notably *Danthonia* and *Stipa*. Those which are confined to higher altitudes, where frosts (or snow) inhibit growth during the winter, grow during the summer months, especially *Agropyron, Agrostis, Deyeuxia, Dichelachne, Echinopogon, Festuca, Glyceria, Hierochloë* and those species of *Poa* confined to higher altitudes.

A large number of grasses occur in the dry inland and the genera are divisible into two segments on the basis of seasonal rainfall incidence. In the north, with a higher summer-rainfall incidence, *Plectrachne* and *Triodia* are the conspicuous grasses of the arid zone. The remaining genera which occur in the northern part of the arid zone are southern extensions or relicts of the wetter tropical and subtropical grasses, of which *Aristida, Chloris, Cymbopogon, Dactyloctenium, Enneapogon, Eragrostis, Iseilema, Sporobolus* and *Themeda* are the most abundant. These drier regions support the endemic genera, *Amphipogon, Astrebla, Neurachne, Paractenium, Plagiosetum, Thellungia* and

Zygochloa. In the southern part of the arid zone, where the rain falls mainly in the winter, the most common genera are *Danthonia* and *Stipa*; *Triodia* is represented by *T. irritans* and *T. scariosa.* The southern area also supports some of the northern genera, as mentioned above.

Soil properties sometimes determine the grass flora at the generic level. The effects of low soil fertility are mentioned above. Also, *Astrebla* is confined to fine textured soils, mainly in those parts of the semi-arid zone which receive a summer rainfall. Several grass genera are restricted to saline environments, viz., *Distichlis, Xerochloa* and *Zoysia* which occur on saline muds around the coast, and a few species of larger genera occur in the same habitat, e.g., *Sporobolus virginicus.* Several grasses occur only around the coast on sands, viz., *Spinifex* spp., *Poa poiformis* and *Stipa teretifolia.*

Several genera occur as aquatics or semi-aquatics in fresh water, e.g. *Oryza* and *Phragmites* and most species of *Ectrosia, Leersia, Potamophila* and *Vetiveria.*

Grasslands occur in areas where some environmental factor, or combination of factors, prevents the development of trees or shrubs. These habitats can be classified as follows:

i. The littoral zone above high tide level where strong winds and/or salinity prevent the growth of woody plants. The littoral grasslands occur into 2 distinct habitats: either on sands commonly blown into dunes, or on mudflats.

ii. Aquatic and waterlogged habitats, such as lagoons or their margins, some swamps (excluding very acid areas which usually support sedgelands), and some areas which are periodically waterlogged; the last occur from the wettest coastal areas to the arid zone.

iii. Areas of fine-textured soils, mainly in the semi-arid zone but extending into the arid zone, where soil texture restricts water-availability, preventing the growth of trees, the factor probably acting in conjunction with deep soil cracking or with temporary seasonal waterlogging.

iv. Those parts of the arid zone where mean annual rainfall alone prevents the development of woody plants. The grasslands are either perennial and of the Hummock-type, or ephemeral and dominated by annuals or short-lived perennials which appear only after rain.

v. Areas subject to very severe frosts, with or without snow mantles in the winter, either on the highest mountains in the south-east or in frost pockets in valleys in mountainous areas.

The grasslands are described below under these categories. Since genera are mostly restricted to specific habitats, this ecological classification is also a floristic one, with a few exceptions. The main exceptions are: (a) *Poa,* of which most of the grassland species occur at higher altitudes, but *P. poiformis* forms grasslands on sands in the littoral zone and *P. labillardieri* and *sieberana* form associations with *Stipa, Danthonia* and *Themeda* on soils of fine texture in the semi-arid zone; (b) *Themeda australis* which appears as a grassland dominant in a variety of habitats.

5. Descriptions of the Alliances

5.1. Littoral Grasslands

Small patches or narrow belts of grasslands occur along the coastline on sand dunes and sandy areas or on mud-flats above high-tide level around the entire coast. The species, with a couple of exceptions, are confined to the littoral zone.

5.1.1. *Spinifex* spp. Alliance

Spinifex, a strictly littoral genus, is represented by two species, *S. longifolius* and *S. hirsutus.* The former grows on coastal dunes along the north and west coasts. The plants are rhizomatous and produce dense tussocks which sometimes build hummocks. The flowers are dioecious, male and female plants commonly growing together in the one clump (Fig. 20.3). The rhizomes spread laterally for several metres if the dune is wide. More often the frontal dune is no more than a couple of metres wide at its summit where grass forms hummocks. Since the species has a wide latitudinal range the associated species vary from north to south, but in no area they are abundant. In the tropics *Canavalia obtusifolia* and *Ipomoea pes-caprae* are almost constant. * *Euphorbia atoto* occurs from north to south as an occasional species. South of the Tropic halophytic shrubs, characteristic of neighbouring zones of vegetation, commonly occur in the grassland, especially *Atriplex isatidea* and *Nitraria billardieri.* Associated

Fig. 20.3. Male (centre) and female (left and right) plants of *Spinifex longifolia*. Bush of *Nitraria billardieri* behind. Carnarvon beach. W. A.

species are more abundant in the Swan R. district (see Chap. 23).

S. hirsitus occurs on the east and south coasts and northward along the west coast to the Swan R. area. The plants produce stolons up to c. 6 m long on loose coastal sands, with tufts of leaves at the nodes. The flowers are dioecious, male and female plants intermixing. On relatively flat areas swards of grass are produced by the interlacing stolons, but in areas subject to local erosion the stolons commonly occupy the crests of residual sand deposits and form hummocks.

The grassland occurs close to high water mark and extends inland on raw sand, meeting and mixing with a zone of mat plants. The associated species are determined by latitude. In the warmer north *Canavalia obtusifolia* and *Ipomoea pes-caprae* are the most constant. In the south, characteristic associates include *Festuca littoralis*, occasional Compositae (*Senecio* and *Sonchus*) and *Carex pumila*. *Distichlis distichophylla* and *Zoysia macrantha* are locally common, and in some areas introduced littoral grasses have become abundant, especially * *Ammopila arenaria*, * *Lagurus ovatus* and * *Ehrharta* spp. (see Chap. 23).

5.1.2. *Sporobolus virginicus* Alliance
Salt Couch

S. virginicus is a stoloniferous perennial forming mats, the flowering culms reaching a height of c. 20 cm. It is the characteristic species of a littoral alliance which occurs on patches of saline mud dotted around the whole coastline. The soils are solonetzic, sometimes with a slight gilgai microrelief and they crack at the surface on drying. The normal position of the grassland is the landward side of the samphire meadows (*Salicornia* and *Arthrocnemum*), or, if samphire is absent, as is sometimes the case in the tropics, the grasslands adjoin mangrove communities or the *Xerochloa* grasslands (Fig. 20.4 and next section).

The floristic composition of the grassland varies with height above high tide mark and with latitude. At all latitudes the *Sporobolus* grasslands nearest the high tide mark are usually monospecific, *S. virginicus* tolerating inundation by sea water, whereas the other species in the alliance do not.

In the tropics, *Xerochloa imberbis* or *X. barbata* occur as ecotonal occurrences in the grassland and *Trianthema turgidifolia* is recorded on the north-west coast. Other species which may be present, chiefly in depressions where salt concentrations are reduced by rain, are: *Diplachne fusca, Fimbristylis littoralis, F. polytrichoides, Panicum maximum* and *Paspalum vaginatum* (Perry, 1953; Pedley, 1971).

In the south (southern Qld to Tas.) a completely different set of associated species occur on the landward, less saline parts of the grassland. These are the annual grasses *Agrostis aemula* and *A. billardieri*. In additon, three perennial grasses become important constituents in the eastern stands *Cynodon dactylon* and *Zoysia macrantha* (southern Qld, through N.S.W. Vic. and eastern S.A.), and *Distichlis distichophylla* (south coast of N.S.W., Vic., Tas. and eastern S.A.). These grasses form mats, which sometimes become spongy, or hummocks, either in association with *S. virginicus* or in pure stands on the landward side of the grassland. These species occur both on clays and on low-lying areas of sand, less commonly among rocks.

5.1.3. *Xerochloa* spp. Alliance

Two closely related species of *Xerochloa, X. imberbis* and *X. barbata*, constitute a small fragmented alliance on saline and subsaline Solonetz soils on mudflats in the tropics. Vegetatively the species are more or less identical; *X. barbata* is distinguished by its having tufts of wool at the bases of the spikelets which are enclosed in a spathe, whereas *X. imberbis* is glabrous. The species, both perennial, are sometimes mixed in the same community. The grasslands occur in small patches or in elongated strips on the landward side of the *Sporobolus virginicus* grasslands or of mangroves (Fig. 20.4).

The grasslands have been described by Perry and Christian (1954) Perry and Lazarides (1964) and Perry (1970), and may reach a height of c. 40 cm, the tussocks being usually scattered. Constants in the communities are *Bassia astrocarpa, Salsola kali* and *Gomphrena* sp. Other species include *Brachyachne convergens, Chloris divaricata, C. pumilio* and *Dactyloctenium radulans*. Swampy, less saline areas on the landward side are sometimes dominated by *Vetiveria elongata* and in these swamps, which dry out during the winter, *X. imberbis* and *Sporobolus virginicus* are subordinate species (Fig. 20.5).

5.1.4. *Aristida hirta* – *A. superpendens* Alliance

This alliance occurs on flats or in depressions on the delta of the Gilbert R. on the Gulf of Carpentaria on Solodized-Solonetz soils which are salt-affected but not necessarily of high salinity (Perry and Lazarides, 1964). The grasslands are c. 15 cm high and the main subsidiary species are *Chloris scariosa, C. pumilio* and *Eriachne armittii. Chrysopogon fallax* occurs as an emergent in small monospecific patches and the grasslands are surrounded by woodlands of *Melaleuca*.

5.1.5. *Themeda australis* on Headlands

T. australis dominates small patches of grassland on coastal headlands overlooking the sea in central New South Wales. The soils are fine-textured, with a nutty structure and are dark grey-brown or red-brown in colour. They are derived from bands of shale or from basalt. The grassland usually contains no or few associated species and the patches are surrounded by or adjoin wind-shorn scrubs of *Banksia integrifolia*.

It has been suggested that these treeless areas were induced by the aboriginals, since many of them were middens. Usage by aboriginals implies soil enrichment. However, since they have not been colonised by woody plants since white occupation, it is possible that they have always been treeless and were chosen by the aboriginals for cooking and eating because the grassland was comfortable to lie on and at the same time provided a pleasant view.

5.2. Grasslands in Fresh-water Aquatic Habitats or on Soils which are Periodically Flooded

These grasslands are dominated by aquatic grasses or by grasses which tolerate or require temporary waterlogging of the soil during their growing period. The grasses are not closely related taxonomically, mostly belonging to different genera, and several tribes are represented (Oryzeae, Arundineae, Eragrostideae, Andropo-

Fig. 20.4. *Sporobolus virginicus* (foreground, beside mud channel) and *Xerochloa barbata* grassland on slightly more elevated areas. Behind (and at a lower level) the dwarf mangrove, *Aegiceras corniculatum*, and *Avicennia marina* (slightly taller). Near Broome, W.A.

goneae, Paniceae and Maydeae). *Phragmites* is noteworthy because two species occur, one in the tropics *(P. karka)*, the other in the south *(P. australis)*.

Ecologically the communities are sometimes zoned, especially in the tropics. The *Oryza*, *Phragmites karka*, and *Themeda – Eriachne* communities occur in the same area, being zoned with decreasing water-level in the order named. Furthermore, some of the understory species of the tropical grasslands extend south of the tropics and become dominants or common

Fig. 20.5. Depression in a mudflat supporting a *Vetiveria elongata* grassland. In the more saline foreground tussocks of *Xerochloa imberberis* and *Sporobolus virginicus* occur. Tree in the centre is *Eucalyptus pruinosa* and those in the background *E. polycarpa*. c. 16 km south of Karumba, Qld.

in small areas either as aquatic or mat-grassland, especially *Pseudoraphis* and *Leersia*.

The aquatic grasslands are related floristically by their understory associates, though mainly at the generic level, e.g., *Eleocharis* and *Marsilea* occur in most of these grasslands, even those in the arid zone.

Most of the grasslands make their growth during the summer months and, with few exceptions, notably *Phragmites australis,* they dry out and become dormant during the dry (usually winter) season. During the drying-out period subsidiary herbaceous species, mostly dicotyledons, become established, completing their life cycles before the soil reaches wilting point.

5.2.1. *Oryza australiensis* Alliance
Native Rice

O. australiensis occurs only in the tropics in coastal lagoons in shallow water (Fig. 20.6). The grass grows to a height of 1–1½ m and is usually mixed with *Eleocharis dulcis, E. sphacelata* and *E. spiralis* or *O. rufipogon* (Perry and Lazarides, 1964; Perry, 1970). *Leersia hexandra* is a constant subsidiary species. Other species of occasional or local occurrence are *Cyperus retzii, Hymenachne acutigluma, Panicum paludosum* and *Pseudoraphis spinescens*. Local associations of *Leersia* and *Pseudoraphis* (both trailing – floating grasses) are common, the taller component being absent. *Marsilea* sp., with floating leaf-blades is often abundant. The grasslands are often dotted with trees of *Melaleuca leucadendron*.

5.2.2. *Phragmites* spp. Alliance

5.2.2.1. *P. karka* Suballiance

This species is confined to the tropics, occurring either at the margins of lagoons or along the banks of rivers (with an isolated record in a pool in Central Australia, Chippendale 1972). Although usually a fringing species, it occasionally dominates small patches of grassland in shallow water on the landward side of the *Oryza* grasslands (Perry, 1970). The grass reaches a height of 1–2 m and has an understory of *Scleria poiformis* and *Pseudoraphis spinescens*. The drier margins adjoin grassland swards dominated by *Imperata cylindrica*.

Fig. 20.6. Grassland dominated by *Oryza australiensis* in a swamp near Darwin, N.T. Photo: George Wray.

5.2.2.2. *Phragmites australis* Suballiance
(Common Reed)

P. australis occurs in the east and south-east along the coast from the Tropic to South Australia and Tasmania. It occurs also in the inland rivers and lakes as far west as the Darling. The species grows in fresh or slightly brackish water and occurs along the coast in river estuaries and in lakes or lagoons (Aston, 1973).

The plants are rhizomatous and amphibious, the culms reaching a height of c. 2 m (rarely to 3½). The culms, which resemble bamboo, form dense monospecific stands at the water's edge. *Typha* spp. are sometimes occasional in the stands (see Chap. 23).

5.2.3. *Themeda australis – Eriachne burkittii* Alliance

This alliance occurs in the tropics in Northern Territory and Queensland close to the coast in areas which are subject to minor waterlogging during the wet summer growing season (Christian and Stewart, 1952; Story, 1969, 1970; Perry, 1970). The grasslands are often dotted with winter-deciduous trees which occupy the less waterlogged areas.

The grasslands, mostly 1 m high, are patterned according to microtopography. *Eriachne burkittii* occupies the lower areas and *Themeda australis* the more elevated ones. Local variations are brought about by the occurrence of wetter depressions in which one or other of the following are dominant: *Sorghum plumosum* (all areas), *Vetiveria pauciflora* and *Sclerandium truncatiglume* (Qld.). Subsidiary constants are species of *Aristida, Bothriochloa, Eriachne, Eragrostis, Ischaemum* and *Panicum*. Cyperaceae are common in some areas, especially *Fimbristylis* spp. and a few dwarf forbs may be present, e.g., *Bulbostylis barbata, Stylidium* spp. and *Xyris complanata*.

The shrub and tree species which are dotted in the grassland are *Pandanus* sp., *Eugenia bleeseri, Eucalyptus papuana, E. polycarpa, Melaleuca nervosa, M. symphyocarpa* and *Tristania lactiflua*. The savannah tends to have a somewhat different grass flora, commonly a mixture of *Sorghum plumosum, Vetiveria, Coelorachis* and *Heteropogon triticeus*.

5.2.4. *Chionachne cyathopoda* Alliance
Coastal Cane-grass

This grass, which reaches a height of 3–4 m occurs along the banks of some of the tropical rivers as an understory species in woodlands, but is dominant in depressions in the tropical *Dichanthium* grasslands in the east (Perry, 1953). The plants have stout rhizomes which produce a dense reed-like thicket of culms during the summer wet season. The understory in the grassland consists of various assemblages of *Cyperus retzii, Eleocharis philippinensis, Elytrophorus spicatus, Eragrostis japonica, Leptochloa brownii* and *Sesbania benthamii*. The soft forbs, *Ludwigia parviflora, Marsilea brownii* and *Ammania multiflora* form a lowermost layer and persist on the mud as the water dries up. The community is desiccated during the dry winter.

5.2.5. *Pseudoraphis spinescens* Alliance

P. spinescens is a subsidiary species in the tropical aquatic grasslands or sometimes a local dominant in shallow water (see 5.2.1, above). Southward, where the taller species do not occur, *P. spinescens* is a common dominant in shallow, usually still waters along the coast to central New South Wales. Its stolons float and form a tangled mass, and it sometimes forms mats on soil at the edge of the water, where *Juncus* spp. commonly occur. Other common associates in water are *Leersia hexandra* and (in N.S.W.) *Potamophila parviflora*. In subtropical areas *Paspalum distichum* is sometimes abundant, both in water and on the adjacent soil, with *Hemarthria uncinata* as a common associate on soil.

5.2.6. *Leptochloa digitata* Alliance
Cane-grass

This species is characteristic of lowlying areas of fine textured soil which are periodically waterlogged. It occurs mainly in the semi-arid and arid eastern half of the continent in the summer rainfall zone along watercourses, in depressions in some woodlands and in the *Astrebla* grasslands where, in particular, it is a local dominant. It is also common in the Channel country of Queensland, where it is often associated with or occurs near thickets of *Muehlenbeckia cunninghamii*. Its abundance has probably been increased in some cases by the creation of wet habitats by man, especially tabledrains along roads. Its associates are "swamp" species, especially *Marsilea drummondii*, Cyperaceae (especially *Eleocharis* spp.), *Minuria integerrima* and, as the water recedes, *Cynodon dactylon*.

5.2.7. *Eragrostis australasica* Alliance
Inland Cane Grass

E. australasica occurs on some arid-zone claypans and some playas (Chap. 21) which are almost impervious to water. After rain, or if irrigated by run-off or overflow water from streams, the claypans become miniature temporary lakes. *E. australasica* is confined to these claypans and forms tussock grasslands up to c. 2 m high (Fig. 20.7). Associated species are few in number. *Marsilea drummondii*, which may have become established as an aquatic before the cane-grass occupied the area, is a constant. Small forbs are usually present, becoming established in mud, rather than in free water, especially *Centipida cunninghamii*, *Minuria cunninghamii* and sometimes *Eragrostis setifolia*. Annual species may invade the community as the water disappears from the surface, especially *Atriplex inflata*, *A. spongiosa* and *Mollugo hirta*.

5.2.8. Other Species of Limited Occurrence

i. *Diplachne fusca* (Beetle grass)

D. fusca is widely distributed in the northern part of the continent, usually in areas of low frost incidence. It is a perennial tussock grass, reaching a height of c. 1 m. It is confined to wet habitats and may be amphibious. It is tolerant to subsaline conditions and is recorded from saline coastal muds, but is occurs also in non-saline areas along rivers and in and around swamps. It sometimes forms small patches of grassland in depressions in other grasslands on fine textured soils. It is abundant today as a dominant of an induced community in bore drains leading from artesian bores to provide water for stock. It is most commonly associated with *Eleocharis pallens*.

ii. *Cynodon dactylon* (Couch Grass)

C. dactylon is a cosmopolitan grass, probably native in Australia, and commonly cultivated as a lawn grass. It occurs in a variety of habitats from saline coastal muds to the interior, mainly in warmer areas. It is summer-growing and the leaves are killed by frost. The grass is stoloniferous and forms lawn-like mats, usually on fine-

Fig. 20.7. *Eragrostis australasica* on a claypan; understory of herbaceous *Minuria integerrima*. Grass in foreground is *Enneapogon avenaceus*, with a plant of *Maireana sedifolia* at the left. Trees behind are *Casuarina cristata*. Kinchega National Park, near Menindee, N.S.W.

textured soils. Its abundance has possibly increased since white settlement and it is now common in many open situations, especially near watercourses, on river flats or in temporarily swampy ground in the interior. It often occurs in areas known to have been dominated by taller grasses which have now been removed by overgrazing, and possibly it occurred as an understory species in these taller grasslands. Blake (1938) records it a common species along the banks of bore drains in western Queensland, sometimes with *Sporobolus benthamii* on lower ground.

5.3. Grasslands Occurring on Fine-textured Soils mainly in the Semi-arid Zone

5.3.1. Distribution of the Grasslands and their Interrelationships
(see Fig. 20.8)

These grasslands, which are the most extensive tussock grasslands in the continent, occur mainly in the semi-arid zone, with minor extensions into the arid zone. They stretch across the Carpentaria Plains in the north-east and follow the clay soils westward across the Barkly Tableland, and extend southward in patches to southern Victoria. Small areas of the same or related grasslands occur to the west in the arid zone, but they cover only small areas in Western Australia where soils suitable for their development are scantily represented.

The northern grasslands grow during the summer and there is a simple climatic zonation, the *Dichanthium* grasslands occurring in the wetter areas, the *Astrebla* grasslands in the drier south and west. The two types of grassland intergrade, the latter extending into the arid zone and mixing (through *A. pectinata*) with the *Eragrostis xerophila* Alliance in Central Australia. *E. xerophila* occurs in patches westward to the Indian Ocean and it mixes with the *E. setifolia* Alliance, the latter extending to the south across the Nullarbor Plain to the Southern Ocean.

The southern grasslands, which make their growth during the cooler months, are usually sharply segregated from the northern grasslands, especially *Astrebla* spp. and *Stipa aristiglumis* which may grow in the same district, without mixing. However, *S. aristiglumis* forms a significant winter-growing component of the *Dichanthium sericeum* grasslands on the Darling Downs, and the latter species extends south as a summer-growing species in some of the southern grasslands.

In the southern grasslands *Stipa* spp. form the main dominants with *Danthonia* spp. becoming increasingly important towards the south. The most northerly member of the series, the *S. aristiglumis* Alliance, is replaced in the south by the *S. scabra* – *S. bigeniculata* Alliance. The latter occurs on fine textured soils and from it two series of related grasslands can be traced. Firstly, in the drier west by other species of *Stipa*, which in sequence from wet to dry are *S. falcata, S. nitida* and *S. eremophila,* the last two extending on to the Nullarbor Plain. A species of *Danthonia* is often a co-dominant in these grasslands. Secondly, the *S. scabra* – *S. bigeniculata* Alliance is related to the grasslands of higher altitudes. This relationship is first seen by the gradual increase in prominence of *Themeda australis* and *Poa sieberana* in moister and/or cooler areas. Also *Poa labillardieri* appears in still moister areas and *Stipa* disappears. *Poa* spp. (Section 8.3.8, below) finally become the dominants at higher altitudes.

5.3.2. *Dichanthium* spp. Alliance
Bluegrasses

Several species of *Dichanthium* dominate grasslands on dark grey to black clays in the tropics and subtropics mainly between the 500 and 800 mm isohyets (Perry, 1953; Perry and Christian, 1954; Perry and Lazarides, 1964). Tiny patches may occur in tropical Western Australia, but suitable soils in this region are almost absent in the appropriate rainfall zone. The grasslands occupy flat areas, less commonly undulating downs which are sometimes dotted with trees of which *Eucalyptus microtheca, Atalaya hemiglauca* and *Bauhinia cunninghamii* are the most common, and (confined to the east) *Eucalyptus brownii* and *E. orgadophila. Acacia cambagei* is sometimes present in the drier areas. The grasslands are mostly 1–1½ m high and make their growth after summer rains. They occur mostly adjacent to *Eucalyptus* woodlands which occupy soils of coarser texture, or, in the drier areas on fine textured soils, they merge

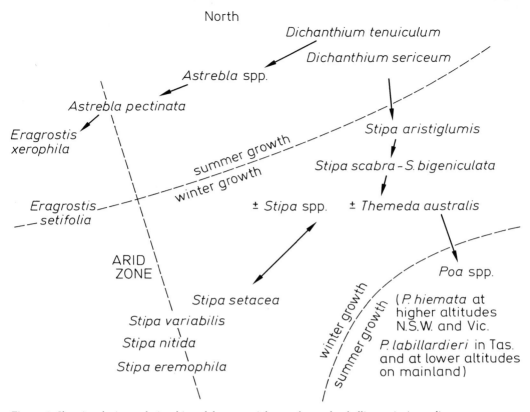

Fig. 20.8. Showing the interrelationships of the perennial tussock grassland alliances in Australia.

with the *Astrebla* grasslands and form extensive ecotones with them.

The *Dichanthium* grasslands vary in floristic composition according to latitude, which determines the proportion of the annual rain falling in the summer. In the north, where 80–90% of the rain falls between November and March, *D. tenuiculum* (formerly *D. superciliatum*) is dominant or codominant. To the south, where 60–70% of the rain falls during the same period, the dominant is usually *D. sericeum*. Two suballiances are identified; they are very alike in general appearance and many of the associated species are common to the two.

5.3.2.1. *Dichanthium tenuiculum* Suballiance

This species and *D. fecundum*, with subsidiary *D. annulatum* and *D. aristatum*, occur in the wettest areas in the north (Fig. 20.9). Distinctive association are: *D. tenuiculum* in pure stands; *D. tenuiculum* – *Ophiurus exaltatus* (Carpentaria Plains and eastward); *D. fecundum* – *Eulalia fulva* (Barkly Tableland). These dominants form an upper story c. 1–1½ m high. A second layer of smaller grasses is present, including *Bothriochloa ewartiana*, *B. bladhii* (formerly *B. intermedia*), *Brachyachne convergens* and *Panicum decompositum*. A third layer of small legumes and forbs is present, including *Aeschynomene indica*, *Alysicarpus rugosus*, *Crotalaria crassipes*, *Psoralea* spp. and *Rhynchosia minima*.

Variations (minor association) include:

i. The local occurrence of certain taller grasses, viz. *Chrysopogon fallax* (west), *Themeda australis* (highest rainfall areas), and *Heteropogon contortus* (wettest areas, east). The last is not usually a grassland dominant but rather an understory species in woodlands. However, small areas of grassland dominated by *H. contortus* are recorded by Story (1970) on the

Fig. 20.9. General view of the *Dichanthium* spp. grassland on the Carpentaria Plains south of Normanton, Qld.

southern part of Cape York Peninsula on basalt in association with *Themeda australis*.

ii. In gilgai depressions where water accumulates several minor grassland communities occur, including the *Oryza* aquatic grassland (see 5.2.1, above), and in the east near the coast on alluvial soils stands of the Coastal Cane Grass *Chionachne cyathopoda* (see 5.2.4, above).

In drier areas ecotones occur with the *Astrebla* grasslands and species commonly associated with the latter are sometimes found, especially *Aristida latifolia*, the annual *Iseilema vaginiflorum*, and several dwarf shrubs and forbs, especially species of *Sida* and *Hibiscus*, *Boerhavia diffusa*, *Convolvulus erubescens* and *Euphorbia drummondii*.

5.3.2.2. *Dichanthium sericeum* Suballiance

These grasslands, mostly $1-1^1/_2$ m tall, occur mainly in south-eastern Queensland, extending from the Darling Downs (rainfall c. 800 mm) inland to c. the 500 mm isohyet, where they mix with the *Astrebla* grasslands. The soils throughout are deep, cracking clays with a high base status, black in the east and becoming progressively paler in colour to grey brown towards the west. Much of the grassland has been altered by grazing and large areas have been ploughed for cultivation. The grasslands are sometimes dotted with trees, *E. albens* and *E. orgadophila* in the east, *E. populnea* in the west. These grasslands differ from the more northerly *Dichanthium* grasslands because of the presence in the south of winter-growing species, chiefly species of *Danthonia* and *Stipa aristiglumis*, and many annual forbs (Blake, 1938, Moore, 1970, Tothill and Hacker, 1973).

D. sericeum is the charcteristic species throughout the suballiance. Co-dominants in various parts suggest the following associations: (a) In the north, *D. sericeum* occurs with *D. humilis*, with *Aristida latifolia* and *A. leptopoda* as subsidiary species; (b) In the same area *Heteropogon contortus* is a co-dominant with *D. sericeum* and sometimes locally dominant (Story, 1967); (c) In the wetter south-east *D. sericeum* forms an association with *Bothriochloa erianthioides* with *Danthonia* spp. and *Stipa aristiglumis* appearing in the winter assemblage; (d) In drier areas species of *Astrebla* occur, especially *A. lappacea*, and *Aristida latifolia* and *A. leptopoda* are occasional to common species (Pedley, 1974).

5.3.3. *Astrebla* spp. Alliance

5.3.3.1. General

Astrebla is endemic in Australia, with four species (Jozwik, 1969) which constitute the Mitchell Grass Downs occurring in the semi-arid zone between the 550 and 300 mm isohyets, with small outliers in the arid zone. The four species sometimes occur in the same stretch of grassland and for this reason the communities are described under one alliance. The habitat factors which separate the species in these mosaics is not always clear, but it is suggested that the species have somewhat different water requirements and/or tolerances to waterlogging.

The grasslands occur in areas receiving a summer rainfall and growth usually commences in early December, seed being produced in midsummer. The plants dry off during the autumn and remain dormant during the winter and early spring. In the most northerly portion of the alliance 70–80% of the annual rainfall falls during the four summer months (Dec. to Mar.), whereas in the most southerly areas only 40% of the annual total falls during the same period. Consequently winter rainfall becomes increasingly important from north to south and it governs the annual species which accompany the Mitchell Grasses, but has little effect on the summer-growing species.

The soils are invariably fine-grained grey, grey-brown or red-brown (in driest areas) clays, self-mulching at the surface, grading into heavy clays in the subsoil and cracking to depths of several cm to 1 m on drying. They are derived either from alluviums or from calcareous Cretaceous shales. The profiles are slightly alkaline throughout and lime commonly occurs at or near the surface (pH c. 7–8). Fertility levels are relatively high in the south-east, total P levels varying from 155–305 ppm (7 samples) in New South Wales but are relatively low in western Queensland (42–113 ppm P–6 samples). In the drier western parts the soil surface is often strewn or covered with gibbers.

The grasslands occur on flat or undulating country and topography plays some part not only in the distribution of the *Astrebla* species but also in determining microcommunities of other species within the *Astrebla* grasslands. When all species of *Astrebla* occur in the same area, *A. pectinata* occupies more elevated patches of soil, whereas *A. squarrosa* dominates in depressions. The other two species occupy flatter areas but are often segregated, each dominating patches of grassland, *A. elymoides* possibly occurring where the soil is loose in the surface and the rate of infiltration of water is high, whereas *A. lappacea* occurs on the more compact soils. Furthermore, in areas subject to prolonged waterlogging (possibly several weeks) species other than *Astrebla* may become locally dominant, e.g. the grass *Leptochloa digitata* (see 5.2.6, above), or Cyperaceae.

The grasslands sometimes continue as an understory to certain woodlands on fine-textured soils in better watered sites, especially along watercourses, e.g. *Eucalyptus microtheca, E. orgadophila, E. populnea, Acacia pendula, Atalaya hemiglauca* and *Casuarina cristata* woodlands. The grasslands are highly valued pastorally and are grazed by cattle in the north and by cattle and sheep in the south. Their productivity and carrying capacity have been studied in some detail (Everist, 1964 and references therein; Roe and Allen, 1945; Whalley and Davidson, 1969).

The alliance has been divided into four suballiances each identified by a species of *Astrebla*. The suballiances intergrade and have many associated species in common. The description that follow are based on Blake (1939), Beadle (1948), Davidson (1954), Perry and Christian (1954), Holland and Moore (1962), Perry and Lazarides (1964), Perry (1970), Pedley (1974).

5.3.3.2. *Astrebla squarrosa* Suballiance
Bull Mitchell Grass

A. squarrosa occupies depressions and badly drained areas in the grasslands, mainly in the tropical north, being most abundant on the Carpentaria Plains where the *Astrebla* grasslands adjoin the *Dichanthium* grasslands. It becomes decreasingly abundant towards the south and west, and westward of the 480 mm isohyet it is a rare species in depressions. The grassland attains a height of 1–2 m, the grass tussocks being 15–30 cm diameter and separated by distances of c. 15–30 cm in the wettest areas.

Associated species include *A. elymoides* and *Eulalia fulva* (both common) and the annuals *Iseilema convexum* and *I. calvum*, together with several species found typically in waterlogged depressions, viz., *Chenopodium auricomum, Cyperus gilesii, Eragrostis leptocarpa, Minuria integerrima, Neptunia gracilis* and *Sesbania benthamii*.

5.3.3.3. *Astrebla elymoides* Suballiance
Hoop Mitchell Grass

A. elymoides occurs mainly in the central part of the alliance, being rare in the wetter north and east and rare in the drier west, where it sometimes forms small monospecific stands in the *A. pectinata* grasslands. In the wetter east and north, mostly between the 400 and 500 mm isohyets, it occurs with *A. lappacea* on the more porous soils on flat country; it occurs also as a subsidiary species with *A. squarrosa* in depressions. The species forms tussocks and when the culms mature they curve over to form rounded hummocks.

In those areas where *A. elymoides* forms mosaics with *A. lappacea* the associated species are those which occur with *A. lappacea* (see below). In the drier areas where *A. elymoides* occurs in depressions in the *A. pectinata* grasslands, the assemblages are similar but with some additions or substitutions. On the Barkly Tableland, *Aristida latifolia*, *Eriachne nervosa*, *Panicum decompositum* and *P. whitei* are the main tall associates. There is a lower story of annual grasses or short-lived perennials including *Iseilema membranacea*, *I. vaginiflorum*, *Sporobolus actinocladus* and *S. mitchellii* and the forbs *Psoralea cinerea*, *Boerhavia diffusa*, with *Astrebla squarrosa*, *Cyperus retzii* and *Sesbania aculeata* in wet depressions. In southern Queensland, *A. elymoides* occurs in association with *Dichanthium sericeum*, and both species occur in mixed or pure stands with *Cyperus retzii*, *C. difformis* and *Astrebla squarrosa*. *Dichanthium sericeum* is a constant component and may become an apparent dominant after very heavy summer rains. *Iseilema* spp. are the more abundant summer annual grasses and the legumes mentioned in the next section also occur. Since winter rainfall is significant in these areas heavy growths of annuals occur, especially *Atriplex muelleri* and Compositae.

5.3.3.4. *Astrebla lappaceae* Suballiance
Curly Mitchell Grass

A. lappaceae is the most common of the Mitchell grasses. It reaches a height of c. 1 m and has an upright, tufted habit. In the best stands tussock are c. 15–30 cm diam., spaced at intervals of c. 15–40 cm (Fig. 20, 10). Details of an individual tussock are described by Everist (1964). The plants have short, much branched rhizomes producing large, scaly buds which normally develop into flowering stems. A second set of buds occurs in the leaf-axils of the aerial stems and these produce tufts of leaves at the nodes. After small showers of rain the buds on the aerial stems may grow, but not the buds on the rhizomes. The root system extends almost vertically to a depth of c. 80 cm without branching, though roots near the surface may produce fine rootlets after rain. Below a depth of c. 80 cm the main roots branch and extend laterally

Fig. 20.10. *Astrebla lappacea* grassland. The shrub is *Nitraria billardieri*. North of Brewarrina, N.S.W.

so that the roots of neighbouring tussocks overlap between a depth of 80–120 cm.

A. lappacea is the dominant and most permanent species in the communities, which usually contain c. 100 species, many of which are perennial. Constants include the summer-growing *Eulalia fulva, Panicum decompositum* and *Thellungia advena*. Other important constituents are *Dichanthium tenuiculum* and *D. fecundum* (north) and *D. sericeum* (south), *Aristida latifolia* (north) and *A. leptopoda* (mainly south) and *Bothriochloa ewartiana* (north). *Eragrostis setifolia* occurs over most of the area and is sometimes locally dominant, especially on shallow soils developed where limestones lie near the surface. Several small perennial shrubs and herbs occur as occasional to rare species, including *Boerhavia diffusa, Hibiscus brachysiphonius, Malvastrum americanum* and legumes belonging to the genera *Alysicarpus, Crotalaria, Desmodium, Glycine, Neptunia, Psoralea, Rhynchosia* and *Vigna*. Annual species are numerous. In the north where the rain falls mainly in the summer, winter-growing annuals are rare but these become increasingly important from the Tropic southward. Annual grasses are most abundant after summer rains (except in degenerate areas) and forbs after winter rains, the families Chenopodiaceae, Cruciferae and Compositae becoming increasingly abundant from north to south. The main annual summer-growing grasses are *Iseilema membranaceum* and *I. vaginiflorum* which commonly occupy all bare space between perennial tussocks. Less abundant are *Dactyloctenium radulans, Eriochloa crebra,* and *Panicum buncei*. Summer growing annual forbs are rare or absent but increase in abundance in degenerate swards (see below). Winter-growing forbs include *Atriplex muelleri, A. spongiosa* (south), *Calotis hispidula, Daucus glochidiatus, Helipterum corymbiflorum,* * *Medicago polymorpha* (south) and *Zygophyllum ammophilum*.

5.3.3.5. *Astrebla pectinata* Suballiance
Barley Mitchell Grass

A. pectinata occurs between the 200 and c. 600 mm isohyets but is dominant only south and west of the 450 mm isohyet, becoming increasingly important as a dominant with decreasing rainfall. A. lappacea occurs occasionally in the drier stands and, rarely, A. elymoides and A. squarrosa. The grasslands are best developed on the Barkly Tableland, with smaller areas to the south, extending to north-western N.S.W. They occur on fine textured soils which are grey-brown in wetter areas becoming yellow-brown or red-brown in the drier parts. In the drier areas the soil surface is usually strewn with gibbers and is often gilgaied. In the driest areas *Astrebla* is confined to gilgais, often within the *Atriplex vesicaria* Alliance.

In the wettest areas, A. pectinata forms tussocks 20–30 cm diam., which are spaced at intervals of 50 cm to 1 m; the culms reach a height of c. 80 cm. In drier areas, the tussocks are usually smaller but may be similarly spaced, especially in gilgais. In the northern areas, the main associated tall species are *Aristida latifolia, Chrysopogon fallax* and *Dichanthium fecundum*, with subsidiary *Panicum decompositum* and *Spathia nervosa*, sometimes with *Themeda avenacea* in depressions. Summer annuals are mainly *Iseilema* spp., with subsidiary *Brachyachne convergens, Dactyloctenium radulans, Eragrostis japonica* and *Sorghum* spp. Small legumes, *Neptunia* spp. and *Rhynchosia minima* are present.

In the more southerly, drier stands A. pectinata occurs either in pure stands or in association with *Eragrostis xerophila*. *Panicum decompositum* is often present and occasionally *Themeda avenacea*. The stands are usually strongly influenced by the southern Chenopodiaceae flora, especially *Bassia* spp. In these communities the summer annuals include *Dactyloctenium radulans, Iseilema membranacea* and *Sporobolus actinocladus*. Winter rains produce heavy growths of *Atriplex spongiosa, A. conduplicata* and *A. inflata*, and Cruciferae and Compositae, especially *Helipterum* spp.

5.3.3.6. Effects of Grazing on the *Astrebla* Grasslands

Light grazing has little effect on the grassland since stock eat the softer associated species in preference to the *Astrebla* tussocks. However, if the latter are removed by heavy grazing and/or a succession of very dry years the perennials are removed and annuals dominate. In such cases summer rains produce an annual cover of grasses and forbs in which *Dactyloctenium radulans* is often conspicuous, with *Portulaca oleracea, Tribulus terrestris* and *Boerhavia diffusa*. The last two have underground perennating organs.

Winter rains produce an assemblage containing various mixtures of *Atriplex muelleri, A. spongiosa, A. inflata, Calotis hispidula, Hibiscus trionum* and *Helipterum* spp, and various crucifers (south). Perennial species of *Bassia* sometimes become prominent and locally dominant in these degenerate swards, especially *B. quinquecuspis, B. tetracuspis* and *B. bicornis* (south). Two woody species are invading parts of the grassland, even in areas where the *Astrebla* tussocks are in good condition. *Acacia farnesiana*, which is usually regarded as an occasional component of the grassland though the species in possibly introduced into Australia, is becoming a weed in parts of western New South Wales. *Nitraria billardieri*, a hummock-forming plant usually in saline and subsaline areas has invaded the grasslands in some places. The species occurs mainly in the south and has become a weed in parts of the saltbush country (see Chap. 19).

5.3.4. *Eragrostis xerophila* Alliance

This species dominates tussock grasslands c. 30 cm high on clays in the arid zone, in western Queensland, through Central Australia and westward to the Indian Ocean, adjoining the 80 mile Beach and extending south to Roebourne. The grasslands occur on narrow belts of clay, rarely exceeding 1 km wide, along some watercourses in the interior. On the west coast the clay belt lies close to the ocean and in places reaches a width of a few km. It is separated from the ocean either by a saline grassland dominated by *Xerochloa* (see 5.1.3, above) or by low sand dunes. The *Eragrostis* grassland is commonly flanked by sandy soils which support grassland or savannah in which species of *Triodia* or *Plectrachne* form the bulk of the sward.

In all areas *E. xerophila* is often associated with *E. setifolia* (see next section), and in Central Australia with *Astrebla pectinata* and the lesser species found with this grass (see 5.3.3.5, above). On the western coastal plain the following associated species are found: *Chloris ruderalis, Dichanthium humilis, Enneapogon planifolius, Eragrostis lacunaria, Panicum decompositum, Sporobolus actinocladus* and *Triraphis mollis*. In damper patches, *Eulalia fulva* and *Bothriochloa decipiens* are dominant, and in saline areas *Sporobolus virginicus* and *Xerochloa barbata* become prominent in an ecotone with the adjacent saline grassland. Trees of *Melaleuca leucadendron* or *M. lasiandra* may occur in the grassland (Burbidge, 1944).

5.3.5. *Eragrostis setifolia* Alliance
Neverfail

In addition to its common occurrence in some woody communities, *E. setifolia* forms patches of grassland in depressions in these woody communities, especially the *Acacia aneura* alliance and the halophytic shrublands. It occurs also as a dominant of grassland on shallow soils on limestone in some of the *Astrebla* grasslands. The species is abundant throughout the arid zone and extends on to the Nullarbor Plain where it forms small patches of grassland in some depressions. In summer rainfall areas the main associates are *Astrebla pectinata, Chloris truncata, Dichanthium sericeum, Panicum decompositum, Sporobolus actinocladus, Themeda australis* and forbs or small shrubs, especially *Psoralea cinerea* and *Bassia* spp.; *Marsilea drummondii* is present in some wetter sites. Annual species appear seasonally, especially Compositae. On the Nullabor Plain the grass occurs usually in pure stands with species of *Helipterum* appearing in the cooler months, and after exceptionally good rains *Stipa* spp. (Johnson and Baird, 1970).

5.3.6. *Eragrostis dielsii* Alliance
Mulka Grass

This species is widespread and of occasional to rare occurrence in several communities of the arid zone. It becomes locally dominant on compacted sandy clays around claypans, usually immediately above the waterlogged zone of the pan. The species, which is usually prostrate, has the capacity to colonize compact, sometimes eroded surfaces, and it commonly occurs without associates. Though potentially a perennial, the species usually behaves as an annual in the sites mentioned above.

5.3.7. The *Stipa* Grasslands

5.3.7.1. General

The species of *Stipa* used to identify the alliances are difficult to separate taxonomically, and the genus is presently under revision. Consequently, the specific names applied here must be regarded as tentative.

The grasslands occur in the winter rainfall zone. Those in the wetter east and south occur on clays. In drier areas the soils are usually sandy at the surface and commonly contain lime in the subsoil. In the arid zone the grasslands are usually ephemeral; the *Stipa* tussocks, though potentially perennial, are killed by drought and are replaced by seedlings after the first favourable rain. One or more species of *Danthonia* are commonly associated with *Stipa* and the proportion of *Stipa* to *Danthonia* in the grassland is often determined by the intensity of grazing.

The *Stipa* spp. dealt with in this section all provide good pasturage. The species, however, have the disadvantage of being troublesome to stock when the grain matures. The callus at the base of the lemma is sharp-pointed and may enter the eyes of animals or bore their way through the hide into the flesh.

5.3.7.2. *Stipa aristiglumis* Alliance
Plains Grass

S. aristiglumis forms small patches of tall grassland, mostly to c. 1½ m tall, on Grey and Brown Clays or Black Earths, all of which have a high base status (Fig. 20.11). The patches are often temporarily waterlogged, possibly for a few weeks, which apparently prevents the growth of trees. Patches are commonly circular in outline or elongated beside watercourses, or they represent the bottoms of old lake beds, as at Breeza, where most of the grassland has been removed and the soils cultivated. The grasslands are found mainly between the 400 and 520 mm isohyets in the winter rainfall zone south of Narrabri. To the north, the *Astrebla* grassland occur on similar soils into which the *Stipa* has scarcely infiltrated.

S. aristiglumis extends to the Darling Downs where it forms the winter-growing component of the *Dichanthium sericeum* grassland. It occurs also in Victoria where it is now an occasional to rare species with other species of *Stipa* and *Themeda australis* (see next section). It is commonly the dominant understory species in some *Eucalyptus* woodlands, especially *E. albens*.

S. aristiglumis usually occurs in pure stands. Native species are almost entirely excluded during the autumn-winter growing period. Summer rains may produce a small growth of *Dichanthium sericeum* with subsidiary *Eriochloa pseudoacrotricha*, *Panicum decompositum* and *P. whitei*, which become decreasingly important towards the south. Forbs and dwarf shrubs are likewise of rare occurrence, except in the drier stands where the *Stipa* tussocks are more distant. These include *Atriplex semibaccata*, *Alternanthera triandra*, *Psoralea patens*, *Rumex*

Fig. 20.11. *Stipa aristiglumis* grassland. Trees behind are *Eucalyptus melliodora*. Breeza, N.S.W.

brownii and *Vittadinia triloba,* which are permanent members of drier stands. In the most northerly stands, summer annual grasses may appear, especially *Dactyloctenium radulans, Iseilema membanaceum* and *Sporobolus caroli.*

The subsidiary species increase in abundance when the grasslands are grazed and trampled by stock and, as well, exotic species enter the stands, the most common of which are * *Medicago* spp., * *Lolium temulentum,* and many alien weeds, such as species of * *Argemone,* * *Carthamus,* * *Centaurea,* * *Sisymbrium* and * *Silybum,* and some native species, especially of *Helipterum.*

5.3.7.3. *Stipa scabra* –
S. *bigeniculata* Alliance
Spear Grasses

This alliance occurs on flat or undulating treeless plains on the southern Tablelands of New South Wales and a similar grassland occurs on the basalt plains in Victoria, extending from near Melbourne almost to the South Australian border, and smaller areas appear to occur as far west as Lake Alexandrina in South Australia.

The grasslands occur under a mean annual rainfall of 400–625 mm, with a slight winter predominance, on clays derived mainly from basalts and shales. At higher altitudes on fine-textured soils the *Stipa* grasslands are replaced by *Poa* grasslands (Costin, 1954). Costin recognizes 11 associations in New South Wales, of which the *S. scabra* association is the most widespread and it becomes admixed with or is replaced by *S. variabilis* in drier areas. Patton (1935) and Willis (1964) record the following species of *Stipa* on the Victorian basalt plains: *S. setacea, S. semibarbata, S. aristiglumis, S. blackii, S. variabilis, S. compacta* and *S. eremophila.* In all areas, one or more species of *Danthonia* occur in association with species of *Stipa.* Willis records 9 species of *Danthonia.* Of these *D. caespitosa* appears to be the most abundant. The summer growing *Themeda australis* appears to be a common associate in wetter climates or in depressions in drier areas.

The subsidiary species are too numerous to be listed here. Costin records 120 species and a larger flora is recorded by Willis (1964) for Victoria. Other grass genera include *Agropyron, Agrostis, Chloris, Enneapogon, Eragrostis* and *Panicum.* The Compositae are well represented, especially *Brachycome, Calocephalus, Craspedia, Helichrysum, Leptorhynchos* and *Podolepsis.*

5.3.7.4. *Stipa falcata* –
Danthonia caespitosa
Disclimax Grassland

The grassland occurs at the eastern end of the Riverine Plain in New South Wales in an area with a mean annual rainfall of c. 350–400 mm, with a winter predominance but a significant summer component. It occurs on soils which exhibit the puff-shelf-depression (gilgai) complex, and on Red-brown Earths.

The composition of the grassland over its whole area is variable as a result of heavy grazing in the past, seasonal incidence and intensity of rain, and present grazing pressures. The component grasses have soil preferences. *Danthonia caespitosa* appears to be dominant on the Red Brown Earths, sometimes with *Stipa falcata* as a co-dominant on the more sandy soils. The latter may also occur on the clayey shelves after heavy rains. *Danthonia* is dominant on the shelves and in the depressions; on the puffs *Stipa aristiglumis* occurs and was possibly abundant in these sites in the original shrub woodland and/or saltbush communities. Williams (1955 a, b) has studied the changes in specific composition in the various segments of the community in relation to seasonal rainfall and waterlogging.

Heavy grazing of the mosaics of grasslands leads to a decrease in *Stipa falcata,* and *S. aristiglumis,* with an increase in *Danthonia.* On the more clayey sites *Chloris truncata* commonly assumes complete temporary dominance. Annual species likewise become very abundant. *Hordeum leporinum* commonly forms continuous swards after autumn and winter rains and Compositae increase in abundance, especially *Calocephalus sonderi.*

5.3.7.5. Other Species of *Stipa*

i. *Stipa variabilis*

This species is the common Spear Grass in western New South Wales and is possibly conspecific with *S. scabra.* It forms the dominant species in the *Casuarina cristata* and *Heterodendrum oleifolium* woodlands, as well as some of the *Eucalyptus* woodlands. As a natural grassland it occupies open spaces in woodlands of *Casuarina* and *Heterodendrum* (Chap. 18), both of which commonly form groves as a result of root suckering. The species occurs mainly on soils with a sandy surface horizon. Following white settlement and the attendant destruction of certain plant communities, especially the southern portion of the *Acacia aneura* alliance and the *Atriplex – Maireana* shrublands, *S. variabilis* has apparently extended its range or has become more abundant.

S. variabilis forms tussocks up to 8 cm diameter and reaches a height of 80 cm. In the wetter areas the tussocks are perennial and growth of established tussocks usually commences after winter rains, the culms reaching maximum size and flowering in spring. Following dry spells, when the tussocks are dormant, shoots may appear at any time of the year following rains, even during the summer months (Biddiscombe et al., 1954).

The associated species vary considerably, possibly since some of the stands are natural, others induced. In the more southerly stands, which are possibly natural grasslands, there are relatively few associates which occur in abundance. This is possibly due to the high density of the *Stipa* tussocks which may be separated by distances equal to 3–4 times their diameter. Occasional species include *Eragrostis dielsii*, *E. setifolia* (depressions), *Danthonia caespitosa* and *Eragrostis lacunaria*. Semi-herbaceous shrubs and herbs include *Bassia obliquicuspis*, *B. uniflora*, *Daucus glochidiatus*, *Goodenia glauca* and *Sida corrugata*, and many of the species listed in the next paragraph.

In the induced grasslands the grass tussocks are more widely spaced and the associated flora is rich. The ephemeral assemblage varies with the incidence of the first effective rain which promotes growth. Late summer and autumn rains produce an assemblage dominated by Chenopodiaceae, especially *Atriplex conduplicata, A. inflata, Bassia lanicuspis, B. obliquicuspis, B. paradoxa, Maireana lobiflora, m. tomentosa, Salsola kali*, and the grasses *Aristida contorta, Enneapogon avenaceus* and *Eragrostis lacunaria*. Late autumn and winter rains produce a colourful assemblage of forbs, probably some 60 species, which are largely Compositae, e.g. *Helipterum corymbiflorum, H. floribundum, H. moschatum, Craspedia pleiocephala*, or Cruciferae, e.g. *Blennodia lasiocarpa, Lepidium papillosum*, or other forbs, including *Erodium cygnorum, Tetragonia expansa, Trichinium alopecuroideum* and *Zygophyllum ammophilum* (Pidgeon and Ashby, 1940). Heavy grazing removes the *Stipa*, and the grassland is converted to an ephemeral herbfield.

ii. *Stipa nitida*

S. nitida is the common spear grass in arid to semi-arid South Australia, extending to areas with as little as 200 mm of rain per annum. It occurs also in New South Wales, where it is occasionally mixed with *S. variabilis*, and on the Nullarbor Plain. The grass forms the main understory species chiefly in the low woodlands on sandy soils dominated by *Acacia aneura* or *Casuarina – Heterodendrum*.

S. nitida reaches a height of c. 50 cm and forms tussocks 5–8 cm diam.; these are usually widely spaced with bare ground in between occupied by forbs which may be abundant after rain. The main species are the winter-growing annuals: *Brachycome pachyptera, Calandrinia volubilis, Calotis hispidula, Daucus glochidiatus, Erodium cygnorum, Helipterum moschatum, Tetragonia eremaea* and *Zygophyllum apiculatum*. The short lived perennial, *Bassia patenticuspis*, is the most permanent species in the community but apparently does not compete for moisture with *S. nitida* which has a deeper root system (Osborn et al., 1931).

S. nitida is recorded on the Nullarbor Plain after suitable rains. Johnson and Baird (1970) record extensive areas of grassland dominated by this species on flat terrain and in depressions. The plants reach a height of 10–90 cm, the larger ones being recorded in the depressions where *S.* aff. *eremophila* may also occur (see next section). *S. nitida* is usually associated with *Danthonia caespitosa* and an undescribed species of *Stipa*. *Bassia* spp. and other forbs, chiefly Compositae, are subsidiary.

iii. *Stipa eremophila* (= *S. fusca*)

S. eremophila occurs mainly in the southern part of the arid zone, extending from western New South Wales to the western part of the Nullarbor Plain. It occurs as an occasional species in the mallee (Chap. 14). In open areas between stands of mallee, *S. eremophila* is the dominant in a grassland c. 60 cm high, in which it is often co-dominant with *Danthonia caespitosa* (Crocker, 1946). As a grassland-dominant it is found also on the Nullarbor Plain where it has possibly increased in abundance following the degeneration or removal of the *Atriplex – Maireana* shrublands as a result of grazing and drought. Johnson and Baird (1970), record *S.*

eremophila from the deeper parts of depressions on theNullarbor Plain, sometimes in association with *S. nitida* (see preceding section).

5.4. Grasslands and Savannahs of the Arid Zone

5.4.1. General

The grasslands are of two types, the perennial hummock grasslands with dominants of *Plectrachne, Triodia* or *Zygochloa,* and the ephemeral grasslands which appear only after rains, the species being annuals or short-lived perennials. The hummock grasslands have two distinctive forms differing in the gross morphology of the hummock. In *Plectrachne* and *Triodia* the hummock appears as a mass of pungent-pointed leaves attached to concealed short aerial stems, whereas in *Zygochloa* the hummocks are shrub-like and composed of tangled cane-like stems.

Within their total area the grasslands are distributed as follows: In the driest unstable areas (moving sand or truncated profiles) ephemeral grasslands occur. On sand dunes in the climatically drier areas, as in the Simpson Desert, *Zygochloa* is dominant, sometimes with *Triodia* on flat sandy troughs between the dunes. In the less dry segment of the total area *Plectrachne* and *Triodia* dominate.

The grasslands are bounded in the north chiefly by the *Eucalyptus brevifolia* and *E. dichromophloia* alliances, and in the south mainly by the *Acacia aneura* alliance (Chap. 18). The communities, however, are not sharply delimited, but woodlands of *Eucalyptus* or *Acacia* become progressively less dense with increasing aridity, so that woodland merges into savannah and savannah merges into grassland.

The savannahs contain species of *Plectrachne* or *Triodia* and a woody component. The number of woody species is high, greatly enriching the total flora, which amounts to at least 300 spp. (combined lists of Burbidge, 1944, 1945; Chippendale, 1963; Winkworth, 1967). Of these, about one-third are shrubs and trees, one-third are perennial herbs and subshrubs and one-third are annuals.

Many of the woody species, especially the taller ones, occur in communities in slightly wetter climates, and some of them even have forest affinities, e.g. *Bauhinia cunninghamii,* *Clerodendron tomentosum, Ficus platypoda,* and *Owenia reticulata.* The woody plants occur either as scattered individuals without an apparent species-pattern, or in clumps. The latter are commonly monospecific and usually are species of *Acacia.* Such clumps possibly develop after fire, which cracks the hard testas of the seeds which have accumulated over a small area under or near a plant. Other clumps are composed of several species seemingly randomly selected. Others again appear to be fragments of woodland which have persisted in a stunted form in a more favourable site. This is suggested by the presence of both a tree layer and a shrub layer above the hummock grasses. The main association-fragments are *Eucalyptus brevifolia* (mainly rocky areas in the north), *E. dichromophloia* (mainly watercourses in rocky terrain in the north-west), *Acacia linophylla, A. coriacea* and *Grevillea eriostachya* (sand in north-west), *Eucalyptus odontocarpa* and *E. pachyphylla* (rocky areas and sand-ridges in the north), *Acacia aneura* (on sand and rocky ridges in the south). Some of the other more abundant species are given under the alliances. On the other hand, many species are confined to the savannahs, e.g. *Casuarina decaisneana, Eucalyptus gongylocarpa* and a few more eucalypts which are sometimes abundant enough to form mallee communities (Chap. 14).

Many hemi-parasites are recorded in the savannahs. Mistletoes (*Amyema* and *Lysiana* spp.) are not uncommon, but usually only on creek banks and flood plains where birds are present (Chippendale, 1963). Santalaceae are represented by *Anthobolus leptomerioides, Santalum lanceolatum* and *S. acuminatum.* The last produces fruits with an edible pericarp and seed, which were harvested by the aboriginals.

The grasslands and savannahs are grazed by domestic stock, but provide poor pasturage. The leaves of *Plectrachne* and *Triodia* are highly xeromorphic and pungent-pointed (Burbidge, 1946b). However, the inflorescences and grain are both palatable and nutritious, but are not commonly eaten by herbivores, since the associated, softer species are available when the hummock grasses are in flower or fruit. Young leaves of *Plectrachne* and *Triodia* are more acceptable to stock and fires are often lit, in addition to those caused by lightning, to remove spinescent hummocks and to promote new growth of both the hummock grasses and the subsidiary species. The grass hummocks burn readily because of their highly resinous leaves, especially *T. pungens,* the resin of which was collected by aboriginals and used as a cement. Fires usually kill most of the shrubs and many or some of the grass hummocks. Regeneration is rapid after rain, and species of *Acacia* germinate freely, commonly producing thickets.

5.4.2. The *Plectrachne* and *Triodia* Alliances "Spinifex Country"

Plectrachne and *Triodia* (both endemic) are closely related, the former being distinguished by having 3-awned lemmas, whereas those of *Triodia* are 3-toothed. Opinions on their tribal affinities have varied in the past but anatomical features coupled with floral morphology favour affinities with the Eragrostideae (Jacobs, 1971). The grasslands are usually referred to as "spinifex country". They occur under a mean annual rainfall of c. 200 to 300 mm. The rainfall is highly erratic, annual totals varying from about one-fifth to about twice the mean. The number of effective falls (estimated at 10 mm) to benefit the hummock grasslands is probably not less than 2 per annum. Such small falls do not establish the annual species which may appear in abundance once in 3 years, though every year some annuals, possibly stunted, appear between the hummocks. Temperatures are extreme, maximum soil-surface temperatures being of the order of 60°C and minima c. −6°C. A diurnal range of 50°C is possible during the winter.

The soils are usually very sandy or skeletal and are developed on flat, undulating or hilly terrain. Over large areas of Western Australia spinifex occurs on a rugged topography (Fig. 20.12) which has been moulded during a wetter climate, as evidenced particularly by deep ravines, a few of them now containing permanent water in rock-holes. On the other hand, large tracts of spinifex occur over flat sandplains or undulating dune-fields, moulded in a previous and not-so-distant drier climate and now stabilized by spinifex. Soil texture plays an important role in delimiting the alliances (see below).

Soil fertility levels measured by agricultural standards are very low. The nutrient in lowest supply is nitrogen. Winkworth (1967) estimated a maximum total soil-N of 0.022%. Phosphorus levels are variable, Winkworth's analyses indicating 80–150 ppm HCl-soluble P. Some much lower values have been recorded by the writer for *Triodia pungens* (range 28–112 for 23 samples). Phosphate analyses of soils from a mosaic of communities of *Acacia aneura* (in slight depressions) and *Triodia basedowii* (on elevated areas) indicated that no significant differences

Fig. 20.12. General view of *Triodia* grassland. c. 60 km south of Roebourne, W.A.

occurred between the two communities (*Triodia* 63 ± 8; *Acacia* 62 ± 3 ppm P). These few data support the view that water rather than nutrients is the controlling factor in plant distribution on these sandy soils.

The four main alliances described below are distributed as follows: *Plectrachne schinzii*, on sand, is dominant in the north-west segment of the arid region; *Triodia pungens*, on sand or stony areas, stretches across the arid region mainly north of the Tropic; *T. basedowii*, usually on sand, occurs mainly south of the Tropic from south-west Queensland to Western Australia; *T. irritans*, usually in rocky situations occurs mainly on ranges in South Australia.

5.4.2.1. *Plectrachne schinzii* Alliance
Feather-top Spinifex

P. schinzii forms hummocks c. 30 cm high and up to 1 m diameter, and provides a ground cover of c. 25%. This species is the most palatable of the hummock grasses (Nunn and Suijendorp, 1954). As a grassland or savannah *P. schinzii* covers vast areas north of the Tropic from near the coast in north-west Western Australia extending through the Great Sandy Desert into Central Australia (Burbidge, 1944, 1945; Winkworth, 1967). Continuous tracts of *Plectrachne* are common, but sometimes patches of *Triodia pungens* occur within the *Plectrachne*, possibly in areas where the sand is more compact. Conversely *P. schinzii* may occur on patches of loose sand within the *Triodia pungens* alliance. In Central Australia *P. schinzii* is sometimes associated with *Triodia basedowii*. Another variation reported by Burbidge is the occurrence of *T. lanigera* (confined to western W.A.) forming colonies where the sand is mixed with numerous quartz pebbles.

Associated species in the grasslands are commonly *Eragrostis eriopoda* (almost constant) and *Danthonia bipartita* (west), with subsidiary species of *Amphipogon, Aristida, Cymbopogon, Enneapogon, Eriachne, Neurachne* and *Sorghum*. Forbs are common after rain, especially *Calandrinia* spp. and Compositae. In the west, small shrubs, mostly equalling the height of the *Plectrachne* or slightly taller, sometimes occur: *Abutilon walcottii, Adriana tomentosa, Calytrix interstans, Cyanostegia bunnyana, Halgania littoralis, Newcastlia cladotricha* and others.

Shrubs, and occasionally small trees, are commonly dotted throughout the grasslands, either

Fig. 20.13. *Plectrachne schinzii* grassland with low shrubs (scarcely exceeding the grass) of *Acacia translucens*. c. 32 km east of Port Hedland, W.A.

Fig. 20.14. *Owenia reticulata* as an emergent in a depression in hummock grassland. The understory shrubs are *Acacia holosericea* and *A*. aff. *stenophylla*. c. 240 km south-west of Broome, W.A.

as individuals but usually in clumps. Some are constant throughout the area, some are localized. Frequently the clumps contain a single species, especially of *Acacia,* which is probably the result of fire. A common shrub associate in the west, which scarcely overtops the *Plectrachne* is *Acacia translucens* (Fig. 20.13). Other common taller species in the west are *A. pachycarpa* (commonly in thickets 1–2 m high), *A. coriacea, A. holosericea, A. inaequilatera, A. stipuligera, A. trachycarpa, A. tumida, Bauhinia cunninghamii, Cassia glutinosa, Clerodendron tomentosa, Duboisia hopwoodii* and the stunted or mallee-like eucalypts, *E. dichromophloia* and *E. zygophylla.* Tall shrubs or small trees occasionally encountered are *Alalaya hemiglauca, Dolichandrone heterophylla, Grevillea striata* and *Owenia reticulata* (Fig. 20.14). The species in Central Australia (Winkworth, 1967) are mainly *Acacia aneura, Codonocarpus cotonifolius, Eucalyptus papuana* and *E. terminalis* (trees) and numerous shrubs including *Acacia ancistrocarpa, A. cowleana, A. ligulata, A. murrayana, Eucalyptus odontocarpa, E. setosa,* species of *Cassia, Dodonaea, Eremophila, Hakea lorea, H. macrocarpa, Santalum lanceolatum* and others.

5.4.2.2. *Triodia pungens* Alliance
Gummy or Soft Spinifex

This species occurs across the arid interior north of the Tropic from western Queensland to the Indian Ocean. In the north the grasslands are replaced by woodlands chiefly of *Eucalyptus brevifolia* and this species and others occur abundantly in the grasslands, mostly in stunted form, to produce savannahs (Fig. 20.15). Consequently grassland merges into savannah and savannah into woodland. In the north-west *T. pungens* is replaced by *Plectrachne schinzii* on loose sands, though the two alliances sometimes mix, the *Triodia* occurring on more compact sandy areas. In the south, *T. pungens* is replaced mainly by *T. basedowii* or by *Zygochloa* on sand in the east. The grassland communities have been described by Blake (1938), Burbidge (1944, 1945), Chippendale (1958, 1963), Perry and Lazarides (1962), Lazarides (1970).

T. pungens forms spiny hummocks up to c. 150 cm diameter and 30–40 cm high, the inflorescences rising to 90 cm. Ring formation sometimes occurs and stolons, which establish new tussocks, are often formed. The species occurs over a wide range of habitats, classified

Fig. 20.15. *Triodia pungens – Eucalyptus brevifolia* savannah on a laterite ridge. c. 110 km north of Camooweal, Qld.

under three broad headings: (a) Sands, mostly in level or undulating country. The soil surface is absorbent, there being little or no runoff. The *Triodia* hummocks reach their maximum size in these areas and commonly cover 60–80% of the soil surface. (b) Rocky areas with a broken surface either on plateaux or over an undulating landscape. Rock type plays little or no part in the distribution of the species; sandstones, quartzites, limestones or truncated laterite profiles support the same species. The tussocks in these areas occur in pockets of soil or in crevices which are sometimes well watered because they receive irrigation from runoff. The tussocks are usually large but scattered. (c) Sandy areas with a small clay content, occurring mostly on undulating country but sometimes on flats. The small amount of clay is redistributed by water to form a film over the soil surface, reducing the permeability of the surface to water. The clay is possibly derived from ancient laterite profiles and the surfaces are commonly strewn with wind-polished pebbles and stones. *Triodia* hummocks in these sites are small, often scattered but sometimes aggregated in depressed areas which receive local run-off water.

Associated perennial species vary in the habitats mentioned above. In sandy areas, *Eragrostis eriopoda* is often common and *Danthonia bipartita* an occasional associate, with annual species of *Aristida*, *Enneapogon* and *Eriachne*, the assemblage in the west being similar to that in the *Plectrachne schinzii* alliance. On broken rocky areas *Cymbopogon bombycinus* and *Themeda australis* sometimes occur, and small shrubs are often abundant, e.g. *Hibiscus*, *Indigofera* and *Trichinium*. On soils with a clay-covered surface perennial associates are usually absent. In all habitats annual species occur in the spaces between hummocks and a variety of plants occurs on porous surfaces, e.g. *Aristida* spp., *Dactyloctenium radulans*, *Euphorbia australis*, *Pterigeron odoratus*. On less pervious sandy clays annuals are rare, e.g., Blake (1938) records only *Eriochlamys behrii*.

Woody species in the savannahs vary across

the continent. In the west the assemblage is the same as that found in the *Plectrachne schinzii* alliance. In Central Australia the main species are *Eucalyptus terminalis, E. setosa* and *E. pachyphylla, Acacia dictylophleba, A. lysiophloia, A. stipuligera, Hakea lorea* and *H. macrocarpa*. In Queensland: *Acacia costinervis, Hakea acacioides, H. lorea, Eucalyptus terminalis, Sterculia australis* and *Terminalia aridicola*.

5.4.2.3. *Triodia basedowii* Alliance
Hard Spinifex

T. basedowii dominates the hummock grassland in the south, mostly between lat. 22° and 27°S., and in this areas it commonly forms mosaics with the *Acacia aneura* alliance in which the grass sometimes occurs as an understory dominant. In the north, *T. basedowii* often adjoins both the *Plectrachne schinzii* and the *T. pungens* alliances. The different alliances are usually sharply delimited, mixtures of the dominant species being uncommon.

T. basedowii forms hummocks up to c. 1 m diam. and 25 cm high and ring formation sometimes occurs. It is dominant over large areas of sand plain and dune fields and sometimes occurs in rocky habitats. It is most abundant as a grassland in the driest areas, especially in the Victoria and parts of the Simpson Deserts, and in the latter it occupies some flat, interdunal areas between loose sand-ridges on which *Zygochloa* is dominant (see 5.4.3, below). In these drier areas emergent shrubs and trees are usually absent.

The quantitive data of Winkworth (1967) for four grassland stands in Central Australia show that living hummocks of *T. basdowii* provide a soil cover of 20–24%, with an additional 1–15% being provided by dead culms. A few associated perennial species occur, of which *Aristida browniana* and *Eragrostis eriopoda* are the most abundant; these together provide less than 1% cover, and annuals provide only 0.2 to 1.2%. Large numbers of annual species are reported, the southern assemblage consisting of *Calandrinia balonensis, C. polyandra, Cleome viscosa, Erodium crinitum, Euphorbia drummondii, E. eremophila, Haloragis gossei, Helipterum* spp., *Lepidium rotundum, Podolepis canescens, Ptychosema trifoliatum, Senecio gre-*

Fig. 20.16. *Triodia basedowii* on red sand. Shrubs are mainly *Hakea ivoryi*. c. 20 km west of Windorah, southwestern Qld.

gorii, Trachymene glaucifolia and *Waitzia acuminata*.

In savannahs the most common tall shrub emergent is *Acacia aneura*. A few other species are found locally, these usually forming identifiable alliances described elsewhere, especially *Casuarina decaisneana* and *Eucalyptus gongylocarpa* (Chap. 14). The total number of species is less than in the northern alliances, and most belong to *Acacia* (*A. adsurgens, A. brachystachya, A. estrophiolata, A. kempeana, A. leursenii*). *Hakea intermedia* occurs in Central Australia and *H. ivoryi* in Queensland (Fig. 20.16). *Cassia* spp., *Eremophila* spp., *Grevillea juncifolia* and *G. striata* occur from east to west.

5.4.2.4. *Triodia irritans* var. *irritans* Alliance
Porcupine Grass

T. irritans, with three variable varieties, and the closely related *T. scariosa* are found south of lat. 26½°S (Burbidge, 1953). Most of the taxa occur as understory species in woodlands or in the mallee, and often they are referred to as *T. irritans* because of the difficulty in separating the varieties and forms.

T. irritans var. *irritans* is confined to rocky ranges and hills where it roots in pockets of soil or in rock-ledges. It occurs mainly in South Australia from Elbow Hill on Eyre's Peninsula, eastward to the Barrier Range in western New South Wales and northward on ranges into southern Northern Territory (Mt. Olga and Gosses Range). Possibly the largest tract of grassland occurs in the Musgrave Ranges in South Australia (Fig. 20.17). The various localities are isolated from each other by distances up to c. 100 km and the distribution of the taxon is difficult to explain.

The grass forms hummock up to 120 cm diam. and 60 cm high, the inflorescences rising to c. 90 cm. It does not form rings. Associated herbaceous species are commonly absent, since in these exposed situations rock dominates most of the area and the *Triodia* hummocks spread on to the rock surface. *Cymbopogon bombycinus* occurs in some areas and the only other herbaceous perennial species are rock-crevice plants, such as *Cheilanthes distans, C. tenuifolia* and *Ophioglossum lusitanicum* (ferns) and *Isotoma petraea*.

The hills commonly support some woody species which occasionally form groves in elongated soil pockets in clefts, especially *Acacia aneura, A. kempeana, Cassia desolata, C. sturtii, Eremophila freelingii, E. longifolia, Ficus platypoda* (north) and *Pittosporum phylliraeoides*. It is of interest that on the sandy soils at the bases of these isolated ranges *T. basedowii* is the common grass, occurring either in grassland-savan-

Fig. 20.17. Hummock grassland of *Triodia irritans* var. *irritans* on skeletal soils. Shrubs of *Cassia sturtii* and taller *Pittosporum phylliraeoides*. Musgrave Ranges, northern S.A.

nah formation or as an understory dominant in an *Acacia* scrub.

5.4.2.5. *Triodia mitchellii* Savannahs
Eastern Spinifex

This species is confined to semi-arid Queensland and northern N.S.W., occurring mostly between the 400 and 500 mm isohyets and extending on to the main Dividing Range in the east. In the south and east it is usually an understory dominant in woodlands (Blake, 1938). In the drier north and west the woodlands are more open and in many areas can be classed as savannahs.

T. mitchellii occurs on sandy or skeletal soils (Perry and Lazarides, 1964; Pedley, 1967). In the extreme north *T. hostilis* is sometimes an associate. On sandstone plateaux patches of *Triodia* alternate with patches of tussock grassland dominated by *Aristida* and *Bothriochloa*, and on these plateaux xeromorphic shrubs sometimes occur. A few perennial grasses may be present between the *Trodia* hummocks, *Aristida* and *Eriachne* spp. and *Cymbopogon bombycinus* being the most common.

The tree species in the savannahs are *Eucalyptus similis* and less often *E. crebra* and *Acacia shirleyi*.

5.4.3. *Zygochloa paradoxa* Alliance
Sandhill Cane Grass

This hummock grass occurs on the crests of sand dunes under a mean annual rainfall of 150–250 mm, with odd occurrences in wetter areas. Its main centre of abundance is in the Simpson Desert (Crocker, 1946) where it occupies the crests of the parallel dunes (Fig. 20.18). It occurs also on some desert dunes and sandy alluviums deposited in creeks or around salt lakes in Central Australia (Perry and Lazarides, 1962; Chippendale 1963). Its most easterly occurrence is near the Darling River between Menindee and Pooncarie where it occupies the loose sandy summits of some of the lunettes (Chap. 21).

Z. paradoxa, which is dioecious and perennial, forms shrublike hummocks mostly $1-1^{1}/_{2}$ m tall and 1–2 m diam. The hummock consists of the tangled many-noded stems, which are photosynthetic, the leaves being only c. 5 cm long. The hummocks are usually spaced at distances of 2–6 m apart in the driest sites, and are rarely clumped and contiguous. Sand movement in the open spaces between hummocks sometimes occurs, resulting in the development of blowouts and unstable patches.

Over most of its areas the species has no

Fig. 20.18. Parallel sand dunes in the Simpson Desert supporting a hummock grassland of *Zygochloa paradoxa*. Trough in foreground supports scattered Chenopodiaceae growing on grey clay. c. 80 km north-east of Oodnadatta, S.A.

associated plants which equal it in height. Occasional species recorded in some localities are *Acacia ligulata, Atalaya hemiglauca, Cassia pleurocarpa, Crotalaria cunninghamii* in Central Australia, *Acacia dictyophleba* in Queensland, and *Dodonaea attenuata* and *Nicotiana glauca* in some of the stands near the Darling R. On sandy alluviums in creeks it occasionally forms an understory layer to *E. camaldulensis* or *E. microtheca*.

The herbaceous flora is entirely ephemeral (Eardley, 1946; Jessup, 1951; Boyland, 1970), the most common species being: *Aristida arenaria, A. browniana, Atriplex spongiosa, Blennodia pterosperma, Citrullus lanatus, Crotalaria dissitiflora, Euphorbia drummondii, E. wheeleri, Myriocephalus stuartii, Plagiosetum refractum, Salsola kali, Senecio gregorii, Tribulus hystrix, Tetragonia expansa, Trichinium alopecuroideum, Trichodesma zeylanicum* and *Zygophyllum howittii*.

5.4.4. The Ephemeral Grasslands

5.4.4.1. General

The ephemeral grasslands appear mostly after rains during the warmer months and they occur on sands, with one exception. Before white settlement unstable sands were probably of rare occurrence in the arid zone, but their extent has been increased by overgrazing and the death of woody species. Also, the truncation of soil profiles by wind erosion has produced bare areas which are sometimes colonized by grasses. All the grasses which constitute the ephemeral grasslands, whether these are natural or induced, are present as understory plants in woodlands or shrublands. The exception mentioned above is an ephemeral grassland of *Echinochloa turneriana* in the channel country in Queensland (Chap. 21).

The grasses are either annual or short-lived perennials which behave as annuals because of the arid conditions imposed at the end of their growing season. Two sets of grasslands occur, the one in the northern summer rainfall zone, which are dealt with here, the other in the south (chiefly *Stipa* spp., see 5.3.7.5, above).

The ephemeral communities follow a similar pattern of development. Seeds germinate within a few days after rain, and in about a week the bare area is tinged with green. If the rain has not been sufficient to effect establishment, the seedlings may be desiccated. In many cases only one species germinates and completely dominates the area. Later rains, if they occur, produce further crops of annuals unless the first-established species is so abundant that it suppresses the growth of the second crop of seedlings. Summer-growing grasslands usually have a winter component of forbs and the establishment of these is most successful when the summer crop has dried and fragmented. When this occurs surface litter assists in the germination and establishment of the forbs. Seasonal and annual rhythms consequently occur and when two or more favourable seasons follow in succession, perennial grasses and shrubs may become established in the ephemeral grassland. When this happens it is suspected that the ephemeral grassland has been induced by degeneration of a climax community, as indicated below.

5.4.4.2. *Sporobolus australasicus – Enneapogon* spp. Grassland

This grassland (Perry and Christan, 1954; Perry, 1970) occurs within the 300 and 600 mm isohyets either in elevated, raised patches of gravelly clay soil within the *Astrebla* Alliance on the Barkly Tableland, or on skeletal soils on limestone or basic volanic rocks in the Ord R. area within the *Eucalyptus terminalis – E. argillacea* woodlands. The soils are relatively dry because they shed water rapidly.

The grasses are annuals, reaching a height of 15 cm and forming a patchy cover, with much bare ground between the tussocks. *Sporobolus actinocladus* is constant in the community, with varying co-dominants of *Enneapogon pallidus, E. avenaceus* and *E. polyphyllus*. The main subsidiary species are *Aristida arenaria, Brachyachne convergens, Chloris pectinata, Dactyloctenium radulans* and *Tragus australianus*. The small legumes, *Indigofera dominii* and *Neptunia* spp. occur, and the forbs *Cleome viscosa* and *Portulaca oleracea*.

5.4.4.3. *Aristida* spp. Grasslands
Kerosene Grasses

Two closely related species of *Aristida, A. contorta, A. browniana* occur as summer growing ephemerals on loose sand in the arid zone. The germination of the former has been studied by Mott (1974) who found that two dormancy mechanisms are evident in the young grain. One is an after-ripening requirement of the embryo; the second is associated with the "hull" (presumably the involute lemma) and requires exposure to prolonged periods of high temperature before it permits sufficient oxygen to pass to the embryo. The layer inhibiting the passage of oxygen consists of lipid-like material on the inner epidermis of the "hull" directly over the embryo (Mott and Tynan, 1974).

After an adequate summer rain (mostly in excess of c. 70 mm of rain entering the soil in the growing period) these two species, mixed or in monospecific stands may form a complete coverage to the sand. The grasses reach a height of 20–30 cm and the grains mature within a few weeks after the plants become established. Usually there are no associated grass species, though *Plagiosetum refractum,* a hummock forming annual up to 50 cm high, occurs in some areas but is not abundant. Occasional summer annuals are *Calandrinia balonensis* (valued as fodder because of its succulent leaves) and *Salsola kali.* Winter annuals may establish under the dead culms, especially *Myriocephalus stuartii* and *Senecio gregorii.*

The grasslands are mostly induced, following the widespread death of *Acacia aneura,* but natural areas of the ephemeral grassland probably occurred in some of the driest sandy areas, e.g. on the western side of the Simpson Desert.

5.4.4.4. *Enneapogon avenaceus* Grasslands

This species is usually an annual, though potentially a short-lived perennial. It is a fairly constant subsidiary species in most of the *Acacia* and *Atriplex-Maireana* communities of the arid zone. Following the degeneration of the climax communities, *E. avenaceus* has increased enormously in abundance and now commonly covers large areas as a monospecific grassland c. 20 cm high (Fig. 20.19). It has the capacity to colonize truncated profiles in which the residual surfaces (A2 or B1 horizons) are compact and sandy. Also, it commonly migrates on to the gibber-strewn downs (bare areas in the *Atriplex vesicaria* Alliance), if wind-blown sand accumulates among the gibbers.

The ephemeral grasslands have become a semi-permanent feature of the inland in the eastern portion of the arid zone. Even if the tussocks

Fig. 20.19. *Enneapogon avenaceus* grassland. No associated species. Bushes in background are *Maireana pyramidata.* Kinchega National Park, near Menindee, N.S.W.

of *Enneapogon* do not grow during the second year, the dead tussocks remain and, in these, either new plants of *E. avenaceus* or another annual species become established. Summer annuals are rare, the more abundant being *Atriplex spongiosa*, *Dactyloctenium radulans* and *Boerhavia diffusa*. Winter annuals are often abundant and include *Atriplex angulata*, *A. spongiosa*, *Bassia brachyptera*, *B. divaricata*, *B. patenticuspis*, *B. ventricosa* (the last 4 may be perennial), *Brachycome ciliaris*, *Helipterum floribundum*, *Tetragonia expansa* and *Zygophyllum iodocarpum*. This grass-forb assemblage is probably a seral stage in the redevelopment of the climax. Small perennials such as *Eragrostis dielsii* and the legumes, *Swainsona* spp. and *Lotus coccineus*, establish in this community and are followed by *Cassia*, especially *C. artemisioides*, *C. nemophila* and *C. phyllodinea*. The bushes lead to a reduction in density of *Enneapogon* and its associates, and by the time they are 1–2 m high they have accumulated a layer of litter below their crowns which prevents the establishment of smaller plants. Some succulent-fruited chenopods, especially *Enchylaena tomentosa* and *Rhagodia spinescens*, become established, apparently from seed dropped by birds which perch in the branches of the bushes.

5.4.4.5. Other Species

i. *Triraphis mollis* is a common dominant in ephemeral assemblages on loose sands, especially on and around lunettes or on sandpatches which have developed in woody communities as a result of disturbance. This species, which reaches a height of 40 cm, grows mainly after late summer and autumn rains. It is often accompanied by one or more of the following forbs: *Citrullus lanatus*, *Melothria micrantha* (late summer rains) and *Myriocephalus stuartii* (winter rains).

ii. *Tripogon loliiformis*, a dwarf annual to c. 12 cm high, which can withstand almost complete desiccation, commonly forms small local societies after late summer rains in the eastern parts of the semi-arid and arid zones. It is sometimes locally dominant in bare areas in woody communities. Pedley (1974) records it in western Queensland with *Sporobolus actinocladus* as the main associate, and *Bassia diacantha* and *B. quinquecuspis*.

iii. *Echinochloa turneriana*, often with *Panicum whitei* as a co-dominant, forms ephemeral grasslands following flooding in the Channel country of western Queensland. Some perennial species may become established in these annual swards, especially *Eulalia fulva* and *Leptichloa digitata*. The habitats and floristic composition of these irrigated grasslands is discussed in more detail in Chap. 21.

5.5. The *Poa* Grasslands

5.5.1. General

Grasslands dominated by *Poa* occur from south-eastern Queensland to the southern highlands, where they are best developed on the Kosciusko Plateau and the Bogong High Plains in Victoria. Small patches occur also in Tasmania and along the south coast.

In most ecological literature the grasslands have been described as *Poa caespitosa* (= *australis*) associations, and ecologists have recognized the existence of several forms of this complex taxon (Costin, 1954; Carr and Turner, 1960). The genus *Poa* has now been reviewed by Vickery (1970) who recognizes a total of 34 indigenous species, 13 of which were formerly included under *P. caespitosa*. Vickery discusses the difficulties in defining the species. She states that climatic barriers do not appear to break up the large and variable populations, distribution being controlled by soil type and aspect, rather than altitude. That the populations Vickery describes as species are genetically distinct is supported by transplant experiments done by Dr. Marie E. Phillips at Cooma (in Vickery, 1970). Phillips cultured clones and established that each population (species) maintained the same relative differences which distinguished the population in the field.

The 13 species formerly included under *P. caespitosa* occur in *Eucalyptus* forests and woodlands and several of them are dominants of grasslands, some at the highest altitudes (especially *P. hiemata*), others at lower altitudes (especially *P. labillardieri*). These two species have been used to designate the major alliances and an attempt has been made in the account that follows to identify the "forms" of *P. caespitosa* with the species defined by Vickery.

The *P. hiemata* alliance includes at least 6 species of *Poa*, of which only *P. hiemata* is confined to this grassland alliance. The remaining species occur at lower altitudes also, either with *P. labillardieri* which dominates the *Poa* grasslands, or in *Eucalyptus* communities. *P. hiemata* does not occur in Tasmania, where the high altitude grasslands contain only three species of *Poa*, *P. gunnii* (endemic to Tasmania), *P. labillardieri* and *P. saxicola*, the last two occurring also on the mainland.

The *Poa* grasslands at lower altitudes are dominated either by *P. poiformis* (near the coast), or by *P. labillardieri*. The latter is often found in association with *P. sieberana* occurring mainly in drier sites and extending into the *Stipa* grasslands. The floristic relationship between the various grasslands is strengthened by *Themeda australis*, which occurs sparingly in the *P. hiemata* grasslands at lower levels and becomes increasingly abundant at lower altitudes through the *P. labillardieri* grasslands to the *Stipa* grasslands.

5.5.2. *Poa hiemata* Alliance
Snow Grass

P. hiemata is the common soft Snow Grass which is restricted to higher altitudes on the Kosciusko plateau and the Bogong high plains in Victoria. Its altitudinal range is 2200 to c. 1420 m. At the highest altitudes the grasslands occur above the tree-line, formed by *E. niphophila* (Chap. 11), and on these mountain plateaux *P. hiemata* is sometimes associated with the composite, *Celmisia longifolia*, as as constituent of the Herbfields (Chap. 17). The *P. hiemata* grasslands extend below the tree-line in frost pockets through the *E. niphophila* woodlands which flank the grasslands on the more elevated and better drained soils. The downward extensions may cover an altitudinal drop of 200–300 m. At these lower altitudes the *P. hiemata* grasslands are replaced by the *P. labillardieri* grasslands. In both the alpine and subalpine tracts the grasslands commonly adjoin bogs or fens which occupy sites with a more or less permanent water table near the surface (Chap. 17).

The grasslands are of the sod tussock type and occur on level or undulating, often rounded plateaux in the alpine tract and along broad valleys in the subalpine tract. The climate of the area is cool to cold, with mean minimum winter temperatures of c. $-6°C$. (lowest minimum c. $-20°C$). However, the grassland is protected by snow for one to six months of the year and the snow blanket limits the freezing of the soil. Mean annual rainfall varies from 750 to 2250 mm, the rain being fairly evenly distributed throughout the year, but with a winter maximum. The soils are highly organic, slightly to very acidic and are classified as Alpine Humus Soils and Gleyed Podzolic Soils. The last are developed under conditions of impeded drainage. The soils are usually moist, especially in depressions, but soil desiccation is possible during hot dry summers and the surface may freeze during the winter.

About 100 species are recorded in the alliance. However, since *Poa* spp. make up at least 90% of the biomass, except in disturbed stands, the subsidiary species are inconspicuous, except when certain ones flower in spring. The species belong mostly to cosmopolitan temperate genera, among them the grasses *Agropyron*, *Agrostis*, *Chionochloa*, *Danthonia*, *Deschampsia*, *Deyeuxia*, *Festuca*, *Trisetum*, and one species of *Stipa* (*S. nivicola*, recorded in Victoria). The other monocotyledonous families recorded are Cyperaceae (*Carex*), Restionaceae (*Calostrophus*), Juncaceae (*Luzula*), and Orchidaceae (*Caladenia*, *Chiloglottis*, *Diurus*, *Eriochilus* and *Prasophyllum*). Cold-climate dicotyledonous families include Caryophyllaceae (*Scleranthus*, *Stellaria*), Ranunculaceae (*Ranunculus*), Rosaceae (*Acaena*), Scrophulariaceae (*Euphrasia*), Umbelliferae (*Hydrocotyle*, *Oreomyrrhis* and *Trachymene*) and several Compositae including *Brachycome* spp., *Celmisia longifolia*, *Craspedia uniflora*, *Microseris scapigera* and *Senecio* spp.

A few small shrubs sometimes occur in the grassland, usually on shallow soils near rocky outcrops. *Bossiaea foliosa* is the most constant. Others include *Epacris serpyllifolia*, *Hovea longifolia*, *Grevillea australis*, and *Orites lancifolia*.

The associations described by Costin (1954), and Carr and Turner (1959) are renamed using Vickery's (1970) terminology.

i. *Poa hiemata* association is the most widespread and occurs on the relatively well drained Alpine Humus Soils in both the alpine and subalpine tracts.

ii. *Danthonia nudiflora* association occurs in exposed situations in the alpine tract where the soils are shallow. This species forms an ecotonal association with *P. hiemata*.

iii. *Poa costiniana* association (Big Horny Grass in Vic., identified by its rigid, "prickly" leaves) occurs in badly drained areas and forms an ecotonal association with *P. hiemata* on better drained areas and with the following association in wetter areas.

iv. *P. costiniana – Calorophus minor* in wet areas, commonly adjoining bogs.

v. *P. helmsii* association (the species is recognized by its rather light-green foliage) occurs along mountain streams or soakage areas in both the alpine and subalpine tracts.

vi. *P. fawcettiae* association (Horny Grass in Vic. and Smooth-blue Snow Grass in N.S.W., recognized by its bluish, rather stiff-leaved tussocks) is confined to creeks between altitudes of 1650 and 1500 m.

vii. *P. saxicola* association occurs among rocks in the alpine and subalpine tracts.

viii. *P. hothamensis* association (Ledge Grass) is confined to Victoria where it dominates certain grasslands which appear to be induced by fire.

ix. *P. hiemata – Themeda australis* association occurs at lower altitudes and merges into the *P. labillardieri* alliance.

5.5.3. *Poa labillardieri* Alliance

5.5.3.1. Associations on the Mainland

P. labillardieri occurs mainly between altitudes of 700 and 1400 m and on the mainland it mixes with *P. hiemata* at its upper limit. *P. sieberana* often occurs in the same area as *P. labillardieri* and in such cases the former occupies the drier sites, the latter the wetter ones. *Themeda australis* occurs throughout the alliance (Fig. 20.20) and this species and *P. sieberana* extend into the *Stipa* grasslands (5.3.7, above) at lower altitudes in drier climates.

The alliance, as it occurs on the Kosciusko plateau, is described in some detail by Costin (1954) and the associations in Tasmania are listed by Jackson (in Specht et al., 1974). Small patches occur elsewhere, as mentioned below. The grasslands occur in areas with a mean annual rainfall of 550–1500 mm, but precipitation is no guide to distribution since the grasslands usually occur on flats or in depressions, often along watercourses which receive additional irrigation water as run-off from higher areas.

The grasslands commonly adjoin forests or woodlands on the higher side and the elongated strip of grasslands is usually dissected by swamps which adjoin or occur in the watercourse. Costin states that ground water is essential to the grass at least during the summer growing season, and that relatively free movement of ground water must occur, otherwise more acidophilous heath communities assume dominance. The soils are commonly Wiesenboden (Meadow Soil) and Gleyed Podzolic Soils, or Prairie Soils in the drier sites. They develop on any rock type.

The grasslands, which are c. 60 cm high (inflorescence to c. 1 m), are almost closed, but the *Poa* tussocks are not basally contiguous. Smaller grasses form a discontinuos understory and forbs are scattered in a third story. *Poa* makes up at least 90% of the biomass in the densest stands, but less in wetter areas where Cyperaceae and Juncaceae become conspicuous in the tallest stratum. The total flora contains c. 60 species, including additional species of *Poa* (see "associations", below). Other grasses are represented by the genera *Agropyron, Danthonia, Deyeuxia, Festuca, Hemarthria, Hierochloë* and *Themeda*. Other monocotyledons include Xanthorrhoeaceae (*Lomandra* – Tasmania only), Juncaceae (*Juncus, Luzula*) and Orchidaceae (*Diurus*). Herbaceous dicotyledons are mainly species of *Brachycome, Calotis, Epilobium, Haloragis, Hypericum, Leptorhynchos, Oreomyrrhis, Plantago* and *Ranunculus*. Small shrubs are sometimes present, especially *Hakea microcarpa* and *Olearia myrsinoides* (Fig. 20.20).

Three main associations with dominants or co-dominants of grasses occur on the Southern Tablelands:

i. *Poa labillardieri* association is the most widespread, being characteristic of slightly acid soils which are maintained in a damp, relatively well aerated condition by a freely moving watertable. This merges into ii.

ii. *P. sieberana – Themeda australis* association (Fig. 20.21) (Moore and Williams, 1976) which occurs on better drained soils, the latter becoming progressively more common with decreasing moisture-availability.

iii. *P. labillardieri – Festuca muelleri* association occupies wetter situations.

iv. *P. ensiformis* occurs along creeks.

With increasing acidity and decreasing soil-aeration as a result of waterlogging, *Poa* is associated with (a) *Juncus filicaulis,* (b) *J. poly-*

Fig. 20.20. *Poa labillardieri* grassland with some *Themeda australis*. Shrubs of *Hakea microcarpa*, *Cassinia aculeata* and *Bossiaea foliosa*. Trees in background are mainly *Eucalyptus pauciflora*. Near Kiandra, N.S.W.

Fig. 20.21. *Themeda australis* (taller, foreground), *Poa sieberana* and *P. labillardieri* (depression). *Eucalyptus pauciflora* on hill behind. Near Adaminaby, N.S.W.

Fig. 20.22. *Poa labillardieri* grassland on a "bald" in rainforest. Subsidiary species in grassland are *Sorghum leiocladum, Themeda australis, Sporobolus elongatus, Eragrostis leptostachya* and *Wahlenbergia gracilis*, with the introduced species: *Pennisetum clandestinum, Paspalum dilatatum* and *Verbena bonariensis*. Rainforest species adjoining the grassland are *Acmena smithii, Diploglottis cunninghamii, Alectryon subcinereus* and *Ehretia acuminata*. Bunya Mtn. Qld.

anthemos, and (c) *Carex tereticaulis,* the last becoming dominant in the wettest sites.

Small patches of similar grassland occur on the Northern Tablelands. Fraser and Vickery (1939) describe an elongated patch of grassland from the Barrington Tops area lying within a forest and dissected by a swamp (*Sphagnum* in the wettest parts), from which *Calorophus minor* forms an ecotonal margin with the *Poa* grassland. A few small shrubs occur in the grassland, notably *Exocarpos nanus, Hakea microcarpa* and *Pultenaea fasiculosa*. The northern grasslands, which have been isolated from the southern stands for a significant length of time resemble the southern grasslands floristically, but mainly at the generic level. A few species occur in both grasslands and are confined to higher altitudes, viz., *Epilobium glabellum, Oreomyrrhis andicola* and *Scaevola hookeri*. The others are more widely distributed (*Luzula campestris, Plantago varia* and *Ranunculus lappaceus*).

In south-eastern Queensland and north-eastern New South Wales patches of grassland dominated by *P. labillardieri* occur within Araucarian rainforest (Chap. 7) or *Eucalyptus* forest. The grasslands are known as "balds" and they stand in stark contrast to the surrounding vegetation which is mostly rainforest, or *Eucalyptus* forest, often with a mesomorphic understory. The balds occur mostly on basalt.

Herbert (1938) has explained the balds at Bunya Mts as fire-scars following the activities of aboriginals. Webb (1964) made further observations and pointed out the low frequency of fire-indicators. He suggested that the grassland patches are relicts of a former climate when the southern *Poa* grasslands extended further north. The rapid rate of colonization of burned areas by eucalypts throws further doubt on the fire hypothesis. Also, the sharp delimination of the grassland from the rainforest stand (Fig. 20.22) and the poor performance of those few stands of vegetation which are encroaching on the balds, especially the depauperate *Eucalyptus tereticornis,* suggest that a soil property, e.g. some toxic substance, is preventing the migration of the expected climax (forest) in these areas.

The grasslands are closed and reach a height

Fig. 20.23. *Poa poiformis* grassland with subsidiary *Gahnia trifida*. Near Westernport Bay, Vic.

of c. 1 m (inflorescences taller). *Poa labillardieri* dominates the stands, with subsidiary *Themeda australis, Danthonia longifolia* and *Carex appressa*. The flora is listed by White (1920).

5.5.3.2. Associations in Tasmania

P. labillardieri is usually the dominant in the high altitude grasslands in Tasmania, and it extends to the lowlands as an understory species in woodlands and forests. *P. labillardieri* as well as forming grasslands, occurs also as a constituent of alpine herbfields as a co-dominant with *Celmisia longfolia* (c.f. *P. hiemata*, 8.5.2, above). Small patches of grassland are described by Rodway (1924), Sutton (1928) and Martin (1939) and the associations, using Vickery's terminology, have been listed by Jackson (in Specht et al., 1974). The grasslands occur on poorly drained soils on mountain plateaux and extend into treeless frost-pockets in the subalpine tract. As on the mainland, they are often edged on the upper side by woody communities, e.g. shrublands or forests of *Arthrotaxis* or *Nothofagus*, or on plateaux by herbfields of *Celmisia longifolia*. On the lower sides they usually adjoin swamps.

Jackson lists the following associations which are confined to mountain plateaux or frost-pockets in the subalpine tract:

i. *Poa labillardieri – P. gunnii* association occurs on plateaux as a closed grassland. The two species occur in association with *Celmisia longifolia* in the subalpine tract.

ii. *P. labillardieri – Diplarrena moraea* (Iridaceae) association occurs throughout the island on the subalpine tract.

iii. *P. labillardieri – P. gunnii – Helipterum anthemoides – Helichrysum acuminatum* association occurs in the subalpine tract in the northeast and on the Central Plateau.

iv. *Poa rodwayi – Lomandra longifolia* is found on the Central Plateau and eastern parts of the island.

v. *Themeda australis – Lepidisperma linearis – Lomandra longifolia* association occurs on the Central Plateau and eastern parts of the island.

Poa gunnii appears to be locally dominant in some areas. Howard (1974) describes a *P. gunnii* grassland, with *Agropyron pectinatum* locally abundant on Kraznozems derived from basalt 30 km south of Burnie, at an altitude of 579 m under a mean annual rainfall of 1770 mm. The grassland adjoins *Nothofagus cunninghamii* forest. Shrubs 30 cm to 7 $^1/_2$ m high form clumps through the grassland, the species being *Lissanthe montana, Hakea microcarpa* and *Tasmannia lanceolata*. The presence of charcoal in the soil led her to suggest that the grassland is fire-induced. The grass-cover is continuous, except for occasional hollows which are water-filled in winter and support a thin cover of moss (*Polytrichum* spp.). The main forbs in the grassland belong to the genera *Diurus, Gnaphalium, Hydrocotyle, Luzula, Oxalis* and *Viola*.

5.5.4. *Poa poiformis* Alliance

P. poiformis occurs along the coast in the south from the vicinity of Newcastle, New South Wales to Geraldton, Western Australia, and in Tasmania (Vickery, 1970). It covers relatively large areas on those islands in Bass Strait which are too small to support a large woody flora (Gillham, 1961, see Chap. 23). The species occurs mainly on sandy soils, often behind coastal dunes or on sandy alluvium near the coast (Fig. 20.23).

On the mainland, where woody vegetation can establish, the grasslands are sometimes dotted with shrubs. For example, Tiver and Crocker (1951) describe a savannah on Rendzinas in South Australia, with *Banksia marginata* as the shrub component. *Poa poiformis* in this area is associated with *Gahnia trifida* and *Cladium filum,* these two occurring on poorly drained sites, some of which are sub-saline and support *Distichlis distichophylla*. Minor species associated with *Poa* are *Agrostis aemula, Dichelachne crinita, Pentapogon quadrifidus* and *Stipa* spp.

The small area of grassland described by Hope and Thompson (1971) from Cliffy Island, off Wilson's Promontory, Victoria, probably represents the maximum development of the grassland. Here *P. poiformis* dominantes a closed tussock grassland, the tussocks being c. 40 cm diameter at the base and are separated by distances of 20–40 cm. Plants reach a height of c. 60 cm, the culms interlacing. The grassland has developed on sandy loams or loams up to 20 cm deep derived from granite and are organically enriched by the droppings of mutton birds. In some places boulders occur at the surface. Other vascular plants are rare, but *Pelargonium australe,* **Hypochoeris radicata* and 8 species of moss occur between the tussocks. With decreasing soil depth the tussocks become smaller and more widely spaced, and *Plantago* * *coronopus* becomes a conspicuous member of the community, finally becoming dominant on very shallow soils.

The grasslands in most areas have been severely damaged by man and/or domestic animals. Probably the greatest disturbance has been caused by rabbits, and alien plants are now common in most stands. However, *P. poiformis* rapidly colonizes degenerate communities and bare sands, if the grazing pressure is reduced (Norman, 1970).

21. Communities of the Inland Watercourses, Flood-Plains and Discharge Areas

1. Introduction

1.1. The Watercourses

The river systems of Australia were described briefly in Chap. 1. In the interior of the continent, three main types of watercourses occur:

i. Rivers, with sources in wet areas, which are usually channelled between clay banks cut into wide, clay flood-plains, the waters discharging into the sea, e.g. the Darling-Murray River systems.

ii. Rivers with sources in wet areas, which are channelled between banks in the semi-arid zone and which spill out over more or less flat country in the arid zone, where they irrigate vast areas of land. The flood-waters ultimately converge and discharge into inland lakes (Queensland rivers and the channel country).

iii. Watercourses with sources in the arid zone and which flow mainly through rocky or sandy country. They discharge either into inland lakes (playas) or are dissipated across broad wadis, or in sandy fan-deltas, those in Western Australia reaching the Indian Ocean.

The gradients of the rivers are low, except close to the Great Divide. Where the eastern rivers enter the semi-arid zone, the gradients of the flood-plains do not exceed 30 cm per km, and this gradient is reduced to 6–11 cm per km in the arid zone. The rivers meander considerably, the distance along a river being usually 2–3 times the distance between any two points in a direct line, so that river gradients are as little as 3 cm per km in the arid zone.

Flooding of the rivers is a regular, but not annual, occurrence, the water spreading laterally for many kilometres over the flood-plains. This irrigation appears to be necessary for the maintenance of the *Eucalyptus* woodlands and the swamp communities which occur on the river-flats.

In Queensland, the rivers which flow into the dry interior are usually confined within banks as far inland as the arid zone where the rivers divide and spread over flat claypans which form a network of streams separated by elongated sand dunes. The country is referred to as "channel country". The area of channel country is not known accurately. In Queensland it is not less than 300,000 sq.km (in Anon., 1947), and to this can be added a lesser figure to account for the flood-plain channels of the Paroo and Bulloo Rivers, and the channel in South Australia.

The gradient of the flood-plain is measured at c. 16 cm per km. The periodicity of flooding of the whole of the Cooper flood-plain can be judged from Ogilvie's (in Anon., 1947) comment: "Of the 57 years since 1892, 22 years have given floods so small as to be almost useless, serving only channels and portions of swamps. Five floods have completely inundated the whole of the flood-plain ... reaching far into South Australia". Ogilvie's data infer a flow rate of c. 6 km per day for a high flood. Floods result mainly from summer rains, with the production of swards of summer ephemerals, mainly grasses. A late summer flood (March) usually produces a mixture of summer- and winter-growing annuals. The latter can appear also after a summer flood if sufficient moisture persists in the soils.

In the arid zone, that portion lying west and north-east of Lake Eyre and extending to the Indian Ocean contains the driest parts of the continent and includes the Simpson, Victoria, Gibson and Great Sandy Deserts and, in the south, the Nullarbor Plain. Unlike the eastern segment of the arid zone, no large streams flow into the area from higher rainfall regions. The streams therefore are small or absent and permanent water at the surface is almost unknown, though some rock holes containing water and some elongated lagoons do occur. It was on these that the aborigines depended for water. As well as having a low rainfall, much of the region is sandy and consequently water penetration is rapid and run-off is negligible. In such sandy areas watercourses are absent, e.g. the Great

Sandy Desert. A second area devoid of watercourses is the limestone Nullarbor Plain and the adjoining sandy area to the north. Any run-off that occurs on this limestone plateau drains into local depressions into which the water percolates, sometimes reaching caves that occur sporadically below the surface in the limestone.

Watercourses with sources in the arid zone may carry large volumes of water during and after heavy rainstorms, especially when the catchment is large and the water is ultimately channelled. In such cases walls of water some 10 m high may submerge the trees along the watercourse. However, the torrents recede within hours, as rapidly as they appeared.

There is evidence to indicate that all the inland watercourses were at one time much larger than they are today, and that areas which now lack watercourses, such as the Great Sandy Desert, were once dissected by streams (Mabbutt, 1963; Beard, 1973).

1.2. Water-tables and Underground Waters

Water-tables, or at least free water accessible to plants, are present under the beds of those inland stream-lines which support eucalypts, and permanent water-tables are often associated with the levees of major streams, especially those with sandy beds. In some places the water is pumped to the surface and used for irrigation, or for town supplies.

Artesian and subartesian waters occur over much of the continent. However, since these lie mostly at a depth of a few hundred to 1000 m (in some areas to 2000 m), they are not available to plants. A few cases are recorded of the natural emergence of artesian water at the surface. In the Simpson Desert and the Denison Range, south of Oodnadatta, springs of artesian water produce mounds of salts (Madigan, 1930; Jessup, 1951). At Millstream on the Fortescue River, warm, sulphurous water is emitted in a gorge at the rate of c. 50 million litres per day; the well watered gorge is a refugium for plants, among them the palm, *Livistona alfredii,* known only from this area (Beard, n.d.).

Artesian and subartesian waters are often tapped by bores to provide water for domestic stock. The former gushes to the surface and is usually unharnessed, but the latter has to be pumped. The water is either allowed to run into local depressions or, in more closely settled areas, it is diverted into narrow man-made channels (bore drains), many kilometres long, to provide stock-water in the various paddocks of a grazing property. Subterranean waters contain various amounts of dissolved salts, either sodium carbonate or chloride, with smaller amounts of other ions.

1.3. Discharge Areas

1.3.1. Playas

With the exception of the rivers of the Darling-Murray systems and those flowing into the Indian Ocean, the inland watercourses lead to inland "lakes", which are usually dry and are referred to as "playas". Several hundred playas occur in the dry interior, the smallest being a few hectares in area. The largest, Lake Eyre, with an area of c. 8000 sq. km, lies about 12 m below sea level. Probably it has been a playa only since the end of the Pleistocene when a greater water-flow than today filled the basin, the overflow water entering the ocean to the south (Bonython, 1956).

Most playas have been formed by the pondings of water on former extensive lake beds or broad flood-plains. The smaller playas are sometimes elongated but are often roughly circular in outline, their shape being determined partly by wind action. Since the beds of the playas are usually devoid of vegetation, they absorb and re-radiate heat to a greater extent than the surrounding sand-hills which are usually partly or completely covered by vegetation. Hot air rises from the surface of the playa and sets up eddies which assume a circular motion at the base and spiral upwards. Soil and salt particles are set in motion. Saltating grains at the periphery of the floor fashion the sandy bank, producing a curved, sometimes circular rim to the playa. Also, wind-eddies sweeping the periphery of the floor throw salt and sand outwards and upwards to produce a flattish or concave beach slightly above the floor of the playa. Sand grains and salt blown out of the playa accumulate on the rim or are blown further afield, particularly on the eastern side because of the general wind movement from west to east (Hills, 1940; Stephens and Crocker, 1946;

Campbell, 1968). When a sufficient volume of sand is present, dunes are formed, which are either crescent-shaped (lunettes) or the seif type (longitudinal), e.g. in the Simpson Desert bordering Lake Eyre (Madigan, 1946; King, 1956).

The floors of playas are either clayey or salt-encrusted. Lake Eyre (Madigan, 1930; Bonython, 1956) contains a surface salt deposit c. 42 cm thick, consisting of c. 98% sodium chloride overlying gypseous clay. Of interest is the pinkish colour of the surface. Bonython ascribes this to the presence of iron hydroxide, probably introduced in wind-blown soil particles, though some of the pigment may be organic and produced by *Bacterium halobium* or by the flagellate, *Dunaliella salina*.

Small playas and those which are regularly flushed out by flood-waters from rivers have clay floors. These may be saline or subsaline and those that have been investigated contain gypsum a few to several centimetres below the surface. The nature of the surface varies considerably with regard to cracking, which is determined by the type of clay and the adsorbed ions. Surfaces which crack deeply (montmorillonite-beidellite clays) are common; less common are surfaces without cracks which are usually developed on a sandy substrate which is heavily impregnated with clay.

1.3.2. Wadis, Minor Watercourses and Fan-Deltas

Many of the minor watercourses, which may produce flash floods during and after rainstorms, spread out over flat, elongated or fan-shaped areas of clay or sand (wadis) which may be traversed in wet weather by a small stream. The flat surface of the wadi is sometimes stone-strewn. The areas appear to have been fashioned by larger, prior streams.

The lowest limit of water supply by natural irrigation is seen in the sandy watercourses and fan-deltas which spread out at the end of a low-gradient watercourse. The water may merely soak into sand or it is channelled within sandy banks not more than a few centimetres high.

2. The Plant Communities

2.1. General

All the recognized life-forms of plants occur in the arid zone (Chap. 5) and it follows that many different habitats are available. These habitats can be classified as follows and the communities are described under these headings:

i. Aquatic habitats supporting hydrophytes.

ii. Swamps which are periodically but regularly flooded.

iii. Clays which occur in watercourses or in playas which are occasionally and sporadically flooded.

iv. Flood-plains of watercourses supporting communities dominated by *Eucalyptus*.

v. Minor watercourses and other small irrigated areas.

2.2. Aquatic Communities

Many hydrophytes occur in the inland, being most abundant in the south-east. Apart from permanency of water, there are other factors that determine the presence or abundance of hydrophytes. The more obvious of these are:

i. River flow: Free-floating plants are usually absent from flowing water, but are found in still waters in protected bends of rivers where a river bank acts as a natural groyne, or in billabongs (ox-bow lakes) or reed-beds.

ii. Turbitity of the water: Submerged hydrophytes are much more abundant in the low-turbidity waters, e.g. in the Murray and Murrumbidgee Rivers, than in the high-turbidity waters of the more northerly rivers.

iii. Salinity. In most flowing rivers salinity does not appear to affect the hydrophyte flora. However, in a few of the inland playas which sometimes become lakes, and in a few stagnant streams too saline to support glycophytes, two marine hydrophytes occur, *Ruppia maritima* and *Naias marina*.

Some unusual and rare occurrences in pools in Central Australia are *Isoëtes humilior* (on Ayer's Rock), *Nymphaea violacea* and *Nymphoides geminatum* in the Macdonnell Ranges (Chippendale, 1972).

Communities of the free-floating hydrophytes, *Azolla filiculoides* and *A. pinnata*, occur commonly in still waters along many of the rivers in the south-east, particularly in the Murray-Murrumbidgee systems, where they sometimes completely cover the surface. The introduced Water Hyacinth, *Eichhornia crassipes* has become established in the lower reaches of the Murray and in several of the upper Darling R. tributaries. In the latter area, at least, it has become a noxious weed because it is a threat to the free-flow of the waterways. Members of the Lemnaceae (*Lemna, Spirodela* and *Wolffia*) are reputedly widespread in the continent. They are found, today, with varying frequency, around the permanent rivers in the south-east.

Two species of completely submerged aquatics rooted in the mud occur throughout the eastern half on the arid zone and are most abundant in the low-turbidity waters in the south-east, viz., *Myriophyllum verrucosum* and *Potamogeton tricarinatus*. They occur mainly in gently flowing water up to c. 150 cm deep. Submerged species rooted in the mud and with floating or slightly emergent leaves and flowers are found in some permanent or semipermanent waters in Central Australia and the eastern States. They occur in water mostly less than 1 m deep. The species are: *Ludwigia peploides, Ottelia ovalifolia,* and *Marsilea drummondii* (see 2.3.3, below). Also, *Callitriche verna, Elatine gratioloides* and *Mentha australis,* sometimes half-submerged but sometimes forming mats on wet soil, occur mainly in the south-eastern parts of the arid zone, extending northward almost to the Tropic.

There are several other species, recorded from the inland watercourses in the past, which are rarely, if ever, seen today. This could well be explained by the elimination of many permanent waterholes and billabongs as a result of siltation caused by soil erosion following white settlement. One known case is reported here: some 60 years ago fish used to be caught in Evelyn Creek near Milparinka, N.S.W. The creek has been full of sand for 30 years. The disappearance of such watercourses not only affects the aquatic vegetation but also the surrounding communities and the dependent animal communities, including migratory bird populations.

2.3. "Swamp" Communities

2.3.1. *Phragmites australis* Alliance
Common Reed

This grass forms communities c. 2 m high in shallow water, mostly c. 50 cm deep, in a few small areas in the lower reaches of the Murray R. (The community is well developed along the east coast, see Chap. 22). *Typha domingensis* (next section) is sometimes an associate.

2.3.2. *Typha domingensis* Alliance
Bulrush or Cumbunji

This species forms the most common helophyte community in the inland. It grows in water up to c. 50 cm deep and can tolerate deeper submergence for short periods. It occurs in shallows along rivers, in elongated billabongs, and in "delta-swamps", the most famous of which is the Great Cumbungi swamp into which the Lachlan R. empties itself. The plant has an underground rhizome buried a few centimetres deep in the mud. The substrate is usually clayey, and sometimes highly organic. The community remains green while free water is available. However, when swamps dry out the leaves dry and falls as trash on the surface; the rhizomes survive.

Seed of the species is carried for long distances and the plant may occur in remote arid areas, e.g. in artificial lakes formed around the outlets of hot, artesian bores. The abundance of the species has been increased in the irrigation areas around the south-eastern rivers where it has become a weed in the irrigation canals.

Typha makes up more than 90% of the total biomass of the community, the leaves reaching a length of c. 2 m. Associated species are few and include the free-floating species (above) and *Marsilea drummondii* (next section). At the shallow margins semi-aquatics herbs are often abundant, especially *Polygonum lapathifolium* and/or *Rumex bidens*. *Typha* swamps often adjoin *Muehlenbeckia* swamps (see 2.3.4, below).

2.3.3. *Marsilea drummondii* Alliance
Nardoo

This heterosporous fern is widely distributed throughout the eastern half of the semi-arid and arid zones, being found as an understory species in *Eucalyptus, Acacia* and *Casuarina* woodlands, as well as most of the treeless communities (halophytic shrublands, grasslands and most swamps). It occurs as the dominant herb in these communities over small areas in depressions and becomes most conspicuous when tall shrubs or trees are absent. Its abundance has been somewhat increased by road-making, the species becoming locally dominant in table-drains along some roads constructed through fine-textured soils.

The plant possesses a rhizome and can propagate vegetatively. However, sexual reproduction cannot occur in the absence of free water and consequently the species becomes abundant and locally dominant in small depressions (often in gilgais) which retain free water periodically for a few weeks. Maximum water-depth tolerated for vegetative growth is c. 30–40 cm when the leaf-blades float (Fig. 21.1). The micro- and megasporangia occur together in the same sporocarp and emerge after it has imbibed water, the sporangia being attached to a filamentous mucilaginous sorophore 2–3 cm long. The sporocarps were collected by the aborigines and ground between stones to make a kind of flour.

Monospecific communities of *Marsilea* are common, but usually other species, tolerant of waterlogging, are found as associates. These vary with the adjacent flora. Some of the more constant species are: *Alternanthera nodiflora, Centipida cunninghamii, Goodenia glauca, Isoetopsis graminifolia, Mimulus gracilis, M. repens, Plantago varia, Teucrium racemosum*. In the summer-rainfall areas in the north, the chief additions to this set of species belong to the Cyperaceae and Gramineae, and in the driest parts to the Chenopodiaceae.

Fig. 21.1. Leaf-blades of *Marsilea drummondii* (with four leaflets) floating on water. Plants with sporocarps being held in the hand. Depression beside road, in grassland. 16 km north of Moree, N.S.W.

2.3.4. *Muehlenbeckia cunninghamii* Alliance
Lignum

Lignum is found throughout the arid and semi-arid east. It is recorded for Western Australia around the 250 mm isohyet in the south and also in the tropics on small patches of clay soil adjacent to Sturt's Creek (Perry, 1970). The descriptions below are based on the eastern communities.

The plant reaches a height of 2–3 m and forms more or less globular, tangled bushes, which may be contiguous and form an almost impenetrable thicket (often a heaven for wild pigs). The linear to lanceolate leaves, 2–3 cm long, are shed soon after they reach maximum size; the terete, striate stems are green. The communities occur mainly on the flood-plains of the inland rivers and depend on regular flooding and a deep penetration of water which is ensured by the cracking of the soil to a depth of 1 m or more. The water-requirement of the communities appears to be intermediate between that of the *Typha* swamps, which, if they occur in the same area, lie at a lower level with respect to flooding, and the *Eleocharis pallens* swamps (Fig. 21.3) which require less water. The soils are usually stiff grey to black clays; they are mostly slightly alkaline and commonly flecked with calcium carbonate.

Several shrubs may be associated with lignum. *Chenopodium nitrariaceum* (to 2 m high) is common in the south and extends almost to the Tropic; it may occur in pure stands (Fig. 21.2). Other associated species are numerous, including both monocotyledons and dicotyledons. The former usually occupy the wettest situations, the latter occurring mostly in less heavily flooded areas and many of them form mats on mud at the periphery of the swamp. Common monocotyledons are: *Cyperus irio, C. victoriensis, Eleocharis pallens, Scirpus validus* (Cyperaceae); *Diplachne fusca, Leptochloa digitata* (Gramineae); *Juncus polyanthemos* (Juncaceae). Dicotyledons include several members of the Scrophulariaceae, unusual in the arid zone, viz., *Mimulus gracilis, M. repens, Morgania floribunda*.

2.3.5. *Eleocharis pallens* Alliance

This species covers small areas of dark grey to black, cracking clay, usually in slight depressions in the *Eucalyptus* woodlands on river flood plains (Fig. 21.3). The swamps receive less water than the *Muehlenbeckia* swamps and are

Fig. 21.2. *Chenopodium nitrariaceum* in a depression in a woodland of *Eucalyptus largiflorens*. c. 12 km east of Balranald, N.S.W.

Fig. 21.3. Depression on Darling R. flood-plain supporting a sedgeland of *Eleocharis pallens*. Associated species are *Marsilea drummondii*, and *Myriophyllum verrucosum*. In left background, a clump of *Muehlenbeckia cunninghamii* and *Eucalyptus microtheca* in the distance. 32 km west of Collarenebri, N.S.W.

wetter for shorter periods. The species forms dense sedgelands c. 40 cm high and constitutes almost the entire bulk of the community. There are few associated species; *Marsilea drummondii* and *Alternanthera nodiflora* are usually present, and the bulb-forming *Calostemma luteum*.

2.3.6. *Chenopodium auricomum* Alliance

Bluebush or Golden Goosefoot

This species, which reaches a height of c. 1 m and the same diameter, occurs as the dominant in depressions chiefly in the *Astrebla pectinata* grasslands, and to a less extent in the *A. lappacea* grasslands (Chap. 20) as far south as c. lat. 30°S. It is confined to grey, cracking clays in areas receiving a mean annual rainfall of 250–400 mm, falling chiefly in the summer. The communities receive some additional water from run-off. The largest areas of the community occur on the Barkly Tableland (Perry and Christian, 1954) where the floristic composition is richest.

In the swamps, *Chenopodium* has its highest density in the wettest parts. Grasses form a subsidiary stratum and their density increases as the density of *Chenopodium* decreases from wet to dry. The main species are: *Aristida latifolia, Astrebla elymoides, A. squarrosa, Dichanthium* spp., *Eriachne nervosa* and *Eulalia fulva*, with subsidiary *Panicum decompositum, P. whitei* and the forbs *Cyperus retzii, Salsola kali* and *Sesbania benthamii*. As the water recedes, dense mats of smaller species become abundant, including *Iseilema* spp. and the legumes *Psoralea cinerea* and *P. patens*. The communities are highly regarded as pastures for cattle, since they have a high nutritive value and remain green long after the adjacent *Astrebla pectinata* grassland has dried off.

The swamps are sometimes traversed by small creeks, when *Muehlenbackia cunninghamii* fringes the depression where water is more abundant. *C. auricomum* sometimes forms a discontinuous understory in some *Eucalyptus microtheca* woodlands.

2.3.7. *Eremophila maculata* Alliance

Poison Fuchsia

E. maculata is a shrub 1–1½ m high. It is poisonous to stock since it contains a cyanogenetic glycoside in the leaves. However, the leaves do not contain the necessary enzyme (emulsin) to hydrolyse the glycoside and release the toxin (HCN). The enzyme, however, may be present in other plants which could be eaten by stock at the same time as the *E. maculata* is ingested.

The species usually forms the sole dominant of open low shrublands on flats or slight depres-

Fig. 21.4. *Eremophila maculata* forming an open scrub in a depression in the *Acacia aneura* alliance. Associates are *Eleocharis pallens, Centipeda cunninghamii* and **Inula graveolens*. In the distance, *Cassia nemophila* and *Acacia aneura*. c. 65 km east of Broken Hill, N.S.W.

sions subject to occasional shallow flooding (Fig. 21.4). These occur in scrubs and woodlands, usually the *A. aneura* alliance, from east to west across the southern half of the arid zone. The soils vary from deep cracking clays to sandy loams, the latter usually lying above a more clayey subsoil, and in some cases the soils are subsaline.

The herbaceous species vary with the soil. On cracking clays *Marsilea drummondii* is usually common with subsidiary *Eragrostis setifolia*, the latter becoming increasingly prominent with increasing sandiness. In subsaline areas, shrubs of approximately the same size as the *Eremophila* sometimes occur, including *Maireana pyramidata, Lycium australe* and *Scaevola spinescens* (east), and in the west *Maireana pyramidata, Frankenia cinerea, Cratystylis subspinescens* and *Plagianthus incanus* (Speck, 1963). In all these communities herbaceous species of *Atriplex, Bassia, Zygophyllum*, and sometimes *Goodenia* and *Velleia* are recorded, the species and their abundance varying with the district.

2.3.8. *Eragrostis australasica* Alliance

Cane Grass

E. australasica occur mainly on clay pans which may be natural occurrences or produced by accelerated soil erosion resulting in the exposure of the B horizon. The species forms a natural grassland (see Chap. 20). The "swamps" are often surrounded by at least two distinctive zones of herbaceous plants (Wood, 1938). At the high water level, usually where sand has been thinly deposited, the prostrate *Cressa cretica* forms a mat up to c. 1 m wide, sometimes with subsidiary species, e.g. *Centipida cunninghamii* and *Isoetopsis graminifolia*. This zone is surrounded by a zone of *Eragrostis dielsii* on more sandy soils, at a slightly higher level, and contains several to many annual species, the most usual being *Aristida anthoxanthoides, Atriplex inflata, A. spongiosa, Babbagia acroptera* and *Lythrum hyssopifolium*.

Organic matter accumulated from herbaceous plants may increase the permeability of the claypan sufficiently to established larger, woody species, especially *Chenopodium nitrariaceum, Eremophila maculata* and *Muehlenbeckia cunninghamii*.

2.4. Communities on Clay in Flooded Areas, Mainly Channel Country and Playas

2.4.1. Ephemeral Communities in the Channel Country

The pervious clays of the flood plains on the Channel Country produce two sets of ephemeral species as a result of flooding. The summer assemblage is dominated by annual grasses, whereas the winter communities are dominated by forbs.

Summer communities are mainly the grasslands of *Echinochloa turneriana* ("Sorghum") and *Panicum whitei* (Pepper Grass) which form the bulk of the summer communities in Queensland. The former, an annual 1–2 m high, is dominant in areas of deepest flooding. The latter, mostly 50–80 cm high, is potentially a perennial, but commonly behaves as an annual. Blake (1938) and Skerman (in anon., 1947) list the following as the main species on the Cooper's Creek flood plain: Most heavily watered: *Echinochloa turneriana*, with subsidiary *Eulalia fulva* and *Leptochloa digitata*. The grasslands sometimes surround thickets of *Muehlenbeckia cunninghamii*. In the central Cooper area *Aeschynomene indica* is sometimes co-dominant with *Echinochloa*. Heavily watered: *Panicum whitei, Brachyachne convergens, Chloris acicularis, Dactyloctenium radulans, Eragrostis dielsii, E. japonica, E. setifolia, Eriochloa* spp., *Iseilema membranacea, Sporobolus mitchellii, Tripogon loliiformis*. A few subshrubs and forbs occur: *Alternanthera nodiflora, Amaranthus mitchellii, Atriplex muelleri, Bidens pilosa, Bassia quinquecuspis, Chenopodium auricomum* and *Zaleya decandra*. Lightly watered: *Boerhavia diffusa*, which has a stout carrot-like taproot. This forms a zone of varying width at the periphery of the flooded area.

The winter communities are composed almost entirely of annual species. *Trigonella suavissima,* the only representative of the Trifolieae in Australia, produces the most dramatic changes in the landscape. After heavy flooding this species carpets the clay soils with a sward of bright green, comparable in colour and density with a field of well grown clover. The plants are always heavily nodulated (Beadle, 1964). Other annuals are usually present, their numbers increasing in areas of lesser flooding at the periphery of the *Trigonella* sward. The most common species in New South Wales are: *Atriplex spongiosa, Babbagia acroptera* (possibly perennial), *Calotis hispidula* and *Glinus lotoides*. For the Queensland swards, Skerman (in Anon., 1947) records *Agrostis avenacea, Craspedia* spp., and *Poa fordeana*.

2.4.2. Communities on and around Playas

2.4.2.1. General

The type and amount of vegetation on saline playas is governed by salt concentrations. When the salt-crust on the floor exceeds 2–3 mm in thickness plants are absent. Plants become established usually on the less saline beaches, which are watered partly by direct rainfall, partly by runoff water from the upper beach levels and partly by water which seeps through the sand of the playa-rim. Contributions of water from the three sources vary with each fall of rain. Whatever the rainfall, there is always a movement of salt downslope to the floor of the playa. If saline water encroaches upon the beach as a result of an increase in brine-level on the floor of the playa, or by the blowing of brine on to the beach, the area of soil capable of supporting plants is reduced, or established plants may be killed.

In this section, "saline" refers to areas where white crystalline salt is visible on the surface; "subsaline" refers to areas without surface salt crystals.

The most salt-tolerant plants belong to the Chenopodiaceae *(Arthrocnemum, Pachycornia, Atriplex, Bassia* and *Maireana)*, Frankeniaceae *Frankenia)*, Malvaceae *(Plagianthus)*, Aizoaceae *(Aizoon, Dysphyma)*, Zygophyllaceae *(Zygophyllum)* and Goodeniaceae *(Scaevola)*. A few larger shrubs or small trees *(Melaleuca, Eucalyptus)* occur in some areas of lower salinity. The species include perennials (most of those above), and annuals (mainly Chenopodiaceae and Compositae). Few quantitative data are available on the physiology of the species so that the account that follows is based mainly on field observations.

The cell saps of the cladodes of *Arthrocnemum* and *Pachycornia* and the leaves of *Frankenia* and *Disphyma* commonly exceed molar concentration. Hellmuth (1971a,b) measured osmotic potentials in *Arthrocnemum bidens* of -52.7 and -59.6 atm. at $20°C$ under optimal water conditions (cooler months), and -80.0 and -94.7 atm. under stress (summer). These values approximate a twice molar concentration. The high values occur in mature vegetative organs and are much higher than those tolerated by germinating seeds.

Malcolm (1964) found that the seed of *Arth-*

rocnemum halocnemoides suffered a 50% reduction in germination in a 0.14 molar solution of NaCl. The seeds possess hard coats, and optimum germination occurred at an alternating temperature range of 5–35°C. Seeds of annual species of *Atriplex* germinate best at low salt concentrations (Chap. 18). It is probable that germination and establishment of all species occurs when salt concentrations are minimal, following rains. Flowering and fruiting do not appear to be inhibited by high salt concentrations and it is almost inevitable that these phases coincide with maximum salt-concentration in both soil and plant.

Apart from high osmotic potentials in the photosynthetic organs, other factors may assist in the survival of the perennials. Root-shedding under conditions of water-stress possibly occurs (Chap. 18). In the case of *Frankenia*, several species secrete salty sap through salt glands and the salts then crystallize on the leaf-surfaces.

Microtopography also plays some part in determining vegetal patterns, e.g. hummocks of sand which are leached by the rain protect plants against high salt-concentrations. Conversely slight depressions into which brine can flow are usually devoid of vegetation.

The plant communities are described below in order of decreasing salt-tolerances.

2.4.2.2. *Arthrocnemum* spp. –
Pachycornia tenuis Alliance
Samphires

A. halocnemoides, *A. leiostachyum* and *P. tenuis* occur in the most saline situations. The first is the most widespread and usually the most abundant, and each species may be locally dominant. The samphires occur mostly on clays which, when dry, are covered by a surface deposit of salt up to c. 2 mm thick (Fig. 21.5) and gypsum occurs in some areas. *Arthrocnemum* spp. occur also in the littoral zone (Chap. 23).

Associated perennial species are rare or absent. Species of *Frankenia* are often abundant or co-dominant, and *Plagianthus glomeratus* and *P. helmsii* occur in Western Australia. Ephemerals sometimes appears on hummocks when salt is partially removed. Common occurrences are: *Eragrostis dielsii*, *Maireana carnosa* (W.A.), *Ptilotus exaltatus* and a few Compositae, of which *Myriocephalus stuartii* is fairly constant. A noteworthy occurrence is the liverwort *Monocarpus sphaerocarpus* (Marchantiaceae), monotypic and endemic, recorded by Carr (1956) on saltpans in Victoria; the thallus is a single-lobed cup-like structure firmly affixed to the soil by rhizoids.

Fig. 21.5. Saline playa, with no vegetation on the floor. On the salt-encrusted, sandy bank the dominant shrub is *Arthrocnemum halocnemoides*. The smaller bush is *Frankenia pauciflora*. A mallee stand in the background. Lake King, W.A.

Fig. 21.9. Savannah woodland of *Eucalyptus camaldulensis*. Herbaceous layer of introduced pasture species, especially lucerne. c. 70 km south of Keith, S.A.

Nauclea orientalis, Terminalia platyphylla (Fig. 21.8) and *Tristania suaveolens,* but they are not all found in the same stand.

iii. Eastern Rivers with clay flood-plains. Along these streams *E. camaldulensis* occurs in narrow bands at the tops of the banks (Fig. 1.9) or as woodlands in flooded areas, less often as forests on the Murray and Murrumbidgee Rivers. The woodlands are flaked either by the *E. microtheca* (north) or the *E. largiflorens* (south) alliances. The understory species are often the same as in these alliances. Certain species, however, are more abundant in the *E. camaldulensis* woodlands, including the small trees *Acacia salicina* and *A. stenophylla*, which are rarely abundant enough to form a lower stratum. The following grasses may be locally abundant: *Leptochloa digitata, Paspalidium jubiflorum* and *Stipa verticillata* and the sedges *Cyperus gunnii* and *C. victoriensis*. The forests attain a height of c. 60 m, the canopies of the trees interlacing. An understory of small trees and saplings of *E. camaldulensis* occurs. The ground is strewn with debris, some of which is brought in by floodwaters; the herbaceous flora is scanty or lacking.

iv. South, remote from rivers. In South Australia and in parts of Victoria between the 1000 and 600 mm isohyets *E. camaldulensis* forms woodlands to 20 m high (Fig. 21.9) on Podzols (often with gley in the wetter areas) and Red Brown Earths. These soils are developed on river flats, broad valleys and on slopes and hills (Wood, 1937; Incoll, 1946; Jessup, 1948; Specht, 1951; Boomsma, 1969). In the wettest areas the woodlands contain a discontinuous tall shrub/small tree layer of *Acacia melanoxylon* and *Banksia marginata*. The herbaceous layer is dominated by *Carex gaudichaudiana* and *Schoenus apogon*, which frequently form a turf. Scattered grasses include *Imperata cylindrica, Microlaena stipoides, Poa caespitosa* and *Themeda australis*. Native forbs occur in small numbers but the most abundant today are introduced species of *Erodium, *Trifolium* and

Compositae. Wet depressions support an almost unique flora consisting of the pteridophytes, *Isoëtes drummondii* and *Phylloglossum drummondii*, with *Drosera* spp. and *Hypoxis glabella*. Lagoons support hydrophyte-helophyte communities (see above). In areas of lesser rainfall on Red Brown Earths, the woodlands, except those adjacent to rivers, show a reduction in height and density. The shrub stratum becomes xeromorphic and consists of *Acacia pycnantha, A. rhetinoides, Astroloma humifusum, Daviesia ulicina, Dillwynia floribunda, Eutaxia microphylla, Pimelea glauca* and *Thomasia petalocalyx*. The herbaceous layer is dominated by species of *Danthonia* and *Stipa*. Small depressions occur containing the aquatics *Potamogenton tricarinatus* and *Triglochin procera*, surrounded by a zone of Cyperaceae and Juncaceae.

v. Dry, sandy watercourses. *E. camaldulensis* forms pure stands, with few or no woody understory species and most of these are found in the adjacent communities where they occur at a lower density. Eastern occurrences include: Tall shrubs, *Acacia oswaldii, A. victoriae, Myoporum montanum, Nicotiana glauca;* shrubs c. 1 m high, *Enchylaena tomentosa, Rhagodia spinescens, Senecio anethifolius, S. magnificus;* subshrubs and herbs, *Cymbopogon exaltatus* (often abundant), *Chenopodium murale, Chloris acicularis, Lavatera plebeia, Malvastrum spicatum*. In Central Australia, where *Acacia estrophiolata* is a common tree associate, Perry and Lazarides (1962) list several kinds of understory, dominated by *Chloris acicularis, Zygochloa paradoxa, Aristida browniana, Triodia pungens* (rocky areas), or *Bothriochloa ewartiana* and *Eulalia fulva* (north).

vi. The beds of the dry watercourses. These are usually bare or dotted with plants, but occasionally small, temporary communities develop which may be wiped out when water flows again. Clay beds, exposed after prolonged flooding or waterlogging, support plants characteristic of swamps, particularly species of *Alternanthera, Ammania, Centipida, Marsilea, Mimulus, Morgania* and *Sporobolus*. Sandy and gravelly-stony beds present unexpectedly dry habitats. The beds may receive a copious water supply during flood-time, but the water either flows rapidly downstream or percolates rapidly into the porous substratum. The surface sand quickly dries, usually before germination and establishment can be affected. If the sand is maintained in a wet condition, annual species grow either as scattered individuals or isolated societies. The species vary with the district and in any district seasonal variations occur. In and near the tropics, with a summer rainfall, Blake (1938) records a mainly Cyperaceae-Gramineae assemblage consisting of *Brachiaria piligera, B. miliiformis, Bulbostylis barbata, Chloris virgata, Fimbristylis microcarpa, F. miliacea, Polanisia viscosa* and *Setaria surgens*. To the south in the winter rainfall zone, the species are forbs, with relatively few grasses (one being the introduced *Schismus barbatus*). Some of the common forbs are *Erodium cygnorum, Heliotropium europaeum, Lotus coccineus, Salsola kali, Senecio gregorii, Tetragonia expansa* and the introduced *Echium plantagineum*. For Central Australia, Chippendale (1963) reports a completely different assemblage: *Trichodesma zeylanicum, Stemodia viscosa, Evolvulus alsinoides, Alternanthera nana, Amaranthus grandiflorus* and others.

2.5.2. *Eucalyptus rudis* Alliance
Western Australia Flooded Gum

E. rudis occurs mainly in the wetter areas of the southwest (mean annual rainfall 375–1000 mm, mainly in the winter) but, because of its close taxonomic relationship with *E. camaldulensis,* it is included in this chapter. Over a distance of some 300 km (north and south of Geraldton) the two species intergrade. At its northern limit *E. rudis* is smooth-barked and in the south it is rough-barked. From north to south, it shows continuous variation in bark, bud and fruit characters.

In the drier north, *E. rudis* forms woodlands 15–25 m high fringing watercourses. There are no tree associates, but a sparse shrub layer of *Acacia pycnantha* 3–4 m high may be present (Fig. 21.10). In the wetter south the woodlands, up to 30 m high, occur along some watercourses and extend on to flats and even over the crests of low hills in areas which have a clay subsoil and are wet or waterlogged for part of the year (c.f. *E. camaldulensis* in South Australia). Under the most favourable conditions it forms forests to 35 m tall. Most of the areas are now used for grazing and the composition of the understory has not been recorded. Ecotonal associations include *E. rudis* – *E. loxophleba* in the drier north; *E. rudis* – *E.* – *calophylla* in the south

Fig. 21.10. *Eucalyptus rudis* woodland with scattered *Acacia saligna* below. Herbaceous layer of introduced species. 102 km south of Geraldton, W.A.

(Speck, 1958), and at the margins of swamps, *E. rudis – Melaleuca raphiophylla*.

2.5.3. *Eucalyptus microtheca* Alliance
Coolibah

E. microtheca is distributed mainly over the northern half of the continent, usually along watercourses. It occurs between the 150 and 1000 mm isohyets and in drier areas it is dependent more on flooding than on precipitation. For example, it edges some claypans in the Simpson Desert at the bases of sand-dunes from which seepage water enhances the meagre supply accumulated in the claypans. Several forms of the species occur. The eastern populations are rough-barked, the branches usually smooth; in the west the trees are smooth-barked. *E. microtheca* usually forms pure stands as open woodlands (Fig. 21.11) (Blake, 1938; Beadle, 1948; Perry and Christian, 1954; James, 1960). It often adjoins the *E. camaldulensis* Alliance which edges streams and, along the Darling R., it is intermixed with *E. largiflorens* which replaces *E. microtheca* in the south (next section). *E. ochrophloia* (Napunyah) with a limited distribution is included in this alliance. It forms associations with *E. microtheca* or occurs in pure stands (Fig. 21.12) usually in the drier areas of flood plains in western Queensland, extending to New South Wales along the Paroo R. and its tributaries.

The soils are usually deeply cracking Grey or Brown Clays which are base-saturated and alkaline, often containing lime. Sandy surface layers sometimes occur in the soil, apparently as the result of the deposition of sand from adjoin-

Fig. 21.11. *Eucalyptus microtheca* woodland. Grass layer contains *Paspalidium jubiflorum, Eriochloa pseudacrotricha, Chloris acicularis, Panicum decompositum, Salsola kali* and *Bassia quinquecuspis.* c. 40 km west of Moree, N.S.W.

ing areas in depressions. The sand usually modifies the herbaceous layer.

Well defined shrub layers in the woodlands are rare, but many tall shrub or small tree species occur between the *E. microtheca* trees, viz., *Bauhinia cunninghamii, Excoecaria par-*

Fig. 21.12. *Eucalyptus ochrophloia* in a shallow watercourse. *Bassia birchii* below. 6 km south of Augathella, Qld.

vifolia and *Terminalia* spp. (mainly tropics), and to the south *Acacia salicina, A. stenophylla, A. sutherlandii, Atalaya hemiglauca, Casuarina cristata, Eremophila bignoniflora* and *E. polyclada*. Shrubs 1–2 m high are *Carissa lanceolata* and, in swampy areas *Chenopodium auricomum* and *Muehlenbeckia cunninghamii.*

The herbaceous layers are dominated by grasses, especially in the north. The grass assemblages occur also as treeless grasslands (for species lists, see Chap. 20). The assemblages arranged in north to south order are:

i. *Chrysopogon – Dichanthium – Eulalia – Sorghum;*

ii. *Astrebla lappacea – A. squarrosa – A. elymoides;*

iii. *A. pectinata* (drier);

iv. *Eragrostis xerophila* (Central Austr. and W.A.);

v. *Eragrostis setifolia – Dichanthium sericeum.* In the south, mainly where *Eucalyptus largiflorens* becomes co-dominant and in Central Australia, *Atriplex nummularia* and/or *A. vesicaria* from a discontinuous shrub layer, sometimes with *Triodia* spp. below.

In Western Australia *E. microtheca* (smooth-barked form) is far less abundant than in the east and few communities have been described. Burbidge (1945) records *E. microtheca* in association with *E. dichromophloia* along the De Grey R., with an understory of grasses (*Eulalia, Chrysopogon, Sorghum, Themeda* and others), and Cyperaceae in wet patches. Speck (1963) states that in the Meekatharra area *E. microtheca* forms an open fringe to calcrete platforms with *Acacia tetragonophylla* as an associate and bushes of *Scaevola spinescens, Atriplex* sp., *Eremophila* spp. and *Rhagodia* sp.; *Eragrostis leptocarpa* is moderately abundant along channels.

2.5.4. *Eucalyptus largiflorens* Alliance
Black Box

E. largiflorens occurs along watercourses or on lowlying areas subject to flooding and replaces *E. microtheca* in the south-east. Along parts of the Darling River, in particular, the two species are mixed. *E. largiflorens* is most abundant in New South Wales (Beadle, 1948; James, 1960) but occurs also in Victoria (Zimmer, 1937; Connor, 1966) and in South Australia (Wood, 1937; Specht, 1972). The species occurs mostly between the 200 and 400 mm isohytes, with a definite winter predominance in the south. It occurs mainly on grey clays which commonly crack deeply on drying; the soils are mostly alkaline throughout the profile (pH 7–7.5), sometimes containing traces of lime and sometimes with gypsum in the lower horizons. Small patches of soil are sometimes moderately saline, which is indicated by the presence of a halophytic understory. In heavy clay areas, mostly subject to flooding for weeks or even months, gilgais develop. The species occurs also on red-brown, sandy deposits (possibly wind-accumulated) which are mostly 1–2 m deep and lie above the clays of the river flats.

E. largiflorens forms open to very open woodlands 10–20 m high (Fig. 21.13). Ecotonal associations with other watercourse eucalypts may occur, viz. *E. camaldulensis* and *E. microtheca*. Other ecotonal associations are rare and the ecotones are narrow, e.g. with *E. woollsiana* in the south. Associated tall shrubs are also rare and do not form a continuous stratum, except, perhaps, locally. The species are mainly

Fig. 21.13. *Eucalyptus largiflorens* woodland with a grass layer of *Hordeum leporinum*. Murrumbidgee R., near Balranald, N.S.W.

watercourse acacias – *A. salicina* and *A. stenophylla*. Near community junctions, understory tree species from adjacent communities may occur as an understory, e.g., *Acacia pendula*, *Casuarina cristata* and *C. luehmannii*. Shrubs c. 3 m high are rare and include *A. oswaldii* and *Exocarpos aphylla* and in the north *Eremophila bignoniflora* and *E. polyclada*.

In the virgin condition the halophytic shrubs *Atriplex nummularia* (c. 2 m tall) and/or *A. vesicaria* (60–80 cm tall) dominated the lower layers over much of the area, with some 100 associated herbaceous species, which occur also in the treeless *Atriplex* shrublands (Chap. 19). Towards the north, where the rain falls mainly in the summer, grasses were more abundant and may have dominated the ground layer to the exclusion of *Atriplex*. The main grasses are: *Dichanthium sericeum*, *Eragrostis setifolia*, *Eulalia fulva*, *Paspalidium jubiflorum* and, in waterlogged areas, the semi-aquatic *Diplachne fusca*. This sward has been modified considerably by heavy grazing, leading to the dominance of the less permanent *Eragrostis parviflora*. Noteworthy plants in the communities are *Swainsona greyana*, a shrub c. 1 m high which is highly toxic to animals, and the Darling lily, *Crinum flaccidum*, the bulb of which is buried to a depth of c. 80 cm.

Local variations in microtopography lead to variations in the understory species:

i. In gilgai depressions, which are periodically waterlogged, societies of *Marsilea drummondii*, *Scleroblitum atriplicinum*, *C. auricomum* (north) and *Muehlenbeckia cunninghamii* occur.

ii. Saline patches, identified by the occurrence of salt crystals on the soil surface, support *societes* of *Disphyma australe* and/or *Pachycornia tenuis* or of *Nitraria billardieri*.

Since the woodlands lie near permanent or semi-permanent water, they have been heavily grazed by domestic stock. The perennial salt bushes have been killed out over large areas; the dominants now are commonly ephemeral species belonging to the families Chenopodiaceae, Compositae, Cruciferae and Gramineae. Common seasonal dominants are *Atriplex inflata* (often with *Bassia* spp.), *Blennodia lasiocarpa*, *Hordeum leporinum* (Fig. 21.13) and *Tribulus terrestris*. The perennial *Bassia quinquecuspis* often becomes dominant following heavy grazing.

2.6. Communities of Minor Watercourses and other Small Irrigated Areas

Minor watercourses are best defined as stream lines which are too dry to support *Eucalyptus camaldulensis*. Other small irrigated areas include rock-holes which contain semi-permanent (rarely permanent) water, rock-crevices and the bases of boulders.

2.6.1. Minor Watercourses

In most cases minor watercourses extend climax communities of somewhat wetter climates into more arid areas. For example:

i. *Acacia cambagei* fringes clayey watercourses in the eastern half of the arid zone.

ii. *Acacia aneura* extends into sandy or rocky watercourses and is often accompanied by *A. estrophiolata* (Central Austr.), *Grevillea striata* and *Pittosporum phylliraeoides*.

iii. Species of *Eucalyptus*, especially *E. brevifolia*, *E. dichromophloia*, *E. intertexta*, *E. lucasii*, *E. oleosa*, and *E. terminalis*.

iv. *Callitris "glauca"* occurs both along some sandy creeks and in large soil pockets in bare sedimentary rocks, especially sandstones and conglomerates, where it is irrigated by run-off.

Wadis and minor watercourses are sometimes dominated by shrubs normally found in the subsidiary stratum of taller communities, particularly the *Acacia aneura* alliance. These form the dominants of scrubs 2–3 m high along the minor watercourse. The main species are: *Acacia victoriae* (Fig. 21.14), *Eremophila duttonii*, *E. freelingii* and *Myoporum montanum*. An herbaceous plant commonly associated with these is the Lemon-Scented Grass *Cymbopogon exaltatus*.

Watercourses with the lowest water supply support a few perennials or herbaceous annuals only, the former usually being short-lived. Two common species are *Nicotiana glauca* (east only) which reaches a height of c. 3 m and *Lavatera plebeia* (c. 1½ m). Species of *Maireana* sometimes occur, and *Cymbopogon exaltatus* may be present. The annual species are not distinctive but are those which occur in other habitats after rain. However, in the Tanami Desert a remarkable assemblage of species is reported by Chippendale (1963) in sandy watercourses. It con-

Fig. 21.14. Minor watercourse with a sandy bed fringed by *Acacia victoriae* and *Cymbopogon exaltatus*. Bushes of *Rhagodia spinescens* and *Enchylaena tomentosa* visible on left and in background. 40 km north of Broken Hill, N.S.W.

sists mainly of species of genera typical of low-nutrient sandy swamps in wetter areas, viz., *Drosera indica, D. burmannii, Stylidium floribundum,* and the fern *Cheilanthes tenuifolia*.

2.6.2. Semi-permanent Rock Holes and Deep Gorges

Permanent water in the gorges of the Macdonnell Ranges in Central Australia account for the relic occurrence of several species normally found in or near rainforests, or of species of genera typical of wet climates (Chap. 5). The species include several ferns, the cycad, *Macrozamia macdonellii* and a few angiosperms, notably the palm, *Livistona mariae*.

2.6.3. Rock-crevice and Boulder Communities

Rock crevices and the bases of bare rock surfaces which receive run-off water often produce distinctive communities. These differ from the gorge communities in that the plants are subjected to full sunlight, high temperatures and low humidities.

i. Rock-crevices commonly support monospecific communities of the ferns *Cheilanthes tenuifolia* (most common and widespread), *C. distans, Gymnogramma reynoldsii* or *Pleurosorus rutifolius,* or of *Isotoma petraea* (Lobeliaceae). Rarely do two of the species occur together. All species have rhizomes deeply rooted in the crevices. The aerial parts of the plants are usually desiccated and killed each year and are replaced from the rhizome after rain.

ii. Boulder communities are defined as those formed over a rock surface by plants which are rooted in the soil at the base of the rocky area. The seed germinates in soil wetted by runoff water and the continued irrigation produces plants of extraordinary large dimensions. The best example is *Ficus platypoda* which occurs in the ranges of the arid zone in the summer rainfall areas of the centre and west. Commonly only one plant forms the "community" and it may cover an area of 100 sq. metres or more (Fig. 5.4).

22. Communities in Fresh or Brackish Water, Mainly on the Coastal Lowlands. Including Lagoons, Lakes, Rivers, Swamps and Flooded Areas

1. Introduction

1.1. General

This chapter deals with communities associated with open water or with soils which are periodically or permanently waterlogged in the climatically wetter parts of the coastal lowlands with a few minor extensions on to the highlands in the east. Similar communities occur also in the drier inland (Chap. 21) and at higher altitudes (Chap. 17) and a few species are found throughout these three climatic zones.

The habitats encountered on the coast include open waters of lakes and lagoons, some of which are brackish, the flowing waters of rivers, the swamp habitat where the soils are permanently or periodically waterlogged and flat areas subject to seasonal flooding. The life-forms of the dominant species provide a system for the classification of the communities into zones which are indicated in the various tables below. The zones are determined by the depth of open water or the height of the water table in the soil.

The zonation is repeated throughout the continent, if all the habitats occur. However, in many cases one or more zones may be absent, according to local topographic conditions; for example, along some stretches of a river with steep banks the open water zone with hydrophytes adjoins the woodland or forest zone on well drained soils.

1.2. The Flora

The flora is an extensive one, all the major plant groups being represented, except the gymnosperms (though *Actinostrobus pyramidalis* is of occasional occurrence in sedgelands in Western Australia). Algae are abundant but cannot be dealt with here, except to mention that of the larger ones, Characeae *(Chara* and *Nitella)* are often abundant; *Spirogyra* and *Zygnema* and often visible and *Cladophora* is fairly common in brackish water. The bryophytes are represented mainly by *Riccia fluitans* (aquatic) and *Sphagnum* is recorded in a few districts; the latter is more abundant at higher altitudes (Chap. 17).

The vascular aquatics, dealt with most adequately by Aston (1973), are mainly angiosperms, but the pteridophytes are well represented by the genera *Azolla, Ceratopsis, Isoëtes* and *Marsilea*. Also, several pteridophytes occur with the helophytes, some of them being local dominants. The aquatic angiosperms, numbering c. 100, are equally divided between the dicotyledons and monocotyledons and they occur in many unrelated families. A list is not provided here but the most common species are mentioned below.

The general distribution of the aquatic flora is determined largely by temperature, two main regions being apparent, tropical and temperate, and about the same number of species occurs in each zone. Many genera are confined to one region, e.g. *Aldrovanda, Caldesia, Nelumbo* and the fern *Ceratopsis* to the tropics. A few tropical genera extend along the eastern coastal lowlands into New South Wales, e.g., *Aponogeton* and *Nymphaea*. Conversely, several genera are confined to the cooler south, some occurring west and east, e.g., *Isoëtes, Damasonium, Elatine, Vallisneria, Villarsia* and the aquatic species of *Ranunculus*. On the other hand, a few are restricted to the south-east, notably *Alisma*. Some genera are widely distributed over the continent (especially the mainland) and many of these are represented by different species in different regions, especially *Myriophyllum, Ottelia, Potamogeton, Triglochin* and *Utri-*

cularia. A few species occur throughout the continent, e.g. *Lemna minor.*

The two regions with the lowest number of hydrophytes are Tasmania and the south-west. Low temperatures combined with isolation may account for the paucity of species in Tasmania which lacks the otherwise common genera *Marsilea, Ceratophyllum, Hydrilla, Ludwigia, Spirodela* and *Wolffia.* On the other hand, isolation of the south-west by the arid zone possibly accounts for the paucity of species in this region.

The tropical flora is related to the flora of Asia and the tropics in general, many species in Australia occurring also in Asia. The temperate flora is related to the flora of New Zealand and to a less extent to South America. Only one genus is endemic to Australia, *Maundia* (Juncaginaceae). *Liparophyllum* (Menyanthaceae) is confined to Tasmania and New Zealand.

A few species show curious extra-Australian disjunctions. The floating Araceae, *Pistia stratiotes,* occurs only on the Arnhem Peninsula (N.T.), being first recorded in 1946 (see Aston 1973); it is a common tropical species. Again, *Hydrocharis dubia* (Hydrocharitaceae) is recorded in south-east Queensland and northern New South Wales; it occurs also in Asia and the islands north of Australia (Aston, 1973). Whether these are recent introductions by man or the result of distant dispersal by birds cannot be stated.

Several aquatics have been purposefully introduced by man as ornamentals for ponds or aquaria and some of these have become weeds in waterways or in lagoons, especially the water Hyacinth (** Eichhornia crassipes*), the fern ** Salvinia auriculata,* the edible Water Cress ** Nasturtium officinale,* and a few others (Aston, 1973).

Noteworthy among the species in aquatic and semi-aquatic habitats are the numerous carnivorous plants (Chap. 2, Table 2.7). These include *Cephalotus* (restricted to the south-west and growing in peaty swamps near Albany), many species of *Drosera* (with the highest concentration in the south-west), *Aldrovanda* (tropics) (both Droseraceae), *Byblis* (Byblidaceae, one species in tropics, one in the south-west), and *Polypompholyx* and *Utricularia* (Lentibulariaceae).

The helophyte flora is largely monocotyledonous, the species belonging mainly to Cyperaceae, Typhaceae (widely distributed), Juncaceae, Restionaceae and Xyridaceae (confined to the south).

The woody species belong to many families and some of them are extensions of rainforest species (Chap. 7) along watercourses or around swamps. Other than these, the most important are *Casuarina cunninghamiana* see 3.4.2, below), and many species of *Melaleuca* on which a general statement is made here.

Melaleuca with c. 140 species in Australia is near-endemic; a few species extend to the islands to the north and north-east of Australia. Most of the species occur in areas which are temporarily or permanently waterlogged, and many species are salt-tolerant. When a zonation of communities occurs, developed as a result of decreasing height of a water-table, it is usual for a species of *Melaleuca* to dominate the zone of vegetation where the water-table is lowest. The species vary from region to region and the *Melaleuca* communities (scrubs, woodlands and forests) are most extensive in the tropics where the intense summer rains lead to waterlogging over considerable areas.

In the account that follows the communities are described under regions, as indicated above by the distribution of the flora.

2. Communities in the Tropics

2.1. General

In the tropics, where the rain falls mainly in the summer, flooding of flat country is an annual event. Water which does not drain away remains at the surface to waterlog the soil or/and to form swamps or lagoons. The permanency of the latter varies, but relatively few permanent lakes occur, notably among these being the crater lakes in Queensland (see below). The lagoons, swamps and waterlogged soils support two distinct groups of communities. In areas where free water lies for a few months, herbaceous communities dominated by hydrophytes and helophytes develop during the wet summer months; these dry out during the dry winter and spring. On waterlogged soils, which may be submerged by water for a few weeks, the vegetation is dominated by woody plant of which *Pandanus,* palms and species of *Melaleuca, Tristania, Eucalyptus* and a few others are the most abundant.

Fig. 22.1. *Barringtonia acutangula* edging a swamp (on right). *Pandanus* sp. on left. Near Darwin, N.T.

The rivers which drain the areas are flooded after rain but they subside rapidly; many of them carry permanent water which provides for a distinctive set of woody and herbaceous species. Many of the rivers spread out over broad deltas and it is on these that most of the swamps occur. On the deltas, two types of alluvium have been deposited, one an "acid alluvium" containing 80–90% sand, the other fine textured material with much organic matter deposited in drowned valleys (Specht, 1958). The type of alluvium has some effect in selecting the flora.

The communities of the lagoons and swamps, described below in order of decreasing water level, are as follows:

i. *Nymphaea-Nelumbo* Alliance (Waterlilies) in the deepest water followed by

ii. aquatic grasslands dominated by species of *Oryza, Phragmites, Sclerandrium* and others (see Chap. 20) or sedgelands dominated by *Leptocarpus* and/or *Eleocharis;*

iii. Scrubs of *Pandanus* or palm *(Livistona);*

iv. Scrubs, woodlands or forests of *Tristania* or *Melaleuca.*

Along rivers where the gradient is such that swamps do not develop, some members of the series usually occur in a contracted sequence,

Fig. 22.8. Open low woodland of *Melaleuca minutiflora*. Herbaceous layer dominated by *Eragrostis browniana*, with subsidiary *Xerochloa barbata*. c. 60 km east of Kununurra (W. A.) but in Northern Territory.

2.7.3. *Melaleuca viridiflora* Alliance

This species is widely distributed from east to west across the tropics in areas with a mean annual rainfall of c. 500–700 mm. It occurs on Yellow and Red Earths which are waterlogged for a few weeks during the summer. *M. viridiflora* is a common dominant of woodlands 5–7 m tall (Fig. 22.7) occurring mostly in pure stands. Other species of *Melaleuca* may be present, either as associates or in the shrub understory. In Queensland, Perry and Lazarides (1964) and Story (1969, 70) recorded *M. acacioides*, *M. foliolosa*, *M. nervosa*, *M. stenos-*

Fig. 22.9. A. brackish lake enclosed on the seaward side by a sand dune. Zonation on nearer side of lake: *Phragmites australis*; sedgeland (scarcely visible), *Melaleuca* scrub (*M. thymifolia* on left, *M. armillaris* on right). Dune behind with scattered plants of *Chrysanthemoides monilifera*, *Leucopogon parviflorus*, *Hibbertia scandens* and *Scaevola calendulacea*. Dee Why, near Sydney.

tachya and *M. symphocarpa,* and each of these species may be locally dominant. Other tall species recorded are *Alphitonia excelsa, Erythrophleum chlorostachys* and *Petalostigma* spp. In the west, *M. viridiflora* has fewer associates (Speck and Lazarides, 1964; Speck, 1965), viz. *M. acacioides, Erythrophleum* and *Terminalia platyptera*. Shrubs are usually absent. The herbaceous layer is dominated by grasses.

2.7.4. *Melaleuca minutiflora* Alliance

This species is dominant over small areas in the Ord River area and westward into the Kimberleys (Speck and Lazarides, 1964). The scrubs and woodlands are open and 5–7 m tall, developing on Yellow Earths subject to minor waterlogging (mean annual rainfall 600–700 mm). The species forms pure stands (Fig. 22.8). Shrubs are usually absent and the herbaceous layer contains *Plectrachne pungens, Chrysopogon* spp., *Eragrostis browniana,* and *Xerochloa barbata* (on saline patches).

3. Communities in the East from South-eastern Queensland to South Australia and Tasmania

3.1. Communities of Coastal, Brackish Lakes and Estuaries

Brackish lakes occur along the coast of New South Wales, Victoria and south-eastern South Australia, being enclosed on the seaward side by sandy alluvium or a sand dune (Fig. 22.9). The lakes are commonly open to the ocean, either permanently through a narrow channel, or temporarily by the local breaking of the dune by floodwaters. The lakes are usually fed by rivers, and consequently the waters vary in salinity, perhaps even daily, according to rainfall and the tides. Maximum salinity is c. 3% (close to sea water), dropping to an average of c. 0.5% (Davis, 1941), and is high enough in places to support the marine angiosperms.

Fig. 22.10. Watercourse and swamp showing a zone of *Phragmites australis* edging the open water. Other species are *Baumea articulata* and *Scirpus nodosus*. Wilson's Promontory, Vic.

Fig. 22.11. A pure stand of *Typha domingensis* edging open water. Main herbaceous species on landward side is *Rumex bidens*. Lake Alexandrina, S.A.

Table 22.1. Zonation in brackish lakes along the coast of New South Wales. Based on Hamilton (1919), Osborn and Robertson (1939), Pidgeon (1940), Davis (1941).

Zone 1	Free-floating hydrophytes
	Lemna minor
Zone 2	Submerged hydrophytes
	Naias marina, Ruppia maritima, Zostera capricorni, Potamogeton pectinatus (more saline waters); *Myriophyllum elatinoides, M. verrucosum, Potamogeton perfoliatus, Triglochin procera, Vallisneria spiralis*
Zone 3a	Helophytes (deeper water)
	Scirpus vallidus, Phragmites australis, Typha domingensis, Scirpus litoralis, Baumea articulata, B. procera, Alisma plantago-aquatica, Triglochin procera, Villarsia exaltata
Zone 3b	Helophytes and herbs forming a sedgeland (see 3.3, below)
	Juncus maritimus, Baumea juncea, Schoenus brevifolius. Small herbs: *Selliera radicans, Cotula coronopifolia, C. reptans, Apium prostratum, Hydrocotyle vulgaris, Samolus repens, Lobelia anceps, Mimulus repens, Dichondra repens, Wilsonia backhousii*
Zone 4	Scrubs, woodland or forests
4a	*Casuarina glauca*, sometimes with epiphytic *Dendrobium teretifolium* and *Platycerium bifurcatum* (rare). Understory of: *Juncus maritimus* and/or *Baumea juncea* and herbs, as in Zone 3b
4b	*Melaleuca quinquenervia* (north), *M. squarrosa* (south)

Table 22.2. Zonation in brackish lakes and river estuaries in Victoria, South Australia (Eardley, 1943; Dodson and Wilson, 1975), and Tasmania. * indicates introduced.

Zone 1	Free-floating hydrophytes
	* *Nasturtium officinale, Lemna minor*
Zone 2	Submerged hydrophytes
	Potamogeton pectinatus, Ruppia maritima (more saline water); *Myriophyllum elatinoides, Lepilaena preissii, Ranunculus rivularis, Potamogeton perfoliatus, Lilaeopsis polyantha* (Vic.), *L. brownii* (Tas.), *Vallisneria spiralis*
Zone 3a	Helophytes (deeper water)
	Phragmites australis, Typha domingensis (rare in Tas.), *Baumea articulata, Eleocharis acuta, E. sphacelata, Cladium procerum, Triglochin procera*
Zone 3b	Helophytes and herbs forming a sedgeland (see 3.3, below)
	Juncus maritimus, Baumea juncea, Scipus spp., *Sium latifolium* and small herbs
Zone 4	Scrubs or thickets
	Melaleuca squarrosa, M. squamea, Leptospermum lanigerum

The various zones of vegetation, which are usually regarded as successional, are shown in Tables 22.1 and 22.2. A division into subregions is necessary because of differences between the floras of the east coast, and the south coast and Tasmania. The main difference is that *Casuarina glauca* which forms a very distinct zone in the east (Fig. 22.12) is absent from the south. Minor differences occur also in the less common species, but the dominants are similar throughout.

All of the species listed do not always occur in the one lake. Of the helophytes, *Scirpus vallidus* is often the dominant in deepest water, though *Phragmites* (Fig. 22.10) is usually the most common, and *Typha domingensis* is dominant in some areas (Fig. 22.11).

The species of *Melaleuca* adjoining the swamp vary from north to south. North of Sydney, *M. quinquenervia* is usually dominant (Fig. 22.13; see also 3.2.3, below); south of Sydney other species occur, especially *M. thymoides* and *M. squarrosa*, the latter extending into South Australia and Tasmania. A swamp forest of *Eucalyptus robusta* (Chap. 9) sometimes adjoins the *Melaleuca* community on the east coast. In South Australia the scrub or thicket adjoining the swamp is dominated either by *M. squarrosa* and/or *Leptospermum lanigerum* in the least saline areas (with a scattered understory of *Gahnia grandis* and *G. trifida*), or in more saline areas by *M. lanceolata* and *M. halmaturorum*. The former, often in association with *Bursaria spinosa* forms scrub-thickets c. 5 m tall in slightly saline habitats (Crocker, 1944). *M. halmaturorum,* of rare occurence, occurs on saline soils as a gnarled shrub up to 6 m tall, with an understory of the salt-tolerant herbs *Apium australe, Centrolepis polygyna, Triglochin mucronata* and *Polygonum maritimum* (Crocker, 1944).

3.2. Communities of Fresh Water Lakes and Lagoons

Fresh water lakes and lagoons are common on the lowlands, but they are rarely more than a few hectares in area. Some occur behind coastal dunes, others near rivers and many have been formed for the supply of water for humans or grazing animals. In all cases the water is of low salinity and consequently salt-requiring species are absent, notably *Casuarina glauca* and the marine angiosperms. Otherwise, a very large number of the species which grow in the brackish lakes occur also in the fresh waters. The zonations in fresh water are similar to those in brackish water. *Phragmites australis* and *Typha domingensis* are the more common species edging large bodies of open water. Smaller lagoons are sometimes fringed with *Eleocharis acuta* (Fig. 22.14). The fresh water swamps are usually surrounded by a sedgeland (see 3.3, below) which is often fringed by a *Melaleuca* swamp.

Fig. 22.12. Pure stand of *Casuarina glauca* with a narrow zone of *Phragmites australis* edging the water. Myall Lake, N.S.W.

Fig. 22.13. Low forest of *Melaleuca quinquenervia* with scattered *M. thymifolia* below. Herbaceous layer contains *Eleocharis sphacelata, Cyperus* sp., *Gahnia grandis, Blechnum indicum* and *Philydrum lanuginosum* Bribie Is., Qld.

Fig. 22.14. *Eleocharis acuta* dominating a sedgeland surrounding free water in which *Ottelia ovalifolia* (leaf-blades floating; the plants are flowering) is dominant. Near Bathurst, N.S.W.

The communities vary floristically from north to south and a division into a mainland (Table 22.3) and a Tasmanian assemblage (Table 22.4) is necessary. On the mainland there is a change of flora from north to south, chiefly through the diminution in abundance of species with large, floating leaves, notably *Nymphaea*, *Nymphoides* (Fig. 22.15) and *Ottelia ovalifolia* (Fig. 22.16). On the other hand, a few species, apart from the rushes and reeds, are almost invariably present, especially *Triglochin procera* (Fig. 22.17). In Tasmania relatively few hydrophytes occur and there are few species with floating leaf-blades (only *Nymphoides*). *Typha* is rare and *Phragmites* is not abundant as a dominant of reed swamps; on the contrary, the open water is more often edged with species of *Eleocharis*, *Juncus* or *Scirpus*, or by small herbaceous species, of which *Neopaxia australasica*, *Hydrocotyle* spp. and *Limosella* are the most common (Jackson, in Specht et al. 1974).

The *Melaleuca* communities adjoining the swamps or sedgelands are those which occur also around the least saline of the brackish lakes. *M. quinquenervia* (Fig. 22.13) is dominant north of Sydney, sometimes with *M. glomerata* forming a shrub layer. The herbaceous layer contains many species from adjacent sedgelands, including *Baumea juncea*, *Schoenus brevifolius* and *Restio* spp. Other species dominating scrubs or thickets in the east are *M. decora* and *M. lineariifolia* (clay soils), *M. glomerata*, *M. armillaris* (Fig. 22.9), *M. genistifolia*, *M. squamea*, *M. styphelioides*, *M. ericifolia* and, towards the south, *M. squarrosa* (Fig. 22.18). The last is the common species in Victoria, South Australia and Tasmania, where it forms impenetrable thickets with only fragile, scattered herbs below (usually *Drosera* and *Utricularia* spp.). On the mainland the thickets often adjoin thickets of *Leptospermum lanigerum* of similar density (Fig. 22.19) and the former species sometimes follows drainage lines through the swamps with scattered plants of *Gahnia clarkei* in the understory (Willis, 1964). In Tasmania, *M. squarrosa* occurs in similar sequence with *Leptospermum juniperinum* and it is sometimes associated with *M. squamea* and *M. gibbosa*.

3.3. The Sedgelands

3.3.1. General

The sedgelands are herbaceous communities dominated by Cyperaceae and Restionaceae. They occur on soils which are waterlogged for long periods or even permanently, and are found mainly near the coast around fresh water or

Fig. 22.18. Swampy flat dissected by a creek (visible on left) supporting a reed-swamp of *Phragmites australis* (paler) and *Melaleuca squarrosa* on both sides of the reed swamp. Foreground with scrub of *Leptospermum laevigatum*. Wilson's Promontory, Vic.

highly organic and acid, developed on siliceous parent materials; *Sphagnum* is sometimes present and adds to the organic accumulations.

Button Grass has a tussock habit, the tussocks reaching a height of c. 2 m, individual plants being elevated on "stools" separated by distances of 50 cm to 1 m. In perhumid climates the communities are not confined to flats but extend up hillsides by marginal creep. In less wet climates the community is restricted to flats or

Fig. 22.19. Thicket of *Leptospermum lanigerum* with scattered *Gahnia trifida* in the understory. The cleared area in the foreground adjoins the road which accounts for the sward of grass. Near Princetown, Vic.

Fig. 22.20. *Calorophus minor* forming a continuous sward on peaty sand. Shrubs of *Callistemon pachyphyllus*. Forest of *Melaleuca quinquenervia* behind. Near Forster, N.S.W.

Fig. 22.21. *Blechnum indicum* as a subsidiary species in a *Calorophus minor* sedgeland. Young plants of *Melaleuca quinquenervia* behind. Near Forster, N.S.W.

	Cyperaceae and Restionaceae	Herbs	Shrubs
South-east Qld and north coast N.S.W.	Restionaceae: *Calorophus minor, Hypolaena fastigiata, Leptocarpus tenax, Lepyrodia interrupta, L. muelleri, L. scariosa, Restio complanatus, R. pallens, R. tetraphyllus* ssp. *meiostachys* Cyperaceae: *Gahnia clarkei, Lepidosperma filiforme, L. longitudinale, Ptilanthelium deustum, Schoenus brevifolius, S. imberbis, S. paludosus*	Pteridophytes: *Blechnum indicum, Gleichenia dicarpa, Lindsaea linearis, Lycopodium laterale, Selaginella uliginosa* Angiosperms: *Blandfordia grandiflora, Drosera binata, D. spathulata, Haloragis micrantha, Sowerbaea juncea, Stackhousia viminea, Utricularia dichotoma, Xanthorrhoea resinosa, Xyris complanata, X. gracilis, X. operculata*	*Banksia latifolia, Boronia parviflora, Callistemon pachyphyllus, Epacris obtusifolia, E. pulchella, Hakea teretifolia, Leptospermum liversidgei, Melaleuca nodosa, M. thymifolia, Sphaerolobium vimineum, Sprengelia sprengelioides, Viminaria juncea*
Sydney region and south-east N.S.W.	Restionaceae: *Calorophus minor, Hypolaena fastigiata, Leptocarpus tenax, Lepyrodia gracilis, L. scariosa, Restio complanatus. R. fastigiatus, R. gracilis, R. tetraphyllus* ssp. *meiostachys* Cyperaceae: *Chorizandra sphaerocephala, Gymnoschoenus sphaerocephalus, Lepidosperma flexuosum, L. longitudinale, Ptilanthelium deustum, Schoenus brevifolius, Tricostularia pauciflora*	Pteridophytes: *Gleichenia dicarpa, Lindsaea linearis, Lycopodium laterale, Selaginella uliginosa* Angiosperms: *Blandfordia grandiflora, Drosera binata, D. pymaea, D. spathulata, Haloragis micrantha, Sowerbaea juncea, Stackhousia viminea, Utricularia dichotoma, U. lateriflora, Xanthorrhoea resinosa, Xyris complanata, X. gracilis, X. operculata*	*Banksia latifolia, Bauera rubioides, Epacris obtusifolia, E. pulchella, Hakea teretifolia, Melaleuca nodosa, M. squarrosa, M. thymifolia, Sphaerolobium vimineum, Sprengelia incarnata, Symphyonema paludosum, Viminaria juncea*
Victoria	Restionaceae: *Calorophus minor, Hypolaena fastigiata, Leptocarpus tenax, Restio complanatus, R. tetraphyllus* ssp. *tetraphyllus* Cyperaceae: *Gahnia radula, G. trifida, Gymnoschoenus sphaerocephalus, Lepidosperma filiforme, L. longitudinale*	Pteridophytes: *Lycopodium laterale, Selaginella uliginosa* Angiosperms: *Drosera binata, D. pygmaea, D. spathulata, Stackhousia viminea, Utricularia australis, Xanthorrhoea resinosa, Xyris gracilis, X. operculata*	*Epacris impressa, E. obtusifolia, Leptospermum lanigerum, Melaleuca gibbosa, M. squamea, M. squarrosa, Sprengelia incarnata*
Tasmania	Restionaceae: *Calorophus minor, Hypolaena fastigiata, Leptocarpus tenax, Restio complanatus, R. monocephalus, R. oligocephalus, R. tetraphyllus* ssp. *tetraphyllus* Cyperaceae: *Baumea acuta, Gahnia radula, Gymnoschoenus sphaerocephalus, Lepidosperma filiforme, L. longitudinale, Schoenus tenuissimus*	Pteridophytes: *Lindsaea linearis, Selaginella uliginosa* Angiosperms: *Drosera pygmaea, Patersonia filiforme, P. fragilis, Xanthorrhoea resinosa, Xyris operculata*	*Banksia marginata, Bauera rubioides, Boronia parviflora, Dillwynia glaberrima, Epacris impressa, E. lanuginosa, Leptospermum scoparium, Melaleuca gibbosa, M. squamea, M. squarrosa, Sprengelia incarnata*

gently sloping country where waterlogged soils occur.

In the most heavily waterlogged areas, *Gymnoschoenus* forms pure stands (Fig. 22.22), with species of *Drosera, Utricularia, Euphrasia* and *Sphagnum* below. In less waterlogged stands larger plants occur, e.g. at Port Davy *Lycopodium densum*, and several ground orchids of the genera *Cryptostylis, Prasophyllum, Pterostylis* and *Thelymitra* (Davis, 1940), and at Cradle Mountain at an altitude of c. 1200 m *Abrotanella fosterioides* and *Astelia alpina*, plants characteristic of the alpine zone (Casson, 1952). With decreased waterlogging *Gymnoschoenus* becomes less abundant and other sedges occur between the stools, especially *Lepidosperma filiforme* and *Restio complanatus* (Curtis and Somerville, 1947) or *Calorophus* and its associates (Table 22.5). In still better drained areas, though periodically waterlogged, shrubs appear in the sedgeland, usually species of *Epacris, Sprengelia, Leptospermum* (Table 22.5) or the Tasmanian endemics, *Cenarrhenes nitida* and *Agastachys odorata*. Streams which flow through the sedgeland provide better aeration along their banks which support woody species, chiefly *Melaleuca squamea, M. squarrosa, Eucalyptus nitida* and *Nothofagus cunninghamii*.

In the Bulli district at an altitude of c. 300 m, Davis (1941) describes a community of *Gymnoschoenus* with *Sphagnum*, with subsidiary *Calorophus minor, Lepidosperma forsythii* and other herbs and shrubs typical of the coastal sedgeland (Table 22.5). On the Northern Tablelands, *Gymnoschoenus* dominates swamps on low fertility granites with *Lepidosperma limicola* as the main associated species and subsidiary *Calorophus minor, Blandfordia grandiflora, Ptilanthelium deustum, Xyris operculata, Schoenus* spp., *Juncus* spp. and *Drosera*. Some areas are dotted with shrubs, of which *Sprengelia incarnata, Callistemon pallidus* and *Leptospermum arachnoides* are the most common.

3.3.5. *Gahnia trifida – G. filum* Alliance
Cutting Grass – Thatching Grass

This alliance, now mutilated or destroyed by agriculture, is described by Crocker (1944) as a treeless assemblage on Rendzina soils, which are subsaline and waterlogged in the winter. A common associate and sometimes a local dominant is the grass *Poa poiformis* (Chap. 20). Some areas were once dotted with or supported dense scrubs of *Banksia marginata*. *G. filum* is more tolerant to salinity than *G. trifida* and dominates more salty patches where *Scirpus antarcticus* and *Distichlis distichophylla* also occur.

Fig. 22.22. *Gymnoschoenus sphaerocephalus* sedgeland. The trees behind are *Eucalyptus gunnii*. Cradle Mt. area, Tas.

Other species present between the *Gahnia* tussocks are *Baumea juncea, Leptocarpus tenax, Centrolepis polygyna* and *Schoenus nitens*.

3.3.6. *Lomandra dura –*
L. effusa Alliance
Iron Grasses

These species occur as dominants in hilly country and across broad valley southward of the Flinders Range in South Australia (Wood, 1937; Jessup, 1948). The mean annual rainfall is 280–600 mm and the soils are Lithosols or alluviums (valleys). The plants are sedgelike, forming tussocks and reaching a height of 30–40 cm. The main associated species are *Danthonia caespitosa* and *Stipa variabilis*. A complete species list is given by Specht (1972).

3.4. Communities Associated with Rivers

3.4.1. General

The herbaceous flora of the open water of rivers is essentially the same as that found in fresh water lakes and lagoons. The abundance of these herbs in rivers is governed partly by the rate of flow of water and partly by the turbidity of the water. In contrast to the inland rivers (Chap. 21) the coastal streams carry relatively little suspended matter and are therefore clear, except in times of flood. Free floating and submerged aquatics are usually absent from swiftly flowing rivers, but they may be abundant in backwaters or among reeds, where they are unlikely to be swept away by currents.

Phragmites australis and *Typha domingensis* are the most common taller helophytes fringing the banks and several species of *Juncus* are often dominant in shallow water, mainly in the south. The most widespread and common of these are *J. bufonius, J. pallidus, J. planifolius* and *J. prismatocarpus*.

In the most humid habitats created by waterfalls the most fragile species are found, especially under conditions of low light intensity. In such places herbaceous species which normally occur in the rainforest understory are usually locally dominant. A very few species are confined to such perhumid climates, e.g. the ferns *Schizaea rupestris* (confined to N.S.W.) and *Leptopteris fraseri* (a filmy tree-fern, occuring in N.S.W. and Qld).

The woody species associated with rivers vary from district to district but relatively few are rigidly confined to or usually associated with flowing water. Some examples of these are *Casuarina cunninghamiana* and a few of its associates (see below), *Tristania laurina, T. neriifolia* (mainland, east), *Leptospermum riparium* (Tas.) and *Acradenia frankliniae* (monotypic endemic genus restricted to western Tas.). In areas where rainforests occur, species belonging to the rainforests commonly fringe water and in extensive rainforest stands certain species are usually associated with watercourses (Chap. 7). *Casuarina cunninghamiana* is regularly associated with streams in the east from southern New South Wales to Cape York Peninsula; it occurs also on the Arnhem Peninsula where it is found on watercourses, often with ribbons of rainforest which edge the stream (Chap. 7). In the east it extends to the tablelands to an altitude of c. 900 m (Fig. 22.23), and westward, where it is replaced by *Eucalyptus camaldulensis* at an altitude of c. 300 m. *Casuarina cunninghamiana* reaches a height of c. 30 m and has no regular associates over its whole range. In the north it is sometimes mixed with *E. tereticornis* on the coastal lowlands; in the south it is sometimes associated with *E. elata*. Because of its preference for wet habitats, rainforest species sometimes occur as an understory and a few xeromorphic shrubs and trees are locally constant associates, especially in the north. Rainforest species in tropical Queensland are *Nauclea orientalis* and *Tristania grandiflora* and xeromorphic species are *Eucalyptus raveretiana* (confined to watercourses), *Melaleuca mimosoides* and *M. saligna* (Perry, 1953). South of the Tropic and extending into northern New South Wales, *Grevillea robusta, Tristania conferta, T. laurina, Angophora subvelutina, Melaleuca bracteata* and *Callistemon viminalis*, with or without rainforest species, are the most common associates. Further south there are no distinctive associated shrubs but *Angophora floribunda* may flank the *Casuarina*, and rainforest species belonging to the *Ceratopetalum apetalum* alliance may be present (Chap. 7).

The herbaceous layer has no distinctive set of species. Blake (1942) records an assemblage containing *Cyperus exaltatus, Carex polyantha,*

Fig. 22.23. *Casuarina cunninghamiana* edging Booralong Creek. Tall trees of *Casuarina* are c. 22 m high. Also present (right) trees of *Angophora floribunda*. Herbaceous plant is *Lomandra longifolia*. Near Armidale, N.S.W.

Pennisetum alopecuroides and *Lomandra longifolia*. The last is probably the most abundant species in the more southerly stands, as in Fig. 22.23, and others of minor importance are *Microlaena stipoides*, *Agrostis avenaceus* and several semi-aquatics, e.g. *Callitriche * stagnalis*.

4. Communities in South-western Western Australia

4.1. General

Swamps occur in the south-west near the coast between the Hill R. and Israelite Bay (Gardner, 1942). Along the west coast many of the swamps surround lakes and lagoons in limestone country. Along the south coast the swamps are located mainly in siliceous sands behind the coastal dunes and rarely surround open water.

The communities in the south-west resemble those in the south-east and many species are common to the two regions; the zonations of the communities are the same. Differences that occur between the two regions involve differences in species in many of the genera, and the occurrence in the west of several endemic genera. In the west, the hydrophytic flora is small in comparison with the east. Furthermore, the area covered by brackish water and mud in the west is very small, being restricted to some river estuaries and a few coastal lakes. In the brackish estuary of the Swan R., Seddon (1972) records *Zostera muelleri* and *Posidonia australis* as submerged aquatics. The zone of helophytes edging the open water is dominated by *Scirpus validus*, with a zone of *Baumea juncea* behind; this is similar to but not identical with conditions in the east. Edging the sedgeland there is a zone of *Casuarina obesa* (c.f. *C. glauca* in the east) which is associated with *Melaleuca raphiophylla* in some areas.

The remaining communities are described under separate headings in order of decreasing water level, viz. aquatic communities, sedgelands, *Melaleuca* and *Banksia* communities.

4.2. Communities in Permanent and Semi-permanent Fresh Water

Few free-floating and submerged aquatics are recorded for this area (Smith and Marchant, 1961) and many which do occur are apparently never abundant. The species recorded are the free-floating *Azolla filiculoides*, *Lemna minor* and *Spirodela oligorrhiza*. The strictly submerged species are *Potamogeton* (4 spp.), *Naias major*, *Triglochin procera*, *T. striata* and *Myriophyllum propinquum*; *Ottelia ovalifolia* has floating leaf-blades.

At Loch McNess, a body of fresh, alkaline water in limestone country in the Yanchep National Park, McComb and McComb (1967) record only algae in the free water, but *Azolla* and *Lemna* are recorded by Smith and Marchant (1961). The main species edging the free water is *Scirpus validus*. Local variations include the dominance of *Typha domingensis* or *Lepidosperma gladiatum*. Subsidiary species are *L. drummondii* and *Baumea juncea*. The last of these dominates the sedgeland which adjoins the swamp in shallower water. In this sedgeland, woody species establish in less waterlogged areas, especially *Viminaria juncea*, *Xanthorrhoea preissii*, *Phebalium anceps*, *Acacia cyanophylla* and *Banksia littoralis*. On more elevated areas, *Melaleuca raphiophylla* forms woodlands to 8 m tall.

4.3. Sedgelands

4.3.1. General

Sedgelands develop on siliceous or peaty sands or on peats which are subject to prolonged waterlogging during the winter. Two alliances are described below though others (not yet reported in the literature) may occur. The communities are superficially similar to those in the east but differ floristically both at the species and generic levels – items which are commented on under the alliances.

4.3.2. *Leptocarpus aristatus* Alliance

This alliance is described for the Cannington area, near Perth (Univ. W.A., 1972), where an extensive sedgeland has developed on flat sandy country crossed by two parallel sandridges supporting *Banksia* woodlands (Chap. 16). The flats are waterlogged during the winter and support a rich flora of c. 350 species, forming a mosaic of communities, which are delimited partly by the height of the water-table, partly by salinity and partly by the proximity of clay to the surface.

Where a surface sandy soil overlies clay in the subsoil, maintaining high moisture levels at the surface, a sedgeland of *Leptocarpus aristatus* occurs, and is sometimes dotted with shrubs, especially *Hakea varia*, *Banksia sphaerocarpa* and *Calothamnus villosus*. In spring and early summer ephemeral species come into flower. These include small geophytes such as orchids (*Caladenia*, *Diurus*, *Prasophyllum*, *Thelymitra*), insectivorous plants (*Byblis*, *Drosera*, *Polypompholyx*, *Utricularia*), the fern allies, *Phylloglossum* and *Isoëtes*, several species of *Stylidium*, some Liliales (*Burchardia*, *Tribonanthes*, *Anigozanthos* and a few Compositae (*Brachycome*, *Cotula*). Mounds of sand on the flats are usually dominated by *Melaleuca fasciculiflora*, *Actinostrobus pyramidalis*, *Banksia sphaerocarpa* and *Calothamnus villosa*.

Saline depressions are sparsely covered by *Arthrocnemum halocnemoides* and *Salicornia australis*, and *Viminaria juncea* dominates areas where clay lies near the surface.

As the terrain rises from the *Leptocarpus* meadow to sand ridges, *Leptospermum ellipticum* dominates scrub communities which give way to *Banksia* woodlands on higher ground.

4.3.3. *Evandra – Anarthria – Lyginia* ssp. Alliance

These three genera, the first Cyperaceae, the other two Restionaceae, are endemic to Australia and confined to the south-west. Con-

Fig. 22.24. Woodland of *Melaleuca raphiophylla* in a swamp. Near Myalup, W. A.

sequently the communities which the various species dominate are unique in the continent. The sedgelands are restricted to small areas near the coast in the climatically wettest areas.

Loneragon (1952) describes a community with dominants of *Anarthria laevis, Lyginia barbata, Lepyrodia muirii, Hypolaena exsulca, Lepidosperma angustatum, Amphipogon laguroides* and *A. debilis*. The ground flora consists of insectivorous plants (*Drosera* and *Polypompholyx*), *Stylidium* spp., *Borya nitida* and *Patersonia occidentalis*. Some shrubs occur – species of *Astartea, Kunzea, Leptospermum* and others. The fringing community is dominated by *Banksia littoralis*.

In the extreme south-west a similar assemblage occurs on black peaty sand behind the coastal dune and extending short distances inland along river flats into the *Eucalyptus marginata* forests (Smith 1972, 1973). Here the characteristic species are *Evandra aristata* (up to 1 m tall), *Anarthria prolifera, A. scabra, Restio applanatus, Leptocarpus scariosus, Lepidosperma persecans* and *Lyginia barbata*. Shrubs occur in the less waterlogged areas, especially *Leptospermum firmum, L. ellipticum, Astartea fascicularis* and *Agonis linearifolia*. On low narrow sand ridges *Banksia ilicifolia* is dominant.

4.4. Communities Dominated by *Melaleuca*

4.4.1. *Melaleuca raphiophylla* Alliance

This species occurs mainly on the western coastal plain where it forms woodlands c. 8–13 m tall around swamps (Fig. 22.24), on flats subject to waterlogging and along streams, mainly within the *Eucalyptus gomphocephala* and *E. marginata* alliances (Chap. 15). It is commonly surrounded by a zone of *E. rudis* which is a typical watercourse species in the south-west (Chap. 21). South of Busselton, *M. raphiophylla* is sometimes associated with *M. acuminata*. Understory shrubs are rare and include *Acacia saligna* and/or *Banksia littoralis*. The herbaceous layer consists almost entirely of helophytes which sometimes form a continuous sward, the most common species being *Baumea juncea* or *Juncus pallidus* (Smith, 1972).

4.4.2. *Melaleuca preissiana* Alliance

This species occurs mainly along the wetter parts of the south coast and it occupies open flats of leached sand subject to seasonal flooding, being most common behind the coastal dunes. It may extend along drainage lines on flats adjacent to streams or lakes (Smith, 1972, 1973). It forms low open scrubs 2–7 m tall, the shrubs sometimes in groups.

Pure stands are common, but mixed stands with other species occur, viz. *Banksia ilicifolia*, stunted *Eucalyptus marginata*, *E. decipiens* *Nuytsia floribunda*, *Xanthorrhoea preissii* or *Agonis juniperina*. The last species is often a local dominant, extending along creeks in forests (especially *E. diversicolor*) where it reaches a height of 10 m. The unterstory in the *Melaleuca* communities is either xeromorphic shrubs in better drained areas or sedges in waterlogged patches. The main shrubs are *Astartea fascicularis*, *Agonis parviceps* and *Leptospermum firmum*. The main sedges are *Lepidosperma longitudinale* and *Mesomalaema tetragona*.

4.4.3. Other Species of *Melaleuca*

Three other species of *Melaleuca* are locally dominant but cover relatively small areas.

i. *M. lanceolata* (which occurs also in the east, see 3.1, above) forms scrubs in patches mainly along the south coast (Smith, 1972) and on the Recherche Islands (Willis, 1953).

ii. *M. thyoides* is recorded in patches along the south coast by Beard (1972); it dominates scrubs c. 3 m tall, with a dense canopy and often no understory species.

iii. *M. polygaloides*, either in pure stands or associated with *Agonis linearifolia* or *A. flexuosa* is recorded by Smith (1973) in the Busselton area.

4.5. Communities Dominated by *Banksia*

Speck (1954) described two associations dominated by species of *Banksia*. These are found in the south-west in areas receiving a mean annual rainfall of 750–1000 mm. They lie adjacent to the *Melaleuca raphiophylla* woodlands. The associations are characterized by:

i. *Banksia ilicifolia*, occurring on flats which, in winter, are covered by a few centimetres of water; the trees, up to 12 m high, are clumped into thickets separated by small more swampy areas.

ii. *B. littoralis*, which is restricted to a narrow bands a few trees deep on slightly higher ground which is not waterlogged for long periods; the species sometimes extends as an understory to the adjacent *Eucalyptus rudis* woodlands.

23. Communities of the Littoral Zone

1. Introduction

1.1. General; Ocean Currents; Climates; Tides; Sea Water

The littoral zone adjoins the coast and includes the strip of substrate lying between tide levels and below low tide level. The coastline has a length of c. 16,000 km and is affected by ocean currents, the tides and the winds, all of which have some influence on topographic features, the climate, the flora and the plant communities. The continent is washed by cold ocean currents from the south and warm currents from the north (Fig. 23.1).

In general terms, the coast line is relatively warm, zero temperatures never occurring close to high tide mark, even in the south, though frosts may be of experienced regularly a few

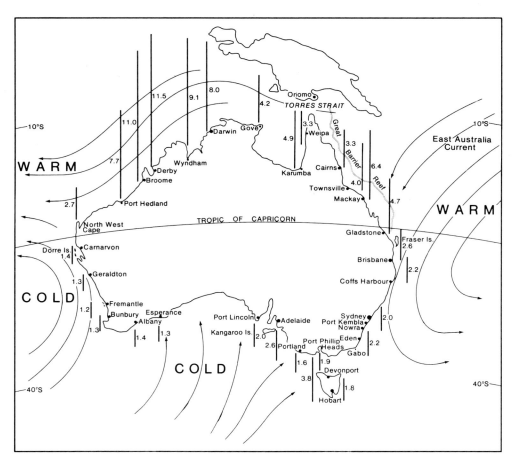

Fig. 23.1. The Australian coast-line and some factors which affect the littoral zone: ocean currents indicated by arrows and "warm" or "cold"; tidal-range (in metres) indicated by vertical lines, which refer to the closest town or city indicated.

hundred metres inland. The north is invariably the warmest, with the lowest mean minimum temperatures above 10°C and absolute minima 5°C. In the south, including Tasmania, mean monthly minima vary between 4°C and 10°C, and absolute minima are not lower than −0.5°C for recording stations located near the ocean. The warming effect of the warm northerly current is more pronounced in the east and accounts for the southern extension of several tropical taxa, especially the mangroves along the east coast well south of the Tropic. However, in river estuaries which extend inland in the south, subzero temperatures may occur and frost damage to littoral communities may follow.

Mean annual rainfall on the coast line varies tremendously and also the seasonal incidence of rain (Chap. 1). Whereas rainfall has a controlling effect on the climax vegetation adjoining the littoral zone, it has a lesser effect on the vegetation along the coast where winds, sea spray and the tides are the main factors governing the plant communities.

The tidal range (the vertical distance between the highest and lowest astronomical tides) varies considerably around the continent, the greatest range (11.5 m) being recorded on the north-west coast, the lowest (1.2–1.3 m) in the south-west (Fig. 23.1). Maximum tidal range can happen only when the sun, earth and moon lie on the same straight line, which is not necessarily an annual event. At other positions of earth, moon and sun, the tidal range is smaller. When the moon is new or full the tides are highest (Spring Tides) and a correspondingly low tide follows a Spring Tide. In the text that follows vegetation zones are referred to the positions of high water at Spring Tide. However, the highest astronomical tides (King Tides) flood the uppermost parts of some littoral and adjoining climax communities, especially when high wind velocities prevail. Under these conditions not only is the salt water at its highest but coastal erosion and the blowing of salt sea spray inland are at their maximum.

Under conditions of high wind velocity salt is carried inland (cyclic salt, see Chap. 1) and the plants of the littoral zone, as well as those growing in or periodically submerged or partially submerged by sea water, are affected in some way by salt from the ocean. Sea water contains c. $3^1/_2$% of inorganic material of which c. 3% is sodium chloride. The remainder consists mainly of sulphate, magnesium, calcium and potassium, all of which are macronutrients for plants. Other nutrients are in much lower supply, especially nitrogen and phosphorus which occur at the very low concentration of no more than 0.02 ppm. However, marine algae build their tissues from these concentrations and supply food for marine animals which cycle the nutrients and add to the supply for the littoral zone angiosperms which are rooted in the mud below low tide mark or between tidal levels. The bodies of marine organisms (plant and animals) and faeces which are carried to the land by the waves or the wind are important sources of nutrients for some terrestrial plants, the significance of which is discussed in more detail below. The introduction of sodium chloride into the terrestrial systems has some effect on the morphology of the plants, which is expressed chiefly in the development of leaf-succulence.

1.2. Habitats

The littoral zone presents a number of habitats classified as: i. Rocky areas; ii. Sand dunes; iii. Mudflats or sandy alluviums between tide levels or below low tide mark. Also, many islands, large and small, occur around the coast on the continental shelf. The communities are described under these four headings.

On many stretches of coast a distinct littoral flora and vegetation do not occur. On the contrary, the climax communities adjoin the sea, the two being separated by a vertical bank of soil or rock. The condition is common in some estuaries and along those parts of the coast where wave action is minimal, e.g. where offshore coral reefs in tropical waters protect the land (Fig. 23.1).

1.3. Flora

The flora is mostly angiospermic. A few pteridophytes are adapted to the saline conditions of the ocean front, notably *Acrostichum speciosum* in some mangrove swamps and *Asplenium obtusatum* on cliffs on the east coast. Of the gymnosperms, *Callitris columellaris* and *Macrozamia communis* occur on some coastal dunes in the east.

The angiosperm members of the littoral flora

have been selected from a large number of families, and a few families are confined to the littoral zone. The latter group includes the marine angiosperms (Posidoniaceae and Zosteraceae) and the mangroves belonging to the Rhizophoraceae. The Chenopodiaceae and Aizoaceae, mostly halophytic, are well represented in the littoral zone but occur elsewhere. These are no endemic genera confined to the littoral zone. Most of the genera and many species which grow submerged in sea water are found on the coast lines of other continents. The terrestrial flora is partly cosmopolitan (at least at the generic level) and partly Australian, the latter including, in particular, the phyllodineous acacias, *Casuarina*, *Melaleuca* and many others. Further information on individual taxa is provided below.

Table 23.1. Most common species occurring on rocks near the sea. Herbs and succulents (usually nearer the sea) are listed before shrubs. Sources of data, see text.

WEST COAST SOUTH OF THE TROPIC		SOUTH-EAST QUEENSLAND AND NEW SOUTH WALES
Carpobrotus *aequilaterus, *Senecio lautus*, *Swainsona* sp. *Atriplex paludosa*, *Frankenia pauciflora*, *Threlkeldia diffusa*, *Acanthocarpus preissii*, *Olearia axillaris*, *Pimelea floribunda*, *Diplolaena dampieri*		*Pandanus pedunculata* (north), *Samolus repens*, *Apium prostratum*, *Lobelia anceps*, *Pelargonium australe*, *Plantago varia*, *Cotula coronopifolia*, *Tetragonia expansa*, *Scirpus nodosus*, *Lomandra longifolia*, *Scaevola calendulacea*, *Rhagodia nutans*, *Asplenium obtusatum*, *Westringia fruticosa*, *Eriostemon buxifolius*, *Pseudanthus orientalis*, *Leptospermum laevigatum*
WETTER WEST COAST (SOUTH) AND WETTER SOUTH COAST W.A.	SOUTH AUSTRALIA AND VICTORIA	TASMANIA
Carpobrotus *aequilaterus, *C. *edulis*, *Cotula coronopifolia*, *Samolus repens*, *Ranunculus muricata*, *Tetragonia implexicoma*, *T. decumbens*, *Wilsonia backhousii*, *Enchylaena tomentosa*, *Nitraria billardieri*, *Rhagodia baccata*, *Westringia rigida*, *Olearia axillaris*, *Alyxia buxifolia*, *Myoporum insulare*	*Arthrocnemum halocnemoides*, *Salicornia quinqueflora*, *Samolus repens*, *Senecio lautus*, *Disphyma australe*, *Carpobrotus rossii*, *Scleranthus pungens*, *Apium prostratum*, *Tetragonia implexicoma*, *Pelargonium australe*, *Asplenium obtusatum*, *Poa poiformis*, *Calocephalus brownii*, *Atriplex cinerea*, *A. paludosa*, *Rhagodia baccata*, *Enchylaena tomentosa*, *Correa alba*, *Myoporum insulare*, *Olearia axillaris*, *Leucopogon parviflorus*, *Casuarina stricta*	*Samolus repens*, *Tetragonia expansa*, *Senecio lautus*, *Calocephalus brownii*, *Dichelachne crinita*, *Senecio odoratus* (north), *Poa poiformis*, *Lomandra longifolia*, *Helichrysum reticulatum*, *Rhagodia billardieri*, *Correa alba*, *Myoporum insulare*, *Helichrysum costatifructum* (south and east), *Cyathodes abietina* (south-west), *Alyxia buxifolia*, *Leucopogon parviflorus*, *Cassinia spectabilis*, *Casuarina stricta*

2. Communities on Rocky Outcrops and Headlands

Rock forms the coast line in many places, the rocks being granitic, gneiss, basalt, dolerite, metamorphics, or sedimentaries especially sandstones and limestones. In the tropics ferruginous fossil laterite is often the main constituent of headlands. The type of rock has some effect on the small communities which develop, chiefly through the method of weathering and fragmentation under the influence of the waves. Sedimentary rocks are most readily colonized, chiefly because plants establish in softer seams or along bedding planes. Harder rocks are colonized partly through the growth of algal mats or lichens on their surfaces and partly by the accumulation of rock fragments and debris among boulders or in clefts.

In the tropics rock outcrops adjoining the ocean do not support a distinctive flora, but a few species found in other littoral communities are haphazardly distributed in pockets of "soil", especially *Sesuvium portulacastrum*, *Ipomoea pes-caprae* and *Pandanus*. South of the Tropic, many species found elsewhere in the littoral zone occur also on rocks, but in addition there are a few species, notably in Tasmania, which are confined to the rocky habitat. The more common species found in various sectors around the continent are listed in Table 23.1. The plants in the various sectors usually occur in distinct zones according to their salt-tolerance but the zones are not necessarily indicative of a successional relationship since, in most cases, the coast line is being degraded by the sea rather than being built up by land plants. Many of the species have succulent or semi-succulent leaves and the more xeromorphic species usually show

Fig. 23.2. *Pandanus pedunculatis* and *Casuarina equisetifolia* on a cliff overlooking the sea. The *Pandanus* is c. 4 m tall; note stilt roots. Near Caloundra, Qld.

some degree of salt-hypertrophy. The amount of salt absorbed by the plants varies from species to species and appears to be related to the thickness of the cuticle (Parsons and Gill, 1968).

The most conspicuous of the subtropical sea-cliff plants in the east is *Pandanus peduncularis* (Fig. 23.2). On the central and south-east coast the shrub growing closest to the salt spray is *Westringia fruticosa* and in Tasmania *Helichrysum reticulatum* (Fig. 23.3), and *Casuarina stricta* is the tallest shrub (3–4 m) growing on rock overlooking the sea (Fig. 23.3); it is common also in estuaries and behind some sand dunes, or even inland (see below).

Two zonations on granite have been described from sites on the south coast. Ashton and Webb (1977) record the following zonation on granite on Wilson's Promontory (mean annual rainfall 1000–1500 mm). The rocks are colonized by the orange crustose lichen, *Caloplaca marina,* and the grey foliose lichen, *Parmelia conspersa*. In the adjacent shallow soil zone, trailing succulents *Disphyma australe* and *Carpobrotus rossii,* and where salt accumulates in the soil, *Samolus repens* and *Salicornia quinqueflora* are occasional. The shrub zone consists of scattered cushions of *Calocephalus brownii*, with *Poa poiformis* and *Correa alba*. On deeper soils a windswept scrub or heath develops with dominants of *Casuarina pusilla, Leptospermum myrsinoides* etc. (Chap. 16) and on slightly less exposed sites woodlands of *Casuarina stricta*. The zonation on Pearson Is. (Specht, 1969) commences with encrustations of blue-green algae followed by: black lichen encrustations (*Lichina-Verrucaria*); orange and grey lichens (as above); *Arthrocnemum halocnemoides – Frankenia pauciflora; Disphyma blackii – Enchylaema tomentosa; Atriplex cinerea – Calocephalus brownii; Atriplex paludosa – Rhagodia crassifolia; Nitraria billardieri; Olearia axillaris – Leucopogon parviflorus – Correa reflexa; Casuarina stricta – Melaleuca lanceolata*.

Several species extend from South Australia westward along the south coast where limestones (Nullarbor region, Fig. 23.4) and gneiss in the south-west form the substrate. Some floral differences are encountered in the south-west where a few of the local endemic genera have produced salt-adapted species. On the drier parts of the west coast, where the rocks are limestones, a relatively small number of species is recorded (Table 23.1) possibly the consequence of the low rainfall.

3. Communities on Sand Dunes

3.1. General

Dunes occur around much of the coast line, being built up from materials washed up from ocean since the Pleistocene glaciation. The sands are sometimes mostly siliceous, the grains being coated with a film of hydrated ferric hydroxide which provides the colour to the grains. The colour varies from snow-white, when no ferric hydroxide is present, to red-brown, the most common colour being yellow. Other materials are usually incorporated into the sand mass, especially shells and coral fragments which are mainly calcium carbonate with some calcium phosphate. The latter is more abundant in the tropics and in some cases coral sand comprises the whole of the dune.

Other minerals may also be present, as determined by the landward rock from which the

Fig. 23.3. Sandstone cliffs and rock platform adjoining the ocean. Species on rock platform are *Lomandra longifolia* and *Dichelachne crinita*. Shrubs at base of cliff are *Helichrysum reticulatum* (lowest), *Correa alba, Myoporum insulare, Helichrysum dendrodium, Leucopogon parviflorus* and *Acacia verticillata*. Taller shrub (3–4 m high) on right is *Casuarina stricta*. Tasman Peninsula, south-east Tas.

Fig. 23.4. The eastern end of the limestone cliffs which extend around the Great Australian Bight. Plants growing in crevices are: *Myoporum insulare, Zygophyllum ammophilum, Disphymia australe, Threlkeldia diffusa* and (at the base of cliff) *Spinifex hirsutus* and *Scirpus nodosus*. The beach of snow-white sand is strewn with marine brown algae. South of Cocklebiddy, W.A.

sand was derived, e.g. fragments of mica, feldspar or hornblende, all of which have some significance with respect to the nutrition of the plants which grow on the dunes. On the central east coast heavy minerals occur below the surface sands and in recent years the economic importance of these minerals has been realized so that "sand mining" is operative on some coastal dunes. These minerals are rutile (titanium dioxide, used in paint), monazite (cerium phosphate), zircon (zirconium silicate) and ilmenite (iron-titanium oxide). Mining companies are obliged to restore the dune to its near-natural condition and this appears usually to be readily effected, though the contour of the dunes is sometimes altered (Broese, 1974; Clark, 1975; Thatcher and Westman, 1975). Broese indicates that on the dunes in northern New South Wales 30 of the 36 original species returned to mined area following restoration. Some alteration to the floristic composition usually results from the introduction of cover crops, e.g. some grasses and lupins, or through the establishment of rapidly growing perennials. The latter include *Acacia saligna* and *Albizia lophantha* both native to the south-west and now established on the coastal dunes in New South Wales.

The dune systems are either simple or complex (Ward and Little, 1975). In all cases there is a frontal (mobile) dune which adjoins the uppermost high tide level (strand). The width of this dune varies considerably and in some cases it measures no more than a few metres, e.g. on some parts of the west coast, where the yellowish sands of the littoral dune meet the red-brown sands of the inland dune systems. Complex dune systems occur mainly along the central-eastern coast where several dunes have been built up over the years, the inner (barrier) dunes often being separated from the frontal dune by an extensive flat which is often swampy, a lake or a creek.

The frontal dune may take on a characteristic shape, gently sloping on the seaward side, abruptly sloping on the landward side, with a height of c. 20 m. Alternatively the frontal dune is almost level, especially in the tropics where the beach is protected from wave action by a coral reef. In most cases, but especially in the south, the frontal dune is subjected to continual change resulting from changes in wind velocity and direction, wave action and ocean currents, the latter altering the location of sand banks. Consequently plant communities (and man-made structures) are often destroyed and new bare areas are being created, which necessitates intensive study relating to the management of the frontal zone (Quilty and Wearne, 1975).

The compositions of the sand vary in the various sectors around the continent. As far as mineral plant nutrition is concerned, the various nutrients needed by plants are readily available from sea water, except nitrogen and phosphorus. The snow-white sands found in the south-west and along part of the south coast (Fig. 23.5) appear to be low in iron but there is no evidence of an iron deficiency in the plants which grow on the sands. An added supply of phosphorus from shells or coral is likely to satisfy the needs of plants; e.g. on the central coast dune sands often contain c. 150 ppm phosphorus, whereas the level in adjacent heath is c. 30 ppm. The supply of nitrogen to the plants is still obscure for species growing on the seaward side of the shrub zone. Plants in the shrub zone (*Acacia* and *Casuarina*) have a nitrogen – fixing symbiont in the roots.

The chemical composition of the sands along a transect at right angles to the beach show changes in certain constituents, which are related to the length of time the sand has been leached and the amount of vegetation which the sand (soil) supports. The first materials to be removed by rain are the soluble ions (Cl^-, SO_4^{2-}, Na^+, K^+, Mg^{2+}) though these are continually being added by spray from the ocean. Before plants colonize the sands, pH value is c. 8.0 and alkalinity is maintained in the medium until calcium carbonate is removed by leaching, this process being assisted by the addition of acid humus from plants. With the development of acidity through the accumulation of organic matter, the iron is finally removed from the profile and a

Fig. 23.5. Snow-white frontal dune. Plants are *Spinifex hirsutus, Rhagodia baccata* (shrub) and on the strand *Arctotheca nivea, Euphorbia *paralias* and *Atriplex isatidea* (in distance). Great Australian Bight, south of Cocklebiddy, W. A.

Fig. 23.6. *Spinifex hirsutus*, a rhizomatous stoloniferous grass binding a sand surface, and a mat of *Scaevola suaveolens*. Port Stephens, N.S.W.

Podzol develops (in the wetter areas). Burges and Drover (1953) estimate that the time required for the development of a Podzol under a mean annual rainfall of 1200 mm is c. 2000 years. In drier areas and in coral sands the calcium carbonate is never removed and the medium remains alkaline.

3.2. Zonation of the Communities

The zonations of plants, usually regarded as successions, are remarkably similar around the coast line though species differ from sector to sector. Furthermore, the biomass of vegetation

Fig. 23.7. A contracted zonation on a frontal dune. At base of dune: *Spinifex hirsutus* and *Ipomoea pes-caprae* (also *Oxalis corniculata* and **Oenothera drummondii*). Top of bank: *Dianella caerulea*, *Lomandra longifolia* and *Imperata cylindrica*. Behind: thicket of *Banksia integrifolia* and *Acacia sophorae*. Bribie Is., Qld.

Table 23.2. Species occurring on coastal sand dunes arranged in sectors around the continent and in zones within the sectors. Zone 1 is the frontal dune where the species are mainly herbaceous but some shrubs also occur. Zone 2 is the shrub zone where the shrubs (rarely trees) form a more or less continuous ground cover. Sources of data: Patton (1934), Wood (1937), Pidgeon (1940), Davis (1941), Eardley (1943), Brass (1953), Herbert (1953), Perry (1953), Smith (1957), Turner et al. (1962), Sauer (1965), Specht (1972), Westman (1975), Beard (1976).

BROOME AREA, W. A.

Zone 1: Spinifex longifolius, Euphorbia *atoto, Ipomoea pes-caprae, Salsola kali, Canavalia obtusifolia, Psoralea martinii, Sesuvium portulacastrum

Zone 2: Acacia bivenosa, Capparis spinosa, Crotalaria cunninghamii, Gardenia pantonii, Terminalia sp., Ficus sp., Pterigenon macrocephalus, Acacia tumida

ADJOINING: Acacia pachycarpa scrub

PORT HEDLAND AREA TO TROPIC, W. A.

Zone 1: Spinifex longifolius, Canavalia obtusifolia, Ipomoea pes-caprae, Euphorbia *atoto, Sporobolus virginicus, *Aerva persica, Salsola kali

Zone 2: Acacia translucens, Crotalaria cunninghamii

ADJOINING: Triodia pungens

TROPIC TO SHARK BAY AREA, W. A.

Zone 1: Spinifex longifolius, Atriplex isatidea, Sesuvium portulacastrum,

ARNHEM PENINSULA, N.T.

Zone 1: Spinifex longifolius, Ipomoea pes-caprae, Sesuvium portulacastrum, Sporobolus virginicus, Euphorbia *atoto

Zone 2: Casuarina equisetifolia, Hibiscus tiliaceus, Pongamia pinnata, Sterculia quadrifida

ADJOINING: Deciduous rainforest; Eucalyptus forest

TROPICAL EAST COAST

Zone 1: Spinifex hirsutus, Ipomoea pes-caprae, Canavalia obtusifolia, Sporobolus virginicus, Sesuvium portulacastrum, Euphorbia *atoto, Lepturus repens

Zone 2: Casuarina equisetifolia, Pandanus, Hibiscus tiliaceus, Terminalia catappa, Thespesia populneoides

ADJOINING: Rainforest (espec. Calophyllum inophyllum); Eucalyptus woodland; scrub or heath (Leptospermum, Melaleuca, Thryptomene)

SOUTH-EAST QLD AND NORTH COAST N.S. W.

Strand: *Cakile edentula

Zone 1: Spinifex hirsutus, Ipomoea pes-caprae, Canavalia obtusifolia, Sporobolus virginicus, Sesuvium portulacastrum, Oxalis corniculata, *Oenothera drummondii, Lepturus repens, Ischaemum triticeum, Dianella caerulea, Scaevola suaveolens

Zone 2: Casuarina equisetifolia,

Zone 2: Carpobrotus glaucescens,
Angianthus cunninghamii,
Euphorbia myrtoides,
Sporobolus virginicus,
Threlkeldia diffusa,
Nitraria billardieri,
Rhagodia preissii,
Scaevola crassifolia,
Myoporum insulare,
Acacia sclerosperma

ADJOINING: Acacia sclerosperma,
A. coriacea scrub

SHARK BAY AREA TO HILL RIVER, W. A.

Strand: *Cakile edentula
Zone 1: Spinifex longifolia,
Atriplex isatidea,
Salsola kali, Scirpus nodosus,
Carpobrotus glaucescens,
Tetragonia decumbens,
Angianthus cunninghamii,
Threlkeldia diffusa
Zone 2: Nitraria billardieri,
Scaevola crassifolia,
Acacia ligulata or
A. rostellifera,
Myoporum insulare,
Olearia axillaris,
Rhagodia baccata
ADJOINING: Banksia scrub or heath

WETTER WEST AND SOUTH COASTS, W. A.

Strand: *Cakile edentula,
*Arctotheca nivea
Zone 1: Spinifex hirsutus (nearest ocean),
Scirpus nodosus,
Tetragonia decumbens,
Festuca littoralis (south),
Calocephalus brownii,
Spinifex longifolius (west),

continued next page.

SOUTH AUSTRALIA AND VICTORIA

Strand: *Cakile edentula,
Salsola kali
Zone 1: Spinifex hirsutus,
Scirpus nodosus,
Calocephalus brownii,
*Ammophila arenaria,
Festuca littoralis,
Sporobolus virginicus,
Cynodon dactylon,
Tetragonia implexicoma,
Lepidosperma gladiatum,
Carpobrotus rossii,
Threlkeldia diffusa,
Poa poiformis,
Sonchus *asper,
*Lagurus ovatus,

Pandanus peduncularis (north),
Acacia sophorae (south),
Banksia integrifolia,
Hibiscus tiliaceus,
Thespesia populnea,
Sophora tomentosa,
Acacia cunninghamii,
Callitris columellaris,
Leucopogon parviflorus,
Hibbertia scandens

ADJOINING: Rainforest
(espec. Cupaniopsis anacardioides);
Eucalyptus forest
(E. tessellaris, E. intermedia)

CENTRAL AND SOUTH COAST N.S.W.

Strand: *Cakile edentula
Zone 1: Spinifex hirsutus,
Festuca littoralis,
*Ammophila arenaria,
Senecio lautus,
Sonchus megalocarpus,
S. *asper,
Sporobolus virginicus,
Pelargonium australe,
Euphorbia *sparrmanii,
Apium prostratum,
Scirpus nodosus,
Carpobrotus glaucescens,
Carex pumila,
Calystegia soldanella,
Dianella caerulea,
Cynodon dactylon,
Stackhousia spathulata,
Lomandra longifolia,
Scaevola suaveolens,
Hibbertia scandens
Zone 2: Acacia sophorae,
Leptospermum laevigatum,
Leucopogon parviflorus,

Table 23.2 continued

```
              Atriplex isatidea,              Pelargonium australe,              Monotoca scoparia,
              Carpobrotus glaucescens,        Scaevola crassifolia,              Myoporum insulare,
              Sporobolus virginicus,          Rhagodia baccata,                  Correa alba,
              Angianthus cunninghamii,        Atriplex cinerea,                  Banksia integrifolia,
              Pelargonium drummondii,         A. paludosa,                       B. serratifolia,
              Sonchus megalocarpus,           Nitraria billardieri,              B. serrata,
              *Ammophila arenaria,            Enchylaena tomentosa               Cupaniopsis anacardioides
              *Oenothera drummondii           Olearia axillaris,     ADJOINING:  Eucalyptus forest
Zone 2:       Rhagodia baccata,      Zone 2:  Leucopogon parviflorus,            (E. pilularis,
              Myoporum insulare,              Acacia ligulata (S. A.)            E. botryoides,
              Scaevola crassifolia,           A. sophorae (mainly Vic.),         E. gummifera,
              Olearia axillaris,              Myoporum insulare,                 Angophora costata);
              Acacia cyclops,                 Alyxia buxifolia,                  rainforest
              A. rostellifera,                Banksia integrifolia (Vic.),
              Lepidosperma gladiatum,         Leptospermum laevigatum (Vic.)
              Spyridium globosum,  ADJOINING: Melaleuca lanceolota (S. A.),   TASMANIA
              Alyxia buxifolia                Casuarina stricta;             Strand:    *Cakile edentula,
ADJOINING:    Banksia scrub;                  Eucalyptus spp.,                          Atriplex billardieri
              Eucalyptus gomphocephala forest;  including mallee in S. A.   Zone 1:     Spinifex hirsutus,
              Agonis flexuosa woodland                                                  Festuca littoralis,
              (Pinjarra area to Point Hood)                                             *Ammophila arenaria,
                                                                                        Scirpus nodosus,
              GREAT AUSTRALIAN BIGHT AREA – W. A. and S. A.                             Lepidosperma gladiatum,
                                                                                        Sonchus megalocarpus,
              Strand:    *Cakile edentula,                                              Senecio lautus,
                         *Arctotheca nivea                                              Carex pumila,
              Zone 1:    Spinifex hirsutus,                                             Pelargonium australe,
                         Scirpus nodosus,                                               Carpobrotus glaucescens,
                         Euphorbia *paralias,                                           Rhagodia baccata
                         Atriplex isatidea,                                 Zone 2:     Leucopogon parviflorus,
                         Senecio lautus                                                 Acacia sophorae,
              Zone 2:    Rhagodia baccata,                                              Correa alba, Myoporum insulare,
                         Acacia rostellifera                                            Helichrysum paralium,
              ADJOINING: Mallee                                                         Leptospermum laevigatum,
                                                                                        Banksia marginata,
                                                                                        Acacia verticillata
                                                                            ADJOINING:  Gymnoschoenus sedgeland
                                                                                        (south-west);
                                                                                        heath;
                                                                                        Eucalyptus spp.
```

on the seaward face of the frontal dunes appears to be independent of mean annual rainfall or its incidence; rather, it appears to be controlled by the intensity of the wind.

In general terms, the following zonation applies to all frontal dunes.

i. Above high tide level (strand) a few scattered plants sometimes occur. These rarely form distinctive communities but the species are often ephemeral and play no part in the succession which leads to the stabilization of the dune.

ii. Rhizomatous or stoloniferous grasses are the first colonizers of the frontal dune and may sometimes form mats. In the more stable sands annual, tufted perennials or mat-forming plants become established. See Fig. 23.6.

iii. Shrubs add further to the stability of the system, often forming thickets which completely cover the soil surface.

The widths of the various zones vary according to the topography of the dune. On low, flattish dunes the zone of rhizomatous grasses is likely to be very wide and in such cases mat plants are sometimes abundant and may even form a distinct zone on the landward side. On the other hand, when the dune rises abruptly from the beach, the zones are contracted and the grass zone may be less than one metre wide, or even absent (Fig. 23.7). The community adjoining the shrub zone in the various sectors varies according to the mean annual rainfall and the soil fertility level (Table 23.2). In some cases it might be regarded as the final stage of the dune succession since the community may develop on dune sand, e.g. *Eucalyptus pilularis* on the central east coast. In other cases the adjoining community may occur on a different parent material, unrelated to the dune sands. The variations encountered (Table 23.2) are: In wetter areas in the north and east the adjoining community on soils of higher fertility is sometimes rainforest; on soils of lower fertility it is usually *Eucalyptus* forest; on soils of lowest fertility it is heath or *Banksia* scrub. Elsewhere it is usually a *Eucalyptus* community (woodland in wetter areas, mallee in drier areas), except in the driest sector on the west coast where the dune vegetation meets *Triodia* grassland.

Fig. 23.8. *Spinifex longifolia* on coral sand. Other herbaceous species are *Canavalia obtusifolia* and *Euphorbia *atoto*. Trees on right are *Hibiscus tiliaceus*, *Pongamia pinnata* and *Acacia auriculiformis*. Mangrove community at rear. Darwin, N.T.

Fig. 23.9. *Casuarina equisetifolia* on a coastal dune. The liana is *Passiflora herbertiana*. Main herbaceous species are *Pteridium esculentum*, *Dianella caerula* and *Imperata cylindrica*. Near Noosa, Qld.

The more important species occurring in the various sectors of the coastline, arranged in zones for each sector, are listed in Table 23.2. The main differences between the sectors are as follows:

i. The species in the tropics and extending southward into New South Wales are different from those in the south.

ii. *Spinifex hirsutus* (Fig. 23.6) is found along the eastern and southern coasts, whereas *S. longifolius* (Fig. 23.8) occurs along the northern and western coasts. The two occur together in the Perth region where the former occupies the frontal position, with *S. longifolius* in more protected sites on the landward side of the dune.

iii. *Casuarina equisetifolia* (Fig. 23.9) dominates the shrub zone on dunes on the Arnhem Peninsula and along the east coast from tropical Queensland to northern New South Wales.

iv. Species of *Acacia* are common or dominant in dune scrubs around that part of the coast where *Casuarina* does not occur, except in south-eastern Queensland and northern New South Wales where *Casuarina* and *Acacia* overlap. Along the east coast, *A. sophorae* is the dominant of dune scrubs, usually with *Banksia* and/or *Leptospermum* spp. *A. sophorae* extends to Tasmania and along most of the southern coast. Towards the west it is replaced by *A. rostellifera*, *A. cyclops* and others, which occur also on the west coast in the south. On the west coast, *A. ligulata, A. sclerosperma, A. translucens* and *A. binervosa* (in that order from south to north) dominate the dune scrubs.

v. In the south-west *Agonis flexuosa* forms woodlands or scrubs behind the coastal dunes, separating the dune system from the *Eucalyptus* forest (See Chap. 16).

vi. *Casuarina stricta* occurs in hind dune areas and is commented on in more detail.

C. stricta, mentioned above as a species on rocky outcrops and in river estuaries, commonly forms scrubs in many hind-dune areas and occasionally elsewhere in Tasmania, Victoria and South Australia. In most areas it is associated with *Bursaria spinosa* (Fig. 23.10) and, near the coast, with *Leptospermum laevigatum* and *Acacia sophorae* and/or *A. longifolia*. In South Australia it extends inland on deep siliceous sands or rocky slopes with different assemblages of understory species. Crocker (1946) records it on Eyre Peninsula on calcareous dunes with an understory of *Eremophila crassifolia, Lasiopetalum discolor* and others. Perry and Specht (1948) record it with *Banksia marginata* and *Callitris "glauca",* and an understory of *Bursaria spinosa* and *Xanthorrhoea australis* in the Mount Lofty Ranges near Adelaide.

4. Communities on Mudflats

4.1. General

Mudflats occur in isolated patches around the coast-line. Some contain a high proportion of coarse sand, whereas others are composed mainly of fine sand, silt and clay, usually with a high proportion of organic matter. The largest areas of these alluviums are found in areas with the highest rainfall in and adjacent to estuaries. The surface of the alluvium is sometimes strewn with rounded stones (shingle) or even large boulders.

In many cases only parts of the mudflats support vegetation, the bare parts lacking vascular plants being of two kinds:

i. Saline, often salt-encrusted areas lying a few centimetres above spring high-tide level (Fig. 23.11). These are washed only by the highest (king) tides when they are wetted by salt water. The highest spring tides do not reach these flats so that salt incrustations develop as the water evaporates. The salt may be partially removed by rains, but, even in high rainfall areas, removal of the salt by percolation into the soil is slow because of the low infiltration rate through the sodium-saturated clay. The accumulation of wind-blown sand, trapped by obstructions on the surface, may lead to local colonization of the surface to produce "islands" of vegetation on the less saline sand.

ii. Areas of mud in the inter-tidal zone lying above low-tide level and the first seaward zone of vegetation. Such bare areas occur either where mangroves are absent (as in Tasmania and most mudflats on the south coast), or where mangroves are present and the tidal range exceeds 2–3 m (as on the north-west coast, in particular). Since mangroves can tolerate immersion to a depth of 2–3 m, any area of mud flooded by more than c. 3 m by tidal action cannot support vascular plants. These bare inter-tidal areas usually support scums of blue-green algae or occasional green algae.

The mudflats support three groups of communities dominated by angiosperms:

i. The sea grasses which form marine meadows mainly in the south in the intertidal zone or below low tide level;

ii. The mangroves, which occur mainly in warmer climates in the inter-tidal zone;

iii. A group of samphire-sedgeland-grassland communities which occur above the mean spring high tide level.

4.2. The Sea-grasses and Marine Meadows

4.2.1. General

In Australian waters there are 22 species of marine herbaceous angiosperms referred to as sea-grasses, belonging to 11 genera. These have been described and illustrated by Aston (1973). They represent a high proportion of the world's sea-grasses (49 species in 12 genera – Den Hartog, quoted in Aston, 1973).

Sea-grasses are found around most of the coast line but only one, *Halophila ovalis* (Sea Wrack) is recorded in both tropical and temperate waters. A few appear to be confined to the tropics, notable *Enhalus acoroides,* and a few to

Fig. 23.10. Scrub of *Casuarina stricta,* with subsidiary, *Acacia longifolia, Leptospermum laevigatum* and *Bursaria spinosa*. Hind-dune area, Wilson's Promontory, Vic.

Fig. 23.11. General view of mudflat at Cambridge Gulf. The dark strips of vegetation at water's edge and along depressions on the landward side are mangal. The bare area above the mangal ist salt-encrusted mudflat. Trees on hill in foreground are *Eucalyptus dichromophloia*. From the Lookout, Wyndham, W. A.

temperate waters, as discussed below. Most information on communities has been collected on the east coast where the dominants change with latitude and, in some cases, there is a zonation of species according to depth of water. The sea-grass meadows occur below low tide level but may be exposed during spring low tides. The meadows are commonly submerged to a depth of 1–2 m but below this the plants become fewer, though some species, especially *Halophila ovalis* and *Posidonia australis* (Marine Fibre Plant) can grow at depths of c. 10 m.

Morphologically the sea grasses, which belong to several monocotyledonous families, have some features in common, especially the rhizomatous habit which restricts their habitat. They require penetrable substrates such as sand or mud and consequently they are absent from rocky areas where algae, with specially adapted holdfasts for securing the plant to a rock, are dominant. However, on sand and mud the sea-grasses are commonly associated with larger algae and, as well, the leaves of the sea-grasses usually support a wealth of epiphytic organisms, chiefly small algae and small animals (Ducker et al. 1977).

Many of the sea-grasses produce extensive communities, often with a single dominant, which cover areas amounting in some cases to hundreds of square kilometres. These meadows have considerable economic importance since they provide food and shelter for marine animals, such as fish, molluscs and crustaceans, which eat the sea-grass leaves or the epiphytes which dwell on the leaves. Some larger animals, especially in warmer waters, are dependent or partially dependent on the sea-grasses, e.g. turtles and dugongs (Heinsohn and Wake, 1976), and many offshore birds include sea-grasses in their diet.

Reproduction of the sea-grasses is both vegetative and by seed. When meadows develop, it is probable that the communities are extended vegetatively on a clonal basis. However, flowers and fruits of all species are known (Aston, 1973), though in some species they are reported as "rare". Many species are dioecious, which implies a precarious sexual life cycle. Pollination in all species but one occurs in water. The exception is *Enhalus acoroides* in which the male flowers become detached, float to the surface and mass around the female flowers.

In addition to the truly marine sea-grasses, *Ruppia maritima* sometimes occurs in saline river estuaries, though its chief occurrences are inland in saline, brackish or fresh waters (Aston, 1973 and Chap. 22).

4.2.2. Communities in the North-east

The subtropical assemblages in southern Queensland are mixtures of tropical and temperate species, plus *Halophila ovalis* which occurs around the continent. The tropical species are *Cymodocea serrulata*, *Syringodium isoetifolium*, *Halodule uninervis* and *Halophila spinulosa*, the first two reaching their southern limit at Stradbroke Is. The temperate species is *Zostera capricorni* (Kirkman, 1975). Kirkman reports that the species may "live in a dynamic climax community of a number of species or as a monospecific meadow"; mixtures of two or three of these species are recorded around Fraser Is. (Olsen, in Lauer, 1978). Where zonation occurs, *Cymodocea* grows as a monospecific meadow at a depth of 150 cm below low tide. On the shoreward side, *Syringodium* forms a dense sward, and in the shallowest water *Zostera capricorni*, *Halodule uninervis* and *Halophila ovalis* adjoin the mangroves.

4.2.3. Communities in the South

Zostera capricorni (Eel Grass), with linear leaves up to 50 cm long, occurs along the whole of the east coast, and in the south it dominates meadows from the low tide mark to a depth of c. 6 m. It is sometimes mixed with *Halophila ovalis* which appears to be associated with muds containing sulphides. *Posidonia australis* is sometimes present in the meadow, but this species dominates a zone in deeper water (to 10 m) and in these deeper waters large algae are usually abundant, especially species of *Caulerpa*, *Codium*, *Dictyota*, *Gracilaria* and others (Wood, 1959).

Along the south coast and in Tasmania, *Zostera muelleri* dominates the meadows, with a zone of *Posidonia australis* in deeper water, sometimes with *Heterozostera tasmanica* (Aston, 1973).

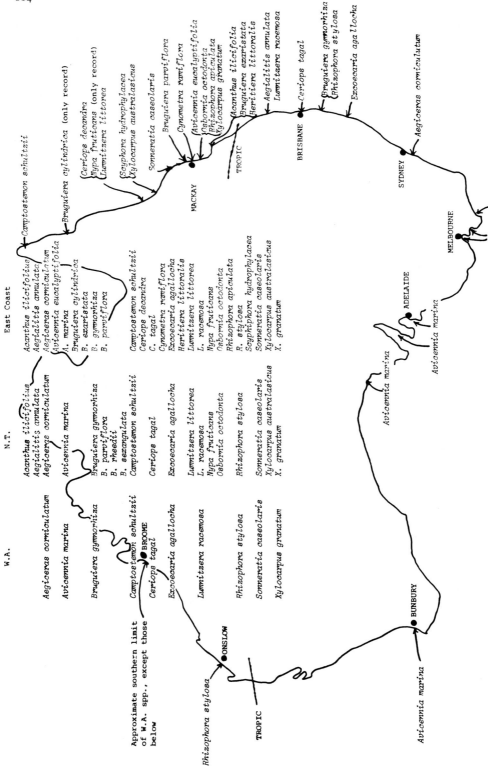

Fig. 23.12. Distribution of the mangroves in Australia. Southern limits of the various species are indicated around the coast-line.

4.3. Mangroves and Mangrove Communities (Mangals)

4.3.1. General

Mangroves are shrubs or trees which grow in the intertidal zone and are regularly flooded by sea water. The mangrove habit appears to have evolved in the early Tertiary, fossil pollens having been identified from Borneo in Eocene and Oligocene deposits (Muller, 1964). In Australia 27 species are classed as mangroves, all being dicotyledons, except *Nypa fruticans* which is a palm.

Mangroves have in common a definite requirement for salt, but salt-tolerance varies considerably within the group. Otherwise there is no single feature which identifies the mangroves as a group. Some mangroves have evolved special adaptations which appear to be beneficial or essential for their existence, viz., pneumatophores and/or vivipary (see below) but these features are not common to all mangroves. Both features occur in all species of Rhizophoraceae, and in *Avicennia* and *Aegiceras*. It is perhaps inconsistent that the species with the largest pneumatophores *(Sonneratia caseolaris)* is not viviparous.

The mangroves have some economic importance. The communities harbour a wealth of animals including fish, crustaceans and molluscs which provide food for man. They are used in New Guinea for creating brackish water ponds for fish culture (Percival and Womersley, 1975). Several species provide useful products, including tannin for tanning leather or toughening ropes, fishing nets and sails. Many provide wood for various purposes, either round poles, firewood, charcoal-making and, in a few cases,

Fig. 23.13. *Rhizophora stylosa* with stilt roots. Note long fruits. Near Darwin, N.T.

milled timber. *Aegiceras corniculatum* produces good honey.

The 27 Australian mangrove species belong to 17 genera in 15 families. Almost all of the species occur in New Guinea (described in detail by Percival and Womersley, 1975) and the flora as a whole is related to the mangrove flora of south-eastern Asia, with most of the genera and some of the species extending to the east coast of Africa and the Red Sea. Two species are endemic in north-eastern Australia, but there are no endemic genera. *Osbornia* (Myrtaceae – Leptospermoideae) possibly originated in Australia. The mangroves are most abundant in the tropics, with the greatest concentration of species the north-east coast of Queensland. The number of species decreases from east to west across the tropics, only 10 species being recorded in the west. The mangrove flora becomes progressively poorer from north to south as a result of the lower winter temperatures in the south. The southern limits of individual species are shown in Fig. 23.12. The extension of several species south of the Tropic in the east is probably due to warmer winter conditions induced by the warm East Australian Current (Fig. 23.1). Only one species, *Avicennia marina*, occurs on the south coast, in isolated patches, but there are none in Tasmania.

4.3.2. Fruits, Seeds and Germination

Many mangroves produce fruits which are characteristic of the family or genus to which they belong, and in such cases the seed and embryo develop normally. In the Rhizophoraceae, *Aegiceras* and *Avicennia*, on the other

Table 23.3. Zonation of mangroves. The zones are governed by tide levels, as indicated. Sources of data, see text.

Zone		Species-groups	Comments
Outer Zone			
Flooded by all high tides	1	*Sonneratia caseolaris*	Deep, soft mud
	2	*Avicennia marina*	Firm substrates
		or *A. eucalyptifolia*	Local. Coral reefs, Q.
	3	*R. stylosa*	
		Acanthus ilicifolius	Understory
		Aegialitis annulata	Understory
Middle Zone			
Flooded by medium high and spring tides	4	*Bruguiera gymnorhiza*	Dominant, common
		B. parviflora	Locally dominant, waterlogged
		Xylocarpus granatum	Occasional to rare
		Aegiceras corniculatum	Subsidiary to locally common
		Avicennia marina	Subsidiary to locally common
		Acanthus ilicifolius	Understory
		Aegialitis annulata	Understory
Flooded by spring tides	5	*Ceriops tagal*	Dominant
		Bruguiera exaristata	Sometimes subdominant. Q. only
		Avicennia marina	Occasional to rare or absent
		Lumnitzera spp.	Higher ground
		Understory as in 4	
Inner Zone			
Flooded by King tides	6	*Camptostemon schultzii*	This zone is of variable
		Cynometra ramiflora	composition and is sometimes
		Excoecaria agallocha	absent, see text
		Heritiera littoralis	
		Osbornia octodonta	
		Scyphiphora hydrophylacea	
		Xylocarpus australasicus	

hand, a condition known as vivipary exists, which involves the development of the zygote into a seedling without the formation of a resting seed. In viviparous species only one ovule per ovary develops into a seedling, though the ovary may contain many ovules.

In *Rhizophora* and *Ceriops* the young seedling, with a greatly elongated hypocotyl, breaks through the pericarp and hangs on the plant in a vertical position, root lowermost (Fig. 23.13). The young seedlings are green and in *R. stylosa* reach a length of 25 cm before they break from the pericarp and fall. Some seedlings may spear themselves into the mud, but most are carried away and are disseminated by the tides, as are the fruits and seeds or seedlings of other mangroves. In the few which have indehiscent fruits, e.g. *Sonneratia*, the pericarp rots before the seeds can germinate.

The seeds germinate and the seedlings can establish in salt water, though there is some evidence to show that young seedlings grow best in diluted sea water (see next section). Many species require either full sunlight or only light shade for establishment, especially *Acanthus* which is incapable of establishment in the shade (Macnae, 1968). In these species a coating of mud over the seed or seedling can prevent or retard establishment. Conversely, species of *Bruguiera, Ceriops* and *Xylocarpus* all develop in the shade of other trees or of themselves.

4.3.3. Tolerance to Flooding; Root Systems; Reactions to Salinity

All mangroves are subjected to flooding by sea water, and in most cases to flooding by fresh water from rain or run-off from the land. Those mangroves which are rooted below the level of all high tides are adapted to diurnal inundations by sea water, involving immersion to a depth of c. 2 m, and in some cases occasional flooding to a depth of 3 or even $3^{1}/_{2}$ m. Mangroves occurring above average high tide levels are flooded

Fig. 23.14. Pneumatophores of *Avicennia marina*. Noosa, Qld.

only by the higher spring tides (Table 23.3). The mangrove species are therefore variously adapted to depth of flooding and to soil-waterlogging and the species are zoned according to depth of immersion.

Since the maximum inter-tidal ranges (Fig. 23.1) over most of the coast-line are within the immersion-limits quoted above, mangroves commonly occupy the whole of the inter-tidal zone. However, where the tidal range is large, e.g. in the north-west, the mangroves occur only near the high-water mark along a strip of mud which is immersed at high tide to a depth of 2–3 m. Consequently, when the tide recedes a zone of bare mud appears, extending, perhaps, hundreds of metres or even a couple of kilometres between the outer fringe of the mangroves and the low-water level (Fig. 23.11).

Adaptations to immersion and soil-waterlogging are manifest mainly in the root systems, including the development of aerial roots, some with apogeotropic branchings referred to as pneumatophores. The tap root of mangroves is soon replaced by a system of lateral roots (cable roots) situated 20–50 cm below the soil surface, and which anchor the plant. The cable roots produce finer roots for absorption. In some mangroves the roots are produced in the air, notably in *Rhizophora* spp. which develop stilt-roots (Fig. 23.13); the first of these roots emerge from the hypocotyl in the seedling stage, later ones from the stem. Again, the aerial roots may occur as a mass above the surface of the mud, as in *Aegialitis*, *Aegiceras* and *Excoecaria*.

In many species the cable roots produce lateral roots which grow more or less vertically above the surface of the mud. These structures (pneumatophores) are periodically immersed in water or exposed to the air according to the rise and fall of the tides. Pneumatophores are of several forms, including massive cones 2–2½ m tall in *Sonneratia*, stumpy cones to 15 cm tall in *Xylocarpus australasicus*, or unbranched finger-like growths up to c. 1 m tall in *Avicennia* and *Lumnitzera racemosa* (Fig. 23.14). For some species (e.g. of *Bruguiera* and *Ceriops*) the cable roots grow to and above the surface, form a loop and return to the mud to produce a knee-pneumatophore; in *Xylocarpus granulatum* the cable roots grow upward above the mud to form a plank- or plate-like pneumatophore (Percival and Womersley, 1975).

Pneumatophores are composed mainly of spongy aerenchymatous tissue which serves to accelerate the passage of oxygen into the root system; the surfaces are covered with numerous lenticels. In addition to acting as aerating organs, the pneumatophores produce vast masses of fine absorbing roots below the surface of the mud.

The fact that mangroves occur only in salt or brackish water indicates that they are obligate halophytes. However, they exhibit different degrees of salt-tolerance (Table 23.4).

Avicennia marina has the highest salinity range, which appears to account for its sporadic distribution from the outer seaward margin in some areas to the inner landward fringe in others, and even in hypersaline patches where the plants are stunted or shrubby.

Salt-accumulation and excretion through glands in the epidermi of the leaves has been reported in *Acanthus*, *Aegialitis*, *Aegiceras* and *Avicennia* spp. (Macnae, 1968).

Experiments done in the laboratory by Clarke and Hannon (1970) on the growth of seedlings

Table 23.4. Salt-tolerance of mangroves. From Macnae (1968).

Species	% Salinity
Avicennia marina var. *resinifera* (dwarfed)	9.0 to <1
Lumnitzera racemosa	9.0
Ceriops tagal var. *tagal*	6.0
Rhizophora mucronata	5.5
Sea water (open ocean)	c. 3.5
Bruguiera gymnorhiza	2.5–1.0
B. parviflora (maximum development)	c. 2.0
B. sexangula	1.0
Sonneratia caseolaris	3.5 to <1.0
Aegiceras corniculatum	? to almost fresh water

of *Aegiceras corniculatum* and *Avicennia marina* var. *resinifera* showed that both mangroves required sea water for growth. However, sea water alone could not provide all the nutrients required, so that nutrients had to be added to the solution cultures. For both species optimum growth occurred when sea water was diluted (solution contained 20–40% sea water plus added nutrients). At higher concentrations of sea water growth was slower and in pure sea water only 33% of the seedlings of *Avicennia* survived. These data indicate that fresh water (in the form of rain or flood water) is at least beneficial and probably essential for these two mangroves in the seedling stage. This is corroborated by the observations that the tallest mangroves occur in areas with the highest rainfall and especially when the winter rainfall component is significant, as in north-east Queensland (Macnae, 1966).

4.3.4. The Mangals: General

Mangrove communities (Mangals) occur mainly in and around estuaries, on deltas, or on coast-lines facing the sea but protected by coral reefs from violent wave action. The communities occupy the inter-tidal zone on alluviums of relatively high fertility (Saenger et al., 1977). Occasionally mangroves may establish in rocky situations, the seedlings taking root in pockets of alluvium accumulated in cracks or crevices.

The floristic composition and structure of the communities is determined mainly by climate, especially winter temperatures. In any area a zonation of species, determined mainly by tolerance to depth of flooding, occurs when two or more species are present. Other factors which cause local or regional variations are:

i. Macro-climate, especially mean annual rainfall and its incidence, and temperature;
ii. Soil parent material and soil, including nutrient levels and the nature of the soil surface;
iii. The fauna.

Optimum mangrove development, both with respect to species diversity and biomass of the communities, occurs in the tropics, especially the north-east coast of Queensland, where rainfall is high. The reduced vigour of the mangroves on the west coast is possibly due to the low rainfall in that area.

Low winter temperatures in the south appear to limit mangrove development. This is supported by the observation that the southern mangroves (*Avicennia* and *Aegiceras*) are sometimes damaged by frosts in river estuaries even a few hundred metres from the ocean.

In all cases the mangrove soils are alluvial deposits. They vary in texture from coarse sands to fine silts which are semi-fluid. Texture is determined partly by the proportion of sand to finer particles, the latter being silt, clay and organic matter. The organic contents vary but few quantitative data are available. Macnae (1968) gives a figure of 5–15% loss on ignition for tropical *Rhizophora* soils, and Clarke and Hannon (1967) state that decomposed organic matter occurs to a depth of 1 m in *Avicennia* soils in New South Wales. It is only in the *Nypa* swamps that the organic content is sufficient to form a type of peat (loss on ignition 80–90% – Macnae, 1968).

The soils are usually alkaline (pH 8.5–9.0) and many soils contains calcium carbonate derived from marine shells. The percentage salt in the soil solution varies from sea water concentration (3½%) which is likely to be reduced considerably during heavy rain. Hypersaline soils are rare, but may occur on the landward fringe in bays cut off from regular tidal flooding. Such soils may support stunted *Avicennia marina* but the plants ultimately die, apparently through salt-toxicity.

All of the soils are regularly waterlogged through tidal inundation and when rain falls or water flows into the swamp freshwater adds to the intensity of waterlogging and may maintain a waterlogged condition even when the tide is out. The depth to which the soil is covered by water decreases from the sea inland and accounts for the zonation of the species (Table 23.3). In the middle and especially the inner zones which are not inundated twice daily by the tides, the soils drain and the water table falls below the surface. Precise depths cannot be stated except for *Avicennia* (Clarke and Hannon, 1969) who measured a water level 66 cm above the soil surface during an exceptionally high tide, and a watertable 28 cm below the surface two hours prior to the inflow of that tide. The main effect of waterlogging is the reduction of oxygen tension in the soil leading to anaerobic conditions. The effects of the high water table are ameliorated to some extent by the digging of holes by animals, and local drainage by gullies and creeks.

The effect of reduced oxygen supply is manifest mainly through the evolution of hydrogen sulphide produced from the sulphate ions in sea water through the activities of bacteria. Further, the sulphide reduces ferric compounds to various hydrated ferrous sulphides which give the mangrove soils their characteristic black colour; also, reducing conditions often lead to the development of gley (greenish ferrous compounds) deep in the profile.

The tallest and floristically richest mangrove communities occur where soil fertility is highest. For example, in north-east Queensland the alluviums derived from basaltic soils support more luxuriant mangrove communities than do the alluviums derived from quartzites (Macnae, 1966).

The surfaces of the inter-tidal zone vary considerably. Smooth surfaces do not exist, even when the inter-tidal zone slopes evenly towards the sea. Small variations in the surface result from the burrowing of marine animals which scoop holes and deposit mud or sand on the surface, in some cases mounds of considerable size being formed. Also, litter from the mangroves falls to the surface, and it is usual for soil material and litter to be redistributed locally by the rain or tides, often against the bases of pneumatophores.

The flow of both fresh and salt water through the mangal results in the formation of rills or gullies, or the development of more or less permanent creeks. The smaller of these watercourses may dry out at low tide and they tend to reduce the level of the watertable in the soils adjacent to the gully. Large watercourses are likely to remain as permanent creeks (brackish or salt) and these drain the adjacent banks at low tide, often permitting the establishment in an otherwise waterlogged zone of a mangrove species which prefers better aerated soil.

The most regular and most severe erosion of the soil surface occurs during the ebb flow along the north and north-western coast-line where the tidal range is c. 10 m. The vast mass of outflowing water scours channels and deposits material, so that anastomosing channels are formed separated by strips of mangal. In such areas the normal zonation is rarely apparent. The materials removed by erosion are deposited closer to or in the sea. Therefore, a succession of mangroves occurs, the outer fringe gradually extending seaward. Islands of alluvium colonized by mangroves are not uncommon both on the ocean front and in river estuaries (see below).

Animals are abundant in mangrove swamps, which are unique in that they provide habitats for two groups of animals, terrestrial and marine (Macnae, 1966, 1968). The mangrove trees harbour flying foxes and a variety of birds and insects, all of which play some part in the system, such as pollination, cycling of nutrients, and the decomposition of organic matter. In the water or mud on the floor of the swamps another set of organisms occur including crocodiles (tropics only), frogs, fish, molluscs (oysters and snails) and crustaceans. These play a similar role in the decomposition of organic matter, the cycling of nutrients and in many cases the accumulation of ions from sea water or from larger algae or microorganisms, the latter probably of considerable significance with regard to phosphorus and nitrogen. The calcium carbonate in the shells, in particular, appears to have some significance in ameliorating the possible deleterious effect of high concentrations of sodium.

In additional to chemical transformations, some of the marine animals have a pronounced effect on the general contour of the surface of the mud-floor. These take the form of holes dug in the mud by crabs, the holes leading to better aeration of the substrate, and to mounds heaped above the surface, in particular by the mud-lobster, *Thallassina anomala*. These mounds, up to 75 cm high, provide a drained soil in an otherwise waterlogged environment and are possibly one of the main habitats for the prothallus of the fern, *Acrostichum* (Macnae, 1966).

Mangals are usually 1-layered, the height of the community varying from c. $1^{1}/_{2}$–30 m. The tallest of the communities, which occur in the tropics, are forests with a closed canopy; the shortest occur along the south coast and are classed as scrubs which may be either closed or open.

Two-layered communities occur in some tropical stands. Layering is brought about either by the development of a lower layer (or layers) of the same species (notably with *Bruguiera parviflora*) or through the occurrence of scattered plants of the shrub-mangrove species (*Acanthus* and *Aegialitis*) or the fern *Acrostichum*.

The mud surface and sometimes the pneumatophores of mangroves usually support an algae flora, containing blue green algae and occasional green algae.

Where mangroves adjoin rainforests, chiefly in the tropics, a few epiphytes and lianas, migrants from the rainforests, have been recorded on mangroves (Macnae, 1966; Jones, 1971). The epiphytes are: *Hydnophytum formicarum* (Smooth Ant Plant), *Myrmecodia antoinii* (Spiny Ant Plant), the orchids *Cymbidium canaliculatum*, *Dendrobium undulatum* and *Dischidia nummularia,* and the ferns *Drynaria rigidula, Platycerium* sp. and *Polypodium acrostichoides*. The lianas are: *Canavalia maritima, Cynanchum carnosum, Derris trifoliata* and *Gymanthera nitida.*

4.3.5. Zonations and Alliances in the Tropics

The normal zonation of the species in the tropics, determined by tidal levels is given in Table 23.3. Three zones have been identified and six species-groups which are regarded as alliances.

4.3.5.1. *Sonneratia caseolaris* Alliance

This species forms the pioneer outer zone over most of the tropical coastlines where mangroves occur. It forms monospecific communities mostly 8–10 m high and it produces massive pneumatophores, up to 2 m tall and 15–20 cm at the base. The communities grow only on soft deep mud which is sometimes stone-strewn and sometimes contains a significant amount of sand mingled throughout the soft matrix of mud. On firmer substrates it is replaced by *Avicennia marina. Sonneratia* tolerates a wide range of salinity, from sea water, which may cover the bases of the trees to 2 m (up to c. $3^{1}/_{2}$ m), to almost fresh water. It may therefore form the outer zone in estuaries. There are no associated vascular species.

4.3.5.2. *Avicennia marina* var. *resinifera* Alliance

This species is found on the west, north and east coasts and it forms the outer fringe in all areas except parts of the tropics where *Sonneratia* may occur on the seaward side of *Avicennia*, and when this happens, the substrate is deep mud. *Avicennia marina* prefers firm substrates and is abundant near coral, but in parts of north-eastern Queensland it is replaced near coral reefs by *A. eucalyptifolia*. The communities are forests to 10 m high and are flooded to a depth of c. 2 m (rarely 3 m) at hightide. There are usually no associated species, except that where there is a strong fresh water influence *Aegiceras corniculatum* may be an associate.

A. marina occurs also in the middle zone, sometimes as an occasional species, sometimes as a local dominant, especially when the mangal adjoins a permanent swamp; in such cases the species typical of the inner zone (Table 23.3) are lacking.

4.3.5.3. *Rhizophora* spp. Alliance

The *Rhizophora* forests, dominated by *R. stylosa* which is readily identified by its stilt roots (Fig. 23.13), mark the inner limit of the outer zone, which is flooded twice daily by the tides. *R. apiculata* is included in the alliance; it is confined to north-eastern Queensland, where it occurs in estuaries where salinity is reduced.

R. stylosa, usually 6–12m high but up to 25 m on the most fertile soils, often forms the outer fringe of the mangal, either on the ocean front or in estuaries, the two outermost zones (above) being absent, usually as the result of an abrupt change in slope at the outermost edge of the mangal.

The *Rhizophora* spp. usually occur in pure stands, except for algae which lodge on the pneumatophores or on the surface of the mud. However, at the inner fringe of the stands an understory of the two shrub mangroves, *Aegialitis annulata* and/or *Acanthus ilicifolius*, may occur, usually as a discontinuous layer of scattered shrubs 1–2 m high. Also a taller understory of *Bruguiera* spp., ecotonal occurrences from the next zone, develops in some areas.

4.3.5.4. *Bruguiera* spp. Alliance

B. gymnorhiza is the common species which dominates the first of the middle zone which lies just above the level of medium high tides and is therefore flooded only by spring tides. The other species of this genus are of local occurrence and their ecological requirements cannot be given, except that Macnae (1966) states that *B. parviflora* prefers waterlogged soils.

B. gymnorhiza in the most favourable habitats (eastern Queensland) forms forests up to 30 m tall, but elsewhere it is rarely more than half so tall, and at its southern limit in the east it is a small tree or tall shrub. In most areas *B. gymnorhiza* forms pure stands, sometimes with *Xylocarpus granatum* as an occasional associate, sometimes with an understory of the fern, *Acrostichum speciosum,* which is often on hummocks formed by lobsters.

Occasional associates are *Avicennia marina, Excoecaria agallocha* and *Heriertera littoralis.* Also, ecotones with the *Ceriops* communities are common, especially in areas which are slightly sloping or almost level.

4.3.5.5. *Ceriops tagal* Alliance

Ceriops tagal dominates the area of relatively well drained soils at the upper limit of the middle zone subject to flooding only by the spring tides (Fig. 23.15). In eastern Queensland *C. decandra* is present in some stands. The communities, mostly c. 6 m high are often very dense, the canopies of *Ceriops* interlacing. In the highest rainfall areas of eastern Queensland the *Ceriops* communities do not occur, presumably as a result of soil-waterlogging induced by the high rainfall rather than the tides.

In the wetter tropics *Ceriops tagal* is often co-dominant with other species, notably *Avicennia marina* and *Lumnitzera racemosa* both of which may assume local dominance (Fig. 23.16). Species from adjacent communities, *Bruguiera* from the outer zone and *Excoecaria* and *Osbornia* from the inner zone, may also be present. An understory may occur, being mostly discontinuous and composed of *Acrostichum speciosum, Osbornia octodonta* (Fig. 23.17) and *Scyphiphora hydrophylacea.*

Towards its southern limit where the typical inner zone trees do not occur *Ceriops tagal* forms the innermost zone of the mangal, which adjoins the climax communities of the area (usually *Eucalyptus* forests). In these areas the plants are shrubs (Fig. 23.15).

4.3.5.6. The Inner Zone (Landward Fringe)

This zone is sometimes well developed, with a variety of species which, at their best develop-

Fig. 23.15. *Ceriops tagal*. Note buttressed trunks. Noosa, Qld.

Fig. 23.16. Interior of a mangal containing *Lumnitzera racemosa* (dark trunks), *Avicennia marina* var. *resinifera* (pale trunks) and *Ceriops tagal* (rare). The fern *Acrostichum speciosum* in foreground and out of focus. Near Darwin, N.T.

ment, form a forest 12 m high (Fig. 23.18). An understory of one or more of *Acanthus*, *Aegialitis* and *Scyphiphora* and *Acrostichum* may occur, and in some areas in the east (adjacent to rainforest) epiphytes may be present (see above). On the other hand, the inner zone is often absent, especially where the mangal adjoins a permanent swamp.

The nature of the soil, especially with regard to soil texture and drainage and the presence or absence of rock (or ferruginous laterite) at or near the surface possibly determine the composition of the stand. In addition to the six species which normally occur in this zone (Table 23.3) *Avicennia marina*, *Lumnitzera littorea* and *L. racemosa* are sometimes present on sandy soils (Macnae, 1966).

A scattered understory of species typical of the more saline mudflats or of dune sands occurs in some areas, especially if the canopy is open. The species are *Arthrocnemum* ssp., *Salicornia australis*, *Sesuvium portulacastrum*, *Sporobolus virginicus*, *Suaeda australis* and *Tecticornia cinerea*, as well as *Aegialitis annulata*.

Several species, sometimes regarded as mangroves, occur either sporadically or commonly in the inner zone, especially *Hibiscus tiliaceus*, *Pemphis acidula* and *Thespesia populnea*.

Fig. 23.17. Five mangrove species: *Aegialitis annulata* (small shrub on left), *Osbornia octodonta* (taller shrub), and three tree species (behind), *Ceriops tagal, Rhizophora stylosa* and *Avicennia marina* var. *resinifera*. At low tide. Near Darwin, N.T.

4.3.5.7. *Nypa fruticans* Alliance

This palm occurs in the east in small patches only along the McIvor and Herbert Rivers. It is possibly more abundant in Northern Territory. In New Guinea it covers relatively large areas and the indigenous people gather several useful products from the palms.

Nypa occurs only along tidal rivers, the palms growing in brackish water which is affected only by the highest spring tides (Fig. 23.19). The palms are almost trunkless and produce crowns of leaves 7–9 m long. The stands are mostly monospecific but tree mangroves may possible occur in the stands, especially *Heritiera litoralis, Xylocarpus granulatum, Bruguiera gymnorhiza, Lumnitzera racemosa* and *Rhizophora apiculata*.

4.3.6. Zonations and Alliances South of the Tropic

4.3.6.1. General

Only eight species are recorded south of the Tropic in the east, and only one species in the west (Fig. 23.12). The mangrove communities become progressively simpler floristically from north to south. The various species maintain the

Fig. 23.18. Inner zone mangroves, *Xylocarpus australasicus*, *Excoecaria agallocha* (right) and *Avicennia marina* var. *resinifera* (left). Near Darwin, N.T.

same relative positions as in the tropics and similar variations occur in relative abundance. For example in south-western Queensland *Avicennia marina* forms the outer zone, followed by *Rhizophora stylosa*, *Bruguiera gymnorhiza*, *Ceriops tagal* and *Excoecaria agallocha*. The main variations include the absence of *Avicennia* on steep slopes, especially estuaries where *Rhizophora* forms the outer zone, the presence of *Avicennia* in the middle zone, the absence or rarity of *Excoecaria* on the landward fringe and the presence of *Aegiceras* in estuaries. Further south along the east coast, where only *Avicennia* and *Aegiceras* occur, the former forms the outer zone (often the only zone) and *Aegiceras* occurs in an inner zone which is influenced by fresh water.

4.3.6.2. *Avicennia marina* var. *australasica* Alliance

On the east coast south of Sydney, this species is the sole mangrove and it is found on a few isolated patches of alluvium along the south coast, with a single occurrence at Bunbury in Western Australia (Fig. 23.12). Two forms of the variety exist, one being a tree 5–8 m tall, the other a shrub $1^1/_2$ m high. The shrub may be a variant stunted by hypersalinity (Hamilton, 1919), though this view is queried by Clarke and

Fig. 23.19. *Nypa fruticans* edging the Wassi Kussa River. Western Province, Papua New Guinea. Photo Donald B. Foreman.

Hannon (1967) who consider that the dwarf form is a genetically stable ecotype.

The mangals lie between high and low tide levels. On the seaward side *Zostera* meadows occur and on the landward side samphire communities (*Salicornia* on the east coast, *Arthrocnemum* in the south-Table 23.5; Fig. 23.20). In South Australia and western Victoria, *Melaleuca halmaturorum* is an occasional associate. Usually, the stands have no vascular understory but occasional plants of *Arthrocnemum* spp. (south coast) and/or *Suaeda australis* may be present.

Fig. 23.20. Shrub form of *Avicennia marina* var. *australasica*. *Salicornia quinqueflora* on right. At spring tide. Port Stephens area, N.S.W.

4.4. Mangrove Islands

Off-shore islands, with an elevation of one to a few metres above spring high tide level sometimes support communities dominated entirely by mangroves, the islands being of two kinds:

i. coral and coral sands, occurring only in the tropics;

ii. alluvial deposits of sand, silt, clay and organic matter built up mainly in river estuaries.

The tropical coral cays (see below) sometimes support mangroves, the presence of which appears to depend on the amount of organic material incorporated in the coral sand. A species which may colonize and persist on such sands is *Aegialitis annulata* and it may even produce thickets up to 2 m high, sometimes with an understory of *Sesuvium portulacastrum* and Chenopodiaceae (Macnae 1966; Jones, 1971).

Avicennia marina var. *resinifera* likewise colonizes some coral islands and, rarely *A. eucalyptifolia* in a restricted area along the Queensland coast. Shrubberies of these species may develop, the success of the colonizers presumably being dependent upon the amount of debris trapped by the young seedlings. *Rhizophora stylosa* has also been recorded either above the *Avicennia* zone or as the first colonizer on coral on the ocean front.

Alluvial deposits in estuaries are colonized by *Avicennia marina* and *Aegiceras corniculatum* and where there is a strong fresh water influence *Aegiceras* is more abundant. Consequently it is common to find *Aegiceras* dominating the mangal on the landward side where the fresh water influence is greatest, whereas *Avicennia* dominates the seaward side of the island. Other mangroves may become established, depending on

Fig. 23.21. *Arthrocnemum leiostachyum* on a mudflat, with subsidiary *Trianthema crystallina* and *Xerochloa imberberis* (behind). c. 20 km south of Karumba, Qld.

the elevation of the island, and an understory of lianas and herbs occurs in these elevated communities (Blake, 1940).

Mangrove islands have become larger and more frequent since white man's occupation of the continent. This results from increased accelerated erosion of the land leading to siltation of the estuaries. In many cases mangrove islands have developed during the last couple of decades on sand-silt spits in shallow water in areas not affected by industrial pollution.

4.5. Communities Adjoining Mangroves on the Landward Side

Many kinds of communities adjoin mangrove swamps, but in few cases are these adjacent communities climax to mangroves in a true succession. True terrestrial climaxes exist only when the fine textured alluviums are built to a level at which they are unaffected by tidal flooding. This may occur and a climax of rainforest (often with palms) adjoins the mangroves, or where certain species of *Eucalyptus* adapted to clays establish behind the mangal, e.g. *E. saligna* in some estuaries on the south-east coast. More often the mangroves adjoin sandy or rocky areas, on which a climax of *Eucalyptus* forest or woodland occurs, but the two communities are unrelated. Again, sandy material, usually in the form of a wind-blown dune, may encroach upon and obliterate the mangrove community in a process possibly best described in successional terms as a deflected succession which may lead to a climax on sand.

4.6. The Samphire, Sedgeland and Grassland Communities

4.6.1. General

This group of communities occurs on the landward side of the mangroves or, if mangroves are absent, as in the south, they form the outermost zone of terrestrial vegetation. In the tropics only two communities occur, samphire and grassland, but in the south several distinctive communities may be present, these being zoned according to the tides (Saenger et al., 1977). The tropical and temperate communities have few species in common, among them *Sporobolus virginicus* and *Suaeda australis*.

4.6.2. Communities in the Tropics

Arthrocnemum halocnemoides and *A. leiostachyum* (Fig. 23.21) occur in varying density on mudflats which are only occasionally inundated by sea water. Such mudflats have the highest salinity resulting from the high evaporation rate following very shallow flooding by the sea. Bare areas may be colonized by *Arthrocnemum*, the seed germinating in cracks. The samphire bushes reach a height of 20–50 cm and a diameter of c. 1 m, and at maximum density lie c. 1 m apart.

Associated species may be lacking but usually a few herbaceous plants are present, especially *Trianthema turgidiflora*, *Sesuvium portulacastrum*, *Tecticornia cinerea* and *Frankenia pauciflora* and the grasses *Sporobolus virginicus* and *Xerochloa* spp. in areas of lower salinity. *Salicornia quinqueflora* has been recorded on the east and west coasts in areas of higher salinity. Very rare species include *Bassia astrocarpa* and *Hemichroa diandra* on the west coast (Sauer, 1965), and *Cressa australis*, *Epaltes australis* and *Statice australis* on the east coast (Perry, 1953). The bushes of *Arthrocnemum* collect wind-blown sand and on the sand hummocks annual species establish, notably *Salsola kali*.

The *Arthrocnemum* communities commonly adjoin sand dunes which are either fixed by plants, e.g., *Acacia translucens* on the west coast, or mobile. In the latter case the sand encroaches on the mudflat, thus permitting the sand-dune species to invade the samphire area (Fig. 23.22). Also, sand dunes may completely enclose a mudflat, cutting the flat off from tidal influence. In such cases the mudflat becomes less saline as rainfall leaches the salt downward; the *Arthrocnemum* then decreases in density and species which are less salt-tolerant invade the area, sometimes developing an almost complete cover to the soil, e.g. *Sporobolus* and *Xerochloa*, and species of *Atriplex*, *Flaveria*, *Pterigeron* and *Threlkeldia* (west coast).

The *Arthrocnemum* communities adjoin either bare highly saline areas of mud flat, mangroves (subject to regular tidal influence), grasslands of *Sporobolus* or *Xerochloa* (less saline soils – see Chap. 20) or *Melaleuca* woodlands.

Fig. 23.22. Mudflat with hummocks of *Arthrocnemum halocnemoides* and subsidiary *Frankenia pauciflora*. The flat is being encroached upon by sand supporting *Spinifex longifolius, Atriplex isatidea* and *Nitraria billardieri*. Carnarvon, W.A.

Fig. 23.23. Pure stand of *Arthrocnemum arbuscula* with a wall of mangrove (*Avicennia marina* var. *australasica*) behind. Westernport Bay, Vic.

Table 23.5. Zonation of mudflat communities in temperate Australia. The dominants are in capital letters. Sources of data, see text.

Tidal levels		South coast, mainland		Tasmania		New South Wales
1. Below low tide	1a. 1b.	POSIDONIA AUSTRALIS ZOSTERA MUELLERI	1a. 1b.	POSIDONIA AUSTRALIS ZOSTERA MUELLERI	1a. 1b.	POSIDONIA AUSTRALIS ZOSTERA CAPRICORNI
2. Intertidal	2.	AVICENNIA MARINA (few areas, S.A. & Vic. Bunbury only in W.A.). Melaleuca balmaturorum in estuaries (S.A. & west Vic.)	2.	Zone absent	2.	AVICENNIA MARINA (seaward), Aegiceras corniculatum (north and landward), rarely Suaeda australis, Atriplex *patula in understory
3. Flooded by highest spring tides	3a.	ARTHROCNEMUM ARBUSCULA (rare W.A. – Mandurah & Swan R.). Sometimes with understory of Salicornia quinqueflora, Suaeda australis, Maireana oppositifolia	3a.	ARTHROCNEMUM ARBUSCULA Understory: Salicornia quinqueflora, Gahnia filum, Wilsonia backhousii	3a.	Zone absent
	3b.	ARTHROCNEMUM HALOCNEMOIDES Frankenia pauciflora, Disphyma blackii	3b.	Zone absent	3b.	Zone absent
4. Flooded by most high tides	4.	SALICORNIA QUINQUEFLORA Zone narrow and mainly estuaries. Suaeda australis, Maireana oppositifolia, Samolus repens, Spergularia rubra	4.	SALICORNIA QUINQUEFLORA Suaeda maritima, Samolus repens, Triglochin striata, Cotula coronopifolia, Wilsonia humilis, Hemichroa pentandra, Gahnia filum	4.	SALICORNIA QUINQUEFLORA Suaeda australis, Samolus repens, Triglochin striata, Cotula coronopifolia, Wilsonia backhousii, Tetragonia expansa, Lobelia anceps, Spergularia rubra
5. Occasionally flooded by highest spring tides	5.	SPOROBOLUS VIRGINICUS Zone narrow or absent and replaced by Atriplex paludosa, A. cinerea, sometimes with Nitraria billardieri	5.	Zone usually absent, see Zone 7	5.	SPOROBOLUS VIRGINICUS Often well developed, sometimes replaced by Zoysia macrantha. Many associates: Cynodon dactylon, Salicornia quinqueflora,

6. Rarely flooded by highest spring tides	6. JUNCUS MARITIMUS Mainly estuaries	6. JUNCUS MARITIMUS Zone narrow, usually incorporated in Zone 7	6. JUNCUS MARITIMUS Often in pure stands or mixed with species of Zone 5 and 7 *Cyperus polystachys, Plantago coronopus, Atriplex *patula, Cotula cornopifolia*
7. Normally beyond tidal influence	7. DISTICHLIS DISTICHOPHYLLA (absent W.A.) *Baumea juncea, Gahnia filum, Juncus maritimus, Spergularia rubra* and alien spp.	7. DISTICHLIS DISTICHOPHYLLA *Juncus maritimus, Apium prostratum, Selliera radicans, Sebaea albidiflora, Gahnia filum, Stipa teretifolia*	7. BAUMEA JUNCEA Usually dense stands with subsidiary *Juncus maritimus*. Extends as an understory in Zone 8
8. Beyond tidal influence	8. MELALEUCA ERICIFOLIA (Vic.) Grassland? (drier S.A.), possibly *Casuarina obesa* (W.A.)	8. MELALEUCA ERICIFOLIA (north)	8. CASUARINA GLAUCA, sometimes with *Melaleuca* spp.

4.6.3. Communities South of the Tropic

Salt marsh communities in the south have been studied by Hamilton (1919), Collins (1921), Pidgeon (1940), Clarke and Hannon (1967, 1969, 1970, 1971) (N.S.W.), Wood (1937) (S.A.). Patton (1942) (Vic.), Curtis and Somerville (1947) (Tas.), and for the whole area Chapman (1960) and Kratochvil et al. (1973). The zonation of the communities is controlled by tidal levels as indicated in Table 23.5. More precise data on the incidence of flooding and the effects of flooding in varying certain soil properties in some communities near Sydney are provided in Table 23.6.

It is of interest that a shrub community of *Arthrocnemum arbuscula* initiates the succession in the zone of deeper flooding adjoining the marine meadow (or mangrove, if it occurs – Fig. 23.23). This is followed by an herbaceous community of *Salicornia* and other herbaceous communities as indicated in Table 23.5. Woody communities of *Casuarina* or *Melaleuca* complete the succession, except in drier parts of South Australia where the mudflats adjoin the arid zone. In such cases the herbaceous communities of the mudflat dominated by *Atriplex paludosa* adjoin the *Atriplex – Maireana* halophytic shrublands of the arid zone (see Chap. 19).

All the zones are not necessarily present in the one transect. On the contrary, a zone, or even two or three zones may be absent, this usually being the result of an abrupt change in the slope of the mudflat; for example, the *Juncus* zone, located on a bank, may adjoin the mangrove zone (Fig. 23.24).

There is some variation in the zonation from sector to sector around the coast and even within sectors. The more important of these are:

i. *Arthrocnemum* does not occur on the east coast.

ii. On the east coast *Zoisia macrantha* (Fig. 23.25) replaces *Sporobolus* in some places.

iii. *Baumea juncea* and *Distichlis distichophylla* are present on both the east and south coasts but *Baumea* is dominant in the east and *Distichlis* in the south (except W.A., where it has not been recorded).

iv. *Atriplex paludosa* or *A. cinerea* (Fig. 23.26) are sometimes locally dominant on somewhat drier areas on mudflats, especially along the south coast.

Pools of brackish water occur on most mud-

Table 23.6. Some characteristics of soils of mudflats in the Sydney district (Woolooware Bay). Data collected from regular observations over periods of one to four years. Osmotic pressures and pH values of subsurface soils only are quoted (surface values also are published by the authors). Maximum osmotic pressures recorded for summer, lower values in cooler months. Data from Clarke and Hannon (1967, 1969).

Community	Number of tidal floodings per month. minimum–maximum	Water table. Highest and lowest recorded. + above mud − below mud surface cm		Osmotic Pressure. Range recorded for subsurface soils. atmospheres	pH Range recorded for subsurface soils
Mangrove	24–60	+70	−58	15.5–32.1	5.5–7.3
Salicornia	0–37	+52	−60	15.7–38.6	5.6–7.4
Juncus	0–7	+12	−52	3.3–31.4	5.4–7.5
Baumea		+8	−38		
Casuarina	0–0	−9	−65	1.4–23.0	5.2–7.4

flats, the water varying in salinity, being least saline or possibly fresh in the *Casuarina* zone. The pools support a distinctive set of species, some of which are *Ruppia maritima* (submerged), *Ludwigia peploides, Paspalum distichum* (floating leaves) and amphibious species of *Scirpus, Phragmites* and *Juncus*.

5. Communities and Flora of Small Islands

5.1. General

Islands are of value to the ecologist and plant geographer for several reasons. In some cases they provide data on migration or interchange of biota between two larger land masses, perhaps continents. In other cases they provide isolated areas with restricted biota on which migration, dispersal and colonization can be studied on different kinds of parent materials in different climates. The closed system, usually with a limited number of taxa, facilitates the study of animal/plant/soil interactions. Again, some islands are the centres of origin of new taxa, these being the products of evolution following isolation; for example, Tasmania has 23 endemic angiosperm genera restricted to the island. The islands dealt with below have been selected to give some information on the items mentioned above.

Many hundreds of islands occur on the continental shelf in climates which vary from tropical to temperate. The islands can be classified into four groups: i. rock-based; ii. coral (cays); iii. sand; iv. mud. The last group, mostly of recent origin and usually occurring in estuaries are often colonized by mangroves and have been dealt with in 4.4, above. The rock-based islands are mostly fragments of the mainland and were connected to the mainland or Tasmania during the Pleistocene glaciation. When sea level rose as the polar ice caps melted, the ancient coast line disappeared under the sea and segments of land were isolated as islands, some in drowned valleys. Sand islands, occurring off the coast of Queensland may have originated prior to the glaciation but the coral islands are undoubtedly a post-glacial development, as discussed below.

The flora of islands is always closely related to the adjacent mainland flora and in the case of the larger rock-based islands may have persisted in its entirety from the glacial period to the present, excepting the newly formed littoral strip. On the contrary, islands of recent origin such as cays and mud islands have been recently colonized from the adjacent mainland. The number of species occurring on islands is governed partly by the size of the island and partly by exposure to wind and the waves. For example, Abbott (1977) studied 121 islands west of Perth, varying in area from 3–160,900 m^2. He concluded that the important factors determining the flora are area, elevation and diversity of habitat and he found linear correlations between species numbers and these three variables. Islands with areas less than 5000 m^2 had an impoverished flora as a result of exposure to sea-spray. Also, inter-island distance plays a

Fig. 23.24. Pure stand of *Juncus maritimus* with *Avicennia marina* var. *australasica* behind. Church Point, Sydney.

negligible part in determining the similarities of island floras, and alien species have higher immigration rates than native species.

Many of the islands are inhabited by man and since white settlement the vegetation of some of these has been altered or destroyed to provide for cultivation or the grazing of domestic stock. Even prior to white settlement the ecology of some islands was disturbed through the introduction of exotic animals, especially goats and rabbits, which were placed on selected islands to provide a meat supply for shipwrecked people or sealers, mainly in the south-east. Many alien plant species have become established on most islands, being most abundant where white man has settled or interfered with the vegetation. However, alien species are sometimes abundant where white man is only a casual visitor and this suggests dispersal of the species from the mainland, probably by wind or birds.

The supply of fresh water on many islands is minimal, there being no pools or springs. This has significance with respect to certain animal populations (and consequently to plant-cover), though some of the larger native animals are specially adapted to surviving on the sap of plants, supplemented by dew.

5.2. Torres Strait Islands

Numerous islands occur in Torres Strait between Australia and New Giunea, being sparated by distances which do not exceed 45 km. They are of interest biologically because they could provide "stepping stones" for the interchange of taxa between Australia and New Guinea.

Stocker (1978) writes that the vegetation of these islands is a disappointment to the botanist, since it is basically similar to that found over much of northern Australia. "Real mingling of the Australian and New Guinea floras takes place still further north along the fresh water reaches of the streams flowing across the Orioma plateau of south-east Papua".

Stocker considers the islands under two main groups: i. the high islands which are the remains of volcanoes; ii. low mud, sand or coral islands.

The high islands have sandy infertile soils, which are very shallow on rocky slopes, though there are a few exceptions where the lava was basic, providing a fertile soil. The vegetation of these islands, especially on those closer to Australia, is dominated by species of *Eucalyptus*, of which 10 have been recorded. *Melaleuca*,

Fig. 23.25. Grassland of *Zoisia macrantha* with *Avicennia marina* var. *australasica* behind. Port Stephens, N.S.W.

Tristania and *Acacia* are also well represented. Deciduous rainforest occurs on the more fertile soils, and the floristic assemblages are similar to those found in northern Australia, containing *Albizia toona, Dysoxylum oppositifolium, Barringtonia calyptrata, Canarium australianum* and *Bombax ceiba.*

The low islands of alluvium are covered by mangroves if they are under the influence of the tides. More elevated areas with a brackish water

Fig. 23.26. *Atriplex cinerea* (plants 60 cm to 1 m high) locally dominant in a well drained site near the ocean. *Scirpus nodosus* forms scattered tufts. Warnambool, Vic.

table support *Pandanus* with occasional *Tristania suaveolens* or *Melaleuca,* and a grass understory of *Imperata cylindrica* and *Themeda.*

Islands with a coral sand beach also occur, and are edged with *Casuarina equisetifolia.* They are similar to those found along the tropical east coast (next section).

5.3. Coral Cays

Cays occur mainly in tropical waters in the east. Most of them are associated with the Great Barrier Reef, which extends for c. 1950 km along the east coast northward of lat. 25°S., and has an estimated area of 250,000 sq. km. The modern reef has developed since the Pleistocene glaciation on an eroded reef which was destroyed when sea level dropped during glaciation. Borings show that the age of the deepest corals is c. 18 million years, and possibly several different reefs were rebuilt as sea level rose. The modern reef has been rebuilt over a period of 8-9000 years, developing on a basal karst of eroded coral situated at a depth of c. 20 m (Davies, 1975).

The various small reefs which make up the Barrier Reef stand at different levels above the sea. Some are completely submerged, some are exposed at low tide, while others have accumulated wind blown, fragmented coral and stand as islands above sea level. Several general accounts of the reef are available (Rougley, 1947; McGregor, 1974; Baglin and Mullins, 1974, copiously illustrated, mainly in colour).

The first plants to appear on the fragmented coral is the strand plant. *Cakile edentula* followed by *Sesuvium portulacastrum, Portulaca oleracea, Ipomoea pes-caprae, Vigna marina,* and some grasses of the genera *Cenchrus, Lepturus, Sporobolus, *Stenotaphrum* and *Thuarea.* The plants are usually scattered but may form a discontinuous cover, and *Sesuvium* and *Portulaca* may form mats. In some cases a herb flat composed of *Boerhavia diffusa, Wedelia biflora* and *Achyranthes aspera* adjoins the first zone. In other cases shrubs or trees form a ring close to the herbaceous species, the main shrubs being *Messerschmidia argentea, Scaevola suaveolens* and *Casuarina equisetifolia* (Heatwole, 1976). On some of the Torres Strait islands Stocker (1978) records *Terminalia catappa* with *Casuarina.*

The more stable and denser community on the more elevated parts of cays vary and appear to be three kinds; *Pandanus* in thickets, *Pisonia grandis* forming low forests (Heatwole, 1976), or a mixed rainforest-like assemblage containing *Manilkara kauki, Hibiscus tiliaceus, Erythrina* sp. and *Cordia subcordata,* with a tangle of spinescent vines (Stocker, 1978). Small patches of deciduous rainforest occur on some of the islands, and some rainforest species may occur with *Pisonia,* especially *Ficus opposita.*

Mangroves sometimes colonize cays, mostly *Aegialitis* and *Avicennia* (see 4.4, above).

5.4. Fraser Island (a sand island)

Fraser Is., the largest of the sand islands lying off the Queensland coast, is 130 km long and 5-22 km wide. It probably originated during the Pleistocene glaciation through the blowing of sand against rock outcrops, the winds at this time being stronger than they are today because of the higher temperature gradients between the equator and the poles. The island consists of a core of cemented yellow-brown and red-brown sands of Pleistocene age which have been partially redistributed and added to under the various climates and sea level positions that have prevailed since the glaciation. Probably the island has been connected to the mainland on more than one occasion permitting the migration of plants and animals from the mainland. The climate is equable frost-free and with a mean annual rainfall of 1200–1400 mm. Noteworthy from the pedological point of view are the deep Podzols in some areas developed on sands and with an A horizon c. 6 m deep. Soil fertility levels vary considerably, from very low (recent sands) to relatively high (supporting rainforests) and it is assumed that the higher fertility levels stem from the accumulation of material from marine organisms transported with the wind-blown sand (various authors in Lauer, 1978).

The island supports communities which are structurally and floristically similar to those on the mainland and include rainforest, tall *Eucalyptus* forest with a rainforest understory, xeromorphic *Eucalyptus* forest on soils of lower fertility, coniferous forest *(Callitris),* swamp vegetation associated with lakes, and littoral zone communities including dune communities

(*Spinifex* and *Casuarina* – *Banksia*), and marine meadows of *Zostera* and *Cymodocea*).

The island supports several species in great abundance and combinations not found elsewhere. Among these are the three conifers (*Agathis robusta* and *Araucaria cunnighamii* – associated with rainforests, and *Callitris columellaris*), and *Syncarpia hillii* which is more abundant on the island than on the mainland. Also the giant fern, *Angiopteris erecta* (along watercourses) is at its southern limit on Fraser Island.

5.5. The Five Islands (rocky)

These islands lie 1/2 to 3 km from the coast near Port Kembla and provide an example of rocky islands, probably once forested, which support species belonging almost entirely to the littoral zone. The plant and animal populations are well documented by Davis et al. (1938). The islands vary in size from 12–2 ha. and receive a mean annual rainfall of c. 1140 mm. Four of the islands are of dolerite (which on the adjacent mainland supports rainforest), the fifth is trachytic-andesite. The difference in the rock type does not lead to a difference in the flora. The percentage cover bears some relationship to the area of the islands, the highest (51% cover) occurring on the largest island, the lowest (3% cover) on the smallest island, which is also the most exposed.

The total number of species on all the islands is 61 (plus 21 aliens), but only four native species are found on all islands. Most of the species are typical of the littoral zone and the tallest community developed is dominated by *Westringia fruticosa* and *Correa alba* which form similar communities on rocky slopes overlooking the sea on the mainland.

5.6. Granite Islands in Southeastern Victoria

These islands lie up to 23 km from the coast of Victoria (Fig. 23.1) and reach altitudes of 20–120 m. The largest is Gabo Is. with an area of c. 170 ha., the others being less than 32 ha. The plant and animal populations have been studied by Gillham (1960, 1961), Norman (1970) and Hope and Thomson (1971). The nine islands have some or much rock exposed and they show various degrees of colonization by plants, there being a zonation on each island into a littoral zone and a more stable zone. The species involved occur elsewhere on the Victorian coast.

Senecio lautus and *Poa poiformis* are the only two occurring on all the islands; the latter dominates climax grasslands on the smaller islands (Chap. 20). Of the shrubs, *Correa alba* is present on most of the islands. On Gabo Is. the tallest vegetation is a scrub containing *Acacia sophorae*, *Banksia integrifolia*, *Leptospermum laevigatum* and a few others, the assemblage being similar to that on the coastal dunes on the mainland.

The chief interest in the islands comes from the controlling effects of vast numbers of sea birds which roost and nest on the islands. The birds probably play some part in the dispersal of seeds, especially gulls which are omnivorous, though most of the other species feed mainly at sea. The birds disturb the habitat by roosting, building nests, burrowing (chiefly mutton birds), and dropping guano. The last of these activities alter the chemical properties of the soil and has the greatest effects on plant populations.

The effects of guano are to increase soil fertility levels, especially with regard to nitrogen and phosphorous. Consequently, xeromorphic species, which are adapted to low levels of nutrients, are almost non-existent in the rookeries where they are replaced by succulents, annuals, biennials and aliens, the last introduced mainly by man. Gillham (1960) concludes: "Guano and sea-spray have parallel effects on vegetation, and the presence of sea-birds leads to a broadening of the coastal belt of salt resistant plants and the elimination of the indigenous, more inland type of flora."

5.7. Kangaroo Island

This island, with an area of 4308 sq. km, is the largest in the south, excluding Tasmania. It supports mainland communities of woodland and forests dominated by eucalypts and shorter communities including littoral zone assemblages typical of the southern coast line (Wood, 1930).

During the Pleistocene glaciation, the island was connected to the mainland, and at this time the strip of coastal lowland, of which Kangaroo Is. was part, formed a corridor for the interchange of taxa between the east and the west. This is supported by the presence on the island of a few eastern species which are recorded in South Australia only on Kangaroo Is., e.g., *Tetratheca ericifolia* and *Hibbertia fascicularis*. Relations with the west are seen through the occurrence of species of western genera (*Lhotskya* and *Conostephium,* see below), which do not occur further east.

The island is unique in the continent in that it supports several endemic taxa confined to the island, including the monotypic genus *Achnophora (A. tatei)*, as well as c. 15 species (listed in Black, 1943–1957) of other Australian endemic genera. Some of these species belong to genera of wide distribution in the south, e.g. *Petrophile multisecta* (the only species of this genus is South Australia), *Pultanaea trifida, Tetratheca halmaturina* and *Cryptandra waterhousei;* others belong to genera which are mainly or entirely western, e.g. *Spyridium halmaturinum, Lhotskya smeatoniana, Conostephium halmaturinum* and *Tetraria halmaturina*.

5.8. Coastal Islands near Fremantle

The three islands, Rottnest, Garden and Carnac (approximate areas 25, 18 and 1.2 sq. km respectively) receive a mean annual rainfall of c. 720 mm with a pronounced winter incidence. They support a flora and plant communities which relate them to the assemblages on the nearby mainland. However, the islands support no eucalypts(c.f. Dorre Is., next section) and the early naturalists who visited the islands commented on the absence of Proteaceae and the paucity of Myrtaceae, Papilionaceae and Epacridaceae. The detailed accounts of McArthur (1957) and Baird (1958 – for Garden Is.) provide information on the communities and an explanation of what might appear to be an aberrant flora.

The total number of native species on the islands is 80, including the parasitic *Orobanche australiana* (all islands). The smallest island, Carnac, supports only 25 species all of which are typical littoral species. The two larger islands each support c. 70 species but only 58 occur on both islands.

The islands are entirely calcareous and the soils are all alkaline. McArthur regards this as the main factor restricting the establishment of xeromorphs, which are mainly adapted to acid soils. He points out that on the mainland xeromorphs of the families mentioned above are not found in areas where conditions approximate those on the islands.

Several woody communities are found on both the larger islands but the communities are dissimilar in several respects. Firstly the proportion of the island covered by the same dominant differs:

i. *Melaleuca lanceolata* forest c. 10 m tall is the most widespread on Rottnest but occurs only in clumps on Garden Is., where the most widespread community is dominated by *Acacia rostellifera.*

ii. Two species locally dominant in scrubs on Rottnest Is. are absent from Garden Is., viz. *Templetonia retusa* and *Acacia cuneata* which indicates direct colonization from the mainland.

McArthur comments on the peculiar structure of the communities. The canopies are closed and dense so that the understory is very sparse or absent.

5.9. Dorre Island

Dorre Is., c. 29 km long and $2^1/_2$ km wide at its widest and with a maximum elevation of c. 45 m lies in or close to the arid zone (mean annual rainfall c. 250 mm). It consists of consolidated calcareous sand with hardpan concretions of calcium carbonate (travertine) and loose sand around its margins. In spite of its exposed situation with respect to winds and the dry climate, it supports no fewer than 118 species (R.D. Royce, quoted in Beard, 1976). Two species of *Eucalyptus* occur; both are recorded also on the mainland, as are the other species on the island.

Royce and Beard describe four zones of vegetation:

i. on white calcareous beach sand, *Spinifex longifolia* and its associates, culminating in

ii. scrubs of *Acacia ligulata* on consolidated dunes;

iii. dwarf scrub 60–90 cm high on crusts of

travertine, containing *Diplolaena dampieri* (usual dominant), *Scaevola crassifolia* and *Westringia rigida*;

iv. flat areas of pink sandy soil overlying travertine and supporting a grassland of *Triodia plurinervata*. The grass hummocks, 40–60 cm high, are interspersed with low, domed shrubs of *Thryptomene baeckeana, Melaleuca cardiophylla, Beyeria cyanescens, Stylobasium spathulatum* and *Brachysema macrocarpum*. In the centre of the island *Eucalyptus oraria* and *E. dongarraensis* form scrubs 1.5 m tall.

References

Abbott, Ian (1977): Species richness, turnover and equilibrium in insular floras near Perth, Western Australia. Aust. J. Bot. 25: 193–208.

Adamson, R.S. and Osborn, T.G.B. (1921): On the ecology of the Ooldea district. Trans. Roy. Soc. S.A. 46: 539–564.

Aldrick, J.M. and Robinson, C.S. (1972): Report on the land units of the Katherine-Douglas area, N.T. 1970. Land Conservation Section. Animal Industry and Agriculture Branch N.T. Administration. Land Conservation Series No. 1. 91 pp.

Allan, H.H. (1961): Flora of New Zealand. Vol 1. Govt. Printer. Wellington.

Anderson, D.J. (1967): Studies on structure in plant communities. V. Pattern in *Atriplex vesicaria* communities in south-eastern Australia. Aust. J. Bot. 15: 451–458.

–, Jacobs, S.W.L. and Malif, A.R. (1969): Studies on structure in plant communities. VI. The significance of pattern evaluation in some Australian dry-land vegetation types. Aust. J. Bot. 17: 315–322.

Anderson, R.H. (1968): The trees of New South Wales. 4th Ed. Govt. Printer. Sydney.

Andrews, E.C. (1913): The development of the Natural Order Myrtaceae. Proc. Linn. Soc. N.S.W. 38: 529–568.

– (1914): The development and distribution of the Natural Order Leguminosae. J. Roy. Soc. N.S.W. 333–407.

– (1916): The geological history of the Australian flowering plants. Amer. J. Sci. Fourth Ser. 42, No. 249: 171–232.

Anon (1947): The Channel Country of south-west Queensland. Bur. of Investigation. Tech. Bull. No. 1. Dept. Public Lands, Brisbane.

Ashby, W.C. and Beadle, N.C.W. (1957): Studies in halophytes. III. Salinity factors in the growth of Australian salt bushes. Ecol. 38: 344–352.

Ashton, D.H. (1956): Studies on the autecology of *Eucalyptus regnans*. Ph. D. thesis, Botany Dept. Univ. Melbourne.

– (1958): The ecology of *Eucalyptus regnans* F. Muell.: The species and its frost resistance. Aust. J. Bot. 6: 154–176.

– and Frankenberg, J. (1976): Ecological studies of *Acmena smithii* with special reference to Wilson's Promontory. Aust. J. Bot. 24: 453–487.

– and Webb, R.N. (1977): The ecology of granite outcrops at Wilson's Promontory, Victoria. Aust. J. Ecol. 2: 269–296.

Aston, Helen I. (1973): Aquatic plants of Australia. Melb. Univ. Press.

Auld, B.A. and Martin, P.M. (1975): Morphology and distribution of *Bassia birchii* (F. Muell.) F. Muell. Proc. Linn. Soc. N.S.W. 100: 167–178.

Austin, M.P. (1975): Vegetation of the south coast study area. MS. C.S.I.R.O. Canberra.

Australian national tide tables (1976): (Dept. Defence,

Baglin, Douglass and Mullins, Barbara (1974): Islands of Australia. Ure Smith. Sydney.

Baird, Alison M. (1958): Notes on the regeneration of vegetation of Garden Island after the 1956 fire. J. Roy. Soc. W.A. 41: 102–107.

– (MS): Vegetation of Yule Brook Reserve, Cannington, W.A. Dept. Botany. Univ. W.A.

Baker, R.T. and Smith, H.G. (1920): A research on the eucalypts and their essential oils. 2nd Ed. Govt. Printer. Sydney.

Baldwin, J.G. and Crocker, R.L. (1941): Soils and vegetation of portion of Kangaroo Island, South Australia. Proc. Roy. Soc. S.A. 65: 263–275.

Balme, Basil E. (1964): The palynological record of Australian pre-Tertiary floras. Tenth Pacif. Sci. Congr. Ser.: Ancient Pacific floras. The pollen story. Ed. L. Cranwell. 49–80.

Bamber, R.K. and Mullette, K.J. (1978): Studies of the lignotubers of *Eucalyptus gummifera*. II. Anatomy. Aust. J. Bot. 26: 15–22.

Barber, H.N. (1955): Adaptive gene substitutions in Tasmanian eucalypts: I. Genes controlling the development of glaucousness. Evolution IX. 1–14.

Barlow, B.A. (1959a): Chromosome numbers in the Casuarinaceae. Aust. J. Bot. 7: 230–237. (b): Polyploidy and apomixis in the *Casuarina distyla* species group. Ibid: 238–251.

– (1966): A revision of the Loranthaceae of Australia and New Zealand. Aust. J. Bot. 14: 421–499.

– (1971): Cytogeography of the genus *Eremophila*. Aust. J. Bot. 19: 295–310.

Barrow, M.D., Costin, A.B. and Lake, P. (1968): Cyclical changes in an Australian fjaeldmark community. J. Ecol. 56: 89–96.

Bates, R. (1976): Terrestrial orchids of South Australia's semi-arid areas. Orchadian 5: 42–48.

Baur, G.N. (1957): Nature and distribution of rain forests in New South Wales. Aust. J. Bot. 5: 190–233.

– (1962): Forest vegetation in north-eastern New South Wales. For. Comm. N.S.W. Res. Note No. 8.

– (1964): The ecological basis of rainforest management. For. Comm. N.S.W.

– (1965): Forest types in New South Wales. For. Comm. N.S.W. Res. Note No. 17.

Beadle, N.C.W. (1940): Soil temperatures during forest fires and their effect on the survival of vegetation. J. Ecol. 28: 180–192.

– (1948): The vegetation and pastures of western New South Wales. Govt. Printer. Sydney.

– (1952): Studies in halophytes. I. The germination of the seed and establishment of the seedlings of five species of *Atriplex* in Australia. Ecol. 33: 49–62.

– (1952): The misuse of climate as an indicator of vegetation and soils. Ecol. 32: 343–345.

– (1953): The edaphic factor in plant ecology with a Commonwealth of Australia). Alexander Bros. Mentone. Vic.

special note on soil phosphates. Ecology 34: 426–428.
- (1954): Soil phosphate and the delimitation of plant communities in eastern Austraila. Ecology 35: 370–375.
- (1962a): Soil phosphate and the delimitation of plant communities in eastern Australia. II. Ecology 43: 281–288.
- (1962b): An alternative hypothesis to account for the generally low phosphate content of Australian soils. Aust. J. Agric. Res. 13: 434–442.
- (1966): Soil phosphate and its role in molding segments of the Australian flora and vegetation with special reference to xeromorphy and sclerophylly. Ecology 47: 991–1007.
- (1968): Some aspects of the ecology and physiology of Australian xeromorphic plants. Aust. J. Sci. 30: 348–355.
- (in press): Origins of the Australian angiosperm flora. Chap. 2 in Biogeography and Ecology of Australia. 2nd Edition. Ed. by A. Keast. Junk, The Hague.
- and Beadle, Lois D. (1971–1972, 1976): Students flora of north-eastern New South Wales. Univ. New England. Armidale.
- and Burges, N.A. (1953): A further note on laterites. Aust. J. Sci. 15: 170–171.
- and Costin, A.B. (1952): Ecological classification and nomenclature. Proc. Linn. Soc. N.S.W. 77: 61–82.
- –, Evans, O.D. and Carolin, R.C, (1972): Flora of the Sydney District. A.H. and A.W. Reid. Sydney.
- –, Whalley, R.D.B. and Gibson, J.B. (1957): Studies in halophytes II. Analytic data on the mineral constituents of three species of *Atriplex* and their accompanying soils in Australia. Ecol. 38: 340–344.

Beard, J.S. (n.d.): Wildflowers in the North-west. Westviews Production. Perth.
- (1967): Some vegetation types of tropical Australia in relation to those of Africa and America. J. Ecol. 55: 271–290.
- (1967): A study of patterns in some West Australian heath and mallee Communities. Aust. J. Bot. 15: 131–139.
- (1969): The Vegetation of the Boorabbin and Lake Johnston areas, Western Australia. Proc. Linn. Soc. N.S.W. 93: 239–268.
- (1970): West Australian plants. 2nd Ed. Surrey Beatty and Sons. Chipping Norton, N.S.W.
- (1972a): The vegetation of the Newdegate and Bremer Bay areas, Western Australia; (b): Southern Cross area; (c): Jackson area; (d): Hyden area; (e) Kalgoorlie area. Vegmap Publications. Sydney.
- (1973): The elucidation of palaeodrainage patterns in Western Australia through vegetation mapping. Vegmap. Public. Occasional Paper No. 1. Applecross., W.A.
- (1973a): Vegetation survey of Western Australia. The vegetation of the Ravensthorpe area; (b): Esperance and Malcolm areas. Vegmap Publications. Perth.
- (1974): Vegetation survey of Western Australia. Great Victoria Desert. Univ. W.A. Press.
- (1975): Vegetation survey of Western Australia. Nullarbor. Univ. W.A. Press.
- (1976): An indigenous term for the Western Australian sandplain and its vegetation. J. Roy. Soc. W.A. 59: 55–57.
- (1976): The monsoon forests of the Admirality Gulf, Western Australia. Vegetatio: 31: 177–192.
- (1976a): Vegetation survey of Western Australia: Shark Bay and Edel Areas; (b): Ajana Area; (c): Geraldton Area (with A.C. Burns); (d): Dongara Area; (e): Perenjori Area. Vegmap Publications. Perth.
- (1977): Tertiary evolution of the Australian flora in the light of latitudinal movements of the continent. J. Biogeography. 4: 111–118.
- and Webb, M.J. (1974): Vegetation survey of Western Australia. Great Sandy Desert. Univ. W.A. Press.

Bentham, George (1863–1878): Flora Australiensis. Lovell Reeve and Co. London. 7 Vols.
- (1875): Revisions of the sub-order Mimoseae. Trans. Linn. Soc. London. 30: 335–664.

Bettenay, E., McArthur, W.M., and Hingston, F.J. (1960): The soil associations of part of the swan Coastal Plain, Western Australia, C.S.I.R.O. Bull. No. 35.

Biddiscombe, E.F. (1963): A vegetation survey in the Macquarie region, N.S.W., C.S.I.R.O. Div. Plant Ind. Tech. Pap. No. 18.
- –, Cuthberston, E.G. and Hutchings, R.J. (1954): Autecology of some natural pasture species at Trangie, N.S.W. Aust. J. Bot. 2: 69–98.

Bird, E.C.F. (1965): Coastal landforms. Aust. Nat. Univ. Press, Canberra.

Bird, E.T. (1975): An investigation into dormancy in selected species of Gramineae. Litt. B. Thesis. Univ. New England. Armidale.

Black, J.M. (1943–1957): Flora of South Australia. (Part IV revised by Enid Robertson). Govt. Printer. Adelaide.

Black, R.F. (1954): The leaf anatomy of Australian members of the genus *Atriplex*. Aust. J. Bot. 2: 269–286.

Blackall, W.E. and Grieve, B.J. (1954, 1956, 1965): How to know Western Australian wilflowers. Parts 1–3. Univ. W.A. Press.

Blake, S.T. (1938): The plant communities of western Queensland and their relationship with special reference to the grazing industry. Proc. Roy. Soc. Qld 49: 156–204.
- (1940): The vegetation of Goat Island and Bird Island in Moreton Bay. Qld Naturalist. 11: 94 to 101.
- (1941): The vegetation of the Lower Stanley River Basin. Proc. Roy. Soc. Qld 52: 61–77.
- (1942): The vegetation of Running Creek Valley,

South-east Queensland, and some neighbouring areas. Qld Naturalist 12: 4–12.
– (1947): The vegetation of Noosa. Qld Naturalist. 13: 47–50.
– (1948): The vegetation of the country surrounding Somerset Dam. Qld Naturalist. 13: 94–100.
– (1953): Botanical contributions to the northern Australia regional survey. 1. Studies of northern Australian species of eucalypts. Aust. J. Bot. 1: 185–352.
– (1968): A revision of *Melaleuca leucadendron* and its allies (Myrtaceae). Contrib. Qld Herb. 1: 1–114.
Blakely, W.F. (1934, 1955): A key to the eucalypts. For. and Timb. Bur. Canberra.
Bonython, C.W. (1956): The salt of Lake Eyre – its occurrence in Madigan Gulf and its possible origin. Trans. Roy. Soc. S.A. 79: 66–92.
Boomsma, C.D. (1946): The vegetation of the southern Flinders Ranges. Trans. Roy. Soc. S.A. 70: 259–276.
– (1969): The forest vegetation of South Australia. S. Aust. Woods and Forests Dept. Bull. No. 18.
Bowen, G.D. (1956): Nodulation of legumes indigenous to Queensland. Qld J. Agric. Sci., 13: 47 to 60.
Bowler, J.M. (1971): Pleistocene salinities and climatic change: evidence from lunettes in south-east Australia. In: Aboriginal man and environment in Australia. A.N.U. Press.
–, Hope, G.S., Jennings, J.N., Singh, G. and Walker, D. (1976): Late Quaternary climates of Australia and New Guinea. Quaternary Research. 6: 359–394.
Boyland, D.E. (1970): Ecological and floristic studies in the Simpson Desert National Park, South Western Queensland. Proc. Roy. Soc. Qld 82: 1–16.
Brass, L.J. (1953): Results of the Archbold expeditions No. 68: Summary of the 1948 Cape York (Australia) expedition. Bull. Amer. Mus. Nat. Hist. 102: 135–205 New York.
Briggs, Barbara G. (1959): *Ranunculus lappaceus* and allied species of the Australian mainland. I. Taxonomy. Proc. Linn. Soc. N.S.W. 84: 295–324.
Broese van Groenou, P.J.R. (1974): Regeneration of native plant communities following mineral sand mining. M. Sc. Thesis, Univ. New England Armidale, N.S.W.
Brooker, M.I.H. (1974a): Notes on *Eucalyptus brachycorys* Blakely and *E. comitae-vallis* Maiden. Nuytsia 1: 294–296.
– (1974b): Six new species of *Eucalyptus* from Western Australia. Nuytsia 1: 297–314.
Brown, D.A., Campbell, K.S.W. and Crook, K.A.W. (1968): The geological evolution of Australia and New Zealand. Pergamon Press.
Browne, W.R. (1945): An attempted post-Tertiary chronology of Australia. Pres. Address. Proc. Linn. Soc. N.S.W. 70.
Brownell, P.F. and Crossland, C.J. (1972): The requirement for sodium as a micronutrient by species having the C_4 dicarboxylic photosynthetic pathway. Plant Physiol. 49: 784–797.
– and Wood, J.G. (1957): Sodium, an essential micronutrient for *Atriplex vesicaria* Heward. Nature. London. 179: 635–636.
Brough, P., McLuckie, J. and Petrie, A.H.K. (1924): An ecological study of the flora of Mount Wilson. Pt. i. The vegetation of the basalt. Proc. Linn. Soc. N.S.W. 49: 475–498.
Burbidge, Nancy T. (1943): Notes on the vegetation of the North-east Goldfields. J. Roy. Soc. W. Aust. 27: 119–132.
– (1944): Ecological succession observed during regeneration of *Triodia pungens* R.Br. after burning. J. Roy. Soc. W.A. 28: 149–156.
– (1944a): Ecological notes on the vegetation of 80-mile Beach. J. Roy. Soc. W.A. 28: 157–164.
– (1945a): Ecological notes on the De Grey – Coongan area, with special reference to physiography. J. Roy. Soc. W.A. 29: 151–161.
– (1945b): Morphology and anatomy of the Western Australian species of *Triodia* R.Br. I. General Morphology. Trans. Roy. Soc. S. Aust. 69: 303–308.
– (1946): Germination studies of Australian Chenopodiaceae with special reference to the conditions necessary for regeneration. Trans. Roy. Soc. S.A. 70: 110–120.
– (1946): Morphology and anatomy of the Western Australian species of *Triodia* R.Br. II. Internal anatomy of leaves. Trans. Roy Soc. S.A. 70: 221–234.
– (1947): Key to the South Australian species of *Eucalyptus* L'Hérit. Trans. Roy. Soc. S.A. 71: 137–167.
– (1952): The significance of the mallee habit in *Eucalyptus*. Proc. Roy. Soc. Qld. 62: 73–78.
– (1953): The genus *Triodia* R.Br. (Gramineae). Aust. J. Bot. 1: 121–184.
– (1959): Notes on plants and plant habitats observed in the Abydos-Woodstock area, Pilbara District, Western Australia. C.S.I.R.O. Tech. Pap. No. 12.
– (1960): Further notes on *Triodia* R.Br. (Gramineae) with description of five new species and one variety. Aust. J. Bot. 8: 381–395.
– (1960): The phytogeography of the Australian region. Aust. J. Bot. 8: 75–212.
– (1963): Dictionary of Australian plant genera. Angus and Robertson. Sydney.
Bureau of Meteorology (1975): Climatic averages, Australia. Aust. Govt. Pub. Service. Canberra.
Burges, A. and Drover, D.P. (1953): The rate of podzol development in the sands of the Woy Woy district, N.S.W. Aust. J. Bot. 1: 83–94.
– and Johnston, R.D. (1953): The structure of a New South Wales subtropical rain forest. J. Ecol. 41: 72–83.
Burrough, P.A., Brown, L. and Morris, E.C. (1977): Variations in vegetation and soil pattern across the Hawkesbury Sandstone plateau from Barren

Grounds to Fitzroy Falls, New South Wales. Aust. J. Ecol. 2: 137–159.

Burrows, W.H. (1973): Regeneration and spatial patterns of *Acacia aneura* in south west Queensland. Trop. Grassl. 7: 57–68.

– and Beale, I.F. (1969): Structure and association in the Mulga *(Acacia aneura)* lands of south-western Queensland. Aust. J. Bot. 17: 539–552.

Byrnes, N.B. (1977): A revision of Combretaceae in Australia. Contrib. Qld Herbarium. No. 20.

–, Everist, S.L., Reynolds, S.T., Specht, A. and Specht, R.L. (1977): The vegetation of Lizard Island, North Queensland. Proc. Roy. Soc. Qld 88: 1–15.

Campbell, E.M. (1968): Lunettes in southern South Australia. Trans. Roy. Soc. S.A. 92: 85–113.

Carey, Gladys (1938): Comparative anatomy of leaves from species in two habitats around Sydney. Proc. Linn. Soc. N.S.W. 63: 439–450.

Carnahan, J.A. (1976): Natural vegetation. Atlas of Australian resources; 2nd series. Divn. Nat. Mapping, Canberra. Map + 26pp.

Carolin, R.C., Jacobs, S.W.L. and Vesk, Maret (1975): Leaf structure in Chenopodiaceae. Bot. Jahrb. Syst. 95: 226–255.

Carr. D.J. (1955): Contributions to Australian Bryology. 1. The structure, development, and systematic affinities of *Monocarpus sphaerocarpus* (Marchantiales). Aust. J. Bot. 4: 175–191.

Carr, S.G.M. (1972): Problems of the geography of the tropical eucalypts. In "Bridge and Barrier: The natural and cultural history of Torres Strait." Dept. Biogeog. A.N.U. Canberra: 153–182.

–, Carr, D.J. and George, A.S. (1970): A new eucalypt from Western Australia. Proc. Roy. Soc. Vic. 83: 159–169.

– and Turner, J.S. (1959): The ecology of the Bogong High Plains. I. The environmental factors and the grassland communities. Aust. J. Bot. 7: 12–33. II. Fencing experiments in Grassland C. Ibid. 34–64.

Carrodus, B.B. and Specht, R.L. (1965): Factors affecting the relative distribution of *Atriplex vesicaria* and *Kochia sedifolia* (Chenopodiaceae) in the arid zone of South Australia. Aust. J. Bot. 13: 419–433.

–, – and Jackman, M.L. (1965): The vegetation of Koonamore Station, South Australia. Trans. Roy. Soc. S.A. 89: 41–57.

Casson, P.B. (1952): The forests of Western Tasmania. Aust. For. 16: 71–86.

Chapman, F. (1921): A sketch of the geological history of Australian plants: The Cainozoic flora. Vic. Nat. 37: 115–133.

– (126): The fossil *Eucalyptus* record. Vic. Nat. 42: 229–231.

– (1937): Descriptions of Tertiary plant remains from Central Australia and from other Australian localities. Trans. Roy. Soc. S.A. 61: 1–16.

Chapman, V.J. (1960): Salt marshes and salt deserts of the world. Leonard Hill Ltd. London.

Charley, J.L. (1959): Soil salinity – vegetation patterns in western N.S.W. and their modification by overgrazing. Ph.D. thesis. Univ. New Engl. Armidale.

– and Cowling, S.W. (1968): Changes in soil nutrient status resulting from overgrazing and their consequences in plant communities in semi-arid zones. Proc. Ecol. Soc. Aust. 3: 25–38.

– and McGarity, J.W. (1964): High soil nitrate-levels in patterned saltbush communities. Nature 201: 1351–1352.

Chinnick, L.J. and Key K.H.L. (1971): Map of soils and timber density in the Bogan-Macquarie outbreak area of the locust *Chortoicetes terminifera*. C.S.I.R.O. Div. Entomol. Tech. Paper 12.

Chippendale, G.M. (1958): Notes on the vegetation of a desert area in Central Australia. Trans. Roy. Soc. S.A. 81: 31–41.

– (1959): Plants of Palm Valley, Central Australia. Vic. Naturalist. 75: 192–200.

– (1963): The relic nature of some Central Australian plants. Trans. Roy. Soc. S.A. 86: 31–34.

– (1963): Ecological notes on the "Western Desert" area of the Northern Territory. Proc. Linn. Soc. N.S.W. 88: 54–66.

– (1971): Check list of Northern Territory plants. Proc. Linn. Soc. N.S.W. 96: 207–267.

– (1973): Eucalypts of the Western Australian Goldfields. Aust. Govt. Printing Service. Canberra.

Christensen, P.E. and Kimber, P.C. (1975): Effect of prescribed burning on the flora and fauna of south-west Australian forests. Proc. Ecol. Soc. Aust. 9: 85–106.

–, –, Noakes, L.C. and Blake, S.T. (1953): General report on Survey of Katherine-Darwin region; 1946. C.S.I.R.O. Land Research Series No. 1.

Churchill, D.M. (1968): The distribution and prehistory of *Eucalyptus diversicolor*, *E. marginata* and *E. calophylla* in relation to rainfall. Aust. J. Bot. 16: 125–151.

– (1973): The ecological significance of tropical mangroves in the early Tertiary floras of southern Australia. Geol. Soc. Aust. Essays in honour of Isabel Cookson. Spec. Publ. 4: 79–86.

Clark, L.R. (1948): Observations of the plant communities at "Bundemar", Trangie district, N.S.W., in relation to *Chortoicetes terminifera* (Walk) and *Austraoicetes cruciata* (Sauss.). C.S.I.R.O. Bull. 236.

Clark, S.S. (1975): The effect of sand mining on coastal heath vegetation in New South Wales. Proc. Ecol. Soc. Aust. 9: 1–16.

Clarke, Lesley D. and Hannon, Nola J. (1967–1971): The mangrove swamp and salt marsh communities of the Sydney district. (1967): I. Vegetation soils and climate. J. Ecol. 55: 753–771; (1969): II. The holocoenotic complex with particular reference to physiography. Ibid. 57: 231–234; (1970): III. Plant growth in relation to salinity and waterlogging. Ibid. 58: 351–369; (1971): IV. The significance of species interaction. Ibid 59: 535–553.

Clayton, W.D. (1975): Chorology of the genera of Gramineae. Kew Bull. 30: 111–132.

Clifford, H.T. (1953): A note on the germination of *Eucalyptus* seed. Aust. For. 17: 17–20.

– (1972): Eucalypts of the Brisbane region. Qld Museum.

Coaldrake, J.E. (1951): The climate, geology, soils and plant ecology of portion of the County of Buckingham (ninety-Mile Plain), South Australia. C.S.I.R.O. Bull. 266.

– (1961): Ecosystems of the coastal lowlands, southern Queensland. C.S.I.R.O. Bull. 283.

– (1971): Variation in some floral, seed, and growth characteristics of *Acacia harpophylla* (Brigalow). Aust. J. Bot. 19: 335–352.

–, Tothill, J.C., McHarg, G.W. and Hargreaves, J.N.G. (1972): Vegetation map of Narayen Research Station, south-east Queensland. Divn. Trop. Pastures. Tech. Paper 12. C.S.I.R.O.

Collins, Marjorie I. (1921): On the mangrove and salt marsh vegetation near Sydney, N.S.W., with special reference to Cabbage Tree Creek, Port Hacking. Proc. Linn. Soc. N.S.W. 46: 376–392.

– (1923–1924): Studies in the vegetation of arid and semi-arid New South Wales. I. The plant ecology of the Barrier district. Proc. Linn. Soc. N.S.W. 48: 229–266. II. The botanical features of the Grey Range and its neighbourhood. Ibid. 49: 1–18.

Condon, R.W. (1960): Scald reclamation experiments at Trida, N.S.W. J. Soil Cons. Serv. N.S.W. 16: 288–303.

Connor, D.J. (1966a): Vegetation studies in north-west Victoria. I. The Beulah-Hopetoun area. Proc. Roy. Soc. Vic. 79: 579–595.

– (1966b): Vegetation studies in north-west Victoria. II. The Horsham area. Proc. Roy. Soc. Vic. 79: 637–653.

– and Wilson, G.L. (1968): Response of a coastal Queensland heath community to fertilizer application. Aust. J. Bot. 16: 117–123.

Cookson, Isabel C. (1945): The pollen content of Tertiary deposits. Aust. J. Sci. 7: 149–150.

– (1946): Pollen of *Nothofagus* Blume from Tertiary deposits in Australia. Proc. Linn. Soc. N.S.W. 71: 49–63.

– (1950): Pollen grains of proteaceous type from Tertiary deposits in Australia. Aust. J. Sci. Res. 3 B: 166–177.

– (1954): A palynological examination of No. 1 Bore, Birregurra, Vic. Proc. Roy. Soc. Vic., 66: 119–128.

– (1954): The occurrence of an older Tertiary microflora in Western Australia. Aust. J. Sci. 17: 37–38.

– (1954): The Cainozoic occurrence of *Acacia* in Australia. Aust. J. Bot. 2: 52–59.

– (1964): Some early angiosperms from Australia: The pollen record. Tenth Pacif. Sci. Congr. Ser.: Ancient Pacific floras. The pollen story. Ed. L. Cranwell: 81–84.

– and Balme, B.E. (1962): *Amosopollis cruciformis* gen. et sp. nov., a pollen tetrad from the Cretaceous of Western Australia. J. Proc. Roy. Soc. W.A. 45: 97–99.

– and Duigan, S.L. (1950): Fossil Banksiae from Yallourn, Victoria, with some notes on the morphology and anatomy of living species. Aust. J. Sci. 3: 133–164.

– and Pike, Kathleen M. (1954): The fossil occurrence of *Phyllocladus* and two other podocarpaceous types in Australia. Aust. J. Bot. 2: 60–68.

Copeland, E.B. (1947): Genera Filicum. Chronica Botanica Co. Waltham, Mass., U.S.A.

Corn, Carolyn A. and Hiesey, W.M. (1973): Altitudinal variation in Hawaiian *Metrosideros*. Amer. J. Bot. 60: 991–1002.

Costin, A.B. (1954): A study of the ecosystems of the Monaro Region of New South Wales. Govt. Printer. Sydney.

– (1957): The high mountain vegetation of Australia. Aust. J. Bot. 5: 173–189.

– (1967): Alpine ecosystems of the Australasian region. Arctic and Alpine Environments: 55–87. Indiana Univ. Press.

– and Polach. (1969): Dating soil organic matter. Atomic Energy in Australia 12: 13–17.

–, Wimbush, D.J., Barrow, M.D. and Lake, P. (1969): Development of soil and vegetation climaxes in the Mount Kosciusko area, Australia. Vegetatio 18: 273–288.

Couper, R.A. (1953): Upper Mesozoic and Cainozoic spores and pollen grains from New Zealand. N.Z. Geol. Surv. Paleontol. Bull. 22.

– (1953): Distribution of Proteaceae, Fagaceae and Podocarpaceae in some Southern Hemisphere Cretaceous and tertiary beds. N.Z. J. Sci. 35: 247–249.

– (1960): Southern Hemisphere Mesozoic and Tertiary Podocarpaceae and Fagaceae and their palaeogeographic significance. Proc. Roy. Soc. B 152 (949): 491–500.

– (1960): New Zealand Mesozoic and Cainozoic plant microfossils. N.Z. Geol. Surv. Paleontolog. Bull. 32.

Cranwell, L.M. (1964): Antarctica: cradle or grave for its *Nothofagus*. In Ancient Pacific Floras: 87–93.

Cremer, K.W. (1960): Eucalypts in Rain Forest. Aust. For. 24: 120–126.

– and Mount, A.B. (1965): Early stages of plant succession following the complete felling and burning of *Eucalyptus regnans* forest in the Florentine Valley, Tasmania. Aust. J. Bot. 13: 303–322.

Crocker, R.L. (1944): Soil and vegetation relationships in the lower South-east of South Australia. Trans. Roy. Soc. S.A. 68: 144–172.

– (1946): The soils and vegetation of the Simpson Desert and its borders. Trans. Roy. Soc. S.A. 70: 235–258.

– (1946): An introduction to the soils and vegetation of Eyre Peninsula, South Australia. Trans. Roy. Soc. S.A. 70: 83–106.

– (1959): Past climatic fluctuations and their influ-

ence upon Australian vegetation. Biogeography and ecology in Australia: 283–290. Junk. The Haque.
- and Eardley, C.M., (1939): A South Australian *Sphagnum* bog. Trans. Roy. Soc. S.A. 63: 210–214.
- and Skewes, H.R. (1941): The principal soil and vegetation relationships on Yudnappina Station, north-west South Australia. Trans. Roy. Soc. S.A. 65: 44–60.
- and Wood, J.G. (1947): Some historical influences on the development of the South Australian vegetation communities. Trans. Roy. Soc. S.A. 71: 91–136.
Cromer, D.A.N. and Pryor, L.D. (1942): A contribution to rainforest ecology. Proc. Linn. Soc. N.S.W. 67: 249–268.
Cronquist, Arthur (1955): Phylogeny and taxonomy of the Compositae. Amer. Midland Naturalist. 53: 478–511.
Croxall, J.P. (1975): The Hymenophyllaceae of Queensland. Aust. J. Bot. 23: 509–547.
Cunningham, T.M. and Cremer, K.W. (1965): Control of the understory in wet eucalypt forests. Aust. For. 29: 4–14.
Curtis, Winifred M. (1963–1967): Student's flora of Tasmania. Parts 2 & 3. Govt. Printer, Tasmania.
- (1969): The Vegetation of Tasmania. Tasmanian Year Book, Vol. 3. 55–59.
- and Somerville, J. (1947): Boomer Marsh – preliminary botanical and historical survey. Pap. Proc. Roy. Soc. Tas.: 151–157.
- and Somerville, Janet (1949): Vegetation. A.N.Z.A.A.S. Handbook for Tasmania. Hobart: 51–57.
- and Morris, Dennis I. (1975). The Student's flora of Tasmania. Part 1, 2nd Ed. Govt. Printer, Tas.
Dallimore, W. and Jackson, A.B. (1966): Revised by S.G. Harrison. A Handbook of the Coniferae and Ginkgoaceae. Edward Arnold. London.
Darlington, P.J. (1965): Biogeography of the southern end of the world. Harvard Univ. Press.
David, T.W.E. and Browne, W.R. (1950): The geology of the Commonwealth of Australia. Vol. 1. Edward Arnold. London.
Davidson, Dorothy (1954): The Mitchell grass association of the Longreach district. Univ. Qld Papers. 3: 45–59.
Davies, P. (1975): The Great Barrier Reef: The geological structure. Habitat. 3: 3–8.
Davis, Consett (1936): Plant ecology of the Bulli district. I. Stratigraphy, physiography and climate; general distribution of plant communities and interpretation. Proc. Linn. Soc. N.S.W. 61: 285–297.
- (1940): Preliminary survey of the vegetation near New Harbour, south-west Tasmania. Pap. Roy. Soc. Tas. 40: 1–10.
- (1941): Plant ecology of the Bulli district II. Plant communities of the plateau and scarp. III. Plant communities of the coastal slopes and plain. Proc. Linn. Soc. N.S.W. 66: 1–19 and 20–32.
-, Day, M.F. and Waterhouse, D.F. (1938): Notes on the terrestrial ecology of the Five Islands. Proc. Linn. Soc. N.S.W. 63: 357–388.
Davis, Gwenda L. (1967): Apomixis in the Compositae. Phytomorphology 17: 270–277.
Dawson, N.M. and Ahern, C.A. (1973): Soils and landscapes of Mulga lands with special reference to south-western Queensland. Trop. Grassl. 7: 23–34.
- and Boyland, D.E. (1971): Western arid region land use study-Part I: Land systems. Div. Land Util. Qld Dept. Prim. Indust. Brisbane.
Deane, Henry (1900): Observations on the Tertiary flora of Australia, with special reference to Ettinghausen's theory of the Tertiary cosmopolitan flora. Proc. Linn. Soc. N.S.W. 25: 463–475.
- (1902a): Preliminary report on the fossil flora of Pitfield, Mornington, Sentinel Rock (Otway Coast), Berwick, and Wonwron. Geol. Surv. Vic. 1: 13–32.
- (1902b): Notes on fossil leaves from the Tertiary doposits of Wingello and Bungonia. Geol. Surv. N.S.W. 7: 59–65.
- (1925a): Tertiary fossil fruits from Deep Head, Foster, South Gippsland. Geol. Surv. Vic. 4: 489–492.
- (1925b): Fossil leaves from the Open Cut, State Brown Coal Mine, Morwell. Geol. Surv. Vic. 4: 492–498.
Dettmann, Mary E. (1963): Upper Mesozoic microfloras from south-eastern Australia. Proc. Roy. Soc. Vic. 77: 1–148.
- (1973): Angiospermous pollen from Albian to Turonian sediments of Eastern Australia. Geol. Soc. Austr. Essays in honour of Isabel Cookson. Spec. Publ. 4: 3–34.
- and Playford, G. (1968): Taxonomy of some Cretaceous spores and pollen grains from eastern Australia. Proc. Roy. Soc. Vic. 81: 69–94.
- and Playford, G. (1968a). Palynology of the Australian Cretaceous: A review. Stratigraphy and Palaeontology: Essays in honour of Dorothy Hill. Ed. K.W.S. Campbell: 174–210.
Diels, L. (1906): Die Pflanzenwelt von West-Australien südlich des Wendekreises. In "Vegetation der Erde, VII". Engelmann. Leipzig.
Dockrill, A.W. (1972): In: Ryan, P. (Ed.) "Encyclopaedia of Papua and New Guinea". Melb. Univ. Press.
Dodson, J.R. (1974a): Vegetation history and water fluctuations at Lake Leake, south-eastern South Australia. I. 10,000 B.P. to Present. Aust. J. Bot. 22: 719–741. II. 50,000 B.P. to 10,000 B.P. Ibid. 23: 815–831.
- (1974b): Vegetation and climatic history near Lake Keilambete, Western Victoria. Aust. J. Bot. 22: 709–717.
- and Wilson, I.B. (1975): Past and present vegetation of Marshes Swamp in south-eastern South Australia. Aust. J. Bot. 23: 123–150.
Dorman, F.H. (1966): Australian Tertiary paleotemperatures. J. Geol. 71: 49–61.

Douglas, J.G. (1973): Spore-plant relationships in Victorian Mesozoic cryptogams. Geol. Soc. Aust. Essays in honour of Isabel Cookson. Spec. Publ. 4: 119–126.

Ducker, Sophie C., Foord, N.J. and Knox, R.B. (1977): Biology of Australian seagrasses: The genus *Amphibolis*. Aust. J. Bot. 25: 67–95.

Duffy, F.A. (1972): The gymnosperm flora of Australia, and its relationships, past and present, to other southern gymnosperm floras. Litt. B. Thesis. Bot. Dept., Univ. New. England. Armidale.

Duigan, Suzanne L. (1951): A catalogue of the Australian Tertiary flora. Proc. Roy. Soc. Vic. 63: 41–56.

Dury, G.H. and Langford-Smith, T. (1969): A minimum age for the Duricrust. Aust. J. Sci. 31: 362–363.

Dwyer, L.J. (1957): 50 years of weather – Western Australia 1908–1957. Commonwealth Bur. Meteorology.

Eardley, C.M. (1943): An ecological study of the vegetation of Eight Mile Creek Swamp: a natural South Australian fen formation. Trans. Roy. Soc. S.A. 67: 200–223.

– (1946): The Simpson desert expedition, 1939. Scientific reports No. 7, Botany, Part 1: Catalogue of plants. Trans. Roy. Soc. S.A. 70(2): 145–174.

– (1948): The Simpson Desert expedition 1939. Scientific Reports 7: Botany, Part II. The phytogeography of some important sandridge deserts compared with that of the Simpson Desert. Trans. Roy. Soc. S.A. 72: 33–68.

Ecos (1978): A destructive fungus in Australian forests. Ecos 15: 3–14.

Eichler, Hansjoerg (1965): Supplement to J.M. Black's Flora of South Australia. Govt. Printer. Adelaide.

Eldridge, K.G. (1968): Physiological studies of altitudinal variation in *Eucalyptus regnans*. Proc. Ecol. Soc. Aust. 3: 70–76.

Etheridge, Robert (1888): Contributions to the Tertiary flora of Australia by Dr. Constantin, Baron von Ettinghausen. (Translation). Mem. Geol. Surv. N.S.W. Palaeontol. No. 2. Govt Printer. Sydney.

Ettinghausen, C. (1883, 1886): Beiträge zur Kenntniss der Tertiärflora Australiens. Staatsdruckerei. Wien. pp. 101–147 and 1–62.

– (1895): Beiträge zur Kenntnis der Kreideflora Australiens. Staatsdruckerei. Wien. pp. 1–56.

Evans, J.W. (1941): Concerning the Peloridiidae. Austr. J. Sci. 4: 95–97.

Evans, L.T. and Knox, R.B. (1969): Environmental control of reproduction in *Themeda australis*. Aust. J. Bot. 17: 375–389.

Everist, S.L. (1949): Mulga (*Acacia aneura* F. Muell.) in Queensland. Qld J. Agric. Sci. 6: 87–139.

– (1964): The Mitchell Grass country. Qld Naturalist. 45–50.

– (1966): Lessons from Arcadia Valley brigalow. Qld Dept. Prim. Indust. Leaflet 867.

Ewart, Alfred J. (1908): On the longevity of seeds. Proc. Roy. Soc. Vic. 21 (N.S.): 1–203.

Fitzpatrick, E.A. and Nix, H.A. (1970): The climatic factor in Australian grassland ecology. In Moore, R.M. (1970): 3–26.

Florence, R.G. (1963): Vegetational pattern in east coast forests. Proc. Linn. Soc. N.S.W. 88: 164–179.

– and Crocker, R.L. (1962): Analysis of Blackbutt (*Eucalyptus pilularis* Sm.) seedling growth in a blackbutt forest soil. Ecology. 43: 670–679.

Floyd, A.G. (1962): Investigations into the natural regeneration of Blackbutt – *E. pilularis*. Research Note 10. Forestry Comm. N.S.W.

Forest Focus (1971): Focus on the jarrah forest. No. 6: 3–8. Forests Dept. W.A. Perth.

Francis, W.D. (1970): Australian rain forest trees. 3rd Ed. Dept. Nat. Devel. Canberra.

Fraser, Lilian and Vickery, Joyce W. (1937): The ecology of the upper Williams River and Barrington Tops district. I. Introduction. Proc. Linn. Soc. N.S.W. 62: 269–283. (1938): II. The rainforest formations. Ibid. 63: 139–184. (1939): III. The *Eucalyptus* forests and general discussion. Ibid. 64: 1–33.

Frenguelli, Joaquín (1943): Restos de *Casuarina* en el Mioceno de el Mirador Patagonia Central. Notos del Museo. 8. Paleontologia 56: 349–354. Buenos Aires.

Gaff, D.F. and Churchill, D.M. (1976): *Borya nitida* – An Australian species in the Liliaceae with desiccation-tolerant leaves. Aust. J. Bot. 24: 209–224.

Gardner, C.A. (1944): The vegetation of Western Australia with special reference to the climate and soils. J. Roy. Soc. W.A. 28: 11–87.

– (1952): Flora of Western Australia. Vol. 1. Part 1. Gramineae. Govt. Printer, Perth.

– (1952–1966): Trees of Western Australia. Bulletins reprinted from J. Agric. W.A. Vols. 1–7.

Gentilli, J. (1972): Australian climate patterns. Thos. Nelson (Australia).

Gibbs, L.S. (1917): A contribution to the phyto-geography of Bellenden-Ker. J. Bot. 55: 297–310.

– (1920): Notes on the phyto-geography and flora of the mountain summit plateaux of Tasmania. J. Ecol. 8: 89–117.

Gilbert, J.M. (1959): Forest succession in the Florentine Valley, Tasmania. Papers and Proc. Roy. Soc. Tas. 93: 129–151.

Gillham, Mary E. (1960): Destruction of indigenous heath vegetation in Victorian sea-bird colonies. Aust. J. Bot. 8: 278–317.

– (1961): Plants and seabirds of granite islands in south-east Victoria. Proc. Roy. Soc. Vic. 74: 21 to 35.

Good, R.O. (1960): On the geographical relationships of the angiosperm flora of New Guinea. Bull. Brit. Mus. (Nat Hist.). 2. No. 8. 205–226.

Goodall, David W., (1966): The nature of the mixed community. Proc. Ecol. Soc. Aust. 1: 84–96.

Gordon, H.D. (1949): The problem of sub-antarctic

plant distribution A.N.Z.A.A.S. Hobart. 27: 142–149.
Gould, R.E. (1975): The succession of Australian Pre-Tertiary megafossil floras. Bot. Rev. 41: 453–483.
Green, J.W. (1964): Discontinuous and presumed vicarious plant species in southern Australia. J. Roy. Soc. W.A. 47: 25–32.
– (1969a): Taxonomic problems associated with continuous variation in *Eucalyptus pauciflora* (Snow Gum) (Myrtaceae). Taxon 18: 209–276.
– (1969b): Temperature responses in altitudinal populations of *Eucalyptus pauciflora* Sieb. ex Spreng. New Phytol. 68: 399–410.
– (1971): Variation in *Eucalyptus obliqua* L'Herit. New Phytol. 70: 897–909.
Green, R. (1975): The basis of plate tectonics. In: Australia's Past. Divn. Ecol. Seminars. Univ. New England, Armidale.
– and Pitt, R.P.B. (1967): Suggested rotation of New Guinea. J. Geomagnetism and Geoelectricity. 19: 317–321.
Greene, S.W. and Walton, D.W.H. (1975): An annotated check list of the sub-Antarctic and Antarctic vascular flora. Polar Record. 17: 473–484.
Grieve, B.J. (1956): Studies in the water relations of plants. I. Transpiration of Western Australian (Swan Plain) sclerophylls. J. Roy. Soc. W.A. 40: 15–30.
– and Blackall, W.E. (1975): How to know Western Australian wildflowers. Part 4. Univ. W.A. Press.
– and Hellmuth, E.O. (1970): Eco-physiology of Western Australian plants. Oecol. Plant. Gauthier-Villars. V: 33–68.
Guillaumin, A. (1948): Flore de la Nouvelle-Caledonie. Paris.
Gunn, R.H. and Nix, H.A. (1977): Land units of the Fitzroy region, Queensland Land Res. Ser. 39: C.S.I.R.O.
Guymer, G.P. (1978): A Monograph of the genus *Brachychiton* (Sterculiaceae). Ph.D. Thesis. Univ. New England.
Hall, Elizabeth A.A., Specht, R.L. and Eardley, Constance (1964): Regeneration of the vegetation on Koonamore Vegetation Reserve, 1926–1962. Aust. J. Bot. 12: 205–264.
Hall, Norman, Brooker, Ian and Gray, A.M. (1970–1977): Forest Tree Series. Dept. Primary Industry. Canberra.
–, Johnston, R.D. and Chippendale, G.M. (1970): Forest trees of Australia. Aust. Govt. Publ. Serv. Canberra.
Hallam, N.D. and Chambers, T.C. (1970): The leaf waxes of the genus *Eucalyptus* L'Héritier. Aust. J. Bot. 18: 335–386.
Hallsworth, E.G. and Beckmann, G.G. (1969): Gilgai in the Quaternary. Soil Sci. 107: 409–420.
–, Colwell, J.D. and Gibbons, F.R. (1953): Studies in pedogenesis in New South Wales. V. The Euchrozems. Aust. J. Agric. Res. 4: 305–325.
– and Costin, A.B. (1953): Studies in pedogenesis in New South Wales. IV. The ironstone soils. J. Soil Sci. 4: 24–47.
Hamilton, A.A. (1919): An ecology study of the salt marsh vegetation in the Port Jackson District. Proc. Linn. Soc. N.S.W. 44: 464–513.
Hannon, Nola J. (1956–1958): The status of nitrogen in the Hawkesbury sandstone soils and their plant communities in the Sydney district. (1956): I. The significance and level of nitrogen. Proc. Linn. Soc. N.S.W. 81: 199–143; 1958: II. The distribution and circulation of nitrogen. Ibid. 83: 65–85.
Harden, G.J. (1976): Provisional flora list – Warrumbungle National Park Nat. Parks and Wildlife Serv. Sydney.
Harland, W.B. et. al. (1967): The Fossil Record: Geol. Soc. London.
Harrington, H.J. (1974): The Tasman geosyncline in Australia. Geol. Soc. Aust. Symposium in honour of Professor Dorothy Hill. Univ. Qld: 383–407.
– (1975): The progressive separation of Australia from surrounding landmasses. In: Australia's Past. Divn. Ecol. Seminars. Univ. New England. Armidale.
Harris, Thistle Y. (1970): Alpine plants of Australia. Angus and Robertson. Sydney.
Harris, W.K. (1965a): Basal Tertiary microfloras from the Princetown area, Victoria, Australia. Palaeontographica. 115: B: 75–106.
– (1965b): Tertiary microfloras from Brisbane, Queensland. Rep. Geol. Surv. Qld 10: 1–7.
Hartley, W. (1958–1973): (1958): Studies on the origin, evolution, and distribution of the Gramineae. I. Andropogoneae. Aust. J. Bot. 6: 116–128; II. Paniceae. Ibid. 6: 343–357; (1961): IV. The genus *Poa*. Ibid 9: 152–161; (1973): V. Subfamily Festucoideae. Ibid 21: 201–234.
– (1964): The distribution of the grasses. In Grasses and Grasslands. C. Barnard, Ed. pp. 26–46. Macmillan & Co.
– and Slater, Christine (1960): Studies on the orgin, evolution and distribution of the Gramineae. III.Tribes of subfamily Eragrostideae. Aust. J. Bot. 8: 256–276.
Hatch, E.D. (1963): Brief comment on the orchids of New Zealand. Orchadian, 1 (1): 3–4.
Havel, J.J. (1968): The potential of the northern Swan coastal plain for *Pinus pinaster* Ait. plantations. For. Dept. Perth, W.A. Bull. 76: 1–73.
Hayman, D.L. (1960): The distribution and cytology of the chromosome races of *Themeda australis* in southern Australia. Aust. J. Bot. 8: 58–68.
Heatwole, Harold (1976): The ecology and biogeography of coral cays. Biol. and Geol. Coral Reefs. 3: 369–387.
Heddle, E.M. and Specht, R.L. (1975): Dark Island Heath (Ninety-Mile Plain, South Australia). VIII. The effect of fertilizers on composition and growth, 1950–1972. Aust. J. Bot. 23: 151–164.
Heinsohn, G.E. and Wake, J.A. (1976): The importance of the Fraser Island region to dugongs. Operculum. March: 15–18.

Helby, R. (1966): Triassic plant microfossils from a shale within the Wollar Sandstone, N.S.W. Proc. Roy. Soc. N.S.W. 100: 61–73.

Hellmuth, E.O. (1971): Eco-physiological studies on plants in arid and semi-arid regions in Western Australia. III. J. Ecol. 59: 225–259.

Herbert, D.A. (1929): The major factors in the present distribution of the genus *Eucalyptus*. Proc. Roy. Soc. Qld 49: 165–193.

– (1935): The climatic sifting of the Australian vegetation. Aust. and N.Z. Assoc. for Advancement of Science. Melbourne. 349–370.

– (1936): An advancing Antarctic Beech forest. Queensland Naturalist. 10: 8–10.

– (1938): The upland savannahs of the Bunya Mts, South Queensland. Proc. Roy. Soc. Qld 49: 145–149.

– (1950): Present day distribution and the geological past. Vict. Nat. 66: 227–232.

– (1960): Tropical and sub-tropical rainforest in Australia. Aust. J. Sci. 22: 283–290.

Hill, A.W. (1918): The genus *Caltha* in the Southern Hemisphere. Ann. Bot. 32: 421–435.

Hills, E.S. (1939): The physiography of north-western Victoria. Proc. Roy. Soc. Vic. 51: 297–320.

– (1940): The lunette, a new landform of aeolian origin. Aust. Geogr. 3: 15–21.

Hnatiuk, R.J. (1977): Population structure of *Livistona eastonii*, Mitchell Plateau, Western Australia. Aust. J. Ecol. 2: 461–466.

Holland, A.A. and Moore, C.W.E. (1962): The vegetation and soils of the Bollon district in south-west Queensland. C.S.I.R.O. Div. Plant Ind. Tech. Pap. No. 17.

Hoogland, R.D. (1958): The alpine flora of Mount Wilhelm (New Guinea). Blumea. Supp. IV. 2. x: 220–238.

Hooker, J.D. (1859): The flora of Australia, its origins, affinities and distribution. Lovell Reeve. London.

– (1872–1897): Flora of British India. 7 vols. Reeve. London.

Hope, G.S. and Thompson, G.K. (1971): The vegetation of Cliffy Island, Victoria, Australia. Proc. Roy. Soc. Vic. 84: 121–127.

Hopper, Stephen D. and Maslin, Bruce R. (1978): Phytogeography of *Acacia* in Western Australia. Aust. J. Bot. 26: 63–78.

Howard, Truda M. (1973): Accelerated tree death in mature *Nothofagus cunninghamii* Oerst forests in Tasmania. Vic. Naturalist 90: 343–345.

– (1973a): Studies in the ecology of *Nothofagus cunninghamii* Oerst. I. Natural regeneration on the Mt. Donna Buang massif, Victoria. Aust. J. Bot., 21: 67–78; II. Phenology. Ibid. 79–92; III. Two limiting factors: light intensity and water stress. Ibid. 93–102.

– (1974): *Nothofagus cunninghamii* ecotonal stages. Buried viable seed in north west Tasmania. Proc. Roy. Soc. Vic. 86: 137–142.

– and Ashton, D.H. (1973): The distribution of *Nothofagus cunninghamii* rainforest. Proc. Roy. Soc. Vic. 86: 47–75.

Hubble, G.D. (1970): Soils. pp. 44–58 in Australian Grasslands. Ed. R.M. Moore. Aust. Nat. Univ. Press. Canberra.

Hughes, N.F. (Ed.) (1973): Organisms and continents through time. Palaeontological Association. London. Special Papers No. 12.

Hutchinson, J. (1959); The families of flowering plants. 2 vols. Clarendon Press. Oxford.

Hutton, J.T. (1976): Chloride in rainwater in relation to distance from the ocean. Search 7: 207–208.

– and Leslie, T.I. (1958): Accession of non-nitrogenous ions dissolved in rain water to soils in Victoria. Aust. J. Agric. Res. 9: 492–507.

Hyland, B.P.M. (1971): A key to common rainforest trees between Townsville and Cooktown based on leaf and bark features. Dept. Forestry, Brisbane, Qld.

– (in prep.). A card key to the rain forest trees of north Queensland. 2nd Ed. C.S.I.R.O. Canberra.

Incoll, F.S. (1946): The red gum *(Eucalyptus camaldulensis)* forests of Victoria. Aust. For. 10: 47–51.

Isbell, R.F. (1962): Soils and vegetation of the Brigalow lands, eastern Australia. Soils and Land Use Series No. 43. C.S.I.R.O. Melbourne.

– and Murtha, G.G. (1972): Burdekin-Townsville Region. Qld Resource Series. Vegetation. Dept. Nat. Devel. Canberra.

Ising, Ernest H. (1922): Ecological notes on South Australia plants. Part 1. Proc. Roy. Soc. S.A. 46: 583–606.

Ives, Alan (1973): The mallee of south-eastern Australia. A short bibliography. Dept. Geography. Monash Univ. Melbourne.

Jackson, W.D. (1965): Vegetation. Atlas of Tasmania: 30–35. A.N.Z.A.A.S. Hobart.

– (1968): Fire, air, water and earth – an elemental ecology of Tasmania. Proc. Ecol. Soc. Aust. 3: 9–16.

– (1973): Vegetation. In "The Lake Country of Tasmania". Ed. M.R. Banks. Roy. Soc. Tas. Hobart. 61–86.

Jacobs, S. (1971): Systematic position of the genera *Triodia* R. Br. and *Plectrachne* Henr. (Gramineae). Proc. Linn. Soc. N.S.W. 96: 175–185.

James, J.W. (1959–1960): Erosion survey of the Paroo-Upper Darling region. Parts I–IV. J. Soil Conser. Ser. N.S.W. Vols. 15–16.

Jarrett, Phyllis H. and Petrie, A.H.K. (1929): The vegetation of the Black's Spur region. A study in the ecology of some Australian mountain *Eucalyptus* forests. II. Pyric succession. J. Ecol. 17: 249–281.

Jensen, H.I. (1914): The soils of New South Wales. Govt. Printer. Sydney.

Jessup, R.W. (1946): The ecology of the area adjacent to Lakes Alexandrina and Albert. Trans. Roy. Soc. S.A. 70: 3–34.

– (1948): A vegetation and pasture survey of Coun-

ties Eyre, Burra and Kimberley, South Australia. Trans. Roy. Soc. S.A. 72: 33–68.
- (1951): The soils, geology and vegetation of north-western South Australia. Trans. Roy. Soc. S.A. 74: 189–273.
- (1960): The lateritic soils of the south-eastern portion of the Australian arid zone. J. Soil Sci. 11: 106–113.
- (1961): A Tertiary-Quaternary pedological chronology for the south-eastern portion of the Australian arid zone. J. Soil Sci. 12: 199–213.

Johnson, E.R.L. and Baird, A.M. (1970): Notes on the flora and vegetation of the Nullarbor Plain at Forrest, W.A. J. Roy. Soc. W.A. 53: 46–61.

Johnson, L.A.S. (1962): Taxonomic notes on Australian plants. Contrib. Nat. Herb. N.S.W. 3: 93–102.
- and Briggs, Barbara G. (1975): On the Proteaceae – the evolution and classification of a southern family. Bot. J. Linn. Soc. 70: 83–182.

Johnson, R.W. (1966): Where the Brigalow scrubs grow. Qld Dept. Prim. Indust. Leaflet 895.
- (1976): Brigalow clearing and regrowth control. Qld Agric. J. 84: 40–56.

Jones, D.L. and Gray, B. (1977): Australian climbing plants. Reed.

Jones, J.B. and Segnit, E.R. (1966): The occurrence and formation of opal at Coober Pedy and Andamooka. Aust. J. Sci. 29: 129–133.

Jones, O.A. and de Jersey, N.J. (1947): The flora of the Ipswich coal measures – Morphology and floral succession. Univ. Qld Papers III (New Series). 1–88.

Jones, R. (1968): The leaf area of an Australian heathland with reference to seasonal changes and the contribution of individual species. Aust. J. Bot. 16: 579–588.
–, Groves, R.H. and Specht, R.L. (1969): Growth of heath vegetation III. Growth curves for heaths in southern Australia: A reassessment. Aust. J. Bot. 17: 309–314.

Jones, R.M. (1969): Scald reclamation studies in the Hay district, Part IV – Scald soils: their properties and changes with reclamation. J. Soil Cons. Serv. N.S.W. 24: 271–278.

Jones, W.T. (1971): The field identification and distribution of mangroves in eastern Australia. Qld Naturalist. 20: 35–51.

Joplin, Germaine A. (1963): Chemical analyses of Australian rocks. Part I. Igneous and metamorphic. Bureau Mineral Resources. Bull. 65. Dept. Nat. Devel. Canberra.

Jozwik, F.X. (1969): Some systematic aspects of Mitchell grasses (*Astrebla* F. Muell.). Aust. J. Bot. 17: 359–374.

Keast, A. (1973): Contemporary biotas and the separation sequence of the southern continents. In "Implications of continental drift to the earth sciences. Vol. 1. Academic Press: 309–343.

Kemp, Elizabeth M. (1972): Reworked palynomorphs from the West Ice Shelf area, East Antarctica, and their possible geological and palaeoclimatological significance. Marine Geol. 13: 145–157.
- (1976): Early Tertiary pollen from Napperby, central Australia. Bur. Min. Res. J. Aust. Geol. Geophys. 1: 109–114.
- and Barrett, P.J. (1975): Antarctic glaciation and early Tertiary vegetation. Nature, 258: 507–508.
- and Harris, Wayne K. (1975): The vegetation of Tertiary islands on the Ninetyeast Ridge. Nature, 258: 303–307.

Kenley, P.R. (1954): The occurrence of Cretaceous sediments in south-western Victoria. Proc. Roy. Vic. 66: 1–16.

Kerr, Lesley R. (1925): The lignotubers of *Eucalyptus* seedlings. Proc. Roy. Soc. Vic. 37: 79–97.

Kershaw, A.P. (1970): A pollen diagram from Quincan Crater, north-east Queensland, Australia. New Phytol. 70: 669–681.
- (1978): The analysis of aquatic vegetation on the Atherton Tableland, north-east Queensland, Australia. Aust. J. Ecol. 3: 23–42.

Khu, K.L. (1969): The distribution of rhizobia and indigenous legumes along rivers and lakes and in native vegetation communities in the Western Division of New South Wales, with special reference to their ecology. M. Sc. Thesis Univ. New England. Armidale.

King, D. (1956): The Quaternary stratigraphic record at Lake Eyre North and the evolution of existing topographic forms. Trans. Roy. Soc. S.A. 79: 93–103.

Kirkman, H. (1975): A description of the sea-grass communities of Stradbroke Island. Proc. Roy. Soc. Qld 86: 129–131.

Kirkpatrick, J.B. (1977): The disappearing heath. Tas. Conserv. Trust. Inc.

Knowles, G.H. (1954): Scald reclamation in the Hay district. J. Soil Cons. Serv. N.S.W. 10: 149–156.

Kratochvil, M., Hannon, Nola J. and Clarke, Lesley D. (1973): Mangrove swamp and salt marsh communities in southern Australia. Proc. Linn. Soc. N.S.W. 97: 262–274.

Lacey, C.J. (1974): Rhizomes in tropical eucalypts and their role in recovery from fire damage. Aust. J. Bot. 22: 29–38.

Ladiges, Pauline Y. and Ashton, D.H. (1977): A comparison of some populations of *Eucalyptus viminalis* Labill. growing on calcareous and acid soils in Victoria, Australia. Aust. J. Ecol. 2: 161–178.

Lang, Gerhard (1970): Die Vegetation der Brindabella Range bei Canberra. Akad. Wissensch. u. Lit. Mainz.

Lange, R.T. (1961): Nodule bacteria associated with the indigenous Leguminosae of south-western Australia. J. Gen. Microbiol. 61: 351–359.
- (1966): Vegetation in the Musgrave Range, South Australia. Trans. Roy. Soc. S.A. 90: 57–64.
- (1966): Sampling for association analysis. Aust. J. Bot. 14: 373–378.
- (1978): Carpological evidence for fossil *Eucalyptus*

and other Leptospermeae from a Tertiary deposit in the South Australian arid zone. Aust. J. Bot. 26: 221–233.
– and Purdie, Rosemary (1976): Western Myall *(Acacia sowdenii)*, its survival prospects and management needs. Aust. Rangel. J. 1: 64–69.
Lauer, Peter K. (Editor) (1978): Fraser Island. Occas. Papers in Anthropology. No. 8. Univ. Qld. Brisbane.
Lavarack, P. (1976): The taxonomic affinities of the Australian Neottioideae. Taxon. 25: 289–296.
– and Stanton, J.P. (1977): Vegetation of the Jardine River catchment and adjacent coastal areas. Proc. Roy. Soc. Qld 88: 39–48.
Lazarides, M. (1970): The grasses of Central Australia. Aust. Nat. Univ. Press. Canberra.
Learmonth, A. and Learmonth, Nancy. (1973): Encyclopaedia of Australia. Hicks, Smith and Sons. Sydney.
Leeper, G.W. (1957): Introduction to soil science. Melb. Univ. Press.
– (Editor) (1973): The Australian environment, C.S.I.R.O. Melbourne.
Leigh, J.H. and Mulham, W.E. (1965): Pastural plants of the Riverine Plain, Jacaranda Press.
Litchfield, W.H. (1956): Species distribution over part of the Connalpyn Downs, South Australia. Aust. J. Bot. 4: 68–116.
Loneragon, W.A. (1952): An ecological survey of Mersea Lake. B. Sc. (Honours) Thesis. Botany Dept. Univ. W.A.
Lyons, M.T., Brooks, R.R., and Craig, D.C. (1974): The influence of soil composition on the vegetation of the Coolac serpentine belt in New South Wales. Proc. Roy. Soc. N.S.W. 107: 65–75.
Mabbutt, J.A. (1965): The weathered land surface in Central Australia. Z. Geomorph. 9: 82–113.
– and Sullivan, M.E. (1970): Landforms and structure. In Australian grasslands. Ed. R.M. Moore. 27–43.
Macnae, W. (1966): Mangroves in eastern and southern Australia. Aust. J. Bot. 14: 67–104.
– (1968): A general account of the fauna and flora of mangrove swamps and forests in the Indo-West Pacific region. Adv. Marine Biol. 6: 73–270.
MacPhail, M.K. and Peterson, J.A. (1975): New deglaciation dates from Tasmania. Search. 6: 127–130.
Madigan, C.T. (1930): Lake Eyre, South Australia. Geograph. J. 76: 215–240.
– (1946): The Simpson Desert Expedition, 1939. Scientific Reports: No. 6 – The sand formation. Trans. Roy. Soc. S.A. 70: 45–63.
Maiden, J.H. (1903–1931): Forest flora of New South Wales. Part IV. Govt. Printer, Sydney.
Malcolm, C.V. (1964): Effects of salt, temperature and seed scarification on germination of two varieties of *Arthrocnemum halocnemoides*. J. Proc. Roy. Soc. W.A. 47: 72–74.
– (1972): Establishing shrubs in saline environments. Tech. Bull. 14. Dept. Agric. W.A. Perth. 1–37.

Martin, A.R.H. (1973): Reappraisal of some palynomorphs of supposed Proteaceous affinity. I. Spec. Publ. Geol. Soc. Aust. 4: 73–78.
Martin, D. (1940): The vegetation of Mount Wellington, Tasmania. Roy. Soc. Tas. 39: 97–124.
Martin, Helene A. (1973a): The palynology of some Tertiary and Pleistocene deposits, Lachlan, River Valley, New South Wales. Aust. J. Bot. Supplementary Series: Supplement 6: 1–57.
– (1973b): Upper Tertiary palynology in southern New South Wales. Geol. Soc. Aust. Essay in honour of Isabel Cookson. Spec. Publ. 4: 35–54.
– (1974): The indentification of some Tertiary pollen belonging to the Family Euphorbiaceae. Aust. J. Bot. 22: 271–291.
– (1977): The history of *Ilex* (Aquifoliaceae) with special reference to Australia. Evidence from pollen. Aust. J. Bot. 25: 655–673.
McArthur, W.M. (1957): Plant Ecology of the coastal islands near Fremantle, W.A. J. Roy. Soc. W.A. 40: 46–64.
– and Bettenay, E. (1958): The soils of the Busselton area, Western Australia. C.S.I.R.O. Div. Report 3/58.
– and – (1960): The development and distribution of the soils of the Swan coastal plain, Western Australia. C.S.I.R.O. Austr. Soils Public. No. 16.
McComb, J.A. and McComb, A.J. (1967): A preliminary account of the vegetation of Loch McNess, a swamp and fen formation in Western Australia. J. Roy. Soc. W.A. 50: 105–112.
McGregor, Craig (1974): The Great Barrier Reef. Time/Life Books. Amsterdam.
McLuckie, J. (1923): Studies in Symbiosis. IV. The Root-nodules of *Casuarina cunninghamiana* and their physiological significance. Proc. Linn. Soc. N.S.W. 48: 194–205.
– and Petrie, A.H.K. (1926): An ecological study of the flora of Mount Wilson. Pt. iii. The vegetation of the valleys. Proc. Linn. Soc. N.S.W. 51: 94–113.
– and – (1927): The Vegetation of the Kosciusko Plateau. Proc. Linn. Soc. N.S.W. 52: 187–221.
McWhae, J.R.H., Playford, P.E., Lindner, A.W., Glenister, B.F., Balme, B.E. (1958): The stratigraphy of Western Australia. Melb. Univ. Press 161 pp.
Medwell, Lorna M. (1952): A review and revision of the flora of the Victorian lower Jurassic. Proc. Roy. Soc. Vic. 65: 63–111.
– (1954): Fossil plants from Killara near Casterton, Victoria. Proc. Roy. Soc. Vic. 66: 17–24.
Melville, R. (1966): Continental drift, Mesozoic continents and the migrations of the angiosperms. Nature 211: 116–120.
Menendez, C.A. (1969): Die fossilen Floren Südamerikas. Biogeography and Ecology in South America. Junk. The Hague.
Miles, J.M., Kenneally, K.F. and George, A.S. (1975): The Prince Regent River reserve environment. Wildlife Research Bull. No. 3: 17–30. W.A. Wildlife Research Centre. Perth.

Millington, R. J. (1954): *Sphagnum* bogs of the New England plateau, New South Wales. J. Ecol. 42: 328–344.

Moewus, Liselotte (1953): About the occurrence of fresh-water algae in the semi-desert round Broken Hill (New South Wales, Australia). Botaniska Notiser 4. Lund. 399–416.

Moore, A. W., Russell, J. S. and Coaldrake, J. E. (1967): Dry matter and nutrient content of a subtropical forest of *Acacia harpophylla* F. Muell. (Brigalow). Aust. J. Bot. 15: 11–24.

Moore, C. W. E. (1953): The Vegetation of the southeastern Riverina, New South Wales. I. The communities. Aust. J. Bot. 1: 485–547. II. The disclimax communities. Ibid. 1: 548–567.

– (1959): The nutrient status of the soils of some plant communities on the Southern Tablelands of New South Wales. Ecology 40: 339–349.

– (1964): Distribution of grasslands. In Barnard (1964): 182–205.

Moore, Lucy B. and Edgar, Elizabeth (1970): Flora of New Zealand. Vol. II. Govt. Printer. Wellington N. Z.

Moore, R. Milton (Editor) (1970): Australian grasslands. Aust. Nat. Univ. Press. Canberra.

– and Williams, J. D. (1976): A study of a subalpine woodland – grassland boundary. Aust. J. Ecol. 1: 145–153.

Morland, R. T. (1949): Preliminary investigations in Hume catchment area. J. Soil Conserv. Serv. N. S. W. 5: 44–54.

– (1951): Notes on the snow lease section of Hume catchment area. J. Soil Conserv. Serv. N. S. W. 7: 5–29.

Morris, M. (1939): Plant regeneration in the Broken Hill district. Aust. J. Sci. 2: 32–48.

Mott, J. J. (1973): Temporal and spacial distribution of an annual flora in an arid region of Western Australia. Trop. Grassl. 7: 89–97.

– (1974): Mechanisms controlling dormancy in the arid zone grass *Aristida contorta*. I. Physiology and mechanisms of dormancy. Aust. J. Bot. 22: 635–645.

– and Tynan, P. W. (1974): Mechanisms controlling dormancy in the arid zone grass *Aristida contorta*. II. Aust. J. Bot. 22: 647–653.

Mount, A. B. (1964): Three Studies in forest ecology. M. Sc. Thesis. Univ. Tasmania.

Müller, F. (1879–1884): Eucalyptographia. Govt. Printer. Melbourne.

Mulcahy, M. J. (1960): Laterites and lateritic soils in south-western Australia. J. Soil. Sci. 11: 206 to 225.

Muller, J. (1964): A palynological contribution to the history of the mangrove vegetation in Borneo. Ancient Pacific floras (ed. L. M. Cranwell): 33–42. Univ. Hawaii Press. Honolulu.

– (1970): Palynological evidence on early differentiation of angiosperms. Biol. Reviews 45: 417–450.

Mullette, K. J. (1978): Studies of the lignotubers of *Eucalyptus gummifera*. I. The nature of the lignotuber. Aust. J. Bot. 26: 9–13.

– and Bamber, R. K. (1978): Studies of the lignotubers of *Eucalyptus gummifera*. III. Inheritance and chemical composition. Aust. J. Bot. 26: 23–28.

–, Hannon, Nola J., and Elliott, A. G. L. (1974): Insoluble phosphorus usage by *Eucalyptus*. Plant and Soil. 41: 199–205.

Murray, B. J. (1931): A study of the vegetation of the Lake Torrens plateau, South Australia. Trans. Roy. Soc. S. A. 55: 91–112.

Myers, Amy (1942): Germination of seed of Curly Mitchell Grass (*Astrebla lappacea* Domin). J. Aust. Instit. Agr. Sci. 8: 31–32.

Nicholls, A. O. (1966): A report on the vegetation of the coastal sands of the Iluka region, North Coast. N. S. W. B. Sc. (Hons) Thesis. Univ. New England.

Nix, H. A. and Austin, M. P. (1973): Mulga: A bioclimatic analysis. Trop. Grassl. 7: 9–21.

Noakes, L. C. (1966): Commentary accompanying map-sheet 'Geology'. Bur. Mineral Resources. Canberra.

Noble, J. C. and Whalley, R. D. B. (1978): The biology and autecology of *Nitraria* L. in Australia. I. Its distribution, morphology, and potential utilization. Aust. J. Ecol. 3: 141–163. II. Seed germination, seedling establishment and response to salinity. Ibid. 3: 165–177.

Norman, F. I. (1970): Ecological effects of rabbit reduction on Rabbit Island, Wilsons Promontory, Victoria. Proc. Roy. Soc. Vic. 83: 235–251.

Nunn, W. M., and Suijdendorp. H. (1954): The value of deferred grazing. J. Dept. Agric. W. A. 3: 585–587.

Ogilvie, C. (1947): The hydrology of Cooper's Creek. See Anon., 1947.

Olsen, H. F. (1978): Great Sandy Strait Fisheries habitat reserves. in Lauer (1978): 223–230.

Osborn, T. G. B. and Robertson, R. N. (1939): A reconnaissance survey of the vegetation of the Myall Lakes. Proc. Linn. Soc. N. S. W. 64: 279 to 296.

–, Wood, J. G. and Paltridge, T. B. (1931): On the autecology of *Stipa nitida*. Proc. Linn. Soc. N. S. W. 56: 299–324.

Osmond, C. B. (1961): Some aspects of calcium and magnesium absorption in the genus *Atriplex*. B. Sc. Honours thesis. Univ. New England. Armidale.

– (1970): Carbon metabolism in *Atriplex* leaves. In: The biology of *Atriplex*. C.S.I.R.O. Canberra.

– (1974): Leaf anatomy of Australian saltbushes in relation to photosynthetic pathways. Aust. J. Bot. 22: 39–44.

Parbery, I. H. (1970): A survey of the algal flora of some semi-arid and arid soils of N. S. W. Litt. B. thesis. Botany Dept. Univ. New England. Armidale.

Parsons, R. F. (1969): Distribution and paleography of two mallee species of *Eucalyptus* in southern Australia. Aust. J. Bot. 17: 323–330.

– (1970): Mallee vegetation of the southern Nullar-

bor and Roe Plains, Australia. Trans. Roy. Soc. S.A. 94: 227–242.
— and Gill, A.M. (1968): The effects of salt spray on coastal vegetation at Wilsons Promontory, Victoria, Australia. Proc. Roy. Soc. Vic. 81: 1–10.
— and Rowan, J.N. (1968): Edaphic range and cohabitation of some mallee eucalypts in south-eastern Australia. Aust. J. Bot. 16: 109–116.
Paton, T.R. (1974): Origin and terminology for gilgai in Australia. Geoderma 11: 221–242.
Patton, R.T. (1933): The Cheltenham Flora. Proc. Roy. Soc. Vic. 45: 205–218.
— (1933): Ecological Studies in Victoria – Pt. II. The Fern Gully. Proc. Roy. Soc. Vic. 46: 117–129.
— (1934): Ecological Studies in Victoria. Part III. – Coastal Sand Dunes. Proc. Roy. Soc. Vic. 47: 135–152.
— (1935): Flora of Victoria. Aust. Assoc. Adv. Sci. Handbooks: 54–65. Melbourne.
— (1935): Ecological studies in Victoria. IV. Basalt plains association. Proc. Roy. Soc. Vic. 48: 172–190.
— (1937): Ecological studies in Victoria. Part V. Red Box – Stringybark association. Proc. Roy. Soc. Vic. 49: 293–307.
— (1942): Ecological studies in Victoria. VI. Salt Marsh. Proc. Roy. Soc. Vic. 54: 131–144.
— (1944): Ecological studies in Victoria. VII. Box-Ironbark association. Proc. Roy. Soc. Vic. 61: 35–51.
— (1951): The mallee. Vict. Nat. 68: 57–62.
Pedley, L. (1967): Vegetation of the Nogoa – Belyando area, Queensland. C.S.I.R.O. Land Res. Ser. No. 18: 138–169.
— (1969): Intermediates between *Eucalyptus populnea* F. Muell. and *E. brownii* Maid. and Cambage. Contrib. Qld Herb. 5: 1–6.
— (1973): Taxonomy of the *Acacia aneura* complex. Trop. Grasslands. 7: 3–8.
— (1974): Vegetation of the Balonne – Maranoa area. Land Res. Ser. No. 34: 180–203. C.S.I.R.O. Melbourne.
— and Isbell, R.F. (1971): Plant communities of Cape York Peninsula. Proc. Roy. Soc. Qld 82: 51–74.
Penfold, A.R. and Willis, J.L. (1961): The Eucalypts. Interscience Publishers, Inc. New York.
Percival, Margaret and Womersley, J.S. (1975): Floristics and ecology of the mangrove vegetation of Papua New Guinea. Papua New Guinea National Herbarium. Bot. Bull 8: 96 pp.
Perry, R.A. (1953): The vegetation communities of the Townsville-Bowen region. In survey of the Townsville-Bowen region. C.S.I.R.O. Land Res. Ser. No. 2: 44–54.
— (1970): Vegetation of the Ord-Victoria area. In Lands of the Ord-Victoria Area, W.A. and N.T. Part VII. C.S.I.R.O. Land Res. Ser. No. 28: 104–119.
— and Christian, C.S. (1954): Vegetation of the Barkly region. C.S.I.R.O. Land Res. Ser. No. 3: 78–112.

— and Lazarides, M. (1962): Vegetation of the Alice Springs area. In Lands of the Alice Springs Area, Northern Territory, 1956–1957. C.S.I.R.O. Land Res. Ser. No. 6: 208–236.
— and — (1964): Vegetation of the Leichhardt-Gilbert area. C.S.I.R.O. Land Res. Ser. No. 11: 152–191.
Petrie, A.H.K. (1925): An ecological study of the flora of Mount Wilson. II. The *Eucalyptus* forests. Proc. Linn. Soc. N.S.W. 50: 145–166.
—, Jarrett, P.H. and Patton, R.T. (1929): The vegetation of the Black Spur region. I. The mature plant communities. J. Ecol. 17: 223–248.
Pettigrew, C.J. and Watson, Leslie (1975): On the classification of Australian acacias. Aust. J. Bot. 23: 833–847.
Philippi, F. (1881): Catalogus plantarum vascularium chilensium ex annalibus Universitatis Chilensis Santiago de Chile. Imprenta Nacional, Calle de la Bandera. Núm. 29.
Phillips, Marie E. (1947): The vegetation of the Wianamatta Shales and associated soil types. M. Sc. Thesis. Univ. Sydney.
Pidgeon, I.M. (1937–1941): The ecology of the central coastal area of New South Wales. (1937): I. The environment and general features of the vegetation. Proc. Linn. Soc. N.S.W. 62: 315–340. II. (1938): Plant succession on the Hawkesbury Sandstone. Ibid. 63: 1–26. III. (1940): Types of Primary succession. Ibid. 65: 221–249. IV. (1941): Forest types on soils from Hawkesbury sandstone and Wianamatta shale. Ibid. 66: 113–137.
— and Ashby, Eric (1940): Studies in applied ecology. I. A statistical analysis of regeneration following protection from grazing. Proc. Linn. Soc. N.S.W. 65: 123–143.
Playford, G. (1965): Plant microfossils from Triassic sediments near Poatina. Tasmania. Geol. Soc. Austr. 12: 173–210.
— and Cornelius, K.D. (1967): Palynological and lithostratigraphic features of the Razorback beds, Mount Morgan district, Queensland. Univ. Qld Papers. Dept. Geology. 6. No. 3: 81–94.
— and Dettmann, Mary E. (1965): Rhaeto-Liassic plant microfossils from the Leigh Creek coal measures, South Australia. Senchenbergiana Lethaea. 46: 127–180.
Plumstead, Edna P. (1962): Trans-antarctic Expedition 1955–1958. Scientific Reports No. 9. Geology. 2: Fossil Floras of Antarctica.
Pook, E.W., Costin, A.B. and Moore, C.W.E. (1966): Water stress in native vegetation during the drought of 1965. Aust. J. Bot. 14: 257–267.
— and Moore, C.W.E. (1966): The influence of aspect on the composition and structure of dry sclerophyll forest on Black Mountain, Canberra, A.C.T. Aust. J. Bot. 14: 223–242.
Preece, P.B. (1971): Contributions to the biology of mulga. I. Flowering. Aust. J. Bot. 19: 21–38. II. Germination. Ibid. 19: 39–49.
Prescott, J.A. (1931): The soils of Australia in relation

to vegetation and climate. Bull. C.S.I.R.O. No. 52. Melbourne.
— and Pendleton, R.L. (1952): Laterite and lateritic soils. Commonwealth Bur. Soil. Tech. Comm. 47.
Price, Doreen (1963): Calendar of flowering times of some plants of the Sydney district, possibly associated with pollinosis. Contrib. Nat. Herb. N.S.W. 3: 171–194.
Prider, R.T. (1966): The lateritized surface of Western Australia. Aust. J. Sci. 28: 443–451.
Pryor, L.D. (1976): Biology of eucalypts. Edward Arnold.
— and Johnson, L.A.S (1971): A classification of the euclypts. Austr. Nat. Univ., Canberra.
Quilty, J.A and Wearne, A.H. (1975): Evaluation of a phase of extreme coastal erosion. J. Soil Conserv. Serv. N.S.W. 31: 179–192.
Randell, Barbara R. (1970): Adaptations in the genetic system of Australian arid zone *Cassia* species. Aust. J. Bot. 18: 77–97.
— and Symon, D.E. (1977): Distributions of *Cassia* and *Solanum* species in arid regions of Australia. Search. 8: 206–207.
Ratkowsky, D.A. and A.V. (1977): Plant communities of the Mount Wellington Range, Tasmania. Aust. J. Ecol. 2: 435–445.
Raunkiaer, C. (1934): The life forms of plants and statistical plant geography. Oxford Univ. Press, New York.
Raven, P.H. and Axelrod, D.I. (1974): Angiosperm biogeography and past continental movements. Ann. Missouri Bot. Gardens. 61: 539–673.
Rayson, Patricia (1957): Dark Island heath (Ninetymile Plain, South Australia). II. The effects of microtopography on climate, soils, and vegetation. Aust. J. Bot. 5: 86–102.
Richards, P.W. (1968): The tropical rain forest. Cambridge Univ. Press.
Robertson, J.A. and Moore, R.M. (1972): Thinning *Eucalyptus populnea* woodlands by injecting trees with chemicals. Trop. Grasslands. 6: 141–150.
Rodway, L. (1903): The Tasmanian flora. Govt. Printer. Hobart.
— (1924): Some ecological features in Tasmania. Rep. Aust. Assoc. Adv. Sci. 17: 730–738. Sydney.
Roe, R. and Allen, G.H. (1945): Studies on the Mitchell Grass association in south-west Queensland. C.S.I.R. Bull. 185.
Rogers, R.W. (1972): Soil surface lichens in arid and subarid south-eastern Australia. II. Phytosociology and geographic zonation. Aust. J. Bot. 20: 215–227.
— and Lange, R.T. (1972): Soil surface lichens in arid and subarid south-eastern Australia. I. Introduction and floristics. Aust. J. Bot. 20: 197–213.
Rose, A.B. and Evans, O.D. (1974): A list of plant species in Ku-ring-gai Chase National Park. Roneod. pp. 40.
Roughley, T.C. (1947): Wonders of the Great Barrier Reef. Angus and Robertson. Sydney.

Russell, T.S., Moore, A.W. and Coaldrake, J.E. (1967): Relationships between subtropical semiarid forest of *Acacia harpophylla* (Brigalow), microrelief, and chemical properties of associated gilgai soil. Aust. J. Bot. 15: 481–498.
Saenger, Peter, Specht, Marion M., Specht, R.L. and Chapman, V.J. (1977): Mangal and coastal saltmarsh communities in Australasia. Ecosystems of the world. I: 293–345. Elsevier Sci. Publ. Co.
Sands, Valerie E. (1975): The cytoevolution of the Australian Papilionaceae. Proc. Linn. Soc. N.S.W. 100: 118–155.
Sanford, W.W. (1974): The Ecology of orchids. In Withner, C.L. (Ed.) The orchids. Scientific Studies: 1–100. John Wiley & Sons.
Sauer, J. (1965): Geographic reconnaissance of Western Australia seashore vegetation. Aust. J. Bot. 13: 39–69.
Schimper, A.F. (1898): Pflanzen-geographie auf physiologischer Grundlage. Jena. (English Translation, 1903).
Schuster, R.M. (1972): Continental movements, "Wallace's Line" and Indomalayan – Australasian dispersal of land plants: some eclectic concepts. Bot. Rev. 38: 3–86.
Seddon, G. (1972): Sense of Place. Univ. W.A. Press.
— (1974): Xerophytes, xeromorphs and sclerophylls: the history of some concepts in ecology. Biol. J. Linn. Soc. 6: 65–87.
Sharma, M.L. (1976): Soil water regimes and water extraction patterns under two semi-arid shrub (*Atriplex* spp.) communities. Aust. J. Ecol. 1: 249–258.
— and Tongway, D.L. (1973): Plant induced soil salinity patterns in two saltbush (*Atriplex* spp.) communities. J. Range Management. C.S.I.R.O. 26: 121–125.
Siddiqi, M.Y., Carolin, R.C. and Anderson, D.J. (1973): Studies in the ecology of coastal heath in New South Wales. Proc. Linn. Soc. N.S.W. 97: 211–224.
— and — (1976): Studies in the ecology of coastal heath in New South Wales. II. The effects of water supply and phosphorus uptake on the growth of *Banksia serratifolia, B. aspleniifolia* and *B. ericifolia*. Proc. Linn. Soc. N.S.W. 101: 38 to 52.
— , — and Myerscough, P.J. (1976): Studies in the ecology of coastal heath in New South Wales. III. Regrowth of vegetation after fire. Proc. Linn. Soc. N.S.W. 101: 53–63.
Siddiqi, P.J., Myerscough, P.J., and Carolin, R.C. (1976): Studies in the ecology of coastal heath in New South Wales. IV. Seed survival, germination and seedling establishment and early growth in *Banksia serratifolia, B. aspleniifolia* and *B. ericifolia* in relation to fire: Temperature and nutritional effects. Aust. J. Ecol. 1: 175–183.
Silcock, R.G. and Williams, Lynn M. (1975): Changes with age in the germinability of seed of native pas-

ture species from south-western Queensland. Aust. Seed Sci. Newsletter. 1: 9–11.

Simon, B.K. (1978): A preliminary check-list of Australian grasses. Tech. Bull. 3. Dept. Prim. Industr. Brisbane.

Sinclair, R. and Thomas, D.A. (1970): Optical properties of leaves of some species in arid South Australia. Aust. J. Bot. 18: 261–273.

Skerman, P.J. (1947): II. The soils of the Cooper Country. IV. The vegetation of the Cooper Country. V. Soil erosion in the Cooper Country. See Anon 1947.

– (1953): The brigalow country and its importance to Queensland. J. Instit. Agric. Sci. 19: 167–176.

Slade, M.J. (1964): A stratigraphic and palaeobotanical study of the Lower Tertiary sediments of the Armidale district, New South Wales. M. Sc. Thesis. Dept. of Geology, Univ. of New England.

Slatyer, R.O. (1965): Measurements of precipitation interception by an arid plant community (*Acacia aneura* F. Muell.). Unesco Arid Zone Res. 25: 181–192.

– (1967): Plant-water relationships. Acad. Press. London and New York.

Smith, A. Gilbert, Briden, J.C. and Drewry, G.E. (1973): Phanerozoic world maps. In Hughes, N.F. (1973): 1–42.

Smith, D.F. (1963): The plant ecology of Lower Eyre Peninsula, South Australia. Trans. Roy. Soc. S.A. 87: 93–118.

Smith, F.G. (1972): Vegetation survey of Western Australia. Vegetation map of Pemberton and Irwin Inlet. W.A. Dept. Agric.: 1–31.

– (1973): Vegetation survey of Western Australia. Vegetation map of Busselton and Augusta. W.A. Dept. Agric.: 1–32.

Smith, G.G. (1957): A guide to sand dune plants of south-western Australia. W.A. Naturalist. 6: 1–18.

– (1962): The flora of granite rocks of the Porongurup Range, South Western Australia. J. Roy. Soc. W.A. 45: 18–23.

– (1966): A census of Pteridophyta of Western Australia. J. Roy. Soc. W.A. 49: 1–12.

– and Marchant, N.G. (1961): A census of aquatic plants of Western Australia. W.A. Naturalist 8: 5–17.

Smith, J.M.B. (1974): Southern biogeography on the basis of continent drift: A review. J. Aust. Mammal Soc. 1: 213–230.

Smith-White, S. (1959): Cytological evolution in the Australian flora. Cold Spr. Harb. Symp. Quant. Biol. 24: 273–289.

–, Carter, C.R. and Stace, Helen M. (1970): The cytology of *Brachycome*. I. The subgenus *Eubrachycome*: A general survey. Aust. J. Bot. 18: 99–125.

Specht, R.L. (1951): A reconnaissance survey of the soils and vegetation of the Hundreds of Tatiara, Wirrega and Stirling of the Country Buckingham. Trans. Roy. Soc. S.A. 74: 79–106.

– (1957): Dark Island Heath (Ninety-mile Plain, South Australia). IV. Soil moisture patterns produced by rainfall interception and stem-flow. Aust. J. Bot. 5: 137–150: V. Water relationships in heath vegetation and pastures on the Makin sand. Ibid: 151–172.

– (1958): see Specht and Mountford, 1958.

– (1963): Dark Island Heath (Ninety-mile Plain, South Australia). VII. The effects of fertiliser and composition and growth, 1950–1960. Aust. J. Bot. 11: 67–94.

– (1966): The growth and distribution of mallee-broombush (*Eucalyptus incrassata* – *Melaleuca uncinata* association) and heath vegetation near Dark Island Soak, Ninety-mile Plain South Australia. Aust. J. Bot. 14: 361–371.

– (1969): The vegetation of the Pearson Islands, South Australia. A re-examination. February, 1960. Trans. Roy. Soc. S.A. 93: 143–152.

– (1972): The vegetation of South Australia. Govt. Printer. Adelaide.

– and Jones, R. (1971): A comparison of the water use by heath vegetation at Frankston, Victoria, and Dark Island Soak, South Australia. Aust. J. Bot. 19: 311–326.

– and Mountford, C.P. (1958): Records of the Americal-Australian scientific expedition to Arnhem Land. Melb. Univ. Press.

– and Perry, R.A. (1948): Plant ecology of part of the Mount Lofty Ranges. Trans. Roy. Soc. S.A. 72: 91–132.

– and Rayson, P. (1957): Dark Island Heath (Ninety-mile Plain, South Australia). I. Definition of the ecosystem. Aust. J. Bot. 5: 52–85.; III. The root systems. Ibid. 4: 103–114.

–, Roe, Ethel M. and Boughton, Valerie H. (1974): Conservation of major plant communities in Australia and Papua New Guinea. Aust. J. Bot. Supplementary Ser. No. 7.

–., Salt, R.B. and Reynolds, S.T. (1977): Vegetation in the vicinity of Weipa, North Queensland. Proc. Roy. Soc. Qld 88: 17–38.

Speck, N.H. (1958): The vegetation of the Darling-Irwin botanical districts and an investigation of the distribution patterns of Family Proteaceae. South Western Australia. Ph. D. Thesis. Botany Dept. Univ. Western Australia.

– (1960): Vegetation of the North Kimberley area, W.A. Lands and pastoral resources of the North Kimberley Area, W.A. C.S.I.R.O. Land Res. Ser. No. 4: 41–63.

– (1963): Vegetation of the Wiluna-Meekatharra area. In: Lands of the Wiluna-Meekatharra Area, Western Australia, 1958. Part VII. C.S.I.R.O. Land Res. Ser. No. 7: 143–161.

– (1965): Vegetation and pastures of the Tipperary area (Northern Territory). C.S.I.R.O. Land Res. Ser. No. 13: 81–98.

(1968): Vegetation of the Dawson-Fitzroy area. Lands of the Dawson-Fitzroy area, Queensland. C.S.I.R.O. Land Res. Ser. No. 21: 157–173.

– and Lazarides, M (1964): Vegetation and pastures of the West Kimberley area. C.S.I.R.O. Land Res. Ser. No. 9: 140–174.

Stace, H.C.T., Hubble, G.D., Brewer, R., Northcote, K.H., Sleeman, J.R., Mulcahy, M.J. and Hallsworth, E.G. (1968): A handbook of Australian soils. Rellim Tech. Public. Glenside, S. Aust. 435 pp.

Stannard, M.E. (1958): Erosion survey of the south-west Cobar Peneplain. J. Soil Conserv. Serv. N.S.W. 14: 1–15 and 137–156.

– (1962–1963): Erosion survey of the central East-Darling region of N.S.W. I. (1962): Climate and land forms. J. Soil Conserv. Serv. N.S.W. 18: 143–153. II. (1962): Soils. Ibid.: 18: 173–182. III. (1963): Vegetation. Ibid.: 19: 1–12.

Stephens, C.G. (1962): A manual of Australian soils. Third Ed. C.S.I.R.O. Melbourne. 61 pp. (First Ed. 1953).

– and Crocker, R.L. (1946): Composition and genesis of lunettes. Trans. Roy. Soc. S.A. 70: 302–312.

Stewart, G.A. (1959): Some aspects of soil ecology. "Biogeography and Ecology in Australia": 303–314. Junk. The Hague.

Stocker, G.C. (1961): Soils. In: Scientific report on expedition to Northern Queensland. Exploration Soc. Rep. No. 2. Univ. New England. Armidale.

– (1968): The plant communities of Karslake Peninsula, Melville Island, Northern Territory. M.Sc. Thesis. Univ. New England. Armidale.

– (1972): The environmental background to forestry in the Northern Territory of Australia. Forest Res. Instit. Canberra. Leaflet 113.

– (1974): Report on cyclone damage to natural vegetation in the Darwin area after Cyclone Tracey 25 December 1974. Forestry and Timber Bur. Leaflet 127. Canberra.

– (1978): The floral patchwork. Aust. Nat. Hist. 19: 186–189.

– (1978): The natural vegetation of the Torres Strait Islands. In: Aust. Nat. Hist. Torres Strait Islands.

– and Mott, J.J. (In Preparation). Fire in the forests and woodlands of north Australia. In Gill, A.M. (Ed.): Fire and the Australian biota.

Story, R. (1963): Vegetation of the Hunter Valley. In General report on the Lands of the Hunter Valley. C.S.I.R.O. Land Res. Ser. No. 8: 136–150.

– (1967): Vegetation of the Isaac-Comet area. Queensland. C.S.I.R.O. Land Res. Ser. No. 19: 108–128.

– (1969): The vegetation of the Queanbeyan-Shoalhaven area. C.S.I.R.O. Land Res. Ser. No. 24: 113–133.

– (1969): Vegetation of the Adelaide-Alligator area. In: Lands of the Adelaide-Alligator area, Northern Territory. C.S.I.R.O. Land Res. Ser. No. 25: 114–130.

– (1970): Vegetation of the Mitchell, Normanby area, Queensland C.S.I.R.O. Land Res. Ser. No. 26: 75–88.

– (1976): Vegetation of the Alligator Rivers area. C.S.I.R.O. Land Res. Ser. No. 38: 89–111.

Stover, L.E. and Evans, P (1973): Upper Cretaceous-Eocene spore-pollen zonation, offshore Gippsland Basin, Australia. Geol. Soc. Austr. Essays in honour of Isabel Cookson. Spec. Publ. 4: 55–72.

Sutton, C.S. (1911–1913): Notes on the Sandringham flora. Vic. Nat. 28: 5–20 and 29: 79–96.

– (1916): A sketch of the Keilor Plains flora. Vic. Nat. 33: 112–123 and 128–143.

– (1928): A sketch of the vegetation of Cradle Mt. and a census of its plants. Pap. Roy. Soc. Tas.: 132–159. Hobart.

Sutton, J. (1970): Continental drift. Sigma Library of Science Surveys. Francis Hodgson Ltd. Guernsey.

Swain, E.H.F. (1928): The forest conditions of Queensland. Govt. Printer, Brisbane.

Tarling, D.H. and Tarling, M.P. (1971): Continental drift. Bell and Sons, London.

Taylor, B.W. (1955): The flora, vegetation and soils of Macquarie Island. A.N.A.R.E. Reports. Series B. Vol. II, Botany. Antarctic Division, Melbourne. 1–192.

Tchan, Y.T. and Beadle, N.C.W. (1955): Nitrogen economy in semi-arid plant communities. II. The non-symbiotic nitrogen-fixing organisms. Proc. Linn. Soc. N.S.W. 80: 97–104.

Teakle, L.J.H. (1937): Saline soils of Western Australia. J. Dept. Agric. W.A. 14: 313–324.

Thatcher, A.C. and Westman, W.E. (1975): Succession following mining on high dunes of coastal southeast Queensland. Proc. Ecol. Soc. Aust. 9: 17–31.

Thiele, Colin (1975): The Little Desert. Rigby.

Thomas, D.A. and Barber, H.N. (1974): Studies on leaf characteristics of a cline of *Eucalyptus urnigera* from Mount Wellington, Tasmania. Aust. J. Bot. 22: 501–512.

Thorne, R.F. (1972): Major disjunctions in the geographic ranges of seed plants. Quart. Rev. Biol. 47: 365–411.

Tindale, Mary D. (1961a): Studies in Australian pteridophytes. No. 3. Contrib. Nat. Herb. N.S.W. 3: 88–92.

– (1961b): Pteridophyta of south-eastern Australia. Contrib. Nat. Herb. Flora Series 208–211: 1–78.

– (1963): Pteridophyta of south-eastern Australia. Hymenophyllaceae. Contrib. Nat. Herb. N.S.W. Flora Series 201, 1–49.

– (1965): A Monograph of the genus *Lastreopsis*. Contrib. Nat. Herb. N.S.W. 3: 249–339.

Tiver, N.S. and Crocker, R.L. (1951): The grasslands of south-east South Australia in relation to climate, soils and developmental history, J. Brit. Grassl. Soc. 6: 29–80.

Tothill, J.C. and Hacker, J.B. (1973): The grasses of southeast Queensland. Univ. Qld Press.

Tracey, J.G. (1969): Edaphic differentiation of some forest types in eastern Australia. 1. Soil physical factors. J. Ecol. 57: 805–816.

– and Webb, L.J. (1975): The vegetation of the humid tropical region of north Queensland. 15 map sheets 1:100,000. C.S.I.R.O. Brisbane, Qld.
Turner, B.L. (1977): Fossil history and geography. In: The biology and chemistry of the Compositae. Editors V.H. Heywood, J.B. Harborne and B.C. Turner. Academic Press.
Turner, J.C. (1976): An altitudinal transect in rain forest in the Barrington Tops area, New South Wales. Aust. J. Ecol. 1: 155–174.
Turner, J.S., Carr, S.G.M. and Bird, E.C.F. (1962): The dune succession at Corner Inlet, Victoria. Proc. Roy. Soc. Vic. 75: 17–33.
University of Western Australia (1963): Botany Dept. Excursion, Yanchep. Species list. Through courtesy of Miss Alison Baird.
– (1970): Botany Dept. Porongurups Field Camp.
– (1971): Botany Dept. Botany Field Camp. Yanchep and Jurien Bay. Through courtesy of Miss Alison Baird.
– (1972): Biology 15 field day to Cannington Swamp and Darling scarp. Through courtesy of Miss Alison Baird.
Van Loon, A.P. (1966): Investigations is regenerating the Tallowwood – Blue Gum forest type. Res. Note No. 19 Forestry Comm. N.S.W.
Vickery, J.W. (1970): A taxonomic study of the genus *Poa* L. in Australia. Contrib. Nat. Herb. N.S.W. 4: 145–243.
Volck, H.E. (1968): Silvicultural research and management in north Queensland rain forests. Prepared for ninth Commonwealth Forestry Conference. Qld For. Dept. Brisbane, Qld 20pp.
– (1975): Problems in the silvicultural treatment of the tropical rainforests of Queensland. Dept. of Forestry, Brisbane.
Wace, N.M. (1965): Vascular plants. Biogeography and ecology in Antarctica. Junk. The Hague: 201–266.
Wakefield, N.A. (1955): Ferns of Victoria and Tasmania. Field Nat. Club. of Vic.
Walker, D. (Editor) (1972): Bridge and barrier. The natural and cultural history of Torres Strait. A.N.U. Canberra. Publ. BG/3.
Walker, J., Moore, R.M. and Robertson, J.A. (1971): Herbage response to tree and shrub thinning in *Eucalyptus populnea* shrub woodlands. Aust. J. Agric. Res. 23: 405–410.
Walker, P.H. (1960): A soil survey of the Country of Cumberland, Sydney Region, New South Wales. Govt. Printer Sydney. 109pp.
Walkom, A.B. (1915): Mesozoic floras of Queensland. Part 1. The flora of the Ipswich and Walloon Series. (a) Introduction (b) Equisitales. Qld Geol. Survey. Public. No. 252: 1–50. Govt. Printer. Brisbane.
– (1917): Mesozoic floras of Queensland. Part 1 – continued. (c) Filicales etc. Qld Geol. Survey. Public. No. 257: 1–66. Govt. Printer. Brisbane.
– (1918): Mesozoic floras of Queensland. Part II. The flora of the Maryborough (Marine) Series. Qld Geol. Survey. Public. No. 262: 1–20. Govt. Printer. Brisbane.
– (1919): Mesozoic floras of Queensland Parts III and IV. The floras of the Burrum and Styx River Series. Qld Geol. Survey. Public. No. 263: 1–76. Govt. Printer. Brisbane.
Wallace, B.J. (1974): Studies in the Orchidales. B. Sc. Honours Thesis. Univ. New England. Armidale. 144pp.
Walter, H., Harnickell, E. und Mueller-Dombois, D. (1974): Klimadiagramm-Karten der einzelnen Großräume und ökologische Klimagliederung der Erde (Bd. X der Vegetationsmonographien der einzelnen Großräume). Stuttgart.
– und Lieth, H. (1967): Klimadiagramm-Weltatlas. Jena.
Ward, W.T. and Little, I.P. (1975): Times of coastal sand accumulation in south-east Queensland. Proc. Ecol. Soc. Aust. 9: 313–317.
Warren, J.F., Newman, J.C., Jones, R.M., and Unwin, B. (1962): Dew in western New South Wales. J. Soil Conservation Serv. N.S.W. 18: 1–7.
Webb, L.J. (1958): Cyclones as an ecological factor in tropical lowland rainforest, North Queensland. Aust. J. Bot. 6: 220–228.
– (1959): A physiognomic classification of Australian rain forests. J. Ecol. 47: 551–570.
– (1963): The influence of soil parent materials on the nature and distribution of rain forests in south Queensland. Sympos. Ecol. Res. in Humid Trop. Veget. Kuching, Sarawak. 3–13.
– (1964): An historical interpretation of the grass balds of the Bunya Mountains, South Queensland. Ecol. 45: 159–162.
– (1966): The identification and conservation of habitat-types in wet tropical lowlands of north Queensland. Proc. Roy. Soc. Qld 78: 59–86.
– (1968): Environmental relationships of the structural types of Australian rain forest vegetation. Ecol. 49: 296–311.
– (1969): Edaphic differentiation of some forest types in eastern Australia. 11. Soil chemical factors. J. Ecol. 57: 817–830.
–, Tracey, J.G. and Williams, W.T. (1976): The value of structural features in tropical forest typology. Aust. J. Ecol. 1: 3–28.
–, –, –, and Lance, G.N. (1967a): Studies in the numerical analysis of complex rainforest communities. I. A comparison of methods applicable to site/species data. J. Ecol. 55: 171–191.
–, –, –, and – (1967b): Studies in the numerical analysis of complex rainforest communities. II. The problem of species sampling. J. Ecol. 55: 525–538.
–, –, –, and – (1970): Studies in the numerical analysis of complex rainforest communities. V. A comparison of the properties of floristic and physiognomic-structural data. J. Ecol. 58: 203–232.
Welbourn, R.M.E. and Lange, R.T. (1968): An analysis of vegetation on stranded coastal dunes

between Robe and Naracoorte, South Australia. Trans. Roy. Soc. S.A. 92: 19–24.

Westman, W.E. (1975): Pattern and diversity in swamp and dune vegetation, North Stradbroke Island. Aust. J. Bot. 23: 339–354.

– and Rogers, R.W. (1977): Nutrient stocks in a subtropical eucalypt forest, North Stradbroke Island. Aust. J. Ecol. 2: 447–460.

– and – (1977): Biomass and structure of a subtropical eucalypt forest, North Stradbroke Island. Aust. J. Bot. 25: 171–191.

Whalley, R.D.B. (1970): Exotic or native species – the orientation of pasture research in Australia. J. Aust. Inst. Agr. Sci. 36: 111–118.

– and Davidson, A.A. (1969): Drought dormancy in *Astrebla lappacea, Chloris acicularis*, and *Stipa aristiglumis*. Aust. J. Agric. Res. 20: 1035–1042.

White, C.T. (1920): Flora of the Bunya Mts. Qld. Agric. J. 13: 25–31.

Whitehouse, F.W. (1940): Studies in the late geological history of Queensland. Univ. Qld Papers Dept. Geol. No. 2.

– (1947): The Geology of the Channel Country of south-western Queensland. I. See Anon., 1947.

Whitmore, T.C. (1975): Tropical rainforests of the Far East. Clarendon Press, Oxford.

Wiedemann, A.M. (1971): Vegetation Studies in the Simpson Desert, N.T. Aust. J. Bot. 19: 99–124.

Wild, A. (1958): The phosphate content of Australian soils. Austr. J. Agric. Res. 9: 193–204.

Wilkins, Jean (1967): The effects of high temperatures on certain root-nodule bacteria. Austr. J. Agric. Res. 18: 299–304.

Williams, J.B. (1963): The vegetation of northern New South Wales from the eastern scarp to the western slopes – a general transect. New England Essays. Univ. New England, Armidale. 41–50.

– (MS): Miscellaneous notes and lists on the flora of north-eastern New South Wales. Botany Dept. Univ. New England. Armidale.

– and Hore-Lacy, I. (1963): Point Lookout Survey. New England National Park. Preliminary botanical survey. Aust. J. Sci. 26: 10–12.

Williams, O.B. (1955): Studies in the ecology of the Riverine Plain. i. The gilgai microrelief and the associated flora. Aust. J. Bot. 3: 99–112. ii. Plant-soil relationships in three semi-arid grasslands. Aust. J. Agric. Res. 7: 127–139.

Williams, R.F. (1932): An ecological analysis of the plant communities of the 'Jarrah' region occurring on a small area near Darlington. J. Roy. Soc. W.A. 18: 105–124.

– (1948): An ecological study near Beraking Forest Station. J. Roy. Soc. W.A. 31: 19–31.

Williams, W.T., Lance, G.N., Webb, L.T., Tracey, J.G. and Connell, J.H. (1969): Studies in the numerical analysis of complex rainforest communities. IV. A method for the elucidation of small scale forest pattern. J. Ecol. 57: 635–654.

Willis, E. (1972): Early history of Jarrahdale. Forest Focus 7: 3 and 10–14. Forests Dept. W.A.

Willis, J.C. and Airy Shaw, H.K. (1973): A dictionary of the flowering plants and ferns. 8th Ed. Camb. Univ. Press.

Willis, J.H. (1943): Statistical notes on the mallee flora. Vic. Nat. 59: 176–177.

– (1953): Plants of the Recherche Archipelago, W.A. Muelleria 1: 92–96.

– (1953): The archipelago of the Recherche. Aust. Geog. Soc. Report 1. Melbourne.

– (1959): Notes on the vegetation of Eucla district, W.A. Muelleria 1: 92–96.

– (1964): Vegetation of the basalt plains in western Victoria. Proc. Roy. Soc. Vic. 77: 397–418.

– (1970, 1972): A handbook to plants in Victoria. 2 Vols. Melbourne Univ. Press.

Wilson, Janet M. (1937): The structure of galls formed by *Cyttaria septentrionalis* on *Fagus moorei*. Proc. Linn. Soc. N.S.W. 62: 1–8.

Wilson, P.G. (1975): A taxonomic revision of the genus *Maireana* (Chenopodiaceae). Nuytsia 2: 2–83.

Winkworth, R.E. (1967): The composition of several arid spinifex grasslands of Central Australia in relation to rainfall, soil water relations and nutrients. Aust. J. Bot. 15: 107–130.

Womersley, J.S. (1969): A dictionary of the generic and family names of flowering plants. New Guinea and South West Pacific Region. Bot. Bull. No. 3. Dept. Forests. Admin. Papua and New Guinea. Port Moresby. 124 pp.

Wood, E.J.F. (1959): Some east-Australian sea-grass communities. Proc. Linn. Soc. N.S.W. 84: 218–226.

Wood, J.G. (1925): Selective absorption of chlorine ions and absorption of water by leaves in the genus *Atriplex*. Aust. J. Exp. Biol. and Med. Sci. 2: 45–56.

– (1929): Floristics and ecology of the mallee. Trans. Roy. Soc. S.A. 53: 359–378.

– (1930): An analysis of the vegetation of Kangaroo Island and the adjacent peninsulas. Trans. Roy. Soc. S.A. 54: 105–139.

– (1937): The vegetation of South Australia. Govt. Printer. Adelaide. 164 pp.

Zimmer, W.J. (1937): Flora of the far north-west of Victoria. For. Comm. Vic. Bull. No. 2.

– (1940): Plant invasions in the mallee. Vict. Nat. 56: 143–147.

Index

Family names (abbreviated) of Angiosperms are recorded in brackets after generic names.
For families of pteridophytes and gymnosperms, see Tables 2.1 (pp. 34–36) and 2.2 (p. 40).
F indicates an illustration.

A

Abarema (Mimos.)
– grandiflora 148
– monilifera 195
– sapindoides 148
Aborigines 32, 68, 517
Abrophyllum (Escallon.) 78
– ornans 146
Abrotanella (Compos.) 66
– fosterioides 445, 447 F, 450, 456
Abrus (Papil.) 85
– precatorius 181
Abutilon (Malv.) 104, 108
– halophilum 504
– walcottii 534
Acacia (Mimos.) 52, 67, 73, 87, 102, 103, 104, 105, 110, 130, 461 (Chap. 18)
– acuaria 393
– acuminata 382, 383, 385 F, 386, 425, 462
– – alliance 483
– adsurgens 474, 538
– Africa 67, 102
– alpina 87
– ancistrocarpa = A. pachycarpa
– aneura 105, 216, 307, 321, 327, 328, 462, 492, 508, 532, 559, 568
– – alliance 469, 470 F, 472 F, 475, 538
– – autecology 469
– armata 311, 334, 338
– aulacocarpa 173, 180, 181, 201, 211
– auriculiformis 181, 609
– baileyana 87, 313
– baxteri 393
– beauverdiana 393
– bidwillii 187
– biflora 393
– binervata 230, 231, 235, 236
– bivenosa 481, 606, 611
– blakelyi 382, 425
– botrycephala 238, 239, 255, 260
– brachystachya 473, 474, 538
– burkittii 322, 474, 486, 491
– buxifolia 280, 282, 304, 313, 332, 334, 335, 350
– calamifolia 309, 343, 348
– calcicola 461, 462
– – alliance 485
– calyculata 399
– cambagei 461, 462, 497, 559, 568
– – alliance 467, 468 F, 469
– cana 469

– carnei 121, 473 F, 474
– catenulata 461, 462
– – alliance 477
– cheelii 410, 462
– chisholmii 224, 474
– cochlearis 383, 430, 432
– colletioides 328, 357
– complanata 305
– concurrens 224, 306, 462
– conferta 306
– coriacea 307, 482, 483, 487, 532, 535, 607
– costinervis 537
– cowleana 535
– craspedocarpa 474
– crassiuscula 355
– cultriformis 336, 340
– cuneata 393, 420, 430
– cunninghamii 239, 303, 607
– cupularis 393
– cyanophylla = saligna
– cyclops 345, 432, 608, 610
– dealbata 87, 152, 153, 261, 262, 268, 273, 275, 279, 280, 285, 287, 288, 291
– deanei 304, 305, 306, 326, 336
– decipiens 377, 435
– – alliance 433
– decora 336
– decurrens 236
– delibrata 181
– dictyophleba 474, 537
– difformis 313
– dilatata 393
– diptera 393
– doratoxylon 282, 313, 410, 462
– elata 87, 148, 230
– elongata 254
– ericifolia 425
– ericoides 404
– estrophiolata 216, 479, 538, 568
– excelsa 305, 307, 321, 327, 462
– – alliance 478 F
– extensa 375
– falcata 330
– falciformis 271
– farnesiana 67, 68, 87
– filicifolia 271
– fimbriata 330
– flavescens 203, 305
– floribunda 241
– fossil record 67
– fragilis 393
– georginae 469
– gladiiformis 313
– gonophylla 393

– grasbyi 461, 462, 483
– – alliance 483, 484 F
– hakeoides 304, 313, 335, 350
– harpophylla 136, 183, 184, 301, 325, 461, 462, 478, 490
– – alliance 464, 465 F
– hilliana 224
– holosericea 220, 481, 535
– homalophylla 321, 326, 328, 336, 478, 490, 491
– humifusa 213
– implexa 231, 238, 279, 285, 330, 334
– impressa 479
– inaequilatera 535
– inophloia 474
– irrorata 230
– ixiophylla 348
– jucunda 305
– kempeana 364, 473, 474, 492, 538
– lasiocalyx 362, 393, 434, 436
– lasiocarpa 393
– leptocarpa 203, 211
– leptostachya 309
– leursenii 224, 538
– ligulata 339, 393, 421, 424, 425, 462, 483, 491, 535, 607, 608, 611, 637
– – alliance 487
– linearis 232
– lineata 304, 350
– linophylla 462, 482, 483, 492, 532
– loderi 107, 121, 461, 462, 485, 486 F, 491
– longifolia 235, 240, 244, 255, 259, 260, 400, 409, 611, 612
– longispicata 301
– longispinea 357, 393, 424, 425
– lysiphloia 214, 224, 364, 399, 474, 537
– maidenii 148, 229
– mangium 181
– mearnsii 279
– – Melaleuca and Casuarina thickets (W.A.) 425
– melanoxylon 148, 152, 153, 155, 268, 273, 274, 275, 277, 279, 283, 287, 291, 336, 339, 563
– microcarpa 348
– microsperma 462
– monticola 481
– mountfordii 213
– mucronata 261, 263, 264, 405
– – – Phebalium squameum alliance 407
– multilineata 393

Acacia (Mimos.)
- multispicata 393
- murrayana 393, 483, 535
- myrtifolia 99, 236, 238, 255, 261, 371, 402, 404
- neriifolia 282, 286, 288
- neurophylla 393
- orthocarpa 214, 224
- oswaldii 325, 347, 466, 490, 508, 564
- oxycedrus 405
- pachycarpa 206, 461, 462, 535
- – alliance 479 F, 480 F
- patens 492
- pendula 462, 490, 580
- – alliance 466, 460 F
- penninervis 239, 282, 288, 375
- pentadenia 370 F, 371, 372
- peuce 107
- pilligaensis 312
- pilosa 393
- pravissima 410
- pruinocarpa 473
- pruinosa 286
- pulchella 371, 375, 382, 393
- pycnantha 278, 289, 334, 335, 337, 338, 343, 564
- quadrimarginea 484
- quadrisulcata 393
- ramulosa 357, 393, 424, 462, 483, 487
- – – A. linophylla alliance 482 F
- resinomarginea 393, 428, 474
- restiacea 434
- rhodoxylon 307, 462
- rigens 326, 335, 348
- root suckering 462
- rossii 393
- rostellifera 353, 358, 380, 418, 422, 430, 483, 487, 607, 608, 610, 637
- – thickets 425
- rothii 393
- salicina 332, 336, 382, 393, 437, 466, 492, 563, 566, 568
- saligna 380, 383, 425, 565, 597, 603
- scirpifolia 393, 425
- sclerosperma 461, 462, 482, 483, 508, 607, 611
- – – A. tetragonophylla – Eremophila pterocarpa – Atriplex spp. edaphic complex 481 F
- sections of genus 67, 87
- sericata 213, 220
- shirleyi 461, 462, 539
- – alliance 463 F
- signata 393
- sophorae 403, 409 F, 605, 607, 608, 610, 611, 636
- sorophylla 352
- sowdenii 107, 461, 462, 485 F, 486
- – – A. loderi alliance 485
- spathulata 382, 393
- spectabilis 304, 336
- spinescens 345
- stenophylla 466, 563, 566, 568
- stereophylla 393, 425
- stipuligera 399, 535, 537
- stricta 261, 281
- strigosa 371, 393
- suaveolens 236, 239, 254, 409, 411
- sutherlandii 566
- tetragonophylla 424, 473, 474, 481, 482, 483, 484
- trachycarpa 535
- translucens 462, 479, 480 F, 483, 534, 535, 606, 611, 628
- – alliance 480
- triptera 336
- tumida 213, 220, 221 F, 481, 535, 606
- ulicina 425
- urophylla 371
- verniciflua 336, 337 F
- verticillata 261, 413, 608
- victoriae 482, 483, 491, 564, 568, 569 F
- villosa 404
- vomeriformis 261
- xanthina 393, 421, 425
- xerophila 224
- xiphophylla 461, 462
- – alliance 483, 484 F
Acaena (Rosac.) 66
- ovina 280
Acanthaceae 45
Acanthocarpus (Xanthorr.)
- preissii 600
Acanthus (Acanth.)
- ilicifolius 614, 621
Aceratium (Elaeocarp.)
- doggrellii 172
Acetosa (Polygon.)
- rosea 103
Achnophora (Compos.)
- tatei 637
Achyranthes (Amaranth.)
- aspera 635
Acianthus (Orchid.) 92, 98
- amplexicaulis 92
Acidonia (Prot.) 97
Ackama (Cunon.)
- paniculata 147, 230
Acmena (Myrt.) 81
- hemilampra 149
- smithii 149, 158, 194, 232, 241, 279, 546
Acrachne (Gramin.) 511
Acradenia (Rut.)
- frankliniae 593
Acronychia (Rut.)
- chooreechillum 178
- laevis 240
- oblongifolia 149
- simplicifolia 149
Acrostichum 36
- speciosum 37
Acrotriche (Epacrid.) 98
- cordata 432
- fasciculiflora 343
- serrulata 402
Actinidiaceae 45
Actinodium (Myrt.) 81, 98
Actinomycetes 33

Actinostrobus 40, 41
- arenarius 356, 367, 417, 421
- – alliance 422, 423 F
- pyramidalis 595
Actinotus (Umbell.) 77
- helianthi 77 F, 282
- minor 411
- paddisonii 309
Adansonia (Bombac.) 181
- gregorii 186, 187 F, 188, 211, 214
- – alliance 186
Adaptations – soil fertility 77
Adenanthos (Prot.) 85, 86, 98
- cuneata 395, 426
- cygnorum 395, 417
- dobsonii 395
- flavidiflora 395
- sericea 334, 362, 420, 421, 422
- stricta 395
- terminalis 334, 414
Adenochilus (Orchid.) 98
Adiantaceae 35
Adiantum 35
- aethiopicum 37, 238
- hispidulum 101
Adriana (Euphorb.) 88, 108
- quadripartita 339, 432
- tomentosa 483, 534
Aegialitis (Plumbag.)
- annulata 614, 621, 623, 624 F, 627
Aegiceras (Aegicerat.)
- corniculatum 518, 614, 621, 627, 630
Aegicerataceae 45
Aeolian deposits 22
Aerva (Amaranth.)
- persica 606
Aeschynomene (Papil.)
- indica 523
Agapetes (Eric.) 69, 88
- meiniana 399
Agastachys (Prot.) 87
- odorata 405
Agathis 40, 41, 58, 76, 142
- atropurpurea 172, 179
- robusta 146, 172, 237, 636
Agavaceae 48, 90
Aglaia (Meliac.) 70
Agonis (Myrt.) 81, 98
- flexuosa 375, 376, 380, 383, 415 F, 429, 430, 432, 433, 597, 608
- – alliance 431
- juniperina 371, 433, 597
- linariifolia 371, 596, 597
- marginata 383
- obtusissima 426
- parviceps 371, 372, 377, 597
- scortechiniana 410
Agropyron (Gramin.) 511
- pectinatum 547
- repens 269
- retrofractum 271
- scabrum 280, 285, 333, 334, 335
Agrostis (Gramin.) 511
- aemula 517
- avenaceus 332, 559, 594
- billardieri 519
Ailanthus (Simaroub.)
- malibaricum 150, 185, 194

Aira (Gramin.)
— caryophyllea 264
Aizoon (Aizo.) 45
— quadrifidum 559
— zygophylloides 559
Akania (Akan.)
— lucens 146
Akaniaceae 45
Alangiaceae 45
Alangium (Alang.)
— villosum 146
Albizia (Mimos.)
— distachya 371
— lebbek 180, 181
— lophantha 383, 603
— toona 634
Alchemilla (Ros.)
— novae-hollandiae 449
Aldrovanda (Droser.) 55, 571
Alectryon (Sapind.) 70
— subcinereus 149, 228, 546
— subdentatus 183
Aleurites (Euphorb.)
— moluccana 172
Algae 33, 579
Alisma (Alismat.)
— plantago-aquatica 581, 585
Alismataceae 48
Allania (Lil.) 90
Alloteropsis (Gramin.) 511
— semialata 202, 215, 246, 487
Alluvial deposits 22
Alogyne (Malv.) 112
— cuneiformis 487
Alphitonia (Rhamn.)
— excelsa 148, 164, 180, 183, 184, 201, 211, 228, 231, 234, 239, 248, 299, 301, 302, 303, 305, 313, 319, 320, 324, 464
— petrei 173
— whitei 173
Alpine communities 131, 438 (Chap. 17)
— endemic genera 442
— flora 441
— soils 440
Alstonia (Apocyn.) 45, 180
— actinophylla 180
— constricta 146
— muelleriana 172
— scholaris 172
Alternanthera (Amaranth.)
— nodiflora 535, 555
— triandra 529
Alysicarpus (Papil.)
— rugosus 523
Alyxia (Apocyn.)
— buxifolia 600, 608
— ruscifolia 146
Amaranthaceae 45
Amaranthus (Amaranth.) 104
— viridis 103
Amaryllidaceae 48
Ammania (Lythr.)
— multiflora 520
Ammobium (Compos.)
— alatum 280
Ammophila (Gramin.)
— arenaria 516, 607, 608

Amorphophallus (Arac.)
— glabra 180
Ampelopteris 36
Amperea (Euphorb.)
— xiphoclada 247
Amphibromus (Gramin.) 67, 511
— neesii 270, 332
Amphipogon (Gramin.) 381, 511
— laguroides 596
— strictus 313
Amyema (Loranth.) 55, 532
— quandong 466, 470
Amylotheca (Loranth.) 55
Anabaena 33
Anacardiaceae 45
Anacolosidites 60
Anarthria (Restion.) 93, 595
— laevis 596
— prolifera 596
— scabra 596
Ancana (Annon.) 53
Ancistrachne (Gramin.) 511
— uncinulatum 478
Andersonia (Epacrid.) 98
— aristata 392
— brevifolia 392
— caerulea 392
— parvifolia 392, 435
— simplex 392
— sprengelioides 435
Angianthus (Compos.) 113
— brachypappus 508
— cunninghamii 607, 608
— drummondii 559
— plumigera 505
Angiopteridaceae 34
Angiopteris 34
— evecta 142, 636
Angiosperms 44, 65
— carnivorous 54, 55
— Cretaceous 59, 65
— endemic families 49
— endemic genera 49
— families 44
— fossil pollen 65
— genera 49, 50
— Gondwanaland assemblage 65
— heterotrophic 52, 54, 55
— history in Australia 65
— naturalized 56
— parasitic 52, 54, 55
— primitive 44, 51, 53
— rainforest assemblage 65
— species 49, 51
Angophora (Myrt.) 82, 83, 98
— bakeri 252, 254 F, 300, 303
— cordifolia 411
— costata 228, 239, 250, 252, 253, 299, 300, 302, 306 F, 608
— — — Eucalyptus spp. suballiance 304
— floribunda 234, 281, 307, 313, 593, 594 F
— melanoxylon 301, 302, 307
— — suballiance 309, 310 F
— subvelutina 234, 248
Anguillaria (Lil.) 90, 92
— dioica 92 F, 338

659

Anigozanthos (Haemodor.) 91
— humilis 421
— rufa 426
Anisopogon (Gramin.) 511
Annonaceae 45, 53
Anodopetalum (Cunon.) 78
— biglandulosum 151, 156
Anoectochilus (Orchid.) 93
Anogramma 35
— leptophylla 37
Anopterus (Escallon.) 78
— glandulosus 152, 295
— macleayanus 146
Antarctica
— fossil flora 62, 68
— islands 68
Anthobolus (Santal.) 55, 115
— leptomerioides 532
Anthocercis (Solan.)
— littoralis 432
Anthotroche (Solan.)
— myoporoides 357
Antiaris (Mor.)
— toxicaria 173
Ant Plant, Smooth = Hydnophytum formicarum
— Spiny = Myrmecodia antoinii
Antrophyum 36
Aotus (Papil.) 85, 98
Aphananthe (Ulm.)
— philippinensis 149, 174, 228
Aphanopetalum (Cunon.)
— resinosum 147
Aphyllorchis (Orchid.) 54, 93
Apium (Umbell.)
— prostratum 581, 600, 607, 631
Apocynaceae 45
Aponogeton 63
Aponogetonaceae 48
Apophyllum (Capparid.) 112
— anomalum 190, 325, 469, 490
Apostasia (Orchid.) 93
Apostasiaceae 48
Apple = Angophora
— Coolabah = A. melanoxylon
— Rough-barked = A. floribunda
— Smooth-barked = A. costata
Aptopteris 34
Aquatic communities 551, 570 (Chap. 22)
Aquifoliaceae 45
Araceae 48, 139
Arachniodes 35
Araliaceae 45
Araucaria 41, 76, 142
— bidwillii 146
— cunninghamii 146, 164, 184, 237, 636
Araucariaceae 40
Araucariacites 58
Archonotophoenix (Palm.)
— alexandrae 169, 175 F, 177, 178, 195
— cunninghamiana 150, 163, 169, 229, 230 F, 234
Arctotheca (Compos.)
— nivea 607, 608
Argemone (Papaver.) 530

Argyrodendron (Stercul.) 70
- actinophyllum 150, 174
- peralatum 174
- polyandrum 174
- spp. alliance 162
- trifoliatum 150
Arid zone 14, 16, 125
- Acacia spp. 111. See Chap. 18
- algae 101
- annuals 116, 117
- apomixis 118
- apospory 118
- Compositae 112
- development 75
- endemic genera 111, 112
- floral disjunctions 106
- deciduous habit 117
- distribution of flora 103, 105
- Eucalyptus 363
- flora 100
- flowering 118
- fruit characters 118
- gymnosperms 101
- habitats 100, 104
- helophytes 115
- hydrophytes 115
- leaf characters 116, 117
- lichens 101
- life-forms 114
- littoral taxa 111
- liverworts 101
- microorganisms 100
- naturalized flora 102
- nitrogen fixation 101
- origin of flora 107
- osmotic potentials 118
- pollination 118
- polyploidy 118
- pteridophytes 101
- rainforest derivatives 109
- refuges 107
- reproduction 118
- root parasites 115
- root suckering 121
- seed dispersal 120
- seed germination 118, 119, 120
- seeds 118, 120
- species abundance 104, 105
- stem succulent 114
- succulence 117
- vegetative features 114
- vegetative reproduction 121
- xeromorphs 110
Arillastrum (Myrt.) 84
Aristida (Gramin.) 52, 104, 106, 511
- anthoxanthoides 504, 556
- arenaria 101 F, 187, 349, 475, 540
- browniana 187, 215, 221, 308, 475, 479, 483, 487, 537, 540
- caput-medusae 303, 464, 478
- contorta 120, 531, 540
- echinata 303, 306, 307, 391
- glumaris 309
- hirta-A. superpendens alliance 517
- hygrometrica 202, 209
- jerichoensis 305, 313, 321, 322, 326, 328, 336, 479
- latifolia 524, 526

- leichhardtiana 240
- leptopoda 524, 527
- pruinosa 187, 222, 223, 224, 319, 399
- ramosa 236, 240, 280, 285, 286, 303, 306, 313, 334, 335
- spp. grasslands 540
- superpendens 517
- vagans 234
- warburgii 332, 409
Aristolochiaceae 45
Aristotelia (Elaeocarp.) 66
- australasica 155
- peduncularis 152
Arnhem plateau 131, 213
Arnocrinum (Lil.) 90
Artesian basin 10
- water 550
Arthraxon (Gramin.) 511
Arthrochilus (Orchid.) 98
Arthrocnemum 114, 493, 623, 626
- arbuscula 388, 629 F, 630, 631
- bidens 557
- halocnemoides 485, 494, 557, 558 F, 560, 595, 600, 602, 628, 629, 630
- leiostachya 504, 627 F, 628
Arthropteris 35
Arthrostylis (Cyper.) 91
Arundinella (Gramin.) 551
- nepalensis 410
Arytera (Sapind.)
- foveolata 149
Ascarina (Chloranth.) 60
Asclepiadaceae 45
Ash (inorganic) 84, 293
Ash (Plants)
- Alpine = Eucalyptus delegatensis
- Black = E. sieberi
- Mountain = E. regnans
Asparagus (Lil.)
- racemosus 180
Asperula (Rub.)
- conferta 270, 289, 291
Aspidiaceae 35
Aspleniaceae 35
Asplenium 35, 140
- australasicum 176
- flabellifolium 37
- nidus 36, 163
- obtusatum 37
Association 123
Astartea (Myrt.) 98
- ambigua 436
- fascicularis 379, 436, 596, 597
Astelia (Lil.) 67
- alpina 442, 448 F, 454 F
- alliance 456, 457 F
Asterolasia (Rut.) 85, 98
- floccosa 279
- phebalioides 419
Astrebla (Gramin.) 106, 112, 320, 475, 511, 523
- elymoides 525
- suballiance 525
- lappacea 120, 325, 466, 467 F, 490, 524
- suballiance 526 F

- pectinata 504, 523, 528
- suballiance 527
- spp. alliance 524
- squarrosa 526
- suballiance 525
Astroloma (Epacrid.) 98, 105
- candolliana 392
- ciliatum 392
- conostephioides 259, 343
- humifusum 337, 343, 402
- pallidum 392
- serratifolium 392
Astrotricha (Aral.)
- pterifolia 305
Atalaya (Sapind.)
- hemiglauca 121, 185, 191, 192 F, 215, 217, 219, 222, 223, 320, 325, 328, 332, 364, 466, 522, 469, 479, 490, 566
- - - Grevillea striata - Ventilago viminalis alliance 191
- multiflora 149, 187, 188
- variifolia 181, 299
Atherosperma (Monim.) 53
- moschatum 148, 151, 152 F, 153, 155, 273, 278
Athertonia (Prot.) 87
Athrixia (Compos.)
- chaetopoda 559
Athrotaxis 40, 43, 58, 142
- cupressoides 294, 454 F
- selaginoides 43 F, 154
Athyriaceae 35
Athyrium 35
Atkinsonia (Loranth.) 55
Atriplex (Chenopod.) 103, 104, 106, 493, 507
- angulata 366, 542
- australasica 495
- billardieri 608
- bunburyana 484
- campanulata 348
- C4 - pathway 495
- cinerea 495, 600, 602, 608, 630, 631, 634 F
- conduplicata 527, 531
- hymenotheca 386, 387, 389, 481, 482, 484, 498, 559
- suballiance 497, 504, 505 F
- inflata 366, 499, 502, 504, 521, 527, 528, 531, 556
- isatidea 495, 515, 606, 607, 608, 629
- leaf anatomy 495
- muelleri 526, 527, 528
- nummularia 387, 388, 389, 467, 469, 484, 496, 507, 568
- alliance 496, 497, 499 F
- paludosa 559, 600, 602, 608, 630, 631
- patula 630, 631
- rhagodioides 501
- alliance 497, 504
- seed germination 119
- semibaccata 332, 529
- spinebractea 350
- spongiosa 502, 504, 521, 527, 528, 542, 556

Atriplex (Chenopod.)
- stipitata 348, 476, 483, 486
- vesicaria 105, 348, 366, 467, 469, 474, 483, 485, 486, 494, 496, 498F, 500, 504, 508, 559, 568
- – alliance 497, 499
- – – Bassia spp. suballiance 497, 501F, 502F
- – – Ixiolaena leptolepis suballiance 497, 503F
- – – Maireana astrotricha suballiance 504, 506F
Austrobaileya (Austrobailey.) 53
- scandens 177
Austrobaileyaceae 45, 53
Austrobuxus (Euphorb.)
- swainii 146
Austromuellera (Prot.) 86
- trinervia 173
Austromyrtus (Myrt.) 81
- acmenoides 149
- bidwillii 149
- dulcis 400
- hillii 149
Avena (Gramin.)
- fatua 103
Avicennia (Verben.) 60
- eucalyptifolia 614
- marina 518, 614, 617F, 630
- – var. australasica alliance 625, 626F, 629F, 633F
- – var. resinifera alliance 621, 624F, 627
Ayer's Rock 7F
Azolla 36
- filiculoides 37, 551, 585, 587, 595
- pinnata 37, 551, 585
Azollaceae 36
Azotobacter 33

B

Babbagia (Chenopod.) 113, 493
- acroptera 336, 502, 556, 559
Backhousia (Myrt.) 81, 82
- angustifolia 183
- anisata 149
- bancroftii 173
- hughesii 173
- myrtifolia 149, 231, 237, 240
- sciadophora 149
Bacterium halobium 551
Baeckea (Myrt.) 52, 80, 81, 98, 103, 105
- ambigua 430
- behrii 345
- brevifolia 411
- camphorosmae 393
- crispiflora 393
- densiflora 349
- dimorphandra 393
- fumana 393
- grandibractea 393
- grandiflora 393
- gunniana 445, 446F, 452, 453, 460
- imbricata 405
- leptocaulis 407
- leptospermoides 393
- ovalifolia 436
- pentagonantha 393
- preissiana 393
- robusta 393
- uncinella 393, 426
Balanophora (Balanophor.) 54
Balanophoraceae 45, 54
Balanops (Balanopsid.)
- australiana 172
Balanopsidaceae 45
Balaustion (Myrt.) 81, 105, 112, 428
Balds 544, 546
Baloghia (Euphorb.) 70
- lucida 146
Bambusa (Gramin.) 511
- arnhemica 195, 512F, 562, 573
- moreheadiana 169, 170F, 177
Bangalay = Eucalyptus botryoides
Banksia (Prot.) 84, 86, 94, 98, 105
- ashbyi 395, 421, 425
- – alliance 424F
- aspleniifolia 244, 247, 252, 254, 391, 408, 409
- – alliance 401, 402F
- attenuata 376, 380, 395, 417, 418, 419, 421, 422, 431, 433
- audax 395, 427, 428
- baueri 361, 395, 433
- baxteri 382, 395, 433
- candollei 395
- collina 271, 391
- dentata 391, 576
- dryandroides 395, 435
- elderana 355, 395, 427, 428
- ericifolia 391, 411, 412
- – alliance 401, 403F
- grandis 371, 375, 376, 381, 431, 432F, 433
- hookerana 418
- ilicifolia 418, 419, 433, 596, 597
- integrifolia 166, 239, 245, 271, 274, 287, 288, 391, 400, 409, 517, 605, 607, 608, 636
- laevigata 395, 427
- latifolia 591
- lehmanniana 362
- littoralis 377, 427, 595, 596, 597
- lullfitzii 395
- marginata 291, 263, 269, 278, 289, 338, 339, 344, 391, 411, 413, 414, 548, 563, 591, 592, 608, 611
- – alliance 402, 404F
- media 395, 426, 427
- menziesii 380, 395, 417, 421, 422
- – – B. attenuata alliance 417F, 418
- – – B. attenuata – Casuarina fraserana – Eucalyptus todtiana alliance 418
- nutans 395
- occidentalis 345
- ornata 338, 391, 413, 414
- petiolaris 395, 427
- prionotes 356, 395, 418, 422
- prostrata 362, 395
- pulchella 395
- repens 395
- robur 391
- – alliance 401, 403F
- sceptrum 356, 395, 422
- scrub-heaths 418
- serrata 232, 239, 391, 411, 608
- serratifolia 239, 259, 391, 608
- – alliance 400, 401F
- speciosa 395, 427
- sphaerocarpa 382, 395, 595
- spinulosa 255, 391, 402
- tricuspis 395, 419
Banksieaephyllum 84
Banksieidites 60, 84
Baobab = Adansonia gregorii
Baragwanathia 57
Barklya (Papil.) 85
- syringifolia 147
Barren Ranges (W.A.) 361, 434
Barringtonia (Barrington.) 60
- acutangula 195, 562, 572F, 573F, 574
- calyptrata 628
- gracilis 195
Basalt 23, 74
Bassia (Chenopod.) 103, 113, 348, 493, 497, 542
- astrocarpa 517, 628
- bicornis 528
- birchii 350, 495
- brachyptera 501, 502, 504, 542
- diacantha 382, 484, 542
- divaricata 500, 501, 504, 542
- lanicuspis 476, 502, 531
- limbata 501
- luehmannii 560
- obliquecuspis 486F, 502, 508, 531
- paradoxa 348, 473, 486F, 531
- patenticuspis 508, 542
- quinquecuspis 325, 467, 484, 495, 500, 528, 542, 566
- sclerolaenoides 508
- tetracuspis 528
- uniflora 350, 508, 531
- ventricosa 501, 502F, 503, 504, 506, 542
Bass Strait 74, 125
Bauera (Bauer.)
- rubioides 152, 254, 257, 263, 264, 295, 405, 407, 411, 454, 591
Baueraceae 45
Bauhinia (Caesalpin.) 108
- carronii 109, 184, 190, 490
- cunninghamii 186, 188F, 189, 192, 202, 211, 217, 220, 319, 479, 481, 522, 532, 535, 562
- – – suballiance 188
Baumea (Cyper.)
- acuta 591
- articulata 574, 580, 581, 582
- juncea 240, 631
- – alliance 587
- procera 581
- teretifolia 585
- tetragona 586
Beaufortia (Myrt.) 81, 98, 105
- dampieri 393
- heterophylla 393

Beaufortia (Myrt.)
- micrantha 393
- squarrosa 382, 393
Beauprea 84
Beaupreaidites 60, 84
Bedfordia (Compos.)
- salicina 152, 153, 273, 274, 291, 293
Beech, Antarctic = Nothofagus cunninghamii
- Negrohead = N. moorei
Beijerinckia 33
Beilschmiedia (Laur.) 53
- bancroftii 172
- obtusifolia 147, 172
Belah = Casuarina cristata
Bellendena (Prot.) 87
- montana 295, 297, 439F, 453
Bellida (Compos.) 113
Belvisia 36
Benthamina (Loranth.) 55
Bertya (Euphorb.) 88
Beyeria (Euphorb.) 88
- canescens 359
- cyanescens 638
- leschenaultii 339
- viscosa 261, 292
Bignoniaceae 45
Bipolar taxa 57
Blackbutt = Eucalyptus pilularis
- Albany = E. staeri
- Dundas = E. dundasii
- Goldfields = E. lesouefii
- Large – fruited = E. pyrocarpa
- New England = E. campanulata
- Swan River = E. patens
Blandfordia (Lil.) 90
- grandiflora 591
- nobilis 252
Blechnaceae 35
Blechnum 35, 287
- cartilagineum 231, 238, 279
- procerum 151
- indicum 37, 240, 405, 581, 590F, 591
- minus 39F
- penna-marina 449
Blennodia (Crucif.) 328
- lasiocarpa 531
- pterosperma 540
Blepharocarya (Blepharocar.)
- involucrigera 172
Blindia robusta 457
Bloodwood 84
- Brittle = Eucalyptus aspera
- Broome = E. zygophylla
- Fan-leaved = E. foelschiana
- Inland = E. terminalis
- Long-fruited = E. polycarpa
- Pink = E. intermedia
- Red = E. ptychocarpa or E. gummifera
- Round-leaved = E. latifolia
- Rusty = E. ferruginea
- Silver-leaved = E. collina
- Swamp = E. ptychocarpa
- Variable-barked = E. dichromophloia
- Yellow = E. eximia

Bluebush = Maireana astrotricha, M. pyramidata, M. sedifolia or Chenopodium auricomum
Boerhavia (Nyctagin.)
- diffusa 524, 526, 527, 635
Bog 124, 455, 456, 457
- raised 457, 459
- valley 457
Bogong High Plains (Vic.) 459
Bolbitis 35
Bolster plant 445
Bombacaceae 45
Bombax (Bombac.)
- ceiba 185, 634
Bonewood = Macropteranthes leichhardtii
Boraginaceae 45, 56
Boree = Acacia pendula
Boronia (Rut.) 52, 98, 105
- bipinnata 306, 321
- bowmannii 399
- citriodora 442, 448F, 455
- coerulescens 345, 397
- crenulata 432
- falcifolia 401
- glabra 306
- parviflora 591
- pinnata 404
- rosmarinifolia 306
- spathulata 397
- ternata 397
Borya (Lil.) 397
- nitida 434, 435, 436, 596
- septentrionalis 409
Bosistoa (Rut.)
- euodiformis 149
Bossiaea (Papil.) 52, 85, 98
- aquifolia 371, 372, 374, 375
- biloba 394
- cinerea 404
- eriocarpa 394
- foliolosa 271, 542, 543, 545
- heterophylla 244
- linophylla 373, 375
- neo-anglica 271
- phylloclada 200
- rhombifolia 247
- scolopendria 257, 411
- spinosa 394
- walkeri 348
Bothriochloa (Gramin.) 511
- bladhii 523
- decipiens 528
- erianthioides 524
- ewartiana 301, 309, 318, 320, 325, 475, 479, 523, 527
- intermedia 234, 465, 523, 578
- macera 234, 244, 285, 288, 330, 332, 333, 336
Botrychium 34
Bottle Tree Scrub 137
Bowenia 40, 41, 142
Bowgada = Acacia ramulosa
Box 84, 315
- Barlee = Eucalyptus lucasii
- Bimble = E. populnea
- Black = E. largiflorens
- Black Mallee = E. porosa

- Brush = Tristania conferta
- Georgetown = E. microneura
- Grey = E. molluccana or E. woollsiana
- Gum-leaved = E. leptophleba
- Gum-topped = E. orgadophila
- Normanton = E. normantonensis
- Northern = E. tectifica
- Pilliga = E. pilligaensis
- Poplar = E. populnea
- Reid River = E. brownii
- Silver-leaved = E. pruinosa
- Western = E. argillacea
- White = E. albens
- woodlands 315
- Yellow = E. melliodora
Brachiaria (Gramin.) 511
- miliiformis 564
- piligera 564
Brachyachne (Gramin.) 511
- convergens 523, 540
Brachychiton (Stercul.) 70, 108, 213, 562
- acerifolius 150, 174
- australis 183, 184, 185
- discolor 150, 165F
- grandiflorus 183, 185
- – suballiance 185
- gregorii 117, 345, 482, 483, 492
- populneus 282, 307, 313, 315, 320, 321, 328, 330, 332, 333
- rupestris 183, 184F
- spp. alliance 183
- viridiflorus 181
Brachycome (Compos.) 52, 103
- ciliaris 118, 476
- nivalis-Danthonia alpicola alliance 448
- pachyptera 531
- stolonifera 451
Brachyloma (Epacrid.) 98, 105
- ciliatum 404
- daphnoides 232, 282, 285, 286, 287, 312, 313, 334
Brachysema (Papil.) 85, 98, 105, 107
- bossiaeoides 213
- chambersii 394
- daviesioides 394
- macrocarpum 359
- preissii 394
- tomentosum 394
Brasenia (Cabomb.)
- schreberi 585
Breutelia 459
Breynia (Euphorb.)
- oblongifolia 146, 232, 233, 235, 244
Bridelia (Euphorb.)
- exaltata 146
Brigalow = Acacia harpophylla
Bromus (Gramin.) 104, 511
- arenarius 503, 510
Broombush = Melaleuca uncinata
Brown Barrel = Eucalyptus fastigata
Bruguiera (Rhizophor.)
- cylindrica 614
- exaristata 614
- gymnorhiza 614, 621, 624, 625

Bruguiera (Rhizophor.)
- parviflora 614
- rheedii 614
- sexangulata 614
Brunonia (Brunon.)
- australis 309, 357, 482
Brunoniaceae 45
Bryum 456, 457
- hypnoides 446
Bubbia (Winter.) 53
Buchanania (Anacard.)
- arborescens 180
- obovata 180, 186, 200, 207
Buckinghamia (Prot.) 86
Budda = Eremophila michellii
Buellia subalbula 507
Bulbine (Lil.) 115
- bulbosa 338
Bulbophyllum (Orchid.) 93
Bulbostylis (Cyper.)
- barbata 564
Bullich = Eucalyptus megacarpa
Buloke = Casuarina luehmannii
Bulrush = Typha domingensis
Bulwaddy = Macropteranthes keckwickii
Burchardia (Lil.) 90
- umbellata 338, 376, 417
Burmanniaceae 48
Burnettia (Orchid.) 98
Burr
- Copper = Bassia spp.
- Electric = B. quinquecuspis
- Galvanized = B. birchii
Bursaria (Pittos.)
- longisepala 336
- multisepala 309 F
- spinosa 234, 278, 285, 289, 332, 334, 339, 611, 612
Burseraceae 45
Burtonia (Papil.) 98, 105, 107
- scabra 375
Butomaceae 48
Button Grass = Gymnoschoenus sphaerocephalus
Button Grass Plains 588
Byblidaceae 45, 55
Byblis (Byblid.) 55
- liniflora 399, 574

C

Cabombaceae 45
Cadellia (Simaroub.) 181
- pentastylis 182 F, 466
- - - alliance 183
Caesalpiniaceae 45
Cakile (Crucif.)
- edentula 353, 606, 607, 608, 635
Caladenia (Orchid.) 52, 92, 98, 115
- carnea 92 F
- flava 431
Calamus (Palm) 150, 178
Calandrinia (Portul.)
- balonensis 505, 537, 541
- calyptrata 413

- polyandra 537, 559
- volubilis 531
Calanthe (Orchid.) 92, 93, 98
Caldcluvia (Cunon.) 62
Caleana (Orchid.) 98
- minor 71
Calectasia (Xanthorrh.)
- cyanea 426
Callicoma (Cunon.) 78
- serratifolia 147, 155, 230, 232
Callistemon (Myrt.) 81, 98
- citrinus 402
- linearis 411
- pachyphyllus 590, 591
- pallidus 413, 452
- sieberi 459, 460
- viminalis 194 F, 234
- viridiflorus 296 F, 297, 407
Callitrichaceae 45
Callitriche (Callitrich.)
- stagnalis 586, 594
- verna 552
Callitris 40, 43
- columellaris 43, 239, 247, 599, 607, 636
- endlicheri 282, 284, 286, 301, 305, 306, 307, 311, 313, 330, 336
- "glauca" 42 F, 101, 282, 302, 303, 306, 307, 309, 325, 336, 346, 464, 466, 473, 478, 483, 491, 568, 611
- intratropica 43, 205, 213
- macleayana 43, 142, 146
- muelleri 282
- preissii 101, 339, 345, 346 F, 347, 354, 355, 383, 427, 428, 435, 437
- rhomboidea 252
- roei 352, 354
Calocephalus (Compos.) 113
- brownii 600, 602, 607
- sonderi 500, 530
Calochilus (Orchid.) 92, 98
Calophyllum (Guttif.)
- australianum 181
- inophyllum 606
- sil 172
Caloplaca 439
- marina 602
Calorophus (Restion.) 92
- minor 401, 403, 404, 405, 408, 460, 544, 591
- - - Leptocarpus tenax alliance 587, 590 F
Calostemma (Amaryll.) 155
Calothamnus (Myrt.) 81, 98, 105
- blepharospermus 393
- chrysantherus 393, 421, 424
- gilesii 436
- gracilis 393
- pinifolius 362, 393
- quadrifidus 361, 393, 425, 427, 435, 436
- sanguineus 393, 422, 423, 430
- validus 436
- villosus 393, 595
Calotis (Compos.)
- hispidula 528, 531
- lappulacea 118, 307, 332, 333, 336

Caltha (Ranunc.) 77
- introloba 451
- phylloptera 451
Calycomis (Cunon.) 78
Calycopeplus (Euphorb.) 88
Calyptochloa (Gramin.) 511
Calystegia (Convolvul.)
- soldanella 607
Calythropsis (Myrt.) 81
Calytrix (Myrt.) 52, 81, 98, 104, 110, 213, 224
- achaeta 399
- brachychaeta 221
- brachyphylla 393
- brevifolia 393, 424
- breviseta 393
- Complex 205, 213
- decandra 393
- flavescens 393
- interstans 534
- laricina 213
- longiflora 437
- microphylla 200, 213, 221, 399
- muricata 483
- tenuifolia 393
- tetragona 239, 303, 313, 343, 345, 349, 350, 352, 404, 410, 411, 413, 414, 452
Campanulaceae 45
Camptostemon (Malv.)
- schultzii 614
Campylopus bicolor 413, 434
Cananga (Annon.) 53
Canarium (Burser.)
- australianum 180, 181, 634
- baileyanum 172
- muelleri 172
Canavalia (Papil.)
- obtusifolia 515, 516, 606, 609, 621
Candlebark = Eucalyptus rubida
Canthium (Rubi.) 108
- coprosmoides 148
- latifolium 492
- odoratum 148, 183
- oleifolium 328, 466, 473, 489
Capillipedium (Gramin.) 511
Capparidaceae 45
Capparis (Capparid.) 108
- arborea 147
- lasiantha 190, 320, 466, 469, 474
- mitchellii 182, 183, 190, 191, 328, 331, 332, 469
- spinosa 606
- umbonata 180, 192
Caprifoliaceae 45
Capusiaceae 45
Carallia (Rhizoph.)
- brachiata 181
Carbeen = Eucalyptus tessellaris
- Broad-leaved = E. confertiflora
Cardamine (Crucif.)
- hirsuta 449, 452
Cardaria (Crucif.)
- draba 330
Cardiopteridaceae 45
Carduus (Compos.)
- tenuiflorus 330

Cardwellia (Prot.) 86
– sublimis 173
Carex (Cyper.)
– appressa 544, 586
– gaudichaudiana 459, 460, 563
– – alliance 455F, 456
– – – Sphagnum cristatum alliance 457
– inversa 350
– polyantha 593
– pumila 516, 607, 608
– tereticaulis 547
Carissa (Apocyn.) 108
– lanceolata 109, 217, 219, 223, 224, 319, 320, 325, 492, 563
– ovata 183, 331
Carnarvonia (Prot.) 86
– araliifolia 173
Carnivorous plants 54
Carpentaria (Palm.)
– acuminata 180
Carpentaria Plains 125, 131
Carpha (Cyper.) 67, 91
– – Uncinnia – Oreobolis alliance 456
– alpina 71, 297, 447
– nivicola 459
Carpobrotus (Aizo.)
– aequilaterus 600
– edulis 600
– glaucescens 607, 608
– rossii 600, 602, 607
Carthamus (Compos.)
– lanatus 103, 330, 530
Cartonema (Cartonemat.)
– parviflorum 574
Cartonemataceae 48
Caryophyllaceae 45, 56
Cassia (Caesalpin.) 104, 108, 233
– artemisioides 321, 328, 335, 336, 350, 474, 476F, 477, 478, 483, 542
– circinata 466
– desertorum 482
– desolata 474, 483, 538
– glutinosa 535
– nemophila 336, 348, 358, 366, 367, 424, 474, 476F, 477, 478, 483, 491, 492, 542
– oligoclada 217
– oligophylla 469, 474
– phyllodinea 474, 486, 542
– planitiicola 107
– pleurocarpa 540
– sturtii 217, 348, 469, 474, 486, 538
Cassinia (Compos.)
– aculeata 263, 276, 285, 303, 312, 313, 336, 545
– laevis 313, 335
– longifolia 277, 279, 291, 334
– spectabilis 600
Cassytha (Cassyth.) 55
– glabella 409
– micrantha 430
– pubescens 232, 339
Cassythaceae 55
Castanospermum (Papil.)
– australe 166, 173, 194F, 234

Castanospora (Sapind.)
– alphandii 149
Casuarina (Casuarin.) 33, 76, 89
– acutivalvis 354, 355, 356, 357, 392, 425, 427, 428
– campestris 392, 425, 426F, 436
– chromosome numbers 89
– corniculata 392, 427, 428
– cristata 121, 184, 190, 320, 325, 328, 334, 357, 465, 466, 473, 483, 486, 488, 492, 507, 508, 559
– – alliance 489F
– cunninghamiana 89, 194F, 195, 228, 241, 571, 593, 594F
– decaisneana 364, 488, 491F, 532, 538
– – alliance 492
– decussata 370, 371
– dielsiana 392, 427
– distyla 313, 349, 391, 400, 402, 410, 411
– – alliance 411, 412F
– – – Jacksonia stackhousei alliance 408
– equisetifolia 601F, 606, 610F, 635
– fossil (Eocene) 60, 61, 89
– fraserana 371, 375, 380, 381, 382, 417, 418, 431F
– glauca 232, 233, 240, 581, 583F
– helmsii 392
– huegeliana 434, 436, 483
– humilis 359, 362, 376, 392, 419, 421, 423, 426, 430, 433
– inophloia 305, 306, 307, 311
– littoralis 228, 240, 245, 259, 271, 288, 289, 313, 391, 399
– – – Banksia aspleniifolia alliance 408, 409F
– luehmannii 303, 305, 307, 319, 326, 332, 334, 335F
– – alliance 487, 488F
– microstachya 392
– monilifera 261, 391, 402, 404
– muelleriana 261, 345, 391, 414
– nana 258, 391, 412F
– – alliance 411
– – – Leptospermum lanigerum alliance 452
– nitrogen-fixing nodules 89
– obesa 595, 631
– paludosa 391, 413
– pinaster 392
– polyploidy 89
– pusilla 259, 391, 405, 406, 413, 414, 602
– rigida 312, 314, 391
– striata 26
– stricta 278, 289, 313, 330, 337, 339, 600, 602, 608, 611, 612F
– torulosa 230, 231, 234, 235, 236, 237, 238, 239, 285, 287
Casuarinaceae 45, 89
Cauliflory 140
Caustis (Cyper.) 91
– blakei 246
– flexuosa 256, 288, 405
Cayratia (Vit.)
– clematidea 149

Ceiba (Bombac.)
– malabaricum 69
Celastraceae 45
Celastrophyllum 59
Celmisia (Compos.)
– longifolia 270, 271, 450F, 543, 544
– – – Poa alliance 451
Celtis (Ulm.)
– paniculata 149
– philippinensis 186
Cenarrhenes (Prot.) 87
– nitida 152, 264, 405, 454
Cenchrus (Gramin.) 511
– ciliaris 103
Centaurea (Compos.) 530
Centella (Umbell.)
– cordifolia 586
Centipida (Compos.)
– cunninghamii 490, 521, 555, 556
Centotheca (Gramin.) 511
Centrolepidaceae 48
Centrolepis (Centrolepid.) 407
– exserta 574
– fascicularis 264
– monogyna 446, 450
– polygyna 593
– strigosa 71
Cephalaralia (Aral.)
– cephalobotrys 146
Cephalomanes 34
Cephalotaceae 45
Cephalotus (Cephalot.) 54, 78, 571
Ceratodon purpureus 293
Ceratopetalum (Cunon.) 76
– apetalum 147, 156, 159F, 231, 241, 593
– – alliance 157
– – – Acmena – Tristania suballiance 157
– – – Argyrodendron suballiance 161
– – – Diploglottis suballiance 161
– – – Dorophora suballiance 157
– – – Schizomeria suballiance 161
– corymbosum 178
– gummiferum 76, 239
– succirubrum 172
Ceratophyllum (Ceratophyll.)
– demersum 574
Ceratopteris 36
– thalicroides 37
Cerbera (Apocyn.)
– inflata 171
Ceriops (Rhizophor.)
– decandra 614
– tagal 614, 623, 624, 625
– – alliance 622F
Chaetanthus (Restion.) 92
Chamaeraphis (Gramin.) 511
Chamelaucioideae 80
Chamelaucium (Myrt.) 81, 98, 105
– axillare 393
– megalopetalum 393
– pauciflorum 393
– virgatum 393
Channel Country 6, 520, 542, 549
Cheilanthes 36, 101

Cheilanthes
- distans 314, 410, 538, 569
- tenuifolia 37, 286, 305, 334, 350, 409, 410, 434, 538, 569
Cheirostylis (Orchid.) 93
Chenopodiaceae 45, 493, 494
Chenopodium (Chenopod.) 104, 493
- auricomum 525, 566, 568
- - alliance 555
- divaricata 517
- nitrariaceum 366, 503, 554 F, 556
Chilocarpus (Apocyn.)
- australis 146
Chiloglottis (Orchid.) 92, 98
- formicifera 71
Chionachne (Gramin.) 511
- cyathopoda 412, 524
- - alliance 520
Chionochloa (Gramin.) 511
- frigida 452
Chisocheton (Mel.) 70
Chlaenosciadium (Umbell.) 78
Chloanthaceae 45
Chloris (Gramin.) 104, 325, 469, 511
- acicularis 303, 305, 566
- pectinata 540
- pumilio 517
- ruderalis 528
- scariosa 517
- truncata 325, 332, 334, 335, 350, 467, 489, 501, 528, 530
- ventricosa 306, 324, 332, 333, 334, 335, 336
- virgata 564
Chondrilla (Compos.)
- juncea 330
Choretrum (Santal.) 55
- candollei 255
Choricarpia (Myrt.) 81, 82
Choriceras (Euphorb.)
- tricorne 399
Chorilaena (Rut.) 98
- quercifolia 371
Chorizandra (Cyper.) 91
- cymbaria 585
- enodis 585
- sphaerocephala 591
Chorizema (Papil.) 98
- aciculare 394
- cytisoides 394
- ilicifolia 371
Chrysanthemoides (Compos.)
- monilifera 409, 579
Chrysopogon (Gramin.) 217, 219, 511, 580
- fallax 189, 200, 210, 211, 215, 216, 303, 319, 329, 479, 487, 517, 523, 535
- latifolius 199
- pallidus 220, 223
Chthonocephalus (Compos.) 113
Cinnamomum (Laur.) 53, 70
- camphora 160
- oliveri 147
- propinquum 178
- virens 147
Cinnamomum flora 60, 61, 64, 65, 76

Cissus (Vit.)
- antarctica 149, 155
- hypoglauca 149, 163
- opaca 183
Citriobatus (Pittos.)
- pauciflorus 148
- spinescens 183
Citronella (Icacin.)
- moorei 147
Citrullus (Cucurb.)
- lanatus 103, 540, 542
Cladonia 434
Cladium (Cyper.)
- filum 548
- procerum 582
Claoxylon (Euphorb.)
- australe 146
Claytonia (Portulac.)
- australis 456
Cleistanthus (Euphorb.)
- cunninghamii 146
Cleistocalyx (Myrt.) 81, 83
- operculata 180 F
Cleistochloa (Gramin.) 511
- subjuncea 240, 464
Clematis (Ranunc.) 148
- aristata 233, 241, 286
- microphylla 381
- pubescens 371, 375
Cleome (Capparid.)
- viscosa 537
Clerodendrum (Verben.) 108
- floribundum 109
- tomentosum 149, 232, 532, 535
Clianthus (Papil.)
- formosus 120
Climate
- dew 18
- diagrams 21 F, 22
- drought 17
- factors 125
- fog 18
- frost 13 F
- isohyets 15 F
- precipitation 14–15 F
- rainfall extremes 17
- rainfall reliability 17
- seasonal patterns 15 F
- solar radiation 11, 12 F
- temperature 14
- types 20
- winds 19
Clostridium 33
Clusiaceae 45
Cochlospermaceae 45
Cochlospermum
- fraseri 180, 181, 186, 188 F, 207, 220, 223
- gillivrayi 185
Codonocarpus (Gyrostemon.) 108
- cotonifolius 109, 321, 328, 347, 348, 474, 478, 485, 535
Coelachne (Gramin.) 511
Coelebogyne (Euphorb.)
- ilicifolia 146
Coelorachis (Gramin.) 511
- rottboellioides 199, 210
Coleanthera (Epacrid.) 98

Colobanthus (Caryophyll.) 66
- affinis 71
- apetalus 71, 450
- benthamianus 450
Colocasia (Arac.)
- esculenta 576
Colysis 36
Combretaceae 45
Commelinaceae 48
Community(ies)
- arid 134
- classification 122, 123
- distribution 122, 130
- eastern coastal lowlands 131
- eastern highlands 132
- eastern inland lowlands 132
- patterning 123
- south-western 134
- structure 124
- Tasmania 133
- wetter tropics 131
Compositae 45, 56, 76, 321
Coniferales 40
Connariaceae 45
Conospermum (Prot.) 86, 98
- acerosum 935
- amoenum 395
- brachyphyllum 395
- caeruleum 395
- crassinervium 395
- densiflorum 395
- distichum 395
- ericifolium 255
- glumaceum 395
- incurvum 395
- leianthum 395
- nervosum 395
- stoechadis 395, 419, 421, 422 F, 432 F, 424
- taxifolium 239
- teretifolium 395
- triplinervium 395, 420 F
Conostephium (Epacrid.) 98, 637
- halmaturinum 637
Conostylis (Haemodor.) 91
- aculeata 421
- seorsiflora 430, 432
- setigera 376
Conothamnus (Myrt.) 81, 98
Continental drift 2
Convolvulaceae 45
Convolvulus (Convolvul.) 104
- erubescens 289, 332, 524
Coolibah = Eucalyptus microtheca
Copper 130
Coprosma (Rub.) 66
- billardieri 151
- pumila 71
- - - Colobanthus benthamianus alliance 450
- quadrifida 148, 153
Cordia (Boragin.)
- subcordata 635
Cornaceae 45
Correa (Rut.) 85, 98, 105
- alba 600, 602, 608, 636
- lawrenciana 291
- reflexa 286, 405

Correa (Rut.)
- rubra 339
Corybas (Orchid.) 92, 98
- abellianus 92
- dilatatus 92
Corymborchis (Orchid.) 93
Corynocarpaceae 45
Cosmelia (Epacrid.) 98
Cotula (Compos.)
- coronopifolia 581, 585, 586, 600, 630
- repens 581
Crab hole 30
Craspedia (Compos.) 113
- alpina 452
- chrysantha 333
- pleiocephala 531
- uniflora 71
Crassula (Crassul.) 104
- sieberana 413
Crassulaceae 45
Cratystylis 113
- conocephalus 348, 387, 389, 486, 504
- subspinescens 366, 556, 559
Crepidomanes 34
Cressa (Convolvul.) 104
- australis 628
Cretaceous 3, 10, 73
- angiosperms 58, 59
- Araucaria 59
- gymnosperms 58
- Leptospermum 59
- Metrosideros 59
- Microcachridites 59
- Nothofagus 59
- Podocarpus 59
- pteridophytes 58
Crinum (Amaryll.) 115
- flaccidum 121, 568
Crosslandia (Cyper.) 91
Crotalaria (Papil.) 104
- crassipes 523
- cunninghamii 105, 106 F, 437, 483, 540, 606
- dissitiflora 483, 540
Croton (Euphorb.)
- arnhemicus 207
- insularis 146, 172
- phebalioides 182, 183
- verreauxii 146
Crowea (Rut.) 98
- dentata 371
Cruciferae 45, 56, 500
Cryptandra (Rhamn.)
- amara 254, 306, 410, 411
- waterhousei 637
Crypthanthemis (Orchid.) 54
Cryptocarya (Laur.) 53
- angulata 172
- corrugata 172
- erythroxylon 147
- foveolata 147
- glaucescens 147
- hypoglauca 172
- laevigata 147
- mackinnoniana 172
- meissneri 147
- microneura 147

- obovata 147
- rigida 147, 172, 230
- triplinervis 147
Cryptostylis (Orchid.) 98
Ctenopteris 35
Cucurbitaceae 45
Cudrania (Mor.)
- javanensis 148
Culcita 35
- dubia 37, 38, 40, 230, 233, 234, 235, 237, 238, 287, 409, 411
Cumbunji = Typha domingensis
Cunoniaceae 46, 75, 76, 156
Cupanieidites 60
Cupaniopsis (Sapind.) 70, 237
- anacardioides 149, 232, 240, 607, 608
- - alliance 166
- serrata 149
Cuphonotus (Crucif.) 113
Cupressaceae 40, 76
Cuscuta (Cuscut.) 55
Cuscutaceae 46, 55
Cushion mosaics 450
Cushion plant 445
Cuttsia (Escallon.) 79
- viburnea 146
Cyanostegia (Verben.)
- bunnyana 534
Cyathea 35, 40, 145
- australis 38, 141, 158, 163 F, 230, 283, 291
Cyatheaceae 35
Cyathochaete (Cyper.) 91
Cyathodes (Epacrid.) 89
- abietina 600
- dealbata 453
- glauca 89
- parvifolia 269, 273, 295, 297
- straminea 454
Cycadaceae 40
Cycads 41
Cycas 40
- media 41 F, 199, 201
Cyclic salt 20
Cyclones 16, 19, 169
Cyclone Tracey 19
Cyclone Vine Forest 19
Cyclosorus 36
- gongylodes 101, 574
Cymbidium (Orchid.) 92
- canaliculatum 92, 114, 115 F, 183, 199, 332, 621
- suave 141
Cymbonotus (Compos.)
- lawsonianus 333
Cymbopogon (Gramin.) 511
- bombycinus 221, 223, 399, 536, 538, 539
- difformis 526
- exaltatus 324, 568, 569 F
- refractus 244, 245, 247, 280, 285, 302, 303, 306, 311, 330, 332, 401, 464
Cymodocea (Najad.)
- serrulata 613
Cynanchum (Asclepiad.) 108
- carnosum 621

Cynodon (Gramin.) 511
- dactylon 517, 521, 607, 630
Cynoglossum (Borag.) 104
- suaveolens 307
Cynometra (Caesal.)
- ramiflora 614
Cyperaceae 48, 75, 91
Cyperus (Cyper.) 52, 91, 103
- exaltatus 593
- gilesii 525
- gunnii 563
- irio 554
- lucidus 586
- polystachys 631
- ramosii 180
- retzii 517, 520, 526
- victoriensis 554, 568
Cyrtococcum 511
Cystopteris 36
Cyttaria
- gunnii 144
- septentrionalis 154

D

Dacrydium 40, 43, 58, 75, 142
- franklandii 151
Dactylis (Gramin.)
- glomerata 292
Dactyloctenium (Gramin.) 511
- radulans 120, 328, 517, 527, 530, 536, 540, 542
Damasonium (Alismat.)
- minus 585
Dampiera (Gooden.) 52, 103, 105
- alata 392
- juncea 392
- lavandulacea 392, 428
- oligophylla 392
- sacculata 392
- spicigera 392
- stricta 246, 312, 314, 413
- wellsiana 392
Danthonia (Gramin.) 307, 313, 330, 333, 350, 489, 511
- alpicola 448
- bipartita 321, 328, 474, 534, 536
- caespitosa 326, 336, 339, 489, 531, 593
- carphoides 337
- laevis 280, 285, 286, 289
- linkii 286
- nudiflora 271, 456, 543
- pallida 240, 287
- paradoxa 287
- pauciflora 447
- penicillata 337, 338
- pilosa 287
- purpurascens 332
- racemosa 271, 332
- semiannularis 325, 326, 339
- setacea 559
Daphnandra (Monim.) 53
- micrantha 148
Darlingia (Prot.) 86
- darlingiana 173
Darling R. 9

Darling Scarp (W.A.) 416
Darwinia (Myrt.) 81, 98, 105
- diosmoides 393, 435
- fascicularis 411, 412
- helichrysoides 393, 419
- meissneri 436
- neildiana 393
- vestita 393
- virescens 393
Dasypogon (Xanthorr.)
- hookeri 376
Daucus (Umbell.) 104
- glochidiatus 527, 531
Davallia 35
Davalliaceae 35
Davidsonia (Davidson.)
- pruriens 147
Daviesia (Papil.) 52, 85, 98, 103, 105, 314
- brevifolia 394
- colletioides 394
- croniana 394
- divaricata 430
- genistifolia 343
- horrida 394, 430, 431, 432
- incrassata 394
- juncea 394
- latifolia 261, 271, 280, 282
- obtusifolia 394
- pectinata 394
- rhombifolia 394
- teretifolia 394
- ulicifolia 247
Dawsonia superba 162F
Decaisnina (Loranth.) 55
Decaspermum (Myrt.) 81
Dendrobium (Orchid.) 52, 92, 93, 140
- aemulum 141
- dicuphum 180
- falcorostrum 141, 155
- linguiforme 411
- speciosum 411
- striolatum 411
- teretifolium 581
- undulatum 621
Dendrocnide (Urtic.)
- excelsa 149, 165F, 229
- photinophylla 149
Dendrophthoe (Loranth.) 55
Denhamia (Celastr.)
- pittosporoides 147
Dennstaedtia 35
- davallioides 235
Dennstaedtiaceae 35
Derris (Papil.) 85
- scandens 69, 147
- trifoliata 621
Desert
- Gibson 549
- Great Sandy 364, 379, 492, 549
- Little 414
- Simpson 104, 109, 467, 479, 537, 549
- Tanami 568
- Victoria 537, 549
Desmodium (Papil.)
- varians 287

Devil's Twine = Cassytha spp.
Deyeuxia (Gramin.) 511
- gunnii 459
Dianella (Lil.) 90
- caerulea 411, 605, 606, 607, 610
- longifolia 408
- revoluta 245, 345, 350
Dichanthium (Gramin.) 106, 215, 469, 511
- annulatum 523
- aristatum 523
- fecundum 187, 189, 320, 324, 523, 527
- humilis 528
- sericeum 325, 326, 328, 330, 332, 333, 465, 466, 490, 523, 526, 527, 529, 568
- - suballiance 524
- spp. alliance 522, 524F
- tenuiculum 523, 527
- - suballiance 523
Dichapetalaceae 46
Dichalachne (Gramin.) 511
- crinita 71, 548, 600, 602
- sciurea 71, 234, 280, 285, 286, 288, 332, 334
Dichondra (Convolvul.)
- repens 333, 581, 586
Dichopogon (Lil.)
- strictus 338
Dichosciadium (Umbell.) 77, 80
Dicksonia 35, 40, 142
- antarctica 37F, 151, 158, 274, 275, 278, 279, 291
Dicksoniaceae 35
Dicranium billardieri 450
Dicranopteris 34
Dicrastylis (Verben.) 105
Dictymia 36
Didymanthus (Chenopod.) 493
Didymocheton (Mel.)
- rufum 147
Didymoglossum 34
Dielsia (Restion.) 92
Digitaria (Gramin.) 104, 511
- brownii 336
- divaricatissima 350
Dillenia 53
Dilleniaceae 46, 53
Dillwynia (Papil.) 85, 98
- ericifolia 239, 256, 350, 406
- floribunda 313
- glaberrima 591
- retorta 282, 285, 311
- sericea 286
Dimeria (Gramin.) 511
Dimorphochloa (Gramin.) 511
Dioscorea (Dioscor.)
- transversa 150
Dioscoreaceae 48
Diospyros (Eben.)
- australis 147
- fasciculosa 147
- pentamera 147
Diplachne (Gramin.) 511
- fusca 517, 521, 554, 559, 568
Diplarrena (Irid.)
- moraea 546

Diplasium 35
Diplaspis (Umbell.) 77, 80
Diplatia (Loranth.) 55
Diploglottis (Sapind.)
- australis 149, 237
Diplolaena (Rut.) 98
- dampieri 487, 600, 638
Diplopogon (Gramin.) 511
Diplopterygium 34
Dipodium (Orchid.) 93, 98
Dipteridaceae 35
Dipteris 35
Dipterocarpaceae 170
Discaria (Rhamn.) 66
Dischidia (Asclep.) 140
- nummularia 621
Disclimax 123, 125
Diselma 40, 43
- archeri 453
Disjunction 57, 107, 343
Disphyma (Aizo.)
- australe 494, 559, 600, 602, 603
- blackii 602, 630
Dissiliaria (Euphorb.) 88
Distance dispersal 68, 78, 93
Distichlis (Gramin.) 511
- distichophylla 516, 517, 548, 592, 631
Diuris (Orchid.) 98
- filifolia 429
Dodonaea (Sapind.) 52, 103, 104
- adenophora 474
- attenuata 305, 307, 313, 321, 328, 366, 474, 479, 487, 492, 509
- boroniifolia 410
- cuneata 303, 334, 335
- inaequifolia 483
- oblongifolia 436
- oxyptera 224
- petiolaris 471
- triquetra 231, 232, 238, 239, 400, 401, 408, 409
- vestita 306
- viscosa 234, 237, 261, 278, 286, 292, 312, 326, 335, 336, 339, 350
Dolichandrone (Bignon.)
- heterophylla 188, 220, 319, 479
Donatia (Donat.) 63, 66
- novae-hollandiae 442, 446, 448F, 450
Donatiaceae 46
Donga 507
Doodia 35
- aspera 38, 236
Doryanthes (Agav.) 90
- excelsa 90F
Doryphora (Monim.) 53
- aromatica 173
- sassafras 148, 154, 155, 159F, 279
- - - Acacia suballiance 158
Drabastrum (Crucifer.) 113
Dracaena (Agav.)
- angustifolia 142F, 180
Dracophyllum (Epacrid.) 60, 98
- minimum 446, 450
- sayeri 178, 399
- secundum 411
Drakaea (Orchid.) 98

Drapetes (Thymel.) 66
Drimys (Winter.) 62, 67, 76
Drosera (Droser.) 37, 52, 55, 116, 411
- binata 591
- burmannii 569
- indica 569, 574
- pygmaea 591
- spathulata 591
Droseraceae 46, 55
Dryandra (Prot.) 52, 84, 86, 94, 98, 105
- arctotidis 395
- armata 362, 395
- bipinnatifida 395
- carlinoides 395
- cirsioides 360, 395
- cuneata 395
- erythrocephala 395
- floribunda 381, 395
- formosa 395
- fraseri 395
- kippistiana 395
- longifolia 352, 361
- mucronulata 395
- multiflora 420
- nivea 378, 379, 381, 430
- nobilis 380
- patens 396
- polycephala 396
- preissii 395
- pteridifolia 395, 435
- quercifolia 395, 426
- runcinata 395
- sclerophylla 396
- senecifolia 396
- sessilis 382, 395, 420, 421, 425, 432, 433
- shuttleworthiana 395, 396
- tridentifera 395, 422
Drymoanthus (Orchid.) 93
Drymophila (Lil.) 90
Drynaria 36, 140
- quercifolia 180
- rigidula 621
Dryopoa (Gramin.) 511
Dryopteris 36
Drypetes (Euphorb.) 70
- australasica 146
- - - Araucaria – Brachychiton – Flindersia alliance 164
Dry rainforest 184
Duboisa (Solan.) 108
- hopwoodii 109, 348, 355, 424
- myoporoides 150, 237
Dunaliella salina 551
Dunes 414, 416, 602
Duriala (Chenopod.) 493
Dust 19
Dysoxylum (Mel.) 60, 70
- decandrum 173
- fraseranum 147, 194
- muelleri 147
- oppositifolium 634
- pettigrewianum 173, 178
- schnifferi 170 F
Dysphania (Chenopod.) 113, 493

E

Ebenaceae 46
Ecdeiocolea (Restion.) 92
Echinochloa (Gramin.) 511
- turneriana 540, 542, 557
Echinopogon (Gramin.) 511
- caespitosus 234, 247, 270, 286, 289, 311
- ovatus 71, 332
Ectrosia (Gramin.) 511
Edaphic factors 126
Ehretia (Ehret.) 108
- acuminata 146
- membranifolia 183, 332
- saligna 192
Ehretiaceae 46
Ehrharta (Gramin.) 516
Eichhornia (Ponteder.)
- crassipes 512, 571, 574, 585
Elaeagnaceae 46
Elaeocarpaceae 46
Elaeocarpus (Elaeocarp.)
- bancroftii 172
- cyaneus 237
- ferruginiflorus 172
- foveolata 172
- grandis 146, 194
- holopetalus 153, 155, 278
- kirtonii 146
- largiflorens 172
- obovatus 146, 228
- reticulatus 146, 230, 232, 233, 235
Elaeodendron (Celastr.)
- australe 147, 183
Elaphoglossum 35
Elatinaceae 46
Elatine (Elatin.)
- gratioloides 552, 586
Elattostachys (Sapind.)
- nervosum 149
- xylocarpum 149, 184
Eleocharis (Cyper.) 91, 521
- acuta 582, 584 F, 585, 586
- dulcis 574
- equisetina 574, 585
- pallens 325, 521, 556
- - - alliance, 554, 555 F
- philippinensis 520
- sphacelata 519, 582, 583, 585, 586 F
- spiralis 519
Elionurus (Gramin.) 511
Elythranthera (Orchid.) 98
Elytranthe (Loranth.) 60
Elytrophorus (Gramin.) 511
- spicatus 520
Embelia (Myrsin.)
- australasica 148
Emex (Polygon.)
- australis 103
Emmenosperma (Rhamn.)
- alphitonioides 148
Enchylaena (Chenopod.) 493
- tomentosa 118, 120, 348, 350, 542, 569, 600, 602, 608
Endemism 78
- families 78
- genera 78, 79, 80

Endiandra (Laur.) 53, 70
- cowleyana 172
- discolor 147
- introrsa 147
- muelleri 147
- palmerstonii 172
- pubescens 147
- sieberi 147, 228, 230, 237, 240
- virens 147
Enhalus (Hydrocharit.)
- acoroides 611, 613
Enneapogon (Gramin.) 104, 349, 475, 511, 540
- avenaceus 321, 322, 475, 476 F, 486, 501, 504, 509, 521, 540
- - grassland 541
- pallidus 540
- planifolius 528
- polyphyllus 475, 501, 504, 508, 540
Entada (Mimos.)
- scandens 177
Enteropogon (Gramin.) 511
Entolasia (Gramin.) 511
- marginata 255
- stricta 235, 240, 288, 314
Epacridaceae 46, 75, 88, 89
Epacris (Epacrid.) 98
- breviflora 450
- - - Blindia robusta alliance 459
- impressa 259, 260, 343, 404, 405, 406, 408, 413, 591
- lanuginosa 404, 408, 591
- microphylla 257, 401, 402, 403, 408, 409, 411, 412, 413
- obtusifolia 401, 403, 408, 591
- paludosa 459
- - - Sphagnum cristatum alliance 459
- petrophila 453
- - - Veronica densiflora alliance 449
- pulchella 239, 405, 408, 411, 591
- serpillifolia 454, 459, 543
- - - Sphagnum Raised Bog 456
Epaltes (Compos.)
- australis 628
Ephedra 60, 76
Epiblema (Orchid.) 98
Epilobium (Onagr.)
- billardierianum 289
- confertifolium 450
Epipogium (Orchid.) 54, 93
Eragrostiella (Gramin.) 511
Eragrostis (Gramin.) 103, 104, 511
- australasica 500, 504, 512
- - alliance 521 F, 556
- brownii 234
- cilianensis 103
- dielsii 509, 531, 542, 556, 558, 559
- - alliance 528
- eriopoda 216, 321, 328, 366, 474, 479, 483, 536, 537
- japonica 103
- lacunaria 306, 309, 321, 322, 326, 336, 486, 488, 528, 531
- leptocarpa 525

Eragrostis (Gramin.)
- leptostachya 234, 285, 288, 332
- parviflora 324, 325, 326, 328, 336
- setifolia 467, 475, 501, 504, 521, 523, 527, 531, 568
- - alliance 528
- xerophila 483, 523, 527
- - alliance 528
Eremaea (Myrt.) 81, 98
- beaufortioides 393, 425
- fimbriata 417, 426
- pauciflora 393
Eremochloa (Gramin.) 511
Eremocitrus (Rut.) 85, 112
- glauca 113F, 121, 190, 325, 331, 332
Eremophila (Myopor.) 52, 80, 102, 103, 105
- abietina 474
- alternifolia 474
- bignoniflora 111, 112, 113F, 466, 469, 566
- bowmanii 474
- brownii 474
- chromosome numbers 112
- clarkei 484
- compacta 474
- crassifolia 611
- cuneifolia 474
- dielsiana 474
- duttonii 474, 568
- foliosissima 474
- forrestiana 474
- fraseri 474
- freelingii 474, 482, 538, 568
- gilesii 474
- glabra 307, 309, 474
- goodwinnii 474
- granitica 474
- latrobei 217, 367, 469, 474
- leucophylla 366, 474, 481F
- longifolia 309, 326, 328, 350, 474, 478, 489, 491, 492, 538
- macmillaniana 474
- maculata 467, 556
- - alliance 555, 556F
- maitlandii 484
- margarethae 474
- miniata 474, 492
- mitchellii 111, 112, 183, 184, 185, 190, 191, 307, 320, 321, 324, 326, 328, 331, 332, 336, 465, 466, 469, 474, 478, 490
- paisleyi 474
- platycalyx 474, 483
- polyclada 112, 465, 566
- pterocarpa 481F, 482
- scoparia 348, 358, 387, 388F, 389, 474, 486, 491, 492
- spathulata 474
- sturtii 321, 466, 474, 478, 491
Eremosynaceae 46
Eriachne (Gramin.) 511
- armittii 517
- burkittii 520
- ciliata 219
- mucronata 221, 302, 307, 367
- nervosa 526, 555
- obtusa 187, 213, 302

Ericaceae 46, 69, 88
Erichsenia (Papil.) 112
Erigeron (Compos.)
- floribundus 103
- pappochroma 452
Eriocaulaceae 48
Eriochilus (Orchid.) 98
Eriochlamys (Compos.) 113
- behrii 536
Eriochloa (Gramin.) 511
- crebra 527
- pseudoacrotricha 332, 529, 566
Erioglossum (Sapind.) 70
Eriostemon (Rut.) 85, 98, 105
- brucei 397
- buxifolius 600
- lanceolatus 401, 402
- coccineus 397
- spicatus 397
Erodiophyllum (Compos.) 113
Erodium (Geran.) 104
- cicutarium 103
- cygnorum 321, 326, 328, 501, 531
Eryngium (Umbell.) 104
- paniculatum 376
Erythranthera (Gramin.) 511
Erythrina (Papil.) 108, 635
- vespertilio 109, 117, 147, 188, 191
Erythrophleum (Caesalpin.)
- chlorostachys 188, 201, 202, 203, 207, 208, 210, 211, 215, 218, 299, 301, 319, 324, 464, 479, 481
- - alliance 186F
Erythroxylaceae 46
Escalloniaceae 46, 75, 76
Eucalyptus (Myrt.) 52, 60, 73, 80, 81, 82, 98, 102, 103, 130
- abergiana 196, 242
- acaciiformis 265, 267, 272, 284
- accedens 369, 378
- - suballiance 379F, 380
- acmenoides 196, 226, 227, 228, 232, 242, 244F, 284, 300
- agglomerata 228, 250, 282
- aggregata 265, 267
- alba 196, 199, 205, 227, 242
- - alliance 210, 212F
- albens 300, 316, 323
- - alliance 332, 333F
- albida 342, 361
- alpina 259, 265
- amplifolia 227, 265
- - suballiance 234
- amygdalina 258, 259, 274
- - alliance 261, 262F, 266
- anceps 342
- andrewsii 265, 266, 272, 284, 301
- - alliance 287
- angophoroides 228, 241
- angulosa 344
- angustissima 342
- annulata 342, 353
- anther types 83
- apiculata 250
- apodophylla 196, 198, 202, 205
- approximans 265
- aquilina 342, 362
- archeri 265, 293

- argillacea 196, 205, 217, 219F
- - - E. terminalis alliance 216, 218F
- arid zone 110
- aspera 196, 206, 216
- - suballiance 221
- astringens 368, 378
- - suballiance 380
- baeuerlenii 250, 265
- baileyana 243, 247
- - - E. planchoniana – E. bancroftii alliance 246, 248F
- bancroftii 246, 265, 284
- banksii 265, 284
- barberi 259
- bark types 84
- baueriana 227, 241, 265
- baxteri 258, 259, 274, 281, 282, 414
- - alliance 259, 261F
- behriana 342
- benthamii 227
- beyeri 300
- bicostata 291F, 265, 266, 274
- - - E. globulus alliance 293
- bigalerita 196, 210
- blakleyi 281, 284, 316, 328, 329F
- blaxlandii 250, 265, 282
- bleeseri 196, 199, 205
- bloxsomei 300
- bosistoana 227, 228, 282
- botryoides 226, 227, 228, 250, 259, 608
- - alliance 232
- brachyandra 196, 206
- brachycalyx 342
- brachycorys 342
- brassii 196
- brevifolia 196, 205, 216, 364, 469, 535, 568, 536, 568
- - alliance 223F
- - ssp. brevifolia 206, 226, 227F
- - ssp. confluens 206, 224
- bridgesiana 259, 275, 281, 284
- brockwayi 369, 389
- brownii 196, 316, 323
- - suballiance 324
- bupestrium 342, 361
- burdettiana 342, 362
- burracoppinensis 342
- - - Casuarina acutivalvis alliance 354F, 355
- caesia 342, 363, 436
- caleyi 265, 284, 301
- caliginosa 265, 266, 272, 275, 284
- - suballiance 285F
- calophylla 99, 361, 368, 369, 372, 374F, 375, 376, 377, 418, 433
- calycogona 342
- camaldulensis 9F, 105, 195, 196, 205, 565, 593
- - alliance 561, 562F, 563F
- cambageana 196, 466
- cameronii 274, 284, 287
- camfieldii 250, 411
- campanulata 227, 265, 266, 274, 284
- - alliance 287F

Eucalyptus (Myrt.)
- campaspe 369, 388
- camphora 265, 267
- capitellata 250, 252
- carnea 227, 228, 284, 300
- carnei 242, 342, 367
- celastroides 342, 369
- chloroclada 307 F, 306
- cinerea 259, 265
- – ssp. cephalocarpa 259, 260 F
- citriodora 196, 242, 244 F, 300, 464
- cladocalyx 317, 323
- – alliance 339
- clavigera 196, 199, 205, 207, 209 F
- clelandii 369
- cliftoniana 196, 205
- clinal variation 293
- cloëziana 242, 300
- – alliance 225, 226, 229 F
- cneorifolia 342
- coccifera 265, 266, 272, 441
- – – E. subcrenulata alliance 293, 294 F
- collina 196, 205, 216, 219
- comitae-vallis 342, 358
- concinna 342
- – alliance 358
- confertiflora 196, 199, 204 F, 205, 206, 208, 463, 575
- conglobata 342, 426
- conglomerata 391
- conica 316, 329
- consideniana 252, 259
- cooperana 342
- – alliance 352
- cordata 259
- cornuta 368, 369, 380
- – alliance 382
- coronata 342, 362
- corrugata 369
- cosmophylla 259
- crebra 196, 205, 227, 228, 242, 250, 281, 298, 299, 300, 301, 302, 464, 478
- – suballiance 302, 305 F
- crenulata 265
- crucis 342, 363, 436
- cullenii 196, 205, 208, 298, 300, 302
- – alliance 299
- cupularis 196, 206
- curtisii 242, 244
- cylindrocarpa 342
- cypellocarpa 250, 259, 260, 265, 266, 272, 274
- – – E. muellerana – E. maidenii alliance 282
- dalrympleana 259, 265, 266, 267, 271, 272, 274, 275, 284, 457
- – ssp. dalrympleana 272 F
- – ssp. heptantha 271, 272
- dawsonii 317, 323
- – alliance 336, 337 F
- dealbata 281, 284, 311 F, 300, 301, 302, 410
- deanei 167 F, 227, 231, 232, 250, 284
- debeuzevillei 267, 269
- decipiens 369, 373, 380, 597
- decorticans 228, 298, 300
- decurva 342, 361, 362
- delegatensis 259, 265, 266, 272, 274
- – – E. dalrympleana alliance 271, 272 F
- desmondensis 342, 362
- dichromophloia 187, 196, 198, 205, 206, 216, 218, 220 F, 300, 463, 469, 479, 480 F, 568, 612
- – alliance 217
- – – E. herbertiana – E. collina suballiance 219
- – – E. zygophylla – Acacia spp. suballiance 220, 221 F
- dielsii 342, 353
- diptera 342, 353
- diversicolor 368, 369, 434
- – alliance 370 F
- diversifolia 259, 342, 413
- – alliance 343, 344 F
- dives 265, 281
- dongarrensis 342, 357, 418, 638
- – – E. oraria alliance 358
- doratoxylon 342, 362
- drepanophylla 196, 242, 298, 300, 301, 302
- – – E. crebra alliance 299, 303 F
- dumosa 342, 345, 348 F
- dundasii 368, 369
- – suballiance 388, 389 F
- dunnii 227, 231
- dwyeri 301
- ebbanoensis 342, 357 F
- economic 82
- elata 226, 228, 593
- – alliance 241 F
- eremophila 342
- – alliance 352
- erythrocorys 368, 369, 421
- – alliance 381
- erythronema 342
- eudesmoides 342, 357, 361, 418, 425
- eugenioides 227, 228, 250, 252, 284, 300
- ewartiana 342
- eximia 250, 252
- – – E. punctata suballiance 255, 256 F
- exserta 196, 227, 228, 242, 244 F, 300, 464
- falcata 342, 369
- fasciculosa 274, 317, 323, 336
- fastigata 259, 265, 266, 274, 275
- ferruginea 196, 205, 206, 212, 213, 214 F, 220 F
- fibrosa 196, 228
- – ssp. fibrosa 298, 300, 316, 330
- – ssp. nubila 298, 312 F, 300, 301, 302
- – – – suballiance 313
- ficifolia 415, 431 F, 432
- fitzgeraldii 196, 205
- floctoniae 342, 369, 385, 389
- foecunda 342, 344, 353, 358, 361
- foelscheana 196, 204, 205, 206 F, 207, 208 F
- formanii 342
- forrestiana 342, 353
- fossils 83
- fraxinoides 265, 274, 282
- froggattii 342
- gamophylla 342, 364, 365 F, 473
- gardneri 342, 352, 369
- gilbertensis 196
- gillii 342, 366 F, 367
- glaucescens 265, 266, 267, 271
- glaucina 227, 233
- globoidea 227, 228, 250, 258, 259, 282
- – alliance 258, 260 F
- globulus 265, 266, 275, 292 F, 293
- gomphocephala 368, 369, 421, 608
- – alliance 380, 381 F
- gongylocarpa 342, 365, 366, 437, 492, 532, 538
- goniantha 342, 353, 361, 362
- goniocalyx 281
- gracilis 342, 345, 348 F, 353 F
- grandifolia 196, 199, 204, 205, 206, 575
- – suballiance 208, 211 F
- grandis 196, 225, 226, 227
- – alliance 229, 230 F
- grasbyi 342
- griffithsii 342, 369
- grossa 342, 362
- guilfoylei 369, 370, 372
- gummifera 99, 166, 225, 227, 228, 238 F, 243, 250, 252, 411, 608
- – – E. micrantha – E. sieberi alliance 247, 252
- – – E. racemosa – Angophora costata suballiance 253, 254 F
- gunnii 265, 266, 267, 272, 293, 457, 592
- – alliance 295 F, 296
- haemastoma 252
- haematoxylon 250, 369, 375 F
- herbertiana 196, 206, 216, 219
- houseana 196, 205
- howittiana 196, 464
- huberana suballiance 278 F
- hybrids 82
- incrassata 342, 353, 360, 426
- – – E. foecunda alliance 344, 345 F
- insularis 342, 362
- intermedia 196, 225, 227, 228, 233, 237, 243, 247, 300, 401, 402, 409, 607
- – – E. acmenoides – E. signata – E. nigra alliance 242, 243, 245 F
- intertexta 107, 301, 315, 316, 469, 478, 479, 568
- – alliance 320
- – – Acacia spp. suballiance 321 F
- – – Callitris glauca suballiance 322 F
- jacksonii 368, 369
- – alliance 371 F, 372
- jacobsiana 196, 205, 206

Eucalyptus (Myrt.)
- jensenii 196, 205, 208, 298
- johnstonii 265, 293
- jucunda 342, 357
- jutsonii 342, 357
- kingsmillii 342, 366, 473
- kochii 342
- kondininensis 369, 560 F
- kruseana 342, 363, 436
- kybeanensis 265, 267, 269, 271
- laeliae 369, 380
- laevopinea 265, 266, 274, 275, 284
- – – E. caliginosa – E. youmanii suballiance 284
- lane-poolei 369, 373
- lansdowniana 342
- largeana 227, 231
- largiflorens 469, 497, 499, 559, 563
- – alliance 567 F
- latifolia 196, 205
- – suballiance 209
- lehmannii 342, 362, 363, 369, 383
- leptocalyx 342, 361, 426
- leptophleba 196, 205, 300, 315
- – – E. microneura – E. normantonensis alliance 317, 318 F
- leptopoda 342, 355 F, 356, 361
- lesouefii 368, 369, 387 F, 388
- – – E. dundasii alliance 387
- leucoxylon 274, 317, 323
- ligulata 342, 362
- lingustrina 250, 265, 284
- lirata 196, 206
- longicornis 369, 385
- longifolia 227, 228
- loxophleba 342, 354, 368, 369, 436, 483
- – alliance 382, 383 F
- lucasii 342, 366, 367, 426, 568
- luehmanniana 250, 257 F, 411
- macarthurii 265
- macrocarpa 342, 361
- macrorhyncha 265, 275, 284, 300, 301
- – – E. rossii alliance 280, 281 F
- maculata 225, 226, 227, 228, 242, 300
- – alliance 239, 240 F
- maidenii 228, 265, 266, 282, 283
- major 227, 233
- mannensis 342
- mannifera 250, 252, 265, 267, 281
- – alliance 279
- – ssp. elliptica 280
- – ssp. maculosa 280
- – ssp. mannifera 280 F
- marginata 368, 369, 373 F, 375, 376, 597
- – – E. calophylla alliance 372
- mckieana 265, 284
- megacarpa 369
- – association 377 F
- megacornuta 342, 362
- melanophloia 196, 298, 300, 301, 302, 307, 478
- – alliance 306, 308 F
- melliodora 275, 284, 316, 323
- – – E. blakelyi alliance 328, 329 F
- merrickae 342
- michaeliana 265
- micrantha 342, 353 F
- microcorys 227, 229, 236, 274
- microneura 196, 315, 316, 319 F
- – suballiance 319
- microtheca 114, 189, 196, 205, 466, 469, 497, 499, 559, 563
- – alliance 565, 566 F
- miniata 196, 198, 200 F, 201 F, 205, 206, 213
- mitchelliana 265, 267, 269
- moluccana 227, 228, 300, 316, 332
- – alliance 330 F, 331 F
- mooreana 196, 206
- moorei 250, 265, 267
- – – E. glaucescens suballiance 271
- morrisbyi 259
- morrisii 107, 342
- muellerana 228, 265, 274, 275, 282
- neglecta 265
- nesophila 196, 205
- nicholii 265, 284
- nigra 242, 244 F
- niphophila 265, 266, 267, 439 F, 543
- – suballiance 8 F, 267, 270 F, 271
- nitens 265, 274
- nitida 258, 259, 264, 274
- – alliance 263 F
- normantonensis 196, 315, 316, 464
- – suballiance 319
- nortonii 265
- notabilis 250, 284
- nova-anglica 265, 266, 267, 275, 284
- – alliance 288 F
- nutans 342
- – – E. gardneri alliance 352
- obliqua 259, 265, 266, 272, 273 F, 275, 281, 282, 284
- – – E. fastigata alliance 274
- oblonga 250, 252
- obtusiflora 250
- occidentalis 358, 360, 368, 369, 426, 436
- – alliance 383, 384 F
- ochrophloia 565, 566
- odontocarpa 223, 342, 363 F, 364, 532
- odorata 317, 323
- – – E. porosa alliance 338 F
- oldfieldii 342, 355, 356 F, 361, 367, 422, 483
- oleosa 342, 345, 353, 354, 356, 367, 369, 382, 425, 483, 507, 568
- oligantha 196, 202, 205
- operculum 82
- oraria 342, 358, 638
- orbifolia 342
- oreades 250, 252, 256 F, 265, 282
- orgadophila 196, 300, 315, 316
- – alliance 320
- ovata 259, 265, 266, 267, 275
- – alliance 289 F
- ovularis 342, 358
- oxymitra 342
- pachyloma 342, 361
- pachyphylla 223, 342, 364, 492, 535
- panda 300
- paniculata 227, 228, 237, 242, 250
- papuana 196, 205, 213, 214, 300
- – alliance 214, 215 F
- parramattensis 227
- parvifolia 265, 267
- pastoralis 210
- patellaris 196, 205, 207, 210 F
- patens 369
- – association 377
- pauciflora 265, 266, 267, 272, 274, 275, 276 F, 284, 455
- – alliance 267, 268 F, 269 F
- pellita 196, 227, 228
- peltata 196, 300
- perfoliata 196, 206, 220
- perianth 83
- perriniana 265, 267, 271
- phaeotricha 196, 227, 228
- phoenicea 196, 198, 205, 206, 213
- – alliance 211
- – – E. ferruginea suballiance 212
- pileata 342
- pilligaensis 317
- – alliance 335, 336 F
- pilularis 158, 225, 226, 227, 228, 250, 252, 608, 609
- – alliance 234, 236 F
- – – E. microcorys suballiance 236
- – – E. intermedia – E. siderophloia suballiance 237
- – – E. saligna – E. paniculata suballiance 237
- – – E. gummifera suballiance 238 F
- – – E. piperita suballiance 239
- – – Angophora costata suballiance 239
- pimpiniana 342
- piperita 228, 239, 250, 252, 253, 258
- planchoniana 242, 246
- platypus 342, 351
- – alliance 351 F
- polyanthemos 259, 265, 281
- polybractea 342
- polycarpa 196, 197, 198, 199, 200, 201, 202 F, 205, 206, 227, 228, 300
- – – E. apodophylla suballiance 202
- – – E. tessellaris suballiance 202
- populnea 196, 300, 301, 316, 322, 323, 466, 469, 478, 479
- – – Acacia ssp. suballiance 327, 328 F
- – – Callitris glauca suballiance 325, 327 F
- – – Casuarina cristata suballiance 325
- – – Eremophila mitchellii suballiance 325, 326 F

Eucalyptus populnea
- – – Geijera parviflora – Flindersia maculosa suballiance 327
- – suballiance 324 F, 325
- porosa 317, 338
- porrecta 196, 205
- preissiana 342, 376, 436
- – alliance 362
- propinqua 227, 232
- pruinosa 196, 205, 206, 211, 216, 298
- – alliance 222 F
- ptychocarpa 196, 199, 205
- – alliance 209
- pulchella 259
- punctata 227, 228, 250, 252, 255, 300
- pyriformis 342, 356
- – – E. oldfieldii – E. leptopoda alliance 355
- pyrocarpa 227
- – suballiance 237
- quadrangulata 227
- racemosa 228, 243, 250, 252, 253, 254 F
- radiata 250, 252, 259, 265, 272, 274, 275, 282
- – alliance 279 F
- rameliana 342
- raveretiana 195, 196
- redunca 342, 361, 369
- – – E. uncinata alliance 350
- regnans 44, 84, 152, 265, 274
- – alliance 290 F
- remota 342
- resinifera 227, 228, 250
- – – E. acmenoides – E. propinqua alliance 232, 233 F
- rigidula 342
- risdonii 259
- robusta 225, 226, 228
- – alliance 240
- rodwayi 265, 267, 457
- – alliance 296 F, 297
- rossii 266, 280, 281 F
- roycei 342, 357
- rubida 259, 265, 266, 267, 276 F, 281
- rudis 369
- – alliance 564, 565 F
- rugosa 342
- rummeryi 228
- rupicola 250
- saligna 158, 163 F, 255, 226, 227, 228, 237, 250, 274, 284
- – alliance 229, 231 F
- salmonophloia 368, 369, 385 F, 497
- – – E. salubris alliance 384
- salubris 368, 369, 384, 386 F, 497
- sargentii 369, 560
- scoparia 265
- seeana 227
- – suballiance 234, 235 F
- sepulcralis 342, 362
- sessilis 342
- setosa 196, 205, 214, 300, 301, 304 F, 479, 480 F, 481
- sheathiana 342
- – – E. loxophleba – E. oleosa alliance 354
- shirleyi 196, 298, 300, 302
- – suballiance 308, 309 F
- siderophloia 227, 228, 237, 242, 298, 300
- sideroxylon 254, 259, 281, 284, 298, 300, 301
- – alliance 310, 311 F
- – – E. dealbata suballiance 312
- sieberi 228, 250, 252, 258, 259, 266, 282
- – – E. piperita – E. racemosa suballiance 253, 255 F
- signata 242, 243, 246 F
- similis 196, 300, 301, 302, 539
- smithii 265, 282
- socialis 107, 342
- – – E. dumosa – E. gracilis – E. oleosa alliance 345, 346 F, 347 F, 348 F, 349 F
- spathulata 342, 353, 361
- squamosa 250
- staeri 369, 373
- staigerana 196, 298, 300, 302
- stellulata 265, 266, 267, 275, 284
- – suballiance 269, 270 F
- stoatei 342, 362
- stowardii 342
- striaticalyx 342, 366, 473
- stricklandii 342, 369
- stricta 250, 252, 258 F, 411
- subcrenulata 265, 266, 272, 293, 294 F, 441
- subgenera 83, 197
- tall forests 225 (Chap. 9)
- tannin 378, 380
- tectifica 196, 199, 207 F, 208
- – – E. confertiflora – E. grandifolia alliance 204
- – – E. confertiflora – E. foelschiana suballiance
- tenuipes 196, 300
- tenuiramis 259, 404
- tereticornis 185, 196, 205, 227, 228, 241, 242, 250, 410, 593
- – alliance 233
- terminalis 196, 205, 206, 216, 217, 218 F, 469, 568
- tessellaris 196, 198, 202, 203 F, 205, 227, 228, 242, 300, 607
- tetragona 342, 358 F, 362, 376, 421, 426
- – – E. incrassata alliance 359 F, 360
- tetrapleura 228
- tetraptera 342, 360 F, 361
- tetrodonta 196, 198, 201, 205, 213, 300, 399
- – – E. miniata – E. polycarpa alliance 197
- thozetiana 107, 300
- todtiana 361, 376, 418, 419 F
- torelliana 196, 203, 205, 229
- torquata 369, 388 F
- trachyphloia 196, 228, 242, 300, 301, 312 F, 410
- transcontinentalis 369, 385
- triflora 250
- trivalvis 342, 367
- tropical alliances 197
- umbra 250, 252
- umbrawarrensis 196
- uncinata 342, 350, 361
- urnigera 265
- vernicosa 265, 293, 452
- viminalis 228, 241, 259, 265, 266, 267, 272, 274, 277 F, 282, 284
- – – E. rubida alliance 275
- viridis 342
- – alliance 349, 350 F
- wandoo 368, 369, 378 F, 421
- – – E. accedens – E. astringens alliance 378
- watsoniana 300
- whitei 196, 298, 300
- woollsiana 301, 316, 323, 488, 489
- – alliance 334 F
- – ssp. microcarpa 316, 334
- – ssp. woollsiana 316, 335
- yarraensis 265
- youmanii 265, 284
- – suballiance 286 F
- zygophylla 196, 206, 216

Eucarya (Santal.) 55, 115
- acuminata 119, 313, 347, 354, 355, 387, 491
- spicata 385, 386

Euchylopsis (Papil.) 98
Eucla Basin 9, 10
Eucryphia (Eucryph.) 63, 66
- lucida 151, 295
- millinganii 151
- moorei 154 F, 158
- sp. 178

Eucryphiaceae 46, 75
Eugenia (Myrt.)
- bleeseri 207
- gustavioides 113
- hemilampra 173
- kuranda 173
- spp. 173
- suborbicularis 180, 195
- striata 451

Eulalia (Gramin.) 511
- fulva 320, 324, 325, 475, 479, 504, 525, 527, 528, 542, 555, 568

Eulophia (Orchid.) 93
Euodia (Rut.)
- bonwickii 173
- micrococca 147
- vitiflora 173

Euphorbia (Euphorb.) 104
- atoto 515, 606, 609
- australis 536
- drummondii 524, 537, 540
- eremophila 537
- myrtoides 607
- paralias 608
- sparrmannii 607
- wheeleri 540

Euphorbiaceae 46, 88
Euphrasia (Scroph.) 55
- striata 451

Eupomatia (Eupomat.) 53
- laurina 54 F, 147, 237
Eupomatiaceae 46, 53
Eurabbie = Eucalyptus bicostata
Euroschinus (Anacard.)
- falcatus 146, 172, 230
Eustrephus (Philes.)
- latifolius 150, 238, 287
Eutaxia (Papil.) 98
- microphylla 489
- obovata 436
Evandra (Cyper.) 91
- - Anarthria - Lyginia spp. alliance 595
- aristata 596
Ewartia (Compos.) 80
- meridithae 442 F, 445, 450
Excaecaria (Euphorb.)
- agallocha 614, 622, 625
- dallachyana 146, 184
- latifolia alliance 189 F
- parvifolia 223, 566
Exocarpos (Santal.) 55, 115
- aphyllus 185, 191, 344, 347, 491
- cupressiformis 231, 233, 235, 238, 239, 240, 260, 261, 262, 271, 274, 279, 281, 287, 304, 313, 330
- humifusus 445, 452
- latifolius 149, 183
- nanus 546
- sparteus 487, 492
Exocarya (Cyper.) 91

F

Fagaceae 46
Fan-delta 551
Faradaya (Verben.)
- splendida 177
Feldmark 447
Fellfield 446, 447
Fen 124, 455, 456
Fenzlia (Myrt.) 81
- - Melaleuca - Leptospermum - Sinoga alliance 398
- obtusa 399
Fern Forest 136
- Thicket 136
Ferns. See Pteridophytes
Festuca (Gramin.) 511
- littoralis 516, 607, 608
- muelleri 456
Ficus (Mor.) 108, 140, 215, 234
- benjamina 180
- coronata 148, 240, 562
- coronulata 180
- destruens 173
- eugenioides 140 F
- henneana 148
- lacor 180
- macrophylla 148
- obliqua 148, 228
- opposita 186, 635
- platypoda 109, 110 F, 532, 538, 569
- pleurocarpa 173
- racemosa 180, 195, 562
- rubiginosa 286
- watkinsiana 148, 173

Fieldia (Gesner.)
- australis 140, 141 F, 147
Fig, Port Jackson = Ficus rubiginosa
Fimbristylis (Cyper.) 52, 91, 574
- dichotoma 211, 244, 246
- littoralis 517
- microcarpa 564
- miliacea 564
- polytrichoides 517
Fire 32, 138, 232, 239, 292, 293, 372, 376, 420, 479, 480, 533, 636
Fireweed = Senecio minimus
Fissistigma (Annon.) 53
Fitzalania (Annon.) 53
Fjaeldmark 447
Flacourtiaceae 46
Flagellaria (Flagellar.)
- indica 177, 180, 181, 195, 241
Flagellariaceae 48
Flindersia (Rut.) 70, 108, 168
- acuminata 174
- australis 149, 169, 184
- bennettiana 149
- brayleyana 174
- collina 149
- dissosperma 332
- laevicarpa 174
- maculosa 190, 321, 327, 328, 466, 469, 490
- pimenteliana 174
- schottiana 149, 193 F, 237
- unifoliolata 178
- xanthoxyla 149
Flinders Range (S.A.) 73
Flood waters 17, 549
Floristic units 123
Flowering times 131
Floydia (Prot.) 87
Forbland 124
Forest 124
Forstera (Stylid.) 63, 66
- bellidifolia 456
Fossil
- angiosperms 60
- gymnosperms 60
- laterite 23, 29, 31
- mangroves 61
- Tertiary 60
Fowl, Jungle 181
Frankenia (Franken.) 103, 104, 105
- cinerea 556, 559
- cordata 559
- fecunda 559
- pauciflora 388, 502, 505, 558 F, 600, 602, 628, 629, 630
- serpyllifolia 503, 559
- spp. alliance 559
Frankeniaceae 46, 111
Franklandia (Prot.) 86, 98
Freycinetia (Pandan.) 178
Fuchsia (Onagr.) 62
- Poison = Eremophila maculata
Funaria hygrometrica 293, 372
Fusanus = Eucarya

G

Gahnia (Cyper.) 91
- clarkei 247, 591
- filum 630, 631
- grandis 240, 586
- psittacorum, 264, 273, 291, 297, 407
- radula 591
- sieberana 402, 405
- trifida 289, 549, 589, 591
- - - G. filum alliance 592
Gaimardia (Centrolepid.) 63, 67
- fitzgeraldii 456
Galeola (Orchid.) 54, 93, 98
Garcinia (Guttif.)
- brassii 178
Gardenia (Rub.) 213
- megasperma 207
- pantonii 606
Garnotia (Gramin.) 511
Gastrodia (Orchid.) 54, 93, 98
Gastrolobium (Papil.) 98, 105
- bidens 395
- calycinum 379 F, 380, 395
- oxylobioides 395
- pycnostachyum 395
- spinosum 395
Gaultheria (Eric.) 66, 85
- appressa 153, 268
- depressa 71
- hispida 295
Geijera (Rut.) 70, 108
- linearifolia 339, 347
- paniculata 184
- parviflora 182, 183, 185, 190 F, 315, 321, 325, 326, 327, 328, 332, 336, 466, 478, 490
- - - Flindersia-Heterodendrum alliance 190
- salicifolia 149, 320
Geissois (Cunon.)
- benthamii 147
- biagiana 171
Geitonoplesium (Philes.)
- cymosum 150, 232, 285
Genoplesium (Orchid.) 98
Gentiana (Gentian.)
- diamensis 451
Gentianaceae 46
Geodorum (Orchid.) 93
- pictum 180
Geological time-scale 2
Geology 7
Geomorphology 7
Geraniaceae 46
Geranium (Geran.)
- potentilloides 289
Germainia (Gramin.) 511
Gesneriaceae 46
Gevuina (Prot.) 63, 66, 86
Gibbers 29, 100, 498
Gidgee = Acacia cambagei or A. xiphophylla
Gilesia (Stercul.) 112
Gilgai 30, 466, 489, 491, 499, 501,, 502, 568
Gimlet, Common = Eucalyptus salubris

673

Ginkgoales 57
Glaciation 10, 440
Glasshouse Mountains (Qld) 10 F, 409
Gleichenia 34, 37
– alpina 406, 407, 456, 457 F
– dicarpa 37, 38, 39 F, 591
– microphylla 411
Gleicheniaceae 34
Glochidion (Euphorb.) 70
– ferdinandii 146, 193 F, 228, 230, 232, 235
Gloeocapsa 411, 413
Glossodia (Orchid.) 98
– emarginata 431
– major 92 F
Glyceria 511
– australis 585
Glycine (Papil.)
– clandestina 287
– sericea 559
– tabacina 270, 332, 334
Glycosmis (Rut.) 70
Glycyrrhiza (Papil.) 104
– acanthocarpa 559
Gmelina (Verben.)
– dalrympleana 180
– leichardtii 149, 169, 237
Gnaphalium (Compos.) 104
– luteo-album 559
Gnaphalodes (Compos.) 113
Gnephosis (Compos.) 113
– brevifolia 505
– skirrophora 508
Gompholobium (Papil.) 85, 98
– baxteri 395
– huegelii 404
– latifolium 239
– marginatum 395
– polymorphum 395
Gondwanaland 1, 2, 3, 4, 57
Gonocormus 34
Goodenia (Gooden.) 52, 103
– bellidifolia 413
– concinna 393
– filiformis 559
– glauca 553
– hederacea 247, 286
– pinifolia 393
– pinnatifida 508
– pterigosperma 393
– strophiolata 393
Goodeniaceae 46
Goodia (Papil.) 85, 98
Goodyera (Orchid.) 93
Goosefoot, Golden = Chenopodium auricomum
Gramineae 48, 56, 510, 511
– aquatic 513
– genera listed 511
– habits 512
Grammitidaceae 35
Grammitis 35
Granite 23
Grass
– Barley Mitchell = Astrebla pectinata
– Beetle = Diplacne fusca
– Big Horny = Poa costiniana
– Blue = Dichanthium spp.
– Bull Mitchell = Astrebla squarrosa
– Cane = Leptochloa digitata
– Coastal Cane = Chionachne cyathopoda
– Couch = Cynodon dactylon
– Curly Mitchell = Astrebla lappacea
– Cutting = Gahnia trifida
– Hoop Mitchell = Astrebla elymoides
– Horny = Poa fawcettiae
– Inland Cane = Eragrostis australasica
– Iron = Lomandra dura or L. effusum
– Kerosene = Aristida browniana or A. contorta
– Ledge = Poa hothamensis
– Mitchell = Astrebla spp.
– Mulka = Eragrostis dielsii
– Plains = Stipa aristiglumis
– Porcupine = Triodia irritans
– Sandhill Cane = Zygochloa paradoxa
– Salt Couch = Sporobolus virginicus
– Snow = Poa hiemata or P. labillardieri
– Spear = Stipa spp.
– Thatching = Gahnia filum
Grassland(s) 123, 235, 446, 452, 510 (Chap. 20)
– aquatic 515, 517
– arid zone 515, 532
– Astrebla 524
– disclimax 530
– distribution 514, 522, 523 F
– ephemeral 540
– habitat factors 515
– hummock 515
– littoral 515, 628
– mat 515
– Poa 542
– – Sedgeland communities 574
– "Spinifex" 533
– Stipa 528
– tussock 512
Gratiola (Scrophular.)
– latifolia 586
– pedunculata 585
Great Divide 6, 549
Great Soil Groups 31
Grevillea (Prot.) 52, 85, 86, 98, 105, 207, 220, 479
– acanthifolia 460
– acerosa 396
– agrifolia 213
– albiflora 309, 310 F
– alpiciloba 396
– angulata 213
– annulifera 396, 422
– asparagoides 396
– australis 270, 443, 450 F, 453
– biformis 396, 422
– concinna 427
– crithmifolia 396
– cunninghamii 213
– didymobotrya 427, 428
– dielsiana 396, 423
– drysandra 213
– eriostachya 356, 396, 421, 422, 425, 481, 483
– eryngioides 396
– excelsior 354, 396, 427, 428
– fasciculata 396
– floribunda 313
– glauca 201
– gordoniana 396, 424
– haplantha 396
– heliosperma 211 F
– hilliana 148
– hookerana 361, 396, 426, 428
– huegelii 348, 487
– integrifolia 396, 427
– juncifolia 307, 309, 364, 474, 492
– juniperina 254
– lanigera 281, 313
– leucopteris 396, 422
– longistyla 305
– nematophylla 508
– nudiflora 396
– obliquistigma 396
– paniculata 396
– parallela 324
– pauciflora 396
– pinnatifida 173
– pteridifolia 201, 210, 211 F, 213, 223, 399, 576
– pterosperma 309, 396, 427
– pulchella 396
– pungens 213
– punicea 257
– robusta 166, 194 F, 593
– rogersiana 396
– rudis 396
– rufa 396, 427
– saccata 396
– sericea 287, 402
– shuttleworthiana 369
– stenobotrya 396, 425, 437, 483
– stenomera 421
– striata 111, 188, 192 F, 193, 222 F, 223, 301, 321, 328, 332, 364, 469, 427 F, 473, 478, 489
– tenuiflora 396
– teretifolia 396
– thelemanniana 381, 396, 426
– thyrsoides 396
– tridentifera 396
– triternata 314
– vestita 381, 396
– wickhamii 214 F, 224
Grey billy 23
Grimmia laevigata 413
Growth régimes 20
Guano 636
Guilfoylia (Simaroub.)
– monostylis 150
Guioa (Sapind.) 70
– semiglauca 149
Gulubia (Palm.)
– costata 169
Gum
– Alpine Snow = Eucalyptus niphophila

Gum
- Bendemeer White = E. mannifera ssp. elliptica
- Black = E. ovata
- Blakely's Red = E. blakelyi
- Brittle = E. mannifera
- Cabbage = E. amplifolia
- Cider = E. gunnii
- Coolabah = E. intertexta
- Flooded = E. grandis (east), E. rudis (W.A.)
- Forest Red = E. tereticornis
- Ghost = E. papuana (Central Aus.) or E. laeliae (W.A.)
- Grey = E. punctata
- Kalumburu = E. herbertiana
- Kopi = E. striaticalyx
- Large-leaved Cabbage = E. grandifolia
- Maiden's = E. maidenii
- Manna = E. viminalis
- Marble = E. gongylocarpa
- Monkey = E. cypellocarpa
- Mountain = E. cypellocarpa (east) or E. haematoxylon (W.A.)
- Mountain White = E. dalrympleana
- Northern White = E. brevifolia
- Orange = E. bancroftii
- Pink = E. fasciculosa
- Poplar = E. alba
- Ribbon = E. viminalis
- River Red = E. camaldulensis
- Rough-barked Manna = E. huberana
- Salmon = E. salmonophloia
- Scarlet = E. phoenicea
- Scribbly = E. haemastoma, E. racemosa or E. rossii
- Slaty = E. dawsonii
- Small-fruited Grey = E. propinqua
- Snappy = E. brevifolia
- Snow = E. pauciflora
- Spotted = E. maculata
- Sugar = E. cladocalyx
- Swamp = E. ovata or E. regnans (Tas.)
- Sydney Blue = E. saligna
- Tasmanian Blue = E. globulus
- Tasmanian Snow = E. coccifera or E. subcrenulata
- Tingiringi = E. glaucescens
- Tumbledown Red = E. dealbata
- Victorian Blue = E. bicostata
- Weeping = E. sepulcralis
- White = E. mannifera
- Yellow = E. leucoxylon
- York = E. loxophleba

Gunnera (Halorag.) 66
- cordifolia 456

Gunngunnu = Eucalyptus caesia
Gymnanthera (Asclepiad.)
- nitida 621

Gymnogramma 35
- reynoldsii 101, 569,

Gymnogrammaceae 35
Gymnoschoenus (Cyper.) 600
- sphaerocephalus 263, 297, 408, 586, 591

- - alliance 588, 592F

Gymnosperms 40, 41
- habitats 41
- naturalized 44

Gymnostoma (Casuarin.) 89
Gypsum 129
Gyrocarpus (Combret.) 53
- americanus 181, 185, 186, 189, 219, 479

Gyrostemon (Gyrostemon.)
- ramulosus 437
- sheathii 432
- tepperi 492

Gyrostemonaceae 46

H

Habenaria (Orchid.) 93
Hackelochloa (Gramin.) 511
Haemodoraceae 48, 91
Haemodorum (Haemodor.) 91
- planifolium 246, 247

Hakea (Prot.) 52, 86, 98, 103, 105, 110, 220
- acacioides 537
- adnata 396, 426, 427
- ambigua 396
- arborescens 219, 223, 301
- auriculata 396, 425
- baxteri 396, 422
- bipinnatifida 396
- buculenta 356
- candollei 396
- ceratophylla 396
- cinerea 361, 396
- clavata 436
- conchifolia 396, 423
- corymbosa 359, 396
- costata 382
- crassifolia 359, 360, 362, 396
- cucullata 396
- cycloptera 339, 345
- dactyloides 244, 246, 255, 256, 271, 413
- eriantha 274
- falcata 396, 427
- ferruginea 396
- intermedia 473, 479
- ivoryi 474, 537F, 538
- laurina 362, 396, 436
- lehmanniana 396
- leucoptera 121, 326, 473, 478, 560
- lissocarpa 297, 396, 456
- lorea 223, 301, 303F, 366, 469, 473, 535, 537
- macrocarpa 535, 537
- marginata 435
- megalosperma 219
- microcarpa 297, 456, 545F, 547
- multilineata 366, 396, 425, 427, 428
- neurophylla 419
- nitida 396
- obliqua 396
- oleifolia 432
- pandanicarpa 396
- platysperma 396, 426

- prostrata 362, 379, 381, 396, 430, 434, 435
- purpurea 307
- pycnoneura 425
- rhombalis 366
- roei 396
- rostrata 413
- rugosa 413
- ruscifolia 375, 396
- salicifolia 238
- scoparia 391
- sericea 254, 257
- smilacifolia 396
- stenophylla 396, 424
- strumosa 396
- subsulcata 396
- sulcata 396
- teretifolia 257, 402, 404, 408, 409, 411, 412, 413, 591
- trifurcata 396, 421
- ulicina 343
- undulata 396
- varia 396, 595

Halfordia (Rut.) 70
- kendack 149
- scleroxyla 174

Halgania (Borag.)
- littoralis 534

Halodule (Zanichell.)
- uninervis 613

Halophila (Hydrocharit.)
- ovalis 611, 613
- spinulosa 613

Halophytes 111, 127
Halophytic Shrublands 493 (Chap. 19)
Haloragaceae 46
Haloragis (Halorag.) 52, 66, 76, 110
- brownii 586
- gossii 537
- micrantha 71, 240, 591
- tetragyna 404
- teucrioides 232, 255

Hamamelidaceae 46
Hammersley Range (W.A.) 366
Hannafordia (Stercul.) 112
Haplostichanthus (Annon.) 53
Hardenbergia (Papil.) 98
- comptoniana 381, 433
- violacea 239, 245, 272, 287, 338, 371

Harmsiodoxa (Crucif.) 113
Harperia (Restion.) 92
Harpullia (Sapind.)
- pendula 149

Heath(s) 131, 390 (Chap. 16)
- alpine 452
- arid zone 436
- coastal (east) 391, 400
- coastal (west) 416, 426
- defined 390
- flora 398, 416
- Glasshouse Mts (Qld) 391, 409
- granite (Vic.) 391, 413
- habitats 391
- headland (east) 408
- headland (west) 430, 435
- hills (W.A.) 434

Heath(s)
- inland 427, 436
- laterite 421
- limestone 420
- location 390, 391
- mallee 341, 358 F, 359, 391
- Northern Territory 399
- rock exposures (east) 409
- rock exposures (west) 436
- sandplain (W.A.) 421
- sandstones (N.S.W.) 391, 411
- scrub 418, 427
- sedge 407
- soils 416
- South Aust. 413
- south-eastern 400
- sandstones (east) 411
- tropical 391, 398
- Warrumbungle Mts (N.S.W.) 391, 410
- Western Aust. 414, 415 F
Heavy metals 130
Hedeia 57
Hedycarya (Monim.) 53
- angustifolia 148, 153
Helichrysum (Compos.) 52, 103
- apiculatum 280, 286, 307, 409
- argophyllum 240
- baccaroides 295, 453
- backhousei 453
- baxteri 413
- bracteatum 450
- costatifructum 600
- diosmifolium 235, 239, 240, 287, 311, 336
- hookeri 268, 453
- incanum 452
- paralium 608
- reticulatum 600, 602
- scorpioides 337
- semipapposum 332, 350
Helicia (Prot.) 63, 86
- youngiana 148
Heliotropium (Borag.) 104
Helipterum (Compos.) 52, 103, 527, 528, 530, 537
- acuminatum 546
- corymbiflorum 531
- anthemoides 546
- floribundum 531, 541
- moschatum 531
Helleboraceae 46
Helminthostachys 34
Hemarthria (Gramin.) 511
- uncinata 520
Hemichroa (Amaranth.)
- diandra 628
- pentandra 630
Henslowia (Santal.) 55
Hensmannia (Lil.) 90
Herbfield 124
Heritiera (Stercul.)
- littoralis 614, 622, 624
Hernandia 53
Hernandiaceae 46, 53
Herpolirion (Lil.)
- novae-zelandiae 71
Heterachne (Gramin.) 511

Heterodendrum (Sapind.) 108
- diversifolium 182, 183, 465, 466
- floribundum 311, 332
- oleifolium 109, 121, 183, 184, 185, 191, 321, 325, 326, 328, 332, 336, 346, 348, 466, 469, 473, 482, 486, 487, 490 F, 492
Heteropogon (Gramin.) 511
- contortus 208, 209, 211, 215, 224, 234, 302, 303, 307, 308, 309, 320, 325, 332, 523, 524
- triticeus 199, 200, 208, 209, 299, 301, 520
Heterozostera (Zoster.)
- tasmanica 613
Hibbertia (Dillen.) 52, 53
- acerosa 392
- acicularis 337, 404
- amplexicaulis 371
- aurea 392
- conspicua 392
- cuneiformis 435
- dentata 237
- fascicularis 637
- huegelii 392
- hypericoides 378, 380, 381, 382, 392, 417
- linearis 239, 282, 330
- montana 371
- obtusifolia 311
- polystachya 392
- racemosa 381
- scandens 235, 579, 607
- sericea 259, 337, 338, 344
- serpyllifolia 413
- spicata 392
- stricta 272, 282, 306, 338, 404, 414
- tetrandra 371
- tomentosa 392
- uncinata 392
Hibiscus (Malv.) 108
- brachysiphonius 527
- diversifolius 574
- heterophyllus 235
- huegelii 382
- pinonianus 483
- tiliaceus 181, 195, 606, 607, 623, 635
- timorensis 181
- trionum 528
- zonatus 213
Hicksbeachia (Prot.) 86
- pinnatifolia 148
Hierochloë (Gramin.) 511
- pinnatifolia 148
Himantandra 53
Himantandraceae 46, 53
Hippocrateaceae 46
Histiopteris 35, 101
- incisa 293
Hodgkinsonia (Rub.)
- ovatifolia 148
Hohera (Malv.) 60
Holcus (Gramin.)
- lanatus 269
Hollandaea (Prot.) 86
Homalanthus (Euphorb.)

Homalocalyx (Myrt.) 81
Homalosciadium (Umbell.) 78
Homopholis (Gramin.) 511
Homoranthus (Myrt.) 81
Hoop Pine scrub 137
Hopkinsia (Restion.) 92
Hordeum (Gramin.)
- leporinum 103, 467, 530, 559
- murinum 330
Horizontal = Anodopetalum biglandulosum
Horsfieldia (Myristic.) 53
Hovea (Papil.) 85, 98
- chorizemifolia 371
- elliptica 371, 375
- longipes 307, 336, 543
- ovalifolia 372
Hoya (Asclep.)
- australis 141
Humata 35
Hybanthus (Viol.)
- calycinus 382, 430
Hydnophytum (Rub.) 140
- formicarum 621
Hydrilla (Hydrocharit.)
- verticillata 574, 585
Hydrocharitaceae 48
Hydrocharis (Hydrocharit.)
- dubia 571
Hydrocotyle (Umbell.)
- laxiflora 289
- peduncularis 286
- vulgaris 581
Hydrophyllaceae 46
Hymenachne (Gramin.) 511
- acutigluma 519
Hymenophyllaceae 34
Hymenophyllum 34
Hymenosporum
- flavum 148
Hypericaceae 46
Hyparrhenia (Gramin.) 511
- hirta 333
Hypocalymma (Myrt.) 81, 98
- angustifolia 375, 379, 430
- myrtifolia 436
- tetrapterum 392
Hypochoeris (Compos.)
- radicata 548
Hypolaena (Restion.)
- exsulca 596
- fastigiata 252, 254, 260, 401, 405, 408, 591
- lateriflora 71
Hypolepis 35
- rugulosa 293
Hypoxidaceae 48
Hypoxis (Hypoxid.)
- glabella 244, 246, 286, 564
Hypsela (Campanul.) 66

I

Icacinaceae 46
Ichnanthus (Gramin.) 511
Idiospermaceae 46, 53
Idiospermum 53
- australiense 54 F, 172

Ilex (Aquifol.) 59
Iliciphyllum 62
Illecebraceae 46
Illyarrie = Eucalyptus erythrocorys
Ilmenite 400, 603
Impeded drainage 23
Imperata (Gramin.) 511
- cylindrica 211, 228, 229, 231, 233, 234, 235, 237, 239, 240, 246, 248, 287, 401, 513, 519, 575, 586, 605, 610
Indigofera (Papil.) 104
- australis 236, 287
- dominii 540
Inula (Compos.)
- graveolens 556
Ipomoea (Convolvul.)
- pes-caprae 515, 516, 605, 606, 635
Iridaceae 48, 56
Ironbark 84, 298
- Blue-leaved = Eucalyptus fibrosa ssp. nubila
- Cullen's = E. cullenii
- forests and woodlands 298
- Grey = E. paniculata or E. siderophloia
- Narrow-leaved = E. crebra or drepanophylla
- Shirley's Silver-leaved = E. shirleyi
- Silver-leaved = E. melanophloia
Ironwood 186
Isachne (Gramin.) 511
- australis 460
Ischaemum (Gramin.) 511
- triticeum 606
Iseilema (Gramin.) 106, 511, 555
- calvum 525
- convexum 525
- membranaceum 504
- vaginiflorum 524, 526, 527
Island(s) 632
- Carnac (W.A.) 637
- coral 632
- Dorre (W.A.) 637
- Five (N.S.W.) 636
- Fraser (Qld) 635
- Gabo (Vic.) 636
- Garden (W.A.) 637
- granite (Vic.) 636
- Kangaroo (S.A.) 343, 636
- mangrove 627
- rock 632
- Rottnest (W.A.) 637
- Torres Strait 633
Isoëtaceae 34
Isoëtes 34, 434
- drummondii 37, 564
- elatior 37
- gunnii 37, 456
- humilior 37, 551
- muelleri 101
Isoëtopsis (Compos.)
- graminifolia 553, 556
Isopogon (Prot.) 85, 86, 98
- alcicornis 427
- anemonifolius 247, 254, 409, 413
- asper 397
- attenuatus 397
- axillaris 397, 426, 427
- baxteri 397
- buxifolius 361, 397, 435
- ceratophyllus 406
- divergens 397
- dubius 397
- formosus 435
- linearis 397
- petraea 436
- polycephalus 397
- scabiusculus 397, 427
- sphaerocephalus 397
- teretifolius 397
- tridens 397
- trilobus 397
- uncinatus 397
- villosus 397
Isotoma (Campanul.)
- axillaris 314
- petraea 438, 569
Isotropis (Papil.) 98
Ixiolaena (Compos.) 113, 501
- leptolepis 503, 504
Ixora (Rub.)
- beckleri 148

J

Jacksonia (Papil.) 85, 98, 105, 107
- aculeata 479
- argentea 200
- capitata 395
- decumbens 395
- dilatata 200
- floribunda 395
- furcellata 381
- hakeoides 395
- horrida 430, 433
- odontoclada 399
- racemosa 395
- scoparia 234, 236, 237, 239, 244, 248, 271, 303; 306, 330, 410
- stackhousei 408
- sternbergiana 381
- thesioides 200, 399
- umbellata 395
Jagera (Sapind.) 174
- pseudorhus 149, 228
Jam = Acacia acuminata
Jansonia (Papil.) 98
Jarrah = Eucalyptus marginata
Jasminum (Oleac.) 108
- didymum 180
- lineare 326, 332
- simplicifolium 183
- singuliflorum 147
Johnsonia (Lil.) 90
- pubescens 90 F
Juncaceae 48
Juncaginaceae 48
Juncus (Juncac.) 520, 585
- bufonius 593
- maritimus 581, 582, 631, 633 F
- pallidus 593, 597
- planifolius 586, 593
- polyanthemos 325, 490, 554
- prismatocarpus 593

K

Kaolin 23
Karri = Eucalyptus diversicolor
Kennedia (Papil.) 85, 98
- coccinea 433
- macrophylla 371
- prostrata 405
- rubicunda 233, 235, 238, 239, 287
Keraudrenia (Stercul.)
- lanceolata 410
Kimberleys 131
Kingia (Xanthorr.) 91
- australis 91 F, 376, 422, 423 F
Knightia (Prot.) 60, 62
Kochia, see Maireana
Kopi 366
Korthalsella (Visc.) 55
Kosciusko uplift 10
Kranz anatomy 495
Kreysigia (Lil.) 90
Kunkar 385
Kunzea (Myrt.) 81, 98
- ambigua 404, 410, 413
- capitata 409
- preissiana 431
- pulchella 436
- sericea 434, 436
Kurrajong = Brachychiton populneus
Kwongan 418

L

Labiatae 46, 56
Labichea (Caesalpin.)
- cassioides 483
Lachnostachys (Verben.) 105
Lagenophora (Compos.) 66
- stipitata 447, 458
Lagurus (Gramin.)
- ovatus 516, 607
Lake Eyre 6
- basin 5 F
Lakes 582
Lamarchea (Myrt.) 81
- hakeifolia 357, 424, 425, 487
Lambertia (Prot.) 86, 98
- ericifolia 397
- formosa 247, 254, 255, 256, 402, 409
- inermis 361, 397, 426, 427, 433
- multiflora 397, 423
Lamprobium (Papil.) 85
Lancewood = Acacia shirleyi
Land connections 73
Lantana (Verben.)
- camara 160
Lasiopetalum (Stercul.)
- behrii 345
- discolor 339, 611
- schulzenii 339
Lastreopsis 35
Laterite 23, 74, 416
- bauxitic 23
- disintegration 75
- ferruginous 23
- formation 23, 29

Laterite
- manganiferous 29
- phosphate fixation 99
- truncation 29
- weathering 99
Latrobea (Papil.) 98
Lauraceae 46, 53, 76
Laurasia 1
Laurelia (Laur.) 62
Lauriphyllum 62
Lavatera (Malv.)
- plebeia 568
Laxmannia (Lil.) 90
- gracilis 247
Lead 130
Leaf-Size classes 123
Lebetanthus (Epacrid.) 88
Lecidia 413
Lecythidaceae 46
Leersia (Gramin.) 511
- hexandra 519, 520
Legnephora (Menisperm.)
- moorei 148
Leguminosites 62
Lemna (Lemn.) 552
- minor 571, 581, 582, 585, 586, 588, 595
- trisulca 586
Lemnaceae 48
Lentibulariaceae 46, 55
Leopard Tree = Flindersia maculosa
Lepidium (Crucif.) 104
- papillosum 531
- rotundum 537
Lepidosperma (Cyper.) 91
- angustatum 376, 596
- concava 404, 405
- congestum 344
- drummondii 595
- filiforme 71, 591
- flexuosum 591
- gladiatum 421, 431 F, 432, 595, 607, 608
- laterale 71, 246, 345, 411, 412
- lineare 404, 547
- longitudinale 260, 591, 597
- persecans 596
Lepidozamia 40, 41, 142
Lepilaena (Zanichell.)
- cylindrocarpa 586
- preissii 71, 582, 586
Lepironia (Cyper.)
- articulata 585
Leporella (formerly Leptoceras – Orchid.) 98
Leptaspis (Gramin.) 511
Leptocarpus (Restion.) 63, 67, 92
- aristatus alliance 595
- brownii 414
- scariosus 596
- spathaceus 574
- tenax 408, 585, 587, 591, 593
Leptoceras, see Leporella
Leptochloa (Gramin.) 511
- brownii 520
- debilis 465
- digitata 542, 557, 563
- - alliance 520

Leptomeria (Santal.) 55, 115
- acida 256
- cunninghamii 375
Leptophloeum 57
Leptopteris 34
- fraseri 38, 593
Leptorhynchos (Compos.)
- squamatus 289, 337
Leptospermoideae 80
Leptospermum (Myrt.) 59, 60, 80, 81, 82, 98, 105, 610
- attenuatum 239, 247, 256, 285, 288, 305, 391, 400, 405
- arachnoides 391, 411, 412, 459
- brachyandrum 194, 391, 410 F
- brevipes 271
- coriaceum 391, 414
- Cretaceous 80
- ellipticum 595, 596
- erubescens 394, 434, 436
- fabricia 391, 399
- firmum 429, 596, 597
- flavescens 391, 401
- - - L. attenuatum alliance
- glaucescens 391, 405, 407
- grandifolium 153
- grandiflorum 391, 407
- humifusum 445, 452, 453
- juniperinum 338, 391, 413
- laevigatum 345, 400, 406 F, 589 F, 600, 607, 608, 611, 612 F, 636
- lanigerum 263, 264, 295, 296, 297, 391, 406 F, 407, 441, 452, 582, 589 F, 591
- liversidgei 391, 591
- - alliance 405
- myrsinoides 261 F, 337, 338, 391, 402, 413, 414, 602
- - alliance 405, 406 F
- myrtifolium 460
- nitidum 391, 405
- obovatum 259, 260
- parvifolium 313, 349, 391, 411
- pubescens 289
- riparium 593
- roei 394
- scoparium 263, 264, 391, 404, 405, 407, 408, 591
- semibaccatum 391, 401, 403
- spinescens 394
- squarrosum 391, 411
- Tasmania 405
- wooroonooran 178, 399
Lepturus (Gramin.) 511
- parviflorus 606
Lepyrodia (Restion.) 92
- gracilis 591
- interrupta 591
- muelleri 591
- muirii 596
- scariosa 255, 257, 408, 411, 412, 591
Leschenaultia (Gooden.)
- biloba 393
- expansa 393
- floribunda 393
- juncoides 393
- linarioides 393

- stenosepala 393
- tubiflora 393
Leucopogon (Epacrid.) 52, 88, 98, 105, 314
- attenuatus 410
- australis 408, 413, 432
- brownii 582, 586
- capitellatus 375
- collinus 404, 408
- crassifolius 392
- cucullatus 392
- dielsianus 392
- ericoides 404
- fimbriatus 392
- flavescens 392
- gibbosus 392
- gnaphalioides 436
- juniperinus 71, 238
- lanceolatus 235, 238, 271, 274, 286
- microphyllus 411
- mitchelli 309
- muticus 311
- obovatus 439
- oppositifolius 392
- parviflorus 71, 379, 402, 433, 579, 600, 602, 607, 608
- plumuliflorus 392
- propinquus 392
- pulchellus 392
- reflexus 392
- richii 392
- unilaterus 436
- verticillatus 371, 375
- virgatus 405
Leviera (Monim.) 53
Lhotskya (Myrt.) 81
- alpestris 414
- smeatonia 637
Liana 140
Libertia (Irid.) 67
- pulchella 71
Lichens 33, 434
Lichina 602
Licuala (Palm.)
- ramsayi 177 F, 178
Life forms 122, 123, 139
Lignotuber 84
Lignum = Muehlenbeckia cunninghamii
Ligustrum (Oleac.)
- lucidum 160
Lilaeopsis (Umbell.) 66
- polyantha 582
Liliaceae 48, 56, 89
Liliacidites 60
Lily, Pink Lotus = Nelumbo nucifera
Limomium (Plumbag.) 104
Limosella (Scrophul.)
- australis 585, 586
Linaceae 46
Lindsaea 35
- ensifolia 101
- linearis 372, 404, 409, 591
Lindsaeaceae 35
Lindsayomyrtus (Myrt.)
- brachyandrus 173
Linospadix (Palm.)
- monostachys 150

Liparis (Orchid.) 93
- bracteata 93
- coelogynoides 93
- habenarina 93
- nugentae 93
- reflexa 93
- simmondsii 93
Liparophyllum (Gentian.)
- gunnii 71, 456
Lissanthe (Epacrid.) 98
- montana 295, 453, 456
- strigosa 280, 282, 285, 286, 304, 330, 332
Litsea (Laur.) 53, 70
- leefeana 146, 147, 228
- reticulata 147
Littoral Zone 598
- dunes 602
- flora 599
- islands 621, 632
- mangroves 615
- mudflats 611
- rocky outcrops 601
Livistona (Palm.) 108
- australis 150, 160, 240
- benthamii 199, 210
- eastonii 199
- humilis 204, 219
- — alliance 575, 576 F
- mariae 109 F, 569
Lobelia (Lobel.)
- anceps 581, 600, 630
- dentata 247
Lobeliaceae 46
Logania (Logan.)
- vaginalis 371, 374, 375, 381
Lolium (Gramin.)
- temulentum 530
Lomandra (Xanthorr.) 90, 313
- brownii 335
- dura – L. effusa alliance 593
- effusa 593
- glauca 244
- hastilis 424
- longifolia 236, 238, 262, 274, 287, 289, 292, 402, 411, 546, 547, 594, 600, 602, 605, 607
Lomariopsidaceae 35
Lomatia (Prot.) 62, 63, 66, 85, 86, 98
- arborescens 148, 155
- fraseri 148, 291
- ilicifolia 255, 460
- myricoides 148
- polymorpha 454
- silaifolia 246, 254, 408, 460
- tinctoria 273
Lonchocarpus (Papil.) 85
- blackii 147
Lophatherum (Gramin.) 511
Loranthaceae 46, 55
Lotus (Papil.) 104
- coccineus 542
Loxocarya (Restion.) 92
Ludwigia (Onagr.)
- parviflora 520
- peploides 552, 585, 632
Luisia (Orchid.)
- teretifolia 180

Lumnitzera (Combret.)
- littorea 614, 623
- racemosa 614, 622, 623 F, 624
Luzula (Junc.)
- campestris 289, 585, 586
Luzuriaga (Philes.) 63
Lycium (Solan.) 104
- australe 348, 387, 388, 474, 483, 486, 487, 491, 504, 508, 556
- ferrocissimum 103
Lycopodiaceae 34
Lycopodium 34, 37, 142
- cernuum 37, 38 F, 411
- laterale 37, 411, 591
- myrtifolium 140
Lyginia (Restion.) 92, 595
- barbata 596
Lygodium 34
Lyperanthus (Orchid.) 98
Lysiana (Loranth.) 55, 532
Lysicarpus (Myrt.) 81, 82
- angustifolius 301, 302, 305, 478
Lysinema (Epacrid.) 98
Lythraceae 46
Lythrum (Lythr.) 104
- hyssopifolium 556
- salicaria 585, 586

M

Macadamia (Prot.) 86, 148
Macklinlaya (Aral.)
- confusa 213
Maclura (Mor.)
- cochinchinensis 148
Macroglena 34
Macropidia (Haemodor.)
- fuliginosa 422, 423 F
Macropteranthes (Combret.)
- alliance 189
- kekwickii 189
- leichhardtii 189, 466
Macrozamia 40, 41
- communis 239, 240, 599
- heteromera 282, 288
- macdonnellii 41, 101, 107, 569
- moorei 42 F
- reidlei 373, 379, 380, 426, 427, 431, 432
- spiralis 313
Mahogany 84
- Red = Eucalyptus resinifera
- Swamp = E. robusta
- White = E. acmenoides
Maireana (Chenopod.) 103, 105, 106, 113, 493
- aphylla 467, 500, 504, 509
- astrotricha 485, 504, 506 F
- — alliance 497, 508
- brevifolia 559, 560
- carnosa 492, 558
- enchylaenoides 505
- georgei 348, 483, 504
- lobiflora 531
- oppositifolia 630
- pyramidata 192 F, 366, 474, 491, 499, 500 F, 504, 506 F, 507 F, 508, 556, 560 F

- — alliance 497, 508 F
- sedifolia 348, 366, 474, 485, 486, 487, 491, 499, 500, 504, 506, 507 F, 508, 559
- — alliance 497, 506, 507 F
- seed germination 119
- tamariscina 334, 531
- tomentosa 348
- triptera 348, 366, 483, 491, 508
- villosa 504
Malasia (Mor.)
- scandens 148, 180
Malaxis (Orchid.) 93
Mallee 84, 124, 133, 340, 608
- arid zone 340
- Bell-fruited = E. preissiana
- Blue = E. gamophylla
- Burracoppin = E. burracoppinensis
- Congoo = E. dumosa
- Crowned = E. coronata
- Curly = E. gillii
- Desmond = E. desmondensis
- distribution 340
- floristics 341
- Giant = E. socialis
- Green (Box) = E. viridis
- — heath 341, 358 F, 359, 391
- Hook-leaved = E. uncinata
- Kingmill's = E. kingsmillii
- Large-fruited = E. pyriformis
- Lerp = E. incrassata
- Many-flowered = E. cooperana
- Oldfield's = E. oldfieldii
- outliers 340
- Ribbon-barked = E. sheathiana
- soils 340
- southern 340
- Spearwood = E. doratoxylon
- Sturt Creek = E. odontocarpa
- Tall Sand = E. eremophila
- Tammin = E. leptopoda
- Thick-leaved = E. pachyphylla
- Victoria Desert = E. concinna
- White = E. diversifolia
- White-leaved = E. albida
Mallet
- Blue = Eucalyptus gardneri
- Brown = E. astringens
Mallotus (Euphorb.) 108
- nesophilus 109
- philippinensis 146, 228, 230
Malococera (Chenopod.) 493
Malpighiaceae 46
Malvaceae 46
Malvacipollis 60
Malvastrum (Malv.)
- americanum 527
Manilkara (Sapot.)
- kauki 635
Mangals
- distribution 619
- epiphytes 621
- fauna 620
- zonation 621, 622, 624
Mangrove(s) 73, 612 F, 615
- communities, see Mangal
- economic 615
- flooding 617

Mangrove(s)
- fossil 67, 615
- fruits 616
- germination 617
- islands 627
- pneumatophores 617F
- roots 617
- salt-tolerance 617, 618
- Tertiary 67
- vivipary 615
- zonation 616

Marattia 34, 145
Marattiaceae 34
Marine meadows 611
Marlock 340
- Black = Eucalyptus redunca
Marri = Eucalyptus calophylla
Marsdenia (Asclep.) 108
Marsh 124
Marsilea 36, 37, 101
- angustifolia 37
- brownii 520
- drummondii 37, 325, 333, 467, 490, 500, 521, 528, 552, 555, 568
- - alliance 552, 553F
- mutica 37, 574
Marsileaceae 36
Marsupial moors 457
Maytenus (Celastr.)
- cunninghamii 313
Medicago (Papil.) 103, 530
Medicosma (Rut.)
- cunninghamii 149
Meeboldina (Restion.) 92
Melaleuca (Myrt.) 52, 80, 81, 98, 102, 103, 105, 110, 202, 571
- acacioides 299, 579
- acerosa 203, 381, 394
- acuminata 492, 559, 597
- argentea 562
- armillaris 260, 579
- bracteata 194, 562, 593
- cardiophylla 359, 425, 487, 638
- citrina 436
- communities 577
- cordata 394
- dealbata 576F
- eleutherostachya 425
- ericifolia 304, 413, 631
- erubescens 313, 410
- fasciculiflora 595
- foliolosa 579
- fulgens 436
- genistifolia 234
- gibbosa 338, 407, 413, 586, 591
- glomerata 254
- halmaturorum 626, 630
- hamulosa 436, 559
- holosericea 394
- huegelii 394, 424, 432, 433
- lanceolata 338, 343, 347, 433, 597, 602, 637
- lateriflora 386, 559
- leiopyxis 394, 421, 425
- leucadendron 210, 585
- - alliance 574F, 577F
- lineariifolia 234, 332
- megacephala 394, 425

- mimosoides 195, 593
- minor 574
- minutiflora alliance 519F, 580
- nematophylla 425
- nervosa 211, 579
- nesophila 394, 426
- nodosa 402, 403, 405, 409, 591
- parviflora 351
- pauperiflora 386, 388, 559
- pentagona 435
- platycalyx 394
- polygaloides 597
- preissii 377, 379, 427, 430
- - alliance 597
- pubescens 304, 336, 350
- pungens 394
- quinquenervia 240, 400, 402, 403, 581, 583F
- radula 394
- raphiophylla 376, 430, 565, 595
- - alliance 596F, 597
- saligna 195, 399, 593
- scabra 360, 394, 559
- squamea 582, 586
- squarrosa 259, 289, 413, 581, 582, 585, 586, 589F, 591
- stenostachya 201, 579
- striata 352, 394
- styphelioides 234, 332
- suberosa 394
- substrigosa 351, 394
- symphyocarpa 399
- thymoides 351, 375, 394
- thymifolia 579, 591
- thyoides 425, 559, 597
- uncinata 201, 304, 313, 314F, 339, 345, 348, 350, 351, 352, 382, 386, 414, 425, 436, 483, 559
- violacea 435
- viridiflora 201, 202, 203, 211, 212, 222, 299, 319, 489
- - alliance 578F, 579
Melastoma (Melastom.)
- denticulatum 246
Melastomataceae 46
Melia (Mel.) 76
- azedarach 147, 194
Meliaceae 46
Melichrus (Epacrid.)
- urceolaris 286, 288, 306, 311, 330
Melicope (Rut.) 70
- australasica 149
Melon hole 30
Melothria (Cucurb.)
- micrantha 542
Menispermaceae 46
Menkea (Crucif.) 113
Menyanthaceae 46
Merremia (Convolv.)
- dentata 169
Mesomolaena (Cyper.) 91
- tetragona 597
Messerschmidia (Borag.)
- argentea 635
Messmate = Eucalyptus obliqua
- Gympie = E. cloëziana
Metrosideros (Myrt.) 59, 60, 81, 82, 83

- Cretaceous 80
- eucalyptoides 195, 576
Miconiiphyllum 62
Micraira (Gramin.) 511
- subulifolia 409
Microcachryidites 58
Microcachrys 40, 43, 75
- tetragona 295, 453
Microchloa (Gramin.) 511
Microcitrus (Rut.) 85
- australis 149
Microgonium 34
Microlepia 34
Microlaena (Gramin.) 511
- stipoides 71, 332, 594
Micromyrtus (Myrt.) 81, 98, 105
- ciliata 349
- elobata 394
- flaviflora 437
- imbricata 394, 424
- minutiflora 307
- peltigera 357, 394
- racemosa 394
- rosea 394
Microseris (Compos.) 66, 110
Microsorium 36
Microstegium (Gramin.) 511
Microstrobus 40
- fitzgeraldii 43
- niphophilus 43, 453
Microtis (Orchid.) 98
Microtrichomanes 34
Migration barriers 125
Milletia (Papil.)
- megasperma 147
Milligania (Lil.) 90
- densiflora 451
Mimosaceae 46, 85
Mimulus (Scrophular.)
- gracilis 553, 554
- repens 553, 554, 581
Minuria (Compos.) 113, 501
- cunninghamii 503, 521, 559
- integerrima 118, 521, 525
- leptophylla 350, 504
Mirbelia (Papil.) 85, 98, 105
- dilatata 371
- floribunda 395
- pungens 306
- spinosa 395
Mischocarpus (Sapind.) 70
- pyriformis 149
Mistletoe 55
Mitrasacme (Logan.) 574
- archeri 451
- polymorpha 411
Mitrephora (Annon.) 53
Mixed stands 124
Mobilabium (Orchid.) 93
Mollinedia (Monim.)
Mollugo (Aizo.) 104
- hirta 521
Monazite 400, 603
Monimiaceae 46, 53, 76
Monocarpus sphaerocarpus 558
Monochater (Gramin.) 511
Monogramma 36
Monotaxis (Euphorb.) 88

Monotoca (Epacrid.) 98
— elliptica 413
— empetrifolia 453
— scoparia 246, 271, 286, 400, 401, 608
— tamarascina 436
Monsoon forest 177
Montia (Portulac.)
— australasica 451
— fontana 586
Moor 457
Moort = Eucalyptus platypus
— Red-flowered = E. nutans
Moraceae 46, 76
Morgania (Scrophular.)
— floribunda 554
Morinda (Rub.)
— reticulata 399
Mosses 33
Moss forest 136, 140
Mossland 124, 457
Moss thicket 136
Mottlecah = Eucalyptus macrocarpa
Mount
— Connor (N.T.) 7 F
— Kosciusko (N.S.W.) 8 F, 439
— Ossa (Tas.) 6
Mucuna (Papil.) 85
— gigantea 147
Mudflats 611, 632
Muehlenbeckia (Polygon.) 66, 110
— cunninghamii 189, 216, 325, 465, 469, 500, 503, 521, 555 F, 556, 557, 566, 568
— — alliance 554
Muellerina (Loranth.) 55
Mugga = Eucalyptus sideroxylon
Mulga = Acacia aneura
Murchison R. (W.A.) 421
Muriantha (Rut.) 98
Murray Basin 9
Musa (Mus.)
— banksii 139 F
Musaceae 48
Musgravea (Prot.) 86
— heterophylla 173
— stenostachya 173
Myall = Acacia pendula
— Western = A. sowdenii
Mycorrhiza 33
Myoporaceae 47
Myoporum (Myopor.) 63, 108
— acuminatum 309
— debile 332, 334
— deserti 321, 336, 466, 474, 483, 489
— insulare 600, 602, 603, 607, 608
— montanum 215, 336, 564, 568
— platycarpum 328, 348, 349, 486, 491, 492, 507
— serratum 380
Myrica (Myric.) 62
Myriocephalus (Compos.) 113
— guerinae 482
— stuartii 475, 509, 540, 541, 542, 560, 561
Myriophyllum (Halorag.)
— amphibium 586

— elatinoides 456, 581, 582, 586
— pedunculatum 586
— propinquum 585, 586, 595
— verrucosum 552, 555, 581
Myristica (Myristic.) 53
Myristicaceae 47, 53
Myrmecodia (Rub.) 140
Myrsinaceae 47
Myrtaceae 47, 76, 80
— chromosome numbers 82
— polyploidy 82
— subfamilies 80
Myrtaceidites 60
Myrtle = Nothofagus cunninghamii
Myrtoideae 80

N

Nablonium (Compos.) 80
Najadaceae 48
Najas (Najad.)
— major 595
— marina 551, 581
Nardoo = Marsilea drummondii
Nasturtium (Crucif.)
— officinale 571, 582, 585, 586
Nauclea (Rub.)
— orientalis 173, 180, 195, 562, 563, 593
— undulata 173
Needhamia (Epacrid.) 98
Nelia = Acacia loderi
Nelumbo (Nymph.)
— nucifera 573, 574 F
Nematolepis (Rut.) 98
Nemcia (Papil.) 98
Neolitsea (Laur.) 53
— cassia 147
— dealbata 147
Neopaxia (Portulac.)
— australasica 585, 588
Neoroepera (Euphorb.)
— banksii 399
Neosciadium (Umbell.) 78
Nepenthaceae 47, 54
Nepenthes (Nepenth.) 54
— mirabilis 399
Nephrolepis 36
— cordifolia 101
Neptunia (Mimos.)
— gracilis 525
Nertera (Rub.) 66
Neurachne (Gramin.) 112, 511
— alopecuroides 354, 383, 385, 386, 422, 427
— mitchelliana 321, 328, 475, 478
Neurosoria 36
Neverfail = Eragrostis eriopoda or E. setifolia
New Caledonia 1
Newcastelia (Verben.) 105
— cladotricha 534
New Guinea 1, 3, 4, 5
— endemic taxa 68, 69
— flora 68, 69
Nicotiana (Solan.) 436
— glauca 103, 509, 540, 561, 564, 568

Ninetyeast Ridge 61 F
Nitraria (Zygophyll.)
— billardieri 120, 500 F, 509, 515, 516 F, 526 F, 528, 568, 600, 602, 607, 608, 629 F, 630
— schoberi = billardieri
Nostoc 33
Notelaea (Oleac.)
— ligustrina 153
— longifolia 147, 231, 235, 236, 238, 305
— microcarpa 183, 306, 307, 332
Nothofagidites 59
Nothofagus (Fag.) 58, 60, 61, 62, 66, 67
— alliances 143
— assembage 73, 75
— cunninghamii 145, 151 F, 152, 153 F, 157, 263, 273, 292, 295, 413
— — alliance 143
— gunnii 154, 155 F
— moorei 141, 145, 156 F
— — alliance 154
— pollen types 64, 76
Notochloë (Gramin.) 511
Notothixos (Visc.) 55
Nullarbor Plain 7, 100, 104, 343, 496, 497, 504, 505, 507
Nuytsia (Loranth.) 55
— floribunda 360, 362, 376, 418, 419 F, 421, 422, 425, 426, 427, 433, 434, 597
Nyctaginaceae 47
Nymphaea (Nymph.)
— capensis 585
— gigantea 577 F, 585, 586 F
— — — Nelumbo nucifera alliance 573
— violacea 551
Nymphaeaceae 47
Nymphoides (Menyanth.)
— exigua 586
— geminatum 551, 585, 586 F
— indica 574, 585
Nypa (Palm.) 60
— fruticans 614, 615
— — alliance 624, 626 F

O

Oak, Black = Casuarina cristata
— Bull = C. luehmannii
— Desert = C. decaisneana
Ochnaceae 47
Oenothera (Onagrac.)
— drummondii 605, 606, 608
Oenotrichia 35
Olacaceae 47, 55
Olax (Olac.) 55
Olea (Oleac.)
— paniculata 147
Oleaceae 47
Oleandra 36
Oleandraceae 35
Olearia (Compos.) 52, 113
— algida 456
— argophylla 152, 153, 273, 274, 275, 277, 291, 292

Olearia (Compos.)
- axillaris 380, 430 F, 433, 600, 602, 607, 608
- ledifolia 453
- lepidophylla 456, 458
- microphylla 286, 287
- myrsinoides 546
- nerstii 274
- pimelioides 327, 348
- pinifolia 453
- ramulosa 405
- stellulata 413
- teretifolia 343
Oligarrhena (Epacrid.) 98
Oligocene 74
Omalanthus (Euphorb.)
- populifolius 147
Omphacomeria (Santal.) 55
Onagraceae 47
Onychosepalum (Restion.) 92
Ooline = Cadellia pentastylis
Opal 23
Ophioglossaceae 34
Ophioglossum 34
- lusitanicum 37, 101, 410, 538
- pendulum 36, 142
Ophiurus (Gramin.) 511
- exaltatus 523
Opilia (Opil.)
- amentacea 180
Opiliaceae 47
Opisthiolepis (Prot.) 86
- heterophylla 173
Oplismenus (Gramin.) 511
- compositus 180
Orania (Palm.)
- appendiculata 177
Orchidaceae 48, 54, 92
- epiphytes 92, 93
- lithophytes 93
- origins 92
- pseudobulbs 93
- saprophytes 56
Oreobolus (Cyper.) 67, 91
- pumilio 71, 451, 456
Oreocallis (Prot.) 63, 66, 86
- pinnata 148
Oreomyrrhis (Umbell.) 66
- andicola 546
Orites (Prot.) 63, 66, 85, 86
- acicularis 453
- densiflora 454
- excelsa 148, 178, 230
- lancifolia 271, 450 F, 454, 543
- revoluta 453
Orobanchaceae 47, 54
Orobanche (Orobanch.) 54
- australiana 116, 637
Orthoceras (Orchid.) 98
- strictum 71
Oryza (Gramin.) 511, 574
- australiensis
- - alliance 519 F
- rufipogon 519
Osbornia (Myrt.) 80, 81, 82
- octodonta 614, 622, 624 F
Oschatsia (Umbell.) 77, 80
Osmundaceae 34

Ottelia (Hydrocharit.)
- ovalifolia 552, 584 F, 585, 587 F, 595
Ottochloa (Gramin.) 511
Ourisia (Scrophular.) 66
Owenia (Mel.) 108
- acidula 119, 121 F, 190, 192
- cepiodora 147
- reticulata 532, 535 F
Oxalidaceae 47
Oxalis (Oxalid.)
- corniculata 334, 605, 606, 608
Oxylobium (Papil.) 85, 98
- capitatum 381
- ellipticum 271, 295
- - - Podocarpus lawrencei alliance 452
- lanceolata 371
- retusum 436

P

Pachycornia (Chenopod.) 114, 493
- tenuis 502, 504, 558, 568
Pachynema (Dillen.) 53
- dilatata 576
Palmae 48, 76
Palmeria (Monim.)
- scandens 148, 177
Palm Valley 109
Pandanaceae 48
Pandanus (Pandan.) 178, 201, 203, 210, 211, 215, 572, 573, 574, 575 F, 601, 606, 635
- aquaticus 195
- peduncularis 600, 601 F, 602, 607
- spiralis 181, 409
- spp. alliance 574
- whitei 195
Pandorea (Bignon.) 108
- pandorana 147, 155, 238, 282, 285, 286, 313
Pangea 1
Panicum (Gramin.) 104, 490, 511
- buncei 527
- decompositum 500, 504, 523, 526, 527, 528, 529, 555, 566
- effusum 288
- maximum 517
- paludosum 519
- prolutum 466, 489
- queenslandicum 320, 489
- trichoides 180
- whitei 526, 529, 542, 555, 557
Papilionaceae 47, 56, 85, 87
- endemic genera 85
- rhizobia 87
- Tribes 85
Papillilabium (Orchid.) 93
Paracaleana (Orchid.) 98
Paraceterach 35
Paractenium (Gramin.) 511
- novae-hollandiae 475, 482
Paraneurachne (Gramin.) 511
Parantennaria (Compos.) 80
Paratephrosia (Papil.) 85
Parinari (Ros.)

- corymbosus 180
- nonda 201
Parkeriaceae 36
Parmelia conspersa 413, 602
Parsonsia (Apocyn.) 108
- brownii 146, 155
- eucalyptiphylla 183, 313, 326, 332
- induplicata 146
- leichhardtii 146
- straminea 146, 232
- velutina 146, 180
Paspalidium (Gramin.) 511
- caespitosum 478
- gracile 240, 245, 313, 321, 350, 409, 475, 489
- jubiflorum 189, 325, 563, 566, 568
Paspalum (Gramin.) 511
- dilatatum 586
- distichum 520, 632
- vaginatum 517
Passiflora (Passiflor.)
- edulis 160
- foetida 210
- herbertiana 148
Passifloraceae 47
Patersonia (Irid.)
- filiformis 591
- fragilis 591
- occidentalis 424, 430, 431, 596
- sericea 288, 413
Pedaliaceae 47
Pelargonium (Geran.)
- australe 548, 600, 608
- littorale 432
Pellaea 36
- falcata 38
Peltophorum (Caesalpin.)
- ferruginea 181
- pterocarpum 180
Pemphis (Lythr.)
- acidula 623
Pennantia (Icacin.)
- cunninghamii 147
Pennisetum (Gramin.) 511
- alopecuroides 594
Pentaceras (Rut.)
- australis 149
Pentachondra (Epacrid.)
- pumila 71, 452, 453 F
Pentapogon (Gramin.) 511
- quadrifidus 548
Pentatropis (Asclep.) 108
Peperomia (Piper.) 140
Peppermint 84
- = Eucalyptus odorata (S. A.) or Agonis flexuosa (W. A.)
- Black = E. amygdalina
- Narrow-leaved = E. radiata
- New England = E. nova-anglica
- River = E. elata
- Smithton = E. nitida
- Swamp = E. rodwayi
- Sydney = E. piperita
Periplocaceae 47
Peristeranthus (Orchid.) 93
Pernettya (Eric.) 66, 88
- tasmanica 456, 458 F

Perotis (Gramin.) 511
- rara 326
Persoonia (Prot.) 86, 98, 105
- angustifolia 397
- arborea 291
- coriacea 397
- cornifolia 246, 247, 271
- elliptica 375
- falcata 201, 299
- juniperina 261
- lanceolata 403, 409
- laurina 238
- levis 239, 255
- linearis 233, 235, 236, 239, 287
- longifolia 371, 373, 375
- nutans 306
- pinifolia 233, 238
- prostrata 409
- saccata 397
- saundersiana 397
- striata 397
- teretifolia 397
- tortifolia 397
Petalostigma (Euphorb.) 88, 108
- banksii 189, 202, 223, 299
- glabrescens 305
- pubescens 203, 301, 303, 309, 332, 489
- quadriloculare 213, 224, 239, 247
Petalostylis (Caesalpin.) 112
- labichioides 224
Petermanniaceae 48
Petiveraceae 47
Petrophile (Prot.) 86, 98
- brevifolia 397
- chrysantha 397
- circinata 397, 428
- conifera 397, 427
- divaricata 397
- drummondii 397
- ericifolia 397, 427
- fastigiata 359, 397
- fucifolia 255, 411, 413
- heterophylla 397
- latifolia 397
- linearis 397, 426
- media 397
- megalostegia 397
- microstachya 397
- multisecta 344, 637
- phylicoides 397
- semifurcata 397, 427, 428
- seminuda 397, 427
- serruriae 397
- shuttleworthiana 397
- squamosa 397
- striata 397
- teretifolia 397
- trifida 397
Phebalium (Rut.) 85, 98, 105
- anceps 595
- glandulosum 474
- microphyllum 355
- ovatifolium 270, 443
- rude 432
- squameum 263, 264, 401, 402, 407
Pheidochloa (Gramin.) 511

Philesiaceae 48
Philotheca (Rut.) 98, 105
- salsolifolia 409
Philydraceae 48
Philydrum (Philydr.)
- lanuginosum 583, 585
Phlegmatospermum (Crucif.) 113
Phormium (Agav.) 60
Phragmites (Gramin.) 511
- alliance 519, 552
- australis 518, 519, 579F, 581, 582, 583, 585, 586, 589F, 593
- - suballiance 520
- karka 518, 573, 574
- - suballiance 519
Phyllachne (Stylid.) 63, 66
- colensoi 71, 446
Phyllanthus (Euphorb.) 52, 108
- calycinus 375, 380, 429, 432
- filicaulis 473
- thymoides 313, 409, 413
Phyllocladus 40, 43, 58, 75, 142
- aspleniifolius 151, 157, 158F, 295, 454
Phylloglossum 34
- drummondii 434, 564
Phyllota (Papil.) 98
- philicoides 254
Phymatocarpus (Myrt.) 81
Phytophthora infestans 374
Pigface, Round-leaved = Disphyma australis
Pileanthus (Myrt.) 81, 98, 105
Pilidiostigma (Myrt.) 81
Pilostyles (Raffles.) 54
Pilularia 36, 37
- novae-hollandiae 37, 586
Pimelea (Thymel.) 52, 103
- alpina 446, 452, 453F
- argentea 432
- ferruginea 435, 436
- floribunda 600
- lehmanniana 426
- ligustrina 236
- linifolia 232, 244, 246, 254, 401, 409, 410
- microcephala 483
- pygmaea 446, 449F, 450
- rosea 430, 432
Pindan Wattle = Acacia pachycarpa = A. ancistrocarpa
Pine, White = Callitris "glauca"
Pinus radiata 44
Piper (Piper.)
- novae-hollandiae 148, 177
Piperaceae 47
Piptocalyx (Trimen.) 53
- moorei 149
Pisonia (Nyctag.)
- aculeata 149, 180
- brunoniana 149
- grandis 635
- umbellata 71
Pittosporaceae 47
Pittosporum (Pittospor.) 63, 108, 173
- bicolor 140, 151, 153, 291
- phylliraeoides 183, 313, 332, 387, 436, 473, 474, 487, 508, 538F, 568

- revolutum 148, 231, 238
- undulatum 148, 231, 232, 238, 239, 240
Pituri 109
Pityrodia (Verben.) 105
- loxocarpa 483
Placospermum (Prot.) 86
- coriaceum 173
Plagianthus (Malv.) 60
- glomeratus 558, 560
- helmsii 366, 434, 558
- incanus 556, 559
Plagiogyria 36
Plagiogyriaceae 36
Plagiosetum (Gramin.) 511
- refractum 475, 540, 541
Planchonella (Sapot.)
- australis 150
- cotonifolia 184
- euphlebia 174
- macrocarpa 174
- myrsinoides 150
- obovoidea 174
- papyracea 174
- singuliflora 178
Planchonia (Lecythid.)
- careya 203, 211, 213, 215, 299, 301, 562
Plantaginaceae 47
Plantago (Plantagin.) 104
- antarctica 451
- coronopus 548, 631
- gunnii 451
- muelleri 451
- - - Montia australasica alliance 451
- paradoxa 451
- varia 553
Platycerium 36, 140
- superbum 176F
Platylobium (Papil.) 85, 98
- formosum 235, 238, 246, 259
Platysace (Umbell.) 77
- billardieri 401
- linearis 238
Platyzoma 35
Playa 129, 550, 557
Plectorrhiza (Orchid.) 93
Plectrachne (Gramin.) 105, 106, 112, 130, 511
- danthonioides 424
- desertorum 357
- melvillei 366
- plurinervata 487
- pungens 200, 201F, 202, 208, 209, 211F, 213, 217, 219, 220F, 221F, 224, 308, 580
- rigidissima 355, 428
- schinzii 475, 479, 480, 481, 482, 483, 492
- - alliance 534F
Plectranthus (Labiat.)
- parviflorus 274, 409
Pleiococca (Rut.)
- wilcoxiana 149
Pleiogynium (Anacard.)
- solandri 146
Pleistocene glaciation 10, 69, 75
Pleuromanes 34

Pleurosorus 34, 101
- rutifolius 37, 569
Plinthanthesis (Gramin.) 511
Plumbaginaceae 47
Poa (Gramin.) 273, 511, 523
- australis = caespitosa 542
- costiniana 543
- drummondii 559
- fawcettiae 544
- fordeana 350
- gunnii 451, 452, 546, 547
- helmsii 544
- hiemata 271, 450 F, 451, 523
- - alliance 543
- hothamensis 544
- labillardieri 452, 523, 543
- - alliance 544, 545 F
- poiformis 289, 597, 600, 602, 607, 636
- - alliance 548
- rodwayi 546
- saxicola 544
- sieberana 240, 268, 269, 270, 272, 274, 276, 280, 285, 288, 289, 330, 545
Podalyrieae 85, 87
- Cretaceous 85
- cytology 85
Podocarpaceae 40
Podocarpus 40, 58, 75, 76, 142
- alpina = lawrencei
- amarus 43
- drouyniana 43, 371, 375
- elatus 43, 146, 194, 237
- ladei 172
- lawrencei 43, 153, 295, 440, 445, 452, 453
- neriifolius 172
- spinulosus 43
Podolepis (Compos.) 508
- aristata 357
- canescens 482, 537
- capillaris 383
Podopetalum (Papil.) 85
Podostemaceae 47
Pogonatherum (Gramin.) 511
Polyalthia (Annon.) 53
- nitidissima 146
Polygalaceae 47
Polygonaceae 47
Polygonum (Polygon.)
- minus 585
- lapathifolium 552
Polyosma (Escallon.)
- cunninghamii 146
Polyphlebium 34
Polypodiaceae 36
Polypodium acrostichoides 621
Polypompholyx (Lentib.) 55, 571
Polyscias (Aral.)
- bellendenkerensis 178
- elegans 147, 172
- murrayi 147
- sambucifolius 147
- willmottii 178
Polystichum 35, 38
- aculeatum 274, 292, 449
Polytrias (Gramin.) 511

Polytrichum 548
- juniperinum 293
Pomaderris (Rhamn.)
- apetala 71, 152, 273, 275, 277, 279, 291, 292, 293
- elliptica 283
- phylicifolia 71
Pongamia (Papil.)
- pinnata 181, 606, 609
Pontederiaceae 48
Popowia (Annon.) 53
Poranthera (Euphorb.) 88
- microphylla 71, 247
Porongorups (W.A.) 361, 371, 434
Portulaca 104
- oleracea 500, 527, 540, 635
Portulacaceae 47
Posidonia (Posidon.)
- australis 594, 613, 630
Posidoniaceae 48
Potaliaceae 47
Potamogeton (Potamogeton.)
- australiensis 586
- crispus 585
- ochreatus 585, 586
- pectinatus 581, 582
- perfoliatus 581, 582, 586
- tricarinatus 552
Potamogetonaceae 48
Potamophila (Gramin.) 511
- parviflora 520
Pothos (Arac.)
- longipes 177
Pouteria (Sapot.)
- sericea 181
Prasophyllum (Orchid.) 52, 98, 115
- alpinum 451, 452
- suttonii 452
Pratia (Campan.) 66, 110
Pre-Cambrian Shield 9
Primulaceae 47
Prionotes (Epacrid.) 80, 88, 89
- ceritheroides 151
Prostanthera (Labiat.) 52
- chlorantha 350
- lasianthos 154, 277, 291
- nivea 410
- sieberi 236
- striatiflora 474
- suborbicularis 474
Proteaceae 47, 59, 75, 76
- centre of origin 84
- fossils 84
- in rainforests 85
- subfamilies 84
- xeromorphy 85
Proteacidites 59, 60, 62
Prunus (Ros.)
- turnerana 173
Psammagrostis (Gramin.) 511
Pseudanthus (Euphorb.) 88
- orientalis 600
Pseudocarapa (Mel.)
- nitidula 148
Pseudochaetochloa (Gramin.) 511
Pseudomorus (Mor.)
- brunonianus 148
Pseudopogonatherum (Gramin.) 511

Pseudoraphis (Gramin.) 511
- spinescens 513, 519
- - alliance 520
Pseudoweinmannia (Cunon.) 78
- lachnocarpa 147, 172
Pseudowintera (Winter.) 62, 67
Psilotaceae 34
Psilotum 34, 142
- complanatum 36, 140
- nudum 36, 38, 140, 247
Psoralea (Papil.) 104, 523
- cinerea 526, 528, 555
- martinii 606
- patens 529, 555
Psychotria (Rub.)
- daphnoides 149
- loniceroides 149
Pteridaceae 36
Pteridium 35
- esculentum 40, 232, 235, 237, 239, 240, 246, 260, 272, 278, 287, 293, 372, 610
Pteridoblechnum 35
Pteridophytes
- aquatic 33, 37
- community dominants 37
- epiphytic 36
- flora 33
- habits 33
- heterosporous 33
- homosporous 33
- in caves and ravines 38
- in road cuttings 38
- in rocky situations 37
- invasion into cleared land 40
- primitive 33
- regeneration after fire 40
- tree-ferns 39
Pteridosperms 58
Pterigeron (Compos.)
- macrocephalus 606
- odoratus 536
Pterigeropappus (Compos.) 8
- lawrencei 445, 450
Pteris 36
- tremula 101
Pterostylis (Orchid.) 52, 98, 115
- baptistii 92
- mutica 71
- nutans 238
- pedunculata 92
- plumosa 71
Pterygopappus (Compos.) 80
Ptilanthelium (Cyper.)
- deustum 411, 591
Ptilotus (Amaranth.) 52, 80, 103, 112
- alopecuroideus 357
- exaltatus 358, 364, 388, 424, 558
- leucocoma 364
- obovatus 425, 484, 485, 508
Ptychosema (Papil.) 85, 98
Ptychosperma (Palm.)
- bleeseri 180
Puccinellia (Gramin.) 511
Puff-shelf soil surface 30
Pultenaea (Papil.) 52, 85, 98, 105
- acerosa 343
- boormanii 314

Pultenaea (Papil.)
- capitata 395
- cunninghamii 306
- daphnoides 404, 413
- dasyphylla 395
- foliolosa 314
- georgei 395
- juniperina 261, 285, 404
- trifida 637
- villosa 246
Pycnonia (Prot.) 87
Pyrrosia 36
- rupestris 411

Q

Quaking bog 460
Quandong 115
Quartzite 23
Quaternary environment 73
Quintinia (Escallon.) 60, 76
- sieberi 140, 146, 155
- verdonii 146

R

Rafflesiaceae 47, 54
Rainforest 124, 130
- alliances 143, 144, 145
- arborescent monocotyledons 143
- arid zone 181
- assemblage 73
- buttressing 139
- classification 136
- cyclones 167
- distribution 136
- dry 136, 164, 301
- edaphic factors 136
- epiphytes 139, 142, 143
- Filicales 141
- fire 138, 181
- floral relations 69
- - Asia 69, 70
- - New Guinea 69, 70
- floristics 141, 142
- gallery 193
- Indo-Malayan 167, 169
- leaf characters 138
- lianas 139
- life-forms 138
- littoral 166, 181
- margins 136
- monsoon 136
- orchids 143
- origins 69, 70, 71
- palms 143
- relicts 181
- saprophytes 143
- subtropical 136
- taxa 69, 70, 71
- temperate 136
- tropical 136, 167, 171
- woody species 138
Randia (Rub.)
- benthamiana 149
Ranunculaceae 47

Ranunculus (Ranuncul.) 77, 452
- anemoneus 450
- glabrifolius 585
- inundatus 451, 585, 586
- muelleri 450
- muricatus 600
- nanus 451
- rivularis 582, 585, 586, 588
- trichophyllus 586
Rapanea (Myrsin.)
- variabilis 148, 231, 238
Rauwenhoffia (Annon.) 53
- leichhardtii 146
Recent Arid Period 75
Recherche Archipeligo 363, 382, 435
Redheart = Eucalyptus decipiens
Redwood = E. oleosa
Reed, Common = Phragmites australis
Reedia (Cyper.) 91
Reediella 34
Regelia (Myrt.) 81, 98
- velutina 436
Relicts 76, 99
Remirea (Cyper.) 91
Restio (Restion.) 63
- applanatus 596
- australis 459
- complanatus 591
- fastigiatus 591
- gracilis 460, 591
- monocephalus 591
- pallens 591
- tetraphyllus 240, 405, 407, 591
Restionaceae 48, 92, 411
Rhacomitrium 413
- pruinosum 450
Rhagodia (Chenopod.) 493
- baccata 380, 600, 604 F, 607, 608
- billardieri 600
- hastata 325
- nutans 350, 600
- obovata 425
- preissii 607
- spinescens 120, 486, 497, 500, 502, 504, 542, 569
Rhamnaceae 47
Rhaphidophora (Arac.) 170 F
- australasica 177
- pinnata 177
Rhinerrhiza (Orchid) 93
Rhizanthella (Orchid.) 54
Rhizobia 33, 127
Rhizophora (Rhizophor.)
- apiculata 614, 621, 624
- spp. alliance 621
- stylosa 614, 615 F, 621, 624 F, 625, 627
Rhizophoraceae 47
Rhodamnia (Myrt.) 70, 81
- argentea 149
- trinervia 149, 231, 237
Rhododendron (Eric.) 69, 88
- lochae 178
Rhodomyrtus (Myrt.)
- beckleri 149, 231
- psidioides 228
Rhodosphaera
- rhodanthera 146

Rhopalostylis (Palm.) 60
Rhynchosia (Papil.)
- minima 523
Riccia fluitans 570
Rice, Native = Oryza australiensis
Richea (Epacrid.) 80, 89
- continentalis 459
- dracophylla 295
- gunnii 295
- pandanifolia 89, 152, 454
- scoparia 442, 448, 453 F, 454, 455, 457
- sprengelioides 454
Ricinocarpus (Euphorb.) 88
- pinifolius 401, 405, 409
Rimicola (Orchid.) 98
Ripogonum (Smilac.) 150
River
- Darling 9
- gradients 549
- Murray 7, 9 F
Rivers 5, 6, 8 F
Rockinghamia (Euphorb.)
- brevipes 178
Rocks 22
- chemical composition 22
- igneous 22
- metamorphic 22
- phosphorus content 22
- sedimentary 22
Rosaceae 47, 56
Rottboellia (Gramin.) 511
Roxburghiaceae 48
Roycea 493
Rubiaceae 47
Rubus (Ros.)
- gunnianus 451
- procerus 270
- vulgaris 160
Rumex (Polygon.)
- bidens 552, 585, 586
- brownii 270, 334, 335, 529
Rumohra 35
Rupicola (Epacrid.) 98
Ruppia (Rupp.)
- maritima 551, 581, 582, 613, 632
Ruppiaceae 48
Rutaceae 85
- chromosome numbers 85
- disjunction 85
- polyploidy 85
Rutile 400, 603

S

Sacciolepis (Gramin.) 511
Saccopetalum (Annon.) 53
Salicornia (Chenopod.) 104, 493
- australis 595, 623
- quinqueflora 602, 626 F, 628, 630
Sally, Black = Eucalyptus stellulata
- Narrow-leaved = E. moorei
Salsola (Chenopod.) 104, 493
- kali 102, 382, 501, 517, 531, 540, 555, 561, 606, 607, 628
Saltbush 493
- Bladder = Atriplex vesicaria

Saltbush
- Oldman = A. nummularia
- Silver = A. rhagodioides
Salvinia auriculata 571
Samolus (Primul.) 581
- repens 600, 602, 630
Samphire = Arthrocnemum or Salicornia spp.
- communities 628
Sand dunes 5, 6, 29
Sandplain 414, 421
Sandstones (Sydney region) 242
Santalaceae 47, 55
Santalum (Santal.) 55, 108, 115
- acuminatum 425, 427, 474, 532
- lanceolatum 181, 466, 469, 532
- spicatum 474
Santalumidites 60
Sapindaceae 47, 76
Sapotaceae 47
Sarcochilus (Orchid.) 92, 93
- falcata 155
- hillii 141
Sarcopetalum (Menisperm.)
- harveyanum 148
Sarcopteryx (Sapind.)
- stipitata 149
Sarcostemma (Asclepiad.) 114, 115F
- australe 213
Saurauiaceae 47
Savannah 124, 510
Scaevola (Gooden.) 52, 63, 103, 105
- aemula 364
- anchusifolia 393
- canescens 393
- crassifolia 433, 607, 608, 638
- glandulifera 393
- globulifera 393
- platyphylla 371
- spinescens 321, 350, 357, 366, 474, 559
- suaveolens 579, 600, 605, 606, 607, 635
- thesioides 393
Scald 495
Schelhammera (Lil.) 90
Schismus (Gramin.)
- barbatus 103, 564
Schizachyrium (Gramin.) 211, 511
Schizaea 34
- rupestris 38, 593
Schizaeaceae 34
Schizeilema (Umbell.) 66
Schizomeria (Cunon.)
- ovata 147, 230
- - - Doryphora - Ackama - Crytocarya alliance 161
Schlefflera (Aral.)
- actinophylla 140, 177, 188
Schoenia (Compos.)
- cassinioides 354
Schoenus (Cyper.) 52, 91, 414
- apogon 338, 563
- brevifolius 581, 588, 591
- imberbis 259, 591
- nitens 593
- paludosus 591
- sparteus 399

- tenuissimus 591
Scholtzia (Myrt.) 81, 98
Scirpitis 62
Scirpus (Cyper.) 91
- antarcticus 592
- fluitans 456, 585, 586
- inundatus 456, 585
- litoralis 581
- nodosus 580, 600, 603, 607, 608, 634
- validus 554, 581, 594, 595
Sclerandrium (Gramin.) 511
Scleranthus (Caryophyll.)
- pungens 600
Scleria (Cyper.)
- poiformis 519
Scleroblitum (Chenopod.) 493
- atriplicinum 568
Sclerophylly 80, 93
Sclerothamnus (Papil.) 98
Scolopia (Flacourt.)
- brownii 147
Screw Pine = Pandanus spp.
Scrophulariaceae 47, 55, 56
Scrub 124
- -heath 427
- high mountain 399
Scyphiphora (Rub.)
- hydrophylacea 614, 622
Sea-grasses 611
Sebaea (Gentian.)
- albidiflora 631
Secamone (Asclep.)
- elliptica 180
Sedge-heath 407
Sedgeland 114, 456, 574, 584, 595, 628
Sehima (Gramin.) 511
- nervosa 202, 215, 223, 308
Selaginella 34, 37
- uliginosa 404, 405, 408, 591
Selaginellaceae 34
Selenodesmium 34
Selliera (Gooden.) 63, 66
- radicans 71, 581, 631
Senecio (Compos.) 104
- australis 292
- gregorii 475, 537, 540, 541
- lautus 430, 600, 608, 636
- linearifolius 274
- minimus 293
- odoratus 600
Sequences 126
Serpentine 130
Sesbania (Papil.)
- benthamii 525, 555
Sesuvium (Aizo.)
- portulacastrum 601, 606, 623, 627, 628, 635
Setaria (Gramin.) 511
Shrubbery 124
Shrubland 124
Shrub steppe 124, 493
Sida (Malv.) 103, 104
- corrugata 332, 334, 350, 531
- trichopoda 504
- virgata 350, 483, 504
Silcrete 23

Silybum (Compos.) 530
- marianum 330
Simaroubaceae 47
Sinoga (Myrt.) 81
- lysicephala 399
Sinopteridaceae 36
Siphonodon (Celastr.)
- australe 147
Siphula 434
Sisymbrium (Crucif.) 330, 530
- orientale 103
Sium (Umbell.)
- latifolium 582
Sloanea (Elaeocarp.)
- australis 146
- macbrydei 172
- woollsii 146
Smilacaceae 48
Smilax (Smilac.)
- australis 150, 177, 180, 230, 233, 237, 238
- glycyphylla 177, 238
Snow Grass = Poa spp.
Snow Daisy = Celmisia longifolia
Softwood scrub 137, 182, 466
Soil(s)
- acidity 129
- alkalinity 129
- classification 31
- depth 126
- development 23
- fertility 23, 126, 127, 249, 400, 533
- group and vegetation 31, 74
- groups 24 (Table 1.4)
- nitrogen 249
- nutrients 127
- organic matter 477
- parent materials 22
- phosphorus 30, 126, 128, 249, 416
- profiles 23
- properties 31
- salinity 127
- texture 127, 128
- waterlogging 127, 128, 129
Solanaceae 47, 56
Solanum (Solan.) 52, 104, 108
- ferocissimum 107
- lasiophyllum 484
- mauritanum 228
- orbiculatum 492
- quadriloculatum 213
Solonization 20
Sonchus (Compos.)
- asper 607
- megalocarpus 607, 608
- oleraceus 103, 559
Sonneratia (Sonnerat.) 60
- caseolaris 614
- - alliance 621
Sonneratiaceae 47
Sophora (Papil.)
- tomentosa 607
Sorghum (Gramin.) 511
- leiocladum 280, 330
- plumosum 199, 200, 202, 211, 215, 299, 520
Sowerbaea (Lil.) 90
- juncea 252, 413, 591

Spartochloa (Gramin.) 511
Spartothamnella (Verben.) 108
– juncea 181, 182
Sparganiaceae 48
Spathia (Gramin.) 511
– nervosa 527
Spergularia (Caryophyll.)
– rubra 630, 631
Sphaerocionium 35
Sphaerolobium (Papil.) 98
– vimineum 591
Sphagnum 129, 240, 418, 443, 456, 460, 570
– cristatum 457, 459
Sphalmium (Prot.) 86
Sphenotoma (Epacrid.) 98
– drummondii 436
Spiculaea (Orchid.) 98
Spigeliaceae 47
Spinifex (Gramin.) 511
– hirsutus 513, 515, 603, 604, 605 F, 606, 608, 610
– longifolius 515 F, 606, 607, 609 F, 628, F, 637, 637
– spp. alliance 515
Spinifex
– "country" 533
– Eastern = Triodia mitchellii
– Feather-top = Plectrachne schinzii
– Gummy = Triodia pungens
– Hard = T. basedowii
– Soft = T. pungens
Spiranthes (Orchid.) 98
Spirodela (Lemn.) 552
– oligorrhiza 585, 595
Spirogardnera (Santal.) 55
Sporobolus (Gramin.) 104, 511
– actinocladus 325, 469, 504, 526, 527, 528, 540, 542
– australasicus 219
– – Enneapogon spp. grassland 540
– benthamii 522
– caroli 324, 465, 467, 489, 530
– elongatus 320, 332, 546
– mitchellii 121, 465, 526
– virginicus 111, 121, 215, 513, 517, 528, 606, 607, 608, 623, 628, 630
– – alliance 516, 518 F
Sprengelia (Epacrid.) 98
– incarnata 252, 263, 264, 401, 404, 407, 408, 591
– sprengelioides 401, 408, 591
Spyridium (Rhamn.)
– globosum 432, 433, 436, 608
– halmaturinum 637
Stackhousia (Stackhous.)
– spathulata 607
– viminea 591
Stackhousiaceae 47
Stawellia (Lil.) 90
Steganthera (Monim.) 53
Stellaria (Caryoph.)
– flaccida 274
Stenocarpus (Prot.) 86
– reticulatus 173
– salignus 148, 233
– sinuatus 148, 173

Stenochlaena 35
– palustris 178, 180
Stenopetalum (Crucif.) 113
Stephania (Menisperm.)
– japonica 148, 232, 241, 287
– hernandifolia 228
Steppe 493
Sterculia (Stercul.) 70
– australis 537
– laurifolia 174
– quadrifida 180, 182, 606
– shillinglawii 174
Sterculiaceae 47, 76
Stichetus 34
Stigonema 411
Stipa (Gramin.) 52, 103, 104, 111, 348, 511, 528
– aristiglumis 190, 320 F, 325, 330, 332, 334, 467, 490, 523, 524, 530
– – alliance 529 F
– bigeniculata 523, 530
– blackii 530
– compacta 530
– compressa 380, 381
– densiflora 314
– elegantissima 241, 354, 383, 386
– eremophila 337, 339, 475, 489, 523, 530, 531
– falcata 313, 326, 334
– – Danthonia caespitosa disclimax 530
– nitida 492, 501, 508, 523, 531
– scabra 326, 330, 332, 334, 523
– – S. bigeniculata alliance 530
– semibarbata 261, 344, 405, 530
– setacea 326, 530
– teretifolia 631
– variabilis 190, 269, 303, 307, 321, 326, 328, 330, 335, 336, 337, 475, 487, 501, 509, 523, 530, 593
– verticillata 313, 336, 563
Stirlingia (Prot.) 86, 98
– latifolia 397, 417, 419, 422
– simplex 397
– tenuifolia 397
Stirling Range (W. A.) 361, 376, 434
Stony downs 498
Storckiella (Caesalpin.) 172
Strangea (Prot.) 86, 98
– linearis 401, 402
– stenocarpoides 422
Streptothamnus (Flacourt.)
– beckleri 155
Stringybark 84
– Bailey's = Eucalyptus baileyana
– Baxter's = E. baxteri
– Broad-leaved = E. caliginosa
– Darwin = E. tetrodonta
– Red = E. macrorhyncha
– Silver-top = E. laevopinea
– Swamp = E. conglomerata
– White = E. eugenioides or E. globoidea
– Yellow = E. muellerana
– Youman's = E. youmanii
Structural form 124
Strychnaceae 47
Stylidiaceae 47

Stylidium (Stylid.) 52, 63, 210, 520, 596
– floribundum 569
– graminifolium 244, 246
– pseudohirtum 426
– pubigerum 435
Stylobasium (Ros.)
– spathulatum 359, 483, 638
Stypandra (Lil.) 90
– glauca 260, 282, 410, 430
– imbricata 436
Styphelia (Epacrid.) 98
– triflora 282
– viridis 239
Suaeda (Chenopod.) 104, 493
– australis 623, 626, 628, 630
Sulphur dioxide 264, 407
Sundew = Drosera spp.
Swainsona (Papil.) 52, 85, 87, 103, 112, 512, 600
– procumbens 489
– stipularis 504
Swamp 124, 570
Symphionema (Prot.) 85, 87, 98
– paludosa 408, 591
Symplocaceae 47
Symplocus (Symplococc.)
– stawellii 150, 174
Synaphea (Prot.) 86, 98
– favosa 397
– petiolaris 397, 419
– polymorpha 397
Syncarpia (Myrt.) 81, 82
– glomulifera 149, 225, 226, 229, 230 F, 231, 239, 246, 410
– hillii 237, 636
Synoum (Mel.)
– glandulosum 148, 230, 232, 233, 237
Syringodium (Zanichell.)
– isoetifolium 613
Syzygium (Myrt.) 70, 81
– coolminianum 149
– erythrodoxum 178
– floribundum 194
– francisii 149
– luehmannii 149
– paniculatum 149

T

Tacca (Taccac.)
– leontopetaloides 180
Taccaceae 49
Tainia (Orchid.) 93
Tallerach = Eucalyptus tetragona
Tallow-wood = Eucalyptus microcorys
Tanglefoot = Nothofagus gunnii
Tasman Geosyncline 9
Tasmannia (Winter.) 53, 62, 63
– lanceolata 153, 268, 273, 291, 295, 297, 413, 439, 453
– purpurascens 155, 268
Taxodiaceae 40
Tectaria 35

Tecticornia (Chenopod.)
– cinerea 623, 628
Telopea (Prot.) 87, 98
– oreades 153, 278
– speciosissima 258
– truncata 152, 295, 296, 454
Templetonia (Papil.) 85, 98
– egena 486, 487
– retusa 352, 420, 421, 430, 637
Tepualia (Myrt.) 80, 82
Teratophyllum 35
Terminalia (Combret.) 108, 215, 319
– aridicola 187, 188, 537
– – Bauhinia alliance 187
– canescens 186, 187, 188, 202, 210
– catappa 606, 635
– circumalata 109
– erythrocarpa 195
– ferdinandiana 187, 319
– oblongata 191, 466
– platyphylla 186, 188, 195
– platyptera 186, 187, 188 F, 189, 562 F, 563
– sericocarpa 180
Tertiary environment 73, 74
Tetracarpaea (Escallon.) 79
Tetracera (Dillen.) 53
– nordtiana 177
Tetragonia (Aizo.)
– decumbens 600, 608
– eremaea 531
– expansa 473, 502, 506, 540, 542, 600, 630
– implexicoma 600, 608
Tetrameles (Tetramel.)
– nudiflora 169, 174
Tetraria (Cyper.)
– halmaturina 637
Tetrarrhena (Gramin.) 511
– juncea 400, 512
Tetrastigma (Vit.)
– nitens 149
Tetrasynandra (Monim.) 53
Tetratheca (Tremandr.)
– ericifolia 261, 262, 337, 637
– halmaturina 637
Tetricornia (Chenopod.) 493
Teucrium (Lab.)
– racemosum 553
Thaumastochloa (Gramin.) 511
Theaceae 47
Theleophyton (Chenopod.) 493
Thellungia (Gramin.) 511
– advena 527
Thelymitra (Orchid.) 92, 98, 115
– ixioides 71
– longifolia 71
– venosa 71
Thelypteridaceae 36
Thelypteris 36
Themeda (Gramin.) 511
– australis 199, 202, 208, 211, 215, 217, 229, 231, 232, 234, 235, 236, 238, 239, 244, 245, 246, 269, 279, 280, 299, 301, 302, 303 F, 318, 320, 325, 330, 332, 338, 402, 405, 408, 409, 475, 511, 523, 530, 545 F, 546

– – – Eriachne burkittii alliance 520
– – on headlands 517
– avenacea 479, 527
Thesium (Santal.) 55
Thespesia (Malv.)
– populnea 606, 608, 623
Thicket 124, 417
Thismia (Thism.) 52
Thismiaceae 49, 52
Threlkeldia (Chenopod.) 113, 493
– diffusa 600, 603, 608
Thryptomene (Myrt.) 81, 98, 105, 110, 112
– australis 436
– baeckeacea 359, 638
– denticulata 394
– kochii 355, 394, 436
– maisonneuvii 437
– oligandra 399
– racemosa 394
– roei 428
– saxicola 430
– urceolaris 394
Thuarea (Gramin.) 511
Thymelaeaceae 47
Thyridolepis (Gramin.) 511
Thysanotus (Lil.) 115
– multiflorus 417
– tuberosus 408, 409
Tides 598
Tieghemopanax (Aral.)
– sambucifolius 238
Tiliaceae 47
Tilliaepollenites 60
Tingle, Red = Eucalyptus jacksonii
– Yellow = E. guilfoylei
Tinospora (Menisperm.) 108
– smilacina 109, 180, 199
Tmesipteridaceae 34
Tmesipteris 34, 36, 140
Todea 34
– barbara 37, 39 F
Toechima (Sapind.)
– daemeliana 174
Toona (Mel.) 70
– australis 148, 163 F, 165 F, 173, 194, 229
Topography 4 F, 5
Townsownia (Orchid.)
– viridis 92
Toxic substances 129
Trachymene (Umbell.)
– glaucifolia 538
– ornata 383
Trachystylis (Cyper.) 91
Tragia (Euphorb.) 70
Tragus (Gramin.) 511
– australianus 328, 540
Trema (Urtic.) 70
– aspera 109, 149, 228, 241
Tremandraceae 48
Trianthema (Aizo.) 104
– crystallina 627
– turgidiflora 628
Tribulus (Zygophyll.) 104
– hystrix 540
– terrestris 500, 527
Trichodesma (Boragin.)

– zeylanica 424
Trichomanes 35
Tricoryne (Lil.) 90
Tricostularia (Cyper.)
– pauciflora 591
Trifolium (Papil.) 329, 330
Triglochin (Juncagin.)
– procera 581, 582, 585, 586, 588 F, 595
– striata 586, 595
Trigonella (Papil.) 85
– suavissima 557
Trimeniaceae 48, 53
Triodia (Gramin.) 106, 130, 213, 322, 511, 533 F
– basedowii 366, 475, 491, 492, 534
– – alliance 537 F, 538
– clelandii 475
– concinna 366
– inutilis 224
– irritans 475
– – alliance 538 F
– longipes 367
– microstachya 213, 219
– mitchellii 304, 306, 307, 308, 309, 320
– – savannahs 539
– plurinervata 359, 638
– pungens 121, 200, 208, 214 F, 216, 221, 222, 223 F, 224, 363, 364, 366, 399, 469, 475, 479, 492, 606
– – alliance 535, 536 F
– scariosa 345, 347, 348, 349, 354, 358, 428, 513 F
– spicata 475
– stenostachya 226
– wiseana 219, 224, 365 F
Tripogon (Gramin.) 511
– loliiformis 542
Triraphis (Gramin.) 511
– mollis 482, 509, 528, 542
Trisetum (Gramin.) 511
Tristania (Myrt.) 70, 81, 82
– conferta 149, 167 F, 225, 226, 229, 230, 401, 593
– – alliance 166
– grandiflora 181, 195, 593
– lactiflua 204, 277 F, 575
– – – Grevillea pteridifolia – Banksia dentata alliance 576
– laurina 149, 194, 232, 241, 593
– neriifolia 593
– suaveolens 188, 189, 228, 231, 232, 233, 234, 235, 239, 245, 247, 562, 563, 635
Trithuria (Centrolepid.)
– thymifolia 456
Triunia (Prot.) 87
Triuridaceae 49
Trochocarpa (Epacrid.) 89
– gunnii 89, 151, 152
– laurina 89, 146, 178, 231, 238
– thymifolia 454
Trymalium (Rhamn.)
– spathulatum 370, 371, 372
Tuart = Eucalyptus gomphocephala
Tylophora (Asclepiad.)
– barbata 147, 232

Typha (Typh.) 104
- domingensis alliance 552, 581, 582, 593, 595
Typhaceae 49

U

Ulmaceae 48
Ulothrix 413
Umbelliferae 48, 56, 77
Uncinnia (Cyper.) 67, 92
- compacta 71
- flaccida 456
- riparia 71
- tenella 71
Unona (Annon.) 53
Uranthoecium (Gramin.) 511
Urticaceae 48
Utricularia (Lentibular.) 55, 408, 571
- australis 585, 586, 591
- chrysantha 399
- dichotoma 585, 586, 591
- exoleta 574
- flexuosa 574
- lateriflora 591
Uvaria (Annon.) 53

V

Vaginularia 36
Vallisneria (Hydrocharit.)
- exiliflora 574
- spiralis 581, 582, 585, 586
Valvanthera (Hernand.) 53
Vegetation regions 124
Velleia (Gooden.)
- montana 451
Ventilago (Rhamn.) 108
- viminalis 148, 187, 188, 191, 193, 213, 319, 320, 364, 469, 479
Verbenaceae 48
Veronica (Scrophul.)
- calycina 274, 371
- densiflora 449
Verticordia (Myrt.) 52, 81, 103, 105
- acerosa 394, 426
- brownii 394
- chrysantha 394, 428
- cunninghamii 200, 399
- densiflora 394
- etheliana 357
- grandiflora 394
- grandis 394
- habrantha 394
- humilis 394
- insignis 394
- lepidophylla 394
- picta 394
- plumosa 394
- prizelii 394
- roei 394
- serrata 394
- stelligera 394
Verrucaria 413
Vetiveria (Gramin.) 511, 520
- elongata 517, 518 F
- pauciflora 215, 520

Vicariiads 99
Vigna (Papil.)
- marina 635
Villarsia (Menyanth.)
- exaltata 581, 585, 586
- reniformis 240, 585, 586, 588
Viminaria (Papil.) 98
- denudata 99, 254, 418
- juncea 591, 595
Vine Forest
- mesophyll 136
- microphyll 136
- notophyll 136
- semi-evergreen 179
Vine Thicket 183
- microphyll 136
- notophyll 136
Viola (Viol.)
- betonicifolia 270
- hederacea 232
Violaceae 48
Viscaceae 48, 55
Viscum (Visc.) 55
Vitaceae 48
Vitex (Verben.)
- glabrata 180
Vittadinia (Compos.) 110
- australis 71, 336, 350
- triloba 337, 530
Vittaria 36
Vittariaceae 36
Volcanic activity 10, 74

W

Wadi 551, 568
Wahlenbergia (Campan.)
- gracilis 286
Waitzia (Compos.) 113
- acuminata 354, 357, 385, 482, 538
- citrina 432
Wandoo = Eucalyptus wandoo
- Powderbark = E. accedens
- Salmonbark = E. lane-poolei
Watercourse communities 549
Waterlily, Blue = Nymphaea gigantea
Waterlogging 23
Watershed (eastern) 19
Weathering 23
Wedelia (Compos.)
- biflora 635
Weeds 56
Wehlia (Myrt.) 81, 105, 112
Weinmannia (Cunon.) 63
- rubifolia 155
Westringia (Labiat.)
- cheelii 314
- fruticosa 600, 602, 636
- rigida 358, 600, 638
Whiteochloa (Gramin.) 511
Wilga = Geijera parviflora
Wilkiea (Monim.) 53
- huegeliana 148
Wilsonia (Convolvul.)
- backhousei 581, 600, 630
Wind pollination 19
Wingecarrabie Tableland (N.S.W.) 411

Winteraceae 48, 53, 75
Wittsteinia (Eric.) 88
- vacciniacea 153
Wolffia (Lemn.) 552
- arrhiza 44
- australiana 585
Woodland 124
Woollsia (Epacrid.) 98
Woollybutt = Eucalyptus miniata
Wrightia (Apocyn.)
- pubescens 181

X

Xanthium (Compos.)
- spinosum 103
Xanthophyllum (Polygal.)
- octandrum 173
Xanthorrhoea (Xanthorrh.) 91, 224
- australis 91 F, 259, 288, 289, 311, 344, 404, 409, 611
- - - Banksia ornata – Hakea rostrata – Casuarina paludosa alliance 413
- - - Banksia ornata – Casuarina pusilla alliance 414
- johnstonii 201, 299, 302, 399
- macronema 266
- media 246, 401, 405, 411
- minor 404
- preissii 373, 375, 378, 379, 381, 419, 427, 431, 432, 433, 595, 597
- reflexa 422, 423 F
- resinosa 252, 254, 591
- semiplana 338
- tateana 339
Xanthorrhoeaceae 49, 90, 91
Xanthosia (Umbell.) 77
- pilosa 409
Xanthostemon (Myrt.) 70, 81, 82
- chrystanthus 173
- whitei 173
Xerochloa (Gramin.) 511, 628
- barbata 516, 517, 518 F, 528, 579, 580
- imberbis 516, 517, 518 F, 627
- spp. alliance 517
Xeromorphs 73, 80, 93
- ash 95
- in arid zone 96, 99
- centres of abundance 95
- cold climate 96
- effects of aridity 97
- geographic districution 95
- habitat factors 96
- history 96, 99
- leaf size 94, 95
- leaf texture 94, 95
- phosphorus levels 94
- silica content 94
- stomates 95
- taxa 95, 97
- Tertiary taxa 94
- water potential 94
Xylocarpus (Mel.)
- australasicus 614, 625 F
- granatum 614

Xylomelum (Prot.) 85, 86, 98
– angustifolium 418, 425, 426 F
Xyridaceae 49
Xyris (Xyrid.) 408
– complanata 520, 591
– gracilis 591
– operculata 264, 407, 591

Y

Yarravia 57
Yate = Eucalyptus cornuta
– Flat-topped = E. occidentalis
– Swamp = E. occidentalis
– Warted = E. megacornuta
Yorell = Eucalyptus gracilis

Z

Zamiaceae 40
Zannichelliaceae 49
Zanthoxylum (Rut.)
– brachyacanthum 149
Zeuxine (Orchid.) 93
Zieria (Rut.) 85, 98
– arborescens 275
– smithii 236, 238, 291
Zinc 130
Zingiberaceae 49
Zircon 400, 603
Ziziphus (Rhamn.)
– quadrilocularis 181
Zonation 122
Zostera (Zoster.)
– capricorni 581, 613, 630
– muelleri 594, 613, 630
Zosteraceae 49
Zosterophyllum 57
Zoysia (Gramin.) 511
– macrantha 516, 517, 630, 634 F
Zygochloa (Gramin.) 111, 112, 511
– paradoxa 105, 512, 560
– – alliance 539 F
Zygophyllaceae 48, 111
Zygophyllum (Zygophyll.) 103, 104
– ammophilum 527, 531, 603
– apiculatum 348, 432, 531
– compressum 560
– fruticulosum 348
– howittii 540
– iodocarpum 542

Arid-land ecosystems:
their structure, functioning and management

IBP Series, Volumes 16 & 17

Edited by
D. W. GOODALL & R. A. PERRY
Division of Land Resources Management,
CSIRO, Wembley, Western Australia

The history of man's use of arid lands is a sad record of deterioration of the natural resources base and of low and declining living standards for the 300 million people who live in them. One pre-requisite to meeting the challenge of reversing the deterioration and of raising living standards is a sound knowledge of the natural ecosystems.

It is to this end that the IBP and other relevant studies, reported in these two volumes, are addressed. The subject matter of the two volumes is organised into five major topics; two are dealt with in the first volume and the remaining three in the second.

The first section of the first volume describes the structure of arid ecosystems in terms of climate, soils, geomorphology, hydrology, flora and fauna. All continents except South America are covered. In the second section the processes which operate within, and control, the ecosystem are dealt with individually.

The first section of the second volume considers interactions between the physical environment and vegetation, and between the plants and animals of arid environments, culminating in studies on the productivity.

The second section is concerned with following the flow of energy and the cycling of nutrients. This section ends with simulation of plant production and the use of modelling techniques that help in our understanding of arid ecosystems.

In the final section, the emphasis turns to management of arid-land resources and the impact of recreation and tourism.

This comprehensive account will be importance to university teachers and professional ecologists throughout the world.

Some reviewer's comments on the first volume (IBP 16)

The book is really a monumental work on the state-of-the-art, and the editors deserve much credit for producing such a quality volume.
Choice

This book is encyclopedic and will provide an excellent reference for biologists and geographers who are looking for a detailed outline of world-wide, arid land ecology. Most of the chapters are very well written and easy to understand. The editors have done an excellent job in providing *Integration* (synthesis and summary) chapters at the end of the sections on atmospheric, plant, and animal processes.
Quart. Rev. Biol.